MW00837260

Fluid Mechanics, Fourth Edition

Founders of Modern Fluid Dynamics

Ludwig Prandtl
(1875–1953)

G. I. Taylor
(1886–1975)

(Biographical sketches of Prandtl and Taylor are given in Appendix C.)

Photograph of Ludwig Prandtl is reprinted with permission from the *Annual Review of Fluid Mechanics*, Vol. 19, Copyright 1987 by Annual Reviews www.AnnualReviews.org.

Photograph of Geoffrey Ingram Taylor at age 69 in his laboratory reprinted with permission from the AIP Emilio Segrè Visual Archieves. Copyright, American Institute of Physics, 2000.

Fluid Mechanics
Fourth Edition

Pijush K. Kundu

Oceanographic Center
Nova Southeastern University
Dania, Florida

Ira M. Cohen

Department of Mechanical Engineering and
Applied Mechanics
University of Pennsylvania
Philadelphia, Pennsylvania

with contributions by P. S. Ayyaswamy and H. H. Hu

ELSEVIER

AMSTERDAM • BOSTON • HEIDELBERG • LONDON
NEW YORK • OXFORD • PARIS • SAN DIEGO
SAN FRANCISCO • SINGAPORE • SYDNEY • TOKYO

Academic Press is an imprint of Elsevier

Academic Press is an imprint of Elsevier
30 Corporate Drive, Suite 400
Burlington, MA 01803, USA

Elsevier, The Boulevard, Langford Lane
Kidlington, Oxford, OX5 1GB, UK

© 2008 Elsevier Inc. All rights reserved.

No part of this publication may be reproduced or transmitted in any form or by any means, electronic or
mechanical, including photocopying, recording, or any information storage and retrieval system, without
permission in writing from the publisher. Details on how to seek permission, further information about the
Publisher's permissions policies and our arrangements with organizations such as the Copyright Clearance
Center and the Copyright Licensing Agency, can be found at our website: www.elsevier.com/permissions.
This book and the individual contributions contained in it are protected under copyright by the
Publisher (other than as may be noted herein).

Notices

Knowledge and best practice in this field are constantly changing. As new research and experience
broaden our understanding, changes in research methods, professional practices, or medical treatment may
become necessary.

Practitioners and researchers must always rely on their own experience and knowledge in evaluating
and using any information, methods, compounds, or experiments described herein. In using such
information or methods they should be mindful of their own safety and the safety of others, including
parties for whom they have a professional responsibility.

To the fullest extent of the law, neither the Publisher nor the authors, contributors, or editors, assume
any liability for any injury and/or damage to persons or property as a matter of products liability,
negligence or otherwise, or from any use or operation of any methods, products, instructions, or ideas
contained in the material herein.

Library of Congress Cataloging-in-Publication Data
Kundu, Pijush K.
Fluid mechanics / Pijush K. Kundu, Ira M. Cohen. – 4th ed.
p.   cm.
Includes bibliographical references and index.
ISBN 978-0-12-373735-9 (alk. paper) /
1. Fluid mechanics.   I. Cohen, Ira M.   II. Title.
 QA901.K86 2008
  620.1'06–dc22   2007042765
Reprinted January 2010 – ISBN: 978-0-12-381399-2

British Library Cataloguing-in-Publication Data
A catalogue record for this book is available from the British Library.

For information on all Academic Press publications
visit our Web site at *www.elsevierdirect.com*

Printed in the United States
10 11 12 13 14   10 9 8 7 6 5 4 3 2 1

Working together to grow
libraries in developing countries

www.elsevier.com | www.bookaid.org | www.sabre.org

ELSEVIER   BOOK AID
              International   Sabre Foundation

The fourth edition is dedicated to the memory of Pijush K. Kundu and also to my wife Linda and daughters Susan and Nancy who have greatly enriched my life.

"Everything should be made as simple as possible,
but not simpler."
—Albert Einstein

"If nature were not beautiful, it would not be worth studying it.
And life would not be worth living."
—Henry Poincaré

In memory of Pijush Kundu

Pijush Kanti Kundu was born in Calcutta, India, on October 31, 1941. He received a B.S. degree in Mechanical Engineering in 1963 from Shibpur Engineering College of Calcutta University, earned an M.S. degree in Engineering from Roorkee University in 1965, and was a lecturer in Mechanical Engineering at the Indian Institute of Technology in Delhi from 1965 to 1968. Pijush came to the United States in 1968, as a doctoral student at Penn State University. With Dr. John L. Lumley as his advisor, he studied instabilities of viscoelastic fluids, receiving his doctorate in 1972. He began his lifelong interest in oceanography soon after his graduation, working as Research Associate in Oceanography at Oregon State University from 1968 until 1972. After spending a year at the University de Oriente in Venezuela, he joined the faculty of the Oceanographic Center of Nova Southeastern University, where he remained until his death in 1994.

During his career, Pijush contributed to a number of sub-disciplines in physical oceanography, most notably in the fields of coastal dynamics, mixed-layer physics, internal waves, and Indian-Ocean dynamics. He was a skilled data analyst, and, in this regard, one of his accomplishments was to introduce the "empirical orthogonal eigenfunction" statistical technique to the oceanographic community.

I arrived at Nova Southeastern University shortly after Pijush, and he and I worked closely together thereafter. I was immediately impressed with the clarity of his scientific thinking and his thoroughness. His most impressive and obvious quality, though, was his love of science, which pervaded all his activities. Some time after we met, Pijush opened a drawer in a desk in his home office, showing me drafts of several chapters to a book he had always wanted to write. A decade later, this manuscript became the first edition of "Fluid Mechanics," the culmination of his lifelong dream; which he dedicated to the memory of his mother, and to his wife Shikha, daughter Tonushree, and son Joydip.

Julian P. McCreary, Jr.,
University of Hawaii

Contents

Chapter 1

Introduction

Chapter 2

Cartesian Tensors

Chapter 3

Kinematics

Chapter 4

Conservation Laws

Chapter 5
Vorticity Dynamics

Chapter 6
Irrotational Flow

Chapter 7
Gravity Waves

Chapter 8
Dynamic Similarity

Chapter 9
Laminar Flow

Chapter 10
Boundary Layers and Related Topics

Chapter 11
Computational Fluid Dynamics

Chapter 12
Instability

Chapter 13
Turbulence

Chapter 14
Geophysical Fluid Dynamics

Chapter 15

Aerodynamics

Chapter 16

Compressible Flow

Chapter 17
Introduction to Biofluid Mechanics

Appendix A
Some Properties of Common Fluids

Appendix B
Curvilinear Coordinates

Appendix C
Founders of Modern Fluid Dynamics

Appendix D
Visual Resources

New in This Book

We are pleased to include a free copy of the DVD *Multimedia Fluid Mechanics, 2/e*, with this copy of *Fluid Mechanics, Fourth Edition*. You will find it in a plastic sleeve on the inside back cover of the book. If you are purchasing a used copy, be aware that the DVD might have been removed by a previous owner.

Inspired by the reception of the first edition, the objectives in *Multimedia Fluid Mechanics, 2/e*, remain to exploit the moving image and interactivity of multimedia to improve the teaching and learning of fluid mechanics in all disciplines by illustrating fundamental phenomena and conveying fascinating fluid flows for generations to come.

The completely new edition on the DVD includes the following:

- Twice the coverage with new modules on turbulence, control volumes, interfacial phenomena, and similarity and scaling

- Four times the number of fluids videos, now more than 800

- Now more than 20 Virtual labs and simulations

- Dozens of new interactive demonstrations and animations

Additional *new* features:

- Improved navigation via side bars that provide rapid overviews of modules and guided browsing

- Media libraries for each chapter that give a snapshot of videos, each with descriptive labels

- Ability to create movie playlists, which are invaluable in teaching

- Higher-resolution graphics, with full or part screen viewing options

- Operates on either a PC or a Mac OSX

Preface

Fluid mechanics has a vast scope and touches every aspect of our lives. Just look at the contents of the 39 volumes of the *Annual Review of Fluid Mechanics* (1969–2007) for validation of that statement. We cover only a tiny fraction of that scope in this book.

This Fourth Edition continues to evolve due to the kindness of readers and users who write to me suggesting corrections. Specifically, Roger Berlind of Columbia University is responsible for the revisions to the **Thermal Wind** subsection of Chapter 14. Howard Hu has revised, streamlined, and updated his chapter on Computational Fluid Dynamics, and P. S. Ayyaswamy has contributed a new chapter (17) on **Introduction to Biofluid Mechanics**. It is an excellently written contribution and unique in that its level is appropriate for this book. It is between the advanced treatises and overly simplified treatments available elsewhere. I have tried to update much of the remaining material, particularly on turbulence, where so many new papers appear each year. Reference to the collection of the National Committee for Fluid Mechanics Films, now available for viewing via the Internet, is made in a new Appendix D. These films may be old but remain an excellent resource for visualization of flows.

On a more personal note, the bladder cancer (transitional cell carcinoma) diagnosed in the Fall of 2003, and attacked by a sequence of surgeries and a regimen of chemotherapy, was never completely killed and grew back to visibility in the Fall of 2006. The minimum visible spot on MRI can contain 70,000 TCC. A new and harsher regimen of chemotherapy was prescribed through the Spring of 2007, during the period when the updates were prepared. The irony of the fact that all the chemical poisons infused into my veins are fluids is not lost on me. Because of the fatigue, I accomplished less than I had hoped. Since radiologists cannot distinguish viable living TCC from those that have been killed and remain in place, the only means of discerning living cancer is to image again and see if there is new growth. The image in early June confirmed the message that my body had already sent me: the cancer was growing back and causing pain. Radiation was tried for a while to shrink the painful tumor but that was unsuccessful. In late July a new regimen of chemotherapy began, to last perhaps through the end of the year and beyond. The initial response was positive and, although I have come to realize that my condition is incurable, I was ever hopeful. After three infusions, however, the toxic effects of the chemotherapy were more destructive than the aggressiveness of the cancer. I was left with no lung capacity, no muscle strength, and was very ill and weak when page proofs came back for checking. I remain so at the time of this writing.

I appreciate all the help provided by Ms. Susan Waddington in the preparation of the final stages of this manuscript.

I am very grateful to my family, friends, and colleagues for their support throughout this ordeal.

Ira M. Cohen

Preface to Third Edition

This edition provided me with the opportunity to include (almost) all of the additional material I had intended for the Second Edition but had to sacrifice because of the crush of time. It also provided me with an opportunity to rewrite and improve the presentation of material on jets in Chapter 10. In addition, Professor Howard Hu greatly expanded his CFD chapter. The expansion of the treatment of surface tension is due to the urging of Professor E. F. "Charlie" Hasselbrink of the University of Michigan.

I am grateful to Mr. Karthik Mukundakrishnan for computations of boundary layer problems, to Mr. Andrew Perrin for numerous suggestions for improvement and some computations, and to Mr. Din-Chih Hwang for sharing his latest results on the decay of a laminar shear layer. The expertise of Ms. Maryeileen Banford in preparing new figures was invaluable and is especially appreciated.

The page proofs of the text were read between my second and third surgeries for stage 3 bladder cancer. The book is scheduled to be released in the middle of my regimen of chemotherapy. My family, especially my wife Linda and two daughters (both of whom are cancer survivors), have been immensely supportive during this very difficult time. I am also very grateful for the comfort provided by my many colleagues and friends.

Ira M. Cohen

Preface to Second Edition

My involvement with Pijush Kundu's *Fluid Mechanics* first began in April 1991 with a letter from him asking me to consider his book for adoption in the first year graduate course I had been teaching for 25 years. That started a correspondence and, in fact, I did adopt the book for the following academic year. The correspondence related to improving the book by enhancing or clarifying various points. I would not have taken the time to do that if I hadn't thought this was the best book at the first-year graduate level. By the end of that year we were already discussing a second edition and whether I would have a role in it. By early 1992, however, it was clear that I had a crushing administrative burden at the University of Pennsylvania and could not undertake any time-consuming projects for the next several years. My wife and I met Pijush and Shikha for the first time in December 1992. They were a charming, erudite, sophisticated couple with two brilliant children. We immediately felt a bond of warmth and friendship with them. Shikha was a teacher like my wife so the four of us had a great deal in common. A couple of years later we were shocked to hear that Pijush had died suddenly and unexpectedly. It saddened me greatly because I had been looking forward to working with Pijush on the second edition after my term as department chairman ended in mid-1997. For the next year and a half, however, serious family health problems detoured any plans. Discussions on this edition resumed in July of 1999 and were concluded in the Spring of 2000 when my work really started. This book remains the principal work product of Pijush K. Kundu, especially the lengthy chapters on Gravity Waves, Instability, and Geophysical Fluid Dynamics, his areas of expertise. I have added new material to all of the other chapters, often providing an alternative point of view. Specifically, vector field derivatives have been generalized, as have been streamfunctions. Additional material has been added to the chapters on laminar flows and boundary layers. The treatment of one-dimensional gasdynamics has been extended. More problems have been added to most chapters. Professor Howard H. Hu, a recognized expert in computational fluid dynamics, graciously provided an entirely new chapter, Chapter 11, thereby providing the student with an entree into this exploding new field. Both finite difference and finite element methods are introduced and a detailed worked-out example of each is provided.

I have been a student of fluid mechanics since 1954 when I entered college to study aeronautical engineering. I have been teaching fluid mechanics since 1963 when I joined the Brown University faculty, and I have been teaching a course corresponding to this book since moving to the University of Pennsylvania in 1966. I am most grateful to two of my own teachers, Professor Wallace D. Hayes (1918–2001), who expressed fluid mechanics in the clearest way I have ever seen, and Professor Martin D. Kruskal, whose use of mathematics to solve difficult physical problems was developed to a high art form and reminds me of a Vivaldi trumpet concerto. His codification of rules of applied limit processes into the principles of "Asymptotology" remains with me today as a way to view problems. I am grateful also to countless students who asked questions, forcing me to rethink many points.

The editors at Academic Press, Gregory Franklin and Marsha Filion (assistant) have been very supportive of my efforts and have tried to light a fire under me. Since this edition was completed, I found that there is even more new and original material I would like to add. But, alas, that will have to wait for the next edition. The new figures and modifications of old figures were done by Maryeileen Banford with occasional assistance from the school's software expert, Paul W. Shaffer. I greatly appreciate their job well done.

Ira M. Cohen

Preface to First Edition

This book is a basic introduction to the subject of fluid mechanics and is intended for undergraduate and beginning graduate students of science and engineering. There is enough material in the book for at least two courses. No previous knowledge of the subject is assumed, and much of the text is suitable in a first course on the subject. On the other hand, a selection of the advanced topics could be used in a second course. I have not tried to indicate which sections should be considered advanced; the choice often depends on the teacher, the university, and the field of study. Particular effort has been made to make the presentation clear and accurate and at the same time easy enough for students. Mathematically rigorous approaches have been avoided in favor of the physically revealing ones.

A survey of the available texts revealed the need for a book with a balanced view, dealing with currently relevant topics, and at the same time easy enough for students. The available texts can perhaps be divided into three broad groups. One type, written primarily for applied mathematicians, deals mostly with classical topics such as irrotational and laminar flows, in which analytical solutions are possible. A second group of books emphasizes engineering applications, concentrating on flows in such systems as ducts, open channels, and airfoils. A third type of text is narrowly focused toward applications to large-scale geophysical systems, omitting small-scale processes which are equally applicable to geophysical systems as well as laboratory-scale phenomena. Several of these geophysical fluid dynamics texts are also written primarily for researchers and are therefore rather difficult for students. I have tried to adopt a balanced view and to deal in a simple way with the basic ideas relevant to both engineering and geophysical fluid dynamics.

However, I have taken a rather cautious attitude toward mixing engineering and geophysical fluid dynamics, generally separating them in different chapters. Although the basic principles are the same, the large-scale geophysical flows are so dominated by the effects of the Coriolis force that their characteristics can be quite different from those of laboratory-scale flows. It is for this reason that most effects of planetary rotation are discussed in a separate chapter, although the concept of the Coriolis force is introduced earlier in the book. The effects of density stratification, on the other hand, are discussed in several chapters, since they can be important in both geophysical and laboratory-scale flows.

The choice of material is always a personal one. In my effort to select topics, however, I have been careful not to be guided strongly by my own research interests. The material selected is what I believe to be of the most interest in a book on general fluid mechanics. It includes topics of special interest to geophysicists (for example, the chapters on *Gravity Waves* and *Geophysical Fluid Dynamics*) and to engineers (for example, the chapters on *Aerodynamics* and *Compressible Flow*). There are also chapters of *common* interest, such as the first five chapters, and those on *Boundary Layers*, *Instability*, and *Turbulence*. Some of the material is now available only in specialized monographs; such material is presented here in simple form, perhaps sacrificing some formal mathematical rigor.

Throughout the book the convenience of tensor algebra has been exploited freely. My experience is that many students feel uncomfortable with tensor notation in the beginning, especially with the permutation symbol ε_{ijk}. After a while, however, they like it. In any case, following an introductory chapter, the second chapter of the book explains the fundamentals of *Cartesian Tensors*. The next three chapters deal with standard and introductory material on *Kinematics*, *Conservation Laws*, and *Vorticity Dynamics*. Most of the material here is suitable for presentation to geophysicists as well as engineers.

In much of the rest of the book the teacher is expected to select topics that are suitable for his or her particular audience. Chapter 6 discusses *Irrotational Flow*; this material is rather classical but is still useful for two reasons. First, some of the results are used in later chapters, especially the one on *Aerodynamics*. Second, most of the ideas are applicable in the study of other potential fields, such as heat conduction and electrostatics. Chapter 7 discusses *Gravity Waves* in homogeneous and stratified fluids; the emphasis is on linear analysis, although brief discussions of nonlinear effects such as hydraulic jump, Stokes's drift, and soliton are given.

After a discussion of *Dynamic Similarity* in Chapter 8, the study of viscous flow starts with Chapter 9, which discusses *Laminar Flow*. The material is standard, but the concept and analysis of similarity solutions are explained in detail. In Chapter 10 on *Boundary Layers*, the central idea has been introduced intuitively at first. Only after a thorough physical discussion has the boundary layer been explained as a singular perturbation problem. I ask the indulgence of my colleagues for including the peripheral section on the dynamics of sports balls but promise that most students will listen with interest and ask a lot of questions. *Instability* of flows is discussed at some length in Chapter 12. The emphasis is on linear analysis, but some discussion of "chaos" is given in order to point out how deterministic nonlinear systems can lead to irregular solutions. Fully developed three-dimensional *Turbulence* is discussed in Chapter 13. In addition to standard engineering topics such as wall-bounded shear flows, the theory of turbulent dispersion of particles is discussed because of its geophysical importance. Some effects of stratification are also discussed here, but the short section discussing the elementary ideas of two-dimensional geostrophic turbulence is deferred to Chapter 14. I believe that much of the material in Chapters 8–13 will be of general interest, but some selection of topics is necessary here for teaching specialized groups of students.

The remaining three chapters deal with more specialized applications in geophysics and engineering. Chapter 14 on *Geophysical Fluid Dynamics* emphasizes the linear analysis of certain geophysically important wave systems. However, elements of barotropic and baroclinic instabilities and geostrophic turbulence are also included. Chapter 15 on *Aerodynamics* emphasizes the application of potential theory to flow around lift-generating profiles; an elementary discussion of finite-wing theory is also given. The material is standard, and I do not claim much originality or innovation, although I think the reader may be especially interested in the discussions of propulsive mechanisms of fish, birds, and sailboats and the material on the historic controversy between Prandtl and Lanchester. Chapter 16 on *Compressible Flow* also contains standard topics, available in most engineering texts. This chapter is included with the belief that all fluid dynamicists should have some familiarity with such topics as shock waves and expansion fans. Besides, very similar phenomena also occur in other nondispersive systems such as gravity waves in shallow water.

The appendices contain conversion factors, properties of water and air, equations in curvilinear coordinates, and short bibliographical sketches of *Founders of Modern Fluid Dynamics*. In selecting the names in the list of founders, my aim was to come up with a very short list of historic figures who made truly fundamental contributions. It became clear that the choice of Prandtl and G. I. Taylor was the only one that would avoid all controversy.

Some problems in the basic chapters are worked out in the text, in order to illustrate the application of the basic principles. In a first course, undergraduate engineering students may need more practice and help than offered in the book; in that case the teacher may have to select additional problems from other books. Difficult problems have been deliberately omitted from the end-of-chapter exercises. It is my experience that the more difficult exercises need a lot of clarification and hints (the degree of which depends on the students' background), and they are therefore better designed by the teacher. In many cases answers or hints are provided for the exercises.

Acknowledgments

I would like to record here my gratitude to those who made the writing of this book possible. My teachers Professor Shankar Lal and Professor John Lumley fostered my interest in fluid mechanics and quietly inspired me with their brilliance; Professor Lumley also reviewed Chapter 13. My colleague Julian McCreary provided support, encouragement, and careful comments on Chapters 7, 12, and 14. Richard Thomson's cheerful voice over the telephone was a constant reassurance that professional science can make some people happy, not simply competitive; I am also grateful to him for reviewing Chapters 4 and 15. Joseph Pedlosky gave very valuable comments on Chapter 14, in addition to warning me against too broad a presentation. John Allen allowed me to use his lecture notes on perturbation techniques. Yasushi Fukamachi, Hyong Lee, and Kevin Kohler commented on several chapters and constantly pointed out things that may not have been clear to the students. Stan Middleman and Elizabeth Mickaily were especially diligent in checking my solutions to the examples and end-of-chapter problems. Terry Thompson constantly got me out of trouble with my personal computer. Kathy Maxson drafted the figures. Chuck Arthur and Bill LaDue, my editors at Academic Press, created a delightful atmosphere during the course of writing and production of the book.

Lastly, I am grateful to Amjad Khan, the late Amir Khan, and the late Omkarnath Thakur for their music, which made working after midnight no chore at all. I recommend listening to them if anybody wants to write a book!

Pijush K. Kundu

Author's Notes

Both indicial and boldface notations are used to indicate vectors and tensors. The comma notation to represent spatial derivatives (for example, $A_{,i}$ for $\partial A/\partial x_i$) is used in only two sections of the book (Sections 5.6 and 13.7), when the algebra became cumbersome otherwise. *Equal to by definition* is denoted by $\equiv$; for example, the ratio of specific heats is introduced as $\gamma \equiv C_p/C_v$. *Nearly equal to* is written as $\simeq$, *proportional to* is written as $\propto$, and *of the order* is written as $\sim$.

Plane polar coordinates are denoted by (r, θ), cylindrical polar coordinates are denoted by either (R, φ, x) or (r, θ, x), and spherical polar coordinates are denoted by (r, θ, φ) (see Figure 3.1). The velocity components in the three Cartesian directions (x, y, z) are indicated by (u, v, w). In geophysical situations the z-axis points upward.

In some cases equations are referred to by a descriptive name rather than a number (for example, "the x-momentum equation shows that . . ."). Those equations and/or results deemed especially important have been indicated by a box.

A list of literature cited and supplemental reading is provided at the end of most chapters. The list has been deliberately kept short and includes only those sources that serve one of the following three purposes: (1) it is a reference the student is likely to find useful, at a level not too different from that of this book; (2) it is a reference that has influenced the author's writing or from which a figure is reproduced; and (3) it is an important work done after 1950. In currently active fields, reference has been made to more recent review papers where the student can find additional references to the important work in the field.

Fluid mechanics forces us fully to understand the underlying physics. This is because the results we obtain often defy our intuition. The following examples support these contentions:

1. Infinitesmally small causes can have large effects (d'Alembert's paradox).
2. Symmetric problems may have nonsymmetric solutions (von Karman vortex street).
3. Friction can make the flow go faster and cool the flow (subsonic adiabatic flow in a constant area duct).
4. Roughening the surface of a body can decrease its drag (transition from laminar to turbulent boundary layer separation).
5. Adding heat to a flow may lower its temperature. Removing heat from a flow may raise its temperature (1-dimensional adiabatic flow in a range of subsonic Mach number).
6. Friction can destabilize a previously stable flow (Orr-Sommerfeld stability analysis for a boundary layer profile without inflection point).

7. Without friction, birds could not fly and fish could not swim (Kutta condition requires viscosity).
8. The best and most accurate visualization of streamlines in an inviscid (infinite Reynolds number) flow is in a Hele-Shaw apparatus for creeping highly viscous flow (near zero Reynolds number).

Every one of these counterintuitive effects will be treated and discussed in this text.

This second edition also contains additional material on streamfunctions, boundary conditions, viscous flows, boundary layers, jets, and compressible flows. Most important, there is an entirely new chapter on computational fluid dynamics that introduces the student to the various techniques for numerically integrating the equations governing fluid motions. Hopefully the introduction is sufficient that the reader can follow up with specialized texts for a more comprehensive understanding.

An historical survey of fluid mechanics from the time of Archimedes (ca. 250 B.C.E.) to approximately 1900 is provided in the Eleventh Edition of *The Encyclopædia Britannica* (1910) in Vol. XIV (under "Hydromechanics," pp. 115–135). I am grateful to Professor Herman Gluck (Professor of Mathematics at the University of Pennsylvania) for sending me this article. Hydrostatics and classical (constant density) potential flows are reviewed in considerable depth. Great detail is given in the solution of problems that are now considered obscure and arcane with credit to authors long forgotten. The theory of slow viscous motion developed by Stokes and others is not mentioned. The concept of the boundary layer for high-speed motion of a viscous fluid was apparently too recent for its importance to have been realized.

IMC

Introduction

1. Fluid Mechanics

Fluid mechanics deals with the flow of fluids. Its study is important to physicists, whose main interest is in understanding phenomena. They may, for example, be interested in learning what causes the various types of wave phenomena in the atmosphere and in the ocean, why a layer of fluid heated from below breaks up into cellular patterns, why a tennis ball hit with "top spin" dips rather sharply, how fish swim, and how birds fly. The study of fluid mechanics is just as important to engineers, whose main interest is in the applications of fluid mechanics to solve industrial problems. Aerospace engineers may be interested in designing airplanes that have low resistance and, at the same time, high "lift" force to support the weight of the plane. Civil engineers may be interested in designing irrigation canals, dams, and water supply systems. Pollution control engineers may be interested in saving our planet from the constant dumping of industrial sewage into the atmosphere and the ocean. Mechanical engineers may be interested in designing turbines, heat exchangers, and fluid couplings. Chemical engineers may be interested in designing efficient devices to mix industrial chemicals. The objectives of physicists and engineers, however, are

©2010 Elsevier Inc. All rights reserved.
DOI: 10.1016/B978-0-12-381399-2.50001-0

not quite separable because the engineers need to understand and the physicists need to be motivated through applications.

Fluid mechanics, like the study of any other branch of science, needs mathematical analyses as well as experimentation. The analytical approaches help in finding the solutions to certain idealized and simplified problems, and in understanding the unity behind apparently dissimilar phenomena. Needless to say, drastic simplifications are frequently necessary because of the complexity of real phenomena. A good understanding of mathematical techniques is definitely helpful here, although it is probably fair to say that some of the greatest theoretical contributions have come from the people who depended rather strongly on their unusual physical intuition, some sort of a "vision" by which they were able to distinguish between what is relevant and what is not. Chess player, Bobby Fischer (appearing on the television program "The Johnny Carson Show," about 1979), once compared a good chess player and a great one in the following manner: When a good chess player looks at a chess board, he thinks of 20 possible moves; he analyzes all of them and picks the one that he likes. A great chess player, on the other hand, analyzes only two or three possible moves; his unusual intuition (part of which must have grown from experience) allows him immediately to rule out a large number of moves *without going through an apparent logical analysis*. Ludwig Prandtl, one of the founders of modern fluid mechanics, first conceived the idea of a boundary layer based solely on physical intuition. His knowledge of mathematics was rather limited, as his famous student von Karman (1954, page 50) testifies. Interestingly, the boundary layer technique has now become one of the most powerful methods in applied mathematics!

As in other fields, our mathematical ability is too limited to tackle the complex problems of real fluid flows. Whether we are primarily interested either in understanding the physics or in the applications, we must depend heavily on experimental observations to test our analyses and develop insights into the nature of the phenomenon. Fluid dynamicists cannot afford to think like pure mathematicians. The well-known English pure mathematician G. H. Hardy once described applied mathematics as a form of "glorified plumbing" (G. I. Taylor, 1974). It is frightening to imagine what Hardy would have said of experimental sciences!

This book is an introduction to fluid mechanics, and is aimed at both physicists and engineers. While the emphasis is on understanding the elementary concepts involved, applications to the various engineering fields have been discussed so as to motivate the reader whose main interest is to solve industrial problems. Needless to say, the reader will not get complete satisfaction even after reading the entire book. It is more likely that he or she will have more questions about the nature of fluid flows than before studying this book. The purpose of the book, however, will be well served if the reader is more curious and interested in fluid flows.

2. *Units of Measurement*

For mechanical systems, the units of all physical variables can be expressed in terms of the units of four basic variables, namely, *length*, *mass*, *time*, and *temperature*. In this book the international system of units (Système international d' unités) and commonly referred to as SI units, will be used most of the time. The basic units

TABLE 1.1 SI Units

Quantity	Name of unit	Symbol	Equivalent
Length	meter	m	
Mass	kilogram	kg	
Time	second	s	
Temperature	kelvin	K	
Frequency	hertz	Hz	s^{-1}
Force	newton	N	$kg\,m\,s^{-2}$
Pressure	pascal	Pa	$N\,m^{-2}$
Energy	joule	J	$N\,m$
Power	watt	W	$J\,s^{-1}$

TABLE 1.2 Common Prefixes

Prefix	Symbol	Multiple
Mega	M	10^6
Kilo	k	10^3
Deci	d	10^{-1}
Centi	c	10^{-2}
Milli	m	10^{-3}
Micro	μ	10^{-6}

of this system are *meter* for length, *kilogram* for mass, *second* for time, and *kelvin* for temperature. The units for other variables can be derived from these basic units. Some of the common variables used in fluid mechanics, and their SI units, are listed in Table 1.1. Some useful conversion factors between different systems of units are listed in Section A1 in Appendix A.

To avoid very large or very small numerical values, prefixes are used to indicate multiples of the units given in Table 1.1. Some of the common prefixes are listed in Table 1.2.

Strict adherence to the SI system is sometimes cumbersome and will be abandoned in favor of common usage where it best serves the purpose of simplifying things. For example, temperatures will be frequently quoted in degrees Celsius (°C), which is related to kelvin (K) by the relation °C = K − 273.15. However, the old English system of units (foot, pound, °F) will not be used, although engineers in the United States are still using it.

3. Solids, Liquids, and Gases

Most substances can be described as existing in two states—solid and fluid. An element of solid has a preferred shape, to which it relaxes when the external forces on it are withdrawn. In contrast, a fluid does not have any preferred shape. Consider a rectangular element of solid ABCD (Figure 1.1a). Under the action of a shear force F the element assumes the shape ABC′D′. If the solid is perfectly elastic, it goes back to its preferred shape ABCD when F is withdrawn. In contrast, a fluid deforms

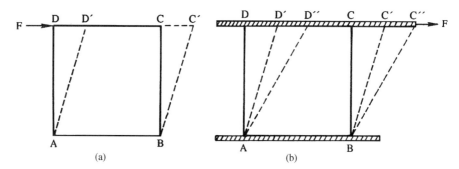

Figure 1.1   Deformation of solid and fluid elements: (a) solid; and (b) fluid.

continuously under the action of a shear force, *however small*. Thus, the element of the fluid ABCD confined between parallel plates (Figure 1.1b) deforms to shapes such as ABC′D′ and ABC″D″ as long as the force F is maintained on the upper plate. Therefore, we say that a fluid flows.

The qualification "however small" in the forementioned description of a fluid is significant. This is because most solids also deform continuously if the shear stress exceeds a certain limiting value, corresponding to the "yield point" of the solid. A solid in such a state is known as "plastic." In fact, the distinction between solids and fluids can be hazy at times. Substances like paints, jelly, pitch, polymer solutions, and biological substances (for example, egg white) simultaneously display the characteristics of both solids and fluids. If we say that an elastic solid has "perfect memory" (because it always relaxes back to its preferred shape) and that an ordinary viscous fluid has zero memory, then substances like egg white can be called *viscoelastic* because they have "partial memory."

Although solids and fluids behave very differently when subjected to shear stresses, they behave similarly under the action of compressive normal stresses. However, whereas a solid can support both tensile and compressive normal stresses, a fluid usually supports only compression (pressure) stresses. (Some liquids can support a small amount of tensile stress, the amount depending on the degree of molecular cohesion.)

Fluids again may be divided into two classes, liquids and gases. A gas always expands and occupies the entire volume of any container. In contrast, the volume of a liquid does not change very much, so that it cannot completely fill a large container; in a gravitational field a free surface forms that separates the liquid from its vapor.

4. *Continuum Hypothesis*

A fluid, or any other substance for that matter, is composed of a large number of molecules in constant motion and undergoing collisions with each other. Matter is therefore discontinuous or discrete at microscopic scales. In principle, it is possible to study the mechanics of a fluid by studying the motion of the molecules themselves, as is done in kinetic theory or statistical mechanics. However, we are generally interested in the *gross behavior* of the fluid, that is, in the *average manifestation* of the molecular motion. For example, forces are exerted on the boundaries of a container due to the constant bombardment of the molecules; the statistical average of this force per unit area is called *pressure*, a macroscopic property. So long as we are not interested in the mechanism of the origin of pressure, we can ignore the molecular motion and think of pressure as simply "force per unit area."

It is thus possible to ignore the discrete molecular structure of matter and replace it by a continuous distribution, called a *continuum*. For the continuum or macroscopic approach to be valid, the size of the flow system (characterized, for example, by the size of the body around which flow is taking place) must be much larger than the mean free path of the molecules. For ordinary cases, however, this is not a great restriction, since the mean free path is usually very small. For example, the mean free path for standard atmospheric air is $\approx 5 \times 10^{-8}$ m. In special situations, however, the mean free path of the molecules can be quite large and the continuum approach breaks down. In the upper altitudes of the atmosphere, for example, the mean free path of the molecules may be of the order of a meter, a kinetic theory approach is necessary for studying the dynamics of these rarefied gases.

5. *Transport Phenomena*

Consider a surface area AB within a mixture of two gases, say nitrogen and oxygen (Figure 1.2), and assume that the concentration C of nitrogen (kilograms of nitrogen per cubic meter of mixture) varies across AB. Random migration of molecules across AB in both directions will result in a *net* flux of nitrogen across AB, from the region of higher C toward the region of lower C. Experiments show that, to a good approximation, the flux of one constituent in a mixture is proportional to its concentration

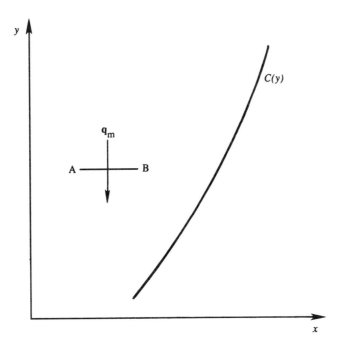

Figure 1.2   Mass flux $\mathbf{q}_m$ due to concentration variation $C(y)$ across AB.

gradient and it is given by

$$\mathbf{q}_m = -k_m \nabla C. \tag{1.1}$$

Here the vector $\mathbf{q}_m$ is the mass flux $(\mathrm{kg\,m^{-2}\,s^{-1}})$ of the constituent, ∇C is the concentration gradient of that constituent, and k_m is a constant of proportionality that depends on the particular pair of constituents in the mixture and the thermodynamic state. For example, k_m for diffusion of nitrogen in a mixture with oxygen is different than k_m for diffusion of nitrogen in a mixture with carbon dioxide. The linear relation (1.1) for mass diffusion is generally known as *Fick's law*. Relations like these are based on empirical evidence, and are called *phenomenological laws*. Statistical mechanics can sometimes be used to derive such laws, but only for simple situations.

The analogous relation for heat transport due to temperature gradient is *Fourier's law* and it is given by

$$\mathbf{q} = -k \nabla T, \tag{1.2}$$

where $\mathbf{q}$ is the heat flux $(\mathrm{J\,m^{-2}\,s^{-1}})$, ∇T is the temperature gradient, and k is the thermal conductivity of the material.

Next, consider the effect of velocity gradient du/dy (Figure 1.3). It is clear that the macroscopic fluid velocity u will tend to become uniform due to the random

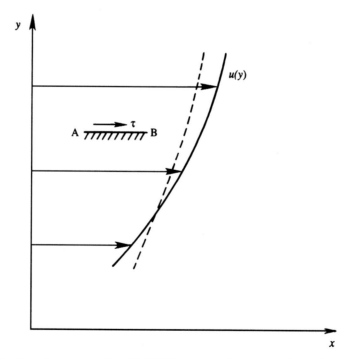

Figure 1.3 Shear stress τ on surface AB. Diffusion tends to decrease velocity gradients, so that the continuous line tends toward the dashed line.

motion of the molecules, because of intermolecular collisions and the consequent exchange of molecular momentum. Imagine two railroad trains traveling on parallel tracks at different speeds, and workers shoveling coal from one train to the other. On the average, the impact of particles of coal going from the slower to the faster train will tend to slow down the faster train, and similarly the coal going from the faster to the slower train will tend to speed up the latter. The net effect is a tendency to equalize the speeds of the two trains. An analogous process takes place in the fluid flow problem of Figure 1.3. The velocity distribution here tends toward the dashed line, which can be described by saying that the x-momentum (determined by its "concentration" u) is being transferred *downward*. Such a momentum flux is equivalent to the existence of a shear stress in the fluid, just as the drag experienced by the two trains results from the momentum exchange through the transfer of coal particles. The fluid above AB tends to push the fluid underneath forward, whereas the fluid below AB tends to drag the upper fluid backward. Experiments show that the magnitude of the shear stress τ along a surface such as AB is, to a good approximation, related to the velocity gradient by the linear relation

$$\tau = \mu \frac{du}{dy}, \tag{1.3}$$

which is called *Newton's law* of friction. Here the constant of proportionality μ (whose unit is $\mathrm{kg\,m^{-1}\,s^{-1}}$) is known as the *dynamic viscosity*, which is a strong function of temperature T. For ideal gases the random thermal speed is roughly proportional to $\sqrt{T}$; the momentum transport, and consequently μ, also vary approximately as $\sqrt{T}$. For liquids, on the other hand, the shear stress is caused more by the intermolecular cohesive forces than by the thermal motion of the molecules. These cohesive forces, and consequently μ for a liquid, decrease with temperature.

Although the shear stress is proportional to μ, we will see in Chapter 4 that the tendency of a fluid to *diffuse* velocity gradients is determined by the quantity

$$\nu \equiv \frac{\mu}{\rho}, \tag{1.4}$$

where ρ is the density $(\mathrm{kg/m^3})$ of the fluid. The unit of ν is $\mathrm{m^2/s}$, which does not involve the unit of mass. Consequently, ν is frequently called the *kinematic viscosity*.

Two points should be noticed in the linear transport laws equations (1.1), (1.2), and (1.3). First, only the *first* derivative of some generalized "concentration" C appears on the right-hand side. This is because the transport is carried out by molecular processes, in which the length scales (say, the mean free path) are too small to feel the curvature of the C-profile. Second, the nonlinear terms involving higher powers of ∇C do not appear. Although this is only expected for small magnitudes of ∇C, experiments show that such linear relations are very accurate for most practical values of ∇C.

It should be noted here that we have written the transport law for momentum far less precisely than the transport laws for mass and heat. This is because we have not developed the *language* to write this law with precision. The transported quantities in (1.1) and (1.2) are scalars (namely, mass and heat, respectively), and the corresponding fluxes are vectors. In contrast, the transported quantity in (1.3) is itself a vector, and the corresponding flux is a "tensor." The precise form of (1.3) will be presented in Chapter 4, after the concept of tensors is explained in Chapter 2. For now, we have avoided complications by writing the transport law for only one component of momentum, using scalar notation.

6. *Surface Tension*

A density discontinuity exists whenever two immiscible fluids are in contact, for example at the interface between water and air. The interface in this case is found to behave as if it were under tension. Such an interface behaves like a stretched membrane, such as the surface of a balloon or of a soap bubble. This is why drops of liquid in air or gas bubbles in water tend to be spherical in shape. The origin of such tension in an interface is due to the intermolecular attractive forces. Imagine a liquid drop surrounded by a gas. Near the interface, all the liquid molecules are trying to pull the molecules on the interface *inward*. The net effect of these attractive forces is for the interface to *contract*. The magnitude of the tensile force per unit length of a line on the interface is called *surface tension* σ, which has the unit N/m. The value of σ depends on the pair of fluids in contact and the temperature.

An important consequence of surface tension is that it gives rise to a pressure jump across the interface whenever it is curved. Consider a spherical interface having a radius of curvature R (Figure 1.4a). If p_i and p_o are the pressures on the two sides of the interface, then a force balance gives

$$\sigma(2\pi R) = (p_i - p_o)\pi R^2,$$

from which the pressure jump is found to be

$$p_i - p_o = \frac{2\sigma}{R}, \tag{1.5}$$

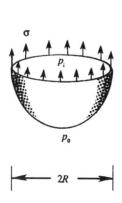
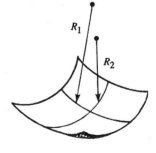

Figure 1.4   (a) Section of a spherical droplet, showing surface tension forces. (b) An interface with radii of curvatures R_1 and R_2 along two orthogonal directions.

showing that the pressure on the concave side is higher. The pressure jump, however, is small unless R is quite small.

Equation (1.5) holds only if the surface is spherical. The curvature of a general surface can be specified by the radii of curvature along two orthogonal directions, say, R_1 and R_2 (Figure 1.4b). A similar analysis shows that the pressure jump across the interface is given by

$$p_i - p_o = \sigma \left(\frac{1}{R_1} + \frac{1}{R_2} \right),$$

which agrees with equation (1.5) if $R_1 = R_2$.

It is well known that the free surface of a liquid in a narrow tube rises above the surrounding level due to the influence of surface tension. This is demonstrated in Example 1.1. Narrow tubes are called *capillary tubes* (from Latin *capillus*, meaning "hair"). Because of this phenomenon the whole group of phenomena that arise from surface tension effects is called *capillarity*. A more complete discussion of surface tension is presented at the end of Chapter 4 (Section 19) as part of an expanded section on boundary conditions.

7. Fluid Statics

The magnitude of the force per unit area in a static fluid is called the *pressure*. (More care is needed to define the pressure in a moving medium, and this will be done in Chapter 4.) Sometimes the ordinary pressure is called the *absolute pressure*, in order to distinguish it from the *gauge pressure*, which is defined as the absolute pressure minus the atmospheric pressure:

$$p_{\text{gauge}} = p - p_{\text{atm}}.$$

The value of the atmospheric pressure is

$$p_{\text{atm}} = 101.3 \, \text{kPa} = 1.013 \, \text{bar},$$

where $1 \, \text{bar} = 10^5 \, \text{Pa}$. The atmospheric pressure is therefore approximately 1 bar.

In a fluid at rest, the tangential viscous stresses are absent and the only force between adjacent surfaces is normal to the surface. We shall now demonstrate that in such a case the surface force per unit area ("pressure") is equal in all directions. Consider a small triangular volume of fluid (Figure 1.5) of unit thickness normal to the paper, and let p_1, p_2, and p_3 be the pressures on the three faces. The z-axis is taken vertically upward. The only forces acting on the element are the pressure forces normal to the faces and the weight of the element. Because there is no acceleration of the element in the x direction, a balance of forces in that direction gives

$$(p_1 \, ds) \sin \theta - p_3 \, dz = 0.$$

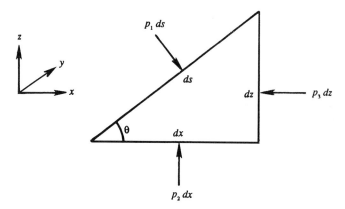

Figure 1.5   Demonstration that $p_1 = p_2 = p_3$ in a static fluid.

Because $dz = ds \sin \theta$, the foregoing gives $p_1 = p_3$. A balance of forces in the vertical direction gives

$$-(p_1 \, ds) \cos \theta + p_2 \, dx - \tfrac{1}{2} \rho g \, dx \, dz = 0.$$

As $ds \cos \theta = dx$, this gives

$$p_2 - p_1 - \tfrac{1}{2} \rho g \, dz = 0.$$

As the triangular element is shrunk to a point, the gravity force term drops out, giving $p_1 = p_2$. Thus, at a point in a static fluid, we have

$$p_1 = p_2 = p_3, \tag{1.6}$$

so that the force per unit area is independent of the angular orientation of the surface. The pressure is therefore a scalar quantity.

We now proceed to determine the *spatial distribution* of pressure in a static fluid. Consider an infinitesimal cube of sides dx, dy, and dz, with the z-axis vertically upward (Figure 1.6). A balance of forces in the x direction shows that the pressures on the two sides perpendicular to the x-axis are equal. A similar result holds in the y direction, so that

$$\frac{\partial p}{\partial x} = \frac{\partial p}{\partial y} = 0. \tag{1.7}$$

This fact is expressed by *Pascal's law*, which states that all points in a resting fluid medium (and connected by the *same* fluid) are at the same pressure if they are at the same depth. For example, the pressure at points F and G in Figure 1.7 are the same.

A vertical equilibrium of the element in Figure 1.6 requires that

$$p \, dx \, dy - (p + dp) \, dx \, dy - \rho g \, dx \, dy \, dz = 0,$$

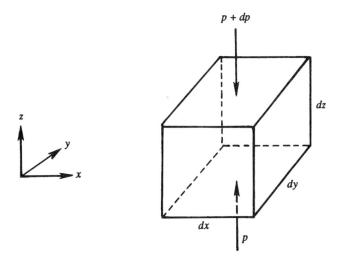

Figure 1.6   Fluid element at rest.

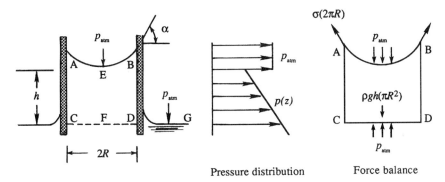

Pressure distribution   Force balance

Figure 1.7   Rise of a liquid in a narrow tube (Example 1.1).

which simplifies to

$$\frac{dp}{dz} = -\rho g. \tag{1.8}$$

This shows that the pressure in a static fluid decreases with height. For a fluid of uniform density, equation (1.8) can be integrated to give

$$p = p_0 - \rho g z, \tag{1.9}$$

where p_0 is the pressure at $z = 0$. Equation (1.9) is the well-known result of *hydrostatics*, and shows that the pressure in a liquid decreases *linearly* with height. It implies that the pressure rise at a depth h below the free surface of a liquid is equal to $\rho g h$, which is the weight of a column of liquid of height h and unit cross section.

Example 1.1. With reference to Figure 1.7, show that the rise of a liquid in a narrow tube of radius R is given by

$$h = \frac{2\sigma \sin \alpha}{\rho g R},$$

where σ is the surface tension and α is the "contact" angle.

Solution. Since the free surface is concave upward and exposed to the atmosphere, the pressure just below the interface at point E is below atmospheric. The pressure then increases linearly along EF. At F the pressure again equals the atmospheric pressure, since F is at the same level as G where the pressure is atmospheric. The pressure forces on faces AB and CD therefore balance each other. Vertical equilibrium of the element ABCD then requires that the weight of the element balances the vertical component of the surface tension force, so that

$$\sigma (2\pi R) \sin \alpha = \rho g h (\pi R^2),$$

which gives the required result.

8. Classical Thermodynamics

Classical thermodynamics is the study of equilibrium states of matter, in which the properties are assumed uniform in space and time. The reader is assumed to be familiar with the basic concepts of this subject. Here we give a review of the main ideas and the most commonly used relations in this book.

A thermodynamic *system* is a quantity of matter separated from the surroundings by a flexible boundary through which the system exchanges heat and work, but no mass. A system in the equilibrium state is free of currents, such as those generated by stirring a fluid or by sudden heating. After a change has taken place, the currents die out and the system returns to equilibrium conditions, when the properties of the system (such as pressure and temperature) can once again be defined.

This definition, however, is not possible in fluid flows, and the question arises as to whether the relations derived in classical thermodynamics are applicable to fluids in constant motion. Experiments show that the results of classical thermodynamics do hold in most fluid flows if the changes along the motion are slow compared to a *relaxation time*. The relaxation time is defined as the time taken by the material to adjust to a new state, and the material undergoes this adjustment through molecular collisions. The relaxation time is very small under ordinary conditions, since only a few molecular collisions are needed for the adjustment. The relations of classical thermodynamics are therefore applicable to most fluid flows.

The basic laws of classical thermodynamics are empirical, and cannot be proved. Another way of viewing this is to say that these principles are so basic that they cannot be derived from anything *more* basic. They essentially establish certain basic *definitions*, upon which the subject is built. The first law of thermodynamics can be regarded as a principle that defines the *internal energy* of a system, and the second law can be regarded as the principle that defines the *entropy* of a system.

First Law of Thermodynamics

The first law of thermodynamics states that the energy of a system is conserved. It states that

$$Q + W = \Delta e, \qquad (1.10)$$

where Q is the heat added to the system, W is the work done on the system, and Δe is the increase of *internal energy* of the system. All quantities in equation (1.10) may be regarded as those referring to unit mass of the system. (In thermodynamics texts it is customary to denote quantities per unit mass by lowercase letters, and those for the entire system by uppercase letters. This will not be done here.) The internal energy (also called "thermal energy") is a manifestation of the random molecular motion of the constituents. In fluid flows, the kinetic energy of the macroscopic motion has to be included in the term e in equation (1.10) in order that the principle of conservation of energy is satisfied. For developing the relations of classical thermodynamics, however, we shall only include the "thermal energy" in the term e.

It is important to realize the difference between heat and internal energy. Heat and work are forms of *energy in transition*, which appear at the *boundary* of the system and are *not contained* within the matter. In contrast, the internal energy resides within the matter. If two equilibrium states 1 and 2 of a system are known, then Q and W depend on the *process* or *path* followed by the system in going from state 1 to state 2. The change $\Delta e = e_2 - e_1$, in contrast, does not depend on the path. In short, e is a thermodynamic property and is a function of the thermodynamic state of the system. Thermodynamic properties are called *state functions*, in contrast to heat and work, which are *path functions*.

Frictionless quasi-static processes, carried out at an extremely slow rate so that the system is at all times in equilibrium with the surroundings, are called *reversible processes*. The most common type of reversible work in fluid flows is by the expansion or contraction of the boundaries of the fluid element. Let $v = 1/\rho$ be the *specific volume*, that is, the volume per unit mass. Then the work done by the body per unit mass in an infinitesimal reversible process is $-p\,dv$, where dv is the increase of v. The first law (equation (1.10)) for a reversible process then becomes

$$de = dQ - p\,dv, \qquad (1.11)$$

provided that Q is also reversible.

Note that irreversible forms of work, such as that done by turning a paddle wheel, are excluded from equation (1.11).

Equations of State

In simple systems composed of a single component only, the specification of two independent properties completely determines the state of the system. We can write relations such as

$$
\begin{aligned}
p &= p(v, T) \quad \text{(thermal equation of state)}, \\
e &= e(p, T) \quad \text{(caloric equation of state)}.
\end{aligned}
\qquad (1.12)
$$

Such relations are called *equations of state*. For more complicated systems composed of more than one component, the specification of two properties is not enough to completely determine the state. For example, for sea water containing dissolved salt, the density is a function of the three variables, salinity, temperature, and pressure.

Specific Heats

Before we define the specific heats of a substance, we define a thermodynamic property called *enthalpy* as

$$h \equiv e + pv. \tag{1.13}$$

This property will be quite useful in our study of compressible fluid flows.

For single-component systems, the specific heats at constant pressure and constant volume are defined as

$$C_p \equiv \left(\frac{\partial h}{\partial T}\right)_p, \tag{1.14}$$

$$C_v \equiv \left(\frac{\partial e}{\partial T}\right)_v. \tag{1.15}$$

Here, equation (1.14) means that we regard h as a function of p and T, and find the partial derivative of h with respect to T, keeping p constant. Equation (1.15) has an analogous interpretation. It is important to note that the specific heats as defined are thermodynamic properties, because they are defined in terms of other properties of the system. That is, we can determine C_p and C_v when two other properties of the system (say, p and T) are given.

For certain processes common in fluid flows, the heat exchange can be related to the specific heats. Consider a reversible process in which the work done is given by $p\,dv$, so that the first law of thermodynamics has the form of equation (1.11). Dividing by the change of temperature, it follows that the heat transferred per unit mass per unit temperature change in a constant volume process is

$$\left(\frac{dQ}{dT}\right)_v = \left(\frac{\partial e}{\partial T}\right)_v = C_v.$$

This shows that $C_v\,dT$ represents the heat transfer per unit mass in a reversible constant volume process, in which the only type of work done is of the pdv type. It is misleading to define $C_v = (dQ/dT)_v$ without any restrictions imposed, as the temperature of a constant-volume system can increase without heat transfer, say, by turning a paddle wheel.

In a similar manner, the heat transferred at constant pressure during a reversible process is given by

$$\left(\frac{dQ}{dT}\right)_p = \left(\frac{\partial h}{\partial T}\right)_p = C_p.$$

Second Law of Thermodynamics

The second law of thermodynamics imposes restriction on the direction in which real processes can proceed. Its implications are discussed in Chapter 4. Some consequences of this law are the following:

(i) There must exist a thermodynamic property S, known as *entropy*, whose change between states 1 and 2 is given by

$$S_2 - S_1 = \int_1^2 \frac{dQ_{rev}}{T}, \tag{1.16}$$

where the integral is taken along any reversible process between the two states.

(ii) For an *arbitrary* process between 1 and 2, the entropy change is

$$S_2 - S_1 \geqslant \int_1^2 \frac{dQ}{T} \quad \text{(Clausius-Duhem)},$$

which states that the entropy of an isolated system ($dQ = 0$) can only increase. Such increases are caused by frictional and mixing phenomena.

(iii) Molecular transport coefficients such as viscosity μ and thermal conductivity k must be positive. Otherwise, spontaneous "unmixing" would occur and lead to a decrease of entropy of an isolated system.

$T dS$ Relations

Two common relations are useful in calculating the entropy changes during a process. For a reversible process, the entropy change is given by

$$T dS = dQ. \tag{1.17}$$

On substituting into (1.11), we obtain

$$\begin{aligned} T dS &= de + p \, dv \\ T dS &= dh - v \, dp \end{aligned} \quad \text{(Gibbs)}, \tag{1.18}$$

where the second form is obtained by using $dh = d(e + pv) = de + p \, dv + v \, dp$. It is interesting that the "$T dS$ relations" in equations (1.18) are also valid for irreversible (frictional) processes, although the relations (1.11) and (1.17), from which equations (1.18) is derived, are true for reversible processes only. This is because equations (1.18) are relations between thermodynamic *state functions* alone and are therefore true for *any* process. The association of $T dS$ with heat and $-p \, dv$ with work does not hold for irreversible processes. Consider paddle wheel work done at constant volume so that $de = T dS$ is the element of work done.

Speed of Sound

In a compressible medium, infinitesimal changes in density or pressure propagate through the medium at a finite speed. In Chapter 16, we shall prove that the square of this speed is given by

$$c^2 = \left(\frac{\partial p}{\partial \rho}\right)_s, \qquad (1.19)$$

where the subscript "s" signifies that the derivative is taken at constant entropy. As sound is composed of small density perturbations, it also propagates at speed c. For incompressible fluids ρ is independent of p, and therefore $c = \infty$.

Thermal Expansion Coefficient

In a system whose density is a function of temperature, we define the thermal expansion coefficient

$$\alpha \equiv -\frac{1}{\rho}\left(\frac{\partial \rho}{\partial T}\right)_p, \qquad (1.20)$$

where the subscript "p" signifies that the partial derivative is taken at constant pressure. The expansion coefficient will appear frequently in our studies of nonisothermal systems.

9. *Perfect Gas*

A relation defining one state function of a gas in terms of two others is called an *equation of state*. A perfect gas is defined as one that obeys the thermal equation of state

$$\boxed{p = \rho R T,} \qquad (1.21)$$

where p is the pressure, ρ is the density, T is the absolute temperature, and R is the *gas constant*. The value of the gas constant depends on the molecular mass m of the gas according to

$$R = \frac{R_u}{m}, \qquad (1.22)$$

where

$$R_u = 8314.36 \, \mathrm{J \, kmol^{-1} \, K^{-1}}$$

is the *universal gas constant*. For example, the molecular mass for dry air is $m = 28.966 \, \mathrm{kg/kmol}$, for which equation (1.22) gives

$$R = 287 \, \mathrm{J \, kg^{-1} \, K^{-1}} \quad \text{for dry air.}$$

Equation (1.21) can be derived from the kinetic theory of gases if the attractive forces between the molecules are negligible. At ordinary temperatures and pressures most gases can be taken as perfect.

The gas constant is related to the specific heats of the gas through the relation

$$R = C_p - C_v, \qquad (1.23)$$

where C_p is the specific heat at constant pressure and C_v is the specific heat at constant volume. In general, C_p and C_v of a gas, including those of a perfect gas, increase with temperature. The ratio of specific heats of a gas

$$\gamma \equiv \frac{C_p}{C_v}, \qquad (1.24)$$

is an important quantity. For air at ordinary temperatures, $\gamma = 1.4$ and $C_p = 1005 \, \text{J kg}^{-1} \, \text{K}^{-1}$.

It can be shown that assertion (1.21) is equivalent to

$$e = e(T)$$
$$h = h(T)$$

and conversely, so that the internal energy and enthalpy of a perfect gas can only be functions of temperature alone. See Exercise 7.

A process is called *adiabatic* if it takes place without the addition of heat. A process is called *isentropic* if it is adiabatic and frictionless, for then the entropy of the fluid does not change. From equation (1.18) it is easy to show that the isentropic flow of a perfect gas with constant specific heats obeys the relation

$$\frac{p}{\rho^\gamma} = \text{const.} \qquad \text{(isentropic)} \qquad (1.25)$$

Using the equation of state $p = \rho R T$, it follows that the temperature and density change during an isentropic process from state 1 to state 2 according to

$$\frac{T_1}{T_2} = \left(\frac{p_1}{p_2}\right)^{(\gamma-1)/\gamma} \quad \text{and} \quad \frac{\rho_1}{\rho_2} = \left(\frac{p_1}{p_2}\right)^{1/\gamma} \quad \text{(isentropic)} \qquad (1.26)$$

See Exercise 8. For a perfect gas, simple expressions can be found for several useful thermodynamic properties such as the speed of sound and the thermal expansion coefficient. Using the equation of state $p = \rho R T$, the speed of sound (1.19) becomes

$$c = \sqrt{\gamma R T}, \qquad (1.27)$$

where equation (1.25) has been used. This shows that the speed of sound increases as the square root of the temperature. Likewise, the use of $p = \rho R T$ shows that the thermal expansion coefficient (1.20) is

$$\alpha = \frac{1}{T},$$

(1.28)

10. *Static Equilibrium of a Compressible Medium*

In an incompressible fluid in which the density is not a function of pressure, there is a simple criterion for determining the stability of the medium in the static state. The criterion is that the medium is stable if the density decreases upward, for then a particle displaced upward would find itself at a level where the density of the surrounding fluid is lower, and so the particle would be forced back toward its original level. In the opposite case in which the density increases upward, a displaced particle would continue to move farther away from its original position, resulting in instability. The medium is in neutral equilibrium if the density is uniform.

For a compressible medium the preceding criterion for determining the stability does not hold. We shall now show that in this case it is not the density but the *entropy* that is constant with height in the neutral state. For simplicity we shall consider the case of an atmosphere that obeys the equation of state for a perfect gas. The pressure decreases with height according to

$$\frac{dp}{dz} = -\rho g.$$

A particle displaced upward would expand adiabatically because of the decrease of the pressure with height. Its original density ρ_0 and original temperature T_0 would therefore decrease to ρ and T according to the isentropic relations

$$\frac{T}{T_0} = \left(\frac{p}{p_0}\right)^{(\gamma-1)/\gamma} \quad \text{and} \quad \frac{\rho}{\rho_0} = \left(\frac{p}{p_0}\right)^{1/\gamma},$$

(1.29)

where $\gamma = C_p/C_v$, and the subscript 0 denotes the original state at some height z_0, where $p_0 > p$ (Figure 1.8). It is clear that the displaced particle would be forced back toward the original level if the new density is larger than that of the surrounding air at the new level. Now if the properties of the *surrounding* air also happen to vary with height in such a way that the entropy is uniform with height, then the displaced particle would constantly find itself in a region where the density is the same as that of itself. Therefore, *a neutral atmosphere is one in which p, ρ, and T decrease in such a way that the entropy is constant with height.* A neutrally stable atmosphere is therefore also called an isentropic or *adiabatic atmosphere*. It follows that a statically stable atmosphere is one in which the density decreases with height *faster* than in an adiabatic atmosphere.

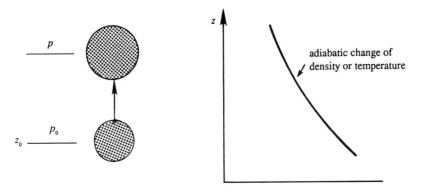

Figure 1.8 Adiabatic expansion of a fluid particle displaced upward in a compressible medium.

It is easy to determine the rate of decrease of temperature in an adiabatic atmosphere. Taking the logarithm of equation (1.29), we obtain

$$\ln T_a - \ln T_0 = \frac{\gamma - 1}{\gamma}[\ln p_a - \ln p_0],$$

where we are using the subscript "a" to denote an adiabatic atmosphere. A differentiation with respect to z gives

$$\frac{1}{T_a}\frac{dT_a}{dz} = \frac{\gamma - 1}{\gamma}\frac{1}{p_a}\frac{dp_a}{dz}.$$

Using the perfect gas law $p = \rho RT$, $C_p - C_v = R$, and the hydrostatic rule $dp/dz = -\rho g$, we obtain

$$\frac{dT_a}{dz} \equiv \Gamma_a = -\frac{g}{C_p} \tag{1.30}$$

where $\Gamma \equiv dT/dz$ is the temperature gradient; $\Gamma_a = -g/C_p$ is called the *adiabatic temperature gradient* and is the largest rate at which the temperature can decrease with height without causing instability. For air at normal temperatures and pressures, the temperature of a neutral atmosphere decreases with height at the rate of $g/C_p \simeq 10\,^\circ$C/km. Meteorologists call vertical temperature gradients the "lapse rate," so that in their terminology the adiabatic lapse rate is $10\,^\circ$C/km.

Figure 1.9a shows a typical distribution of temperature in the atmosphere. The lower part has been drawn with a slope nearly equal to the adiabatic temperature gradient because the mixing processes near the ground tend to form a neutral atmosphere, with its entropy "well mixed" (that is, uniform) with height. Observations show that the neutral atmosphere is "capped" by a layer in which the temperature increases with height, signifying a very stable situation. Meteorologists call this an *inversion*, because the temperature gradient changes sign here. Much of the atmospheric turbulence and mixing processes cannot penetrate this very stable layer. Above this inversion layer the

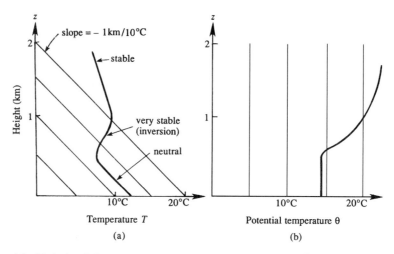

Figure 1.9 Vertical variation of the (a) actual and (b) potential temperature in the atmosphere. Thin straight lines represent temperatures for a neutral atmosphere.

temperature decreases again, but less rapidly than near the ground, which corresponds to stability. It is clear that an isothermal atmosphere (a vertical line in Figure 1.9a) is quite stable.

Potential Temperature and Density

The foregoing discussion of static stability of a compressible atmosphere can be expressed in terms of the concept of *potential temperature*, which is generally denoted by θ. Suppose the pressure and temperature of a fluid particle at a certain height are p and T. Now if we take the particle *adiabatically* to a standard pressure p_s (say, the sea level pressure, nearly equal to 100 kPa), then the temperature θ attained by the particle is called its *potential temperature*. Using equation (1.26), it follows that the actual temperature T and the potential temperature θ are related by

$$T = \theta \left(\frac{p}{p_s} \right)^{(\gamma-1)/\gamma}. \tag{1.31}$$

Taking the logarithm and differentiating, we obtain

$$\frac{1}{T}\frac{dT}{dz} = \frac{1}{\theta}\frac{d\theta}{dz} + \frac{\gamma-1}{\gamma}\frac{1}{p}\frac{dp}{dz}.$$

Substituting $dp/dz = -\rho g$ and $p = \rho R T$, we obtain

$$\frac{T}{\theta}\frac{d\theta}{dz} = \frac{dT}{dz} + \frac{g}{C_p} = \frac{d}{dz}(T - T_a) = \Gamma - \Gamma_a. \tag{1.32}$$

Now if the temperature decreases at a rate $\Gamma = \Gamma_a$, then the potential temperature θ (and therefore the entropy) is uniform with height. It follows that the stability of the atmosphere is determined according to

$$\frac{d\theta}{dz} > 0 \quad \text{(stable)},$$

$$\frac{d\theta}{dz} = 0 \quad \text{(neutral)}, \tag{1.33}$$

$$\frac{d\theta}{dz} < 0 \quad \text{(unstable)}.$$

This is shown in Figure 1.9b. It is the gradient of *potential* temperature that determines the stability of a column of gas, not the gradient of the actual temperature. However, the difference between the two is negligible for laboratory-scale phenomena. For example, over a height of 10 cm the compressibility effects result in a decrease of temperature in the air by only $10 \, \text{cm} \times (10 \, ^\circ\text{C/km}) = 10^{-3} \, ^\circ\text{C}$.

Instead of using the potential temperature, one can use the concept of *potential density* ρ_θ, defined as the density attained by a fluid particle if taken isentropically to a standard pressure p_s. Using equation (1.26), the actual and potential densities are related by

$$\rho = \rho_\theta \left(\frac{p}{p_s} \right)^{1/\gamma}. \tag{1.34}$$

Multiplying equations (1.31) and (1.34), and using $p = \rho RT$, we obtain $\theta \rho_\theta = p_s / R = \text{const}$. Taking the logarithm and differentiating, we obtain

$$-\frac{1}{\rho_\theta} \frac{d\rho_\theta}{dz} = \frac{1}{\theta} \frac{d\theta}{dz}. \tag{1.35}$$

The medium is stable, neutral, or unstable depending upon whether $d\rho_\theta/dz$ is negative, zero, or positive, respectively.

Compressibility effects are also important in the deep ocean. In the ocean the density depends not only on the temperature and pressure, but also on the *salinity*, defined as kilograms of salt per kilogram of water. (The salinity of sea water is $\approx 3\%$.) Here, one defines the potential density as the density attained if a particle is taken to a reference pressure isentropically *and* at constant salinity. The potential density thus defined must decrease with height in stable conditions. Oceanographers automatically account for the compressibility of sea water by converting their density measurements at any depth to the sea level pressure, which serves as the reference pressure.

From (1.32), the temperature of a dry neutrally stable atmosphere decreases upward at a rate $dT_a/dz = -g/C_p$ due to the decrease of pressure with height and the compressibility of the medium. Static stability of the atmosphere is determined by whether the actual temperature gradient dT/dz is slower or faster than dT_a/dz. To determine the static stability of the *ocean*, it is more convenient to formulate the criterion in terms of density. The plan is to compare the density gradient of the actual static state with that of a neutrally stable reference state (denoted here by the subscript "a").

The pressure of the reference state decreases vertically as

$$\frac{dp_a}{dz} = -\rho_a g. \tag{1.36}$$

In the ocean the speed of sound c is defined by $c^2 = \partial p / \partial \rho$, where the partial derivative is taken at constant values of entropy and salinity. In the reference state these variables are uniform, so that $dp_a = c^2 d\rho_a$. Therefore, the density in the neutrally stable state varies due to the compressibility effect at a rate

$$\frac{d\rho_a}{dz} = \frac{1}{c^2}\frac{dp_a}{dz} = \frac{1}{c^2}(-\rho_a g) = -\frac{\rho g}{c^2}, \tag{1.37}$$

where the subscript "a" on ρ has been dropped because ρ_a is nearly equal to the actual density ρ.

The static stability of the ocean is determined by the sign of the *potential density gradient*

$$\frac{d\rho_{pot}}{dz} = \frac{d\rho}{dz} - \frac{d\rho_a}{dz} = \frac{d\rho}{dz} + \frac{\rho g}{c^2}. \tag{1.38}$$

The medium is statically stable if the potential density gradient is negative, and so on. For a perfect gas, it can be shown that equations (1.30) and (1.38) are equivalent.

Scale Height of the Atmosphere

Expressions for pressure distribution and "thickness" of the atmosphere can be obtained by assuming that they are isothermal. This is a good assumption in the lower 70 km of the atmosphere, where the absolute temperature remains within 15% of 250 K. The hydrostatic distribution is

$$\frac{dp}{dz} = -\rho g = -\frac{pg}{RT}.$$

Integration gives

$$p = p_0\, e^{-gz/RT},$$

where p_0 is the pressure at $z = 0$. The pressure therefore falls to e^{-1} of its surface value in a height RT/g. The quantity RT/g, called the *scale height*, is a good measure of the thickness of the atmosphere. For an average atmospheric temperature of $T = 250$ K, the scale height is $RT/g = 7.3$ km.

Exercises

1. Estimate the height to which water at $20\,°C$ will rise in a capillary glass tube 3 mm in diameter exposed to the atmosphere. For water in contact with glass the wetting angle is nearly $90°$. At $20\,°C$ and water-air combination, $\sigma = 0.073$ N/m. (*Answer*: $h = 0.99$ cm.)

2. Consider the viscous flow in a channel of width $2b$. The channel is aligned in the x direction, and the velocity at a distance y from the centerline is given by the parabolic distribution

$$u(y) = U_0 \left[1 - \frac{y^2}{b^2} \right].$$

In terms of the viscosity μ, calculate the shear stress at a distance of $y = b/2$.

3. Figure 1.10 shows a *manometer*, which is a U-shaped tube containing mercury of density ρ_m. Manometers are used as pressure measuring devices. If the fluid in the tank A has a pressure p and density ρ, then show that the gauge pressure in the tank is

$$p - p_{atm} = \rho_m g h - \rho g a.$$

Note that the last term on the right-hand side is negligible if $\rho \ll \rho_m$. (*Hint*: Equate the pressures at X and Y.)

4. A cylinder contains 2 kg of air at 50 °C and a pressure of 3 bars. The air is compressed until its pressure rises to 8 bars. What is the initial volume? Find the final volume for both isothermal compression and isentropic compression.

5. Assume that the temperature of the atmosphere varies with height z as

$$T = T_0 + Kz.$$

Show that the pressure varies with height as

$$p = p_0 \left[\frac{T_0}{T_0 + Kz} \right]^{g/KR},$$

where g is gravity and R is the gas constant.

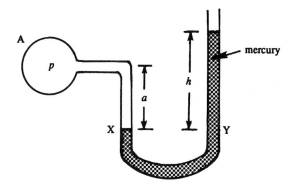

Figure 1.10   A mercury manometer.

6. Suppose the atmospheric temperature varies according to

$$T = 15 - 0.001z$$

where T is in degrees Celsius and height z is in meters. Is this atmosphere stable?

7. Prove that if $e(T, v) = e(T)$ only and if $h(T, p) = h(T)$ only, then the (thermal) equation of state is equation (1.21) or $pv = kT$.

8. For a reversible adiabatic process in a perfect gas with constant specific heats, derive equations (1.25) and (1.26) starting from equation (1.18).

9. Consider a heat insulated enclosure that is separated into two compartments of volumes V_1 and V_2, containing perfect gases with pressures and temperatures of p_1, p_2, and T_1, T_2, respectively. The compartments are separated by an impermeable membrane that conducts heat (but not mass). Calculate the final steady-state temperature assuming each of the gases has constant specific heats.

10. Consider the initial state of an enclosure with two compartments as described in Exercise 9. At $t = 0$, the membrane is broken and the gases are mixed. Calculate the final temperature.

11. A heavy piston of weight W is dropped onto a thermally insulated cylinder of cross-sectional area A containing a perfect gas of constant specific heats, and initially having the external pressure p_1, temperature T_1, and volume V_1. After some oscillations, the piston reaches an equilibrium position L meters below the equilibrium position of a weightless piston. Find L. Is there an entropy increase?

Literature Cited

Taylor, G. I. (1974). The interaction between experiment and theory in fluid mechanics. *Annual Review of Fluid Mechanics* **6**: 1–16.
Von Karman, T. (1954). *Aerodynamics*, New York: McGraw-Hill.

Supplemental Reading

Batchelor, G. K. (1967). "*An Introduction to Fluid Dynamics*," London: Cambridge University Press, (A detailed discussion of classical thermodynamics, kinetic theory of gases, surface tension effects, and transport phenomena is given.)
Hatsopoulos, G. N. and J. H. Keenan (1981). *Principles of General Thermodynamics.* Melbourne, FL: Krieger Publishing Co. (This is a good text on thermodynamics.)
Prandtl, L. and O. G. Tietjens (1934). *Fundamentals of Hydro- and Aeromechanics*, New York: Dover Publications. (A clear and simple discussion of potential and adiabatic temperature gradients is given.)

Cartesian Tensors

1. Scalars and Vectors

In fluid mechanics we need to deal with quantities of various complexities. Some of these are defined by only one component and are called scalars, some others are defined by three components and are called vectors, and certain other variables called tensors need as many as nine components for a complete description. We shall assume that the reader is familiar with a certain amount of algebra and calculus of vectors. The concept and manipulation of tensors is the subject of this chapter.

A *scalar* is any quantity that is completely specified by a magnitude only, along with its unit. It is independent of the coordinate system. Examples of scalars are temperature and density of the fluid. A *vector* is any quantity that has a magnitude and a direction, and can be completely described by its components along three specified coordinate directions. A vector is usually denoted by a boldface symbol, for example, $\mathbf{x}$ for position and $\mathbf{u}$ for velocity. We can take a Cartesian coordinate system x_1, x_2, x_3, with unit vectors $\mathbf{a}^1$, $\mathbf{a}^2$, and $\mathbf{a}^3$ in the three mutually perpendicular directions (Figure 2.1). (In texts on vector analysis, the unit vectors are usually denoted by $\mathbf{i}$, $\mathbf{j}$, and $\mathbf{k}$. We cannot use this simple notation here because we shall use ijk to

©2010 Elsevier Inc. All rights reserved.
DOI: 10.1016/B978-0-12-381399-2.50002-2

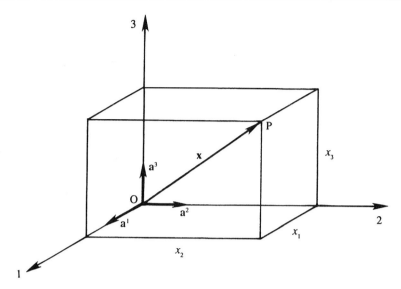

Figure 2.1   Position vector OP and its three Cartesian components (x_1, x_2, x_3). The three unit vectors are $\mathbf{a}^1$, $\mathbf{a}^2$, and $\mathbf{a}^3$.

denote *components* of a vector.) Then the position vector is written as

$$\mathbf{x} = \mathbf{a}^1 x_1 + \mathbf{a}^2 x_2 + \mathbf{a}^3 x_3,$$

where (x_1, x_2, x_3) are the components of $\mathbf{x}$ along the coordinate directions. (The superscripts on the unit vectors $\mathbf{a}$ do *not* denote the components of a vector; the $\mathbf{a}$'s are vectors themselves.) Instead of writing all three components explicitly, we can indicate the three Cartesian components of a vector by an index that takes all possible values of 1, 2, and 3. For example, the components of the position vector can be denoted by x_i, where i takes all of its possible values, namely, 1, 2, and 3. To obey the laws of algebra that we shall present, the components of a vector should be written as a column. For example,

$$\mathbf{x} = \begin{bmatrix} x_1 \\ x_2 \\ x_3 \end{bmatrix}.$$

In matrix algebra, one defines the *transpose* as the matrix obtained by interchanging rows and columns. For example, the transpose of a column matrix $\mathbf{x}$ is the row matrix

$$\mathbf{x}^{\mathrm{T}} = [x_1 \quad x_2 \quad x_3].$$

2. *Rotation of Axes: Formal Definition of a Vector*

A vector can be formally defined as any quantity whose components change similarly to the components of a position vector under the rotation of the coordinate system.

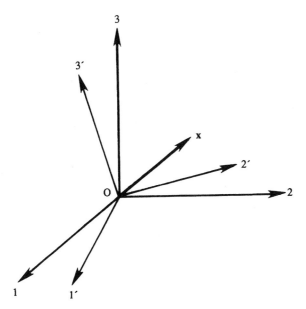

Figure 2.2   Rotation of coordinate system O 1 2 3 to O 1′ 2′ 3′.

Let $x_1\, x_2\, x_3$ be the original axes, and $x'_1\, x'_2\, x'_3$ be the rotated system (Figure 2.2). The components of the position vector **x** in the original and rotated systems are denoted by x_i and x'_i, respectively. The cosine of the angle between the old i and new j axes is represented by C_{ij}. Here, the *first* index of the **C** matrix refers to the *old* axes, and the second index of **C** refers to the new axes. It is apparent that $C_{ij} \neq C_{ji}$. A little geometry shows that the components in the rotated system are related to the components in the original system by

$$x'_j = x_1 C_{1j} + x_2 C_{2j} + x_3 C_{3j} = \sum_{i=1}^{3} x_i C_{ij}. \tag{2.1}$$

For simplicity, we shall verify the validity of equation (2.1) in two dimensions only. Referring to Figure 2.3, let α_{ij} be the angle between old i and new j axes, so that $C_{ij} = \cos \alpha_{ij}$. Then

$$x'_1 = \text{OD} = \text{OC} + \text{AB} = x_1 \cos \alpha_{11} + x_2 \sin \alpha_{11}. \tag{2.2}$$

As $\alpha_{11} = 90° - \alpha_{21}$, we have $\sin \alpha_{11} = \cos \alpha_{21} = C_{21}$. Equation (2.2) then becomes

$$x'_1 = x_1 C_{11} + x_2 C_{21} = \sum_{i=1}^{2} x_i C_{i1}. \tag{2.3}$$

In a similar manner

$$x'_2 = \text{PD} = \text{PB} - \text{DB} = x_2 \cos \alpha_{11} - x_1 \sin \alpha_{11}.$$

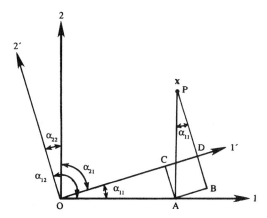

Figure 2.3   Rotation of a coordinate system in two dimensions.

As $\alpha_{11} = \alpha_{22} = \alpha_{12} - 90°$ (Figure 2.3), this becomes

$$x_2' = x_2 \cos\alpha_{22} + x_1 \cos\alpha_{12} = \sum_{i=1}^{2} x_i C_{i2}. \qquad (2.4)$$

In two dimensions, equation (2.1) reduces to equation (2.3) for $j = 1$, and to equation (2.4) for $j = 2$. This completes our verification of equation (2.1).

Note that the index i appears twice in the same term on the right-hand side of equation (2.1), and a summation is carried out over all values of this repeated index. This type of summation over repeated indices appears frequently in tensor notation. A convention is therefore adopted that, *whenever an index occurs twice in a term, a summation over the repeated index is implied, although no summation sign is explicitly written*. This is frequently called the *Einstein summation convention*. Equation (2.1) is then simply written as

$$x_j' = x_i C_{ij}, \qquad (2.5)$$

where a summation over i is understood on the right-hand side.

The free index on both sides of equation (2.5) is j, and i is the repeated or dummy index. Obviously any letter (other than j) can be used as the dummy index without changing the meaning of this equation. For example, equation (2.5) can be written equivalently as

$$x_i C_{ij} = x_k C_{kj} = x_m C_{mj} = \cdots,$$

because they all mean $x_j' = C_{1j} x_1 + C_{2j} x_2 + C_{3j} x_3$. Likewise, any letter can also be used for the free index, as long as the same free index is used on *both* sides of the equation. For example, denoting the free index by i and the summed index by k, equation (2.5) can be written as

$$x_i' = x_k C_{ki}. \qquad (2.6)$$

This is because the set of three equations represented by equation (2.5) corresponding to all values of j is the same set of equations represented by equation (2.6) for all values of i.

It is easy to show that the components of $\mathbf{x}$ in the old coordinate system are related to those in the rotated system by

$$x_j = C_{ji} x_i'. \tag{2.7}$$

Note that the indicial positions on the right-hand side of this relation are different from those in equation (2.5), because the first index of $\mathbf{C}$ is summed in equation (2.5), whereas the second index of $\mathbf{C}$ is summed in equation (2.7).

We can now formally define a Cartesian vector as any quantity that transforms like a position vector under the rotation of the coordinate system. Therefore, by analogy with equation (2.5), $\mathbf{u}$ is a vector if its components transform as

$$\boxed{u_j' = u_i C_{ij}.} \tag{2.8}$$

3. *Multiplication of Matrices*

In this chapter we shall generally follow the convention that 3×3 matrices are represented by uppercase letters, and column vectors are represented by lowercase letters. (An exception will be the use of lowercase τ for the stress matrix.) Let $\mathbf{A}$ and $\mathbf{B}$ be two 3×3 matrices. The product of $\mathbf{A}$ and $\mathbf{B}$ is defined as the matrix $\mathbf{P}$ whose elements are related to those of $\mathbf{A}$ and $\mathbf{B}$ by

$$P_{ij} = \sum_{k=1}^{3} A_{ik} B_{kj},$$

or, using the summation convention

$$\boxed{P_{ij} = A_{ik} B_{kj}.} \tag{2.9}$$

Symbolically, this is written as
$$\mathbf{P} = \mathbf{A} \cdot \mathbf{B}. \tag{2.10}$$

A single dot between $\mathbf{A}$ and $\mathbf{B}$ is included in equation (2.10) to signify that a single index is summed on the right-hand side of equation (2.9). The important thing to note in equation (2.9) is that the elements are summed over the inner or *adjacent* index k. It is sometimes useful to write equation (2.9) as

$$P_{ij} = A_{ik} B_{kj} = (\mathbf{A} \cdot \mathbf{B})_{ij},$$

where the last term is to be read as the "ij-element of the product of matrices $\mathbf{A}$ and $\mathbf{B}$."

In explicit form, equation (2.9) is written as

$$
\begin{bmatrix} P_{11} & P_{12} & P_{13} \\ P_{21} & P_{22} & P_{23} \\ P_{31} & P_{32} & P_{33} \end{bmatrix} = \begin{bmatrix} A_{11} & A_{12} & A_{13} \\ A_{21} & A_{22} & A_{23} \\ A_{31} & A_{32} & A_{33} \end{bmatrix} \begin{bmatrix} B_{11} & B_{12} & B_{13} \\ B_{21} & B_{22} & B_{23} \\ B_{31} & B_{32} & B_{33} \end{bmatrix} \quad (2.11)
$$

Note that equation (2.9) signifies that the ij-element of $\mathbf{P}$ is determined by multiplying the elements in the i-row of $\mathbf{A}$ and the j-column of $\mathbf{B}$, and summing. For example,

$$
P_{12} = A_{11}B_{12} + A_{12}B_{22} + A_{13}B_{32}.
$$

This is indicated by the dotted lines in equation (2.11). It is clear that we can define the product $\mathbf{A} \cdot \mathbf{B}$ only if the number of columns of $\mathbf{A}$ equals the number of rows of $\mathbf{B}$.

Equation (2.9) can be used to determine the product of a 3×3 matrix and a vector, if the vector is written as a column. For example, equation (2.6) can be written as $x_i' = C_{ik}^{\mathrm{T}} x_k$, which is now of the form of equation (2.9) because the summed index k is adjacent. In matrix form equation (2.6) can therefore be written as

$$
\begin{bmatrix} x_1' \\ x_2' \\ x_3' \end{bmatrix} = \begin{bmatrix} C_{11} & C_{12} & C_{13} \\ C_{21} & C_{22} & C_{23} \\ C_{31} & C_{32} & C_{33} \end{bmatrix}^{\mathrm{T}} \begin{bmatrix} x_1 \\ x_2 \\ x_3 \end{bmatrix}.
$$

Symbolically, the preceding is

$$
\mathbf{x}' = \mathbf{C}^{\mathrm{T}} \cdot \mathbf{x},
$$

whereas equation (2.7) is

$$
\mathbf{x} = \mathbf{C} \cdot \mathbf{x}'.
$$

4. Second-Order Tensor

We have seen that scalars can be represented by a single number, and a Cartesian vector can be represented by three numbers. There are other quantities, however, that need more than three components for a complete description. For example, the stress (equal to force per unit area) at a point in a material needs nine components for a complete specification because *two* directions (and, therefore, *two* free indices) are involved in its description. One direction specifies the orientation of the *surface* on which the stress is being sought, and the other specifies the direction of the *force* on that surface. For example, the j-component of the force on a surface whose outward normal points in the i-direction is denoted by τ_{ij}. (Here, we are departing from the convention followed in the rest of the chapter, namely, that tensors are represented by uppercase letters. It is customary to denote the stress tensor by the lowercase τ.) The first index of τ_{ij} denotes the direction of the normal, and the second index denotes the direction in which the force is being projected.

This is shown in Figure 2.4, which gives the normal and shear stresses on an infinitesimal cube whose surfaces are parallel to the coordinate planes. The stresses

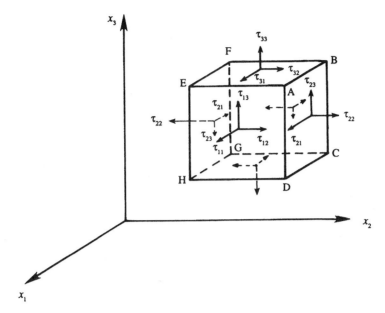

Figure 2.4   Stress field at a point. Positive normal and shear stresses are shown. For clarity, the stresses on faces FBCG and CDHG are not labeled.

are positive if they are directed as in this figure. The sign convention is that, on a surface whose outward normal points in the positive direction of a coordinate axis, the normal and shear stresses are positive if they point in the positive direction of the axes. For example, on the surface ABCD, whose outward normal points in the positive x_2 direction, the positive stresses τ_{21}, τ_{22}, and τ_{23} point toward the x_1, x_2 and x_3 directions, respectively. (Clearly, the normal stresses are positive if they are tensile and negative if they are compressive.) On the opposite face EFGH the stress components have the same value as on ABCD, but their directions are reversed. This is because Figure 2.4 shows the stresses *at a point*. The cube shown is supposed to be of "zero" size, so that the faces ABCD and EFGH are just opposite faces of a plane perpendicular to the x_2-axis. That is why the stresses on the opposite faces are equal and opposite.

Recall that a vector **u** can be completely specified by the three components u_i (where $i = 1, 2, 3$). We say "completely specified" because the components of **u** in any direction other than the original axes can be found from equation (2.8). Similarly, the state of stress at a point can be completely specified by the nine components τ_{ij} (where $i, j = 1, 2, 3$), which can be written as the matrix

$$\boldsymbol{\tau} = \begin{bmatrix} \tau_{11} & \tau_{12} & \tau_{13} \\ \tau_{21} & \tau_{22} & \tau_{23} \\ \tau_{31} & \tau_{32} & \tau_{33} \end{bmatrix}.$$

The specification of the preceding nine components of the stress on surfaces parallel to the coordinate axes completely determines the state of stress at a point, because the stresses on any arbitrary plane can then be determined. To find the stresses on any arbitrary surface, we shall consider a rotated coordinate system $x_1'\, x_2'\, x_3'$ one of whose axes is perpendicular to the given surface. It can be shown by a force balance on a tetrahedron element (see, e.g., Sommerfeld (1964), page 59) that the components of τ in the rotated coordinate system are

$$\tau_{mn}' = C_{im} C_{jn} \tau_{ij}. \qquad (2.12)$$

Note the similarity between the transformation rule equation (2.8) for a vector, and the rule equation (2.12). In equation (2.8) the first index of $\mathbf{C}$ is summed, while its second index is free. The rule equation (2.12) is identical, except that this happens twice. A quantity that obeys the transformation rule equation (2.12) is called a *second-order tensor*.

The transformation rule equation (2.12) can be expressed as a matrix product. Rewrite equation (2.12) as

$$\tau_{mn}' = C_{mi}^{\mathrm{T}} \tau_{ij} C_{jn},$$

which, with adjacent dummy indices, represents the matrix product

$$\tau' = \mathbf{C}^{\mathrm{T}} \cdot \tau \cdot \mathbf{C}.$$

This says that the tensor τ in the rotated frame is found by multiplying $\mathbf{C}$ by τ and then multiplying the product by $\mathbf{C}^{\mathrm{T}}$.

The concepts of tensor and matrix are not quite the same. A matrix is any *arrangement* of elements, written as an array. The elements of a matrix represent the components of a tensor only if they obey the transformation rule equation (2.12).

Tensors can be of any order. In fact, a scalar can be considered a tensor of zero order, and a vector can be regarded as a tensor of first order. The number of free indices correspond to the order of the tensor. For example, $\mathbf{A}$ is a fourth-order tensor if it has four free indices, and the associated 81 components change under the rotation of the coordinate system according to

$$A_{mnpq}' = C_{im} C_{jn} C_{kp} C_{lq} A_{ijkl}. \qquad (2.13)$$

Tensors of various orders arise in fluid mechanics. Some of the most frequently used are the stress tensor τ_{ij} and the velocity gradient tensor $\partial u_i / \partial x_j$. It can be shown that the nine products $u_i v_j$ formed from the components of the two vectors $\mathbf{u}$ and $\mathbf{v}$ also transform according to equation (2.12), and therefore form a second-order tensor. In addition, certain "isotropic" tensors are also frequently used; these will be discussed in Section 7.

5. *Contraction and Multiplication*

When the two indices of a tensor are equated, and a summation is performed over this repeated index, the process is called *contraction*. An example is

$$A_{jj} = A_{11} + A_{22} + A_{33},$$

which is the sum of the diagonal terms. Clearly, A_{jj} is a scalar and therefore independent of the coordinate system. In other words, A_{jj} is an *invariant*. (There are three independent invariants of a second-order tensor, and A_{jj} is one of them; see Exercise 5.)

Higher-order tensors can be formed by multiplying lower tensors. If **u** and **v** are vectors, then the nine components $u_i v_j$ form a second-order tensor. Similarly, if **A** and **B** are two second-order tensors, then the 81 numbers defined by $P_{ijkl} \equiv A_{ij} B_{kl}$ transform according to equation (2.13), and therefore form a fourth-order tensor.

Lower-order tensors can be obtained by performing contraction on these multiplied forms. The four contractions of $A_{ij} B_{kl}$ are

$$
\begin{aligned}
A_{ij} B_{ki} &= B_{ki} A_{ij} = (\mathbf{B} \cdot \mathbf{A})_{kj}, \\
A_{ij} B_{ik} &= A_{ji}^{T} B_{ik} = (\mathbf{A}^{T} \cdot \mathbf{B})_{jk}, \\
A_{ij} B_{kj} &= A_{ij} B_{jk}^{T} = (\mathbf{A} \cdot \mathbf{B}^{T})_{ik}, \\
A_{ij} B_{jk} &= (\mathbf{A} \cdot \mathbf{B})_{ik}.
\end{aligned}
\tag{2.14}
$$

All four products in the preceding are second-order tensors. Note in equation (2.14) how the terms have been rearranged until the summed index is adjacent, at which point they can be written as a product of matrices.

The contracted product of a second-order tensor **A** and a vector **u** is a vector. The two possibilities are

$$
\begin{aligned}
A_{ij} u_j &= (\mathbf{A} \cdot \mathbf{u})_i, \\
A_{ij} u_i &= A_{ji}^{T} u_i = (\mathbf{A}^{T} \cdot \mathbf{u})_j.
\end{aligned}
$$

The doubly contracted product of two second-order tensors **A** and **B** is a scalar. The two possibilities are $A_{ij} B_{ji}$ (which can be written as $\mathbf{A} : \mathbf{B}$ in boldface notation) and $A_{ij} B_{ij}$ (which can be written as $\mathbf{A} : \mathbf{B}^{T}$).

6. Force on a Surface

A surface area has a magnitude and an orientation, and therefore should be treated as a vector. The orientation of the surface is conveniently specified by the direction of a unit vector normal to the surface. If dA is the magnitude of an element of surface and **n** is the unit vector normal to the surface, then the surface area can be written as the vector

$$d\mathbf{A} = \mathbf{n}\, dA.$$

Suppose the nine components of the stress tensor with respect to a given set of Cartesian coordinates are given, and we want to find the force per unit area on a surface of given orientation **n** (Figure 2.5). One way of determining this is to take a rotated coordinate system, and use equation (2.12) to find the normal and shear stresses on the given surface. An alternative method is described in what follows.

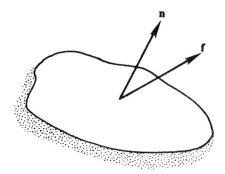

Figure 2.5 Force **f** per unit area on a surface element whose outward normal is **n**.

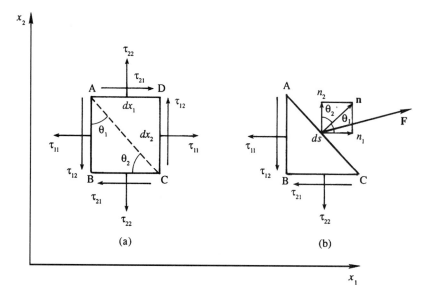

Figure 2.6 (a) Stresses on surfaces of a two-dimensional element; (b) balance of forces on element ABC.

For simplicity, consider a two-dimensional case, for which the known stress components with respect to a coordinate system $x_1 \, x_2$ are shown in Figure 2.6a. We want to find the force on the face AC, whose outward normal **n** is known (Figure 2.6b). Consider the balance of forces on a triangular element ABC, with sides AB $= dx_2$, BC $= dx_1$, and AC $= ds$; the thickness of the element in the x_3 direction is unity. If **F** is the force on the face AC, then a balance of forces in the x_1 direction gives the component of **F** in that direction as

$$F_1 = \tau_{11} \, dx_2 + \tau_{21} \, dx_1.$$

Dividing by ds, and denoting the force per unit area as $\mathbf{f} = \mathbf{F}/ds$, we obtain

$$f_1 = \frac{F_1}{ds} = \tau_{11}\frac{dx_2}{ds} + \tau_{21}\frac{dx_1}{ds}$$
$$= \tau_{11}\cos\theta_1 + \tau_{21}\cos\theta_2 = \tau_{11}n_1 + \tau_{21}n_2,$$

where $n_1 = \cos\theta_1$ and $n_2 = \cos\theta_2$ because the magnitude of $\mathbf{n}$ is unity (Figure 2.6b). Using the summation convention, the foregoing can be written as $f_1 = \tau_{j1}n_j$, where j is summed over 1 and 2. A similar balance of forces in the x_2 direction gives $f_2 = \tau_{j2}n_j$. Generalizing to three dimensions, it is clear that

$$f_i = \tau_{ji}n_j.$$

Because the stress tensor is symmetric (which will be proved in the next chapter), that is, $\tau_{ij} = \tau_{ji}$, the foregoing relation can be written in boldface notation as

$$\boxed{\mathbf{f} = \mathbf{n}\cdot\boldsymbol{\tau}.} \tag{2.15}$$

Therefore, the contracted or "inner" product of the stress tensor $\boldsymbol{\tau}$ and the unit outward vector $\mathbf{n}$ gives the force per unit area on a surface. Equation (2.15) is analogous to $u_n = \mathbf{u}\cdot\mathbf{n}$, where u_n is the component of the vector $\mathbf{u}$ along unit normal $\mathbf{n}$; however, whereas u_n is a scalar, $\mathbf{f}$ in equation (2.15) is a vector.

Example 2.1. Consider a two-dimensional parallel flow through a channel. Take x_1, x_2 as the coordinate system, with x_1 parallel to the flow. The viscous stress tensor at a point in the flow has the form

$$\boldsymbol{\tau} = \begin{bmatrix} 0 & a \\ a & 0 \end{bmatrix},$$

where the constant a is positive in one half of the channel, and negative in the other half. Find the magnitude and direction of force per unit area on an element whose outward normal points at 30° to the direction of flow.

Solution by using equation (2.15):   Because the magnitude of $\mathbf{n}$ is 1 and it points at 30° to the x_1 axis (Figure 2.7), we have

$$\mathbf{n} = \begin{bmatrix} \sqrt{3}/2 \\ 1/2 \end{bmatrix}.$$

The force per unit area is therefore

$$\mathbf{f} = \boldsymbol{\tau}\cdot\mathbf{n} = \begin{bmatrix} 0 & a \\ a & 0 \end{bmatrix}\begin{bmatrix} \sqrt{3}/2 \\ 1/2 \end{bmatrix} = \begin{bmatrix} a/2 \\ \sqrt{3}\,a/2 \end{bmatrix} = \begin{bmatrix} f_1 \\ f_2 \end{bmatrix}.$$

The magnitude of $\mathbf{f}$ is

$$f = (f_1^2 + f_2^2)^{1/2} = |a|.$$

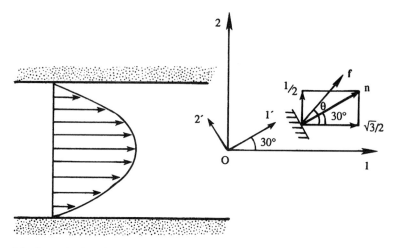

Figure 2.7   Determination of force on an area element (Example 2.1).

If θ is the angle of **f** with the x_1 axis, then

$$\sin\theta = \frac{f_2}{f} = \frac{\sqrt{3}}{2}\frac{a}{|a|} \quad \text{and} \quad \cos\theta = \frac{f_1}{f} = \frac{1}{2}\frac{a}{|a|}.$$

Thus $\theta = 60°$ if a is positive (in which case both $\sin\theta$ and $\cos\theta$ are positive), and $\theta = 240°$ if a is negative (in which case both $\sin\theta$ and $\cos\theta$ are negative).

Solution by using equation (2.12):   Take a rotated coordinate system x_1', x_2', with x_1' axis coinciding with **n** (Figure 2.7). Using equation (2.12), the components of the stress tensor in the rotated frame are

$$\tau_{11}' = C_{11}C_{21}\tau_{12} + C_{21}C_{11}\tau_{21} = \frac{\sqrt{3}}{2}\frac{1}{2}a + \frac{1}{2}\frac{\sqrt{3}}{2}a = \frac{\sqrt{3}}{2}a,$$

$$\tau_{12}' = C_{11}C_{22}\tau_{12} + C_{21}C_{12}\tau_{21} = \frac{\sqrt{3}}{2}\frac{\sqrt{3}}{2}a - \frac{1}{2}\frac{1}{2}a = \frac{1}{2}a.$$

The normal stress is therefore $\sqrt{3}\,a/2$, and the shear stress is $a/2$. This gives a magnitude a and a direction $60°$ or $240°$ depending on the sign of a.

7. *Kronecker Delta and Alternating Tensor*

The *Kronecker delta* is defined as

$$\delta_{ij} = \begin{cases} 1 & \text{if } i = j \\ 0 & \text{if } i \neq j \end{cases}, \tag{2.16}$$

which is written in the matrix form as

$$\delta = \begin{bmatrix} 1 & 0 & 0 \\ 0 & 1 & 0 \\ 0 & 0 & 1 \end{bmatrix}.$$

The most common use of the Kronecker delta is in the following operation: If we have a term in which one of the indices of δ_{ij} is repeated, then it simply replaces the dummy index by the other index of δ_{ij}. Consider

$$\delta_{ij}u_j = \delta_{i1}u_1 + \delta_{i2}u_2 + \delta_{i3}u_3.$$

The right-hand side is u_1 when $i = 1$, u_2 when $i = 2$, and u_3 when $i = 3$. Therefore

$$\delta_{ij}u_j = u_i. \tag{2.17}$$

From its definition it is clear that δ_{ij} is an *isotropic tensor* in the sense that its components are unchanged by a rotation of the frame of reference, that is, $\delta'_{ij} = \delta_{ij}$. Isotropic tensors can be of various orders. There is no isotropic tensor of first order, and δ_{ij} is the only isotropic tensor of second order. There is also only one isotropic tensor of third order. It is called the *alternating tensor* or *permutation symbol*, and is defined as

$$\varepsilon_{ijk} = \begin{cases} 1 & \text{if } ijk = 123, 231, \text{ or } 312 \text{ (cyclic order)}, \\ 0 & \text{if any two indices are equal}, \\ -1 & \text{if } ijk = 321, 213, \text{ or } 132 \text{ (anticyclic order)}. \end{cases} \tag{2.18}$$

From the definition, it is clear that *an index on ε_{ijk} can be moved two places* (*either to the right or to the left*) *without changing its value*. For example, $\varepsilon_{ijk} = \varepsilon_{jki}$ where i has been moved two places to the right, and $\varepsilon_{ijk} = \varepsilon_{kij}$ where k has been moved two places to the left. For a movement of one place, however, the sign is reversed. For example, $\varepsilon_{ijk} = -\varepsilon_{ikj}$ where j has been moved one place to the right.

A very frequently used relation is the *epsilon delta relation*

$$\varepsilon_{ijk}\varepsilon_{klm} = \delta_{il}\delta_{jm} - \delta_{im}\delta_{jl}. \tag{2.19}$$

The reader can verify the validity of this relationship by taking some values for $ijlm$. Equation (2.19) is easy to remember by noting the following two points: (1) The adjacent index k is summed; and (2) the first two indices on the right-hand side, namely, i and l, are the first index of ε_{ijk} and the first *free* index of ε_{klm}. The remaining indices on the right-hand side then follow immediately.

8. Dot Product

The dot product of two vectors $\mathbf{u}$ and $\mathbf{v}$ is defined as the scalar

$$\mathbf{u} \cdot \mathbf{v} = \mathbf{v} \cdot \mathbf{u} = u_1 v_1 + u_2 v_2 + u_3 v_3 = u_i v_i.$$

It is easy to show that $\mathbf{u} \cdot \mathbf{v} = uv\cos\theta$, where u and v are the magnitudes and θ is the angle between the vectors. The dot product is therefore the magnitude of one vector times the component of the other in the direction of the first. Clearly, the dot product $\mathbf{u} \cdot \mathbf{v}$ is equal to the sum of the diagonal terms of the tensor $u_i v_j$.

9. Cross Product

The cross product between two vectors $\mathbf{u}$ and $\mathbf{v}$ is defined as the vector $\mathbf{w}$ whose magnitude is $uv\sin\theta$, where θ is the angle between $\mathbf{u}$ and $\mathbf{v}$, and whose direction is perpendicular to the plane of $\mathbf{u}$ and $\mathbf{v}$ such that $\mathbf{u}$, $\mathbf{v}$, and $\mathbf{w}$ form a right-handed system. Clearly, $\mathbf{u}\times\mathbf{v} = -\mathbf{v}\times\mathbf{u}$, and the unit vectors obey the cyclic rule $\mathbf{a}^1\times\mathbf{a}^2 = \mathbf{a}^3$. It is easy to show that

$$\mathbf{u}\times\mathbf{v} = (u_2 v_3 - u_3 v_2)\mathbf{a}^1 + (u_3 v_1 - u_1 v_3)\mathbf{a}^2 + (u_1 v_2 - u_2 v_1)\mathbf{a}^3, \qquad (2.20)$$

which can be written as the symbolic determinant

$$\mathbf{u}\times\mathbf{v} = \begin{vmatrix} \mathbf{a}^1 & \mathbf{a}^2 & \mathbf{a}^3 \\ u_1 & u_2 & u_3 \\ v_1 & v_2 & v_3 \end{vmatrix}.$$

In indicial notation, the k-component of $\mathbf{u}\times\mathbf{v}$ can be written as

$$\boxed{(\mathbf{u}\times\mathbf{v})_k = \varepsilon_{ijk} u_i v_j = \varepsilon_{kij} u_i v_j.} \qquad (2.21)$$

As a check, for $k = 1$ the nonzero terms in the double sum in equation (2.21) result from $i = 2$, $j = 3$, and from $i = 3$, $j = 2$. This follows from the definition equation (2.18) that the permutation symbol is zero if any two indices are equal. Then equation (2.21) gives

$$(\mathbf{u}\times\mathbf{v})_1 = \varepsilon_{ij1} u_i v_j = \varepsilon_{231} u_2 v_3 + \varepsilon_{321} u_3 v_2 = u_2 v_3 - u_3 v_2,$$

which agrees with equation (2.20). Note that the second form of equation (2.21) is obtained from the first by moving the index k two places to the left; see the remark below equation (2.18).

10. Operator ∇: Gradient, Divergence, and Curl

The vector operator "del"[1] is defined symbolically by

$$\nabla \equiv \mathbf{a}^1 \frac{\partial}{\partial x_1} + \mathbf{a}^2 \frac{\partial}{\partial x_2} + \mathbf{a}^3 \frac{\partial}{\partial x_3} = \mathbf{a}^i \frac{\partial}{\partial x_i}. \qquad (2.22)$$

When operating on a scalar function of position ϕ, it generates the vector

$$\nabla\phi = \mathbf{a}^i \frac{\partial\phi}{\partial x_i},$$

---

[1]The inverted Greek delta is called a "nabla" ($\nu\alpha\beta\lambda\alpha$). The origin of the word is from the Hebrew נֶבֶל (pronounced navel), which means lyre, an ancient harp-like stringed instrument. It was on this instrument that the boy, David, entertained King Saul (Samuel II) and it is mentioned repeatedly in Psalms as a musical instrument to use in the praise of God.

whose i-component is

$$(\nabla \phi)_i = \frac{\partial \phi}{\partial x_i}.$$

The vector $\nabla \phi$ is called the *gradient* of ϕ. It is clear that $\nabla \phi$ is perpendicular to the $\phi =$ constant lines and gives the magnitude and direction of the *maximum* spatial rate of change of ϕ (Figure 2.8). The rate of change in any other direction $\mathbf{n}$ is given by

$$\frac{\partial \phi}{\partial n} = (\nabla \phi) \cdot \mathbf{n}.$$

The *divergence* of a vector field $\mathbf{u}$ is defined as the scalar

$$\nabla \cdot \mathbf{u} \equiv \frac{\partial u_i}{\partial x_i} = \frac{\partial u_1}{\partial x_1} + \frac{\partial u_2}{\partial x_2} + \frac{\partial u_3}{\partial x_3}. \tag{2.23}$$

So far, we have defined the operations of the gradient of a scalar and the divergence of a vector. We can, however, generalize these operations. For example, we can define the divergence of a second-order tensor τ as the vector whose i-component is

$$(\nabla \cdot \tau)_i = \frac{\partial \tau_{ij}}{\partial x_j}.$$

It is evident that the divergence operation *decreases* the order of the tensor by one. In contrast, the gradient operation *increases* the order of a tensor by one, changing

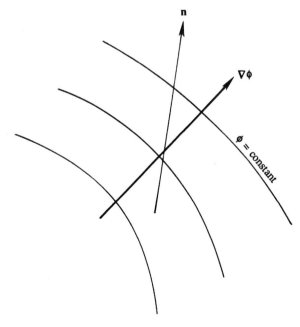

Figure 2.8   Lines of constant ϕ and the gradient vector $\nabla \phi$.

a zero-order tensor to a first-order tensor, and a first-order tensor to a second-order tensor.

The *curl* of a vector field **u** is defined as the vector $\nabla \times \mathbf{u}$, whose i-component can be written as (using equations (2.21) and (2.22))

$$(\nabla \times \mathbf{u})_i = \varepsilon_{ijk} \frac{\partial u_k}{\partial x_j}. \tag{2.24}$$

The three components of the vector $\nabla \times \mathbf{u}$ can easily be found from the right-hand side of equation (2.24). For the $i = 1$ component, the nonzero terms in the double sum in equation (2.24) result from $j = 2, k = 3$, and from $j = 3, k = 2$. The three components of $\nabla \times \mathbf{u}$ are finally found as

$$\left(\frac{\partial u_3}{\partial x_2} - \frac{\partial u_2}{\partial x_3} \right), \quad \left(\frac{\partial u_1}{\partial x_3} - \frac{\partial u_3}{\partial x_1} \right), \quad \text{and} \quad \left(\frac{\partial u_2}{\partial x_1} - \frac{\partial u_1}{\partial x_2} \right). \tag{2.25}$$

A vector field **u** is called *solenoidal* if $\nabla \cdot \mathbf{u} = 0$, and *irrotational* if $\nabla \times \mathbf{u} = 0$. The word "solenoidal" refers to the fact that the magnetic induction **B** always satisfies $\nabla \cdot \mathbf{B} = 0$. This is because of the absence of magnetic monopoles. The reason for the word "irrotational" will be clear in the next chapter.

11. Symmetric and Antisymmetric Tensors

A tensor **B** is called *symmetric* in the indices i and j if the components do not change when i and j are interchanged, that is, if $B_{ij} = B_{ji}$. The matrix of a second-order tensor is therefore symmetric about the diagonal and made up of only six distinct components. On the other hand, a tensor is called *antisymmetric* if $B_{ij} = -B_{ji}$. An antisymmetric tensor must have zero diagonal terms, and the off-diagonal terms must be mirror images; it is therefore made up of only three distinct components. Any tensor can be represented as the sum of a symmetric part and an antisymmetric part. For if we write

$$B_{ij} = \tfrac{1}{2}(B_{ij} + B_{ji}) + \tfrac{1}{2}(B_{ij} - B_{ji})$$

then the operation of interchanging i and j does not change the first term, but changes the sign of the second term. Therefore, $(B_{ij} + B_{ji})/2$ is called the symmetric part of B_{ij}, and $(B_{ij} - B_{ji})/2$ is called the antisymmetric part of B_{ij}.

Every vector can be associated with an antisymmetric tensor, and vice versa. For example, we can associate the vector

$$\boldsymbol{\omega} = \begin{bmatrix} \omega_1 \\ \omega_2 \\ \omega_3 \end{bmatrix},$$

with an antisymmetric tensor defined by

$$\mathbf{R} \equiv \begin{bmatrix} 0 & -\omega_3 & \omega_2 \\ \omega_3 & 0 & -\omega_1 \\ -\omega_2 & \omega_1 & 0 \end{bmatrix}, \tag{2.26}$$

where the two are related as

$$R_{ij} = -\varepsilon_{ijk}\omega_k$$

$$\omega_k = -\tfrac{1}{2}\varepsilon_{ijk}R_{ij}. \tag{2.27}$$

As a check, equation (2.27) gives $R_{11} = 0$ and $R_{12} = -\varepsilon_{123}\omega_3 = -\omega_3$, which is in agreement with equation (2.26). (In Chapter 3 we shall call $\mathbf{R}$ the "rotation" tensor corresponding to the "vorticity" vector $\boldsymbol{\omega}$.)

A very frequently occurring operation is the doubly contracted product of a *symmetric* tensor $\boldsymbol{\tau}$ and any tensor $\mathbf{B}$. The doubly contracted product is defined as

$$P \equiv \tau_{ij}B_{ij} = \tau_{ij}(S_{ij} + A_{ij}),$$

where $\mathbf{S}$ and $\mathbf{A}$ are the symmetric and antisymmetric parts of $\mathbf{B}$, given by

$$S_{ij} \equiv \frac{1}{2}(B_{ij} + B_{ji}) \quad \text{and} \quad A_{ij} \equiv \frac{1}{2}(B_{ij} - B_{ji}).$$

Then

$$\begin{aligned} P &= \tau_{ij}S_{ij} + \tau_{ij}A_{ij} \tag{2.28}\\ &= \tau_{ij}S_{ji} - \tau_{ij}A_{ji} \quad \text{because } S_{ij} = S_{ji} \text{ and } A_{ij} = -A_{ji},\\ &= \tau_{ji}S_{ji} - \tau_{ji}A_{ji} \quad \text{because } \tau_{ij} = \tau_{ji},\\ &= \tau_{ij}S_{ij} - \tau_{ij}A_{ij} \quad \text{interchanging dummy indices.} \tag{2.29} \end{aligned}$$

Comparing the two forms of equations (2.28) and (2.29), we see that $\tau_{ij}A_{ij} = 0$, so that

$$\boxed{\tau_{ij}B_{ij} = \frac{1}{2}\tau_{ij}(B_{ij} + B_{ji}).}$$

The important rule we have proved is that the *doubly contracted product of a symmetric tensor $\boldsymbol{\tau}$ with any tensor $\mathbf{B}$ equals $\boldsymbol{\tau}$ times the symmetric part of $\mathbf{B}$*. In the process, we have also shown that the doubly contracted product of a symmetric tensor and an antisymmetric tensor is zero. This is analogous to the result that the definite integral over an even (symmetric) interval of the product of a symmetric and an antisymmetric function is zero.

12. Eigenvalues and Eigenvectors of a Symmetric Tensor

The reader is assumed to be familiar with the concepts of eigenvalues and eigenvectors of a matrix, and only a brief review of the main results is given here. Suppose $\boldsymbol{\tau}$ is a symmetric tensor with real elements, for example, the stress tensor. Then the following facts can be proved:

(1) There are three real eigenvalues λ^k ($k = 1, 2, 3$), which may or may not be all distinct. (The superscripted λ^k does not denote the k-component of a vector.) The eigenvalues satisfy the third-degree equation

$$\det |\tau_{ij} - \lambda \delta_{ij}| = 0,$$

which can be solved for λ^1, λ^2, and λ^3.

(2) The three eigenvectors $\mathbf{b}^k$ corresponding to distinct eigenvalues λ^k are mutually orthogonal. These are frequently called the *principal axes* of $\boldsymbol{\tau}$. Each $\mathbf{b}$ is found by solving a set of three equations

$$(\tau_{ij} - \lambda \delta_{ij}) b_j = 0,$$

where the superscript k on λ and $\mathbf{b}$ has been omitted.

(3) If the coordinate system is rotated so as to coincide with the eigenvectors, then $\boldsymbol{\tau}$ has a diagonal form with elements λ^k. That is,

$$\boldsymbol{\tau}' = \begin{bmatrix} \lambda^1 & 0 & 0 \\ 0 & \lambda^2 & 0 \\ 0 & 0 & \lambda^3 \end{bmatrix}$$

in the coordinate system of the eigenvectors.

(4) The elements τ_{ij} change as the coordinate system is rotated, but they cannot be larger than the largest λ or smaller than the smallest λ. That is, the eigenvalues are the extremum values of τ_{ij}.

Example 2.2. The strain rate tensor $\mathbf{E}$ is related to the velocity vector $\mathbf{u}$ by

$$E_{ij} = \frac{1}{2} \left(\frac{\partial u_i}{\partial x_j} + \frac{\partial u_j}{\partial x_i} \right).$$

For a two-dimensional parallel flow

$$\mathbf{u} = \begin{bmatrix} u_1(x_2) \\ 0 \end{bmatrix},$$

show how $\mathbf{E}$ is diagonalized in the frame of reference coinciding with the principal axes.

Solution: For the given velocity profile $u_1(x_2)$, it is evident that $E_{11} = E_{22} = 0$, and $E_{12} = E_{21} = \frac{1}{2}(du_1/dx_2) = \Gamma$. The strain rate tensor in the unrotated coordinate system is therefore

$$\mathbf{E} = \begin{bmatrix} 0 & \Gamma \\ \Gamma & 0 \end{bmatrix}.$$

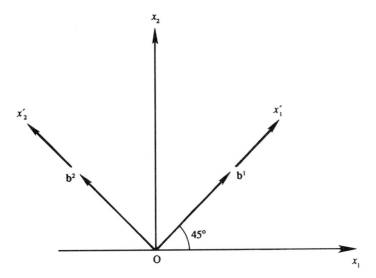

Figure 2.9   Original coordinate system $O\,x_1\,x_2$ and rotated coordinate system $O\,x_1'\,x_2'$ coinciding with the eigenvectors (Example 2.2).

The eigenvalues are given by

$$\det |E_{ij} - \lambda \delta_{ij}| = \begin{vmatrix} -\lambda & \Gamma \\ \Gamma & -\lambda \end{vmatrix} = 0,$$

whose solutions are $\lambda^1 = \Gamma$ and $\lambda^2 = -\Gamma$. The first eigenvector $\mathbf{b}^1$ is given by

$$\begin{bmatrix} 0 & \Gamma \\ \Gamma & 0 \end{bmatrix} \begin{bmatrix} b_1^1 \\ b_2^1 \end{bmatrix} = \lambda^1 \begin{bmatrix} b_1^1 \\ b_2^1 \end{bmatrix},$$

whose solution is $b_1^1 = b_2^1 = 1/\sqrt{2}$, thus normalizing the magnitude to unity. The first eigenvector is therefore $\mathbf{b}^1 = [1/\sqrt{2}, 1/\sqrt{2}]$, writing it in a row. The second eigenvector is similarly found as $\mathbf{b}^2 = [-1/\sqrt{2}, 1/\sqrt{2}]$. The eigenvectors are shown in Figure 2.9. The direction cosine matrix of the original and the rotated coordinate system is therefore

$$\mathbf{C} = \begin{bmatrix} \dfrac{1}{\sqrt{2}} & -\dfrac{1}{\sqrt{2}} \\[2mm] \dfrac{1}{\sqrt{2}} & \dfrac{1}{\sqrt{2}} \end{bmatrix},$$

which represents rotation of the coordinate system by $45°$. Using the transformation rule (2.12), the components of $\mathbf{E}$ in the rotated system are found as follows:

$$E_{12}' = C_{i1}C_{j2}E_{ij} = C_{11}C_{22}E_{12} + C_{21}C_{12}E_{21}$$

$$= \frac{1}{\sqrt{2}}\frac{1}{\sqrt{2}}\Gamma - \frac{1}{\sqrt{2}}\frac{1}{\sqrt{2}}\Gamma = 0$$

$$E'_{21} = 0$$

$$E'_{11} = C_{i1}C_{j1}E_{ij} = C_{11}C_{21}E_{12} + C_{21}C_{11}E_{21} = \Gamma$$

$$E'_{22} = C_{i2}C_{j2}E_{ij} = C_{12}C_{22}E_{12} + C_{22}C_{12}E_{21} = -\Gamma$$

(Instead of using equation (2.12), all the components of $\mathbf{E}$ in the rotated system can be found by carrying out the matrix product $\mathbf{C}^T \cdot \mathbf{E} \cdot \mathbf{C}$.) The matrix of $\mathbf{E}$ in the rotated frame is therefore

$$\mathbf{E}' = \begin{bmatrix} \Gamma & 0 \\ 0 & -\Gamma \end{bmatrix}.$$

The foregoing matrix contains only diagonal terms. It will be shown in the next chapter that it represents a linear stretching at a rate Γ along one principal axis, and a linear compression at a rate $-\Gamma$ along the other; there are no shear strains along the principal axes.

13. Gauss' Theorem

This very useful theorem relates a volume integral to a surface integral. Let V be a volume bounded by a closed surface A. Consider an infinitesimal surface element dA, whose outward unit normal is $\mathbf{n}$ (Figure 2.10). The vector $\mathbf{n} \, dA$ has a magnitude dA and direction $\mathbf{n}$, and we shall write $d\mathbf{A}$ to mean the same thing. Let $Q(\mathbf{x})$ be a scalar, vector, or tensor field of any order. Gauss' theorem states that

$$\int_V \frac{\partial Q}{\partial x_i} \, dV = \int_A dA_i \, Q. \tag{2.30}$$

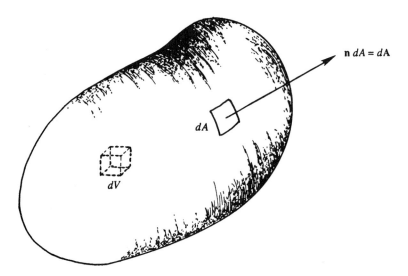

Figure 2.10 Illustration of Gauss' theorem.

The most common form of Gauss' theorem is when $\mathbf{Q}$ is a vector, in which case the theorem is

$$\int_V \frac{\partial Q_i}{\partial x_i}\, dV = \int_A dA_i\, Q_i,$$

which is called the *divergence theorem*. In vector notation, the divergence theorem is

$$\int_V \nabla \cdot \mathbf{Q}\, dV = \int_A d\mathbf{A} \cdot \mathbf{Q}.$$

Physically, it states that the volume integral of the divergence of $\mathbf{Q}$ is equal to the surface integral of the outflux of $\mathbf{Q}$. Alternatively, equation (2.30), when considered in its limiting form for an infintesmal volume, can define a generalized field derivative of Q by the expression

$$\mathcal{D}Q = \lim_{V \to 0} \frac{1}{V} \int_A dA_i\, Q. \qquad (2.31)$$

This includes the gradient, divergence, and curl of any scalar, vector, or tensor Q. Moreover, by regarding equation (2.31) as a definition, the recipes for the computation of the vector field derivatives may be obtained in any coordinate system. For a tensor Q of any order, equation (2.31) as written defines the gradient. For a tensor of order one (vector) or higher, the divergence is defined by using a dot (scalar) product under the integral

$$\operatorname{div} \mathbf{Q} = \lim_{V \to 0} \frac{1}{V} \int_A d\mathbf{A} \cdot \mathbf{Q}, \qquad (2.32)$$

and the curl is defined by using a cross (vector) product under the integral

$$\operatorname{curl} \mathbf{Q} = \lim_{V \to 0} \frac{1}{V} \int_A d\mathbf{A} \times \mathbf{Q}. \qquad (2.33)$$

In equations (2.31), (2.32), and (2.33), A is the closed surface bounding the volume V.

Example 2.3. Obtain the recipe for the divergence of a vector $\mathbf{Q}(\mathbf{x})$ in cylindrical polar coordinates from the integral definition equation (2.32). Compare with Appendix B.1.

Solution: Consider an elemental volume bounded by the surfaces $R - \Delta R/2$, $R + \Delta R/2$, $\theta - \Delta\theta/2$, $\theta + \Delta\theta/2$, $x - \Delta x/2$ and $x + \Delta x/2$. The volume enclosed ΔV is $R\Delta\theta \Delta R \Delta x$. We wish to calculate $\operatorname{div} \mathbf{Q} = \lim_{\Delta V \to 0} \frac{1}{\Delta V} \int_A d\mathbf{A} \cdot \mathbf{Q}$ at the central point R, θ, x by integrating the net outward flux through the bounding surface A of ΔV:

$$\mathbf{Q} = \mathbf{i}_R Q_R(R, \theta, x) + \mathbf{i}_\theta Q_\theta(R, \theta, x) + \mathbf{i}_x Q_x(R, \theta, x).$$

In evaluating the surface integrals, we can show that in the limit taken, each of the six surface integrals may be approximated by the product of the value at the center of the surface and the surface area. This is shown by Taylor expanding each of the scalar products in the two variables of each surface, carrying out the integrations, and applying the limits. The result is

$$
\text{div}\,\mathbf{Q} = \lim_{\substack{\Delta R \to 0 \\ \Delta \theta \to 0 \\ \Delta x \to 0}} \left\{ \frac{1}{R\Delta\theta\,\Delta R\,\Delta x} \left[Q_R\left(R + \frac{\Delta R}{2}, \theta, x\right)\left(R + \frac{\Delta R}{2}\right)\Delta\theta\,\Delta x \right.\right.
$$

$$
- Q_R\left(R - \frac{\Delta R}{2}, \theta, x\right)\left(R - \frac{\Delta R}{2}\right)\Delta\theta\,\Delta x
$$

$$
+ Q_x\left(R, \theta, x + \frac{\Delta x}{2}\right)R\Delta\theta\,\Delta R - Q_x\left(R, \theta, x - \frac{\Delta x}{2}\right)R\Delta\theta\,\Delta R
$$

$$
+ \mathbf{Q}\left(R, \theta + \frac{\Delta\theta}{2}, x\right)\cdot\left(\mathbf{i}_\theta - \mathbf{i}_R\frac{\Delta\theta}{2}\right)\Delta R\,\Delta x
$$

$$
\left.\left.+ \mathbf{Q}\left(R, \theta - \frac{\Delta\theta}{2}, x\right)\cdot\left(-\mathbf{i}_\theta - \mathbf{i}_R\frac{\Delta\theta}{2}\right)\Delta R\,\Delta x \right]\right\},
$$

where an additional complication arises because the normals to the two planes $\theta \pm \Delta\theta/2$ are not antiparallel:

$$
\mathbf{Q}\left(R, \theta \pm \frac{\Delta\theta}{2}, x\right) = Q_R\left(R, \theta \pm \frac{\Delta\theta}{2}, x\right)\mathbf{i}_R\left(R, \theta \pm \frac{\Delta\theta}{2}, x\right)
$$

$$
+ Q_\theta\left(R, \theta \pm \frac{\Delta\theta}{2}, x\right)\mathbf{i}_\theta\left(R, \theta \pm \frac{\Delta\theta}{2}, x\right)
$$

$$
+ Q_x\left(R, \theta \pm \frac{\Delta\theta}{2}, x\right)\mathbf{i}_x.
$$

Now we can show that

$$
\mathbf{i}_R\left(\theta \pm \frac{\Delta\theta}{2}\right) = \mathbf{i}_R(\theta) \pm \frac{\Delta\theta}{2}\mathbf{i}_\theta(\theta), \qquad \mathbf{i}_\theta\left(\theta \pm \frac{\Delta\theta}{2}\right) = \mathbf{i}_\theta(\theta) \mp \frac{\Delta\theta}{2}\mathbf{i}_R(\theta).
$$

Evaluating the last pair of surface integrals explicitly,

$$
\text{div}\,\mathbf{Q} = \lim_{\substack{\Delta R \to 0 \\ \Delta \theta \to 0 \\ \Delta x \to 0}} \left\{ \frac{1}{R\Delta\theta\,\Delta R\,\Delta x} \left[Q_R\left(R + \frac{\Delta R}{2}, \theta, x\right)\left(R + \frac{\Delta R}{2}\right)\Delta\theta\,\Delta x \right.\right.
$$

$$
- Q_R\left(R - \frac{\Delta R}{2}, \theta, x\right)\left(R - \frac{\Delta R}{2}\right)\Delta\theta\,\Delta x
$$

$$+ \left(Q_x \left(R, \theta, x + \frac{\Delta x}{2} \right) - Q_x \left(R, \theta, x - \frac{\Delta x}{2} \right) \right) R \Delta \theta \Delta R$$

$$+ \left(Q_R \left(R, \theta + \frac{\Delta \theta}{2}, x \right) \frac{\Delta \theta}{2} - Q_R \left(R, \theta + \frac{\Delta \theta}{2}, x \right) \frac{\Delta \theta}{2} \right) \Delta R \Delta x$$

$$+ \left(Q_\theta \left(R, \theta + \frac{\Delta \theta}{2}, x \right) - Q_\theta \left(R, \theta - \frac{\Delta \theta}{2}, x \right) \right) \Delta R \Delta x$$

$$- \left(Q_R \left(R, \theta - \frac{\Delta \theta}{2}, x \right) \frac{\Delta \theta}{2} - Q_R \left(R, \theta - \frac{\Delta \theta}{2}, x \right) \frac{\Delta \theta}{2} \right) \Delta R \Delta x \Big] \Big\},$$

where terms of second order in the increments have been neglected as they will vanish in the limits. Carrying out the limits, we obtain

$$\operatorname{div} \mathbf{Q} = \frac{1}{R} \frac{\partial}{\partial R} (R Q_R) + \frac{1}{R} \frac{\partial Q_\theta}{\partial \theta} + \frac{\partial Q_x}{\partial x}.$$

Here, the physical interpretation of the divergence as the net outward flux of a vector field per unit volume has been made apparent by its evaluation through the integral definition.

This level of detail is required to obtain the gradient correctly in these coordinates.

14. Stokes' Theorem

Stokes' theorem relates a surface integral over an open surface to a line integral around the boundary curve. Consider an open surface A whose bounding curve is C (Figure 2.11). Choose one side of the surface to be the outside. Let $d\mathbf{s}$ be an element of the bounding curve whose magnitude is the length of the element and whose direction is that of the tangent. The positive sense of the tangent is such that, when seen from the "outside" of the surface in the direction of the tangent, the interior is on the left. Then the theorem states that

$$\int_A (\nabla \times \mathbf{u}) \cdot d\mathbf{A} = \oint_C \mathbf{u} \cdot d\mathbf{s}, \qquad (2.34)$$

which signifies that the surface integral of the curl of a vector field $\mathbf{u}$ is equal to the line integral of $\mathbf{u}$ along the bounding curve.

The line integral of a vector $\mathbf{u}$ around a closed curve C (as in Figure 2.11) is called the "circulation of $\mathbf{u}$ about C." This can be used to define the curl of a vector through the limit of the circulation integral bounding an infinitesmal surface as follows:

$$\mathbf{n} \cdot \operatorname{curl} \mathbf{u} = \lim_{A \to 0} \frac{1}{A} \oint_C \mathbf{u} \cdot d\mathbf{s}, \qquad (2.35)$$

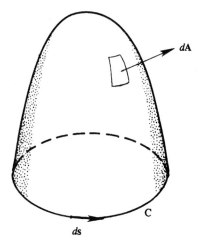

Figure 2.11   Illustration of Stokes' theorem.

where **n** is a unit vector normal to the local tangent plane of A. The advantage of the integral definitions of the field derivatives is that they may be applied regardless of the coordinate system.

Example 2.4.   Obtain the recipe for the curl of a vector **u(x)** in Cartesian coordinates from the integral definition given by equation (2.35).

Solution: This is obtained by considering rectangular contours in three perpendicular planes intersecting at the point (x, y, z). First, consider the elemental rectangle in the $x = $ const. plane. The central point in this plane has coordinates (x, y, z) and the area is $\Delta y\, \Delta z$. It may be shown by careful integration of a Taylor expansion of the integrand that the integral along each line segment may be represented by the product of the integrand at the center of the segment and the length of the segment with attention paid to the direction of integration $d\mathbf{s}$. Thus we obtain

$$(\text{curl } \mathbf{u})_x = \lim_{\substack{\Delta y \to 0 \\ \Delta z \to 0}} \left\{ \frac{1}{\Delta y \Delta z}\left[u_z\left(x, y + \frac{\Delta y}{2}, z\right) - u_z\left(x, y - \frac{\Delta y}{2}, z\right)\right]\Delta z \right.$$
$$\left. + \frac{1}{\Delta y \Delta z}\left[u_y\left(x, y, z - \frac{\Delta z}{2}\right) - u_y\left(x, y, z + \frac{\Delta z}{2}\right)\right]\Delta y \right\}.$$

Taking the limits,

$$(\text{curl } \mathbf{u})_x = \frac{\partial u_z}{\partial y} - \frac{\partial u_y}{\partial z}.$$

Similarly, integrating around the elemental rectangles in the other two planes

$$(\text{curl } \mathbf{u})_y = \frac{\partial u_x}{\partial z} - \frac{\partial u_z}{\partial x},$$

$$(\text{curl } \mathbf{u})_z = \frac{\partial u_y}{\partial x} - \frac{\partial u_x}{\partial y}.$$

15. Comma Notation

Sometimes it is convenient to introduce the notation

$$A_{,i} \equiv \frac{\partial A}{\partial x_i}, \tag{2.36}$$

where A is a tensor of any order. In this notation, therefore, the comma denotes a spatial derivative. For example, the divergence and curl of a vector $\mathbf{u}$ can be written, respectively, as

$$\nabla \cdot \mathbf{u} = \frac{\partial u_i}{\partial x_i} = u_{i,i},$$

$$(\nabla \times \mathbf{u})_i = \varepsilon_{ijk} \frac{\partial u_k}{\partial x_j} = \varepsilon_{ijk} u_{k,j}.$$

This notation has the advantages of economy and that all subscripts are written on one line. Another advantage is that variables such as $u_{i,j}$ "look like" tensors, which they are, in fact. Its disadvantage is that it takes a while to get used to it, and that the comma has to be written clearly in order to avoid confusion with other indices in a term. The comma notation has been used in the book only in two sections, in instances where otherwise the algebra became cumbersome.

16. Boldface vs Indicial Notation

The reader will have noticed that we have been using both boldface and indicial notations. Sometimes the boldface notation is loosely called "vector" or *dyadic* notation, while the indicial notation is called "tensor" notation. (Although there is no reason why vectors cannot be written in indicial notation!). The advantage of the boldface form is that the physical meaning of the terms is generally clearer, and there are no cumbersome subscripts. Its disadvantages are that algebraic manipulations are difficult, the ordering of terms becomes important because $\mathbf{A} \cdot \mathbf{B}$ is not the same as $\mathbf{B} \cdot \mathbf{A}$, and one has to remember formulas for triple products such as $\mathbf{u} \times (\mathbf{v} \times \mathbf{w})$ and $\mathbf{u} \cdot (\mathbf{v} \times \mathbf{w})$. In addition, there are other problems, for example, the order or rank of a tensor is not clear if one simply calls it $\mathbf{A}$, and sometimes confusion may arise in products such as $\mathbf{A} \cdot \mathbf{B}$ where it is not immediately clear *which* index is summed. To add to the confusion, the singly contracted product $\mathbf{A} \cdot \mathbf{B}$ is frequently written as $\mathbf{AB}$ in books on matrix algebra, whereas in several other fields $\mathbf{AB}$ usually stands for the uncontracted fourth-order tensor with elements $A_{ij}B_{kl}$.

The indicial notation avoids all the problems mentioned in the preceding. The algebraic manipulations are especially simple. The ordering of terms is unnecessary because $A_{ij}B_{kl}$ means the same thing as $B_{kl}A_{ij}$. In this notation we deal with *components* only, which are *scalars*. Another major advantage is that one does not have to remember formulas except for the product $\varepsilon_{ijk}\varepsilon_{klm}$, which is given by equation (2.19). The disadvantage of the indicial notation is that the physical meaning of a term becomes clear only after an examination of the indices. A second disadvantage is that the cross product involves the introduction of the cumbersome ε_{ijk}. This, however, can frequently be avoided by writing the *i*-component of the vector product of **u** and **v** as $(\mathbf{u} \times \mathbf{v})_i$ using a mixture of boldface and indicial notations. In this book we shall use boldface, indicial and mixed notations in order to take advantage of each. As the reader might have guessed, the algebraic manipulations will be performed mostly in the indicial notation, sometimes using the comma notation.

Exercises

1. Using indicial notation, show that

$$\mathbf{a} \times (\mathbf{b} \times \mathbf{c}) = (\mathbf{a} \cdot \mathbf{c})\mathbf{b} - (\mathbf{a} \cdot \mathbf{b})\mathbf{c}.$$

[*Hint*: Call $\mathbf{d} \equiv \mathbf{b} \times \mathbf{c}$. Then $(\mathbf{a} \times \mathbf{d})_m = \varepsilon_{pqm}a_p d_q = \varepsilon_{pqm}a_p \varepsilon_{ijq}b_i c_j$. Using equation (2.19), show that $(\mathbf{a} \times \mathbf{d})_m = (\mathbf{a} \cdot \mathbf{c})b_m - (\mathbf{a} \cdot \mathbf{b})c_m$.]

2. Show that the condition for the vectors **a**, **b**, and **c** to be coplanar is

$$\varepsilon_{ijk}a_i b_j c_k = 0.$$

3. Prove the following relationships:

$$\delta_{ij}\delta_{ij} = 3$$
$$\varepsilon_{pqr}\varepsilon_{pqr} = 6$$
$$\varepsilon_{pqi}\varepsilon_{pqj} = 2\delta_{ij}.$$

4. Show that

$$\mathbf{C} \cdot \mathbf{C}^{\mathrm{T}} = \mathbf{C}^{\mathrm{T}} \cdot \mathbf{C} = \delta,$$

where **C** is the direction cosine matrix and δ is the matrix of the Kronecker delta. Any matrix obeying such a relationship is called an *orthogonal matrix* because it represents transformation of one set of orthogonal axes into another.

5. Show that for a second-order tensor **A**, the following three quantities are invariant under the rotation of axes:

$$I_1 = A_{ii}$$
$$I_2 = \begin{vmatrix} A_{11} & A_{12} \\ A_{21} & A_{22} \end{vmatrix} + \begin{vmatrix} A_{22} & A_{23} \\ A_{32} & A_{33} \end{vmatrix} + \begin{vmatrix} A_{11} & A_{13} \\ A_{31} & A_{33} \end{vmatrix}$$
$$I_3 = \det(A_{ij}).$$

[*Hint*: Use the result of Exercise 4 and the transformation rule (2.12) to show that $I_1' = A_{ii}' = A_{ii} = I_1$. Then show that $A_{ij}A_{ji}$ and $A_{ij}A_{jk}A_{ki}$ are also invariants. In fact, *all* contracted scalars of the form $A_{ij}A_{jk}\cdots A_{mi}$ are invariants. Finally, verify that

$$I_2 = \tfrac{1}{2}[I_1^2 - A_{ij}A_{ji}]$$
$$I_3 = A_{ij}A_{jk}A_{ki} - I_1 A_{ij}A_{ji} + I_2 A_{ii}.$$

Because the right-hand sides are invariant, so are I_2 and I_3.]

6. If **u** and **v** are vectors, show that the products $u_i v_j$ obey the transformation rule (2.12), and therefore represent a second-order tensor.

7. Show that δ_{ij} is an isotropic tensor. That is, show that $\delta_{ij}' = \delta_{ij}$ under rotation of the coordinate system. [*Hint*: Use the transformation rule (2.12) and the results of Exercise 4.]

8. Obtain the recipe for the gradient of a scalar function in cylindrical polar coordinates from the integral definition.

9. Obtain the recipe for the divergence of a vector in spherical polar coordinates from the integral definition.

10. Prove that $\text{div}(\text{curl}\,\mathbf{u}) = 0$ for any vector **u** regardless of the coordinate system. [Hint: use the vector integral theorems.]

11. Prove that $\text{curl}(\text{grad}\,\phi) = 0$ for any single-valued scalar ϕ regardless of the coordinate system. [Hint: use Stokes' theorem.]

Literature Cited

Sommerfeld, A. (1964). *Mechanics of Deformable Bodies*, New York: Academic Press. (Chapter 1 contains brief but useful coverage of Cartesian tensors.)

Supplemental Reading

Aris, R. (1962). *Vectors, Tensors, and the Basic Equations of Fluid Mechanics*, Englewood Cliffs, NJ: Prentice-Hall. (This book gives a clear and easy treatment of tensors in Cartesian and non-Cartesian coordinates, with applications to fluid mechanics.)

Prager, W. (1961). *Introduction to Mechanics of Continua*, New York: Dover Publications. (Chapters 1 and 2 contain brief but useful coverage of Cartesian tensors.)

Kinematics

1. Introduction

Kinematics is the branch of mechanics that deals with quantities involving space and time only. It treats variables such as displacement, velocity, acceleration, deformation, and rotation of fluid elements without referring to the forces responsible for such a motion. Kinematics therefore essentially describes the "appearance" of a motion. Some important kinematical concepts are described in this chapter. The forces are considered when one deals with the *dynamics* of the motion, which will be discussed in later chapters.

A few remarks should be made about the notation used in this chapter and throughout the rest of the book. The convention followed in Chapter 2, namely, that vectors are denoted by lowercase letters and higher-order tensors are denoted by uppercase letters, is no longer followed. Henceforth, the number of subscripts will specify the order of a tensor. The Cartesian coordinate directions are denoted by (x, y, z), and the corresponding velocity components are denoted by (u, v, w). When using tensor expressions, the Cartesian directions are denoted alternatively

©2010 Elsevier Inc. All rights reserved.
DOI: 10.1016/B978-0-12-381399-2.50003-4

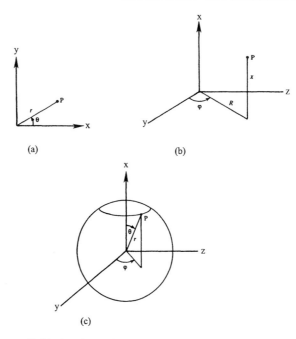

Figure 3.1   Plane, cylindrical, and spherical polar coordinates: (a) plane polar; (b) cylindrical polar;
(c) spherical polar coordinates.

by (x_1, x_2, x_3), with the corresponding velocity components (u_1, u_2, u_3). Plane
polar coordinates are denoted by (r, θ), with u_r and u_θ the corresponding velocity
components (Figure 3.1a). Cylindrical polar coordinates are denoted by (R, φ, x),
with (u_R, u_φ, u_x) the corresponding velocity components (Figure 3.1b). Spherical polar coordinates are denoted by (r, θ, φ), with $(u_r, u_\theta, u_\varphi)$ the corresponding
velocity components (Figure 3.1c). The method of conversion from Cartesian to
plane polar coordinates is illustrated in Section 14 of this chapter.

2.   *Lagrangian and Eulerian Specifications*

There are two ways of describing a fluid motion. In the *Lagrangian description,* one
essentially follows the history of individual fluid particles (Figure 3.2). Consequently,
the two independent variables are taken as time and a label for fluid particles. The label
can conveniently be taken as the position vector $\mathbf{a}$ of the particle at some reference time
$t = 0$. In this description, any flow variable F is expressed as $F(\mathbf{a}, t)$. In particular,
the position vector is written as $\mathbf{r} = \mathbf{r}(\mathbf{a}, t)$, which represents the location at t of a
particle whose position was $\mathbf{a}$ at $t = 0$.

In the *Eulerian* description, one concentrates on what happens at a spatial point
$\mathbf{r}'$, so that the independent variables are taken as $\mathbf{r}'$ and t'. (Here the primes are meant
to distinguish Lagrangian dependent variables from Eulerian independent variables.)
Flow variables are written, for example, as $F(\mathbf{r}', t')$.

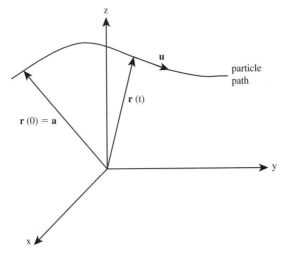

Figure 3.2 Particle—Lagrangian description. Independent variables: $(\mathbf{a}, t)$; dependent variables: $\mathbf{r}(\mathbf{a}, t)$, $\mathbf{u} = (\partial \mathbf{r}/\partial t)_{\mathbf{a}}$, $\rho = \rho(\mathbf{a}, t)$, and so on.

The velocity and acceleration of a fluid particle in the Lagrangian description are simply the partial time derivatives

$$\mathbf{u} = \partial \mathbf{r}/\partial t, \quad \text{acceleration } \mathbf{a} = \partial \mathbf{u}/\partial t = \partial^2 \mathbf{r}/\partial t^2 \tag{3.1}$$

as the particle identity is kept constant during the differentiation. In the Eulerian description, however, the partial derivative $\partial/\partial t'$ gives only the *local* rate of change at a point $\mathbf{r}'$ and is not the total rate of change as seen by the fluid particle. Additional terms are needed to form derivatives following a particle in the Eulerian description, as explained in the next section.

The Eulerian specification is used in most problems of fluid flows. The Lagrangian description is used occasionally when we are interested in finding particle paths of fixed identity; examples can be found in Chapters 7 and 13.

3. Eulerian and Lagrangian Descriptions: The Particle Derivative

Classical mechanics has two alternative descriptions: the field description (Eulerian) and the particle description (Lagrangian), associated with two of the great European mathematical physicists of the eighteenth century [Leonhard Euler (1707–1783) and Joseph Louis, Comte de Lagrange (1736–1813)]. Most of this book is written in the field description (Figure 3.3) but it is frequently very useful to express a particle derivative in the field description. Thus we wish to compare and relate the two descriptions.

Consider any fluid property $F(\mathbf{r}', t') = F(\mathbf{a}, t)$ at the same position and time in the two descriptions. F may be a scalar, vector, or tensor property. We seek to express $(\partial F/\partial t)_{\mathbf{a}}$, which is the rate of change of F as seen by an observer on the fixed particle labeled by coordinate $\mathbf{a} = \mathbf{r}(0)$ at $t = 0$, in field variables. That is,

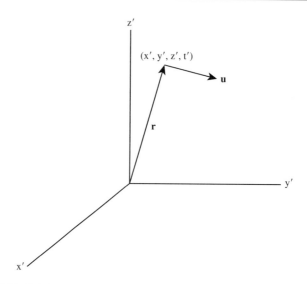

Figure 3.3 Field—Eulerian description. Independent variables: (x', y', z', t'); dependent variables: $\mathbf{u}(\mathbf{r}', t')$, $\rho(\mathbf{r}', t)$, and so on.

we ask what combination of $\mathbf{r}'$, t' field derivatives corresponds to $(\partial F/\partial t)_{\mathbf{a}}$? We do our calculation at $\mathbf{r}' = \mathbf{r}$ and $t' = t$ so we are at the same point and time in the two descriptions. Thus

$$F(\mathbf{a}, t) = F[\mathbf{r}(\mathbf{a}, t), t] = F(\mathbf{r}', t'). \tag{3.2}$$

Differentiating, taking care to differentiate dependent variables with respect to independent variables, and using the chain rule,

$$[\partial F(\mathbf{a}, t)/\partial t]_{\mathbf{a}} = (\partial F/\partial t')_{\mathbf{r}'}(\partial t'/\partial t) + (\partial F/\partial \mathbf{r}')_{t'} \cdot (\partial \mathbf{r}'/\partial \mathbf{r}) \cdot (\partial \mathbf{r}/\partial t)_{\mathbf{a}}. \tag{3.3}$$

Now $\partial t'/\partial t$ is simply the ratio of time scales used in the two descriptions. We take this equal to 1 by measuring the time in the same units (say seconds). Here $\partial \mathbf{r}'/\partial \mathbf{r}$ is the transformation matrix between the two coordinate systems. If $\mathbf{r}'$ and $\mathbf{r}$ are not rotated or stretched with respect to each other, but with parallel axes and with lengths measured in the same units (say meters), then $\partial \mathbf{r}'/\partial \mathbf{r} = \mathbf{I}$, the unit matrix, with elements δ_{ij}.

Since $(\partial \mathbf{r}/\partial t)_{\mathbf{a}} = \mathbf{u}$, we have the result

$$(\partial F/\partial t)_{\mathbf{a}} = \partial F/\partial t' + (\nabla' F) \cdot \mathbf{u} \equiv DF/Dt. \tag{3.4}$$

The total rate of change D/Dt is generally called the *material derivative* (also called the *substantial derivative*, or *particle derivative*) to emphasize the fact that the derivative is taken following a fluid element. It is made of two parts: $\partial F/\partial t$ is the *local* rate of change of F at a given point, and is zero for steady flows. The second part $u_i \, \partial F/\partial x_i$ is called the *advective* derivative, because it is the change in F as a result of advection of the particle from one location to another where the value of F is different. (In this book, the movement of fluid from place to place is called "advection."

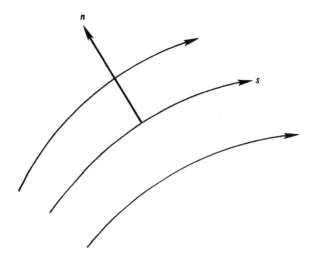

Figure 3.4   Streamline coordinates (s, n).

Engineering texts generally call it "convection." However, we shall reserve the term *convection* to describe heat transport by fluid movements.)

In vector notation, equation (3.4) is written as

$$\frac{DF}{Dt} = \frac{\partial F}{\partial t} + \mathbf{u} \cdot \nabla F. \tag{3.5}$$

The scalar product $\mathbf{u} \cdot \nabla F$ is the magnitude of $\mathbf{u}$ times the component of ∇F in the direction of $\mathbf{u}$. It is customary to denote the magnitude of the velocity vector $\mathbf{u}$ by q. Equation (3.5) can then be written in scalar notation as

$$\frac{DF}{Dt} = \frac{\partial F}{\partial t} + q \frac{\partial F}{\partial s}, \tag{3.6}$$

where the "streamline coordinate" s points along the local direction of $\mathbf{u}$ (Figure 3.4).

4.   *Streamline, Path Line, and Streak Line*

At an instant of time, there is at every point a velocity vector with a definite direction. The instantaneous curves that are everywhere tangent to the direction field are called the *streamlines* of flow. For unsteady flows the streamline pattern changes with time. Let $d\mathbf{s} = (dx, dy, dz)$ be an element of arc length along a streamline (Figure 3.5), and let $\mathbf{u} = (u, v, w)$ be the local velocity vector. Then by definition

$$\frac{dx}{u} = \frac{dy}{v} = \frac{dz}{w}, \tag{3.7}$$

along a streamline. If the velocity components are known as a function of time, then equation (3.7) can be integrated to find the equation of the streamline. It is easy to show that equation (3.7) corresponds to $\mathbf{u} \times d\mathbf{s} = 0$. All streamlines passing through

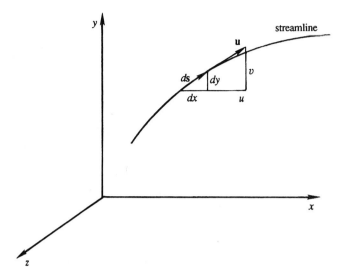

Figure 3.5   Streamline.

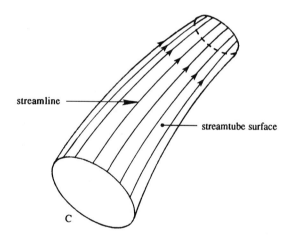

Figure 3.6   Streamtube.

any closed curve C at some time form a tube, which is called a *streamtube* (Figure 3.6). No fluid can cross the streamtube because the velocity vector is tangent to this surface.

In experimental fluid mechanics, the concept of path line is important. The *path line* is the trajectory of a fluid particle of fixed identity over a period of time. The path line or particle path is represented as in Section 2 by $\mathbf{r} = \mathbf{r}(\mathbf{a}, t)$ where $\mathbf{a}$ is the location of the particle at the reference time, say $t = 0$. Then $\mathbf{u} = \partial\mathbf{r}/\partial t$ for fixed particle. Assuming a nonzero Jacobian determinant, we may invert to obtain the reference particle location at $t = 0$, $\mathbf{a} = \mathbf{a}(\mathbf{r}, t)$. Path lines and streamlines are identical in a steady flow, but not in an unsteady flow. Consider the flow around a body moving from right to left in a fluid that is stationary at an infinite distance from the body (Figure 3.7). The flow pattern observed by a stationary observer (that is,

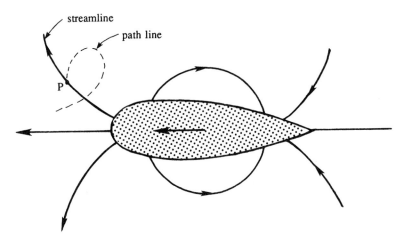

Figure 3.7 Several streamlines and a path line due to a moving body.

an observer stationary with respect to the undisturbed fluid) changes with time, so that to the observer this is an unsteady flow. The streamlines in front of and behind the body are essentially directed forward as the body pushes forward, and those on the two sides are directed laterally. The path line (shown dashed in Figure 3.7) of the particle that is now at point P therefore loops outward and forward again as the body passes by.

The streamlines and path lines of Figure 3.7 can be visualized in an experiment by suspending aluminum or other reflecting materials on the fluid surface, illuminated by a source of light. Suppose that the entire fluid is covered with such particles, and a *brief* time exposure is made. The photograph then shows short dashes, which indicate the instantaneous directions of particle movement. Smooth curves drawn through these dashes constitute the instantaneous streamlines. Now suppose that only a few particles are introduced, and that they are photographed with the shutter open for a *long* time. Then the photograph shows the paths of a few individual particles, that is, their path lines.

A *streak line* is another concept in flow visualization experiments. It is defined as the current location of all fluid particles that have passed through a fixed spatial point at a succession of previous times. It is determined by injecting dye or smoke at a fixed point for an interval of time. Suppose a particle on a streak line passes the location of dye $\boldsymbol{\xi}$ at a time $\tau \leq t$. Then the equation of the streak line is $\mathbf{x} = \mathbf{x}[\mathbf{a}(\boldsymbol{\xi}, \tau), t]$. See Aris (1962), Chapter 4 for more details. In steady flow the streamlines, path lines, and streak lines all coincide.

5. Reference Frame and Streamline Pattern

A flow that is steady in one reference frame is not necessarily so in another. Consider the flow past a ship moving at a steady velocity $\mathbf{U}$, with the frame of reference (that is, the observer) attached to the river bank (Figure 3.8a). To this observer the local flow characteristics appear to change with time, and thus appear to be unsteady. If,

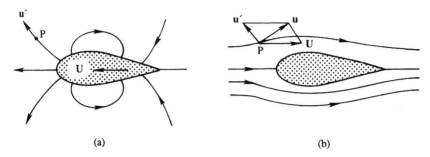

Figure 3.8   Flow past a ship with respect to two observers: (a) observer on river bank; (b) observer on ship.

on the other hand, the observer is standing on the ship, the flow pattern is steady (Figure 3.8b). The steady flow pattern can be obtained from the unsteady pattern of Figure 3.8a by superposing on the latter a velocity $\mathbf{U}$ to the right. This causes the ship to come to a halt and the river to move with velocity $\mathbf{U}$ at infinity. It follows that any velocity vector $\mathbf{u}$ in Figure 3.8b is obtained by adding the corresponding velocity vector $\mathbf{u}'$ of Figure 3.8a and the free stream velocity vector $\mathbf{U}$.

6. Linear Strain Rate

A study of the dynamics of fluid flows involves determination of the forces on an element, which depend on the amount and nature of its deformation, or strain. The deformation of a fluid is similar to that of a solid, where one defines normal strain as the change in length per unit length of a linear element, and shear strain as change of a 90° angle. Analogous quantities are defined in a fluid flow, the basic difference being that one defines strain *rates* in a fluid because it *continues* to deform.

Consider first the *linear* or *normal strain rate* of a fluid element in the x_1 direction (Figure 3.9). The rate of change of length per unit length is

$$\frac{1}{\delta x_1}\frac{D}{Dt}(\delta x_1) = \frac{1}{dt}\frac{A'B' - AB}{AB}$$

$$= \frac{1}{dt}\frac{1}{\delta x_1}\left[\delta x_1 + \frac{\partial u_1}{\partial x_1}\delta x_1\, dt - \delta x_1\right] = \frac{\partial u_1}{\partial x_1}.$$

The material derivative symbol D/Dt has been used because we have implicitly *followed* a fluid particle. In general, the linear strain rate in the α direction is

$$\frac{\partial u_\alpha}{\partial x_\alpha}, \tag{3.8}$$

where *no summation* over the repeated index α is implied. Greek symbols such as α and β are commonly used when the summation convention is violated.

The sum of the linear strain rates in the three mutually orthogonal directions gives the rate of change of volume per unit volume, called the *volumetric strain rate* (also called the *bulk strain rate*). To see this, consider a fluid element of sides δx_1,

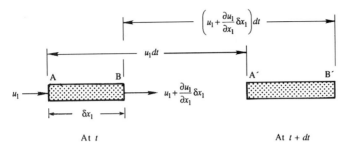

Figure 3.9 Linear strain rate. Here, $A'B' = AB + BB' - AA'$.

δx_2, and δx_3. Defining $\delta \mathcal{V} \equiv \delta x_1\,\delta x_2\,\delta x_3$, the volumetric strain rate is

$$\frac{1}{\delta \mathcal{V}}\frac{D}{Dt}(\delta \mathcal{V}) = \frac{1}{\delta x_1\,\delta x_2\,\delta x_3}\frac{D}{Dt}(\delta x_1\,\delta x_2\,\delta x_3),$$

$$= \frac{1}{\delta x_1}\frac{D}{Dt}(\delta x_1) + \frac{1}{\delta x_2}\frac{D}{Dt}(\delta x_2) + \frac{1}{\delta x_3}\frac{D}{Dt}(\delta x_3),$$

that is,

$$\frac{1}{\delta \mathcal{V}}\frac{D}{Dt}(\delta \mathcal{V}) = \frac{\partial u_1}{\partial x_1} + \frac{\partial u_2}{\partial x_2} + \frac{\partial u_3}{\partial x_3} = \frac{\partial u_i}{\partial x_i}. \tag{3.9}$$

The quantity $\partial u_i/\partial x_i$ is the sum of the diagonal terms of the velocity gradient tensor $\partial u_i/\partial x_j$. As a scalar, it is invariant with respect to rotation of coordinates. Equation (3.9) will be used later in deriving the law of conservation of mass.

7. Shear Strain Rate

In addition to undergoing normal strain rates, a fluid element may also simply deform in *shape*. The shear strain rate of an element is defined as the rate of decrease of the angle formed by two mutually perpendicular lines on the element. The shear strain so calculated depends on the orientation of the line pair. Figure 3.10 shows the position of an element with sides parallel to the coordinate axes at time t, and its subsequent position at $t + dt$. The rate of shear strain is

$$\frac{d\alpha + d\beta}{dt} = \frac{1}{dt}\left\{\frac{1}{\delta x_2}\left(\frac{\partial u_1}{\partial x_2}\delta x_2\,dt\right) + \frac{1}{\delta x_1}\left(\frac{\partial u_2}{\partial x_1}\delta x_1\,dt\right)\right\}$$

$$= \frac{\partial u_1}{\partial x_2} + \frac{\partial u_2}{\partial x_1}. \tag{3.10}$$

An examination of equations (3.8) and (3.10) shows that we can describe the deformation of a fluid element in terms of the *strain rate tensor*

$$\boxed{e_{ij} \equiv \frac{1}{2}\left(\frac{\partial u_i}{\partial x_j} + \frac{\partial u_j}{\partial x_i}\right).} \tag{3.11}$$

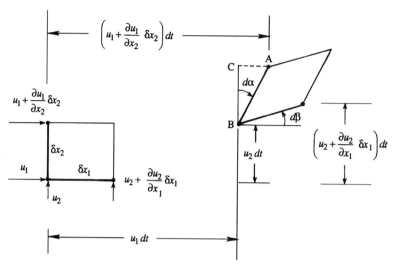

Figure 3.10   Deformation of a fluid element. Here, $d\alpha = CA / CB$; a similar expression represents $d\beta$.

The diagonal terms of **e** are the normal strain rates given in (3.8), and the off-diagonal terms are *half* the shear strain rates given in (3.10). Obviously the strain rate tensor is symmetric as $e_{ij} = e_{ji}$.

8. Vorticity and Circulation

Fluid lines oriented along different directions rotate by different amounts. To define the rotation rate unambiguously, two mutually perpendicular lines are taken, and the *average* rotation rate of the two lines is calculated; it is easy to show that this average is independent of the orientation of the line pair. To avoid the appearance of certain factors of 2 in the final expressions, it is generally customary to deal with *twice* the angular velocity, which is called the *vorticity* of the element.

   Consider the two perpendicular line elements of Figure 3.10. The angular velocities of line elements about the x_3 axis are $d\beta/dt$ and $-d\alpha/dt$, so that the average is $\frac{1}{2}(-d\alpha/dt + d\beta/dt)$. The vorticity of the element about the x_3 axis is therefore twice this average, as given by

$$\omega_3 = \frac{1}{dt}\left\{\frac{1}{\delta x_2}\left(-\frac{\partial u_1}{\partial x_2}\delta x_2\,dt\right) + \frac{1}{\delta x_1}\left(\frac{\partial u_2}{\partial x_1}\delta x_1\,dt\right)\right\}$$

$$= \frac{\partial u_2}{\partial x_1} - \frac{\partial u_1}{\partial x_2}.$$

From the definition of curl of a vector (see equations 2.24 and 2.25), it follows that the vorticity vector of a fluid element is related to the velocity vector by

$$\boxed{\boldsymbol{\omega} = \nabla \times \mathbf{u}} \quad \text{or} \quad \omega_i = \varepsilon_{ijk}\frac{\partial u_k}{\partial x_j}, \tag{3.12}$$

whose components are

$$\omega_1 = \frac{\partial u_3}{\partial x_2} - \frac{\partial u_2}{\partial x_3}, \qquad \omega_2 = \frac{\partial u_1}{\partial x_3} - \frac{\partial u_3}{\partial x_1}, \qquad \omega_3 = \frac{\partial u_2}{\partial x_1} - \frac{\partial u_1}{\partial x_2}. \qquad (3.13)$$

A fluid motion is called *irrotational* if $\boldsymbol{\omega} = \mathbf{0}$, which would require

$$\frac{\partial u_i}{\partial x_j} = \frac{\partial u_j}{\partial x_i} \quad i \neq j. \qquad (3.14)$$

In irrotational flows, the velocity vector can be written as the gradient of a scalar function $\phi(\mathbf{x}, t)$. This is because the assumption

$$u_i \equiv \frac{\partial \phi}{\partial x_i}, \qquad (3.15)$$

satisfies the condition of irrotationality (3.14).

Related to the concept of vorticity is the concept of circulation. The *circulation* Γ around a closed contour C (Figure 3.11) is defined as the line integral of the tangential component of velocity and is given by

$$\boxed{\Gamma \equiv \oint_C \mathbf{u} \cdot d\mathbf{s},} \qquad (3.16)$$

where $d\mathbf{s}$ is an element of contour, and the loop through the integral sign signifies that the contour is closed. The loop will be omitted frequently because it is understood that such line integrals are taken along closed contours called *circuits*. Then *Stokes' theorem* (Chapter 2, Section 14) states that

$$\int_C \mathbf{u} \cdot d\mathbf{s} = \int_A (\text{curl } \mathbf{u}) \cdot d\mathbf{A} \qquad (3.17)$$

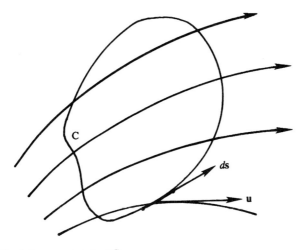

Figure 3.11   Circulation around contour C.

which says that the line integral of **u** around a closed curve C is equal to the "flux" of curl **u** through an arbitrary surface **A** bounded by C. (The word "flux" is generally used to mean the integral of a vector field normal to a surface. [See equation (2.32), where the integral written is the net outward flux of the vector field **Q**.]) Using the definitions of vorticity and circulation, Stokes' theorem, equation (3.17), can be written as

$$\Gamma = \int_A \boldsymbol{\omega} \cdot d\mathbf{A}. \tag{3.18}$$

Thus, the circulation around a closed curve is equal to the surface integral of the vorticity, which we can call the *flux of vorticity*. Equivalently, the *vorticity at a point equals the circulation per unit area.* That follows directly from the definition of curl as the limit of the circulation integral. (See equation (2.35) of Chapter 2.)

9. Relative Motion near a Point: Principal Axes

The preceding two sections have shown that fluid particles deform and rotate. In this section we shall formally show that the relative motion between two neighboring points can be written as the sum of the motion due to local rotation, plus the motion due to local deformation.

Let $\mathbf{u}(\mathbf{x}, t)$ be the velocity at point O (position vector **x**), and let $\mathbf{u} + d\mathbf{u}$ be the velocity at the same time at a neighboring point P (position vector $\mathbf{x} + d\mathbf{x}$; see Figure 3.12). The relative velocity at time t is given by

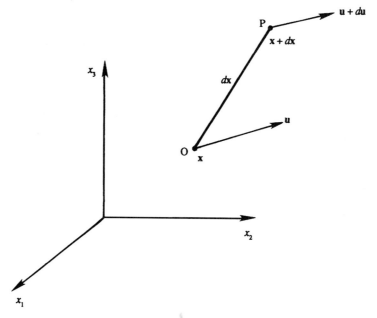

Figure 3.12   Velocity vectors at two neighboring points O and P.

$$du_i = \frac{\partial u_i}{\partial x_j} dx_j, \tag{3.19}$$

which stands for three relations such as

$$du_1 = \frac{\partial u_1}{\partial x_1} dx_1 + \frac{\partial u_1}{\partial x_2} dx_2 + \frac{\partial u_1}{\partial x_3} dx_3. \tag{3.20}$$

The term $\partial u_i / \partial x_j$ in equation (3.19) is the *velocity gradient tensor*. It can be decomposed into symmetric and antisymmetric parts as follows:

$$\frac{\partial u_i}{\partial x_j} = \frac{1}{2}\left(\frac{\partial u_i}{\partial x_j} + \frac{\partial u_j}{\partial x_i}\right) + \frac{1}{2}\left(\frac{\partial u_i}{\partial x_j} - \frac{\partial u_j}{\partial x_i}\right), \tag{3.21}$$

which can be written as

$$\frac{\partial u_i}{\partial x_j} = e_{ij} + \frac{1}{2}r_{ij}, \tag{3.22}$$

where e_{ij} is the strain rate tensor defined in equation (3.11), and

$$r_{ij} \equiv \frac{\partial u_i}{\partial x_j} - \frac{\partial u_j}{\partial x_i}, \tag{3.23}$$

is called the *rotation tensor*. As r_{ij} is antisymmetric, its diagonal terms are zero and the off-diagonal terms are equal and opposite. It therefore has three independent elements, namely, r_{13}, r_{21}, and r_{32}. Comparing equations (3.13) and (3.22), we can see that $r_{21} = \omega_3$, $r_{32} = \omega_1$, and $r_{13} = \omega_2$. Thus the rotation tensor can be written in terms of the components of the vorticity vector as

$$\mathbf{r} = \begin{bmatrix} 0 & -\omega_3 & \omega_2 \\ \omega_3 & 0 & -\omega_1 \\ -\omega_2 & \omega_1 & 0 \end{bmatrix}. \tag{3.24}$$

Each antisymmetric tensor in fact can be associated with a vector as discussed in Chapter 2, Section 11. In the present case, the rotation tensor can be written in terms of the vorticity vector as

$$r_{ij} = -\varepsilon_{ijk}\omega_k. \tag{3.25}$$

This can be verified by taking various components of equation (3.24) and comparing them with equation (3.23). For example, equation (3.24) gives $r_{12} = -\varepsilon_{12k}\omega_k = -\varepsilon_{123}\omega_3 = -\omega_3$, which agrees with equation (3.23). Equation (3.24) also appeared as equation (2.27).

Substitution of equations (3.21) and (3.24) into equation (3.19) gives

$$du_i = e_{ij}\,dx_j - \frac{1}{2}\varepsilon_{ijk}\omega_k\,dx_j,$$

which can be written as

$$du_i = e_{ij}\,dx_j + \tfrac{1}{2}(\boldsymbol{\omega} \times d\mathbf{x})_i. \tag{3.26}$$

In the preceding, we have noted that $\varepsilon_{ijk}\omega_k\,dx_j$ is the i-component of the cross product $-\boldsymbol{\omega} \times d\mathbf{x}$. (See the definition of cross product in equation (2.21).) The meaning of the second term in equation (3.25) is evident. We know that the velocity at a distance $\mathbf{x}$ from the axis of rotation of a body rotating rigidly at angular velocity $\boldsymbol{\Omega}$ is $\boldsymbol{\Omega} \times \mathbf{x}$. The second term in equation (3.25) therefore represents the relative velocity at point P due to rotation of the element at angular velocity $\boldsymbol{\omega}/2$. (Recall that the angular velocity is half the vorticity $\boldsymbol{\omega}$.)

The first term in equation (3.25) is the relative velocity due only to deformation of the element. The deformation becomes particularly simple in a coordinate system coinciding with the principal axes of the strain rate tensor. The components of $\mathbf{e}$ change as the coordinate system is rotated. For a particular orientation of the coordinate system, a symmetric tensor has only diagonal components; these are called the *principal axes* of the tensor (see Chapter 2, Section 12 and Example 2.2). Denoting the variables in the principal coordinate system by an overbar (Figure 3.13), the *first* part of equation (3.25) can be written as the matrix product

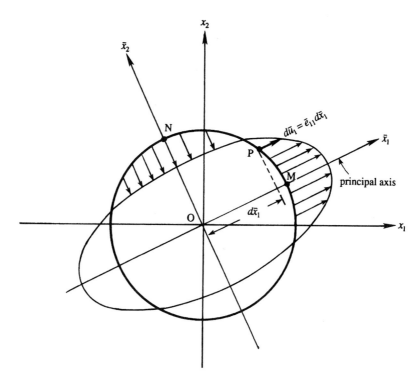

Figure 3.13   Deformation of a spherical fluid element into an ellipsoid.

$$d\bar{\mathbf{u}} = \bar{\mathbf{e}} \cdot d\bar{\mathbf{x}} = \begin{bmatrix} \bar{e}_{11} & 0 & 0 \\ 0 & \bar{e}_{22} & 0 \\ 0 & 0 & \bar{e}_{33} \end{bmatrix} \begin{bmatrix} d\bar{x}_1 \\ d\bar{x}_2 \\ d\bar{x}_3 \end{bmatrix}. \tag{3.27}$$

Here, $\bar{e}_{11}$, $\bar{e}_{22}$, and $\bar{e}_{33}$ are the diagonal components of $\mathbf{e}$ in the principal coordinate system and are called the eigenvalues of $\mathbf{e}$. The three components of equation (3.26) are

$$d\bar{u}_1 = \bar{e}_{11}\, d\bar{x}_1 \qquad d\bar{u}_2 = \bar{e}_{22}\, d\bar{x}_2 \qquad d\bar{u}_3 = \bar{e}_{33}\, d\bar{x}_3. \tag{3.28}$$

Consider the significance of the first of equations (3.27), namely, $d\bar{u}_1 = \bar{e}_{11}\, d\bar{x}_1$ (Figure 3.13). If $\bar{e}_{11}$ is positive, then this equation shows that point P is moving *away* from O in the $\bar{x}_1$ direction at a rate proportional to the distance $d\bar{x}_1$. Considering all points on the surface of a sphere, the movement of P in the $\bar{x}_1$ direction is therefore the maximum when P coincides with M (where $d\bar{x}_1$ is the maximum) and is zero when P coincides with N. (In Figure 3.13 we have illustrated a case where $\bar{e}_{11} > 0$ and $\bar{e}_{22} < 0$; the deformation in the x_3 direction cannot, of course, be shown in this figure.) In a small interval of time, *a spherical fluid element around O therefore becomes an ellipsoid whose axes are the principal axes of the strain tensor* $\mathbf{e}$.

Summary: The relative velocity in the neighborhood of a point can be divided into two parts. One part is due to the angular velocity of the element, and the other part is due to deformation. A spherical element deforms to an ellipsoid whose axes coincide with the principal axes of the local strain rate tensor.

10. *Kinematic Considerations of Parallel Shear Flows*

In this section we shall consider the rotation and deformation of fluid elements in the parallel shear flow $\mathbf{u} = [u_1(x_2), 0, 0]$ shown in Figure 3.14. Let us denote the velocity gradient by $\gamma(x_2) \equiv du_1/dx_2$. From equation (3.13), the only nonzero component of vorticity is $\omega_3 = -\gamma$. In Figure 3.13, the angular velocity of line element AB is $-\gamma$, and that of BC is zero, giving $-\gamma/2$ as the overall angular velocity (half the

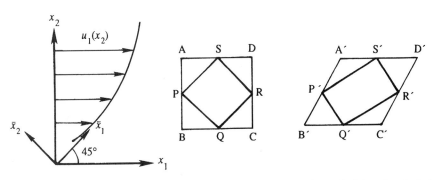

Figure 3.14 Deformation of elements in a parallel shear flow. The element is stretched along the principal axis $\bar{x}_1$ and compressed along the principal axis $\bar{x}_2$.

vorticity). The average value does not depend on *which* two mutually perpendicular elements in the $x_1 x_2$-plane are chosen to compute it.

In contrast, the components of strain rate do depend on the orientation of the element. From equation (3.11), the strain rate tensor of an element such as ABCD, with the sides parallel to the $x_1 x_2$-axes, is

$$
\mathbf{e} =
\begin{bmatrix}
0 & \frac{1}{2}\gamma & 0 \\
\frac{1}{2}\gamma & 0 & 0 \\
0 & 0 & 0
\end{bmatrix},
$$

which shows that there are only off-diagonal elements of $\mathbf{e}$. Therefore, the element ABCD undergoes shear, but no normal strain. As discussed in Chapter 2, Section 12 and Example 2.2, a symmetric tensor with zero diagonal elements can be diagonalized by rotating the coordinate system through 45°. It is shown there that, along these *principal axes* (denoted by an overbar in Figure 3.14), the strain rate tensor is

$$
\bar{\mathbf{e}} =
\begin{bmatrix}
\frac{1}{2}\gamma & 0 & 0 \\
0 & -\frac{1}{2}\gamma & 0 \\
0 & 0 & 0
\end{bmatrix},
$$

so that there is a linear extension rate of $\bar{e}_{11} = \gamma/2$, a linear compression rate of $\bar{e}_{22} = -\gamma/2$, and no shear. This can be understood physically by examining the deformation of an element PQRS oriented at 45°, which deforms to P′Q′R′S′. It is clear that the side PS elongates and the side PQ contracts, but the angles between the sides of the element remain 90°. In a small time interval, a small spherical element in this flow would become an ellipsoid oriented at 45° to the $x_1 x_2$-coordinate system.

Summarizing, the element ABCD in a parallel shear flow undergoes only shear but no normal strain, whereas the element PQRS undergoes only normal but no shear strain. Both of these elements rotate at the same angular velocity.

11. Kinematic Considerations of Vortex Flows

Flows in circular paths are called *vortex flows*, some basic forms of which are described in what follows.

Solid-Body Rotation

Consider first the case in which the velocity is proportional to the radius of the stream-lines. Such a flow can be generated by steadily rotating a cylindrical tank containing a viscous fluid and waiting until the transients die out. Using polar coordinates (r, θ), the velocity in such a flow is

$$
u_\theta = \omega_0 r \qquad u_r = 0, \tag{3.29}
$$

where ω_0 is a constant equal to the angular velocity of *revolution* of each particle about the origin (Figure 3.15). We shall see shortly that ω_0 is also equal to the angular

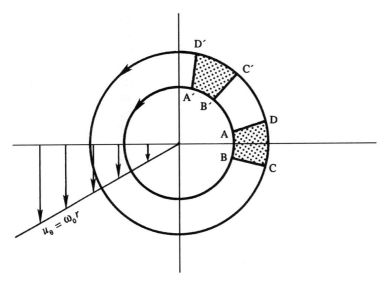

Figure 3.15   Solid-body rotation. Fluid elements are spinning about their own centers while they revolve around the origin. There is no deformation of the elements.

speed of *rotation* of each particle about its *own* center. The vorticity components of a fluid element in polar coordinates are given in Appendix B. The component about the *z*-axis is

$$\omega_z = \frac{1}{r}\frac{\partial}{\partial r}(ru_\theta) - \frac{1}{r}\frac{\partial u_r}{\partial \theta} = 2\omega_0, \tag{3.30}$$

where we have used the velocity distribution equation (3.28). This shows that the angular velocity of each fluid element about its own center is a constant and equal to ω_0. This is evident in Figure 3.15, which shows the location of element ABCD at two successive times. It is seen that the two mutually perpendicular fluid lines AD and AB both rotate counterclockwise (about the center of the element) with speed ω_0. The time period for one *rotation* of the particle about its own center equals the time period for one *revolution* around the origin. It is also clear that the deformation of the fluid elements in this flow is zero, as each fluid particle retains its location relative to other particles. A flow defined by $u_\theta = \omega_0 r$ is called a *solid-body rotation* as the fluid elements behave as in a rigid, rotating solid.

The circulation around a circuit of radius *r* in this flow is

$$\Gamma = \int \mathbf{u} \cdot d\mathbf{s} = \int_0^{2\pi} u_\theta r\, d\theta = 2\pi r u_\theta = 2\pi r^2 \omega_0, \tag{3.31}$$

which shows that circulation equals vorticity $2\omega_0$ times area. It is easy to show (Exercise 12) that this is true of *any* contour in the fluid, regardless of whether or not it contains the center.

Irrotational Vortex

Circular streamlines, however, do not imply that a flow should have vorticity everywhere. Consider the flow around circular paths in which the velocity vector is tangential and is inversely proportional to the radius of the streamline. That is,

$$u_\theta = \frac{C}{r} \qquad u_r = 0. \tag{3.32}$$

Using equation (3.29), the vorticity at any point in the flow is

$$\omega_z = \frac{0}{r}.$$

This shows that the vorticity is zero everywhere except at the origin, where it cannot be determined from this expression. However, the vorticity at the origin can be determined by considering the circulation around a circuit enclosing the origin. Around a contour of radius r, the circulation is

$$\Gamma = \int_0^{2\pi} u_\theta r \, d\theta = 2\pi C.$$

This shows that Γ is constant, independent of the radius. (Compare this with the case of solid-body rotation, for which equation (3.30) shows that Γ is proportional to r^2.) In fact, the circulation around a circuit of *any shape* that encloses the origin is $2\pi C$. Now consider the implication of Stokes' theorem

$$\Gamma = \int_A \boldsymbol{\omega} \cdot d\mathbf{A}, \tag{3.33}$$

for a contour enclosing the origin. The left-hand side of equation (3.32) is nonzero, which implies that $\boldsymbol{\omega}$ must be nonzero somewhere within the area enclosed by the contour. Because Γ in this flow is independent of r, we can shrink the contour without altering the left-hand side of equation (3.32). In the limit the area approaches zero, so that the vorticity at the origin must be infinite in order that $\boldsymbol{\omega} \cdot \delta\mathbf{A}$ may have a finite nonzero limit at the origin. We have therefore demonstrated that the *flow represented by $u_\theta = C/r$ is irrotational everywhere except at the origin, where the vorticity is infinite.* Such a flow is called an *irrotational or potential vortex.*

Although the circulation around a circuit containing the origin in an irrotational vortex is nonzero, that around a circuit *not* containing the origin is zero. The circulation around any such contour ABCD (Figure 3.16) is

$$\Gamma_{\text{ABCD}} = \int_{\text{AB}} \mathbf{u} \cdot d\mathbf{s} + \int_{\text{BC}} \mathbf{u} \cdot d\mathbf{s} + \int_{\text{CD}} \mathbf{u} \cdot d\mathbf{s} + \int_{\text{DA}} \mathbf{u} \cdot d\mathbf{s}.$$

Because the line integrals of $\mathbf{u} \cdot d\mathbf{s}$ around BC and DA are zero, we obtain

$$\Gamma_{\text{ABCD}} = -u_\theta r \, \Delta\theta + (u_\theta + \Delta u_\theta)(r + \Delta r) \, \Delta\theta = 0,$$

where we have noted that the line integral along AB is negative because $\mathbf{u}$ and $d\mathbf{s}$ are oppositely directed, and we have used $u_\theta r = \text{const}$. A zero circulation around ABCD is expected because of Stokes' theorem, and the fact that vorticity vanishes everywhere within ABCD.

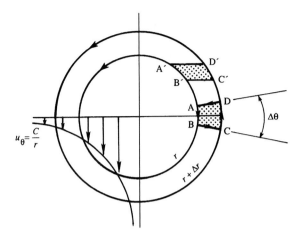

Figure 3.16 Irrotational vortex. Vorticity of a fluid element is infinite at the origin and zero everywhere else.

Rankine Vortex

Real vortices, such as a bathtub vortex or an atmospheric cyclone, have a core that rotates nearly like a solid body and an approximately irrotational far field (Figure 3.17a). A rotational core must exist because the tangential velocity in an irrotational vortex has an infinite velocity jump at the origin. An idealization of such a behavior is called the *Rankine vortex*, in which the vorticity is assumed uniform within a core of radius R and zero outside the core (Figure 3.17b).

12. One-, Two-, and Three-Dimensional Flows

A truly *one-dimensional flow* is one in which all flow characteristics vary in one direction only. Few real flows are strictly one dimensional. Consider the flow in a conduit (Figure 3.18a). The flow characteristics here vary both along the direction of flow and over the cross section. However, for some purposes, the analysis can be simplified by assuming that the flow variables are uniform over the cross section (Figure 3.18b). Such a simplification is called a *one-dimensional approximation*, and is satisfactory if one is interested in the overall effects at a cross section.

A *two-dimensional* or *plane* flow is one in which the variation of flow characteristics occurs in two Cartesian directions only. The flow past a cylinder of arbitrary cross section and infinite length is an example of plane flow. (Note that in this context the word "cylinder" is used for describing any body whose shape is invariant along the length of the body. It can have an *arbitrary* cross section. A cylinder with a *circular* cross section is a special case. Sometimes, however, the word "cylinder" is used to describe circular cylinders only.)

Around bodies of revolution, the flow variables are identical in planes containing the axis of the body. Using cylindrical polar coordinates (R, φ, x), with x along the axis of the body, only two coordinates (R and x) are necessary to describe motion

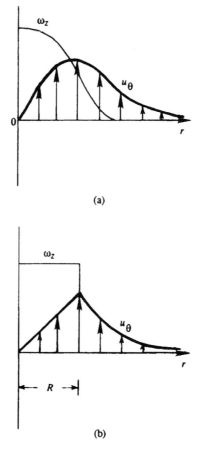

Figure 3.17 Velocity and vorticity distributions in a real vortex and a Rankine vortex: (a) real vortex; (b) Rankine vortex.

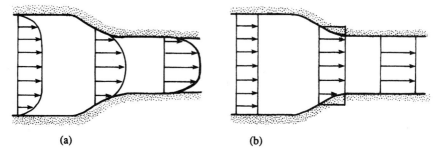

Figure 3.18 Flow through a conduit and its one-dimensional approximation: (a) real flow; (b) one-dimensional approximation.

(see Figure 6.27). The flow could therefore be called "two dimensional" (although not plane), but it is customary to describe such motions as *three-dimensional axisymmetric flows*.

13.   The Streamfunction

The description of incompressible two-dimensional flows can be considerably simplified by defining a function that satisfies the law of conservation of mass for such flows. Although the conservation laws are derived in the following chapter, a simple and alternative derivation of the mass conservation equation is given here. We proceed from the volumetric strain rate given in (3.9), namely,

$$\frac{1}{\delta \mathcal{V}} \frac{D}{Dt} (\delta \mathcal{V}) = \frac{\partial u_i}{\partial x_i}.$$

The *D/Dt* signifies that a specific fluid particle is followed, so that the volume of a particle is inversely proportional to its density. Substituting $\delta \mathcal{V} \propto \rho^{-1}$, we obtain

$$-\frac{1}{\rho} \frac{D\rho}{Dt} = \frac{\partial u_i}{\partial x_i}. \tag{3.34}$$

This is called the *continuity equation* because it assumes that the fluid flow has no voids in it; the name is somewhat misleading because all laws of continuum mechanics make this assumption.

   The density of fluid particles does not change appreciably along the fluid path under certain conditions, the most important of which is that the flow speed should be small compared with the speed of sound in the medium. This is called the Boussinesq approximation and is discussed in more detail in Chapter 4, Section 18. The condition holds in most flows of liquids, and in flows of gases in which the speeds are less than about 100 m/s. In these flows $\rho^{-1} D\rho/Dt$ is much less than any of the derivatives in $\partial u_i/\partial x_i$, under which condition the continuity equation (steady or unsteady) becomes

$$\boxed{\frac{\partial u_i}{\partial x_i} = 0.}$$

In many cases the continuity equation consists of two terms only, say

$$\frac{\partial u}{\partial x} + \frac{\partial v}{\partial y} = 0. \tag{3.35}$$

This happens if w is not a function of z. A plane flow with $w = 0$ is the most common example of such two-dimensional flows. If a function $\psi(x, y, t)$ is now defined such that

$$u \equiv \frac{\partial \psi}{\partial y},$$

$$v \equiv -\frac{\partial \psi}{\partial x}, \tag{3.36}$$

then equation (3.34) is automatically satisfied. Therefore, a streamfunction ψ can be defined whenever equation (3.34) is valid. (A similar streamfunction can be defined for incompressible axisymmetric flows in which the continuity equation involves R and x coordinates only; for compressible flows a streamfunction can be defined if the motion is two dimensional *and* steady (Exercise 2).)

The streamlines of the flow are given by

$$\frac{dx}{u} = \frac{dy}{v}. \tag{3.37}$$

Substitution of equation (3.35) into equation (3.36) shows

$$\frac{\partial \psi}{\partial x}\, dx + \frac{\partial \psi}{\partial y}\, dy = 0,$$

which says that $d\psi = 0$ along a streamline. The instantaneous streamlines in a flow are therefore given by the curves $\psi = $ const., a different value of the constant giving a different streamline (Figure 3.19).

Consider an arbitrary line element $d\mathbf{x} = (dx, dy)$ in the flow of Figure 3.19. Here we have shown a case in which both dx and dy are positive. The volume rate of flow across such a line element is

$$v\, dx + (-u)\, dy = -\frac{\partial \psi}{\partial x}\, dx - \frac{\partial \psi}{\partial y}\, dy = -d\psi,$$

showing that the volume flow rate between a pair of streamlines is numerically equal to the difference in their ψ values. The sign of ψ is such that, facing the direction of motion, ψ increases to the *left*. This can also be seen from the definition equation

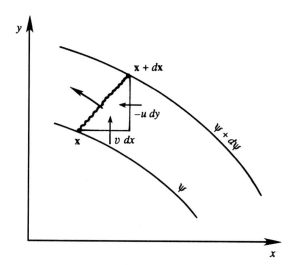

Figure 3.19   Flow through a pair of streamlines.

(3.35), according to which the derivative of ψ in a certain direction gives the velocity component in a direction 90° clockwise from the direction of differentiation. This requires that ψ in Figure 3.19 must increase downward if the flow is from right to left.

One purpose of defining a streamfunction is to be able to plot streamlines. A more theoretical reason, however, is that it decreases the number of simultaneous equations to be solved. For example, it will be shown in Chapter 10 that the momentum and mass conservation equations for viscous flows near a planar solid boundary are given, respectively, by

$$ u\frac{\partial u}{\partial x} + v\frac{\partial u}{\partial y} = \nu\frac{\partial^2 u}{\partial y^2}, \qquad (3.38) $$

$$ \frac{\partial u}{\partial x} + \frac{\partial v}{\partial y} = 0. \qquad (3.39) $$

The pair of simultaneous equations in u and v can be combined into a single equation by defining a streamfunction, when the momentum equation (3.37) becomes

$$ \frac{\partial \psi}{\partial y}\frac{\partial^2 \psi}{\partial x\,\partial y} - \frac{\partial \psi}{\partial x}\frac{\partial^2 \psi}{\partial y^2} = \nu\frac{\partial^3 \psi}{\partial y^3}. $$

We now have a single unknown function and a single differential equation. The continuity equation (3.38) has been satisfied automatically.

Summarizing, a streamfunction can be defined whenever the continuity equation consists of two terms. The flow can otherwise be completely general, for example, it can be rotational, viscous, and so on. The lines $\psi = C$ are the instantaneous streamlines, and the flow rate between two streamlines equals $d\psi$. This concept will be generalized following our derivation of mass conservation in Chapter 4, Section 3.

14. Polar Coordinates

It is sometimes easier to work with polar coordinates, especially in problems involving circular boundaries. In fact, we often select a coordinate system to conform to the shape of the body (boundary). It is customary to consult a reference source for expressions of various quantities in non-Cartesian coordinates, and this practice is perfectly satisfactory. However, it is good to know how an equation can be transformed from Cartesian into other coordinates. Here, we shall illustrate the procedure by transforming the Laplace equation

$$ \nabla^2 \psi = \frac{\partial^2 \psi}{\partial x^2} + \frac{\partial^2 \psi}{\partial y^2}, $$

to plane polar coordinates.

Cartesian and polar coordinates are related by

$$ \begin{aligned} x &= r\cos\theta & \theta &= \tan^{-1}(y/x), \\ y &= r\sin\theta & r &= \sqrt{x^2 + y^2}. \end{aligned} \qquad (3.40) $$

Let us first determine the polar velocity components in terms of the streamfunction. Because $\psi = f(x, y)$, and x and y are themselves functions of r and θ, the chain rule of partial differentiation gives

$$\left(\frac{\partial \psi}{\partial r}\right)_\theta = \left(\frac{\partial \psi}{\partial x}\right)_y \left(\frac{\partial x}{\partial r}\right)_\theta + \left(\frac{\partial \psi}{\partial y}\right)_x \left(\frac{\partial y}{\partial r}\right)_\theta.$$

Omitting parentheses and subscripts, we obtain

$$\frac{\partial \psi}{\partial r} = \frac{\partial \psi}{\partial x} \cos\theta + \frac{\partial \psi}{\partial y} \sin\theta = -v\cos\theta + u\sin\theta. \qquad (3.41)$$

Figure 3.20 shows that $u_\theta = v\cos\theta - u\sin\theta$, so that equation (3.40) implies $\partial\psi/\partial r = -u_\theta$. Similarly, we can show that $\partial\psi/\partial\theta = ru_r$. Therefore, the polar velocity components are related to the streamfunction by

$$u_r = \frac{1}{r}\frac{\partial\psi}{\partial\theta},$$

$$u_\theta = -\frac{\partial\psi}{\partial r}.$$

This is in agreement with our previous observation that the derivative of ψ gives the velocity component in a direction 90° clockwise from the direction of differentiation.

Now let us write the Laplace equation in polar coordinates. The chain rule gives

$$\frac{\partial\psi}{\partial x} = \frac{\partial\psi}{\partial r}\frac{\partial r}{\partial x} + \frac{\partial\psi}{\partial\theta}\frac{\partial\theta}{\partial x} = \cos\theta\frac{\partial\psi}{\partial r} - \frac{\sin\theta}{r}\frac{\partial\psi}{\partial\theta}.$$

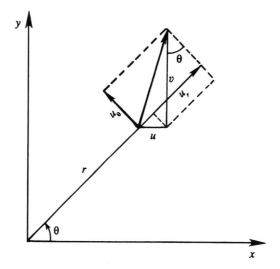

Figure 3.20   Relation of velocity components in Cartesian and plane polar coordinates.

Differentiating this with respect to x, and following a similar rule, we obtain

$$\frac{\partial^2 \psi}{\partial x^2} = \cos\theta \frac{\partial}{\partial r}\left[\cos\theta \frac{\partial\psi}{\partial r} - \frac{\sin\theta}{r}\frac{\partial\psi}{\partial\theta}\right] - \frac{\sin\theta}{r}\frac{\partial}{\partial\theta}\left[\cos\theta \frac{\partial\psi}{\partial r} - \frac{\sin\theta}{r}\frac{\partial\psi}{\partial\theta}\right].$$

(3.42)

In a similar manner,

$$\frac{\partial^2 \psi}{\partial y^2} = \sin\theta \frac{\partial}{\partial r}\left[\sin\theta \frac{\partial\psi}{\partial r} + \frac{\cos\theta}{r}\frac{\partial\psi}{\partial\theta}\right] + \frac{\cos\theta}{r}\frac{\partial}{\partial\theta}\left[\sin\theta \frac{\partial\psi}{\partial r} + \frac{\cos\theta}{r}\frac{\partial\psi}{\partial\theta}\right].$$

(3.43)

The addition of equations (3.41) and (3.42) leads to

$$\frac{\partial^2 \psi}{\partial x^2} + \frac{\partial^2 \psi}{\partial y^2} = \frac{1}{r}\frac{\partial}{\partial r}\left(r\frac{\partial\psi}{\partial r}\right) + \frac{1}{r^2}\frac{\partial^2 \psi}{\partial\theta^2} = 0,$$

which completes the transformation.

Exercises

1. A two-dimensional steady flow has velocity components

$$u = y \qquad v = x.$$

Show that the streamlines are rectangular hyperbolas

$$x^2 - y^2 = \text{const.}$$

Sketch the flow pattern, and convince yourself that it represents an irrotational flow in a 90° corner.

2. Consider a steady axisymmetric flow of a compressible fluid. The equation of continuity in cylindrical coordinates (R, φ, x) is

$$\frac{\partial}{\partial R}(\rho R u_R) + \frac{\partial}{\partial x}(\rho R u_x) = 0.$$

Show how we can define a streamfunction so that the equation of continuity is satisfied automatically.

3. If a velocity field is given by $u = ay$, compute the circulation around a circle of radius $r = 1$ about the origin. Check the result by using Stokes' theorem.

4. Consider a plane Couette flow of a viscous fluid confined between two flat plates at a distance b apart (see Figure 9.4c). At steady state the velocity distribution is

$$u = Uy/b \qquad v = w = 0,$$

where the upper plate at $y = b$ is moving parallel to itself at speed U, and the lower plate is held stationary. Find the rate of linear strain, the rate of shear strain, and vorticity. Show that the streamfunction is given by

$$\psi = \frac{Uy^2}{2b} + \text{const.}$$

5. Show that the vorticity for a plane flow on the xy-plane is given by

$$\omega_z = -\left(\frac{\partial^2 \psi}{\partial x^2} + \frac{\partial^2 \psi}{\partial y^2}\right).$$

Using this expression, find the vorticity for the flow in Exercise 4.

6. The velocity components in an unsteady plane flow are given by

$$u = \frac{x}{1+t} \quad \text{and} \quad v = \frac{2y}{2+t}.$$

Describe the path lines and the streamlines. Note that path lines are found by following the motion of each particle, that is, by solving the differential equations

$$dx/dt = u(\mathbf{x}, t) \quad \text{and} \quad dy/dt = v(\mathbf{x}, t),$$

subject to $\mathbf{x} = \mathbf{x}_0$ at $t = 0$.

7. Determine an expression for ψ for a Rankine vortex (Figure 3.17b), assuming that $u_\theta = U$ at $r = R$.

8. Take a plane polar element of fluid of dimensions dr and $r\,d\theta$. Evaluate the right-hand side of Stokes' theorem

$$\int \boldsymbol{\omega} \cdot d\mathbf{A} = \int \mathbf{u} \cdot d\mathbf{s},$$

and thereby show that the expression for vorticity in polar coordinates is

$$\omega_z = \frac{1}{r}\left[\frac{\partial}{\partial r}(ru_\theta) - \frac{\partial u_r}{\partial \theta}\right].$$

Also, find the expressions for ω_r and ω_θ in polar coordinates in a similar manner.

9. The velocity field of a certain flow is given by

$$u = 2xy^2 + 2xz^2, \qquad v = x^2 y, \qquad w = x^2 z.$$

Consider the fluid region inside a spherical volume $x^2 + y^2 + z^2 = a^2$. Verify the validity of Gauss' theorem

$$\int \nabla \cdot \mathbf{u}\,dV = \int \mathbf{u} \cdot d\mathbf{A},$$

by integrating over the sphere.

10. Show that the vorticity field for *any* flow satisfies

$$\nabla \cdot \boldsymbol{\omega} = 0.$$

11. A flow field on the xy-plane has the velocity components

$$u = 3x + y \qquad v = 2x - 3y.$$

Show that the circulation around the circle $(x - 1)^2 + (y - 6)^2 = 4$ is 4π.

12. Consider the solid-body rotation

$$u_\theta = \omega_0 r \qquad u_r = 0.$$

Take a polar element of dimension $r\, d\theta$ and dr, and verify that the circulation is vorticity times area. (In Section 11 we performed such a verification for a circular element surrounding the *origin*.)

13. Using the indicial notation (and without using any vector identity) show that the acceleration of a fluid particle is given by

$$\mathbf{a} = \frac{\partial \mathbf{u}}{\partial t} + \nabla \left(\frac{1}{2} q^2 \right) + \boldsymbol{\omega} \times \mathbf{u},$$

where q is the magnitude of velocity $\mathbf{u}$ and $\boldsymbol{\omega}$ is the vorticity.

14. The definition of the streamfunction in vector notation is

$$\mathbf{u} = -\mathbf{k} \times \nabla \psi,$$

where $\mathbf{k}$ is a unit vector perpendicular to the plane of flow. Verify that the vector definition is equivalent to equations (3.35).

Supplemental Reading

Aris, R. (1962). *Vectors, Tensors, and the Basic Equations of Fluid Mechanics*, Englewood Cliffs, NJ: Prentice-Hall. (The distinctions among streamlines, path lines, and streak lines in unsteady flows are explained; with examples.)

Prandtl, L. and O. C. Tietjens (1934). *Fundamentals of Hydro- and Aeromechanics*, New York: Dover Publications. (Chapter V contains a simple but useful treatment of kinematics.)

Prandtl, L. and O. G. Tietjens (1934). *Applied Hydro- and Aeromechanics*, New York: Dover Publications. (This volume contains classic photographs from Prandtl's laboratory.)

Conservation Laws

©2010 Elsevier Inc. All rights reserved.
DOI: 10.1016/B978-0-12-381399-2.50004-6

1. Introduction

All fluid mechanics is based on the conservation laws for mass, momentum, and energy. These laws can be stated in the *differential* form, applicable at a point. They can also be stated in the *integral* form, applicable to an extended region. In the integral form, the expressions of the laws depend on whether they relate to a *volume fixed in space*, or to a *material volume*, which consists of the same fluid particles and whose bounding surface moves with the fluid. Both types of volumes will be considered in this chapter; *a fixed region will be denoted by V and a material volume will be denoted by 𝒱*. In engineering literature a fixed region is called a *control volume*, whose surfaces are called *control surfaces*.

The integral and differential forms can be derived from each other. As we shall see, during the derivation surface integrals frequently need to be converted to volume integrals (or vice versa) by means of the divergence theorem of Gauss

$$\int_V \frac{\partial F}{\partial x_i} dV = \int_A dA_i\, F, \qquad (4.1)$$

where $F(\mathbf{x}, t)$ is a tensor of *any* rank (including vectors and scalars), V is either a fixed volume or a material volume, and A is its boundary surface. Gauss' theorem was presented in Section 2.13.

2. Time Derivatives of Volume Integrals

In deriving the conservation laws, one frequently faces the problem of finding the time derivative of integrals such as

$$\frac{d}{dt} \int_{V(t)} F\, dV,$$

where $F(\mathbf{x}, t)$ is a tensor of any order, and $V(t)$ is any region, which may be fixed or move with the fluid. The d/dt sign (in contrast to $\partial/\partial t$) has been written because only a function of time remains after performing the integration in space. The different possibilities are discussed in what follows.

General Case

Consider the general case in which $V(t)$ is neither a fixed volume nor a material volume. The surfaces of the volume are moving, but not with the local fluid velocity. The rule for differentiating an integral becomes clear at once if we consider a one-dimensional (1D) analogy. In books on calculus,

$$\frac{d}{dt} \int_{x=a(t)}^{b(t)} F(x, t)\, dx = \int_a^b \frac{\partial F}{\partial t}\, dx + \frac{db}{dt} F(b, t) - \frac{da}{dt} F(a, t). \qquad (4.2)$$

This is called the *Leibniz theorem*, and shows how to differentiate an integral whose integrand F as well as the limits of integration are functions of the variable with respect to which we are differentiating. A graphical illustration of the three terms on

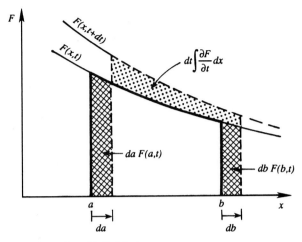

Figure 4.1   Graphical illustration of Leibniz's theorem.

the right-hand side of the Leibniz theorem is shown in Figure 4.1. The continuous line shows the integral $\int F\,dx$ at time t, and the dashed line shows the integral at time $t+dt$. The first term on the right-hand side in equation (4.2) is the integral of $\partial F/\partial t$ over the region, the second term is due to the gain of F at the outer boundary moving at a rate db/dt, and the third term is due to the loss of F at the inner boundary moving at da/dt.

Generalizing the Leibniz theorem, we write

$$\frac{d}{dt}\int_{V(t)} F(\mathbf{x},t)\,dV = \int_{V(t)} \frac{\partial F}{\partial t}\,dV + \int_{A(t)} d\mathbf{A}\cdot \mathbf{u}_A F, \qquad (4.3)$$

where $\mathbf{u}_A$ is the velocity of the boundary and $A(t)$ is the surface of $V(t)$. The surface integral in equation (4.3) accounts for both "inlets" and "outlets," so that separate terms as in equation (4.2) are not necessary.

Fixed Volume

For a fixed volume we have $\mathbf{u}_A = 0$, for which equation (4.3) becomes

$$\boxed{\frac{d}{dt}\int_{V} F(\mathbf{x},t)\,dV = \int_{V} \frac{\partial F}{\partial t}\,dV,} \qquad (4.4)$$

which shows that the time derivative can be simply taken inside the integral sign if the boundary is fixed. This merely reflects the fact that the "limit of integration" V is not a function of time in this case.

Material Volume

For a material volume $\mathcal{V}(t)$ the surfaces move with the fluid, so that $\mathbf{u}_A = \mathbf{u}$, where $\mathbf{u}$ is the fluid velocity. Then equation (4.3) becomes

$$\frac{D}{Dt} \int_{\mathcal{V}} F(\mathbf{x}, t) \, d\mathcal{V} = \int_{\mathcal{V}} \frac{\partial F}{\partial t} \, d\mathcal{V} + \int_A d\mathbf{A} \cdot \mathbf{u} F. \tag{4.5}$$

This is sometimes called the *Reynolds transport theorem*. Although not necessary, we have used the D/Dt symbol here to emphasize that we are following a material volume.

Another form of the transport theorem is derived by using the mass conservation relation equation (3.32) derived in the last chapter. Using Gauss' theorem, the transport theorem equation (4.5) becomes

$$\frac{D}{Dt} \int_{\mathcal{V}} F \, d\mathcal{V} = \int_{\mathcal{V}} \left[\frac{\partial F}{\partial t} + \frac{\partial}{\partial x_j} (F u_j) \right] d\mathcal{V}.$$

Now define a new function f such that $F \equiv \rho f$, where ρ is the fluid density. Then the preceding becomes

$$\frac{D}{Dt} \int \rho f \, d\mathcal{V} = \int \left[\frac{\partial (\rho f)}{\partial t} + \frac{\partial}{\partial x_j} (\rho f u_j) \right] d\mathcal{V}$$

$$= \int \left[\rho \frac{\partial f}{\partial t} + f \frac{\partial \rho}{\partial t} + f \frac{\partial}{\partial x_j} (\rho u_j) + \rho u_j \frac{\partial f}{\partial x_j} \right] d\mathcal{V}.$$

Using the continuity equation

$$\frac{\partial \rho}{\partial t} + \frac{\partial}{\partial x_j} (\rho u_j) = 0.$$

we finally obtain

$$\frac{D}{Dt} \int_{\mathcal{V}} \rho f \, d\mathcal{V} = \int_{\mathcal{V}} \rho \frac{Df}{Dt} \, d\mathcal{V}. \tag{4.6}$$

Notice that the D/Dt operates only on f on the right-hand side, although ρ is variable. Applications of this rule can be found in Sections 7 and 14.

3. Conservation of Mass

The differential form of the law of conservation of mass was derived in Chapter 3, Section 13 from a consideration of the volumetric rate of strain of a particle. In this chapter we shall adopt an alternative approach. We shall first state the principle in an integral form for a fixed region and then deduce the differential form. Consider a

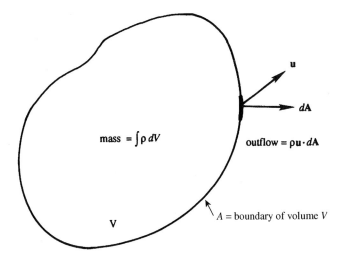

Figure 4.2 Mass conservation of a volume fixed in space.

volume fixed in space (Figure 4.2). The rate of increase of mass inside it is the volume integral

$$\frac{d}{dt} \int_V \rho \, dV = \int_V \frac{\partial \rho}{\partial t} \, dV.$$

The time derivative has been taken inside the integral on the right-hand side because the volume is fixed and equation (4.4) applies. Now the rate of mass flow out of the volume is the surface integral

$$\int_A \rho \mathbf{u} \cdot d\mathbf{A},$$

because $\rho \mathbf{u} \cdot d\mathbf{A}$ is the outward flux through an area element $d\mathbf{A}$. (Throughout the book, we shall write $d\mathbf{A}$ for $\mathbf{n} \, dA$, where $\mathbf{n}$ is the unit outward normal to the surface. Vector $d\mathbf{A}$ therefore has a magnitude dA and a direction along the outward normal.) The law of conservation of mass states that the rate of increase of mass within a fixed volume must equal the rate of inflow through the boundaries. Therefore,

$$\int_V \frac{\partial \rho}{\partial t} \, dV = - \int_A \rho \mathbf{u} \cdot d\mathbf{A}, \tag{4.7}$$

which is the integral form of the law for a volume fixed in space.

The differential form can be obtained by transforming the surface integral on the right-hand side of equation (4.7) to a volume integral by means of the divergence theorem, which gives

$$\int_A \rho \mathbf{u} \cdot d\mathbf{A} = \int_V \nabla \cdot (\rho \mathbf{u}) \, dV.$$

Equation (4.7) then becomes

$$\int_V \left[\frac{\partial \rho}{\partial t} + \nabla \cdot (\rho \mathbf{u}) \right] dV = 0.$$

The forementioned relation holds for *any* volume, which can be possible only if the integrand vanishes at every point. (If the integrand did not vanish at every point, then we could choose a small volume around that point and obtain a nonzero integral.) This requires

$$\frac{\partial \rho}{\partial t} + \nabla \cdot (\rho \mathbf{u}) = 0, \qquad (4.8)$$

which is called the *continuity equation* and expresses the differential form of the principle of conservation of mass.

The equation can be written in several other forms. Rewriting the divergence term in equation (4.8) as

$$\frac{\partial}{\partial x_i} (\rho u_i) = \rho \frac{\partial u_i}{\partial x_i} + u_i \frac{\partial \rho}{\partial x_i},$$

the equation of continuity becomes

$$\frac{1}{\rho} \frac{D\rho}{Dt} + \nabla \cdot \mathbf{u} = 0. \qquad (4.9)$$

The derivative $D\rho/Dt$ is the rate of change of density following a fluid particle; it can be nonzero because of changes in pressure, temperature, or composition (such as salinity in sea water). A fluid is usually called *incompressible* if its density does not change with *pressure*. Liquids are almost incompressible. Although gases are compressible, for speeds $\lesssim 100\,\text{m/s}$ (that is, for Mach numbers <0.3) the fractional change of absolute pressure in the flow is small. In this and several other cases the density changes in the flow are also small. The neglect of $\rho^{-1} D\rho/Dt$ in the continuity equation is part of a series of simplifications grouped under the Boussinesq approximation, discussed in Section 18. In such a case the continuity equation (4.9) reduces to the incompressible form

$$\nabla \cdot \mathbf{u} = 0, \qquad (4.10)$$

whether or not the flow is steady.

4. Streamfunctions: Revisited and Generalized

Consider the steady-state form of mass conservation from equation (4.8),

$$\nabla \cdot (\rho \mathbf{u}) = 0. \tag{4.11}$$

In Exercise 10 of Chapter 2 we showed that the divergence of the curl of any vector field is identically zero. Thus we can represent the mass flow vector as the curl of a vector potential

$$\rho \mathbf{u} = \nabla \times \mathbf{\Omega}, \tag{4.12}$$

where we can write $\mathbf{\Omega} = \chi \nabla \psi + \nabla \phi$ in terms of three scalar functions. We are concerned with the mass flux field $\rho \mathbf{u} = \nabla \chi \times \nabla \psi$ because the curl of any gradient is identically zero (Chapter 2, Exercise 11). The gradients of the surfaces $\chi = $ const. and $\psi = $ const. are in the directions of the surface normals. Thus the cross product is perpendicular to both normals and must lie simultaneously in both surfaces $\chi = $ const. and $\psi = $ const. Thus streamlines are the intersections of the two surfaces, called streamsurfaces or streamfunctions in a three-dimensional (3D) flow. Consider an edge view of two members of each of the families of the two streamfunctions $\chi = a$, $\chi = b$, $\psi = c$, $\psi = d$. The intersections shown as darkened dots in Figure 4.3 are the streamlines coming out of the paper. We calculate the mass per time through a surface A bounded by the four streamfunctions with element $\mathbf{dA}$ having $\mathbf{n}$ out of the paper. By Stokes' theorem,

$$\dot{m} = \int_A \rho \mathbf{u} \cdot \mathbf{dA} = \int_A (\nabla \times \mathbf{\Omega}) \cdot \mathbf{dA} = \int_C \mathbf{\Omega} \cdot \mathbf{ds} = \int_C (\chi \nabla \psi + \nabla \phi) \cdot \mathbf{ds}$$
$$= \int_C (\chi d\psi + d\phi) = \int_C \chi d\psi = b(d - c) + a(c - d) = (b - a)(d - c).$$

Here we have used the vector identity $\nabla \phi \bullet \mathbf{ds} = d\phi$ and recognized that integration around a closed path of a single-valued function results in zero. The mass per time through a surface bounded by adjacent members of the two families of streamfunctions is just the product of the differences of the numerical values of the respective streamfunctions. As a very simple special case, consider flow in a $z = $ constant plane

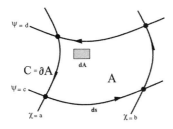

Figure 4.3 Edge view of two members of each of two families of streamfunctions. Contour C is the boundary of surface area $A : C = \partial A$.

(described by x and y coordinates). Because all the streamlines lie in $z = $ constant planes, z is a streamfunction. Define $\chi = -z$, where the sign is chosen to obey the usual convention. Then $\nabla \chi = -\mathbf{k}$ (unit vector in the z direction), and

$$\rho \mathbf{u} = -\mathbf{k} \times \nabla \psi; \quad \rho u = \partial \psi / \partial y, \quad \rho v = -\partial \psi / \partial x,$$

in conformity with Chapter 3, Exercise 14.

Similarly, in cyclindrical polar coordinates as shown in Figure 3.1, flows, symmetric with respect to rotation about the x-axis, that is, those for which $\partial / \partial \phi = 0$, have streamlines in $\phi = $ constant planes (through the x-axis). For those axisymmetric flows, $\chi = -\phi$ is one streamfunction:

$$\rho \mathbf{u} = -\frac{1}{R} \mathbf{i}_\phi \times \nabla \psi,$$

then gives $\rho R u_x = \partial \psi / \partial R, \rho R u_R = -\partial \psi / \partial x$. We note here that if the density may be taken as a constant, mass conservation reduces to $\nabla \cdot \mathbf{u} = 0$ (steady or not) and the entire preceding discussion follows for $\mathbf{u}$ rather than $\rho \mathbf{u}$ with the interpretation of streamfunction in terms of volumetric rather than mass flux.

5. Origin of Forces in Fluid

Before we can proceed further with the conservation laws, it is necessary to classify the various types of forces on a fluid mass. The forces acting on a fluid element can be divided conveniently into three classes, namely, body forces, surface forces, and line forces. These are described as follows:

(1) *Body forces*: Body forces are those that arise from "action at a distance," without physical contact. They result from the medium being placed in a certain *force field*, which can be gravitational, magnetic, electrostatic, or electromagnetic in origin. They are distributed throughout the mass of the fluid and are proportional to the mass. Body forces are expressed either per unit mass or per unit volume. In this book, the body force per unit mass will be denoted by $\mathbf{g}$.

Body forces can be conservative or nonconservative. *Conservative body forces* are those that can be expressed as the gradient of a potential function:

$$\mathbf{g} = -\nabla \Pi, \tag{4.13}$$

where Π is called the *force potential*. All forces directed *centrally* from a source are conservative. Gravity, electrostatic and magnetic forces are conservative. For example, the gravity force can be written as the gradient of the potential function

$$\Pi = gz,$$

where g is the acceleration due to gravity and z points vertically upward. To verify this, equation (4.13) gives

$$\mathbf{g} = -\nabla(gz) = -\left[\mathbf{i}\frac{\partial}{\partial x} + \mathbf{j}\frac{\partial}{\partial y} + \mathbf{k}\frac{\partial}{\partial z}\right](gz) = -\mathbf{k}g,$$

which is the gravity force per unit mass. (Here we have changed our usual convention for unit vectors and used the more standard form.) The negative sign in front of $\mathbf{k}g$ ensures that $\mathbf{g}$ is downward, along the negative z direction. The expression $\Pi = gz$ also shows that the *force potential equals the potential energy per unit mass*. Forces satisfying equation (4.13) are called "conservative" because the resulting motion conserves the sum of kinetic and potential energies, if there are no dissipative processes. Conservative forces also satisfy the property that the work done is independent of the path.

(2) *Surface forces*: Surface forces are those that are exerted on an area element by the surroundings through direct contact. They are proportional to the extent of the area and are conveniently expressed per unit of area. Surface forces can be resolved into components normal and tangential to the area. Consider an element of area dA in a fluid (Figure 4.4). The force $d\mathbf{F}$ on the element can be resolved into a component dF_n normal to the area and a component dF_s tangential to the area. The normal and shear stress on the element are defined, respectively, as,

$$\tau_n \equiv \frac{dF_n}{dA} \qquad \tau_s \equiv \frac{dF_s}{dA}.$$

These are scalar definitions of stress components. Note that the component of force tangential to the surface is a two-dimensional (2D) vector in the surface. The state of stress at a point is, in fact, specified by a stress tensor, which has nine components. This was explained in Section 2.4 and is again discussed in the following section.

(3) *Line forces*: Surface tension forces are called *line forces* because they act along a line (Figure 1.4) and have a magnitude proportional to the extent of the line. They appear at the interface between a liquid and a gas, or at the interface between two immiscible liquids. Surface tension forces do not appear directly in the equations of motion, but *enter only in the boundary conditions*.

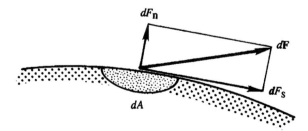

Figure 4.4   Normal and shear forces on an area element.

6. *Stress at a Point*

It was explained in Chapter 2, Section 4 that the stress at a point can be completely specified by the nine components of the stress tensor $\boldsymbol{\tau}$. Consider an infinitesimal rectangular parallelepiped with faces perpendicular to the coordinate axes (Figure 4.5). On each face there is a normal stress and a shear stress, which can be further resolved into two components in the directions of the axes. The figure shows the directions of *positive* stresses on four of the six faces; those on the remaining two faces are omitted for clarity. The first index of τ_{ij} indicates the direction of the normal to the surface on which the stress is considered, and the second index indicates the direction in which the stress acts. The diagonal elements τ_{11}, τ_{22}, and τ_{33} of the stress matrix are the normal stresses, and the off-diagonal elements are the tangential or shear stresses. Although a cube is shown, the figure really shows the stresses on four of the six orthogonal planes passing through a point; the cube may be imagined to shrink to a point.

We shall now prove that the *stress tensor is symmetric*. Consider the torque on an element about a centroid axis parallel to x_3 (Figure 4.6). This torque is generated only by the shear stresses in the $x_1 x_2$-plane and is (assuming $dx_3 = 1$)

$$
T = \left[\tau_{12} + \frac{1}{2} \frac{\partial \tau_{12}}{\partial x_1} dx_1 \right] dx_2 \frac{dx_1}{2} + \left[\tau_{12} - \frac{1}{2} \frac{\partial \tau_{12}}{\partial x_1} dx_1 \right] dx_2 \frac{dx_1}{2}
$$
$$
- \left[\tau_{21} + \frac{1}{2} \frac{\partial \tau_{21}}{\partial x_2} dx_2 \right] dx_1 \frac{dx_2}{2} - \left[\tau_{21} - \frac{1}{2} \frac{\partial \tau_{21}}{\partial x_2} dx_2 \right] dx_1 \frac{dx_2}{2}.
$$

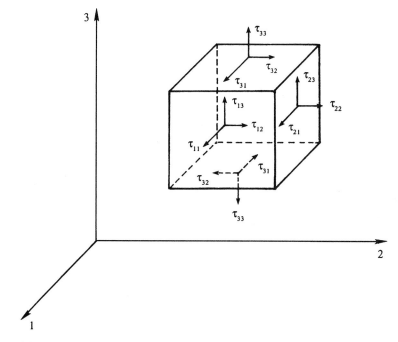

Figure 4.5   Stress at a point. For clarity, components on only four of the six faces are shown.

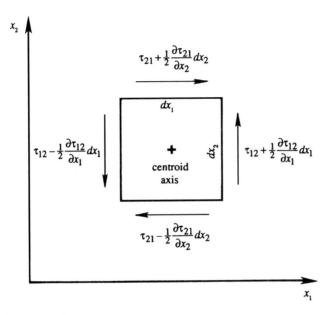

Figure 4.6 Torque on an element.

After canceling terms, this gives

$$T = (\tau_{12} - \tau_{21}) \, dx_1 \, dx_2.$$

The rotational equilibrium of the element requires that $T = I\dot{\omega}_3$, where $\dot{\omega}_3$ is the angular acceleration of the element and I is its moment of inertia. For the rectangular element considered, it is easy to show that $I = dx_1 \, dx_2 (dx_1^2 + dx_2^2)\rho/12$. The rotational equilibrium then requires

$$(\tau_{12} - \tau_{21}) \, dx_1 \, dx_2 = \frac{\rho}{12} \, dx_1 \, dx_2 (dx_1^2 + dx_2^2) \, \dot{\omega}_3,$$

that is,

$$\tau_{12} - \tau_{21} = \frac{\rho}{12} (dx_1^2 + dx_2^2) \, \dot{\omega}_3.$$

As dx_1 and dx_2 go to zero, the preceding condition can be satisfied only if $\tau_{12} = \tau_{21}$. In general,

$$\boxed{\tau_{ij} = \tau_{ji}.} \tag{4.14}$$

See Exercise 3 at the end of the chapter.

The stress tensor is therefore symmetric and has only six independent components. The symmetry, however, is violated if there are "body couples" proportional to the mass of the fluid element, such as those exerted by an electric field on polarized fluid molecules. Antisymmetric stresses must be included in such fluids.

7. *Conservation of Momentum*

In this section the law of conservation of momentum will be expressed in the differential form directly by applying Newton's law of motion to an infinitesimal fluid element. We shall then show how the differential form could be derived by starting from an integral form of Newton's law.

Consider the motion of the infinitesimal fluid element shown in Figure 4.7. Newton's law requires that the net force on the element must equal mass times the acceleration of the element. The sum of the surface forces in the x_1 direction equals

$$\left(\tau_{11} + \frac{\partial \tau_{11}}{\partial x_1} \frac{dx_1}{2} - \tau_{11} + \frac{\partial \tau_{11}}{\partial x_1} \frac{dx_1}{2} \right) dx_2 \, dx_3$$

$$+ \left(\tau_{21} + \frac{\partial \tau_{21}}{\partial x_2} \frac{dx_2}{2} - \tau_{21} + \frac{\partial \tau_{21}}{\partial x_2} \frac{dx_2}{2} \right) dx_1 \, dx_3$$

$$+ \left(\tau_{31} + \frac{\partial \tau_{31}}{\partial x_3} \frac{dx_3}{2} - \tau_{31} + \frac{\partial \tau_{31}}{\partial x_3} \frac{dx_3}{2} \right) dx_1 \, dx_2,$$

which simplifies to

$$\left(\frac{\partial \tau_{11}}{\partial x_1} + \frac{\partial \tau_{21}}{\partial x_2} + \frac{\partial \tau_{31}}{\partial x_3} \right) dx_1 \, dx_2 \, dx_3 = \frac{\partial \tau_{j1}}{\partial x_j} d\mathcal{V},$$

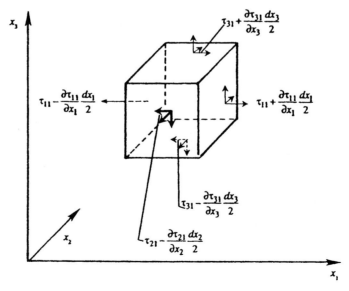

Figure 4.7 Surface stresses on an element moving with the flow. Only stresses in the x_1 direction are labeled.

where $d\mathcal{V}$ is the volume of the element. Generalizing, the i-component of the *surface force per unit volume* of the element is

$$\frac{\partial \tau_{ij}}{\partial x_j},$$

where we have used the symmetry property $\tau_{ij} = \tau_{ji}$. Let $\mathbf{g}$ be the body force per unit mass, so that $\rho \mathbf{g}$ is the body force per unit volume. Then Newton's law gives

$$\rho \frac{Du_i}{Dt} = \rho g_i + \frac{\partial \tau_{ij}}{\partial x_j}. \qquad (4.15)$$

This is the equation of motion relating acceleration to the net force at a point and holds for any continuum, solid or fluid, no matter how the stress tensor τ_{ij} is related to the deformation field. Equation (4.15) is sometimes called *Cauchy's equation of motion*.

We shall now deduce Cauchy's equation starting from an *integral* statement of Newton's law for a material volume $\mathcal{V}$. In this case we do not have to consider the internal stresses within the fluid, but only the surface forces at the boundary of the volume (along with body forces). It was shown in Chapter 2, Section 6 that the surface force per unit area is $\mathbf{n} \cdot \boldsymbol{\tau}$, where $\mathbf{n}$ is the unit outward normal. The surface force on an area element $d\mathbf{A}$ is therefore $d\mathbf{A} \cdot \boldsymbol{\tau}$. Newton's law for a material volume $\mathcal{V}$ requires that the rate of change of its momentum equals the sum of body forces throughout the volume, plus the surface forces at the boundary. Therefore

$$\frac{D}{Dt} \int_{\mathcal{V}} \rho u_i \, d\mathcal{V} = \int_{\mathcal{V}} \rho \frac{Du_i}{Dt} d\mathcal{V} = \int_{\mathcal{V}} \rho g_i \, d\mathcal{V} + \int_A \tau_{ij} \, dA_j, \qquad (4.16)$$

where equations (4.6) and (4.14) have been used. Transforming the surface integral to a volume integral, equation (4.16) becomes

$$\int \left[\rho \frac{Du_i}{Dt} - \rho g_i - \frac{\partial \tau_{ij}}{\partial x_j} \right] d\mathcal{V} = 0.$$

As this holds for any volume, the integrand must vanish at every point and therefore equation (4.15) must hold. We have therefore derived the differential form of the equation of motion, starting from an integral form.

8. Momentum Principle for a Fixed Volume

In the preceding section the momentum principle was applied to a *material* volume of finite size and this led to equation (4.16). In this section the form of the law will be derived for a fixed region in space. It is easy to do this by starting from the differential form (4.15) and integrating over a fixed volume V. Adding u_i times the continuity equation

$$\frac{\partial \rho}{\partial t} + \frac{\partial}{\partial x_j}(\rho u_j) = 0,$$

to the left-hand side of equation (4.15), we obtain

$$\frac{\partial}{\partial t}(\rho u_i) + \frac{\partial}{\partial x_j}(\rho u_i u_j) = \rho g_i + \frac{\partial \tau_{ij}}{\partial x_j}. \tag{4.17}$$

Each term of equation (4.17) is now integrated over a fixed region V. The time derivative term gives

$$\int_V \frac{\partial(\rho u_i)}{\partial t} dV = \frac{d}{dt} \int_V \rho u_i \, dV = \frac{dM_i}{dt}, \tag{4.18}$$

where

$$M_i \equiv \int_V \rho u_i \, dV,$$

is the momentum of the fluid inside the volume. The volume integral of the second term in equation (4.17) becomes, after applying Gauss' theorem,

$$\int_V \frac{\partial}{\partial x_j}(\rho u_i u_j) \, dV = \int_A \rho u_i u_j \, dA_j \equiv \dot{M}_i^{\text{out}}, \tag{4.19}$$

where $\dot{M}_i^{\text{out}}$ is the net rate of outflux of i-momentum. (Here $\rho u_j \, dA_j$ is the mass outflux through an area element $d\mathbf{A}$ on the boundary. Outflux of momentum is defined as the outflux of mass times the velocity.) The volume integral of the third term in equation (4.17) is simply

$$\int \rho g_i \, dV = F_{bi}, \tag{4.20}$$

where $\mathbf{F}_b$ is the net body force acting over the entire volume. The volume integral of the fourth term in equation (4.17) gives, after applying Gauss' theorem,

$$\int_V \frac{\partial \tau_{ij}}{\partial x_j} dV = \int_A \tau_{ij} \, dA_j \equiv F_{si}, \tag{4.21}$$

where $\mathbf{F}_s$ is the net surface force at the boundary of V. If we define $\mathbf{F} = \mathbf{F}_b + \mathbf{F}_s$ as the sum of all forces, then the volume integral of equation (4.17) finally gives

$$\boxed{\mathbf{F} = \frac{d\mathbf{M}}{dt} + \dot{\mathbf{M}}^{\text{out}},} \tag{4.22}$$

where equations (4.18)–(4.21) have been used.

Equation (4.22) is the law of conservation of momentum for a fixed volume. It states that the net force on a fixed volume equals the rate of change of momentum within the volume, plus the net outflux of momentum through the surfaces. The equation has three independent components, where the x-component is

$$F_x = \frac{dM_x}{dt} + \dot{M}_x^{\text{out}}.$$

The momentum principle (frequently called the *momentum theorem*) has wide application, especially in engineering. An example is given in what follows. More illustrations can be found throughout the book, for example, in Chapter 9, Section 4, Chapter 10, Section 11, Chapter 13, Section 10, and Chapter 16, Sections 2 and 3.

Example 4.1. Consider an experiment in which the drag on a 2D body immersed in a steady incompressible flow can be determined from measurement of the velocity distributions far upstream and downstream of the body (Figure 4.8). Velocity far upstream is the uniform flow U_∞, and that in the wake of the body is measured to be $u(y)$, which is less than U_∞ due to the drag of the body. Find the drag force D per unit length of the body.

Solution: The wake velocity $u(y)$ is less than U_∞ due to the drag forces exerted by the body on the fluid. To analyze the flow, take a fixed volume shown by the dashed lines in Figure 4.8. It consists of the rectangular region PQRS and has a hole in the center coinciding with the surface of the body. The sides PQ and SR are chosen far enough from the body so that the pressure nearly equals the undisturbed pressure p_∞. The side QR at which the velocity profile is measured is also at a far enough distance for the streamlines to be nearly parallel; the pressure variation across the wake is therefore small, so that it is nearly equal to the undisturbed pressure p_∞. The surface forces on PQRS therefore cancel out, and the only force acting at the boundary of the chosen fixed volume is D, the force exerted by the body at the central hole.

For steady flow, the x-component of the momentum principle (4.22) reduces to

$$D = \dot{M}^{\text{out}}, \tag{4.23}$$

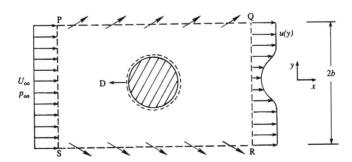

Figure 4.8   Momentum balance of flow over a body (Example 4.1).

where $\dot{M}^{\text{out}}$ is the net outflow rate of x-momentum through the boundaries of the region. There is no flow of momentum through the central hole in Figure 4.8. Outflow rates of x-momentum through PS and QR are

$$\dot{M}^{\text{PS}} = -\int_{-b}^{b} U_\infty (\rho U_\infty \, dy) = -2b\rho U_\infty^2, \tag{4.24}$$

$$\dot{M}^{\text{QR}} = \int_{-b}^{b} u(\rho u \, dy) = \rho \int_{-b}^{b} u^2 \, dy. \tag{4.25}$$

An important point is that there is an outflow of mass and x-momentum through PQ and SR. A mass flux through PQ and SR is required because the velocity across QR is less than that across PS. Conservation of mass requires that the inflow through PS, equal to $2b\rho U_\infty$, must balance the outflows through PQ, SR, and QR. This gives

$$2b\rho U_\infty = \dot{m}^{\text{PQ}} + \dot{m}^{\text{SR}} + \rho \int_{-b}^{b} u \, dy,$$

where $\dot{m}^{\text{PQ}}$ and $\dot{m}^{\text{SR}}$ are the outflow rates of mass through the sides. The mass balance can be written as

$$\dot{m}^{\text{PQ}} + \dot{m}^{\text{SR}} = \rho \int_{-b}^{b} (U_\infty - u) \, dy.$$

Outflow rate of x-momentum through PQ and SR is therefore

$$\dot{M}^{\text{PQ}} + \dot{M}^{\text{SR}} = \rho U_\infty \int_{-b}^{b} (U_\infty - u) \, dy, \tag{4.26}$$

because the x-directional velocity at these surfaces is nearly U_∞. Combining equations (4.22)–(4.26) gives a net outflow of x-momentum of:

$$\dot{M}^{\text{out}} = \dot{M}^{\text{PS}} + \dot{M}^{\text{QR}} + \dot{M}^{\text{PQ}} + \dot{M}^{\text{SR}} = -\rho \int_{-b}^{b} u(U_\infty - u) \, dy.$$

The momentum balance (4.23) now shows that the body exerts a force on the fluid in the negative x direction of magnitude

$$D = \rho \int_{-b}^{b} u(U_\infty - u) \, dy,$$

which can be evaluated from the measured velocity profile.

A more general way of obtaining the force on a body immersed in a flow is by using the Euler momentum integral, which we derive in what follows. We must assume that the flow is steady and body forces are absent. Then integrating (4.17) over a fixed volume gives

$$\int_V \nabla \cdot (\rho \mathbf{uu} - \boldsymbol{\tau}) dV = \int_A (\rho \mathbf{uu} - \boldsymbol{\tau}) \cdot \mathbf{dA}, \qquad (4.27)$$

where A is the closed surface bounding V. This volume V contains *only* fluid particles. Imagine a body immersed in a flow and surround that body with a closed surface. We seek to calculate the force on the body by an integral over a possibly distant surface. In order to apply (4.27), A must bound a volume containing only fluid particles. This is accomplished by considering A to be composed of three parts (see Figure 4.9),

$$A = A_1 + A_2 + A_3.$$

Here A_1 is the outer surface, A_2 is wrapped around the body like a tight-fitting rubber glove with dA_2 pointing outwards from the fluid volume and, therefore, into the body, and A_3 is the connection surface between the outer A_1 and the inner A_2. Now

$$\int_{A_3} (\rho \mathbf{uu} - \boldsymbol{\tau}) \cdot \mathbf{dA}_3 \to 0 \qquad \text{as } A_3 \to 0,$$

because it may be taken as the bounding surface of an evanescent thread. On the surface of a solid body, $\mathbf{u} \cdot \mathbf{dA}_2 = 0$ because no mass enters or leaves the surface. Here $\int_{A_2} \boldsymbol{\tau} \cdot \mathbf{dA}_2$ is the force the body exerts on the fluid from our definition of $\boldsymbol{\tau}$. Then the force the fluid exerts on the body is

$$\mathbf{F_B} = -\int_{A_2} \boldsymbol{\tau} \cdot \mathbf{dA}_2 = -\int_{A_1} (\rho \mathbf{uu} - \boldsymbol{\tau}) \cdot \mathbf{dA}_1. \qquad (4.28)$$

Using similar arguments, mass conservation can be written in the form

$$\int_{A_1} \rho \mathbf{u} \cdot \mathbf{dA}_1 = 0. \qquad (4.29)$$

Equations (4.28) and (4.29) can be used to solve Example 4.1. Of course, the same final result is obtained when $\boldsymbol{\tau} \approx$ constant pressure on all of A_1, $\rho = $ constant, and the x component of $\mathbf{u} = U_\infty \mathbf{i}$ on segments PQ and SR of A_1.

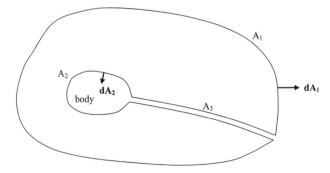

Figure 4.9 Surfaces of integration for the Euler momentum integral.

9. *Angular Momentum Principle for a Fixed Volume*

In mechanics of solids it is shown that

$$\mathbf{T} = \frac{d\mathbf{H}}{dt}, \tag{4.30}$$

where $\mathbf{T}$ is the torque of all external forces on the body about any chosen axis, and $d\mathbf{H}/dt$ is the rate of change of angular momentum of the body about the same axis. The angular momentum is defined as the "moment of momentum," that is

$$\mathbf{H} \equiv \int \mathbf{r} \times \mathbf{u}\, dm,$$

where dm is an element of mass, and $\mathbf{r}$ is the position vector from the chosen axis (Figure 4.10). The angular momentum principle is *not* a separate law, but can be derived from Newton's law by performing a cross product with $\mathbf{r}$. It can be shown that equation (4.30) also holds for a material volume in a fluid. When equation (4.30) is transformed to apply to a *fixed volume*, the result is

$$\mathbf{T} = \frac{d\mathbf{H}}{dt} + \dot{\mathbf{H}}^{\text{out}}, \tag{4.31}$$

where

$$\mathbf{T} = \int_A \mathbf{r} \times (\boldsymbol{\tau} \cdot d\mathbf{A}) + \int_V \mathbf{r} \times (\rho \mathbf{g}\, dV),$$

$$\mathbf{H} = \int_V \mathbf{r} \times (\rho \mathbf{u}\, dV),$$

$$\dot{\mathbf{H}}^{\text{out}} = \int_A \mathbf{r} \times [(\rho \mathbf{u} \cdot d\mathbf{A})\mathbf{u}].$$

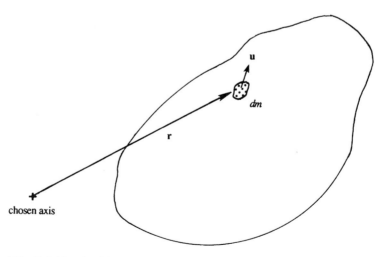

Figure 4.10   Definition sketch for angular momentum theorem.

Here **T** represents the sum of torques due to surface and body forces, $\boldsymbol{\tau} \cdot d\mathbf{A}$ is the surface force on a boundary element, and $\rho \mathbf{g} dV$ is the body force acting on an interior element. Vector **H** represents the angular momentum of fluid inside the fixed volume because $\rho \mathbf{u} dV$ is the momentum of a volume element. Finally, $\dot{\mathbf{H}}^{\text{out}}$ is the rate of outflow of angular momentum through the boundary, $\rho \mathbf{u} \cdot d\mathbf{A}$ is the mass flow rate, and $(\rho \mathbf{u} \cdot d\mathbf{A})\mathbf{u}$ is the momentum outflow rate through a boundary element $d\mathbf{A}$.

The angular momentum principle (4.31) is analogous to the linear momentum principle (4.22), and is very useful in investigating rotating fluid systems such as turbomachines, fluid couplings, and even lawn sprinklers.

Example 4.2. Consider a lawn sprinkler as shown in Figure 4.11. The area of the nozzle exit is A, and the jet velocity is U. Find the torque required to hold the rotor stationary.

Solution: Select a stationary volume V shown by the dashed lines. Pressure everywhere on the control surface is atmospheric, and there is no net moment due to the pressure forces. The control surface cuts through the vertical support and the torque T exerted by the support on the sprinkler arm is the only torque acting on V. Apply the angular momentum balance

$$T = \dot{H}_z^{\text{out}}.$$

Let $\dot{m} = \rho A U$ be the mass flux through each nozzle. As the angular momentum is the moment of momentum, we obtain

$$\dot{H}_z^{\text{out}} = (\dot{m} U \cos \alpha)a + (\dot{m} U \cos \alpha)a = 2a\rho A U^2 \cos \alpha.$$

Therefore, the torque required to hold the rotor stationary is

$$T = 2a\rho A U^2 \cos \alpha.$$

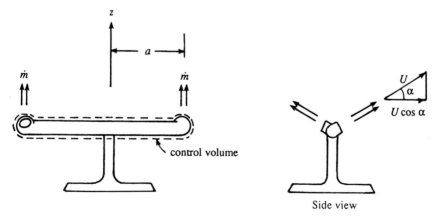

Figure 4.11   Lawn sprinkler (Example 4.2).

When the sprinkler is rotating at a steady state, this torque is balanced by both air resistance and mechanical friction.

10. Constitutive Equation for Newtonian Fluid

The relation between the stress and deformation in a continuum is called a *constitutive equation*. An equation that linearly relates the stress to the rate of strain in a fluid medium is examined in this section.

In a fluid at rest there are only normal components of stress on a surface, and the stress does not depend on the orientation of the surface. In other words, the stress tensor is *isotropic* or spherically symmetric. An isotropic tensor is defined as one whose components do not change under a rotation of the coordinate system (see Chapter 2, Section 7). The only second-order isotropic tensor is the Kronecker delta

$$\delta = \begin{bmatrix} 1 & 0 & 0 \\ 0 & 1 & 0 \\ 0 & 0 & 1 \end{bmatrix}.$$

Any isotropic second-order tensor must be proportional to δ. Therefore, because the stress in a static fluid is isotropic, it must be of the form

$$\tau_{ij} = -p\delta_{ij}, \tag{4.32}$$

where p is the *thermodynamic pressure* related to ρ and T by an equation of state (e.g., the thermodynamic pressure for a perfect gas is $p = \rho RT$). A negative sign is introduced in equation (4.32) because the normal components of τ are regarded as positive if they indicate tension rather than compression.

A moving fluid develops additional components of stress due to viscosity. The diagonal terms of τ now become unequal, and shear stresses develop. For a moving fluid we can split the stress into a part $-p\delta_{ij}$ that would exist if it were at rest and a part σ_{ij} due to the fluid motion alone:

$$\tau_{ij} = -p\delta_{ij} + \sigma_{ij}. \tag{4.33}$$

We shall assume that p appearing in equation (4.33) is still the thermodynamic pressure. The assumption, however, is not on a very firm footing because thermodynamic quantities are defined for equilibrium states, whereas a moving fluid undergoing diffusive fluxes is generally not in equilibrium. Such departures from thermodynamic equilibrium are, however, expected to be unimportant if the relaxation (or adjustment) time of the molecules is small compared to the time scale of the flow, as discussed in Chapter 1, Section 8.

The nonisotropic part σ, called the *deviatoric stress tensor*, is related to the velocity gradients $\partial u_i/\partial x_j$. The velocity gradient tensor can be decomposed into symmetric and antisymmetric parts:

$$\frac{\partial u_i}{\partial x_j} = \frac{1}{2}\left(\frac{\partial u_i}{\partial x_j} + \frac{\partial u_j}{\partial x_i}\right) + \frac{1}{2}\left(\frac{\partial u_i}{\partial x_j} - \frac{\partial u_j}{\partial x_i}\right).$$

The antisymmetric part represents fluid rotation without deformation, and cannot by itself generate stress. The stresses must be generated by the strain rate tensor

$$e_{ij} \equiv \frac{1}{2}\left(\frac{\partial u_i}{\partial x_j} + \frac{\partial u_j}{\partial x_i}\right),$$

alone. We shall assume a linear relation of the type

$$\sigma_{ij} = K_{ijmn}e_{mn}, \tag{4.34}$$

where K_{ijmn} is a fourth-order tensor having 81 components that depend on the thermodynamic state of the medium. Equation (4.34) simply means that *each* stress component is linearly related to *all* nine components of e_{ij}; altogether 81 constants are therefore needed to completely describe the relationship.

It will now be shown that only two of the 81 elements of K_{ijmn} survive if it is assumed that the medium is isotropic and that the stress tensor is symmetric. An isotropic medium has no directional preference, which means that the stress–strain relationship is independent of rotation of the coordinate system. This is only possible if K_{ijmn} is an isotropic tensor. It is shown in books on tensor analysis (e.g., see Aris (1962), pp. 30–33) that all isotropic tensors of even order are made up of products of δ_{ij}, and that a fourth-order isotropic tensor must have the form

$$K_{ijmn} = \lambda\delta_{ij}\delta_{mn} + \mu\delta_{im}\delta_{jn} + \gamma\delta_{in}\delta_{jm}, \tag{4.35}$$

where λ, μ, and γ are scalars that depend on the local thermodynamic state. As σ_{ij} is a symmetric tensor, equation (4.34) requires that K_{ijmn} also must be symmetric in i and j. This is consistent with equation (4.35) only if

$$\gamma = \mu. \tag{4.36}$$

Only two constants μ and λ, of the original 81, have therefore survived under the restrictions of material isotropy and stress symmetry. Substitution of equation (4.35) into the constitutive equation (4.34) gives

$$\sigma_{ij} = 2\mu e_{ij} + \lambda e_{mm}\delta_{ij},$$

where $e_{mm} = \nabla \cdot \mathbf{u}$ is the volumetric strain rate (explained in Chapter 3, Section 6). The complete stress tensor (4.33) then becomes

$$\tau_{ij} = -p\delta_{ij} + 2\mu e_{ij} + \lambda e_{mm}\delta_{ij}. \tag{4.37}$$

The two scalar constants μ and λ can be further related as follows. Setting $i = j$, summing over the repeated index, and noting that $\delta_{ii} = 3$, we obtain

$$\tau_{ii} = -3p + (2\mu + 3\lambda)\,e_{mm},$$

from which the pressure is found to be

$$p = -\frac{1}{3}\tau_{ii} + \left(\frac{2}{3}\mu + \lambda\right)\nabla \cdot \mathbf{u}. \tag{4.38}$$

Now the diagonal terms of e_{ij} in a flow may be unequal. In such a case the stress tensor τ_{ij} can have unequal diagonal terms because of the presence of the term proportional to μ in equation (4.37). We can therefore take the average of the diagonal terms of $\boldsymbol{\tau}$ and define a *mean pressure* (as opposed to thermodynamic pressure p) as

$$\bar{p} \equiv -\frac{1}{3}\tau_{ii}. \tag{4.39}$$

Substitution into equation (4.38) gives

$$p - \bar{p} = \left(\frac{2}{3}\mu + \lambda\right)\nabla \cdot \mathbf{u}. \tag{4.40}$$

For a completely incompressible fluid we can only define a mechanical or mean pressure, because there is no equation of state to determine a thermodynamic pressure. (In fact, the *absolute pressure in an incompressible fluid is indeterminate,* and only its *gradients* can be determined from the equations of motion.) The λ-term in the constitutive equation (4.37) drops out because $e_{mm} = \nabla \cdot \mathbf{u} = 0$, and no consideration of equation (4.40) is necessary. For *incompressible fluids*, the constitutive equation (4.37) takes the simple form

$$\boxed{\tau_{ij} = -p\delta_{ij} + 2\mu e_{ij}} \quad \text{(incompressible)}, \tag{4.41}$$

where p can only be interpreted as the mean pressure. For a compressible fluid, on the other hand, a thermodynamic pressure can be defined, and it seems that p and $\bar{p}$ can be different. In fact, equation (4.40) relates this difference to the rate of expansion through the proportionality constant $\kappa = \lambda + 2\mu/3$, which is called the *coefficient of bulk viscosity*. In principle, κ is a measurable quantity; however, extremely large values of $D\rho/Dt$ are necessary in order to make any measurement, such as within shock waves. Moreover, measurements are inconclusive about the nature of κ. For many applications the *Stokes assumption*

$$\lambda + \frac{2}{3}\mu = 0, \tag{4.42}$$

is found to be sufficiently accurate, and can also be supported from the kinetic theory of monatomic gases. Interesting historical aspects of the Stokes assumption $3\lambda + 2\mu = 0$ can be found in Truesdell (1952).

To gain additional insight into the distinction between thermodynamic pressure and the mean of the normal stresses, consider a system inside a cylinder in which a piston may be moved in or out to do work. The first law of thermodynamics may be written in general terms as $de = dw + dQ = -\bar{p}dv + dQ = -pdv + TdS$, where the last equality is written in terms of state functions. Then $TdS - dQ = (p - \bar{p})dv$. The Clausius-Duhem inequality (see under equation 1.16) tells us $TdS - dQ \geq 0$ for any process and, consequently, $(p - \bar{p})dv \geq 0$. Thus, for an expansion, $dv > 0$, so $p > \bar{p}$, and conversely for a compression. Equation (4.40) is:

$$p - \bar{p} = \left(\frac{2}{3}\mu + \lambda\right) \nabla \cdot \mathbf{u} = -\left(\frac{2}{3}\mu + \lambda\right) \frac{1}{\rho} \frac{D\rho}{Dt} = \left(\frac{2}{3}\mu + \lambda\right) \frac{1}{v} \frac{Dv}{Dt}, \qquad v = \frac{1}{\rho}.$$

Further, we require $(2/3)\mu + \lambda > 0$ to satisfy the Clausius-Duhem inequality statement of the second law.

With the assumption $\kappa = 0$, the constitutive equation (4.37) reduces to

$$\boxed{\tau_{ij} = -\left(p + \tfrac{2}{3}\mu\nabla \cdot \mathbf{u}\right)\delta_{ij} + 2\mu e_{ij}} \qquad (4.43)$$

This linear relation between $\boldsymbol{\tau}$ and $\mathbf{e}$ is consistent with Newton's definition of viscosity coefficient in a simple parallel flow $u(y)$, for which equation (4.43) gives a shear stress of $\tau = \mu(du/dy)$. Consequently, a fluid obeying equation (4.43) is called a *Newtonian fluid*. The fluid property μ in equation (4.43) can depend on the local thermodynamic state alone.

The nondiagonal terms of equation (4.43) are easy to understand. They are of the type

$$\tau_{12} = \mu\left(\frac{\partial u_1}{\partial x_2} + \frac{\partial u_2}{\partial x_1}\right),$$

which relates the shear stress to the strain rate. The diagonal terms are more difficult to understand. For example, equation (4.43) gives

$$\tau_{11} = -p + 2\mu\left[-\frac{1}{3}\frac{\partial u_i}{\partial x_i} + \frac{\partial u_1}{\partial x_1}\right],$$

which means that the normal viscous stress on a plane normal to the x_1-axis is proportional to the *difference* between the extension rate in the x_1 direction and the average expansion rate at the point. Therefore, only those extension rates different from the average will generate normal viscous stress.

Non-Newtonian Fluids

The linear Newtonian friction law is expected to hold for small rates of strain because higher powers of $\mathbf{e}$ are neglected. However, for common fluids such as air and water the linear relationship is found to be surprisingly accurate for most applications. Some liquids important in the chemical industry, on the other hand, display non-Newtonian behavior at moderate rates of strain. These include: (1) solutions containing polymer molecules, which have very large molecular weights and form long chains coiled together in spongy ball-like shapes that deform under shear; and (2) emulsions and slurries containing suspended particles, two examples of which are blood and water containing clay. These liquids violate Newtonian behavior in several ways—for example, shear stress is a *nonlinear* function of the local strain rate. It depends not only on the local strain rate, but also on its *history*. Such a "memory" effect gives the fluid an elastic property, in addition to its viscous property. Most non-Newtonian fluids are therefore *viscoelastic*. Only Newtonian fluids will be considered in this book.

11. Navier–Stokes Equation

The equation of motion for a Newtonian fluid is obtained by substituting the constitutive equation (4.43) into Cauchy's equation (4.15) to obtain

$$\rho \frac{Du_i}{Dt} = -\frac{\partial p}{\partial x_i} + \rho g_i + \frac{\partial}{\partial x_j}\left[2\mu e_{ij} - \frac{2}{3}\mu(\nabla \cdot \mathbf{u})\delta_{ij}\right], \qquad (4.44)$$

where we have noted that $(\partial p/\partial x_j)\delta_{ij} = \partial p/\partial x_i$. Equation (4.44) is a general form of the *Navier–Stokes equation*. Viscosity μ in this equation can be a function of the thermodynamic state, and indeed μ for most fluids displays a rather strong dependence on temperature, decreasing with T for liquids and increasing with T for gases. However, if the temperature differences are small within the fluid, then μ can be taken outside the derivative in equation (4.44), which then reduces to

$$\rho \frac{Du_i}{Dt} = -\frac{\partial p}{\partial x_i} + \rho g_i + 2\mu \frac{\partial e_{ij}}{\partial x_j} - \frac{2\mu}{3}\frac{\partial}{\partial x_i}(\nabla \cdot \mathbf{u})$$

$$= -\frac{\partial p}{\partial x_i} + \rho g_i + \mu\left[\nabla^2 u_i + \frac{1}{3}\frac{\partial}{\partial x_i}(\nabla \cdot \mathbf{u})\right],$$

where

$$\nabla^2 u_i \equiv \frac{\partial^2 u_i}{\partial x_j \partial x_j} = \frac{\partial^2 u_i}{\partial x_1^2} + \frac{\partial^2 u_i}{\partial x_2^2} + \frac{\partial^2 u_i}{\partial x_3^2},$$

is the Laplacian of u_i. For incompressible fluids $\nabla \cdot \mathbf{u} = 0$, and using vector notation, the Navier–Stokes equation reduces to

$$\boxed{\rho \frac{D\mathbf{u}}{Dt} = -\nabla p + \rho \mathbf{g} + \mu \, \nabla^2 \mathbf{u}.} \qquad \text{(incompressible)} \qquad (4.45)$$

If viscous effects are negligible, which is generally found to be true far from boundaries of the flow field, we obtain the *Euler equation*

$$\rho \frac{D\mathbf{u}}{Dt} = -\nabla p + \rho \mathbf{g}. \qquad (4.46)$$

Comments on the Viscous Term

For an incompressible fluid, equation (4.41) shows that the viscous stress at a point is

$$\sigma_{ij} = \mu \left(\frac{\partial u_i}{\partial x_j} + \frac{\partial u_j}{\partial x_i} \right), \tag{4.47}$$

which shows that σ depends only on the deformation rate of a fluid element at a point, and not on the rotation rate ($\partial u_i / \partial x_j - \partial u_j / \partial x_i$). We have built this property into the Newtonian constitutive equation, based on the fact that in a solid-body rotation (that is a flow in which the tangential velocity is proportional to the radius) the particles do not deform or "slide" past each other, and therefore they do not cause viscous stress.

However, consider the net viscous force per unit volume at a point, given by

$$F_i = \frac{\partial \sigma_{ij}}{\partial x_j} = \mu \frac{\partial}{\partial x_j} \left(\frac{\partial u_i}{\partial x_j} + \frac{\partial u_j}{\partial x_i} \right) = \mu \frac{\partial^2 u_i}{\partial x_j \, \partial x_j} = -\mu (\nabla \times \boldsymbol{\omega})_i, \tag{4.48}$$

where we have used the relation

$$
\begin{aligned}
(\nabla \times \boldsymbol{\omega})_i &= \varepsilon_{ijk} \frac{\partial \omega_k}{\partial x_j} = \varepsilon_{ijk} \frac{\partial}{\partial x_j} \left(\varepsilon_{kmn} \frac{\partial u_n}{\partial x_m} \right) \\
&= (\delta_{im} \delta_{jn} - \delta_{in} \delta_{jm}) \frac{\partial^2 u_n}{\partial x_j \, \partial x_m} = \frac{\partial^2 u_j}{\partial x_j \, \partial x_i} - \frac{\partial^2 u_i}{\partial x_j \, \partial x_j} \\
&= -\frac{\partial^2 u_i}{\partial x_j \, \partial x_j}.
\end{aligned}
$$

In the preceding derivation the "epsilon delta relation," given by equation (2.19), has been used. Relation (4.48) can cause some confusion because it seems to show that the net viscous force depends on *vorticity*, whereas equation (4.47) shows that viscous stress depends only on strain rate and is independent of local vorticity. The apparent paradox is explained by realizing that the net viscous force is given by either the spatial *derivative* of vorticity or the spatial *derivative* of deformation rate; both forms are shown in equation (4.48). The net viscous force vanishes when ω is uniform everywhere (as in solid-body rotation), in which case the incompressibility condition requires that the deformation is zero everywhere as well.

12. *Rotating Frame*

The equations of motion given in Section 7 are valid in an inertial or "fixed" frame of reference. Although such a frame of reference cannot be defined precisely, experience shows that these laws are accurate enough in a frame of reference stationary with respect to "distant stars." In geophysical applications, however, we naturally measure positions and velocities with respect to a frame of reference fixed on the surface of the earth, which rotates with respect to an inertial frame. In this section we shall derive the equations of motion in a rotating frame of reference. Similar derivations are also given by Batchelor (1967), Pedlosky (1987), and Holton (1979).

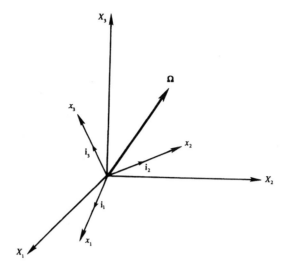

Figure 4.12 Coordinate frame (x_1, x_2, x_3) rotating at angular velocity $\mathbf{\Omega}$ with respect to a fixed frame (X_1, X_2, X_3).

Consider (Figure 4.12) a frame of reference (x_1, x_2, x_3) rotating at a uniform angular velocity $\mathbf{\Omega}$ with respect to a fixed frame (X_1, X_2, X_3). Any vector $\mathbf{P}$ is represented in the rotating frame by

$$\mathbf{P} = P_1 \mathbf{i}_1 + P_2 \mathbf{i}_2 + P_3 \mathbf{i}_3.$$

To a fixed observer the directions of the rotating unit vectors $\mathbf{i}_1$, $\mathbf{i}_2$, and $\mathbf{i}_3$ change with time. To this observer the time derivative of $\mathbf{P}$ is

$$\left(\frac{dP}{dt} \right)_F = \frac{d}{dt} (P_1 \mathbf{i}_1 + P_2 \mathbf{i}_2 + P_3 \mathbf{i}_3)$$

$$= \mathbf{i}_1 \frac{dP_1}{dt} + \mathbf{i}_2 \frac{dP_2}{dt} + \mathbf{i}_3 \frac{dP_3}{dt} + P_1 \frac{d\mathbf{i}_1}{dt} + P_2 \frac{d\mathbf{i}_2}{dt} + P_3 \frac{d\mathbf{i}_3}{dt}.$$

To the rotating observer, the rate of change of $\mathbf{P}$ is the sum of the first three terms, so that

$$\left(\frac{d\mathbf{P}}{dt} \right)_F = \left(\frac{d\mathbf{P}}{dt} \right)_R + P_1 \frac{d\mathbf{i}_1}{dt} + P_2 \frac{d\mathbf{i}_2}{dt} + P_3 \frac{d\mathbf{i}_3}{dt}. \tag{4.49}$$

Now each unit vector $\mathbf{i}$ traces a cone with a radius of $\sin \alpha$, where α is a constant angle (Figure 4.13). The magnitude of the change of $\mathbf{i}$ in time dt is $|d\mathbf{i}| = \sin \alpha \, d\theta$, which is the length traveled by the tip of $\mathbf{i}$. The magnitude of the rate of change is therefore $(d\mathbf{i}/dt) = \sin \alpha \, (d\theta/dt) = \Omega \sin \alpha$, and the direction of the rate of change is perpendicular to the $(\mathbf{\Omega}, \mathbf{i})$-plane. Thus $d\mathbf{i}/dt = \mathbf{\Omega} \times \mathbf{i}$ for any rotating unit vector $\mathbf{i}$. The sum of the last three terms in equation (4.49) is then $P_1 \mathbf{\Omega} \times \mathbf{i}_1$

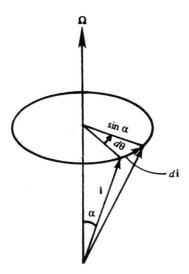

Figure 4.13 Rotation of a unit vector.

$+P_2\boldsymbol{\Omega} \times \mathbf{i}_2 + P_3\boldsymbol{\Omega} \times \mathbf{i}_3 = \boldsymbol{\Omega} \times \mathbf{P}$. Equation (4.49) then becomes

$$\left(\frac{d\mathbf{P}}{dt}\right)_F = \left(\frac{d\mathbf{P}}{dt}\right)_R + \boldsymbol{\Omega} \times \mathbf{P}, \tag{4.50}$$

which relates the rates of change of the vector $\mathbf{P}$ as seen by the two observers.

Application of rule (4.50) to the position vector $\mathbf{r}$ relates the velocities as

$$\mathbf{u}_F = \mathbf{u}_R + \boldsymbol{\Omega} \times \mathbf{r}. \tag{4.51}$$

Applying rule (4.50) on $\mathbf{u}_F$, we obtain

$$\left(\frac{d\mathbf{u}_F}{dt}\right)_F = \left(\frac{d\mathbf{u}_F}{dt}\right)_R + \boldsymbol{\Omega} \times \mathbf{u}_F,$$

which becomes, upon using equation (4.51),

$$\frac{d\mathbf{u}_F}{dt} = \frac{d}{dt}(\mathbf{u}_R + \boldsymbol{\Omega} \times \mathbf{r})_R + \boldsymbol{\Omega} \times (\mathbf{u}_R + \boldsymbol{\Omega} \times \mathbf{r})$$
$$= \left(\frac{d\mathbf{u}_R}{dt}\right)_R + \boldsymbol{\Omega} \times \left(\frac{d\mathbf{r}}{dt}\right)_R + \boldsymbol{\Omega} \times \mathbf{u}_R + \boldsymbol{\Omega} \times (\boldsymbol{\Omega} \times \mathbf{r}).$$

This shows that the accelerations in the two frames are related as

$$\mathbf{a}_F = \mathbf{a}_R + 2\boldsymbol{\Omega} \times \mathbf{u}_R + \boldsymbol{\Omega} \times (\boldsymbol{\Omega} \times \mathbf{r}), \qquad \dot{\boldsymbol{\Omega}} = 0, \tag{4.52}$$

The last term in equation (4.52) can be written in terms of the vector $\mathbf{R}$ drawn perpendicularly to the axis of rotation (Figure 4.14). Clearly, $\boldsymbol{\Omega} \times \mathbf{r} = \boldsymbol{\Omega} \times \mathbf{R}$. Using the

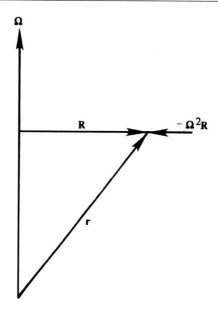

Figure 4.14 Centripetal acceleration.

vector identity $\mathbf{A} \times (\mathbf{B} \times \mathbf{C}) = (\mathbf{A} \cdot \mathbf{C})\mathbf{B} - (\mathbf{A} \cdot \mathbf{B})\mathbf{C}$, the last term of equation (4.52) becomes

$$\mathbf{\Omega} \times (\mathbf{\Omega} \times \mathbf{R}) = -(\mathbf{\Omega} \cdot \mathbf{\Omega})\mathbf{R} = -\Omega^2 \mathbf{R},$$

where we have set $\mathbf{\Omega} \cdot \mathbf{R} = 0$. Equation (4.52) then becomes

$$\mathbf{a}_F = \mathbf{a} + 2\mathbf{\Omega} \times \mathbf{u} - \Omega^2 \mathbf{R}, \tag{4.53}$$

where the subscript "R" has been dropped with the understanding that velocity $\mathbf{u}$ and acceleration $\mathbf{a}$ are measured in a rotating frame of reference. Equation (4.53) states that the "true" or inertial acceleration equals the acceleration measured in a rotating system, plus the Coriolis acceleration $2\mathbf{\Omega} \times \mathbf{u}$ and the centripetal acceleration $-\Omega^2 \mathbf{R}$.

Therefore, Coriolis and centripetal accelerations have to be considered if we are measuring quantities in a rotating frame of reference. Substituting equation (4.53) in equation (4.45), the equation of motion in a rotating frame of reference becomes

$$\frac{D\mathbf{u}}{Dt} = -\frac{1}{\rho}\nabla p + \nu\nabla^2 \mathbf{u} + (\mathbf{g}_n + \Omega^2 \mathbf{R}) - 2\mathbf{\Omega} \times \mathbf{u}, \tag{4.54}$$

where we have taken the Coriolis and centripetal acceleration terms to the right-hand side (now signifying Coriolis and centrifugal *forces*), and added a subscript on $\mathbf{g}$ to mean that it is the body force per unit mass due to (Newtonian) gravitational attractive forces alone.

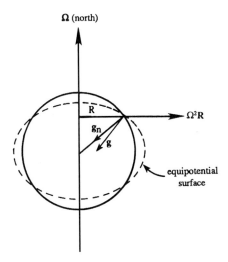

Figure 4.15   Effective gravity **g** and equipotential surface.

Effect of Centrifugal Force

The additional apparent force $\Omega^2\mathbf{R}$ can be added to the Newtonian gravity $\mathbf{g}_n$ to define an *effective gravity force* $\mathbf{g} = \mathbf{g}_n + \Omega^2\mathbf{R}$ (Figure 4.15). The Newtonian gravity would be uniform over the earth's surface, and be centrally directed, if the earth were spherically symmetric and homogeneous. However, the earth is really an ellipsoid with the equatorial diameter 42 km larger than the polar diameter. In addition, the existence of the centrifugal force makes the effective gravity less at the equator than at the poles, where $\Omega^2\mathbf{R}$ is zero. In terms of the effective gravity, equation (4.54) becomes

$$\frac{D\mathbf{u}}{Dt} = -\frac{1}{\rho}\nabla p + \nu\nabla^2\mathbf{u} + \mathbf{g} - 2\mathbf{\Omega} \times \mathbf{u}. \qquad (4.55)$$

The Newtonian gravity can be written as the gradient of a scalar potential function. It is easy to see that the centrifugal force can also be written in the same manner. From Definition (2.22), it is clear that the gradient of a spatial direction is the unit vector in that direction (e.g., $\nabla x = \mathbf{i}_x$), so that $\nabla(R^2/2) = R\mathbf{i}_R = \mathbf{R}$. Therefore, $\Omega^2\mathbf{R} = \nabla(\Omega^2 R^2/2)$, and the centrifugal potential is $-\Omega^2 R^2/2$. The *effective gravity* can therefore be written as $\mathbf{g} = -\nabla\Pi$, where Π is now the potential due to the Newtonian gravity, plus the centrifugal potential. The equipotential surfaces (shown by the dashed lines in Figure 4.15) are now perpendicular to the effective gravity. The average sea level is one of these equipotential surfaces. We can then write $\Pi = gz$, where z is measured perpendicular to an equipotential surface, and g is the effective acceleration due to gravity.

Effect of Coriolis Force

The angular velocity vector $\mathbf{\Omega}$ points out of the ground in the northern hemisphere. The Coriolis force $-2\mathbf{\Omega} \times \mathbf{u}$ therefore tends to deflect a particle to the right of its

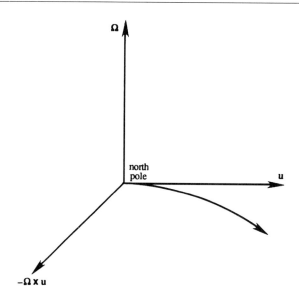

Figure 4.16 Deflection of a particle due to the Coriolis force.

direction of travel in the northern hemisphere (Figure 4.16) and to the left in the southern hemisphere.

Imagine a projectile shot horizontally from the north pole with speed u. The Coriolis force $2\Omega u$ constantly acts perpendicular to its path and therefore does not change the speed u of the projectile. The forward distance traveled in time t is ut, and the deflection is $\Omega u t^2$. The angular deflection is $\Omega u t^2 / u t = \Omega t$, which is the earth's rotation in time t. This demonstrates that the projectile in fact travels in a straight line if observed from the inertial outer space; its apparent deflection is merely due to the rotation of the earth underneath it. Observers on earth need an imaginary force to account for the apparent deflection. A clear physical explanation of the Coriolis force, with applications to mechanics, is given by Stommel and Moore (1989).

It is the Coriolis force that is responsible for the wind circulation patterns around centers of high and low pressure in the earth's atmosphere. Fluid flows from regions of higher pressure to regions of lower pressure, as (4.55) indicates acceleration of a fluid particle in a direction opposite the pressure gradient. Imagine a cylindrical polar coordinate system, as defined in Appendix B1, with the x-axis normal (outwards) to the local tangent plane to the earth's surface and the origin at the center of the "high" or "low." If it is a high pressure zone, u_R is outwards (positive) since flow is away from the center of high pressure. Then the Coriolis acceleration, the last term of (4.55), becomes $-2\mathbf{\Omega} \times \mathbf{u} = -\Omega_z u_r = -u_\theta$ is in the $-\theta$ direction (in the Northern hemisphere), or clockwise as viewed from above. On the other hand, flow is inwards toward the center of a low pressure zone, which reverses the direction of u_r and, therefore, u_θ is counter-clockwise. In the Southern hemisphere, the direction of Ω_z is reversed so that the circulation patterns described above are reversed.

Although the effects of a rotating frame will be commented on occasionally in this and subsequent chapters, most of the discussions involving Coriolis forces are given in Chapter 14, which deals with geophysical fluid dynamics.

13. *Mechanical Energy Equation*

An equation for kinetic energy of the fluid can be obtained by finding the scalar product of the momentum equation and the velocity vector. The kinetic energy equation is therefore not a separate principle, and is not the same as the first law of thermodynamics. We shall derive several forms of the equation in this section. The Coriolis force, which is perpendicular to the velocity vector, does not contribute to any of the energy equations. The equation of motion is

$$\rho \frac{Du_i}{Dt} = \rho g_i + \frac{\partial \tau_{ij}}{\partial x_j}.$$

Multiplying by u_i (and, of course, summing over i), we obtain

$$\rho \frac{D}{Dt} \left(\frac{1}{2} u_i^2 \right) = \rho u_i g_i + u_i \frac{\partial \tau_{ij}}{\partial x_j}, \tag{4.56}$$

where, for the sake of notational simplicity, we have written u_i^2 for $u_i u_i = u_1^2 + u_2^2 + u_3^2$. A summation over i is therefore implied in u_i^2, although no repeated index is explicitly written. Equation (4.56) is the simplest as well as most revealing mechanical energy equation. Recall from Section 7 that the resultant imbalance of the surface forces at a point is $\nabla \cdot \tau$, per unit volume. Equation (4.56) therefore says that the rate of increase of kinetic energy at a point equals the sum of the rate of work done by body force $\mathbf{g}$ and the rate of work done by the net surface force $\nabla \cdot \tau$ per unit volume.

Other forms of the mechanical energy equation are obtained by combining equation (4.56) with the continuity equation in various ways. For example, $u_i^2/2$ times the continuity equation is

$$\frac{1}{2} u_i^2 \left[\frac{\partial \rho}{\partial t} + \frac{\partial}{\partial x_j} (\rho u_j) \right] = 0,$$

which, when added to equation (4.56), gives

$$\frac{\partial}{\partial t} \left(\frac{1}{2} \rho u_i^2 \right) + \frac{\partial}{\partial x_j} \left[u_j \frac{1}{2} \rho u_i^2 \right] = \rho u_i g_i + u_i \frac{\partial \tau_{ij}}{\partial x_j}.$$

Using vector notation, and defining $E \equiv u_i^2/2$ as the kinetic energy per unit volume, this becomes

$$\frac{\partial E}{\partial t} + \nabla \cdot (\mathbf{u} E) = \rho \mathbf{u} \cdot \mathbf{g} + \mathbf{u} \cdot (\nabla \cdot \tau). \tag{4.57}$$

The second term is in the form of divergence of kinetic energy flux $\mathbf{u} E$. Such *flux divergence* terms frequently arise in energy balances and can be interpreted as the net loss at a point due to divergence of a flux. For example, if the source terms on the right-hand side of equation (4.57) are zero, then the local E will increase with time if $\nabla \cdot (\mathbf{u}E)$ is negative. Flux divergence terms are also called *transport* terms because they transfer quantities from one region to another without making a net contribution over the entire field. When integrated over the entire volume, their contribution vanishes if there are no sources at the boundaries. For example, Gauss' theorem transforms the volume integral of $\nabla \cdot (\mathbf{u}E)$ as

$$\int_V \nabla \cdot (\mathbf{u}E)\, dV = \int_A E\mathbf{u} \cdot d\mathbf{A},$$

which vanishes if the flux $\mathbf{u}E$ is zero at the boundaries.

Concept of Deformation Work and Viscous Dissipation

Another useful form of the kinetic energy equation will now be derived by examining how kinetic energy can be lost to internal energy by deformation of fluid elements. In equation (4.56) the term $u_i(\partial \tau_{ij}/\partial x_j)$ is velocity times the net force imbalance at a point due to differences of stress on opposite faces of an element; the net force accelerates the local fluid and increases its kinetic energy. However, this is *not* the total rate of work done by stress on the element, and the remaining part goes into *deforming* the element without accelerating it. The total rate of work done by surface forces on a fluid element must be $\partial(\tau_{ij}u_i)/\partial x_j$, because this can be transformed to a surface integral of $\tau_{ij}u_i$ over the element. (Here $\tau_{ij}\, dA_j$ is the force on an area element, and $\tau_{ij}u_i\, dA_j$ is the scalar product of force and velocity. The total rate of work done by surface forces is therefore the surface integral of $\tau_{ij}u_i$.) The total work rate per volume at a point can be split up into two components:

$$\underbrace{\frac{\partial}{\partial x_j}(u_i \tau_{ij})}_{\substack{\text{total work} \\ \text{(rate/volume)}}} = \underbrace{\tau_{ij}\frac{\partial u_i}{\partial x_j}}_{\substack{\text{deformation} \\ \text{work} \\ \text{(rate/volume)}}} + \underbrace{u_i\frac{\partial \tau_{ij}}{\partial x_j}}_{\substack{\text{increase} \\ \text{of KE} \\ \text{(rate/volume)}}}.$$

We have seen from equation (4.56) that the last term in the preceding equation results in an increase of kinetic energy of the element. Therefore, the rest of the work rate per volume represented by $\tau_{ij}(\partial u_i/\partial x_j)$ can only deform the element and increase its *internal* energy.

The *deformation work* rate can be rewritten using the symmetry of the stress tensor. In Chapter 2, Section 11 it was shown that the contracted product of a symmetric tensor and an antisymmetric tensor is zero. The product $\tau_{ij}(\partial u_i/\partial x_j)$ is therefore equal to τ_{ij} times the *symmetric* part of $\partial u_i/\partial x_j$, namely e_{ij}. Thus

$$\text{Deformation work rate per volume} = \tau_{ij}\frac{\partial u_i}{\partial x_j} = \tau_{ij}e_{ij}. \qquad (4.58)$$

On substituting the Newtonian constitutive equation

$$\tau_{ij} = -p\delta_{ij} + 2\mu e_{ij} - \frac{2}{3}\mu(\nabla \cdot \mathbf{u})\delta_{ij},$$

relation (4.58) becomes

$$\text{Deformation work} = -p(\nabla \cdot \mathbf{u}) + 2\mu e_{ij}e_{ij} - \frac{2}{3}\mu(\nabla \cdot \mathbf{u})^2,$$

where we have used $e_{ij}\delta_{ij} = e_{ii} = \nabla \cdot \mathbf{u}$. Denoting the viscous term by ϕ, we obtain

$$\text{Deformation work (rate per volume)} = -p(\nabla \cdot \mathbf{u}) + \phi, \qquad (4.59)$$

where

$$\phi \equiv 2\mu e_{ij}e_{ij} - \frac{2}{3}\mu(\nabla \cdot \mathbf{u})^2 = 2\mu \left[e_{ij} - \frac{1}{3}(\nabla \cdot \mathbf{u})\delta_{ij} \right]^2. \qquad (4.60)$$

The validity of the last term in equation (4.60) can easily be verified by completing the square (Exercise 5).

In order to write the energy equation in terms of ϕ, we first rewrite equation (4.56) in the form

$$\rho \frac{D}{Dt}\left(\frac{1}{2}u_i^2 \right) = \rho g_i u_i + \frac{\partial}{\partial x_j}(u_i \tau_{ij}) - \tau_{ij}e_{ij}, \qquad (4.61)$$

where we have used $\tau_{ij}(\partial u_i/\partial x_j) = \tau_{ij}e_{ij}$. Using equation (4.59) to rewrite the deformation work rate per volume, equation (4.61) becomes

$$\rho \frac{D}{Dt}\left(\frac{1}{2}u_i^2 \right) = \underset{\substack{\text{rate of work by} \\ \text{body force}}}{\rho \mathbf{g} \cdot \mathbf{u}} + \underset{\substack{\text{total rate of} \\ \text{work by } \tau}}{\frac{\partial}{\partial x_j}(u_i \tau_{ij})} + \underset{\substack{\text{rate of work} \\ \text{by volume} \\ \text{expansion}}}{p(\nabla \cdot \mathbf{u})} - \underset{\substack{\text{rate of} \\ \text{viscous} \\ \text{dissipation}}}{\phi} \qquad (4.62)$$

It will be shown in Section 14 that the last two terms in the preceding equation (representing pressure and viscous contributions to the rate of deformation work) also appear in the *internal* energy equation but with their signs changed. The term $p(\nabla \cdot \mathbf{u})$ can be of either sign, and converts mechanical to internal energy, or vice versa, by volume changes. The viscous term ϕ is always positive and represents a rate of loss of mechanical energy and a gain of internal energy due to deformation of the element. The term $\tau_{ij}e_{ij} = p(\nabla \cdot \mathbf{u}) - \phi$ represents the total deformation work rate per volume; the part $p(\nabla \cdot \mathbf{u})$ is the reversible conversion to internal energy by volume changes, and the part ϕ is the irreversible conversion to internal energy due to viscous effects.

The quantity ϕ defined in equation (4.60) is proportional to μ and represents the rate of *viscous dissipation* of kinetic energy per unit volume. Equation (4.60) shows that it is proportional to the square of velocity gradients and is therefore more important in regions of high shear. The resulting heat could appear as a hot lubricant in a bearing, or as burning of the surface of a spacecraft on reentry into the atmosphere.

Equation in Terms of Potential Energy

So far we have considered kinetic energy as the only form of mechanical energy. In doing so we have found that the effects of gravity appear as work done on a fluid particle, as equation (4.62) shows. However, the rate of work done by body forces can be taken to the left-hand side of the mechanical energy equations and be interpreted as changes in the potential energy. Let the body force be represented as the gradient of a scalar potential $\Pi = gz$, so that

$$u_i g_i = -u_i \frac{\partial}{\partial x_i}(gz) = -\frac{D}{Dt}(gz),$$

where we have used $\partial(gz)/\partial t = 0$, because z and t are independent. Equation (4.62) then becomes

$$\rho \frac{D}{Dt}\left(\frac{1}{2}u_i^2 + gz\right) = \frac{\partial}{\partial x_j}(u_i \tau_{ij}) + p(\nabla \cdot \mathbf{u}) - \phi,$$

in which the function $\Pi = gz$ clearly has the significance of *potential energy* per unit mass. (This identification is possible only for conservative body forces for which a potential may be written.)

Equation for a Fixed Region

An integral form of the mechanical energy equation can be derived by integrating the differential form over either a fixed volume or a material volume. The procedure is illustrated here for a fixed volume. We start with equation (4.62), but write the left-hand side as given in equation (4.57). This gives (in mixed notation)

$$\frac{\partial E}{\partial t} + \frac{\partial}{\partial x_i}(u_i E) = \rho \mathbf{g} \cdot \mathbf{u} + \frac{\partial}{\partial x_j}(u_i \tau_{ij}) + p(\nabla \cdot \mathbf{u}) - \phi,$$

where $E = \rho u_i^2/2$ is the kinetic energy per unit volume. Integrate each term of the foregoing equation over the fixed volume V. The second and fourth terms are in the flux divergence form, so that their volume integrals can be changed to surface integrals by Gauss' theorem. This gives

$$\underbrace{\frac{d}{dt}\int E\,dV}_{\substack{\text{rate of change} \\ \text{of KE}}} + \underbrace{\int E\mathbf{u} \cdot d\mathbf{A}}_{\substack{\text{rate of outflow} \\ \text{across} \\ \text{boundary}}}$$

$$= \underbrace{\int \rho \mathbf{g} \cdot \mathbf{u}\,dV}_{\substack{\text{rate of work} \\ \text{by body} \\ \text{force}}} + \underbrace{\int u_i \tau_{ij}\,dA_j}_{\substack{\text{rate of work} \\ \text{by surface} \\ \text{force}}} + \underbrace{\int p(\nabla \cdot \mathbf{u})\,dV}_{\substack{\text{rate of work} \\ \text{by volume} \\ \text{expansion}}} - \underbrace{\int \phi\,dV}_{\substack{\text{rate of viscous} \\ \text{dissipation}}}$$

$$(4.63)$$

where each term is a time rate of change. The description of each term in equation (4.63) is obvious. The fourth term represents rate of work done by forces at the boundary, because $\tau_{ij}\,dA_j$ is the force in the i direction and $u_i \tau_{ij}\,dA_j$ is the scalar product of the force with the velocity vector.

The energy considerations discussed in this section may at first seem too "theoretical." However, they are very useful in understanding the physics of fluid flows. The concepts presented here will be especially useful in our discussions of turbulent flows (Chapter 13) and wave motions (Chapter 7). It is suggested that the reader work out Exercise 11 at this point in order to acquire a better understanding of the equations in this section.

14. First Law of Thermodynamics: Thermal Energy Equation

The mechanical energy equation presented in the preceding section is derived from the momentum equation and is not a separate principle. In flows with temperature variations we need an independent equation; this is provided by the first law of thermodynamics. Let **q** be the heat flux vector (per unit area), and e the *internal energy per unit mass*; for a perfect gas $e = C_V T$, where C_V is the specific heat at constant volume (assumed constant). The sum $(e + u_i^2/2)$ can be called the "stored" energy per unit mass. The first law of thermodynamics is most easily stated for a material volume. It says that the *rate of change of stored energy equals the sum of rate of work done and rate of heat addition to a material volume*. That is,

$$\frac{D}{Dt} \int_{\mathcal{V}} \rho \left(e + \tfrac{1}{2} u_i^2 \right) d\mathcal{V} = \int_{\mathcal{V}} \rho g_i u_i \, d\mathcal{V} + \int_A \tau_{ij} u_i \, dA_j - \int_A q_i \, dA_i. \quad (4.64)$$

Note that work done by body forces has to be included on the right-hand side if potential energy is *not* included on the left-hand side, as in equations (4.62)–(4.64). (This is clear from the discussion of the preceding section and can also be understood as follows. Imagine a situation where the surface integrals in equation (4.64) are zero, and also that e is uniform everywhere. Then a rising fluid particle ($\mathbf{u} \cdot \mathbf{g} < 0$), which is constantly pulled down by gravity, must undergo a decrease of kinetic energy. This is consistent with equation (4.64).) The negative sign is needed on the heat transfer term, because the direction of $d\mathbf{A}$ is along the outward normal to the area, and therefore $\mathbf{q} \cdot d\mathbf{A}$ represents the rate of heat *outflow*.

To derive a differential form, all terms need to be expressed in the form of volume integrals. The left-hand side can be written as

$$\frac{D}{Dt} \int_{\mathcal{V}} \rho \left(e + \frac{1}{2} u_i^2 \right) d\mathcal{V} = \int_{\mathcal{V}} \rho \frac{D}{Dt} \left(e + \frac{1}{2} u_i^2 \right) d\mathcal{V},$$

where equation (4.6) has been used. Converting the two surface integral terms into volume integrals, equation (4.64) finally gives

$$\rho \frac{D}{Dt} \left(e + \frac{1}{2} u_i^2 \right) = \rho g_i u_i + \frac{\partial}{\partial x_j} (\tau_{ij} u_i) - \frac{\partial q_i}{\partial x_i}. \quad (4.65)$$

This is the first law of thermodynamics in the differential form, which has both mechanical and thermal energy terms in it. A thermal energy equation is obtained if the mechanical energy equation (4.62) is subtracted from it. This gives the *thermal energy equation* (commonly called the *heat equation*)

$$\rho \frac{De}{Dt} = -\nabla \cdot \mathbf{q} - p(\nabla \cdot \mathbf{u}) + \phi, \quad (4.66)$$

which says that internal energy increases because of convergence of heat, volume compression, and heating due to viscous dissipation. Note that the last two terms in equation (4.66) also appear in mechanical energy equation (4.62) with their signs reversed.

The thermal energy equation can be simplified under the *Boussinesq approximation*, which applies under several restrictions including that in which the flow speeds are small compared to the speed of sound and in which the temperature differences in the flow are small. This is discussed in Section 18. It is shown there that, under these restrictions, heating due to the viscous dissipation term is negligible in equation (4.66), and that the term $-p(\nabla \cdot \mathbf{u})$ can be combined with the left-hand side of equation (4.66) to give (for a perfect gas)

$$\rho C_{\mathrm{p}} \frac{DT}{Dt} = -\nabla \cdot \mathbf{q}. \tag{4.67}$$

If the heat flux obeys the Fourier law

$$\mathbf{q} = -k\nabla T,$$

then, if $k = $ const., equation (4.67) simplifies to:

$$\boxed{\frac{DT}{Dt} = \kappa \nabla^2 T.} \tag{4.68}$$

where $\kappa \equiv k/\rho C_{\mathrm{p}}$ is the *thermal diffusivity*, stated in m^2/s and which is the same as that of the momentum diffusivity ν.

The viscous heating term ϕ may be negligible in the thermal energy equation (4.66) if flow speeds are low compared with the sound speed, but not in the mechanical energy equation (4.62). In fact, there must be a sink of mechanical energy so that a steady state can be maintained in the presence of the various types of forcing.

15. Second Law of Thermodynamics: Entropy Production

The second law of thermodynamics essentially says that real phenomena can only proceed in a direction in which the "disorder" of an isolated system increases. Disorder of a system is a measure of the degree of *uniformity* of macroscopic properties in the system, which is the same as the degree of randomness in the molecular arrangements that generate these properties. In this connection, disorder, uniformity, and randomness have essentially the same meaning. For analogy, a tray containing red balls on one side and white balls on the other has more order than in an arrangement in which the balls are mixed together. A real phenomenon must therefore proceed in a direction in which such orderly arrangements decrease because of "mixing." Consider two possible states of an isolated fluid system, one in which there are nonuniformities of temperature and velocity and the other in which these properties are uniform. Both of these states have the same internal energy. Can the system spontaneously go from the state in which its properties are uniform to one in which they are nonuniform? The second law asserts that it cannot, based on experience. Natural processes, therefore, tend to cause mixing due to transport of heat, momentum, and mass.

A consequence of the second law is that there must exist a property called *entropy*, which is related to other thermodynamic properties of the medium. In addition, the second law says that the entropy of an isolated system can only increase; entropy is therefore a measure of disorder or randomness of a system. Let S be the entropy per unit mass. It is shown in Chapter 1, Section 8 that the change of entropy is related to the changes of internal energy e and specific volume $v\ (= 1/\rho)$ by

$$T\, dS = de + p\, dv = de - \frac{p}{\rho^2}\, d\rho.$$

The rate of change of entropy following a fluid particle is therefore

$$T\frac{DS}{Dt} = \frac{De}{Dt} - \frac{p}{\rho^2}\frac{D\rho}{Dt}. \tag{4.69}$$

Inserting the internal energy equation (see equation (4.66))

$$\rho\frac{De}{Dt} = -\nabla \cdot \mathbf{q} - p(\nabla \cdot \mathbf{u}) + \phi,$$

and the continuity equation

$$\frac{D\rho}{Dt} = -\rho(\nabla \cdot \mathbf{u}),$$

the entropy production equation (4.69) becomes

$$\rho\frac{DS}{Dt} = -\frac{1}{T}\frac{\partial q_i}{\partial x_i} + \frac{\phi}{T}$$

$$= -\frac{\partial}{\partial x_i}\left(\frac{q_i}{T}\right) - \frac{q_i}{T^2}\frac{\partial T}{\partial x_i} + \frac{\phi}{T}.$$

Using Fourier's law of heat conduction, this becomes

$$\rho\frac{DS}{Dt} = -\frac{\partial}{\partial x_i}\left(\frac{q_i}{T}\right) + \frac{k}{T^2}\left(\frac{\partial T}{\partial x_i}\right)^2 + \frac{\phi}{T}.$$

The first term on the right-hand side, which has the form (heat gain)/T, is the entropy gain due to reversible heat transfer because this term does not involve heat conductivity. The last two terms, which are proportional to the square of temperature and velocity gradients, represent the *entropy production* due to heat conduction and viscous generation of heat. The second law of thermodynamics requires that the entropy production due to irreversible phenomena should be positive, so that

$$\mu, k > 0.$$

An explicit appeal to the second law of thermodynamics is therefore not required in most analyses of fluid flows because it has already been satisfied by taking positive values for the molecular coefficients of viscosity and thermal conductivity.

If the flow is inviscid and nonheat conducting, entropy is preserved along the particle paths.

16. Bernoulli Equation

Various conservation laws for mass, momentum, energy, and entropy were presented in the preceding sections. The well-known Bernoulli equation is not a separate law, but is derived from the momentum equation for *inviscid* flows, namely, the Euler equation (4.46):

$$\frac{\partial u_i}{\partial t} + u_j \frac{\partial u_i}{\partial x_j} = -\frac{\partial}{\partial x_i}(gz) - \frac{1}{\rho}\frac{\partial p}{\partial x_i},$$

where we have assumed that gravity $\mathbf{g} = -\nabla(gz)$ is the only body force. The advective acceleration can be expressed in terms of vorticity as follows:

$$u_j \frac{\partial u_i}{\partial x_j} = u_j \left(\frac{\partial u_i}{\partial x_j} - \frac{\partial u_j}{\partial x_i} \right) + u_j \frac{\partial u_j}{\partial x_i} = u_j r_{ij} + \frac{\partial}{\partial x_i} \left(\frac{1}{2} u_j u_j \right)$$

$$= -u_j \varepsilon_{ijk} \omega_k + \frac{\partial}{\partial x_i} \left(\frac{1}{2} q^2 \right) = -(\mathbf{u} \times \boldsymbol{\omega})_i + \frac{\partial}{\partial x_i} \left(\frac{1}{2} q^2 \right), \qquad (4.70)$$

where we have used $r_{ij} = -\varepsilon_{ijk} \omega_k$ (see equation 3.23), and used the customary notation

$$q^2 = u_j^2 = \text{twice kinetic energy}.$$

Then the Euler equation becomes

$$\frac{\partial u_i}{\partial t} + \frac{\partial}{\partial x_i} \left(\frac{1}{2} q^2 \right) + \frac{1}{\rho}\frac{\partial p}{\partial x_i} + \frac{\partial}{\partial x_i}(gz) = (\mathbf{u} \times \boldsymbol{\omega})_i. \qquad (4.71)$$

Now assume that ρ is a function of p only. A flow in which $\rho = \rho(p)$ is called a *barotropic flow*, of which isothermal and isentropic ($p/\rho^\gamma = \text{constant}$) flows are special cases. For such a flow we can write

$$\frac{1}{\rho}\frac{\partial p}{\partial x_i} = \frac{\partial}{\partial x_i} \int \frac{dp}{\rho}, \qquad (4.72)$$

where dp/ρ is a perfect differential, and therefore the integral does not depend on the path of integration. To show this, note that

$$\int_{\mathbf{x}_0}^{\mathbf{x}} \frac{dp}{\rho} = \int_{\mathbf{x}_0}^{\mathbf{x}} \frac{1}{\rho}\frac{dp}{d\rho} \, d\rho = \int_{\mathbf{x}_0}^{\mathbf{x}} \frac{dP}{d\rho} \, d\rho = P(\mathbf{x}) - P(\mathbf{x}_0), \qquad (4.73)$$

where $\mathbf{x}$ is the "field point," $\mathbf{x}_0$ is any arbitrary reference point in the flow, and we have defined the following function of ρ alone:

$$\frac{dP}{d\rho} \equiv \frac{1}{\rho}\frac{dp}{d\rho}. \qquad (4.74)$$

The gradient of equation (4.73) gives

$$\frac{\partial}{\partial x_i} \int_{\mathbf{x}_0}^{\mathbf{x}} \frac{dp}{\rho} = \frac{\partial P}{\partial x_i} = \frac{dP}{dp}\frac{\partial p}{\partial x_i} = \frac{1}{\rho}\frac{\partial p}{\partial x_i},$$

where equation (4.74) has been used. The preceding equation is identical to equation (4.72).

Using equation (4.72), the Euler equation (4.71) becomes

$$\frac{\partial u_i}{\partial t} + \frac{\partial}{\partial x_i}\left[\frac{1}{2}q^2 + \int \frac{dp}{\rho} + gz\right] = (\mathbf{u} \times \boldsymbol{\omega})_i.$$

Defining the Bernoulli function

$$B \equiv \frac{1}{2}q^2 + \int \frac{dp}{\rho} + gz = \frac{1}{2}q^2 + P + gz, \qquad (4.75)$$

the Euler equation becomes (using vector notation)

$$\frac{\partial \mathbf{u}}{\partial t} + \nabla B = \mathbf{u} \times \boldsymbol{\omega}. \qquad (4.76)$$

Bernoulli equations are integrals of the conservation laws and have wide applicability as shown by the examples that follow. Important deductions can be made from the preceding equation by considering two special cases, namely a steady flow (rotational or irrotational) and an unsteady irrotational flow. These are described in what follows.

Steady Flow

In this case equation (4.76) reduces to

$$\nabla B = \mathbf{u} \times \boldsymbol{\omega}. \qquad (4.77)$$

The left-hand side is a vector normal to the surface B = constant, whereas the right-hand side is a vector perpendicular to both $\mathbf{u}$ and $\boldsymbol{\omega}$ (Figure 4.17). It follows that surfaces of constant B must contain the streamlines and vortex lines. Thus, an inviscid, steady, barotropic flow satisfies

$$\boxed{\frac{1}{2}q^2 + \int \frac{dp}{\rho} + gz = \text{constant along streamlines and vortex lines}} \qquad (4.78)$$

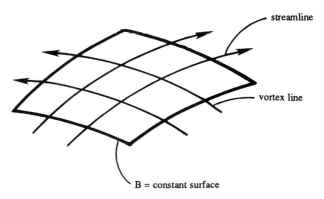

Figure 4.17   Bernoulli's theorem. Note that the streamlines and vortex lines can be at an arbitrary angle.

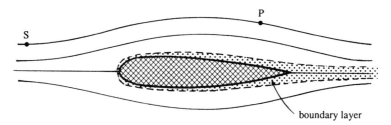

Figure 4.18   Flow over a solid object. Flow outside the boundary layer is irrotational.

which is called *Bernoulli's equation*. If, in addition, the flow is irrotational ($\omega = 0$), then equation (4.72) shows that

$$\frac{1}{2}q^2 + \int \frac{dp}{\rho} + gz = \text{constant everywhere.} \tag{4.79}$$

It may be shown that a sufficient condition for the existence of the surfaces containing streamlines and vortex lines is that the flow be barotropic. Incidentally, these are called Lamb surfaces in honor of the distinguished English applied mathematician and hydrodynamicist, Horace Lamb. In a general, that is, nonbarotropic flow, a path composed of streamline and vortex line segments can be drawn between any two points in a flow field. Then equation (4.78) is valid with the proviso that the integral be evaluated on the specific path chosen. As written, equation (4.78) requires the restrictions that the flow be steady, inviscid, and have only gravity (or other conservative) body forces acting upon it. Irrotational flows are studied in Chapter 6. We shall note only the important point here that, in a nonrotating frame of reference, barotropic irrotational flows remain irrotational if viscous effects are negligible. Consider the flow around a solid object, say an airfoil (Figure 4.18). The flow is irrotational at all points outside the thin viscous layer close to the surface of the body. This is because a particle P on a streamline outside the viscous layer started from some point S, where the flow is uniform and consequently irrotational. The Bernoulli equation (4.79) is therefore satisfied everywhere outside the viscous layer in this example.

Unsteady Irrotational Flow

An unsteady form of Bernoulli's equation can be derived only if the flow is irrotational. For irrotational flows the velocity vector can be written as the gradient of a scalar potential ϕ (called velocity potential):

$$\mathbf{u} \equiv \nabla\phi. \tag{4.80}$$

The validity of equation (4.80) can be checked by noting that it automatically satisfies the conditions of irrotationality

$$\frac{\partial u_i}{\partial x_j} = \frac{\partial u_j}{\partial x_i} \qquad i \neq j.$$

On inserting equation (4.80) into equation (4.76), we obtain

$$\nabla\left[\frac{\partial\phi}{\partial t} + \frac{1}{2}q^2 + \int\frac{dp}{\rho} + gz\right] = 0,$$

that is

$$\boxed{\frac{\partial\phi}{\partial t} + \frac{1}{2}q^2 + \int\frac{dp}{\rho} + gz = F(t),} \tag{4.81}$$

where the integrating function $F(t)$ is independent of location. This form of the Bernoulli equation will be used in studying irrotational wave motions in Chapter 7.

Energy Bernoulli Equation

Return to equation (4.65) in the steady state with neither heat conduction nor viscous stresses. Then $\tau_{ij} = -p\delta_{ij}$ and equation (4.65) becomes

$$\rho u_i \frac{\partial}{\partial x_i}(e + q^2/2) = \rho u_i g_i - \frac{\partial}{\partial x_i}(\rho u_i p/\rho).$$

If the body force per unit mass g_i is conservative, say gravity, then $g_i = -(\partial/\partial x_i)(gz)$, which is the gradient of a scalar potential. In addition, from mass conservation, $\partial(\rho u_i)/\partial x_i = 0$ and thus

$$\rho u_i \frac{\partial}{\partial x_i}\left(e + \frac{p}{\rho} + \frac{q^2}{2} + gz\right) = 0. \tag{4.82}$$

From equation (1.13), $h = e + p/\rho$. Equation (4.82) now states that gradients of $B' = h + q^2/2 + gz$ must be normal to the local streamline direction u_i. Then $B' = h + q^2/2 + gz$ is a constant on streamlines. We showed in the previous section that inviscid, non-heat conducting flows are isentropic (S is conserved along particle paths), and in equation (1.18) we had the relation $dp/\rho = dh$ when $S = $ constant. Thus the path integral $\int dp/\rho$ becomes a function h of the endpoints only if, in the momentum Bernoulli equation, both heat conduction and viscous stresses may be neglected. This latter form from the energy equation becomes very useful for high-speed gas flows to show the interplay between kinetic energy and internal energy or enthalpy or temperature along a streamline.

17. Applications of Bernoulli's Equation

Application of Bernoulli's equation will now be illustrated for some simple flows.

Pitot Tube

Consider first a simple device to measure the local velocity in a fluid stream by inserting a narrow bent tube (Figure 4.19). This is called a *pitot tube*, after the French mathematician Henri Pitot (1695–1771), who used a bent glass tube to measure the velocity of the river Seine. Consider two points 1 and 2 at the same level, point 1 being away from the tube and point 2 being immediately in front of the open end where the fluid velocity is zero. Friction is negligible along a streamline through 1 and 2, so that Bernoulli's equation (4.78) gives

$$\frac{p_1}{\rho} + \frac{u_1^2}{2} = \frac{p_2}{\rho} + \frac{u_2^2}{2} = \frac{p_2}{\rho},$$

from which the velocity is found to be

$$u_1 = \sqrt{2(p_2 - p_1)/\rho}.$$

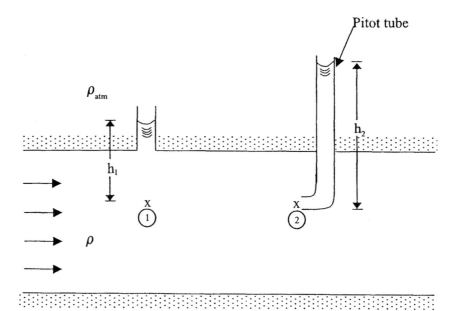

Figure 4.19   Pitot tube for measuring velocity in a duct.

Pressures at the two points are found from the hydrostatic balance

$$p_1 = \rho g h_1 \quad \text{and} \quad p_2 = \rho g h_2,$$

so that the velocity can be found from

$$u_1 = \sqrt{2g(h_2 - h_1)}.$$

Because it is assumed that the fluid density is very much greater than that of the atmosphere to which the tubes are exposed, the pressures at the tops of the two fluid columns are assumed to be the same. They will actually differ by $\rho_{atm} g (h_2 - h_1)$. Use of the hydrostatic approximation above station 1 is valid when the streamlines are straight and parallel between station 1 and the upper wall. In working out this problem, the fluid density also has been taken to be a constant.

The pressure p_2 measured by a pitot tube is called "stagnation pressure," which is larger than the local static pressure. Even when there is no pitot tube to measure the stagnation pressure, it is customary to refer to the local value of the quantity $(p + \rho u^2 / 2)$ as the local *stagnation pressure*, defined as the pressure that would be reached if the local flow is *imagined* to slow down to zero velocity frictionlessly. The quantity $\rho u^2 / 2$ is sometimes called the *dynamic pressure*; stagnation pressure is the sum of static and dynamic pressures.

Orifice in a Tank

As another application of Bernoulli's equation, consider the flow through an orifice or opening in a tank (Figure 4.20). The flow is slightly unsteady due to lowering of the water level in the tank, but this effect is small if the tank area is large as compared to the orifice area. Viscous effects are negligible everywhere away from the walls of the tank. All streamlines can be traced back to the free surface in the tank, where they have the same value of the Bernoulli constant $B = q^2 / 2 + p / \rho + gz$. It follows that the flow is irrotational, and B is constant *throughout* the flow.

We want to apply Bernoulli's equation between a point at the free surface in the tank and a point in the jet. However, the conditions right at the opening (section A in Figure 4.20) are not simple because the pressure is *not* uniform across the jet. Although pressure has the atmospheric value everywhere on the free surface of the jet (neglecting small surface tension effects), it is not equal to the atmospheric pressure *inside* the jet at this section. The streamlines at the orifice are curved, which requires that pressure must vary across the width of the jet in order to balance the centrifugal force. The pressure distribution across the orifice (section A) is shown in Figure 4.20. However, the streamlines in the jet become parallel at a short distance away from the orifice (section C in Figure 4.20), where the jet area is smaller than the orifice area. The pressure across section C is uniform and equal to the atmospheric value because it has that value at the surface of the jet.

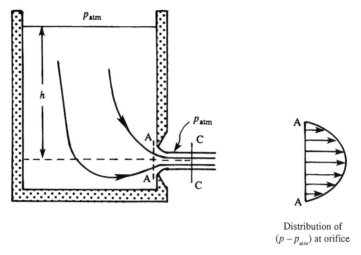

Distribution of
$(p - p_{\text{atm}})$ at orifice

Figure 4.20   Flow through a sharp-edged orifice. Pressure has the atmospheric value everywhere across section CC; its distribution across orifice AA is indicated.

Application of Bernoulli's equation between a point on the free surface in the tank and a point at C gives

$$\frac{p_{\text{atm}}}{\rho} + gh = \frac{p_{\text{atm}}}{\rho} + \frac{u^2}{2},$$

from which the jet velocity is found as

$$u = \sqrt{2gh},$$

which simply states that the loss of potential energy equals the gain of kinetic energy. The mass flow rate is

$$\dot{m} = \rho A_c u = \rho A_c \sqrt{2gh},$$

where A_c is the area of the jet at C. For orifices having a sharp edge, A_c has been found to be $\approx 62\%$ of the orifice area.

If the orifice happens to have a well-rounded opening (Figure 4.21), then the jet does not contract. The streamlines right at the exit are then parallel, and the pressure at the exit is uniform and equal to the atmospheric pressure. Consequently the mass flow rate is simply $\rho A \sqrt{2gh}$, where A equals the orifice area.

18. Boussinesq Approximation

For flows satisfying certain conditions, Boussinesq in 1903 suggested that the density changes in the fluid can be neglected except in the gravity term where ρ is multiplied by g. This approximation also treats the other properties of the fluid (such as μ, k, C_p) as constants. A formal justification, and the conditions under which the Boussinesq approximation holds, is given in Spiegel and Veronis (1960). Here we shall discuss the

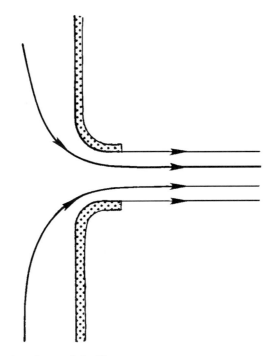

Figure 4.21   Flow through a rounded orifice.

basis of the approximation in a somewhat intuitive manner and examine the resulting simplifications of the equations of motion.

Continuity Equation

The Boussinesq approximation replaces the continuity equation

$$\frac{1}{\rho}\frac{D\rho}{Dt} + \nabla \cdot \mathbf{u} = 0, \qquad (4.83)$$

by the incompressible form

$$\nabla \cdot \mathbf{u} = 0. \qquad (4.84)$$

However, this does not mean that the density is regarded as constant along the direction of motion, but simply that the magnitude of $\rho^{-1}(D\rho/Dt)$ is small in comparison to the magnitudes of the velocity gradients in $\nabla \cdot \mathbf{u}$. We can immediately think of several situations where the density variations cannot be neglected as such. The first situation is a steady flow with large Mach numbers (defined as U/c, where U is a typical measure of the flow speed and c is the speed of sound in the medium). At large Mach numbers the compressibility effects are large, because the large pressure changes cause large density changes. It is shown in Chapter 16 that compressibility effects are negligible in flows in which the Mach number is <0.3. A typical value of c for air at ordinary temperatures is $350 \, \text{m/s}$, so that the assumption is good for speeds $< 100 \, \text{m/s}$. For water $c = 1470 \, \text{m/s}$, but the speeds normally achievable in liquids are much smaller than this value and therefore the incompressibility assumption is very good in liquids.

A second situation in which the compressibility effects are important is unsteady flows. The waves would propagate at infinite speed if the density variations are neglected.

A third situation in which the compressibility effects are important occurs when the vertical scale of the flow is so large that the hydrostatic pressure variations cause large changes in density. In a hydrostatic field the vertical scale in which the density changes become important is of order $c^2/g \sim 10 \, \text{km}$ for air. (This length agrees with the *e*-folding height RT/g of an "isothermal atmosphere," because $c^2 = \gamma RT$; see Chapter 1, Section 10.) The Boussinesq approximation therefore requires that the vertical scale of the flow be $L \ll c^2/g$.

In the three situations mentioned the medium is regarded as "compressible," in which the density depends strongly on pressure. Now suppose the compressibility effects are small, so that the density changes are caused by temperature changes alone, as in a thermal convection problem. In this case the Boussinesq approximation applies when the temperature variations in the flow are small. Assume that ρ changes with T according to

$$\frac{\delta \rho}{\rho} = -\alpha \delta T,$$

where $\alpha = -\rho^{-1}(\partial \rho / \partial T)_p$ is the thermal expansion coefficient. For a perfect gas $\alpha = 1/T \sim 3 \times 10^{-3} \, \text{K}^{-1}$ and for typical liquids $\alpha \sim 5 \times 10^{-4} \, \text{K}^{-1}$. With a temperature difference in the fluid of $10 \, °\text{C}$, the variation of density can be only a few percent at most. It turns out that $\rho^{-1}(D\rho/Dt)$ can also be no larger than a few percent of the velocity gradients in $\nabla \cdot \mathbf{u}$. To see this, assume that the flow field is characterized by a length scale L, a velocity scale U, and a temperature scale δT. By this we mean that the velocity varies by U and the temperature varies by δT, in a distance of order L. The ratio of the magnitudes of the two terms in the continuity equation is

$$\frac{(1/\rho)(D\rho/Dt)}{\nabla \cdot \mathbf{u}} \sim \frac{(1/\rho)u(\partial \rho / \partial x)}{\partial u / \partial x} \sim \frac{(U/\rho)(\delta \rho / L)}{U/L} = \frac{\delta \rho}{\rho} = \alpha \delta T \ll 1,$$

which allows us to replace continuity equation (4.83) by its incompressible form (4.84).

Momentum Equation

Because of the incompressible continuity equation $\nabla \cdot \mathbf{u} = 0$, the stress tensor is given by equation (4.41). From equation (4.45), the equation of motion is then

$$\rho \frac{D\mathbf{u}}{Dt} = -\nabla p + \rho \mathbf{g} + \mu \nabla^2 \mathbf{u}. \tag{4.85}$$

Consider a hypothetical static reference state in which the density is ρ_0 everywhere and the pressure is $p_0(z)$, so that $\nabla p_0 = \rho_0 \mathbf{g}$. Subtracting this state from equation (4.85) and writing $p = p_0 + p'$ and $\rho = \rho_0 + \rho'$, we obtain

$$\rho \frac{D\mathbf{u}}{Dt} = -\nabla p' + \rho' \mathbf{g} + \mu \nabla^2 \mathbf{u}. \qquad (4.86)$$

Dividing by ρ_0, we obtain

$$\left(1 + \frac{\rho'}{\rho_0}\right) \frac{D\mathbf{u}}{Dt} = -\frac{1}{\rho_0} \nabla p' + \frac{\rho'}{\rho_0} \mathbf{g} + \nu \nabla^2 \mathbf{u},$$

where $\nu = \mu / \rho_0$. The ratio ρ'/ρ_0 appears in both the inertia and the buoyancy terms. For small values of ρ'/ρ_0, the density variations generate only a small correction to the inertia term and can be neglected. However, the buoyancy term $\rho' g / \rho_0$ is very important and cannot be neglected. For example, it is these density variations that drive the convective motion when a layer of fluid is heated. The magnitude of $\rho' g / \rho_0$ is therefore of the same order as the vertical acceleration $\partial w / \partial t$ or the viscous term $\nu \nabla^2 w$. We conclude that the density variations are negligible the momentum equation, except when ρ is multiplied by g.

Heat Equation

From equation (4.66), the thermal energy equation is

$$\rho \frac{De}{Dt} = -\nabla \cdot \mathbf{q} - p(\nabla \cdot \mathbf{u}) + \phi. \qquad (4.87)$$

Although the continuity equation is approximately $\nabla \cdot \mathbf{u} = 0$, an important point is that the volume expansion term $p(\nabla \cdot \mathbf{u})$ is *not* negligible compared to other dominant terms of equation (4.87); only for incompressible liquids is $p(\nabla \cdot \mathbf{u})$ negligible in equation (4.87). We have

$$-p\nabla \cdot \mathbf{u} = \frac{p}{\rho} \frac{D\rho}{Dt} \simeq \frac{p}{\rho} \left(\frac{\partial \rho}{\partial T}\right)_{\mathrm{p}} \frac{DT}{Dt} = -p\alpha \frac{DT}{Dt}.$$

Assuming a perfect gas, for which $p = \rho RT$, $C_{\mathrm{p}} - C_{\mathrm{v}} = R$ and $\alpha = 1/T$, the foregoing estimate becomes

$$-p\nabla \cdot \mathbf{u} = -\rho RT\alpha \frac{DT}{Dt} = -\rho(C_{\mathrm{p}} - C_{\mathrm{v}}) \frac{DT}{Dt}.$$

Equation (4.87) then becomes

$$\rho C_{\mathrm{p}} \frac{DT}{Dt} = -\nabla \cdot \mathbf{q} + \phi, \qquad (4.88)$$

where we used $e = C_{\mathrm{v}} T$ for a perfect gas. Note that we would have gotten C_{v} (instead of C_{p}) on the left-hand side of equation (4.88) if we had dropped $\nabla \cdot \mathbf{u}$ in equation (4.87).

Now we show that the heating due to viscous dissipation of energy is negligible under the restrictions underlying the Boussinesq approximation. Comparing the magnitudes of viscous heating with the left-hand side of equation (4.88), we obtain

$$\frac{\phi}{\rho C_{\mathrm{p}}(DT/Dt)} \sim \frac{2\mu e_{ij}e_{ij}}{\rho C_{\mathrm{p}}u_j(\partial T/\partial x_j)} \sim \frac{\mu U^2/L^2}{\rho_0 C_{\mathrm{p}}U\delta T/L} = \frac{\nu}{C_{\mathrm{p}}}\frac{U}{\delta T L}.$$

In typical situations this is extremely small ($\sim 10^{-7}$). Neglecting ϕ, and assuming Fourier's law of heat conduction

$$\mathbf{q} = -k\nabla T,$$

the heat equation (4.88) finally reduces to (if $k = $ const.)

$$\frac{DT}{Dt} = \kappa\nabla^2 T,$$

where $\kappa \equiv k/\rho C_{\mathrm{p}}$ is the *thermal diffusivity*.

Summary: The Boussinesq approximation applies if the Mach number of the flow is small, propagation of sound or shock waves is not considered, the vertical scale of the flow is not too large, and the temperature differences in the fluid are small. Then the density can be treated as a constant in both the continuity and the momentum equations, except in the gravity term. Properties of the fluid such as μ, k, and C_{p} are also assumed constant in this approximation. Omitting Coriolis forces, the set of equations corresponding to the Boussinesq approximation is

$$\nabla \cdot \mathbf{u} = 0$$
$$\frac{Du}{Dt} = -\frac{1}{\rho_0}\frac{\partial p}{\partial x} + \nu\nabla^2 u$$
$$\frac{Dv}{Dt} = -\frac{1}{\mu_0}\frac{\partial p}{\partial y} + \nu\nabla^2 v$$
$$\frac{Dw}{Dt} = -\frac{1}{\rho_0}\frac{\partial p}{\partial z} - \frac{\rho g}{\rho_0} + \nu\nabla^2 w \qquad (4.89)$$
$$\frac{DT}{Dt} = \kappa\nabla^2 T$$
$$\rho = \rho_0[1 - \alpha(T - T_0)],$$

where the z-axis is taken upward. The constant ρ_0 is a reference density corresponding to a reference temperature T_0, which can be taken to be the mean temperature in the flow or the temperature at a boundary. Applications of the Boussinesq set can be found in several places throughout the book, for example, in the problems of wave propagation in a density-stratified medium, thermal instability, turbulence in a stratified medium, and geophysical fluid dynamics.

19. Boundary Conditions

The differential equations we have derived for the conservation laws are subject to boundary conditions in order to properly formulate any problem. Specifically, the Navier-Stokes equations are of a form that requires the velocity vector to be given on all surfaces bounding the flow domain.

If we are solving for an external flow, that is, a flow over some body, we must specify the velocity vector and the thermodynamic state on a closed distant surface. On a solid boundary or at the interface between two immiscible liquids, conditions may be derived from the three basic conservation laws as follows.

In Figure 4.22, a "pillbox" is drawn through the interface surface separating medium 1 (fluid) from medium 2 (solid or liquid immiscible with fluid 1). Here $d\mathbf{A}_1$ and $d\mathbf{A}_2$ are elements of the end face areas in medium 1 and medium 2, respectively, locally tangent to the interface, and separated from each other by a distance l. Now apply the conservation laws to the volume defined by the pillbox. Next, let $l \to 0$, keeping A_1 and A_2 in the different media. As $l \to 0$, all volume integrals $\to 0$ and the integral over the side area, which is proportional to l, tends to zero as well. Define a unit vector $\mathbf{n}$, normal to the interface at the pillbox and pointed into medium 1. Mass conservation gives $\rho_1 \mathbf{u}_1 \cdot \mathbf{n} = \rho_2 \mathbf{u}_2 \cdot \mathbf{n}$ at each point on the interface as the end face area becomes small. (Here we assume that the coordinates are fixed to the interface, that is, the interface is at rest. Later in this section we show the modifications necessary when the interface is moving.)

If medium 2 is a solid, then $\mathbf{u}_2 = 0$ there. If medium 1 and medium 2 are immiscible liquids, no mass flows across the boundary surface. In either case, $\mathbf{u}_1 \cdot \mathbf{n} = 0$ on the boundary. The same procedure applied to the integral form of the momentum equation (4.16) gives the result that the force/area on the surface, $n_i \tau_{ij}$ is continuous across the interface if surface tension is neglected. If surface tension is included, a jump in pressure in the direction normal to the interface must be added; see Chapter 1, Section 6 and the discussion later in this section.

Applying the integral form of energy conservation (4.64) to a pillbox of infinitesimal height l gives the result $n_i q_i$ is continuous across the interface, or explicity, $k_1 (\partial T_1 / \partial n) = k_2 (\partial T_2 / \partial n)$ at the interface surface. The heat flux must be continuous at the interface; it cannot store heat.

Two more boundary conditions are required to completely specify a problem and these are not consequences of any conservation law. These boundary conditions are: no slip of a viscous fluid is permitted at a solid boundary $\mathbf{v}_1 \cdot \mathbf{t} = 0$; and no

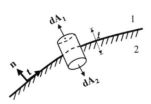

Figure 4.22   Interface between two media; evaluation of boundary conditions.

temperature jump is permitted at the boundary $T_1 = T_2$. Here **t** is a unit vector tangent to the boundary.

Known violations of the no-slip boundary condition occur for superfluid helium at or below $2.17°$ K, which has an immeasurable small (essentially zero) viscosity. On the other hand, the *appearance* of slip is created when water or water-based fluids flow over finely textured "superhydrophobic" (strongly water repellent) coated surfaces. This is described by Gogte et al. (2005). Surface textures must be much smaller than the capillary length for water and were typically about $10\mu m$. The fluid did not slip on the protrusions but did not penetrate the valleys because of the surface tension, giving the appearance of slip. Both slip and temperature jump are known to occur in highly rarefied gases, where the mean distances between intermolecular collisions become of the order of the length scales of interest in the problem. The details are closely related to the manner of gas-surface interaction of momentum and energy. A recent review was given by McCormick (2005).

Boundary condition at a moving, deforming surface

Consider a surface in space that may be moving or deforming in some arbitrary way. Examples may be flexible solid boundaries, the interface between two immiscible liquids, or a moving shock wave, as described in Chapter 16. The first two examples do not permit mass flow across the interface, whereas the third does. Such a surface can be defined and its motion described in inertial coordinates by the equation $f(x, y, z, t) = 0$. We often must treat problems in which boundary conditions must be satisfied on such a moving, deforming interface. Let the velocity of a point that remains on the surface be $\mathbf{u}_s$. An observer that remains on the surface always sees $f = 0$, so for that observer,

$$df/dt = \partial f/\partial t + \mathbf{u}_s \cdot \nabla f = 0 \quad \text{on} \ \ f = 0. \tag{4.90}$$

A fluid particle has velocity **u**. If no fluid flows across $f = 0$, then $\mathbf{u} \cdot \nabla f = \mathbf{u}_s \cdot \nabla f = -\partial f/\partial t$. Thus the condition that there be no mass flow across the surface becomes,

$$\partial f/\partial t + \mathbf{u} \cdot \nabla f \equiv Df/Dt = 0 \quad \text{on} \ \ f = 0. \tag{4.91}$$

If there is mass flow across the surface, it is proportional to the relative velocity between the fluid and the surface, $(u_r)_n = \mathbf{u} \cdot \mathbf{n} - \mathbf{u}_s \cdot \mathbf{n}$, where $\mathbf{n} = \nabla f/|\nabla f|$.

$$(u_r)_n = \mathbf{u} \cdot \nabla f/|\nabla f| + [1/|\nabla f|][\partial f/\partial t] = [1/|\nabla f|]Df/Dt. \tag{4.92}$$

Thus the mass flow rate across the surface (per unit surface area) is represented by

$$[\rho/|\nabla f|]Df/Dt \quad \text{on} \ \ f = 0. \tag{4.93}$$

Again, if no mass flows across the surface, the requirement is $Df/Dt = 0$ on $f = 0$.

Surface tension revisited: generalized discussion

As we discussed in Section 1.6 (p. 8), attractive intermolecular forces dominate in a liquid, whereas in a gas repulsive forces are larger. However, as a liquid-gas phase boundary is approached from the liquid side, these attractive forces are not felt equally because there are many fewer liquid phase molecules near the phase boundary. Thus there tends to be an unbalanced attraction to the interior of the liquid of the molecules on the phase boundary. This is called "surface tension" and its manifestation is a pressure increment across a curved interface. A somewhat more detailed description is provided in texts on physicochemical hydrodynamics. Two excellent sources are Probstein (1994, Chapter 10) and Levich (1962, Chapter VII).

H. Lamb, *Hydrodynamics* (6th Edition, p. 456) writes, "Since the condition of stable equilibrium is that the free energy be a minimum, the surface tends to contract as much as is consistent with the other conditions of the problem." Thus we are led to introduce the Helmoltz free energy (per unit mass) via

$$F = e - TS, \qquad (4.94)$$

where the notation is consistent with that used in Section 1.8. If the free energy is a minimum, then the system is in a state of stable equilibrium. F is called the thermodynamic potential at constant volume [E. Fermi, *Thermodynamics*, p. 80]. For a reversible, isothermal change, the work done on the system is the gain in total free energy F,

$$dF = de - TdS - SdT, \qquad (4.95)$$

where the last term is zero for an isothermal change. Then, from (1.18), $dF = -pdv =$ work done on system. (These relations suggest that surface tension decreases with increasing temperature.)

For an interface of area $= A$, separating two media of densities ρ_1 and ρ_2, with volumes V_1 and V_2, respectively, and with a surface tension coefficient σ (corresponding to free energy per unit area), the total (Helmholtz) free energy of the system can be written as

$$F = \rho_1 V_1 F_1 + \rho_2 V_2 F_2 + A\sigma. \qquad (4.96)$$

If $\sigma > 0$, then the two media (fluids) are immiscible; on the other hand, if $\sigma < 0$, corresponding to surface compression, then the two fluids mix freely. In the following, we shall assume that $\sigma = $ const. Flows driven by surface tension gradients are called Marangoni flows and are not discussed here. Our discussion will follow that given by G. K. Batchelor, *An Introduction to Fluid Dynamics*, pp. 61ff.

We wish to determine the shape of a boundary between two stationary fluids compatible with mechanical equilibrium. Let the equation of the interface surface be given by $f(x, y, z) = 0 = z - \zeta(x, y)$. Align the coordinates so that $\zeta(0, 0) = 0$, $\partial\zeta/\partial x|_{0,0} = 0$, $\partial\zeta/\partial y|_{0,0} = 0$. See Figure 4.23. A normal to this surface is obtained by forming the gradient, $\mathbf{n} = \nabla[z - \zeta(x, y)] = \mathbf{k} - \mathbf{i}\partial\zeta/\partial x - \mathbf{j}\partial\zeta/\partial y$. The (x, y, z) components of $\mathbf{n}$ are $(-\partial\zeta/\partial x, -\partial\zeta/\partial y, 1)$. Now the tensile forces on the bounding

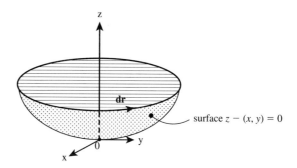

Figure 4.23   Geometry of equilibrium interface with surface tension.

line of the surface are obtained from the line integral

$$= \sigma \oint \mathbf{dr} \times \mathbf{n}$$

$$= \sigma \oint (\mathbf{i}\, dx + \mathbf{j}\, dy + \mathbf{k}\, dz) \times (\mathbf{k} - \mathbf{i}\partial \zeta/\partial x - \mathbf{j}\partial \zeta/\partial y)$$

$$= \sigma \oint [-\mathbf{k}(\partial \zeta/\partial y)dx - \mathbf{j}dx + \mathbf{k}(\partial \zeta/\partial x)dy + \mathbf{i}dy - \mathbf{j}(\partial \zeta/\partial x)dz + \mathbf{i}(\partial \zeta/\partial y)dz].$$

This integral is carried out over a contour C, which bounds the area A. Let that contour C be in a $z = $ const. plane so that $dz = 0$ on C. Then note that

$$\oint (\mathbf{i}dy - \mathbf{j}dx) = -\mathbf{k} \times \oint (\mathbf{i}dx + \mathbf{j}dy) = -\mathbf{k} \times \oint \mathbf{dr} = 0.$$

Then the tensile force acting on the bounding line C of the surface A

$$= \mathbf{k}\sigma \oint [-(\partial \zeta/\partial y)dx + (\partial \zeta/\partial x)dy].$$

Now use Stokes' theorem in the form

$$\oint_{C=\partial A} \mathbf{F} \cdot \mathbf{dr} = \int_A (\nabla \times \mathbf{F}) \cdot d\mathbf{A}, \text{ where here } F = -(\partial \zeta/\partial y)\mathbf{i} + (\partial \zeta/\partial x)\mathbf{j}. \text{ Then}$$

$$\nabla \times \mathbf{F} = (\partial F_y/\partial x - \partial F_x/\partial y)\mathbf{k} = (\partial^2 \zeta/\partial x^2 + \partial^2 \zeta/\partial y^2)\mathbf{k}, \text{ and}$$

$$\sigma \oint_{C=\partial A} [(-\partial \zeta/\partial y)dx + (\partial \zeta/\partial x)dy] = \sigma \int_A (\partial^2 \zeta/\partial x^2 + \partial^2 \zeta/\partial y^2)dA_z. \quad (4.97)$$

We had expanded in a small neighborhood of the origin so the force per surface area is the last integrand $= \sigma(\partial^2 \zeta/\partial x^2 + \partial^2 \zeta/\partial y^2)_{0,0}$, and this is interpreted as a pressure difference across the surface. The curvature of the surface in the $y = 0$ plane $= [\partial^2 \zeta/\partial x^2][1 + (\partial \zeta/\partial x)^2]^{-3/2}$. Since this is evaluated at $(0,0)$ where $\partial \zeta/\partial x = 0$, the curvature reduces to $\partial^2 \zeta/\partial x^2 \equiv 1/R_1$ (defining R_1). Similarly, the curvature in the $x = 0$ plane at (0,0) is $\partial^2 \zeta/\partial y^2 \equiv 1/R_2$ (defining R_2). Thus we say

$$\Delta p = \sigma(1/R_1 + 1/R_2), \tag{4.98}$$

where the pressure is greater on the side with the center of curvature of the interface. Batchelor (loc. cit., p. 64) writes "An unbounded surface with a constant sum of the principal curvatures is spherical, and this must be the equilibrium shape of the surface. This result also follows from the fact that in a state of (stable) equilibrium the energy of the surface must be a minimum consistent with a given value of the volume of the drop or bubble, and the sphere is the shape which has the least surface area for a given volume." The original source of this analysis is Lord Rayleigh (J. W. Strutt), "On the Theory of Surface Forces," *Phil. Mag.* (Ser. 5), Vol. **30**, pp. 285–298, 456–475 (1890).

For an air bubble in water, gravity is an important factor for bubbles of millimeter size, as we shall see here. The hydrostatic pressure for a liquid is obtained from $p_L + \rho gz =$ const., where z is measured positively upwards from the free surface and g is downwards. Thus for a gas bubble beneath the free surface,

$$p_G = p_L + \sigma(1/R_1 + 1/R_2) = \text{const.} - \rho gz + \sigma(1/R_1 + 1/R_2).$$

Gravity and surface tension are of the same order in effect over a length scale $(\sigma/\rho g)^{1/2}$. For an air bubble in water at $288\,°K$, this scale $= [7.35 \times 10^{-2}\,\text{N/m}/(9.81\,\text{m/s}^2 \times 10^3\,\text{kg/m}^3)]^{1/2} = 2.74 \times 10^{-3}\,\text{m}$.

Example 4.3. Calculation of the shape of the free surface of a liquid adjoining an infinite vertical plane wall. With reference to Figure 4.24, as defined above, $1/R_1 = [\partial^2 \zeta/\partial x^2][1 + (\partial \zeta/\partial x)^2]^{-3/2} = 0$, and $1/R_2 = [\partial^2 \zeta/\partial y^2][1 + (\partial \zeta/\partial y)^2]^{-3/2}$. At the free surface, $\rho g\zeta - \sigma/R_2 = $ const. As $y \to \infty$, $\zeta \to 0$, and $R_2 \to \infty$, so const. $= 0$. Then $\rho g\zeta/\sigma - \zeta''/(1 + \zeta'^2)^{3/2} = 0$.

Multiply by the integrating factor ζ' and integrate. We obtain $(\rho g/2\sigma)\zeta^2 + (1 + \zeta'^2)^{-1/2} = C$. Evaluate C as $y \to \infty$, $\zeta \to 0$, $\zeta' \to 0$. Then $C = 1$. We look at $y = 0$, $z = \zeta = h$ to find h. The slope at the wall, $\zeta' = \tan(\theta + \pi/2) = -\cot\theta$. Then $1 + \zeta'^2 = 1 + \cot^2\theta = \csc^2\theta$. Thus we now have $(\rho g/2\sigma)h^2 = 1 - 1/\csc\theta$

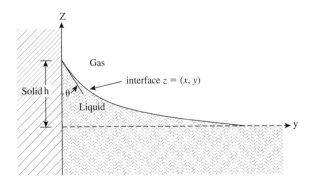

Figure 4.24 Free surface of a liquid adjoining a vertical plane wall.

$= 1 - \sin\theta$, so that $h^2 = (2\sigma/\rho g)(1 - \sin\theta)$. Finally we seek to integrate to obtain the shape of the interface. Squaring and rearranging the result above, the differential equation we must solve may be written as $1 + (d\zeta/dy)^2 = [1 - (\rho g/2\sigma)\zeta^2]^{-2}$. Solving for the slope and taking the negative square root (since the slope is negative for positive y),

$$d\zeta/dy = -\{1 - [1 - (\rho g/2\sigma)\zeta^2]^2\}^{1/2}[1 - (\rho g/2\sigma)\zeta^2]^{-1}.$$

Define $\sigma/\rho g = d^2$, $\zeta/d = \eta$. Rewriting the equation in terms of y/d and η, and separating variables,

$$2(1 - \eta^2/2)\eta^{-1}(4 - \eta^2)^{-1/2}d\eta = d(y/d).$$

The integrand on the left is simplified by partial fractions and the constant of integration is evaluated at $y = 0$ when $\eta = h/d$. Finally

$$\cosh^{-1}(2d/\zeta) - (4 - \zeta^2/d^2)^{1/2} - \cosh^{-1}(2d/h) + (4 - h^2/d^2)^{1/2} = y/d$$

gives the shape of the interface in terms of $y(\zeta)$.

   Analysis of surface tension effects results in the appearance of additional dimensionless parameters in which surface tension is compared with other effects such as viscous stresses, body forces such as gravity, and inertia. These are defined in Chapter 8.   □

Exercises

   1. Let a one-dimensional velocity field be $u = u(x, t)$, with $v = 0$ and $w = 0$. The density varies as $\rho = \rho_0(2 - \cos\omega t)$. Find an expression for $u(x, t)$ if $u(0, t) = U$.

   2. In Section 3 we derived the continuity equation (4.8) by starting from the integral form of the law of conservation of mass for a *fixed* region. Derive equation (4.8) by starting from an integral form for a *material* volume. [*Hint*: Formulate the principle for a material volume and then use equation (4.5).]

   3. Consider conservation of angular momentum derived from the angular momentum principle by the word statement: Rate of increase of angular momentum in volume V = net influx of angular momentum across the bounding surface A of V + torques due to surface forces + torques due to body forces. Here, the only torques are due to the same forces that appear in (linear) momentum conservation. The possibilities for body torques and couple stresses have been neglected. The torques due to the surface forces are manipulated as follows. The torque about a point O due to the element of surface force $\tau_{mk}dA_m$ is $\int \epsilon_{ijk}x_j\tau_{mk}dA_m$, where x is the position vector from O to the element dA. Using Gauss' theorem, we write this as a volume integral,

$$\int_V \epsilon_{ijk}\frac{\partial}{\partial x_m}(x_j\tau_{mk})dV = \epsilon_{ijk}\int_V\left(\frac{\partial x_j}{\partial x_m}\tau_{mk} + x_j\frac{\partial\tau_{mk}}{\partial x_m}\right)dV$$

$$= \epsilon_{ijk}\int_V\left(\tau_{jk} + x_j\frac{\partial\tau_{mk}}{\partial x_m}\right)dV,$$

where we have used $\partial x_j/\partial x_m = \delta_{jm}$. The second term is $\int_V \mathbf{x} \times \nabla \cdot \boldsymbol{\tau} \, dV$ and combines with the remaining terms in the conservation of angular momentum to give $\int_V \mathbf{x} \times (\text{Linear Momentum: equation (4.17)}) \, dV = \int_V \epsilon_{ijk}\tau_{jk} \, dV$. Since the left-hand side $= 0$ for any volume V, we conclude that $\varepsilon_{ijk}\tau_{kj} = 0$, which leads to $\tau_{ij} = \tau_{ji}$.

4. Near the end of Section 7 we derived the equation of motion (4.15) by starting from an integral form for a material volume. Derive equation (4.15) by starting from the integral statement for a *fixed region*, given by equation (4.22).

5. Verify the validity of the second form of the viscous dissipation given in equation (4.60). [*Hint*: Complete the square and use $\delta_{ij}\delta_{ij} = \delta_{ii} = 3$.]

6. A rectangular tank is placed on wheels and is given a constant horizontal acceleration a. Show that, at steady state, the angle made by the free surface with the horizontal is given by $\tan\theta = a/g$.

7. A jet of water with a diameter of 8 cm and a speed of 25 m/s impinges normally on a large stationary flat plate. Find the force required to hold the plate stationary. Compare the average pressure on the plate with the stagnation pressure if the plate is 20 times the area of the jet.

8. Show that the thrust developed by a stationary rocket motor is $F = \rho A U^2 + A(p - p_{\text{atm}})$, where p_{atm} is the atmospheric pressure, and p, ρ, A, and U are, respectively, the pressure, density, area, and velocity of the fluid at the nozzle exit.

9. Consider the propeller of an airplane moving with a velocity U_1. Take a reference frame in which the air is moving and the propeller [disk] is stationary. Then the effect of the propeller is to accelerate the fluid from the upstream value U_1 to the downstream value $U_2 > U_1$. Assuming incompressibility, show that the thrust developed by the propeller is given by

$$F = \frac{\rho A}{2}(U_2^2 - U_1^2),$$

where A is the projected area of the propeller and ρ is the density (assumed constant). Show also that the velocity of the fluid at the plane of the propeller is the average value $U = (U_1 + U_2)/2$. [*Hint*: The flow can be idealized by a pressure jump, of magnitude $\Delta p = F/A$ right at the location of the propeller. Also apply Bernoulli's equation between a section far upstream and a section immediately upstream of the propeller. Also apply the Bernoulli equation between a section immediately downstream of the propeller and a section far downstream. This will show that $\Delta p = \rho(U_2^2 - U_1^2)/2$.]

10. A hemispherical vessel of radius R has a small rounded orifice of area A at the bottom. Show that the time required to lower the level from h_1 to h_2 is given by

$$t = \frac{2\pi}{A\sqrt{2g}}\left[\frac{2}{3}R\left(h_1^{3/2} - h_2^{3/2}\right) - \frac{1}{5}\left(h_1^{5/2} - h_2^{5/2}\right)\right].$$

11. Consider an incompressible planar Couette flow, which is the flow between two parallel plates separated by a distance b. The upper plate is moving parallel to

itself at speed U, and the lower plate is stationary. Let the x-axis lie on the lower plate. All flow fields are independent of x. Show that the pressure distribution is hydrostatic and that the solution of the Navier–Stokes equation is

$$u(y) = \frac{Uy}{b}.$$

Write the expressions for the stress and strain rate tensors, and show that the viscous dissipation per unit volume is $\phi = \mu U^2/b^2$.

Take a rectangular control volume for which the two horizontal surfaces coincide with the walls and the two vertical surfaces are perpendicular to the flow. Evaluate every term of energy equation (4.63) for this control volume, and show that the balance is between the viscous dissipation and the work done in moving the upper surface.

12. The components of a mass flow vector $\rho\mathbf{u}$ are $\rho u = 4x^2y$, $\rho v = xyz$, $\rho w = yz^2$. Compute the net outflow through the closed surface formed by the planes $x = 0, x = 1, y = 0, y = 1, z = 0, z = 1$.

(a)   Integrate over the closed surface.
(b)   Integrate over the volume bounded by that surface.

13. Prove that the velocity field given by $u_r = 0$, $u_\theta = k/(2\pi r)$ can have only two possible values of the circulation. They are (a) $\Gamma = 0$ for any path not enclosing the origin, and (b) $\Gamma = k$ for any path enclosing the origin.

14. Water flows through a pipe in a gravitational field as shown in the accompanying figure. Neglect the effects of viscosity and surface tension. Solve the appropriate conservation equations for the variation of the cross-sectional area of the fluid column $A(z)$ after the water has left the pipe at $z = 0$. The velocity of the fluid at $z = 0$ is uniform at v_0 and the cross-sectional area is A_0.

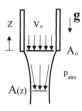

15. Redo the solution for the "orifice in a tank" problem allowing for the fact that in Fig. 4.20, $h = h(t)$. How long does the tank take to empty?

Literature Cited

Aris, R. (1962). *Vectors, Tensors, and the Basic Equations of Fluid Mechanics*, Englewood Cliffs, NJ: Prentice-Hall. (The basic equations of motion and the various forms of the Reynolds transport theorem are derived and discussed.)

Batchelor, G. K. (1967). *An Introduction to Fluid Dynamics*, London: Cambridge University Press. (This contains an excellent and authoritative treatment of the basic equations.)

Fermi, E. (1956). *Thermodynamics*, New York: Dover Publications, Inc.

Gogte, S. P. Vorobieff, R. Truesdell, A. Mammoli, F. van Swol, P. Shah, and C. J. Brinker (2005). "Effective slip on textured superhydrophobic surfaces." *Phys. Fluids* **17**: 051701.

Lamb, H. (1945). *Hydrodynamics*, Sixth Edition, New York: Dover Publications, Inc.

Levich, V. G. (1962). *Physicochemical Hydrodynamics*, Second Edition, Englewood Cliffs, NJ: Prentice-Hall, Chapter VII.

Lord Rayleigh (J. W. Strutt) (1890). "On the Theory of Surface Forces." *Phil. Mag.* (Ser. 5), **30**: 285–298, 456–475.

McCormick, N. J. (2005). "Gas-surface accomodation coefficients from viscous slip and temperature jump coefficients." *Phys. Fluids* **17**: 107104.

Holton, J. R. (1979). *An Introduction to Dynamic Meteorology*, New York: Academic Press.

Pedlosky, J. (1987). *Geophysical Fluid Dynamics*, New York: Springer-Verlag.

Probstein, R. F. (1994). *Physicochemical Hydrodynamics*, Second Edition, New York: John Wiley & Sons, Chapter 19.

Spiegel, E. A. and G. Veronis (1960). On the Boussinesq approximation for a compressible fluid. *Astrophysical Journal* **131**: 442–447.

Stommel H. M. and D. W. Moore (1989) *An Introduction to the Coriolis Force*. New York: Columbia University Press.

Truesdell, C. A. (1952). Stokes' principle of viscosity. *Journal of Rational Mechanics and Analysis* **1**: 228–231.

Supplemental Reading

Chandrasekhar, S. (1961). *Hydrodynamic and Hydromagnetic Stability*, London: Oxford University Press. (This is a good source to learn the basic equations in a brief and simple way.)

Dussan V., E. B. (1979). "On the Spreading of Liquids on Solid Surfaces: Static and Dynamic Contact Lines." *Annual Rev. of Fluid Mech.* **11**, 371–400.

Levich, V. G. and V. S. Krylov (1969). "Surface Tension Driven Phenomena." *Annual Rev. of Fluid Mech.* **1**, 293–316.

Vorticity Dynamics

1. Introduction

Motion in circular streamlines is called vortex motion. The presence of closed stream-lines does not necessarily mean that the fluid particles are rotating about their own centers, and we may have rotational as well as irrotational vortices depending on whether the *fluid particles* have vorticity or not. The two basic vortex flows are the solid-body rotation

$$u_\theta = \frac{1}{2}\omega r, \tag{5.1}$$

and the irrotational vortex

$$u_\theta = \frac{\Gamma}{2\pi r}. \tag{5.2}$$

These are discussed in Chapter 3, Section 11, where also, the angular velocity in the solid-body rotation was denoted by $\omega_0 = \omega/2$. Moreover, the vorticity of an element is *everywhere* equal to ω for the solid-body rotation represented by equation (5.1), so that the circulation around any contour is ω times the area enclosed by the contour.

©2010 Elsevier Inc. All rights reserved.
DOI: 10.1016/B978-0-12-381399-2.50005-8

In contrast, the flow represented by equation (5.2) is irrotational everywhere except at the origin, where the vorticity is infinite. All the vorticity of this flow is therefore concentrated on a line coinciding with the vortex axis. Circulation around any circuit not enclosing the origin is therefore zero, and that enclosing the origin is Γ. An irrotational vortex is therefore called a *line vortex*. Some aspects of the dynamics of flows with vorticity are examined in this chapter.

2. Vortex Lines and Vortex Tubes

A vortex line is a curve in the fluid such that its tangent at any point gives the direction of the local vorticity. A vortex line is therefore related to the vorticity vector the same way a streamline is related to the velocity vector. If ω_x, ω_y, and ω_z are the Cartesian components of the vorticity vector $\boldsymbol{\omega}$, then the orientation of a vortex line satisfies the equations

$$\frac{dx}{\omega_x} = \frac{dy}{\omega_y} = \frac{dz}{\omega_z},\tag{5.3}$$

which is analogous to equation (3.7) for a streamline. *In an irrotational vortex, the only vortex line in the flow field is the axis of the vortex. In a solid-body rotation, all lines perpendicular to the plane of flow are vortex lines.*

Vortex lines passing through any closed curve form a tubular surface, which is called a *vortex tube*. Just as streamlines bound a streamtube, a group of vortex lines bound a vortex tube (Figure 5.1). The circulation around a narrow vortex tube is $d\Gamma = \boldsymbol{\omega} \cdot d\mathbf{A}$, which is similar to the expression for the rate of flow $dQ = \mathbf{u} \cdot d\mathbf{A}$ through a narrow streamtube. The *strength of a vortex tube* is defined as the circulation around a closed circuit taken on the surface of the tube and embracing it just once. From Stokes' theorem it follows that the strength of a vortex tube is equal to the mean vorticity times its cross-sectional area.

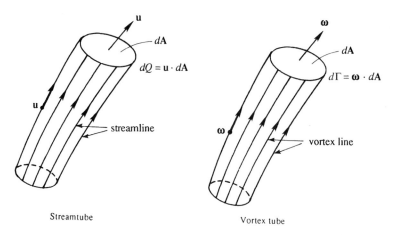

Figure 5.1   Analogy between streamtube and vortex tube.

3. Role of Viscosity in Rotational and Irrotational Vortices

The role of viscosity in the two basic types of vortex flows, namely the solid-body rotation and the irrotational vortex, is examined in this section. Assuming incompressible flow, we shall see that in one of these flows the viscous terms in the momentum equation drop out, although the viscous stress and dissipation of energy are nonzero. The two flows are examined separately in what follows.

Solid-Body Rotation

As discussed in Chapter 3, fluid elements in a solid-body rotation do not deform. Because viscous stresses are proportional to deformation rate, they are zero in this flow. This can be demonstrated by using the expression for viscous stress in polar coordinates:

$$\sigma_{r\theta} = \mu \left[\frac{1}{r} \frac{\partial u_r}{\partial \theta} + r \frac{\partial}{\partial r} \left(\frac{u_\theta}{r} \right) \right] = 0,$$

where we have substituted $u_\theta = \omega r/2$ and $u_r = 0$. We can therefore apply the inviscid Euler equations, which in polar coordinates simplify to

$$-\rho \frac{u_\theta^2}{r} = -\frac{\partial p}{\partial r}$$

$$0 = -\frac{\partial p}{\partial z} - \rho g. \tag{5.4}$$

The pressure difference between two neighboring points is therefore

$$dp = \frac{\partial p}{\partial r} dr + \frac{\partial p}{\partial z} dz = \frac{1}{4} \rho r \omega^2 \, dr - \rho g \, dz,$$

where $u_\theta = \omega r/2$ has been used. Integration between any two points 1 and 2 gives

$$p_2 - p_1 = \frac{1}{8} \rho \omega^2 (r_2^2 - r_1^2) - \rho g(z_2 - z_1). \tag{5.5}$$

Surfaces of constant pressure are given by

$$z_2 - z_1 = \frac{1}{8} (\omega^2/g)(r_2^2 - r_1^2),$$

which are paraboloids of revolution (Figure 5.2).

The important point to note is that viscous stresses are absent in this flow. (The viscous stresses, however, are important during the transient period of *initiating* the motion, say by steadily rotating a tank containing a viscous fluid at rest.) In terms of velocity, equation (5.5) can be written as

$$p_2 - \frac{1}{2} \rho u_{\theta 2}^2 + \rho g z_2 = p_1 - \frac{1}{2} \rho u_{\theta 1}^2 + \rho g z_1,$$

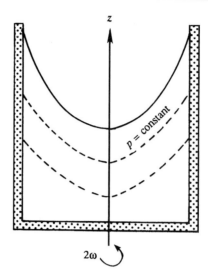

Figure 5.2   Constant pressure surfaces in a solid-body rotation generated in a rotating tank containing liquid.

which shows that the Bernoulli function $B = u_\theta^2/2 + gz + p/\rho$ is *not* constant for points on different streamlines. This is expected of inviscid *rotational* flows.

Irrotational Vortex

In an irrotational vortex represented by

$$u_\theta = \frac{\Gamma}{2\pi r},$$

the viscous stress is

$$\sigma_{r\theta} = \mu \left[\frac{1}{r} \frac{\partial u_r}{\partial \theta} + r \frac{\partial}{\partial r} \left(\frac{u_\theta}{r} \right) \right] = -\frac{\mu \Gamma}{\pi r^2},$$

which is nonzero everywhere. This is because fluid elements do undergo deformation in such a flow, as discussed in Chapter 3. However, the interesting point is that the *net viscous force* on an element again vanishes, just as in the case of solid body rotation. In an incompressible flow, the net viscous force per unit volume is related to vorticity by (see equation 4.48)

$$\frac{\partial \sigma_{ij}}{\partial x_j} = -\mu (\nabla \times \boldsymbol{\omega})_i, \tag{5.6}$$

which is zero for irrotational flows. The viscous forces on the surfaces of an element cancel out, leaving a zero resultant. *The equations of motion therefore reduce to the inviscid Euler equations, although viscous stresses are nonzero everywhere.* The pressure distribution can therefore be found from the inviscid set (5.4), giving

$$dp = \frac{\rho \Gamma^2}{4\pi^2 r^3} dr - \rho g \, dz,$$

where we have used $u_\theta = \Gamma/(2\pi r)$. Integration between any two points gives

$$p_2 - p_1 = -\frac{\rho}{2}(u_{\theta 2}^2 - u_{\theta 1}^2) - \rho g(z_2 - z_1),$$

which implies

$$\frac{p_1}{\rho} + \frac{u_{\theta 1}^2}{2} + g z_1 = \frac{p_2}{\rho} + \frac{u_{\theta 2}^2}{2} + g z_2.$$

This shows that Bernoulli's equation is applicable between any two points in the flow field and not necessarily along the same streamline, as would be expected of inviscid irrotational flows. Surfaces of constant pressure are given by

$$z_2 - z_1 = \frac{u_{\theta 1}^2}{2g} - \frac{u_{\theta 2}^2}{2g} = \frac{\Gamma^2}{8\pi^2 g}\left(\frac{1}{r_1^2} - \frac{1}{r_2^2}\right),$$

which are hyperboloids of revolution of the second degree (Figure 5.3). Flow is singular at the origin, where there is an infinite velocity discontinuity. Consequently, a real vortex such as that found in the atmosphere or in a bathtub necessarily has a rotational core (of radius R, say) in the center where the velocity distribution can be idealized by $u_\theta = \omega r/2$. Outside the core the flow is nearly irrotational and can be idealized by $u_\theta = \omega R^2/2r$; here we have chosen the value of circulation such that u_θ is continuous at $r = R$ (see Figure 3.17b). The strength of such a vortex is given by $\Gamma = $ (vorticity)(core area) $= \pi \omega R^2$.

One way of generating an irrotational vortex is by rotating a solid circular cylinder in an infinite viscous fluid (see Figure 9.7). It is shown in Chapter 9, Section 6 that the steady solution of the Navier–Stokes equations satisfying the no-slip boundary condition ($u_\theta = \omega R/2$ at $r = R$) is

$$u_\theta = \frac{\omega R^2}{2r} \quad r \geqslant R,$$

where R is the radius of the cylinder and $\omega/2$ is its constant angular velocity; see equation (9.15). This flow does not have any singularity in the entire field and is irrotational everywhere. Viscous stresses are present, and the resulting viscous dissipation of kinetic energy is exactly compensated by the work done at the surface of the cylinder. However, there is no *net* viscous force at any point in the steady state.

Discussion

The examples given in this section suggest that *irrotationality does not imply the absence of viscous stresses*. In fact, they must *always* be present in irrotational flows

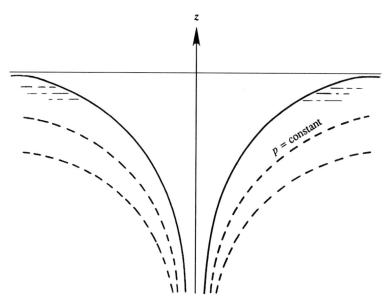

Figure 5.3   Irrotational vortex in a liquid.

of real fluids, simply because the fluid elements deform in such a flow. However the *net* viscous force vanishes if $\omega = 0$, as can be seen in equation (5.6). We have also given an example, namely that of solid-body rotation, in which there is *uniform* vorticity and no viscous stress at all. However, this is the *only* example in which rotation can take place without viscous effects, because equation (5.6) implies that the net force is zero in a rotational flow if ω is *uniform everywhere*. Except for this example, fluid rotation is accomplished by viscous effects. Indeed, we shall see later in this chapter that viscosity is a primary agent for vorticity generation.

4. Kelvin's Circulation Theorem

Several theorems of vortex motion in an inviscid fluid were published by Helmholtz in 1858. He discovered these by analogy with electrodynamics. Inspired by this work, Kelvin in 1868 introduced the idea of circulation and proved the following theorem: *In an inviscid, barotropic flow with conservative body forces, the circulation around a closed curve moving with the fluid remains constant with time,* if the motion is observed from a nonrotating frame. The theorem can be restated in simple terms as follows: At an instant of time take any closed contour C and locate the new position of C by following the motion of all of its fluid elements. Kelvin's circulation theorem states that the circulations around the two locations of C are the same. In other words,

$$\boxed{\frac{D\Gamma}{Dt} = 0,}$$

(5.7)

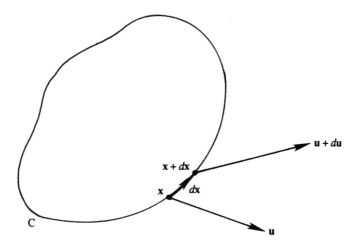

Figure 5.4   Proof of Kelvin's circulation theorem.

where D/Dt has been used to emphasize that the circulation is calculated around a *material contour* moving with the fluid.

To prove Kelvin's theorem, the rate of change of circulation is found as

$$\frac{D\Gamma}{Dt} = \frac{D}{Dt} \int_C u_i \, dx_i = \int_C \frac{Du_i}{Dt} dx_i + \int_C u_i \frac{D}{Dt}(dx_i), \qquad (5.8)$$

where dx is the separation between two points on C (Figure 5.4). Using the momentum equation

$$\frac{Du_i}{Dt} = -\frac{1}{\rho}\frac{\partial p}{\partial x_i} + g_i + \frac{1}{\rho}\sigma_{ij,j},$$

where σ_{ij} is the deviatoric stress tensor (equation (4.33)). The first integral in equation (5.8) becomes

$$\int \frac{Du_i}{Dt} dx_i = -\int \frac{1}{\rho}\frac{\partial p}{\partial x_i} dx_i + \int g_i \, dx_i + \int \frac{1}{\rho}\sigma_{ij,j} \, dx_i$$

$$= -\int \frac{dp}{\rho} + \int g_i \, dx_i + \int \frac{1}{\rho}\sigma_{ij,j} \, dx_i,$$

where we have noted that $dp = \nabla p \cdot dx$ is the difference in pressure between two neighboring points. Equation (5.8) then becomes

$$\frac{D\Gamma}{Dt} = \int_C \mathbf{g} \cdot d\mathbf{x} - \int_C \frac{dp}{\rho} + \int \frac{1}{\rho}(\nabla \cdot \boldsymbol{\sigma}) \cdot d\mathbf{x} + \int_C u_i \frac{D}{Dt}(dx_i). \qquad (5.9)$$

Each term of equation (5.9) will now be shown to be zero. Let the body force be conservative, so that $\mathbf{g} = -\nabla\Pi$, where Π is the force potential or potential energy per unit mass. Then the line integral of $\mathbf{g}$ along a fluid line AB is

$$\int_A^B \mathbf{g} \cdot d\mathbf{x} = -\int_A^B \nabla\Pi \cdot d\mathbf{x} = -\int_A^B d\Pi = \Pi_A - \Pi_B.$$

When the integral is taken around the closed fluid line, points A and B coincide, showing that the first integral on the right-hand side of equation (5.9) is zero.

Now assume that the flow is *barotropic*, which means that density is a function of pressure alone. Incompressible and isentropic ($p/\rho^\gamma = $ constant for a perfect gas) flows are examples of barotropic flows. In such a case we can write ρ^{-1} as some function of p, and we choose to write this in the form of the derivative $\rho^{-1} \equiv dP/dp$. Then the integral of dp/ρ between any two points A and B can be evaluated, giving

$$\int_A^B \frac{dp}{\rho} = \int_A^B \frac{dP}{dp}\, dp = P_B - P_A.$$

The integral around a closed contour is therefore zero.

If viscous stresses can be neglected for those particles making up contour C, then the integral of the deviatoric stress tensor is zero. To show that the last integral in equation (5.9) vanishes, note that the velocity at point $\mathbf{x} + d\mathbf{x}$ on C is given by

$$\mathbf{u} + d\mathbf{u} = \frac{D}{Dt}(\mathbf{x} + d\mathbf{x}) = \frac{D\mathbf{x}}{Dt} + \frac{D}{Dt}(d\mathbf{x}),$$

so that

$$d\mathbf{u} = \frac{D}{Dt}(d\mathbf{x}),$$

The last term in equation (5.9) then becomes

$$\int_C u_i \frac{D}{Dt}(dx_i) = \int_C u_i\, du_i = \int_C d\left(\tfrac{1}{2}u_i^2\right) = 0.$$

This completes the proof of Kelvin's theorem.

We see that the three agents that can create or destroy vorticity in a flow are nonconservative body forces, nonbarotropic pressure-density relations, and viscous stresses. An example of each follows. A Coriolis force in a rotating coordinate system generates the "bathtub vortex" when a filled tank, initially as rest on the earth's surface, is drained. Heating from below in a gravitational field creates a buoyant force generating an upward plume. Cooling from above and mass conservation require that the motion be in cyclic rolls so that vorticity is created. Viscous stresses create vorticity in the neighborhood of a boundary where the no-slip condition is maintained. A short distance away from the boundary, the tangential velocity may be large. Then, because there are large gradients transverse to the flow, vorticity is created.

Discussion of Kelvin's Theorem

Because circulation is the surface integral of vorticity, Kelvin's theorem essentially shows that irrotational flows remain irrotational if the four restrictions are satisfied:

(1) *Inviscid flow*: In deriving the theorem, the inviscid Euler equation has been used, but only along the contour C itself. This means that circulation is preserved if there are no net viscous forces along the path followed by C. If C moves into viscous regions such as boundary layers along solid surfaces, then the circulation changes. The presence of viscous effects causes a *diffusion* of vorticity into or out of a fluid circuit, and consequently changes the circulation.

(2) *Conservative body forces*: Conservative body forces such as gravity act through the center of mass of a fluid particle and therefore do not tend to rotate it.

(3) *Barotropic flow*: The third restriction on the validity of Kelvin's theorem is that density must be a function of pressure only. A homogeneous incompressible liquid for which ρ is constant everywhere and an isentropic flow of a perfect gas for which p/ρ^{γ} is constant are examples of barotropic flows. Flows that are not barotropic are called *baroclinic*. Consider fluid elements in barotropic and baroclinic flows (Figure 5.5). For the barotropic element, lines of constant p are parallel to lines of constant ρ, which implies that the resultant pressure forces pass through the center of mass of the element. For the baroclinic element, the

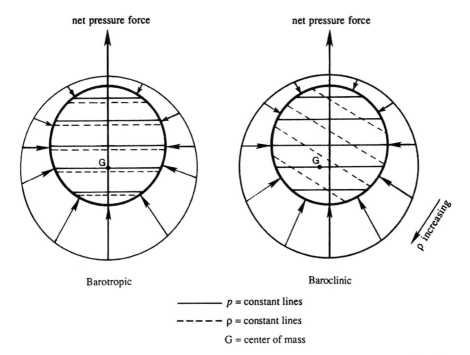

Figure 5.5   Mechanism of vorticity generation in baroclinic flow, showing that the net pressure force does not pass through the center of mass G. The radially inward arrows indicate pressure forces on an element.

lines of constant p and ρ are not parallel. The net pressure force does not pass through the center of mass, and the resulting torque changes the vorticity and circulation.

As an example of the generation of vorticity in a baroclinic flow, consider a gas at rest in a gravitational field. Let the gas be heated locally, say by chemical action (such as explosion of a bomb) or by a simple heater (Figure 5.6). The gas expands and rises upward. The flow is baroclinic because density here is also a function of temperature. A doughnut-shaped ring-vortex (similar to the smoke ring from a cigarette) forms and rises upward. (In a bomb explosion, a mushroom-shaped cloud occupies the central hole of such a ring.) Consider a closed fluid circuit ABCD when the gas is at rest; the circulation around it is then zero. If the region near AB is heated, the circuit assumes the new location A′B′CD after an interval of time; circulation around it is nonzero because $\mathbf{u} \cdot d\mathbf{x}$ along A′B′ is nonzero. The circulation around a material circuit has therefore changed, solely due to the baroclinicity of the flow. This is one of the reasons why geophysical flows, which are dominated by baroclinicity, are full of vorticity. It should be noted that no restriction is placed on the compressibility of the fluid, and Kelvin's theorem is valid for incompressible as well as compressible fluids.

(4) *Nonrotating frame*: Motion observed with respect to a rotating frame of reference can develop vorticity and circulation by mechanisms not considered in our demonstration of Kelvin's theorem. Effects of a rotating frame of reference are considered in Section 7.

Under the four restrictions mentioned in the foregoing, Kelvin's theorem essentially states that *irrotational flows remain irrotational at all times.*

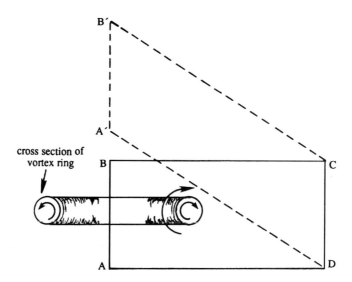

Figure 5.6   Local heating of a gas, illustrating vorticity generation on baroclinic flow.

Helmholtz Vortex Theorems

Under the same four restrictions, Helmholtz proved the following theorems on vortex motion:

(1) Vortex lines move with the fluid.
(2) Strength of a vortex tube, that is the circulation, is constant along its length.
(3) A vortex tube cannot end within the fluid. It must either end at a solid boundary or form a closed loop (a "vortex ring").
(4) Strength of a vortex tube remains constant in time.

Here, we shall prove only the first theorem, which essentially says that fluid particles that at any time are part of a vortex line always belong to the same vortex line. To prove this result, consider an area S, bounded by a curve, lying on the surface of a vortex tube without embracing it (Figure 5.7). As the vorticity vectors are everywhere lying on the area element S, it follows that the circulation around the edge of S is zero. After an interval of time, the same fluid particles form a new surface, say S′. According to Kelvin's theorem, the circulation around S′ must also be zero. As this is true for any S, the component of vorticity normal to every element of S′ must vanish, demonstrating that S′ must lie on the surface of the vortex tube. Thus, vortex tubes move with the fluid. Applying this result to an infinitesimally thin vortex tube, we get the Helmholtz vortex theorem that vortex lines move with the fluid. A different proof may be found in Sommerfeld (*Mechanics of Deformable Bodies*, pp. 130–132).

5. *Vorticity Equation in a Nonrotating Frame*

An equation governing the vorticity in a fixed frame of reference is derived in this section. The *fluid density is assumed to be constant*, so that the flow is barotropic. Viscous effects are retained. Effects of nonbarotropic behavior and a rotating frame

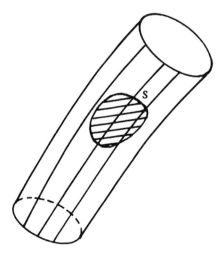

Figure 5.7   Proof of Helmholtz's vortex theorem.

of reference are considered in the following section. The derivation given here uses vector notation, so that we have to use several vector identities, including those for triple products of vectors. Readers not willing to accept the use of such vector identities can omit this section and move on to the next one, where the algebra is worked out in tensor notation without using such identities.

Vorticity is defined as

$$\boldsymbol{\omega} \equiv \nabla \times \mathbf{u}.$$

Because the divergence of a curl vanishes, vorticity for any flow must satisfy

$$\nabla \cdot \boldsymbol{\omega} = 0. \tag{5.10}$$

An equation for rate of change of vorticity is obtained by taking the curl of the equation of motion. We shall see that pressure and gravity are eliminated during this operation. In symbolic form, we want to perform the operation

$$\nabla \times \left\{ \frac{\partial \mathbf{u}}{\partial t} + \mathbf{u} \cdot \nabla \mathbf{u} = -\frac{1}{\rho} \nabla p + \nabla \Pi + \nu \nabla^2 \mathbf{u} \right\}, \tag{5.11}$$

where $\prod$ is the body force potential. Using the vector identity

$$\mathbf{u} \cdot \nabla \mathbf{u} = (\nabla \times \mathbf{u}) \times \mathbf{u} + \frac{1}{2} \nabla (\mathbf{u} \cdot \mathbf{u}) = \boldsymbol{\omega} \times \mathbf{u} + \frac{1}{2} \nabla q^2,$$

and noting that the curl of a gradient vanishes, (5.11) gives

$$\frac{\partial \boldsymbol{\omega}}{\partial t} + \nabla \times (\boldsymbol{\omega} \times \mathbf{u}) = \nu \nabla^2 \boldsymbol{\omega}, \tag{5.12}$$

where we have also used the identity $\nabla \times \nabla^2 \mathbf{u} = \nabla^2 (\nabla \times \mathbf{u})$ in rewriting the viscous term. The second term in equation (5.12) can be written as

$$\nabla \times (\boldsymbol{\omega} \times \mathbf{u}) = (\mathbf{u} \cdot \nabla) \boldsymbol{\omega} - (\boldsymbol{\omega} \cdot \nabla) \mathbf{u},$$

where we have used the vector identity

$$\nabla \times (\mathbf{A} \times \mathbf{B}) = \mathbf{A} \nabla \cdot \mathbf{B} + (\mathbf{B} \cdot \nabla) \mathbf{A} - \mathbf{B} \nabla \cdot \mathbf{A} - (\mathbf{A} \cdot \nabla) \mathbf{B},$$

and that $\nabla \cdot \mathbf{u} = 0$ and $\nabla \cdot \boldsymbol{\omega} = 0$. Equation (5.12) then becomes

$$\frac{D\boldsymbol{\omega}}{Dt} = (\boldsymbol{\omega} \cdot \nabla) \mathbf{u} + \nu \nabla^2 \boldsymbol{\omega}. \tag{5.13}$$

This is the equation governing rate of change of vorticity in a fluid with constant ρ and conservative body forces. The term $\nu \nabla^2 \boldsymbol{\omega}$ represents the rate of change of $\boldsymbol{\omega}$ due to diffusion of vorticity in the same way that $\nu \nabla^2 \mathbf{u}$ represents acceleration due to diffusion of momentum. The term $(\boldsymbol{\omega} \cdot \nabla) \mathbf{u}$ represents rate of change of vorticity due to stretching and tilting of vortex lines. This important mechanism of vorticity generation is discussed further near the end of Section 7, to which the reader can proceed if the rest of that section is not of interest. Note that pressure and gravity terms do not appear in the vorticity equation, as these forces act through the center of mass of an element and therefore generate no torque.

6. Velocity Induced by a Vortex Filament: Law of Biot and Savart

It is often useful to be able to calculate the velocity induced by a vortex filament with arbitrary orientation in space. This result is used in thin airfoil theory. We shall derive the velocity induced by a vortex filament for a constant density flow. (What actually is required is a solenoidal velocity field.) We start with the definition of vorticity, $\boldsymbol{\omega} \equiv \nabla \times \mathbf{u}$. Take the curl of this equation to obtain

$$\nabla \times \boldsymbol{\omega} = \nabla \times (\nabla \times \mathbf{u}) = \nabla(\nabla \cdot \mathbf{u}) - \nabla^2 \mathbf{u}.$$

We shall asume that mass conservation can be written as $\nabla \cdot \mathbf{u} = 0$, (for example, if $\rho = \text{const}$) and solve the vector Poisson equation for $\mathbf{u}$ in terms of $\boldsymbol{\omega}$. The Poisson equation in the form $\nabla^2 \phi = -\rho(r)/\varepsilon$ leads to the solution expressed as $\phi(r) = (4\pi\varepsilon)^{-1} \int_{V'} \rho(r')|r - r'|^{-1} dV'$ where the integration is over all of $V'(r')$ space. Using this form for each component of vorticity, we obtain for $\mathbf{u}$,

$$\mathbf{u} = (4\pi)^{-1} \int_{V'} (\nabla' \times \boldsymbol{\omega})|r - r'|^{-1} dV' \tag{5.14}$$

We take V' to be a small cylinder wrapped around the vortex line C through the point r'. See Figure 5.8. Equation (5.14) can be rewritten in general as

$$\mathbf{u} = (4\pi)^{-1} \int_{V'} \{\nabla' \times [\boldsymbol{\omega}/|r - r'|] - [\nabla'|r - r'|^{-1}] \times \boldsymbol{\omega}\} dV' \tag{5.15}$$

We use the divergence theorem on the first integral in the form $\int_V (\nabla \times \mathbf{F}) dV = \int_{A=\partial V} d\mathbf{A} \times \mathbf{F}$. Then (5.15) becomes

$$\mathbf{u} = (4\pi)^{-1} \left\{ \int_{A'=\partial V'} d\mathbf{A}' \times \boldsymbol{\omega}/|r - r'| + \int_{V'} dV'(\nabla'|r - r'|) \times \boldsymbol{\omega}/|r - r'|^2 \right\} \tag{5.16}$$

Now shrink V' and $A' = \partial V'$ to surround the vortex line segment in the neighborhood of $\mathbf{r}'$. On the two end faces of A', $d\mathbf{A}'||\boldsymbol{\omega}$ so $d\mathbf{A}' \times \boldsymbol{\omega} = 0$. Since, $\nabla \cdot \boldsymbol{\omega} = 0$, $\boldsymbol{\omega}$ is constant along a vortex line, so $\int_{A'_{\text{sides}}} d\mathbf{A}' \times \boldsymbol{\omega} = (\int_{A'_{\text{sides}}} d\mathbf{A}') \times \boldsymbol{\omega} = 0$ and $\int_{A'_{\text{sides}}} d\mathbf{A}' = 0$ because the generatrix of A'_{sides} is a closed curve. For the second integral, $dV' = d\mathbf{A}' \cdot d\mathbf{l}$, where $d\mathbf{A}'$ is an element of end face area and $d\mathbf{l}$ is arc length along the vortex line. Now, by Stokes' theorem, $\int_{\text{end}} \boldsymbol{\omega} \cdot d\mathbf{A}' = \oint_{C=\partial A'} \mathbf{u} \cdot d\mathbf{s} = \Gamma$, where Γ is the circulation around the vortex line C and $d\mathbf{s}$ is an element of arc

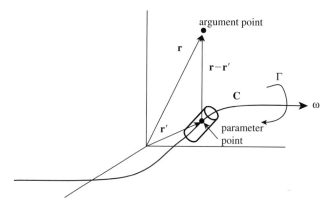

Figure 5.8 Geometry for derivation of Law of Biot and Savart.

length on the generatrix of A'. Then $\omega d\mathbf{A}' \cdot d\mathbf{l} = \Gamma dl$ since ω is parallel to $d\mathbf{l}$. Now $\nabla'|r - r'| = -\mathbf{1}_{\mathbf{r}-\mathbf{r}'}$ (unit vector), so (5.16) reduces to $\mathbf{u} = -(4\pi)^{-1} \int_C (\mathbf{1}_{\mathbf{r}-\mathbf{r}'}/|r - r'|^2) \times (\Gamma d\mathbf{l})$ for any length of vortex line C. For a small segment of vortex line $d\mathbf{l}$,

$$d\mathbf{u} = (\Gamma/4\pi)[d\mathbf{l} \times \mathbf{1}_{\mathbf{r}-\mathbf{r}'}/|r - r'|^2] \tag{5.17}$$

is an expression of the Law of Biot and Savart.

7. *Vorticity Equation in a Rotating Frame*

A vorticity equation was derived in Section 5 for a fluid of uniform density in a fixed frame of reference. We shall now generalize this derivation to include a rotating frame of reference and nonbarotropic fluids. The flow, however, will be assumed nearly incompressible in the Boussinesq sense, so that the continuity equation is approximately $\nabla \cdot \mathbf{u} = 0$. We shall also use tensor notation and not assume any vector identity. Algebraic manipulations are cleaner if we adopt the comma notation introduced in Chapter 2, Section 15, namely, that a comma stands for a spatial derivative:

$$\mathbf{A}_{,i} \equiv \frac{\partial \mathbf{A}}{\partial x_i}.$$

A little practice may be necessary to feel comfortable with this notation, but it is very convenient.

We first show that the divergence of ω is zero. From the definition $\omega = \nabla \times \mathbf{u}$, we obtain

$$\omega_{i,i} = (\varepsilon_{inq} u_{q,n})_{,i} = \varepsilon_{inq} u_{q,ni}.$$

In the last term, ε_{inq} is antisymmetric in i and n, whereas the derivative $u_{q,ni}$ is symmetric in i and n. As the contracted product of a symmetric and an antisymmetric tensor is zero, it follows that

$$\omega_{i,i} = 0 \quad \text{or} \quad \boxed{\nabla \cdot \boldsymbol{\omega} = 0} \tag{5.18}$$

which shows that the vorticity field is nondivergent (solenoidal), even for compressible and unsteady flows.

The continuity and momentum equations for a nearly incompressible flow in rotating coordinates are

$$u_{i,i} = 0, \tag{5.19}$$

$$\frac{\partial u_i}{\partial t} + u_j u_{i,j} + 2\varepsilon_{ijk}\Omega_j u_k = -\frac{1}{\rho}p_{,i} + g_i + \nu u_{i,jj}, \tag{5.20}$$

where $\boldsymbol{\Omega}$ is the angular velocity of the coordinate system and g_i is the effective gravity (including centrifugal acceleration); see equation (4.55). The advective acceleration can be written as

$$\begin{aligned}
u_j u_{i,j} &= u_j(u_{i,j} - u_{j,i}) + u_j u_{j,i} \\
&= -u_j \varepsilon_{ijk}\omega_k + \frac{1}{2}(u_j u_j)_{,i} \\
&= -(\mathbf{u} \times \boldsymbol{\omega})_i + \frac{1}{2}(u_j^2)_{,i},
\end{aligned} \tag{5.21}$$

where we have used the relation

$$\begin{aligned}
\varepsilon_{ijk}\omega_k &= \varepsilon_{ijk}(\varepsilon_{kmn} u_{n,m}) \\
&= (\delta_{im}\delta_{jn} - \delta_{in}\delta_{jm}) u_{n,m} = u_{j,i} - u_{i,j}.
\end{aligned} \tag{5.22}$$

The viscous diffusion term can be written as

$$\nu u_{i,jj} = \nu(u_{i,j} - u_{j,i})_{,j} + \nu u_{j,ij} = -\nu\varepsilon_{ijk}\omega_{k,j}, \tag{5.23}$$

where we have used equation (5.22) and the fact that $u_{j,ij} = 0$ because of the continuity equation (5.19). Relation (5.22) says that $\nu\nabla^2\mathbf{u} = -\nu\nabla \times \boldsymbol{\omega}$, which we have used several times before (e.g., see equation (4.48)). Because $\boldsymbol{\Omega} \times \mathbf{u} = -\mathbf{u} \times \boldsymbol{\Omega}$, the Coriolis term in equation (5.20) can be written as

$$2\varepsilon_{ijk}\Omega_j u_k = -2\varepsilon_{ijk}\Omega_k u_j. \tag{5.24}$$

Substituting equations (5.21), (5.23), and (5.24) into equation (5.20), we obtain

$$\frac{\partial u_i}{\partial t} + \left(\frac{1}{2}u_j^2 + \Pi\right)_{,i} - \varepsilon_{ijk}u_j(\omega_k + 2\Omega_k) = -\frac{1}{\rho}p_{,i} - \nu\varepsilon_{ijk}\,\omega_{k,j}, \tag{5.25}$$

where we have also assumed $\mathbf{g} = -\nabla\Pi$.

Equation (5.25) is another form of the Navier–Stokes equation, and the vorticity equation is obtained by taking its curl. Since $\omega_n = \varepsilon_{nqi} u_{i,q}$, it is clear that we need to operate on (5.25) by $\varepsilon_{nqi}(\quad)_{,q}$. This gives

$$\frac{\partial}{\partial t}(\varepsilon_{nqi} u_{i,q}) + \varepsilon_{nqi}\left(\frac{1}{2}u_j^2 + \Pi\right)_{,iq} - \varepsilon_{nqi}\varepsilon_{ijk}[u_j(\omega_k + 2\Omega_k)]_{,q}$$

$$= -\varepsilon_{nqi}\left(\frac{1}{\rho}p_{,i}\right)_{,q} - \nu\varepsilon_{nqi}\varepsilon_{ijk}\omega_{k,jq}. \tag{5.26}$$

The second term on the left-hand side vanishes on noticing that ε_{nqi} is antisymmetric in q and i, whereas the derivative $(u_j^2/2 + \Pi)_{,iq}$ is symmetric in q and i. The third term on the left-hand side of (5.26) can be written as

$$-\varepsilon_{nqi}\varepsilon_{ijk}[u_j(\omega_k + 2\Omega_k)]_{,q} = -(\delta_{nj}\delta_{qk} - \delta_{nk}\delta_{qj})[u_j(\omega_k + 2\Omega_k)]_{,q}$$

$$= -[u_n(\omega_k + 2\Omega_k)]_{,k} + [u_j(\omega_n + 2\Omega_n)]_{,j}$$

$$= -u_n(\omega_{k,k} + 2\Omega_{k,k}) - u_{n,k}(\omega_k + 2\Omega_k) + u_j(\omega_n + 2\Omega_n)_{,j}$$

$$= -u_n(0+0) - u_{n,k}(\omega_k + 2\Omega_k) + u_j(\omega_n + 2\Omega_n)_{,j}$$

$$= -u_{n,j}(\omega_j + 2\Omega_j) + u_j\,\omega_{n,j}, \tag{5.27}$$

where we have used $u_{i,i} = 0$, $\omega_{i,i} = 0$ and the fact that the derivatives of $\boldsymbol{\Omega}$ are zero.

The first term on the right-hand side of equation (5.26) can be written as follows:

$$-\varepsilon_{nqi}\left(\frac{1}{\rho}p_{,i}\right)_{,q} = -\frac{1}{\rho}\varepsilon_{nqi}\,p_{,iq} + \frac{1}{\rho^2}\varepsilon_{nqi}\rho_{,q}\,p_{,i}$$

$$= 0 + \frac{1}{\rho^2}[\nabla\rho \times \nabla p]_n, \tag{5.28}$$

which involves the n-component of the vector $\nabla\rho \times \nabla p$. The viscous term in equation (5.26) can be written as

$$-\nu\varepsilon_{nqi}\varepsilon_{ijk}\omega_{k,jq} = -\nu(\delta_{nj}\delta_{qk} - \delta_{nk}\delta_{qj})\omega_{k,jq}$$

$$= -\nu\omega_{k,nk} + \nu\omega_{n,jj} = \nu\omega_{n,jj}. \tag{5.29}$$

If we use equations (5.27)–(5.29), vorticity equation (5.26) becomes

$$\frac{\partial\omega_n}{\partial t} = u_{n,j}(\omega_j + 2\Omega_j) - u_j\omega_{n,j} + \frac{1}{\rho^2}[\nabla\rho \times \nabla p]_n + \nu\omega_{n,jj}.$$

Changing the free index from n to i, this becomes

$$\frac{D\omega_i}{Dt} = (\omega_j + 2\Omega_j)u_{i,j} + \frac{1}{\rho^2}[\nabla\rho \times \nabla p]_i + \nu\omega_{i,jj}.$$

In vector notation it is written as

$$\boxed{\frac{D\boldsymbol{\omega}}{Dt} = (\boldsymbol{\omega} + 2\boldsymbol{\Omega})\cdot\nabla\mathbf{u} + \frac{1}{\rho^2}\nabla\rho \times \nabla p + \nu\nabla^2\boldsymbol{\omega}.} \tag{5.30}$$

This is the *vorticity equation* for a nearly incompressible (that is, Boussinesq) fluid in rotating coordinates. Here **u** and **ω** are, respectively, the (relative) velocity and vorticity observed in a frame of reference rotating at angular velocity **Ω**. As vorticity is defined as twice the angular velocity, $2\boldsymbol{\Omega}$ is the *planetary vorticity* and $(\boldsymbol{\omega} + 2\boldsymbol{\Omega})$ is the *absolute vorticity* of the fluid, measured in an inertial frame. In a nonrotating frame, the vorticity equation is obtained from equation (5.30) by setting **Ω** to zero and interpreting **u** and **ω** as the absolute velocity and vorticity, respectively.

The left-hand side of equation (5.30) represents the rate of change of relative vorticity following a fluid particle. The last term $\nu \nabla^2 \boldsymbol{\omega}$ represents the rate of change of **ω** due to molecular diffusion of vorticity, in the same way that $\nu \nabla^2 \mathbf{u}$ represents acceleration due to diffusion of velocity. The second term on the right-hand side is the rate of generation of vorticity due to baroclinicity of the flow, as discussed in Section 4. In a barotropic flow, density is a function of pressure alone, so $\nabla \rho$ and ∇p are parallel vectors. The first term on the right-hand side of equation (5.30) plays a crucial role in the dynamics of vorticity; it is discussed in more detail in what follows.

Meaning of $(\boldsymbol{\omega} \cdot \nabla)\mathbf{u}$

To examine the significance of this term, take a natural coordinate system with s along a vortex line, n away from the center of curvature, and m along the third normal (Figure 5.9). Then

$$(\boldsymbol{\omega} \cdot \nabla)\mathbf{u} = \left[\boldsymbol{\omega} \cdot \left(\mathbf{i}_s \frac{\partial}{\partial s} + \mathbf{i}_n \frac{\partial}{\partial n} + \mathbf{i}_m \frac{\partial}{\partial m}\right)\right]\mathbf{u} = \omega \frac{\partial \mathbf{u}}{\partial s} \qquad (5.31)$$

where we have used $\boldsymbol{\omega} \cdot \mathbf{i}_n = \boldsymbol{\omega} \cdot \mathbf{i}_m = 0$, and $\boldsymbol{\omega} \cdot \mathbf{i}_s = \omega$ (the magnitude of **ω**). Equation (5.31) shows that $(\boldsymbol{\omega} \cdot \nabla)\mathbf{u}$ equals the magnitude of **ω** times the derivative of **u** in the direction of **ω**. The quantity $\omega(\partial \mathbf{u}/\partial s)$ is a vector and has the components $\omega(\partial u_s/\partial s)$, $\omega(\partial u_n/\partial s)$, and $\omega(\partial u_m/\partial s)$. Among these, $\partial u_s/\partial s$ represents the increase of u_s along the vortex line s, that is, the stretching of vortex lines. On the other hand, $\partial u_n/\partial s$ and $\partial u_m/\partial s$ represent the change of the normal velocity components along s and, therefore, the rate of turning or tilting of vortex lines about the m and n axes, respectively.

To see the effect of these terms more clearly, let us write equation (5.30) and suppress all terms except $(\boldsymbol{\omega} \cdot \nabla)\mathbf{u}$ on the right-hand side, giving

$$\frac{D\boldsymbol{\omega}}{Dt} = (\boldsymbol{\omega} \cdot \nabla)\mathbf{u} = \omega \frac{\partial \mathbf{u}}{\partial s} \quad \text{(barotropic, inviscid, nonrotating)}$$

whose components are

$$\frac{D\omega_s}{Dt} = \omega \frac{\partial u_s}{\partial s}, \qquad \frac{D\omega_n}{Dt} = \omega \frac{\partial u_n}{\partial s}, \quad \text{and} \quad \frac{D\omega_m}{Dt} = \omega \frac{\partial u_m}{\partial s}. \qquad (5.32)$$

The first equation of (5.32) shows that the vorticity along s changes due to stretching of vortex lines, reflecting the principle of conservation of angular momentum. Stretching decreases the moment of inertia of fluid elements that constitute a vortex line, resulting in an increase of their angular speed. Vortex stretching plays an especially crucial role in the dynamics of turbulent and geophysical flows The second and third equations

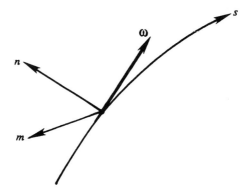

Figure 5.9   Coordinate system aligned with vorticity vector.

of (5.32) show how vorticity along n and m change due to tilting of vortex lines. For example, in Figure 5.9, the turning of the vorticity vector $\boldsymbol{\omega}$ toward the n-axis will generate a vorticity component along n. *The vortex stretching and tilting term $(\boldsymbol{\omega} \cdot \nabla)\,\mathbf{u}$ is absent in two-dimensional flows, in which $\boldsymbol{\omega}$ is perpendicular to the plane of flow.*

Meaning of $2(\boldsymbol{\Omega} \cdot \nabla)\,\mathbf{u}$

Orienting the z-axis along the direction of $\boldsymbol{\Omega}$, this term becomes $2(\boldsymbol{\Omega} \cdot \nabla)\mathbf{u} = 2\Omega(\partial\mathbf{u}/\partial z)$. Suppressing all other terms in equation (5.30), we obtain

$$\frac{D\boldsymbol{\omega}}{Dt} = 2\Omega\frac{\partial\mathbf{u}}{\partial z} \quad \text{(barotropic, inviscid, two-dimensional)}$$

whose components are

$$\frac{D\omega_z}{Dt} = 2\Omega\frac{\partial w}{\partial z}, \qquad \frac{D\omega_x}{Dt} = 2\Omega\frac{\partial u}{\partial z}, \quad \text{and} \quad \frac{D\omega_y}{Dt} = 2\Omega\frac{\partial v}{\partial z}.$$

This shows that stretching of fluid lines in the z direction increases ω_z, whereas a tilting of vertical lines changes the relative vorticity along the x and y directions. Note that merely a stretching or turning of vertical *fluid lines* is required for this mechanism to operate, in contrast to $(\boldsymbol{\omega} \cdot \nabla)\,\mathbf{u}$ where a stretching or turning of *vortex lines* is needed. This is because vertical fluid lines contain "planetary vorticity" $2\boldsymbol{\Omega}$. A vertically stretching fluid column tends to acquire positive ω_z, and a vertically shrinking fluid column tends to acquire negative ω_z (Figure 5.10). For this reason large-scale geophysical flows are almost always full of vorticity, and the change of $\boldsymbol{\Omega}$ due to the presence of planetary vorticity $2\boldsymbol{\Omega}$ is a central feature of geophysical fluid dynamics.

We conclude this section by writing down Kelvin's circulation theorem in a rotating frame of reference. It is easy to show that (Exercise 5) the circulation theorem

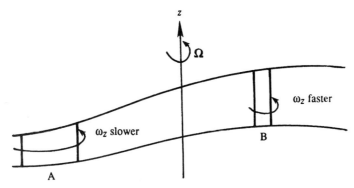

Figure 5.10 Generation of relative vorticity due to stretching of fluid columns parallel to planetary vorticity 2Ω. A fluid column acquires ω_z (in the same sense as Ω) by moving from location A to location B.

is modified to

$$\frac{D\Gamma_a}{Dt} = 0 \tag{5.33}$$

where

$$\Gamma_a \equiv \int_A (\omega + 2\Omega) \cdot d\mathbf{A} = \Gamma + 2 \int_A \Omega \cdot d\mathbf{A}.$$

Here, Γ_a is circulation due to the absolute vorticity $(\omega + 2\Omega)$ and differs from Γ by the "amount" of planetary vorticity intersected by $\mathbf{A}$.

8. Interaction of Vortices

Vortices placed close to one another can mutually interact, and generate interesting motions. To examine such interactions, we shall idealize each vortex by a concentrated line. A real vortex, with a core within which vorticity is distributed, can be idealized by a concentrated vortex line with a strength equal to the average vorticity in the core times the core area. Motion outside the core is assumed irrotational, and therefore inviscid. It will be shown in the next chapter that irrotational motion of a constant density fluid is governed by the linear Laplace equation. The principle of superposition therefore holds, and the flow at a point can be obtained by adding the contribution of all vortices in the field. To determine the mutual interaction of line vortices, the important principle to keep in mind is the Helmholtz vortex theorem, which says that vortex lines move with the flow.

Consider the interaction of two vortices of strengths Γ_1 and Γ_2, with both Γ_1 and Γ_2 positive (that is, counterclockwise vorticity). Let $h = h_1 + h_2$ be the distance between the vortices (Figure 5.11). Then the velocity at point 2 due to vortex Γ_1 is directed upward, and equals

$$V_1 = \frac{\Gamma_1}{2\pi h}.$$

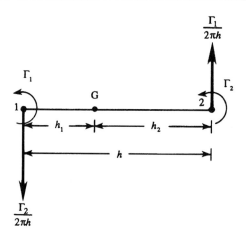

Figure 5.11   Interaction of line vortices of the same sign.

Similarly, the velocity at point 1 due to vortex Γ_2 is downward, and equals

$$V_2 = \frac{\Gamma_2}{2\pi h}.$$

The vortex pair therefore rotates counterclockwise around the "center of gravity" G, which is stationary.

Now suppose that the two vortices have the same circulation of magnitude Γ, but an opposite sense of rotation (Figure 5.12). Then the velocity of each vortex at the location of the other is $\Gamma/(2\pi h)$ and is directed in the same sense. The entire system therefore translates at a speed $\Gamma/(2\pi h)$ relative to the fluid. A pair of counter-rotating vortices can be set up by stroking the paddle of a boat, or by briefly moving the blade of a knife in a bucket of water (Figure 5.13). After the paddle or knife is withdrawn, the vortices do not remain stationary but continue to move under the action of the velocity induced by the other vortex.

The behavior of a single vortex near a wall can be found by superposing two vortices of equal and opposite strength. The technique involved is called the *method of images*, which has wide applications in irrotational flow, heat conduction, and electromagnetism. It is clear that the inviscid flow pattern due to vortex A at distance h from a wall can be obtained by eliminating the wall and introducing instead a vortex of equal strength and opposite sense at "image point" B (Figure 5.14). The velocity at any point P on the wall, made up of V_A due to the real vortex and V_B due to the image vortex, is then parallel to the wall. The wall is therefore a streamline, and the inviscid boundary condition of zero normal velocity across a solid wall is satisfied. Because of the flow induced by the image vortex, vortex A moves with speed $\Gamma/(4\pi h)$ parallel to the wall. For this reason, vortices in the example of Figure 5.13 move apart along the boundary on reaching the side of the vessel.

Now consider the interaction of two doughnut-shaped vortex rings (such as smoke rings) of equal and opposite circulation (Figure 5.15a). According to the method of images, the flow field for a single ring near a wall is identical to the flow of two rings

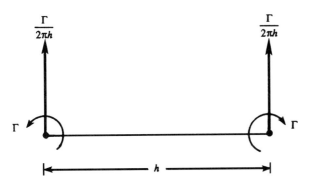

Figure 5.12 Interaction of line vortices of opposite spin, but of the same magnitude. Here Γ refers to the *magnitude* of circulation.

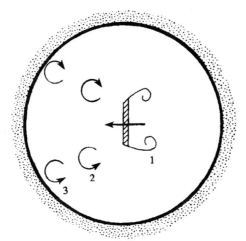

Figure 5.13 Top view of a vortex pair generated by moving the blade of a knife in a bucket of water. Positions at three instances of time 1, 2, and 3 are shown. (After Lighthill (1986).)

of opposite circulations. The translational motion of each element of the ring is caused by the induced velocity of each element of the same ring, plus the induced velocity of each element of the other vortex. In the figure, the motion at A is the resultant of V_B, V_C, and V_D, and this resultant has components parallel to and toward the wall. Consequently, the vortex ring increases in diameter and moves toward the wall with a speed that decreases monotonically (Figure 5.15b).

Finally, consider the interaction of two vortex rings of equal magnitude and similar sense of rotation. It is left to the reader (Exercise 6) to show that they should both translate in the same direction, but the one in front increases in radius and therefore slows down in its translational speed, while the rear vortex contracts and translates faster. This continues until the smaller ring passes through the larger one, at which point the roles of the two vortices are reversed. The two vortices can pass through each other forever in an ideal fluid. Further discussion of this intriguing problem can be found in Sommerfeld (1964, p. 161).

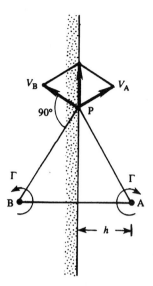

Figure 5.14   Line vortex A near a wall and its image B.

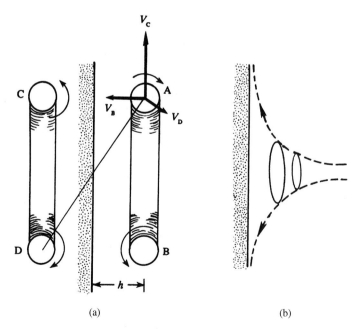

Figure 5.15   (a) Torus or doughnut-shaped vortex ring near a wall and its image. A section through the middle of the ring is shown. (b) Trajectory of vortex ring, showing that it widens while its translational velocity toward the wall decreases.

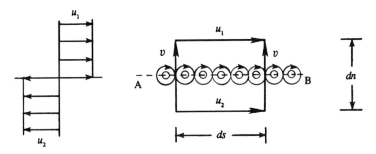

Figure 5.16   Vortex sheet.

9.   *Vortex Sheet*

Consider an infinite number of infinitely long vortex filaments, placed side by side on a surface AB (Figure 5.16). Such a surface is called a *vortex sheet*. If the vortex filaments all rotate clockwise, then the tangential velocity immediately above AB is to the right, while that immediately below AB is to the left. Thus, a discontinuity of tangential velocity exists across a vortex sheet. If the vortex filaments are not infinitesimally thin, then the vortex sheet has a finite thickness, and the velocity change is spread out.

In Figure 5.16, consider the circulation around a circuit of dimensions dn and ds. The normal velocity component v is continuous across the sheet ($v = 0$ if the sheet does not move normal to itself), while the tangential component u experiences a sudden jump. If u_1 and u_2 are the tangential velocities on the two sides, then

$$d\Gamma = u_2\,ds + v\,dn - u_1\,ds - v\,dn = (u_2 - u_1)\,ds,$$

Therefore the circulation per unit length, called the *strength of a vortex sheet*, equals the jump in tangential velocity:

$$\gamma \equiv \frac{d\Gamma}{ds} = u_2 - u_1.$$

The concept of a vortex sheet will be especially useful in discussing the flow over aircraft wings (Chapter 15).

Exercises

1. A closed cylindrical tank 4 m high and 2 m in diameter contains water to a depth of 3 m. When the cylinder is rotated at a constant angular velocity of 40 rad/s, show that nearly 0.71 m^2 of the bottom surface of the tank is uncovered. [*Hint*: The free surface is in the form of a paraboloid. For a point on the free surface, let h be the height above the (imaginary) vertex of the paraboloid and r be the local radius of the paraboloid. From Section 3 we have $h = \omega_0^2 r^2/2g$, where ω_0 is the angular velocity of the tank. Apply this equation to the two points where the paraboloid cuts the top and bottom surfaces of the tank.]

2. A tornado can be idealized as a Rankine vortex with a core of diameter 30 m. The gauge pressure at a radius of 15 m is $-2000\,\mathrm{N/m^2}$ (that is, thc absolute pressure is $2000\,\mathrm{N/m^2}$ below atmospheric). (a) Show that the circulation around any circuit surrounding the core is $5485\,\mathrm{m^2/s}$. [*Hint*: Apply the Bernoulli equation between infinity and the edge of the core.] (b) Such a tornado is moving at a linear speed of $25\,\mathrm{m/s}$ relative to the ground. Find the time required for the gauge pressure to drop from -500 to $-2000\,\mathrm{N/m^2}$. Neglect compressibility effects and assume an air temperature of $25\,^\circ\mathrm{C}$. (Note that the tornado causes a sudden decrease of the local atmospheric pressure. The damage to structures is often caused by the resulting excess pressure on the inside of the walls, which can cause a house to explode.)

3. The velocity field of a flow in cylindrical coordinates (R, φ, x) is

$$u_R = 0 \qquad u_\varphi = aRx \qquad u_x = 0$$

where a is a constant. (a) Show that the vorticity components are

$$\omega_R = -aR \qquad \omega_\varphi = 0 \qquad \omega_x = 2ax$$

(b) Verify that $\boldsymbol{\nabla} \cdot \boldsymbol{\omega} = 0$. (c) Sketch the streamlines and vortex lines in an Rx-plane. Show that the vortex lines are given by $xR^2 = $ constant.

4. Consider the flow in a 90° angle, confined by the walls $\theta = 0$ and $\theta = 90^\circ$. Consider a vortex line passing through (x, y), and oriented parallel to the z-axis. Show that the vortex path is given by

$$\frac{1}{x^2} + \frac{1}{y^2} = \text{constant}.$$

[*Hint*: Convince yourself that we need three image vortices at points $(-x, -y)$, $(-x, y)$ and $(x, -y)$. What are their senses of rotation? The path lines are given by $dx/dt = u$ and $dy/dt = v$, where u and v are the velocity components at the location of the vortex. Show that $dy/dx = v/u = -y^3/x^3$, an integration of which gives the result.]

5. Start with the equations of motion in the rotating coordinates, and prove Kelvin's circulation theorem

$$\frac{D}{Dt}(\Gamma_a) = 0$$

where

$$\Gamma_a = \int (\boldsymbol{\omega} + 2\boldsymbol{\Omega}) \cdot d\mathbf{A}$$

Assume that the flow is inviscid and barotropic and that the body forces are conservative. Explain the result physically.

6. Consider the interaction of two vortex rings of equal strength and similar sense of rotation. Argue that they go through each other, as described near the end of Section 8.

7. A constant density irrotational flow in a rectangular torus has a circulation Γ and volumetric flow rate Q. The inner radius is r_1, the outer radius is r_2, and the height is h. Compute the total kinetic energy of this flow in terms of only ρ, Γ, and Q.

8. Consider a cylindrical tank of radius R filled with a viscous fluid spinning steadily about its axis with constant angular velocity $\boldsymbol{\Omega}$. Assume that the flow is in a steady state. (a) Find $\int_A \boldsymbol{\omega} \cdot d\mathbf{A}$ where A is a horizontal plane surface through the fluid normal to the axis of rotation and bounded by the wall of the tank. (b) The tank then stops spinning. Find again the value of $\int_A \boldsymbol{\omega} \cdot d\mathbf{A}$.

9. In Figure 5.11, locate point G.

Literature Cited

Lighthill, M. J. (1986). *An Informal Introduction to Theoretical Fluid Mechanics*, Oxford, England: Clarendon Press.
Sommerfeld, A. (1964). *Mechanics of Deformable Bodies*, New York: Academic Press. (This book contains a good discussion of the interaction of vortices.)

Supplemental Reading

Batchelor, G. K. (1967). *An Introduction to Fluid Dynamics*, London: Cambridge University Press.
Pedlosky, J. (1987). *Geophysical Fluid Dynamics*, New York: Springer-Verlag. (This book discusses the vorticity dynamics in rotating coordinates, with application to geophysical systems.)
Prandtl, L. and O. G. Tietjens (1934). *Fundamentals of Hydro- and Aeromechanics*, New York: Dover Publications. (This book contains a good discussion of the interaction of vortices.)

Chapter 6

Irrotational Flow

1. *Relevance of Irrotational Flow Theory*

The vorticity equation given in the preceding chapter implies that the irrotational flow (such as the one starting from rest) of a barotropic fluid observed in a nonrotating frame remains irrotational if the fluid viscosity is identically zero and any body forces

©2010 Elsevier Inc. All rights reserved.
DOI: 10.1016/B978-0-12-381399-2.50006-X

are conservative. Such an ideal flow has a nonzero tangential velocity at a solid surface (Figure 6.1a). In contrast, a real fluid with a nonzero ν must satisfy a no-slip boundary condition. It can be expected that viscous effects in a real flow will be confined to thin layers close to solid surfaces if the fluid viscosity is small. We shall see later that the viscous layers are thin not just when the viscosity is small, but when a non-dimensional quantity $\mathrm{Re} = UL/\nu$, called the *Reynolds number*, is much larger than 1. (Here, U is a scale of variation of velocity in a length scale L.) The thickness of such *boundary layers*, within which viscous diffusion of vorticity is important, approaches zero as $\mathrm{Re} \rightarrow \infty$ (Figure 6.1b). In such a case, the vorticity equation implies that fluid elements starting from rest, or from any other irrotational region, remain irrotational unless they move into these boundary layers. The flow field can therefore be divided into an "outer region" where the flow is inviscid and irrotational and an "inner region" where viscous diffusion of vorticity is important. The outer flow can be approximately predicted by ignoring the existence of the thin boundary layer and applying irrotational flow theory around the solid object. Once the outer problem is determined, viscous flow equations within the boundary layer can be solved and matched to the outer solution.

An important exception in which this method would not work is where the solid object has such a shape that the boundary layer separates from the surface, giving rise to eddies in the wake (Figure 6.2). In this case viscous effects are not confined to thin layers around solid surfaces, and the real flow in the limit $\mathrm{Re} \rightarrow \infty$ is quite different

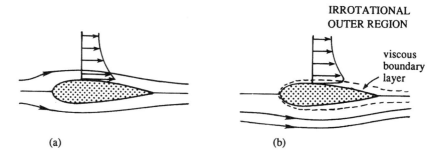

(a)   **(b)**

Figure 6.1 Comparison of a completely irrotational flow and a high Reynolds number flow: (a) ideal flow with $\nu = 0$; (b) flow at high Re.

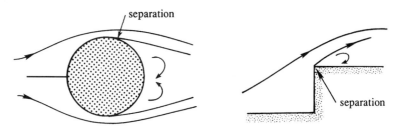

Figure 6.2 Examples of flow separation. Upstream of the point of separation, irrotational flow theory is a good approximation of the real flow.

from the ideal flow ($\nu = 0$). Ahead of the point of separation, however, irrotational flow theory is still a good approximation of the real flow (Figure 6.2).

Irrotational flow patterns around bodies of various shapes is the subject of this chapter. Motion will be assumed inviscid and incompressible. Most of the examples given are from two-dimensional plane flows, although some examples of axisymmetric flows are also given later in the chapter. Both Cartesian (x, y) and polar (r, θ) coordinates are used for plane flows.

2. Velocity Potential: Laplace Equation

The two-dimensional incompressible continuity equation

$$\frac{\partial u}{\partial x} + \frac{\partial v}{\partial y} = 0, \tag{6.1}$$

guarantees the existence of a stream function ψ, from which the velocity components can be derived as

$$u \equiv \frac{\partial \psi}{\partial y} \qquad v \equiv -\frac{\partial \psi}{\partial x}. \tag{6.2}$$

Likewise, the condition of irrotationality

$$\frac{\partial v}{\partial x} - \frac{\partial u}{\partial y} = 0, \tag{6.3}$$

guarantees the existence of another scalar function ϕ, called the *velocity potential*, which is related to the velocity components by

$$u \equiv \frac{\partial \phi}{\partial x} \qquad \text{and} \qquad v \equiv \frac{\partial \phi}{\partial y}. \tag{6.4}$$

Because a velocity potential must exist in all irrotational flows, such flows are frequently called *potential flows*. Equations (6.2) and (6.4) imply that the derivative of ψ gives the velocity component in a direction 90° clockwise from the direction of differentiation, whereas the derivative of ϕ gives the velocity component in the direction of differentiation. Comparing equations (6.2) and (6.4) we obtain

$$\frac{\partial \phi}{\partial x} = \frac{\partial \psi}{\partial y}$$

$$\text{Cauchy–Riemann conditions} \tag{6.5}$$

$$\frac{\partial \phi}{\partial y} = -\frac{\partial \psi}{\partial x}$$

from which one of the functions can be determined if the other is known. Equipotential lines (on which ϕ is constant) and streamlines are orthogonal, as equation (6.5) implies that

$$\nabla \phi \cdot \nabla \psi = \left(\mathbf{i} \frac{\partial \phi}{\partial x} + \mathbf{j} \frac{\partial \phi}{\partial y} \right) \cdot \left(\mathbf{i} \frac{\partial \psi}{\partial x} + \mathbf{j} \frac{\partial \psi}{\partial y} \right) = \frac{\partial \phi}{\partial x} \frac{\partial \psi}{\partial x} + \frac{\partial \phi}{\partial y} \frac{\partial \psi}{\partial y} = 0.$$

This demonstration fails at *stagnation points* where the velocity is zero.

The streamfunction and velocity potential satisfy the Laplace equations

$$\nabla^2 \phi = \frac{\partial^2 \phi}{\partial x^2} + \frac{\partial^2 \phi}{\partial y^2} = 0, \tag{6.6}$$

$$\nabla^2 \psi = \frac{\partial^2 \psi}{\partial x^2} + \frac{\partial^2 \psi}{\partial y^2} = 0, \tag{6.7}$$

as can be seen by cross differentiating equation (6.5). Equation (6.7) holds for two-dimensional flows only, because a single streamfunction is insufficient for three-dimensional flows. As we showed in Chapter 4, Section 4, two streamfunctions are required to describe three-dimensional steady flows (or, if density may be regarded as constant, three-dimensional unsteady flows). However, a velocity potential ϕ can be defined in three-dimensional irrotational flows, because $\mathbf{u} = \nabla \phi$ identically satisfies the irrotationality condition $\nabla \times \mathbf{u} = 0$. A three-dimensional potential flow satisfies the three-dimensional version of $\nabla^2 \phi = 0$.

A function satisfying the Laplace equation is sometimes called a *harmonic function*. The Laplace equation is encountered not only in potential flows, but also in heat conduction, elasticity, magnetism, and electricity. Therefore, solutions in one field of study can be found from a known analogous solution in another field. In this manner, an extensive collection of solutions of the Laplace equation have become known. The Laplace equation is of a type that is called *elliptic*. It can be shown that solutions of elliptic equations are smooth and do not have discontinuities, except for certain singular points on the boundary of the region. In contrast, hyperbolic equations such as the wave equation can have discontinuous "wavefronts" in the middle of a region.

The boundary conditions normally encountered in irrotational flows are of the following types:

(1) *Condition on solid surface*—Component of fluid velocity normal to a solid surface must equal the velocity of the boundary normal to itself, ensuring that fluid does not penetrate a solid boundary. For a stationary body, the condition is

$$\frac{\partial \phi}{\partial n} = 0 \qquad \text{or} \qquad \frac{\partial \psi}{\partial s} = 0 \tag{6.8}$$

where s is direction along the surface, and n is normal to the surface.

(2) *Condition at infinity*—For the typical case of a body immersed in a uniform stream flowing in the x direction with speed U, the condition is

$$\frac{\partial \phi}{\partial x} = U \qquad \text{or} \qquad \frac{\partial \psi}{\partial y} = U \tag{6.9}$$

However, solving the Laplace equation subject to boundary conditions of the type of equations (6.8) and (6.9) is not easy. Historically, irrotational flow theory was developed by finding a function that satisfies the Laplace equation and then determining what boundary conditions are satisfied by that function. As the Laplace equation is linear, superposition of known harmonic functions gives another harmonic function satisfying a new set of boundary conditions. A rich collection of solutions has thereby emerged. We shall adopt this "inverse" approach of studying irrotational flows in this chapter; numerical methods of finding a solution under given boundary conditions are illustrated in Sections 16 and 21.

After a solution of the Laplace equation has been obtained, the velocity components are then determined by taking derivatives of ϕ or ψ. Finally, the pressure distribution is determined by applying the Bernoulli equation

$$p + \tfrac{1}{2}\rho q^2 = \text{const.},$$

between any two points in the flow field; here q is the magnitude of velocity. Thus, a solution of the nonlinear equation of motion (the Euler equation) is obtained in irrotational flows in a much simpler manner.

For quick reference, the important equations in polar coordinates are listed in the following:

$$\frac{1}{r}\frac{\partial}{\partial r}(ru_r) + \frac{1}{r}\frac{\partial u_\theta}{\partial \theta} = 0 \qquad \text{(continuity)}, \tag{6.10}$$

$$\frac{1}{r}\frac{\partial}{\partial r}(ru_\theta) - \frac{1}{r}\frac{\partial u_r}{\partial \theta} = 0 \qquad \text{(irrotationality)}, \tag{6.11}$$

$$u_r = \frac{\partial \phi}{\partial r} = \frac{1}{r}\frac{\partial \psi}{\partial \theta}, \tag{6.12}$$

$$u_\theta = \frac{1}{r}\frac{\partial \phi}{\partial \theta} = -\frac{\partial \psi}{\partial r}, \tag{6.13}$$

$$\nabla^2 \phi = \frac{1}{r}\frac{\partial}{\partial r}\left(r\frac{\partial \phi}{\partial r}\right) + \frac{1}{r^2}\frac{\partial^2 \phi}{\partial \theta^2} = 0, \tag{6.14}$$

$$\nabla^2 \psi = \frac{1}{r}\frac{\partial}{\partial r}\left(r\frac{\partial \psi}{\partial r}\right) + \frac{1}{r^2}\frac{\partial^2 \psi}{\partial \theta^2} = 0, \tag{6.15}$$

3. Application of Complex Variables

In this chapter z will denote the complex variable

$$z \equiv x + iy = r\,e^{i\theta}, \tag{6.16}$$

where $i = \sqrt{-1}$, (x, y) are the Cartesian coordinates, and (r, θ) are the polar coordinates. In the Cartesian form the complex number z represents a point in the xy-plane whose real axis is x and imaginary axis is y (Figure 6.3). In the polar form, z represents the position vector $0z$, whose magnitude is $r = (x^2 + y^2)^{1/2}$ and whose angle with the x-axis is $\tan^{-1}(y/x)$. The product of two complex numbers z_1 and z_2 is

$$z_1 z_2 = r_1 r_2\, e^{i(\theta_1 + \theta_2)}.$$

Therefore, the process of multiplying a complex number z_1 by another complex number z_2 can be regarded as an operation that "stretches" the magnitude from r_1 to $r_1 r_2$ and increases the argument from θ_1 to $\theta_1 + \theta_2$.

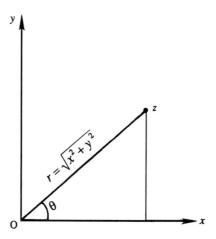

Figure 6.3   Complex z-plane.

When x and y are regarded as variables, the complex quantity $z = x + iy$ is called a *complex variable*. Suppose we define another complex variable w whose real and imaginary parts are ϕ and ψ:

$$w \equiv \phi + i\psi. \tag{6.17}$$

If ϕ and ψ are functions of x and y, then so is w. It is shown in the theory of complex variables that w is a function of the combination $x + iy = z$, and in particular has a finite and "unique derivative" dw/dz when its real and imaginary parts satisfy the pair of relations, equation (6.5), which are called *Cauchy–Riemann conditions*. Here the derivative dw/dz is regarded as unique if the value of $\delta w/\delta z$ does not depend on the *orientation* of the differential δz as it approaches zero. A single-valued function $w = f(z)$ is called an *analytic function* of a complex variable z in a region if a finite dw/dz exists everywhere within the region. Points where w or dw/dz is zero or infinite are called *singularities*, at which constant ϕ and constant ψ lines are not orthogonal. For example, $w = \ln z$ and $w = 1/z$ are analytic everywhere except at the singular point $z = 0$, where the Cauchy–Riemann conditions are not satisfied.

The combination $w = \phi + i\psi$ is called *complex potential* for a flow. Because the velocity potential and stream function satisfy equation (6.5), and the real and imaginary parts of *any* function of a complex variable $w(z) = \phi + i\psi$ also satisfy equation (6.5), it follows that *any analytic function of z represents the complex potential of some two-dimensional flow*. The derivative dw/dz is an important quantity in the description of irrotational flows. By definition

$$\frac{dw}{dz} = \lim_{\delta z \to 0} \frac{\delta w}{\delta z}.$$

As the derivative is independent of the orientation of δz in the xy-plane, we may take δz parallel to the x-axis, leading to

$$\frac{dw}{dz} = \lim_{\delta x \to 0} \frac{\delta w}{\delta x} = \frac{\partial w}{\partial x} = \frac{\partial}{\partial x}(\phi + i\psi),$$

which implies

$$\frac{dw}{dz} = u - iv. \tag{6.18}$$

It is easy to show that taking δz parallel to the y-axis leads to an identical result. The derivative dw/dz is therefore a complex quantity whose real and imaginary parts give Cartesian components of the local velocity; dw/dz is therefore called the *complex velocity*. If the local velocity vector has a magnitude q and an angle α with the x-axis, then

$$\frac{dw}{dz} = qe^{-i\alpha}. \tag{6.19}$$

It may be considered remarkable that any twice differentiable function $w(z), z = x + iy$ is an identical solution to Laplace's equation in the plane (x, y). A general function of the two variables (x, y) may be written as $f(z, z^*)$ where $z^* = x - iy$ is the complex conjugate of z. It is the very special case when $f(z, z^*) = w(z)$ alone that we consider here.

As Laplace's equation is linear, solutions may be superposed. That is, the sums of elemental solutions are also solutions. Thus, as we shall see, flows over specific shapes may be solved in this way.

4. Flow at a Wall Angle

Consider the complex potential

$$w = Az^n \qquad (n \geqslant \frac{1}{2}), \tag{6.20}$$

where A is a real constant. If r and θ represent the polar coordinates in the z-plane, then

$$w = A(re^{i\theta})^n = Ar^n(\cos n\theta + i \sin n\theta),$$

giving

$$\phi = Ar^n \cos n\theta \qquad \psi = Ar^n \sin n\theta. \tag{6.21}$$

For a given n, lines of constant ψ can be plotted. Equation (6.21) shows that $\psi = 0$ for all values of r on lines $\theta = 0$ and $\theta = \pi/n$. As any streamline, including the $\psi = 0$ line, can be regarded as a rigid boundary in the z-plane, it is apparent that equation (6.20) is the complex potential for flow between two plane boundaries of included angle $\alpha = \pi/n$. Figure 6.4 shows the flow patterns for various values of n. Flow within a certain sector of the z-plane only is shown; that within other sectors can be found by symmetry. It is clear that the walls form an angle larger than $180°$ for $n < 1$ and an angle smaller than $180°$ for $n > 1$. The complex velocity in terms of $\alpha = \pi/n$ is

$$\frac{dw}{dz} = nAz^{n-1} = \frac{A\pi}{\alpha}z^{(\pi-\alpha)/\alpha},$$

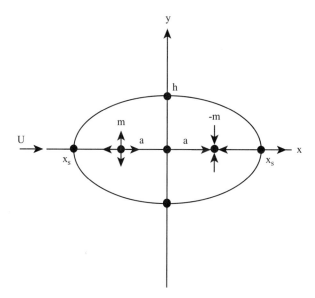

Figure 6.4   Irrotational flow at a wall angle. Equipotential lines are dashed.

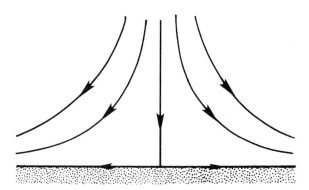

Figure 6.5   Stagnation flow represented by $w = Az^2$.

which shows that at the origin $dw/dz = 0$ for $\alpha < \pi$, and $dw/dz = \infty$ for $\alpha > \pi$. Thus, the *corner is a stagnation point for flow in a wall angle smaller than 180°; in contrast, it is a point of infinite velocity for wall angles larger than 180°.* In both cases the origin is a singular point.

The pattern for $n = 1/2$ corresponds to flow around a semi-infinite plate. When $n = 2$, the pattern represents flow in a region bounded by perpendicular walls. By including the field within the second quadrant of the z-plane, it is clear that $n = 2$ also represents the flow impinging against a flat wall (Figure 6.5). The streamlines and equipotential lines are all rectangular hyperbolas. This is called a *stagnation flow* because it represents flow in the neighborhood of the stagnation point of a blunt body.

Real flows near a sharp change in wall slope are somewhat different than those shown in Figure 6.4. For $n < 1$ the irrotational flow velocity is infinite at the origin, implying that the boundary streamline ($\psi = 0$) accelerates before reaching this point and decelerates after it. Bernoulli's equation implies that the pressure force downstream of the corner is "adverse" or against the flow. It will be shown in Chapter 10 that an adverse pressure gradient causes separation of flow and generation of stationary eddies. A real flow in a corner with an included angle larger than $180°$ would therefore separate at the corner (see the right panel of Figure 6.2).

5. Sources and Sinks

Consider the complex potential

$$w = \frac{m}{2\pi} \ln z = \frac{m}{2\pi} \ln (re^{i\theta}). \tag{6.22}$$

The real and imaginary parts are

$$\phi = \frac{m}{2\pi} \ln r \qquad \psi = \frac{m}{2\pi} \theta, \tag{6.23}$$

from which the velocity components are found as

$$u_r = \frac{m}{2\pi r} \qquad u_\theta = 0. \tag{6.24}$$

This clearly represents a radial flow from a two-dimensional line source at the origin, with a volume flow rate per unit depth of m (Figure 6.6). The flow represents a line sink if m is negative. For a source situated at $z = a$, the complex potential is

$$w = \frac{m}{2\pi} \ln (z - a). \tag{6.25}$$

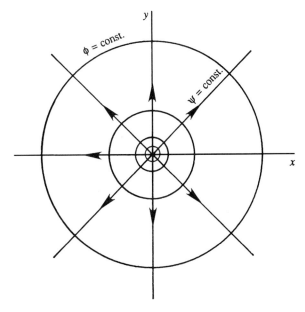

Figure 6.6   Plane source.

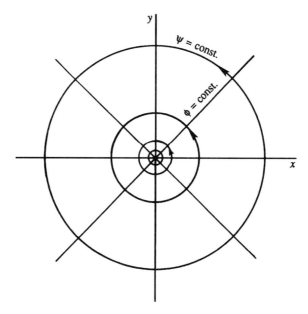

Figure 6.7   Plane irrotational vortex.

6. *Irrotational Vortex*

The complex potential

$$w = -\frac{i\Gamma}{2\pi} \ln z. \tag{6.26}$$

represents a line vortex of counterclockwise circulation Γ. Its real and imaginary parts are

$$\phi = \frac{\Gamma}{2\pi}\theta \qquad \psi = -\frac{\Gamma}{2\pi} \ln r, \tag{6.27}[1]$$

from which the velocity components are found to be

$$u_r = 0 \qquad u_\theta = \frac{\Gamma}{2\pi r}. \tag{6.28}$$

The flow pattern is shown in Figure 6.7.

7. *Doublet*

A *doublet* or *dipole* is obtained by allowing a source and a sink of equal strength to approach each other in such a way that their strengths increase as the separation

---

[1] The argument of transcendental functions such as the logarithm must always be dimensionless. Thus a constant must be added to ψ in equation (6.27) to put the logarithm in proper form. This is done explicitly when we are solving a problem as in Section 10 in what follows.

distance goes to zero, and that the product tends to a finite limit. The complex potential for a source-sink pair on the x-axis, with the source at $x = -\varepsilon$ and the sink at $x = \varepsilon$, is

$$w = \frac{m}{2\pi} \ln(z + \varepsilon) - \frac{m}{2\pi} \ln(z - \varepsilon) = \frac{m}{2\pi} \ln\left(\frac{z + \varepsilon}{z - \varepsilon}\right),$$

$$\simeq \frac{m}{2\pi} \ln\left(1 + \frac{2\varepsilon}{z} + \cdots\right) \simeq \frac{m\varepsilon}{\pi z}.$$

Defining the limit of $m\varepsilon/\pi$ as $\varepsilon \to 0$ to be μ, the preceding equation becomes

$$w = \frac{\mu}{z} = \frac{\mu}{r} e^{-i\theta}, \tag{6.29}$$

whose real and imaginary parts are

$$\phi = \frac{\mu x}{x^2 + y^2} \qquad \psi = -\frac{\mu y}{x^2 + y^2}. \tag{6.30}$$

The expression for ψ in the preceding can be rearranged in the form

$$x^2 + \left(y + \frac{\mu}{2\psi}\right)^2 = \left(\frac{\mu}{2\psi}\right)^2.$$

The streamlines, represented by $\psi = $ const., are therefore circles whose centers lie on the y-axis and are tangent to the x-axis at the origin (Figure 6.8). Direction of flow at the origin is along the negative x-axis (pointing outward from the *source* of the limiting source-sink pair), which is called the *axis* of the doublet. It is easy to show that (Exercise 1) the doublet flow equation (6.29) can be equivalently defined by superposing a clockwise vortex of strength $-\Gamma$ on the y-axis at $y = \varepsilon$, and a counterclockwise vortex of strength Γ at $y = -\varepsilon$.

The complex potentials for concentrated source, vortex, and doublet are all singular at the origin. It will be shown in the following sections that several interesting flow patterns can be obtained by superposing a uniform flow on these concentrated singularities.

8. Flow past a Half-Body

An interesting flow results from superposition of a source and a uniform stream. The complex potential for a uniform flow of strength U is $w = Uz$, which follows from integrating the relation $dw/dz = u - iv$. Adding to that, the complex potential for a source at the origin of strength m, we obtain,

$$w = Uz + \frac{m}{2\pi} \ln z, \tag{6.31}$$

whose imaginary part is

$$\psi = Ur \sin\theta + \frac{m}{2\pi}\theta. \tag{6.32}$$

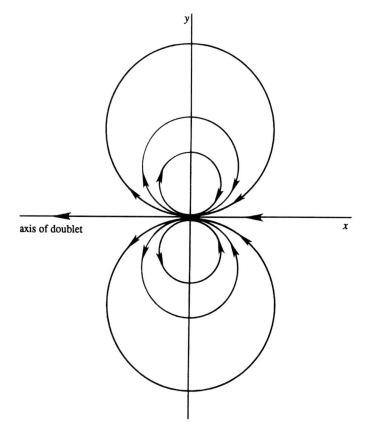

Figure 6.8   Plane doublet.

From equations (6.12) and (6.13) it is clear that there must be a stagnation point to the left of the source (S in Figure 6.9), where the uniform stream cancels the velocity of flow from the source. If the polar coordinate of the stagnation point is (a, π), then cancellation of velocity requires

$$U - \frac{m}{2\pi a} = 0,$$

giving

$$a = \frac{m}{2\pi U}.$$

(This result can also be found by finding dw/dz and setting it to zero.) The value of the streamfunction at the stagnation point is therefore

$$\psi_{\mathrm{s}} = U r \sin\theta + \frac{m}{2\pi}\theta = U a \, \sin\pi + \frac{m}{2\pi}\pi = \frac{m}{2}.$$

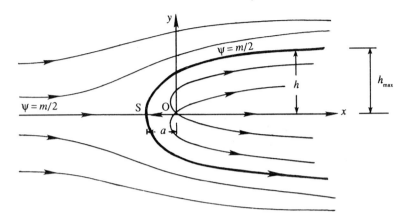

Figure 6.9 Irrotational flow past a two-dimensional half-body. The boundary streamline is given by $\psi = m/2$.

The equation of the streamline passing through the stagnation point is obtained by setting $\psi = \psi_s = m/2$, giving

$$Ur\,\sin\theta + \frac{m}{2\pi}\theta = \frac{m}{2}. \tag{6.33}$$

A plot of this streamline is shown in Figure 6.9. It is a semi-infinite body with a smooth nose, generally called a *half-body*. The stagnation streamline divides the field into a region external to the body and a region internal to it. The internal flow consists entirely of fluid emanating from the source, and the external region contains the originally uniform flow. The half-body resembles several practical shapes, such as the front part of a bridge pier or an airfoil; the upper half of the flow resembles the flow over a cliff or a side contraction in a wide channel.

The half-width of the body is found to be

$$h = r\sin\theta = \frac{m(\pi - \theta)}{2\pi U},$$

where equation (6.33) has been used. The half-width tends to $h_{\max} = m/2U$ as $\theta \to 0$ (Figure 6.9). (This result can also be obtained by noting that mass flux from the source is contained entirely within the half-body, requiring the balance $m = (2h_{\max})U$ at a large downstream distance where $u = U$.)

The pressure distribution can be found from Bernoulli's equation

$$p + \frac{1}{2}\rho q^2 = p_\infty + \frac{1}{2}\rho U^2.$$

A convenient way of representing pressure is through the nondimensional excess pressure (called *pressure coefficient*)

$$C_p \equiv \frac{p - p_\infty}{\frac{1}{2}\rho U^2} = 1 - \frac{q^2}{U^2}.$$

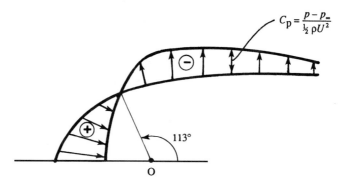

Figure 6.10 Pressure distribution in irrotational flow over a half-body. Pressure excess near the nose is indicated by $\oplus$ and pressure deficit elsewhere is indicated by $\ominus$.

A plot of C_p on the surface of the half-body is given in Figure 6.10, which shows that there is pressure excess near the nose of the body and a pressure deficit beyond it. It is easy to show by integrating p over the surface that the net pressure force is zero (Exercise 2).

9. Flow past a Circular Cylinder without Circulation

Combination of a uniform stream and a doublet with its axis directed against the stream gives the irrotational flow over a circular cylinder. The complex potential for this combination is

$$w = Uz + \frac{\mu}{z} = U\left(z + \frac{a^2}{z}\right), \tag{6.34}$$

where $a \equiv \sqrt{\mu/U}$. The real and imaginary parts of w give

$$
\begin{aligned}
\phi &= U\left(r + \frac{a^2}{r}\right)\cos\theta \\
\psi &= U\left(r - \frac{a^2}{r}\right)\sin\theta.
\end{aligned} \tag{6.35}
$$

It is seen that $\psi = 0$ at $r = a$ for all values of θ, showing that the streamline $\psi = 0$ represents a circular cylinder of radius a. The streamline pattern is shown in Figure 6.11. Flow inside the circle has no influence on that outside the circle. Velocity components are

$$u_r = \frac{\partial \phi}{\partial r} = U\left(1 - \frac{a^2}{r^2}\right)\cos\theta.$$

$$u_\theta = \frac{1}{r}\frac{\partial \phi}{\partial \theta} = -U\left(1 + \frac{a^2}{r^2}\right)\sin\theta,$$

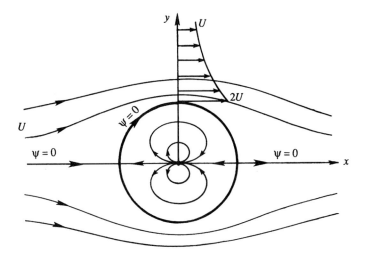

Figure 6.11   Irrotational flow past a circular cylinder without circulation.

from which the flow speed on the surface of the cylinder is found as

$$q|_{r=a} = |u_\theta|_{r=a} = 2U \sin\theta, \tag{6.36}$$

where what is meant is the positive value of $\sin\theta$. This shows that there are stagnation points on the surface, whose polar coordinates are $(a, 0)$ and (a, π). The flow reaches a maximum velocity of $2\,U$ at the top and bottom of the cylinder.

Pressure distribution on the surface of the cylinder is given by

$$C_p = \frac{p - p_\infty}{\frac{1}{2}\rho U^2} = 1 - \frac{q^2}{U^2} = 1 - 4\sin^2\theta.$$

Surface distribution of pressure is shown by the continuous line in Figure 6.12. The symmetry of the distribution shows that there is no net pressure drag. In fact, a general result of irrotational flow theory is that a steadily moving body experiences no drag. This result is at variance with observations and is sometimes known as *d'Alembert's paradox.* The existence of tangential stress, or "skin friction," is not the only reason for the discrepancy. For blunt bodies, the major part of the drag comes from separation of the flow from sides and the resulting generation of eddies. The surface pressure in the wake is smaller than that predicted by irrotational flow theory (Figure 6.12), resulting in a pressure drag. These facts will be discussed in further detail in Chapter 10.

The flow due to a cylinder moving steadily through a fluid appears unsteady to an observer at rest with respect to the fluid at infinity. This flow can be obtained by superposing a uniform stream along the negative x direction to the flow shown in Figure 6.11. The resulting instantaneous flow pattern is simply that of a doublet, as is clear from the decomposition shown in Figure 6.13.

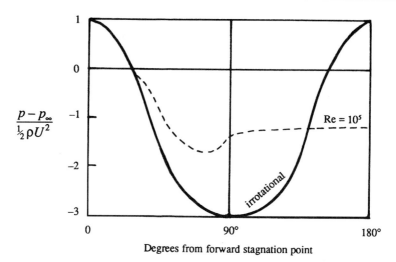

Figure 6.12   Comparison of irrotational and observed pressure distributions over a circular cylinder. The observed distribution changes with the Reynolds number Re; a typical behavior at high Re is indicated by the dashed line.

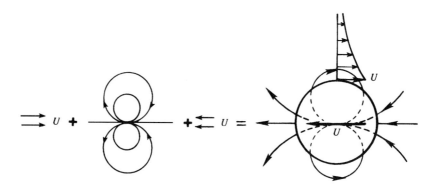

Figure 6.13   Decomposition of irrotational flow pattern due to a moving cylinder.

10. Flow past a Circular Cylinder with Circulation

It was seen in the last section that there is no net force on a circular cylinder in steady irrotational flow without circulation. It will now be shown that a lateral force, akin to a lift force on an airfoil, results when circulation is introduced into the flow. If a *clockwise* line vortex of circulation $-\Gamma$ is added to the irrotational flow around a circular cylinder, the complex potential becomes

$$w = U\left(z + \frac{a^2}{z}\right) + \frac{i\Gamma}{2\pi}\ln(z/a), \qquad (6.37)$$

whose imaginary part is

$$\psi = U \left(r - \frac{a^2}{r} \right) \sin \theta + \frac{\Gamma}{2\pi} \ln (r/a), \qquad (6.38)$$

where we have added to w the term $-(i\Gamma/2\pi) \ln a$ so that the argument of the logarithm is dimensionless, as it must be always.

Figure 6.14 shows the resulting streamline pattern for various values of Γ. The close streamline spacing and higher velocity on top of the cylinder is due to the addition of velocity fields of the clockwise vortex and the uniform stream. In contrast, the smaller velocities at the bottom of the cylinder are a result of the vortex field counteracting the uniform stream. Bernoulli's equation consequently implies a higher pressure below the cylinder and an upward "lift" force.

The tangential velocity component at any point in the flow is

$$u_\theta = -\frac{\partial \psi}{\partial r} = -U \left(1 + \frac{a^2}{r^2} \right) \sin \theta - \frac{\Gamma}{2\pi r}.$$

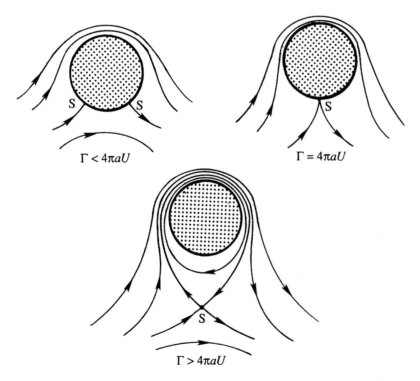

$$\Gamma < 4\pi a U \qquad\qquad \Gamma = 4\pi a U$$

$$\Gamma > 4\pi a U$$

Figure 6.14 Irrotational flow past a circular cylinder for different values of circulation. Point S represents the stagnation point.

At the surface of the cylinder, velocity is entirely tangential and is given by

$$u_\theta \mid_{r=a} = -2\,U\sin\theta - \frac{\Gamma}{2\pi a}, \tag{6.39}$$

which vanishes if

$$\sin\theta = -\frac{\Gamma}{4\pi aU}. \tag{6.40}$$

For $\Gamma < 4\pi aU$, two values of θ satisfy equation (6.40), implying that there are two stagnation points on the surface. The stagnation points progressively move down as Γ increases (Figure 6.14) and coalesce at $\Gamma = 4\pi aU$. For $\Gamma > 4\pi aU$, the stagnation point moves out into the flow along the y-axis. The radial distance of the stagnation point in this case is found from

$$u_\theta \mid_{\theta=-\pi/2} = U\left(1 + \frac{a^2}{r^2}\right) - \frac{\Gamma}{2\pi r} = 0.$$

This gives

$$r = \frac{1}{4\pi U}\,[\Gamma \pm \sqrt{\Gamma^2 - (4\pi aU)^2}],$$

one root of which is $r > a$; the other root corresponds to a stagnation point inside the cylinder.

Pressure is found from the Bernoulli equation

$$p + \rho q^2/2 = p_\infty + \rho U^2/2.$$

Using equation (6.39), the surface pressure is found to be

$$p_{r=a} = p_\infty + \tfrac{1}{2}\rho\left[U^2 - \left(-2U\sin\theta - \frac{\Gamma}{2\pi a}\right)^2\right]. \tag{6.41}$$

The symmetry of flow about the y-axis implies that the pressure force on the cylinder has no component along the x-axis. The pressure force along the y-axis, called the "lift" force in aerodynamics, is (Figure 6.15)

$$L = -\int_0^{2\pi} p_{r=a}\sin\theta\,a\,d\theta.$$

Substituting equation (6.41), and carrying out the integral, we finally obtain

$$L = \rho U\Gamma, \tag{6.42}$$

where we have used

$$\int_0^{2\pi}\sin\theta\,d\theta = \int_0^{2\pi}\sin^3\theta\,d\theta = 0.$$

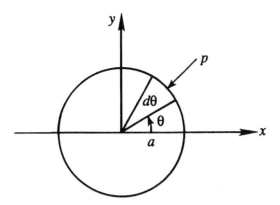

Figure 6.15   Calculation of pressure force on a circular cylinder.

It is shown in the following section that equation (6.42) holds for irrotational flows around *any* two-dimensional shape, not just circular cylinders. The result that lift force is proportional to circulation is of fundamental importance in aerodynamics. Relation equation (6.42) was proved independently by the German mathematician, Wilhelm Kutta (1902), and the Russian aerodynamist, Nikolai Zhukhovsky (1906); it is called the *Kutta–Zhukhovsky lift theorem.* (Older western texts transliterated Zhukhovsky's name as Joukowsky.) The interesting question of how certain two-dimensional shapes, such as an airfoil, develop circulation when placed in a stream is discussed in chapter 15. It will be shown there that fluid viscosity is responsible for the development of circulation. The magnitude of circulation, however, is independent of viscosity, and depends on flow speed U and the shape and "attitude" of the body.

For a circular cylinder, however, the only way to develop circulation is by rotating it in a flow stream. Although viscous effects are important in this case, the observed pattern for *large* values of cylinder rotation displays a striking similarity to the ideal flow pattern for $\Gamma > 4\pi aU$; see Figure 3.25 in the book by Prandtl (1952). For lower rates of cylinder rotation, the retarded flow in the boundary layer is not able to overcome the adverse pressure gradient behind the cylinder, leading to separation; the real flow is therefore rather unlike the irrotational pattern. However, even in the presence of separation, observed speeds are higher on the upper surface of the cylinder, implying a lift force.

A second reason for generating lift on a rotating cylinder is the asymmetry generated due to delay of separation on the upper surface of the cylinder. The resulting asymmetry generates a lift force. The contribution of this mechanism is small for two-dimensional objects such as the circular cylinder, but it is the only mechanism for side forces experienced by spinning three-dimensional objects such as soccer, tennis and golf balls. The interesting question of why spinning balls follow curved paths is discussed in Chapter 10, Section 9. The lateral force experienced by rotating bodies is called the *Magnus effect.*

The nonuniqueness of solution for two-dimensional potential flows should be noted in the example we have considered in this section. It is apparent that solutions for various values of Γ all satisfy the *same* boundary condition on the solid surface (namely, no normal flow) and at infinity (namely, $u = U$), and there is no way to determine the solution simply from the boundary conditions. A general result is that solutions of the Laplace equation in a *multiply connected region* are nonunique. This is explained further in Section 15.

11. Forces on a Two-Dimensional Body

In the preceding section we demonstrated that the drag on a circular cylinder is zero and the lift equals $L = \rho U \Gamma$. We shall now demonstrate that these results are valid for cylindrical shapes of *arbitrary* cross section. (The word "cylinder" refers to any plane two-dimensional body, not just to those with *circular* cross sections.)

Blasius Theorem

Consider a general cylindrical body, and let D and L be the x and y components of the force exerted on it by the surrounding fluid; we refer to D as "drag" and L as "lift." Because only normal pressures are exerted in inviscid flows, the forces on a surface element dz are (Figure 6.16)

$$dD = -p\,dy,$$
$$dL = p\,dx.$$

We form the complex quantity

$$dD - i\,dL = -p\,dy - ip\,dx = -ip\,dz^*,$$

where an asterisk denotes the complex conjugate. The total force on the body is therefore given by

$$D - iL = -i \oint_C p\,dz^*, \qquad (6.43)$$

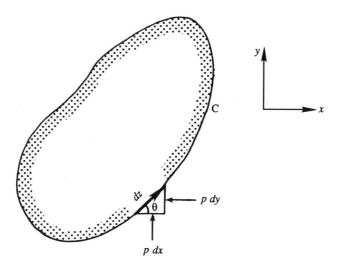

Figure 6.16   Forces exerted on an element of a body.

where C denotes a counterclockwise contour coinciding with the body surface. Neglecting gravity, the pressure is given by the Bernoulli equation

$$p_\infty + \frac{1}{2}\rho U^2 = p + \frac{1}{2}\rho(u^2 + v^2) = p + \frac{1}{2}\rho(u + iv)(u - iv).$$

Substituting for p in equation (6.43), we obtain

$$D - iL = -i \oint_C [p_\infty + \tfrac{1}{2}\rho U^2 - \tfrac{1}{2}\rho(u + iv)(u - iv)]\,dz^*, \qquad (6.44)$$

Now the integral of the constant term $(p_\infty + \frac{1}{2}\rho U^2)$ around a closed contour is zero. Also, on the body surface the velocity vector and the surface element dz are parallel (Figure 6.16), so that

$$u + iv = \sqrt{u^2 + v^2}\, e^{i\theta},$$
$$dz = |dz|\, e^{i\theta}.$$

The product $(u + iv)\, dz^*$ is therefore real, and we can equate it to its complex conjugate:

$$(u + iv)\, dz^* = (u - iv)\, dz.$$

Equation (6.44) then becomes

$$D - iL = \frac{i}{2}\rho \oint_C \left(\frac{dw}{dz}\right)^2 dz, \qquad (6.45)$$

where we have introduced the complex velocity $dw/dz = u - iv$. Equation (6.45) is called the *Blasius theorem*, and applies to any plane steady irrotational flow. The integral need not be carried out along the contour of the body because the theory of complex variables shows that *any contour surrounding the body can be chosen*, provided that there are no singularities between the body and the contour chosen.

Kutta–Zhukhovsky Lift Theorem

We now apply the Blasius theorem to a steady flow around an arbitrary cylindrical body, around which there is a clockwise circulation Γ. The velocity at infinity has a magnitude U and is directed along the x-axis. The flow can be considered a superposition of a uniform stream and a set of singularities such as vortex, doublet, source, and sink.

As there are no singularities outside the body, we shall take the contour C in the Blasius theorem at a very large distance from the body. From large distances, all singularities appear to be located near the origin $z = 0$. The complex potential is then of the form

$$w = Uz + \frac{m}{2\pi}\ln z + \frac{i\Gamma}{2\pi}\ln z + \frac{\mu}{z} + \cdots.$$

The first term represents a uniform flow, the second term represents a source, the third term represents a clockwise vortex, and the fourth term represents a doublet. Because the body contour is closed, the mass efflux of the sources must be absorbed by the sinks. It follows that the sum of the strength of the sources and sinks is zero, thus we should set $m = 0$. The Blasius theorem, equation (6.45), then becomes

$$D - iL = \frac{i\rho}{2} \oint \left[U + \frac{i\Gamma}{2\pi z} - \frac{\mu}{z^2} + \cdots \right]^2 dz. \tag{6.46}$$

To carry out the contour integral in equation (6.46), we simply have to find the coefficient of the term proportional to $1/z$ in the integrand. The coefficient of $1/z$ in a power series expansion for $f(z)$ is called the *residue* of $f(z)$ at $z = 0$. It is shown in complex variable theory that the contour integral of a function $f(z)$ around the contour C is $2\pi i$ times the sum of the residues at the singularities within C:

$$\oint_C f(z)\, dz = 2\pi i \,[\text{sum of residues}].$$

The residue of the integrand in equation (6.46) is easy to find. Clearly the term μ/z^2 does not contribute to the residue. Completing the square $(U + i\Gamma/2\pi z)^2$, we see that the coefficient of $1/z$ is $i\Gamma U/\pi$. This gives

$$D - iL = \frac{i\rho}{2} \left[2\pi i \left(\frac{i\Gamma U}{\pi} \right) \right],$$

which shows that

$$\boxed{\begin{aligned} D &= 0, \\ L &= \rho U \Gamma. \end{aligned}} \tag{6.47}$$

The first of these equations states that there is no drag experienced by a body in steady two-dimensional irrotational flow. The second equation shows that there is a lift force $L = \rho U \Gamma$ perpendicular to the stream, experienced by a two-dimensional body of arbitrary cross section. This result is called the *Kutta–Zhukhovsky lift theorem*, which was demonstrated in the preceding section for flow around a circular cylinder. The result will play a fundamental role in our study of flow around airfoil shapes (Chapter 15). We shall see that the circulation developed by an airfoil is nearly proportional to U, so that the lift is nearly proportional to U^2.

The following points can also be demonstrated. First, irrotational flow over a finite three-dimensional object has no circulation, and there can be no net force on the body in steady state. Second, in an *unsteady* flow a force *is* required to push a body, essentially because a mass of fluid has to be accelerated from rest.

Let us redrive the Kutta–Zhukhovsky lift theorem from considerations of vector calculus without reference to complex variables. From equations (4.28) and (4.33), for steady flow with no body forces, and with **I** the dyadic equivalent of the Kronecker delta δ_{ij}

$$\mathbf{F}_B = -\int_{A_1} (\rho \mathbf{u}\mathbf{u} + p\mathbf{I} - \sigma) \cdot d\mathbf{A}_1.$$

Assuming an inviscid fluid, $\boldsymbol{\sigma} = 0$. Now additionally assume a two-dimensional constant density flow that is uniform at infinity $\mathbf{u} = U\mathbf{i}_x$. Then, from Bernoulli's theorem, $p + \rho q^2/2 = p_\infty + \rho q^2/2 = p_0$, so $p = p_0 - \rho q^2/2$. Referring to Figure 6.17, for two-dimensional flow $\mathbf{dA_1} = \mathbf{ds} \times \mathbf{i}_z dz$, where here z is the coordinate out of the paper. We will carry out the integration over a unit depth in z so that the result for $\mathbf{F_B}$ will be force per unit depth (in z).

With $\mathbf{r} = x\mathbf{i}_x + y\mathbf{i}_y$, $\mathbf{dr} = dx\mathbf{i}_x + dy\mathbf{i}_y = \mathbf{ds}$, $\mathbf{dA_1} = \mathbf{ds} \times \mathbf{i}_z \cdot 1 = -\mathbf{i}_y\,dx + \mathbf{i}_x\,dy$. Now let $\mathbf{u} = U\mathbf{i}_x + \mathbf{u}'$, where $\mathbf{u}' \to 0$ as $r \to \infty$ at least as fast as $1/r$. Substituting for $\mathbf{uu}$ and q^2 in the integral for $\mathbf{F_B}$, we find

$$\begin{aligned}
\mathbf{F_B} = -\rho \int_{\mathbf{A}_1} &\{ UU\mathbf{i}_x\mathbf{i}_x + U\mathbf{i}_x(u'\mathbf{i}_x + v'\mathbf{i}_y) + (u'\mathbf{i}_x + v'\mathbf{i}_y)\mathbf{i}_x U \\
&+ \mathbf{u}'\mathbf{u}' + (\mathbf{i}_x\mathbf{i}_x + \mathbf{i}_y\mathbf{i}_y)[p_0/\rho - U^2/2 - Uu' \\
&- (u'^2 + v'^2)/2] \} \cdot (-\mathbf{i}_y\,dx + \mathbf{i}_x\,dy).
\end{aligned}$$

Let $r \to \infty$ so that the contour C is far from the body. The constant terms U^2, p_0/ρ, $-U^2/2$ integrate to zero around the closed path. The quadratic terms $\mathbf{u}'\mathbf{u}'$, $(u'^2 + v'^2)/2 \lesssim 1/r^2$ as $r \to \infty$ and the perimeter of the contour increases only as r. Thus the quadratic terms $\to 0$ as $r \to \infty$. Separating the force into x and y components,

$$\mathbf{F_B} = -\mathbf{i}_x\rho U \oint_c [(u'\,dy - v'\,dx) + (u'\,dy - u'\,dy)] - \mathbf{i}_y\rho U \oint_c (v'\,dy + u'\,dx).$$

We note that the first integrand is $\mathbf{u}' \cdot \mathbf{ds} \times \mathbf{i}_z$, and that we may add the constant $U\mathbf{i}_x$ to each of the integrands because the integration of a constant velocity over a

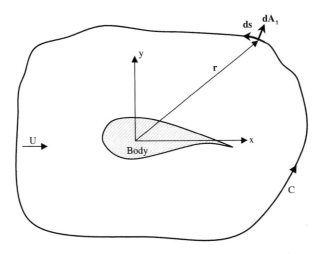

Figure 6.17 Domain of integration for the Kutta–Zhukhovsky theorem.

closed contour or surface will result in zero force. The integrals for the force then become

$$\mathbf{F}_B = -\mathbf{i}_x \rho U \int_{A_t} (U\mathbf{i}_x + \mathbf{u}') \cdot d\mathbf{A}_1 - \mathbf{i}_y \rho U \oint_c (U\mathbf{i}_x + \mathbf{u}') \cdot d\mathbf{s}.$$

The first integral is zero by equation (4.29) (as a consequence of mass conservation for constant density flow) and the second is the circulation Γ by definition. Thus,

$$\mathbf{F}_B = -\mathbf{i}_y \rho U \Gamma \quad \text{(force/unit depth)},$$

where Γ is positive in the counterclockwise sense. We see that there is no force component in the direction of motion (drag) under the assumptions necessary for the derivation (steady, inviscid, no body forces, constant density, two-dimensional, uniform at infinity) that were believed to be valid to a reasonable approximation for a wide variety of flows. Thus it was labeled a paradox—d'Alembert's paradox (Jean Le Rond d'Alembert, 16 November 1717–29 October 1783).

Unsteady Flow

The Euler momentum integral [(4.28)] can be extended to unsteady flows as follows. The extension may have some utility for constant density irrotational flows with moving boundaries; thus it is derived here.

Integrating (4.17) over a fixed volume V bounded by a surface A ($A = \partial V$) containing within it only fluid particles, we obtain

$$d/dt \int_V \rho \mathbf{u} \, dV = -\int_{A=\partial V} \rho \mathbf{u}\mathbf{u} \cdot d\mathbf{A} + \int_{A=\partial V} \tau \cdot d\mathbf{A}$$

where body forces $\mathbf{g}$ have been neglected, and the divergence theorem has been used. Because the immersed body cannot be part of V, we take $A = A_1 + A_2 + A_3$, as shown in Figure 4.9. Here A_1 is a "distant" surface, A_2 is the body surface, and A_3 is the connection between A_1 and A_2 that we allow to vanish. We identified the force on the immersed body as

$$\mathbf{F}_B = -\int_{A_2} \tau \cdot d\mathbf{A}_2$$

Then,

$$\mathbf{F}_B = -\int_{A_1} (\rho \mathbf{u}\mathbf{u} - \tau) \cdot d\mathbf{A}_1 - \int_{A_2} \rho \mathbf{u}\mathbf{u} \cdot d\mathbf{A}_2 - d/dt \int_V \rho \mathbf{u} \, dV \qquad (6.48)$$

If the flow is unsteady because of a moving boundary (A_2), then $\mathbf{u} \cdot d\mathbf{A}_2 \neq 0$, as we showed at the end of Section 4.19. If the body surface is described by $f(x, y, z, t) = 0$, then the condition that no mass of fluid with local velocity $\mathbf{u}$ flow across the boundary is (4.92): $Df/Dt = \partial f/\partial t + \mathbf{u} \cdot \nabla f = 0$. Since ∇f is normal to the boundary (as is $\mathbf{dA_2}$), $\mathbf{u} \cdot \nabla f = -\partial f / \partial t$ on $f = 0$. Thus $\mathbf{u} \cdot \mathbf{dA_2}$ is in general $\neq 0$ on the body surface. Equation (6.48) may be simplified if the density $\rho = $ const. and if viscous effects can be neglected in the flow. Then, by Kelvin's theorem the flow is circulation preserving. If it is initially irrotational, it will remain so. With $\nabla \times \mathbf{u} = 0, \mathbf{u} = \nabla \phi$ and $\rho = $ const., the last integral in (6.48) can be transformed by the divergence theorem

$$d/dt \int_V \rho \mathbf{u} \, dV = \rho d/dt \int_V \nabla \phi dV = \rho d/dt \int_{A = \partial V} \phi \mathbf{I} \cdot \mathbf{dA}$$

With $A = A_1 + A_2 + A_3$ and $A_3 \to 0$, the A_1 and A_2 integrals can be combined with the first two integrals in (6.48) to yield

$$\mathbf{F_B} = -\int_{A_1} (\rho \mathbf{uu} + p\mathbf{I} + \rho \mathbf{I} \partial \phi \partial t) \cdot \mathbf{dA_1} - \int_{A_2} (\rho \mathbf{uu} + \rho \mathbf{I} \partial \phi / \partial t) \cdot \mathbf{dA_2}, \quad (6.49)$$

where $\tau = -p\mathbf{I} + \sigma$ and $\sigma = 0$ with the neglect of viscosity. The Bernoulli equation for unsteady irrotational flow [(4.81)], $\rho \partial \phi / \partial t + p + \rho u^2 / 2 = 0$, where the function of integration $F(t)$ has been absorbed in the ϕ, can be used if desired to achieve a slightly different form.

12.   Source near a Wall: Method of Images

The method of images is a way of determining a flow field due to one or more singularities near a wall. It was introduced in Chapter 5, Section 8, where vortices near a wall were examined. We found that the flow due to a line vortex near a wall can be found by omitting the wall and introducing instead a vortex of opposite strength at the "image point." The combination generates a straight streamline at the location of the wall, thereby satisfying the boundary condition.

Another example of this technique is given here, namely, the flow due to a line source at a distance a from a straight wall. This flow can be simulated by introducing an image source of the same strength and sign, so that the complex potential is

$$w = \frac{m}{2\pi} \ln(z - a) + \frac{m}{2\pi} \ln(z + a) - \frac{m}{2\pi} \ln a^2,$$

$$= \frac{m}{2\pi} \ln(x^2 - y^2 - a^2 + i2xy) - \frac{m}{2\pi} \ln a^2. \quad (6.50)$$

We know that the logarithm of any complex quantity $\zeta = |\zeta| \exp(i\theta)$ can be written as $\ln \zeta = \ln |\zeta| + i\theta$. The imaginary part of equation (6.50) is therefore

$$\psi = \frac{m}{2\pi} \tan^{-1} \frac{2xy}{x^2 - y^2 - a^2},$$

from which the equation of streamlines is found as

$$x^2 - y^2 - 2xy \cot\left(\frac{2\pi \psi}{m}\right) = a^2.$$

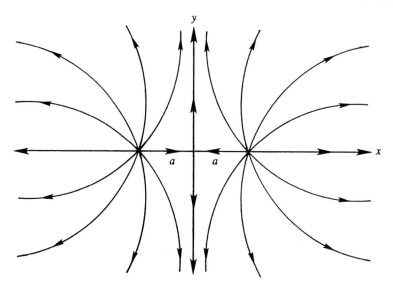

Figure 6.18   Irrotational flow due to two equal sources.

The streamline pattern is shown in Figure 6.18. The x and y axes form part of the streamline pattern, with the origin as a stagnation point. It is clear that the complex potential equation (6.48) represents three interesting flow situations:

(1)   flow due to two equal sources (entire Figure 6.18);
(2)   flow due to a source near a plane wall (right half of Figure 6.18); and
(3)   flow through a narrow slit in a right-angled wall (first quadrant of Figure 6.18).

13. Conformal Mapping

We shall now introduce a method by which complex flow patterns can be transformed into simple ones using a technique known as *conformal mapping* in complex variable theory. Consider the functional relationship $w = f(z)$, which maps a point in the w-plane to a point in the z-plane, and vice versa. We shall prove that infinitesimal figures in the two planes preserve their geometric similarity if $w = f(z)$ is analytic. Let lines C_z and C'_z in the z-plane be transformations of the curves C_w and C'_w in the w-plane, respectively (Figure 6.19). Let δz, $\delta' z$, δw, and $\delta' w$ be infinitesimal elements along the curves as shown. The four elements are related by

$$\delta w = \frac{dw}{dz}\delta z, \tag{6.51}$$

$$\delta' w = \frac{dw}{dz}\delta' z. \tag{6.52}$$

If $w = f(z)$ is analytic, then dw/dz is independent of orientation of the elements, and therefore has the same value in equation (6.51) and (6.52). These two equations

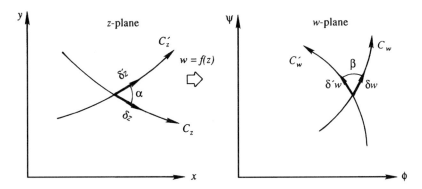

Figure 6.19 Preservation of geometric similarity of small elements in conformal mapping.

then imply that the elements δz and $\delta' z$ are rotated by the *same amount* (equal to the argument of dw/dz) to obtain the elements δw and $\delta' w$. It follows that

$$\alpha = \beta,$$

which demonstrates that infinitesimal figures in the two planes are geometrically similar. The demonstration fails at singular points at which dw/dz is either zero or infinite. Because dw/dz is a function of z, the amount of magnification and rotation that an element δz undergoes during transformation from the z-plane to the w-plane varies. Consequently, *large* figures become distorted during the transformation.

In application of conformal mapping, we always choose a rectangular grid in the w-plane consisting of constant ϕ and ψ lines (Figure 6.20). In other words, we define ϕ and ψ to be the real and imaginary parts of w:

$$w = \phi + i\psi.$$

The rectangular net in the w-plane represents a uniform flow in this plane. The constant ϕ and ψ lines are transformed into certain curves in the z-plane through the transformation $w = f(z)$. *The pattern in the z-plane is the physical pattern under investigation*, and the images of constant ϕ and ψ lines in the z-plane form the equipotential lines and streamlines, respectively, of the desired flow. We say that $w = f(z)$ transforms a uniform flow in the w-plane into the desired flow in the z-plane. In fact, all the preceding flow patterns studied through the transformation $w = f(z)$ can be interpreted this way.

If the physical pattern under investigation is too complicated, we may introduce intermediate transformations in going from the w-plane to the z-plane. For example, the transformation $w = \ln(\sin z)$ can be broken into

$$w = \ln \zeta \qquad \zeta = \sin z.$$

Velocity components in the z-plane are given by

$$u - iv = \frac{dw}{dz} = \frac{dw}{d\zeta}\frac{d\zeta}{dz} = \frac{1}{\zeta}\cos z = \cot z.$$

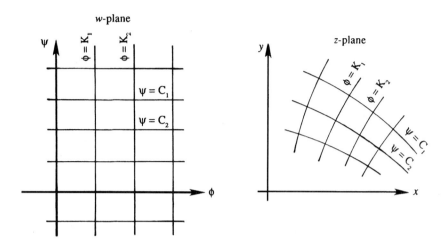

Figure 6.20   Flow patterns in the *w*-plane and the *z*-plane.

A simple example of conformal mapping is given immediately below as a special case of the flow discussed in Section 4 above. Consider the transformation, $w = \phi + i\psi = z^2 = x^2 - y^2 + 2ixy$. Streamlines are $\psi = const = 2xy$, rectangular hyperbolae. See one quadrant of the flow depicted in Fig. 6.5. Uniform flow in the *w*-plane has been mapped onto flow in a right hand corner in the *z*-plane by this transformation. A more involved example is shown in the next section. Additional applications are discussed in Chapter 15.

14. Flow around an Elliptic Cylinder with Circulation

We shall briefly illustrate the method of conformal mapping by considering a transformation that has important applications in airfoil theory. Consider the following transformation:

$$z = \zeta + \frac{b^2}{\zeta},\tag{6.53}$$

relating z and ζ planes. We shall now show that a circle of radius b centered at the origin of the ζ-plane transforms into a straight line on the real axis of the z-plane. To prove this, consider a point $\zeta = b\exp(i\theta)$ on the circle (Figure 6.21), for which the corresponding point in the z-plane is

$$z = be^{i\theta} + be^{-i\theta} = 2b\cos\theta.$$

As θ varies from 0 to π, z goes along the x-axis from $2b$ to $-2b$. As θ varies from π to 2π, z goes from $-2b$ to $2b$. The circle of radius b in the ζ-plane is thus transformed into a straight line of length $4b$ in the z-plane. It is clear that the region *outside* the circle in ζ-plane is mapped into the *entire* z-plane. It can be shown that the region inside the circle is also transformed into the entire z-plane. This, however, is

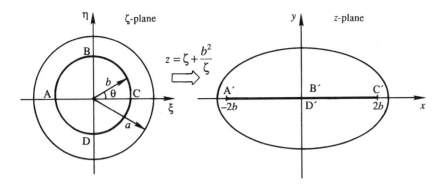

Figure 6.21 Transformation of a circle into an ellipse by means of the Zhukhovsky transformation $z = \zeta + b^2/\zeta$.

of no concern to us because we shall not consider the interior of the circle in the ζ-plane.

Now consider a circle of radius $a > b$ in the ζ-plane (Figure 6.21). Points $\zeta = a \exp(i\theta)$ on this circle are transformed to

$$z = a e^{i\theta} + \frac{b^2}{a} e^{-i\theta}, \tag{6.54}$$

which traces out an ellipse for various values of θ. This becomes clear by elimination of θ in equation (6.54), giving

$$\frac{x^2}{(a + b^2/a)^2} + \frac{y^2}{(a - b^2/a)^2} = 1. \tag{6.55}$$

For various values of $a > b$, equation (6.55) represents a family of ellipses in the z-plane, with foci at $x = \pm 2b$.

The flow around one of these ellipses (in the z-plane) can be determined by first finding the flow around a circle of radius a in the ζ-plane, and then using the transformation equation (6.53) to go to the z-plane. To be specific, suppose the desired flow in the z-plane is that of flow around an elliptic cylinder with clockwise circulation Γ, which is placed in a stream moving at U. The corresponding flow in the ζ-plane is that of flow with the same circulation around a circular cylinder of radius a placed in a stream of the same strength U for which the complex potential is (see equation (6.37))

$$w = U \left(\zeta + \frac{a^2}{\zeta} \right) + \frac{i\Gamma}{2\pi} \ln \zeta - \frac{i\Gamma}{2\pi} \ln a. \tag{6.56}$$

The complex potential $w(z)$ in the z-plane can be found by substituting the inverse of equation (6.53), namely,

$$\zeta = \frac{1}{2} z + \frac{1}{2} (z^2 - 4b^2)^{1/2}, \tag{6.57}$$

into equation (6.56). (Note that the negative root, which falls inside the cylinder, has been excluded from equation (6.57).) Instead of finding the complex velocity in the z-plane by directly differentiating $w(z)$, it is easier to find it as

$$u - iv = \frac{dw}{dz} = \frac{dw}{d\zeta}\frac{d\zeta}{dz}.$$

The resulting flow around an elliptic cylinder with circulation is qualitatively quite similar to that around a circular cylinder as shown in Figure 6.14.

15. *Uniqueness of Irrotational Flows*

In Section 10 we saw that plane irrotational flow over a cylindrical object is nonunique. In particular, flows with *any* amount of circulation satisfy the same boundary conditions on the body and at infinity. With such an example in mind, we are ready to make certain general statements concerning solutions of the Laplace equation. We shall see that the topology of the region of flow has a great influence on the uniqueness of the solution.

Before we can make these statements, we need to define certain terms. A *reducible circuit* is any closed curve (lying wholly in the flow field) that can be reduced to a point by continuous deformation without ever cutting through the boundaries of the flow field. We say that a region is *singly connected* if *every* closed circuit in the region is reducible. For example, the region of flow around a finite body of revolution is reducible (Figure 6.22a). In contrast, the flow field over a cylindrical object of infinite length is multiply connected because certain circuits (such as C_1 in Figure 6.22b) are reducible while others (such as C_2) are not reducible.

To see why solutions are nonunique in a multiply connected region, consider the two circuits C_1 and C_2 in Figure 6.22b. The vorticity everywhere within C_1 is zero, thus Stokes' theorem requires that the circulation around it must vanish. In contrast, the circulation around C_2 can have any strength Γ. That is,

$$\oint_{C_2} \mathbf{u} \cdot d\mathbf{x} = \Gamma, \tag{6.58}$$

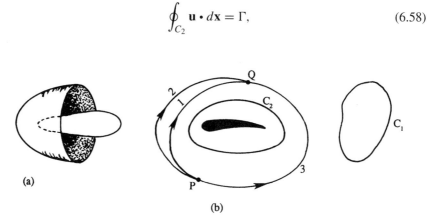

(a)

(b)

Figure 6.22   Singly connected and multiply connected regions: (a) singly connected; (b) multiply connected.

where the loop around the integral sign has been introduced to emphasize that the circuit C_2 is closed. As the right-hand side of equation (6.58) is nonzero, it follows that $\mathbf{u} \cdot d\mathbf{x}$ is not a "perfect differential," which means that the line integral between any two points *depends on the path followed* ($\mathbf{u} \cdot d\mathbf{x}$ is called a *perfect differential* if it can be expressed as the differential of a function, say as $\mathbf{u} \cdot d\mathbf{x} = df$. In that case the line integral around a *closed* circuit must vanish). In Figure 6.22b, the line integrals between P and Q are the same for paths 1 and 2, but not the same for paths 1 and 3. The solution is therefore nonunique, as was physically evident from the whole family of irrotational flows shown in Figure 6.14.

In singly connected regions, circulation around *every* circuit is zero, and the solution of $\nabla^2 \phi = 0$ is unique when values of ϕ are specified at the boundaries (the *Dirichlet problem*). When normal *derivatives* of ϕ are specified at the boundary (the *Neumann problem*), as in the fluid flow problems studied here, the solution is unique within an arbitrary additive constant. Because the arbitrary constant is of no consequence, we shall say that the solution of the irrotational flow in a singly connected region is unique. (Note also that the solution depends only on the *instantaneous* boundary conditions; the differential equation $\nabla^2 \phi = 0$ is independent of t.)

Summary: Irrotational flow around a plane two-dimensional object is nonunique because it allows an arbitrary amount of circulation. Irrotational flow around a finite three-dimensional object is unique because there is no circulation.

In Sections 4 and 5 of Chapter 5 we learned that vorticity is solenoidal ($\nabla \cdot \boldsymbol{\omega} = 0$), or that vortex lines cannot begin or end anywhere in the fluid. Here we have learned that a circulation in a two dimensional flow results in a force normal to an oncoming stream. This is used to simulate lifting flow over a wing by the following artifice, discussed in more detail in our chapter on Aerodynamics. Since Stokes' theorem tells us that the circulation about a closed contour is equal to the flux of vorticity through any surface bounded by that contour, the circulation about a thin airfoil section is simulated by a continuous row of vortices (a vortex sheet) along the centerline of a wing cross-section (the mean camber line of an airfoil). For a (real) finite wing, these vortices must bend downstream to form trailing vortices and terminate in starting vortices (far downstream), always forming closed loops. Although the wing may be a finite three dimensional shape, the contour cannot cut any of the vortex lines without changing the circulation about the contour. Generally, the circulation about a wing does vary in the spanwise direction, being a maximum at the root or centerline and tending to zero at the wingtips.

Additional boundary conditions that the mean camber line be a streamline and that a real trailing edge be a stagnation point serve to render the circulation distribution unique.

16. *Numerical Solution of Plane Irrotational Flow*

Exact solutions can be obtained only for flows with simple geometries, and approximate methods of solution become necessary for practical flow problems. One of these approximate methods is that of building up a flow by superposing a distribution of sources and sinks; this method is illustrated in Section 21 for axisymmetric flows.

Another method is to apply perturbation techniques by assuming that the body is thin. A third method is to solve the Laplace equation numerically. In this section we shall illustrate the numerical method in its simplest form. No attempt is made here to use the most efficient method. It is hoped that the reader will have an opportunity to learn numerical methods that are becoming increasingly important in the applied sciences in a separate study. See Chapter 11 for introductory material on several important techniques of computational fluid dynamics.

Finite Difference Form of the Laplace Equation

In finite difference techniques we divide the flow field into a system of *grid points*, and approximate the derivatives by taking differences between values at adjacent grid points. Let the coordinates of a point be represented by

$$x = i \, \Delta x \qquad (i = 1, 2, \ldots,),$$

$$y = j \, \Delta y \qquad (j = 1, 2, \ldots,).$$

Here, Δx and Δy are the dimensions of a grid box, and the integers i and j are the indices associated with a grid point (Figure 6.23). The value of a variable $\psi(x, y)$ can be represented as

$$\psi(x, y) = \psi(i \, \Delta x, j \, \Delta y) \equiv \psi_{i,j},$$

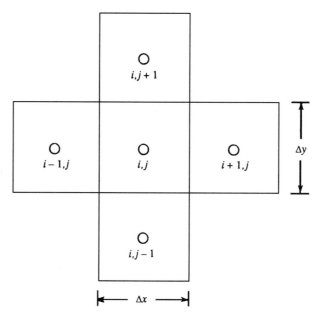

Figure 6.23   Adjacent grid boxes in a numerical calculation.

where $\psi_{i,j}$ is the value of ψ at the grid point (i, j). In finite difference form, the first derivatives of ψ are approximated as

$$\left(\frac{\partial \psi}{\partial x}\right)_{i,j} \simeq \frac{1}{\Delta x}\left(\psi_{i+\frac{1}{2},j} - \psi_{i-\frac{1}{2},j}\right),$$

$$\left(\frac{\partial \psi}{\partial y}\right)_{i,j} \simeq \frac{1}{\Delta y}\left(\psi_{i,j+\frac{1}{2}} - \psi_{i,j-\frac{1}{2}}\right).$$

The quantities on the right-hand side (such as $\psi_{i+1/2,j}$) are half-way between the grid points and therefore undefined. However, this would not be a difficulty in the present problem because the Laplace equation does not involve first derivatives. Both derivatives are written as first-order centered differences.

The finite difference form of $\partial^2 \psi / \partial x^2$ is

$$\left(\frac{\partial^2 \psi}{\partial x^2}\right)_{i,j} \simeq \frac{1}{\Delta x}\left[\left(\frac{\partial \psi}{\partial x}\right)_{i+\frac{1}{2},j} - \left(\frac{\partial \psi}{\partial x}\right)_{i-\frac{1}{2},j}\right],$$

$$\simeq \frac{1}{\Delta x}\left[\frac{1}{\Delta x}(\psi_{i+1,j} - \psi_{i,j}) - \frac{1}{\Delta x}(\psi_{i,j} - \psi_{i-1,j})\right],$$

$$= \frac{1}{\Delta x^2}[\psi_{i+1,j} - 2\psi_{i,j} + \psi_{i-1,j}]. \tag{6.59}$$

Similarly,

$$\left(\frac{\partial^2 \psi}{\partial y^2}\right)_{i,j} \simeq \frac{1}{\Delta y^2}[\psi_{i,j+1} - 2\psi_{i,j} + \psi_{i,j-1}] \tag{6.60}$$

Using equations (6.59) and (6.60), the Laplace equation for the streamfunction in a plane two-dimensional flow

$$\frac{\partial^2 \psi}{\partial x^2} + \frac{\partial^2 \psi}{\partial y^2} = 0,$$

has a finite difference representation

$$\frac{1}{\Delta x^2}[\psi_{i+1,j} - 2\psi_{i,j} + \psi_{i-1,j}] + \frac{1}{\Delta y^2}[\psi_{i,j+1} - 2\psi_{i,j} + \psi_{i,j-1}] = 0.$$

Taking $\Delta x = \Delta y$, for simplicity, this reduces to

$$\psi_{i,j} = \frac{1}{4}[\psi_{i-1,j} + \psi_{i+1,j} + \psi_{i,j-1} + \psi_{i,j+1}], \tag{6.61}$$

which shows that ψ satisfies the Laplace equation if its value at a grid point equals the average of the values at the four surrounding points.

Simple Iteration Technique

We shall now illustrate a simple method of solution of equation (6.61) when the values of ψ are given in a simple geometry. Assume the rectangular region of Figure 6.24, in which the flow field is divided into 16 grid points. Of these, the values of ψ are known at the 12 boundary points indicated by open circles. The values of ψ at the four interior points indicated by solid circles are unknown. For these interior points, the use of equation (6.61) gives

$$\psi_{2,2} = \tfrac{1}{4}\left[\psi_{1,2}^{B} + \psi_{3,2} + \psi_{2,1}^{B} + \psi_{2,3}\right],$$

$$\psi_{3,2} = \tfrac{1}{4}\left[\psi_{2,2} + \psi_{4,2}^{B} + \psi_{3,1}^{B} + \psi_{3,3}\right],$$

$$\psi_{2,3} = \tfrac{1}{4}\left[\psi_{1,3}^{B} + \psi_{3,3} + \psi_{2,2} + \psi_{2,4}^{B}\right],$$

$$\psi_{3,3} = \tfrac{1}{4}\left[\psi_{2,3} + \psi_{4,3}^{B} + \psi_{3,2} + \psi_{3,4}^{B}\right].$$

$$(6.62)$$

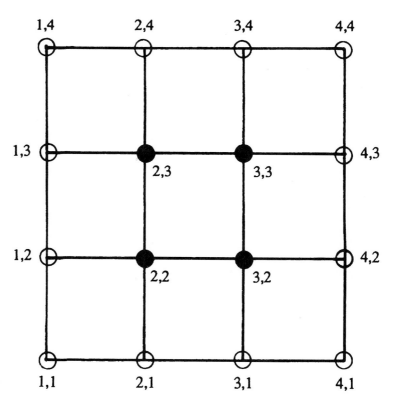

Figure 6.24 Network of grid points in a rectangular region. Boundary points with known values are indicated by open circles. The four interior points with unknown values are indicated by solid circles.

In the preceding equations, the known boundary values have been indicated by a superscript "B." Equation set (6.62) represents four linear algebraic equations in four unknowns and is therefore solvable.

In practice, however, the flow field is likely to have a large number of grid points, and the solution of such a large number of simultaneous algebraic equations can only be performed using a computer. One of the simplest techniques of solving such a set is the *iteration method*. In this a solution is initially assumed and then gradually improved and updated until equation (6.61) is satisfied at every point. Suppose the values of ψ at the four unknown points of Figure 6.24 are initially taken as zero. Using equation (6.62), the first estimate of $\psi_{2,2}$ can be computed as

$$\psi_{2,2} = \frac{1}{4}\left[\psi_{1,2}^{B} + 0 + \psi_{2,1}^{B} + 0\right].$$

The old zero value for $\psi_{2,2}$ is now replaced by the preceding value. The first estimate for the next grid point is then obtained as

$$\psi_{3,2} = \frac{1}{4}\left[\psi_{2,2} + \psi_{4,2}^{B} + \psi_{3,1}^{B} + 0\right],$$

where the *updated* value of $\psi_{2,2}$ has been used on the right-hand side. In this manner, we can sweep over the entire region in a systematic manner, *always using the latest available value at the point*. Once the first estimate at every point has been obtained, we can sweep over the entire region once again in a similar manner. The process is continued until the values of $\psi_{i,j}$ do not change appreciably between two successive sweeps. The iteration process has now "converged."

The foregoing scheme is particularly suitable for implementation using a computer, whereby it is easy to replace old values at a point as soon as a new value is available. In practice, a more efficient technique, for example, the successive over-relaxation method, will be used in a large calculation. The purpose here is not to describe the most efficient technique, but the one which is simplest to illustrate. The following example should make the method clear.

Example 6.1. Figure 6.25 shows a contraction in a channel through which the flow rate per unit depth is $5\,\mathrm{m^2/s}$. The velocity is uniform and parallel across the inlet and outlet sections. Find the flow field.

Solution: Although the region of flow is plane two-dimensional, it is clearly singly connected. This is because the flow field *interior* to a boundary is desired, so that every fluid circuit can be reduced to a point. The problem therefore has a unique solution, which we shall determine numerically.

We know that the difference in ψ values is equal to the flow rate between two streamlines. If we take $\psi = 0$ at the bottom wall, then we must have $\psi = 5\,\mathrm{m^2/s}$ at the top wall. We divide the field into a system of grid points shown, with $\Delta x = \Delta y = 1\mathrm{m}$. Because $\Delta\psi/\Delta y\,(= u)$ is given to be uniform across the inlet and the outlet, we must have $\Delta\psi = 1\,\mathrm{m^2/s}$ at the inlet and $\Delta\psi = 5/3 = 1.67\,\mathrm{m^2/s}$ at the outlet. The resulting values of ψ at the boundary points are indicated in Figure 6.25.

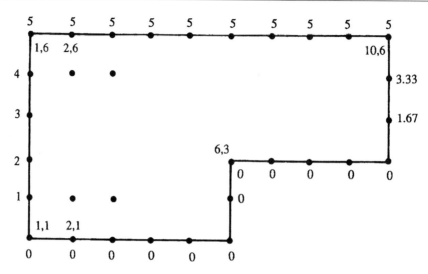

Figure 6.25 Grid pattern for irrotational flow through a contraction (Example 16). The boundary values of ψ are indicated on the outside. The values of i, j for some grid points are indicated on the inside.

The FORTRAN code for solving the problem is as follows:

```
DIMENSION S(10, 6)

      DO 10 I = 1, 6
 10 S(I, 1) = 0.
      DO 20 J = 2, 3
 20 S(6, J) = 0.
      DO 30 I = 7, 10        Set ψ = 0 on top and bottom walls
 30 S(I, 3) = 0.
      DO 40 I = 1, 10
 40 S(I, 6) = 5.

      DO 50 J = 2, 6
 50 S(1, J)  = J - 1.        Set ψ at inlet

      DO 60 J = 4, 6
 60 S(10, J) = (J - 3) * (5. / 3.)   Set ψ at outlet

      DO 100 N = 1, 20
      DO 70 I = 2, 5
      DO 70 J = 2, 5
 70 S(I, J) = (S(I, J+1) + S(I, J-1) + S(I+1, J) + S(I-1, J)) / 4.
      DO 80 I = 6, 9
      DO 80 J = 4, 5
 80 S(I, J) = (S(I, J+1) + S(I, J-1) + S(I+1, J) + S(I-1, J)) / 4.
100 CONTINUE
      PRINT 1, ((S(I, J), I = 1, 10), J = 1, 6)
  1 FORMAT (' ', 10 E 12.4)
      END
```

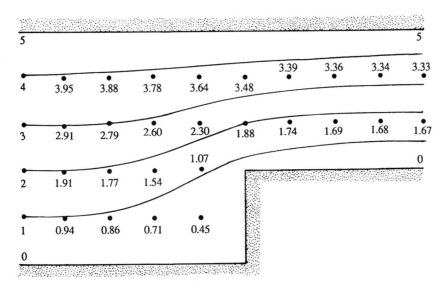

Figure 6.26   Numerical solution of Example 6.1.

Here, S denotes the stream function ψ. The code first sets the boundary values. The iteration is performed in the N loop. In practice, iterations will not be performed arbitrarily 20 times. Instead the convergence of the iteration process will be checked, and the process is continued until some reasonable criterion (such as less than 1% change at every point) is met. These improvements are easy to implement, and the code is left in its simplest form.

The values of ψ at the grid points after 50 iterations, and the corresponding streamlines, are shown in Figure 6.26.

It is a usual practice to iterate until successive iterates change only by a prescribed small amount. The solution is then said to have "converged." However, a caution is in order. To be sure a solution has been obtained, all of the terms in the equation must be calculated and the satisfaction of the equation by the "solution" must be verified.

17. Axisymmetric Irrotational Flow

Several examples of irrotational flow around plane two-dimensional bodies were given in the preceding sections. We used Cartesian (x, y) and plane polar (r, θ) coordinates, and found that the problem involved the solution of the Laplace equation in ϕ or ψ with specified boundary conditions. We found that a very powerful tool in the analysis was the method of complex variables, including conformal transformation.

Two streamfunctions are required to describe a fully three-dimensional flow (Chapter 4, Section 4), although a velocity potential (which satisfies the three-dimensional version of $\nabla^2 \phi = 0$) can be defined if the flow is irrotational.

If, however, the flow is symmetrical about axis, one of the streamfunctions is known because all streamlines must lie in planes passing through the axis of symmetry. In cylindrical polar coordinates, one streamfunction, say, χ, may be taken as $\chi = -\varphi$. In spherical polar coordinates (see Figure 6.27), the choice $\chi = -\varphi$ is also appropriate if all streamlines are in $\varphi =$ const. planes through the axis of symmetry. Then $\rho\mathbf{u} = \nabla\chi \times \nabla\psi$. We shall see that the streamfunction for these axisymmetric flows does *not* satisfy the Laplace equation (and consequently the method of complex variables is not applicable) and the lines of constant ϕ and ψ are *not* orthogonal. Some simple examples of axisymmetric irrotational flows around bodies of revolution, such as spheres and airships, will be given in the rest of this chapter.

In axisymmetric flow problems, it is convenient to work with both cylindrical and spherical polar coordinates, often going from one set to the other in the same problem. In this chapter cylindrical coordinates will be denoted by (R, φ, x), and spherical coordinates by (r, θ, φ). These are illustrated in Figure 6.27a, from which their relation to Cartesian coordinates is seen to be

cylindrical	spherical	
$x = x$	$x = r\cos\theta$	(6.63)
$y = R\cos\varphi$	$y = r\sin\theta\cos\varphi$	
$z = R\sin\varphi$	$z = r\sin\theta\sin\varphi$	

Note that r is the distance from the origin, whereas R is the radial distance from the x-axis. The bodies of revolution will have their axes coinciding with the x-axis (Figure 6.27b). The resulting flow pattern is independent of the azimuthal coordinate φ, and is identical in all planes containing the x-axis. Further, the velocity component u_φ is zero.

Important expressions for curvilinear coordinates are listed in Appendix B. For axisymmetric flows, several relevant expressions are presented in the following for quick reference.

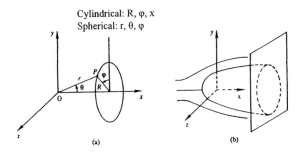

Figure 6.27   (a) Cylindrical and spherical coordinates; (b) axisymmetric flow. In Fig. 6.27, the coordinate axes are not aligned according to the conventional definitions. Specifically in (a), the polar axis from which θ is measured is usually taken to be the z-axis and φ is measured from the x-axis. In (b), the axis of symmetry is usually taken to be the z-axis and the angle θ or φ is measured from the x-axis.

Continuity equation:

$$\frac{\partial u_x}{\partial x} + \frac{1}{R}\frac{\partial}{\partial R}(Ru_R) = 0 \quad \text{(cylindrical)} \tag{6.64}$$

$$\frac{1}{r}\frac{\partial}{\partial r}(r^2 u_r) + \frac{1}{\sin\theta}\frac{\partial}{\partial\theta}(u_\theta \sin\theta) = 0 \quad \text{(spherical)} \tag{6.65}$$

Laplace equation:

$$\nabla^2\phi = \frac{1}{R}\frac{\partial}{\partial R}\left(R\frac{\partial\phi}{\partial R}\right) + \frac{\partial^2\phi}{\partial x^2} = 0 \quad \text{(cylindrical)} \tag{6.66}$$

$$\nabla^2\phi = \frac{1}{r^2}\left[\frac{\partial}{\partial r}\left(r^2\frac{\partial\phi}{\partial r}\right)\right] + \frac{1}{r^2\sin\theta}\frac{\partial}{\partial\theta}\left(\sin\theta\frac{\partial\phi}{\partial\theta}\right) = 0 \quad \text{(spherical)} \tag{6.67}$$

Vorticity:

$$\omega_\varphi = \frac{\partial u_R}{\partial x} - \frac{\partial u_x}{\partial R} \quad \text{(cylindrical)} \tag{6.68}$$

$$\omega_\varphi = \frac{1}{r}\left[\frac{\partial}{\partial r}(ru_\theta) - \frac{\partial u_r}{\partial\theta}\right] \quad \text{(spherical)} \tag{6.69}$$

18. Streamfunction and Velocity Potential for Axisymmetric Flow

A streamfunction can be defined for axisymmetric flows because the continuity equation involves two terms only. In cylindrical coordinates, the continuity equation can be written as

$$\frac{\partial}{\partial x}(Ru_x) + \frac{\partial}{\partial R}(Ru_R) = 0 \tag{6.70}$$

which is satisfied by $\mathbf{u} = -\nabla\varphi \times \nabla\psi$, yielding

$$u_x \equiv \frac{1}{R}\frac{\partial\psi}{\partial R} \qquad \text{(cylindrical)}, \tag{6.71}$$

$$u_R \equiv -\frac{1}{R}\frac{\partial\psi}{\partial x}.$$

The axisymmetric stream function is sometimes called the *Stokes streamfunction*. It has units of m^3/s, in contrast to the streamfunction for plane flow, which has units of m^2/s. Due to the symmetry of flow about the x-axis, constant ψ surfaces are surfaces of revolution. Consider two streamsurfaces described by constant values of ψ and $\psi + d\psi$ (Figure 6.28). The volumetric flow rate through the annular space is

$$dQ = -u_R(2\pi R\,dx) + u_x(2\pi R\,dR) = 2\pi\left[\frac{\partial\psi}{\partial x}dx + \frac{\partial\psi}{\partial R}dR\right] = 2\pi\,d\psi,$$

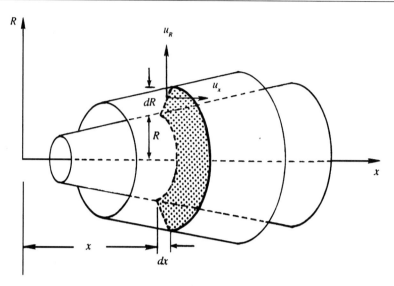

Figure 6.28 Axisymmetric streamfunction. The volume flow rate through two streamsurfaces is $2\pi\,\Delta\psi$.

where equation (6.71) has been used. The form $d\psi = dQ/2\pi$ shows that the difference in ψ values is the flow rate between two concentric streamsurfaces per unit radian angle around the axis. This is consistent with the extended discussion of streamfunctions in Chapter 4, Section 4. The factor of 2π is absent in plane two-dimensional flows, where $d\psi = dQ$ is the flow rate per unit depth. The sign convention is the same as for plane flows, namely, that ψ increases toward the left if we look downstream.

If the flow is also irrotational, then

$$\omega_\varphi = \frac{\partial u_R}{\partial x} - \frac{\partial u_x}{\partial R} = 0. \tag{6.72}$$

On substituting equation (6.71) into equation (6.72), we obtain

$$\frac{\partial^2 \psi}{\partial R^2} - \frac{1}{R}\frac{\partial \psi}{\partial R} + \frac{\partial^2 \psi}{\partial x^2} = 0, \tag{6.73}$$

which is different from the Laplace equation (6.66) satisfied by ϕ. It is easy to show that lines of constant ϕ and ψ are not orthogonal. This is a basic difference between axisymmetric and plane flows.

In spherical coordinates, the streamfunction is defined as $\mathbf{u} = -\nabla\varphi \times \nabla\psi$, yielding

$$u_r = \frac{1}{r^2 \sin\theta}\frac{\partial \psi}{\partial \theta}$$

$$\text{(spherical)}, \tag{6.74}$$

$$u_\theta = -\frac{1}{r \sin\theta}\frac{\partial \psi}{\partial r},$$

which satisfies the axisymmetric continuity equation (6.65).

The velocity potential for axisymmetric flow is defined as

cylindrical	spherical	
$u_R = \dfrac{\partial \phi}{\partial R}$	$u_r = \dfrac{\partial \phi}{\partial R}$	(6.75)
$u_x = \dfrac{\partial \phi}{\partial x}$	$u_\theta = \dfrac{1}{r}\dfrac{\partial \phi}{\partial \theta}$	

which satisfies the condition of irrotationality in a plane containing the x-axis.

19. Simple Examples of Axisymmetric Flows

Axisymmetric irrotational flows can be developed in the same manner as plane flows, except that complex variables cannot be used. Several elementary flows are reviewed briefly in this section, and some practical flows are treated in the following sections.

Uniform Flow

For a uniform flow U parallel to the x-axis, the velocity potential and streamfunction are

cylindrical	spherical	
$\phi = Ux$	$\phi = Ur\cos\theta$	(6.76)
$\psi = \frac{1}{2}UR^2$	$\psi = \frac{1}{2}Ur^2\sin^2\theta$	

These expressions can be verified by using equations (6.71), (6.74), and (6.75). Equipotential surfaces are planes normal to the x-axis, and streamsurfaces are coaxial tubes.

Point Source

For a point source of strength Q (m^3/s), the velocity is $u_r = Q/4\pi r^2$. It is easy to show (Exercise 6) that in polar coordinates

$$\phi = -\frac{Q}{4\pi r} \qquad \psi = -\frac{Q}{4\pi}\cos\theta. \qquad (6.77)$$

Equipotential surfaces are spherical shells, and streamsurfaces are conical surfaces on which $\theta = \mathrm{const}$.

Doublet

For the limiting combination of a source–sink pair, with vanishing separation and large strength, it can be shown (Exercise 7) that

$$\phi = \frac{m}{r^2}\cos\theta \qquad \psi = -\frac{m}{r}\sin^2\theta, \qquad (6.78)$$

where m is the strength of the doublet, directed along the negative x-axis. Streamlines in an axial plane are qualitatively similar to those shown in Figure 6.8, except that they are no longer circles.

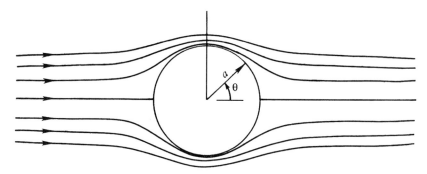

Figure 6.29 Irrotational flow past a sphere.

Flow around a Sphere

Irrotational flow around a sphere can be generated by the superposition of a uniform stream and an axisymmetric doublet opposing the stream. The stream function is

$$\psi = -\frac{m}{r}\sin^2\theta + \frac{1}{2}Ur^2\sin^2\theta. \qquad (6.79)$$

This shows that $\psi = 0$ for $\theta = 0$ or π (any r), or for $r = (2m/U)^{1/3}$ (any θ). Thus all of the x-axis and the spherical surface of radius $a = (2m/U)^{1/3}$ form the streamsurface $\psi = 0$. Streamlines of the flow are shown in Figure 6.29. In terms of the radius of the sphere, velocity components are found from equation (6.79) as

$$
\begin{aligned}
u_r &= \frac{1}{r^2\sin\theta}\frac{\partial\psi}{\partial\theta} = U\left[1 - \left(\frac{a}{r}\right)^3\right]\cos\theta, \\[2mm]
u_\theta &= -\frac{1}{r\sin\theta}\frac{\partial\psi}{\partial r} = -U\left[1 + \frac{1}{2}\left(\frac{a}{r}\right)^3\right]\sin\theta.
\end{aligned}
\qquad (6.80)
$$

The pressure coefficient on the surface is

$$C_p = \frac{p - p_\infty}{\frac{1}{2}\rho U^2} = 1 - \left(\frac{u_\theta}{U}\right)^2 = 1 - \frac{9}{4}\sin^2\theta, \qquad (6.81)$$

which is symmetrical, again demonstrating zero drag in steady irrotational flows.

20. *Flow around a Streamlined Body of Revolution*

As in plane flows, the motion around a closed body of revolution can be generated by superposition of a source and a sink of equal strength on a uniform stream. The closed surface becomes "streamlined" (that is, has a gradually tapering tail) if, for example, the sink is *distributed* over a finite length. Consider Figure 6.30, where there is a point source Q (m^3/s) at the origin O, and a line sink distributed on the x-axis

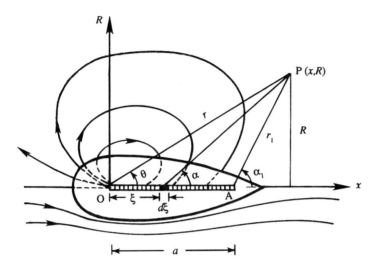

Figure 6.30 Irrotational flow past a streamlined body generated by a point source at O and a distributed line sink from O to A.

from O to A. Let the volume absorbed per unit length of the line sink be k (m^2/s). An elemental length $d\xi$ of the sink can be regarded as a point sink of strength $k\,d\xi$, for which the streamfunction at any point P is [see equation (6.77)]

$$d\psi_{\text{sink}} = \frac{k\,d\xi}{4\pi}\cos\alpha.$$

The total streamfunction at P due to the entire line sink from O to A is

$$\psi_{\text{sink}} = \frac{k}{4\pi}\int_0^a \cos\alpha\,d\xi. \tag{6.82}$$

The integral can be evaluated by noting that $x - \xi = R\cot\alpha$. This gives $d\xi = R\,d\alpha/\sin^2\alpha$ because x and R remain constant as we go along the sink. The streamfunction of the line sink is therefore

$$\psi_{\text{sink}} = \frac{k}{4\pi}\int_\theta^{\alpha_1}\cos\alpha\,\frac{R}{\sin^2\alpha}\,d\alpha = \frac{kR}{4\pi}\int_\theta^{\alpha_1}\frac{d(\sin\alpha)}{\sin^2\alpha},$$

$$= \frac{kR}{4\pi}\left[\frac{1}{\sin\theta} - \frac{1}{\sin\alpha_1}\right] = \frac{k}{4\pi}(r - r_1). \tag{6.83}$$

To obtain a closed body, we must adjust the strengths so that the efflux from the source is absorbed by the sink, that is, $Q = ak$. Then the streamfunction at any point P due to the superposition of a point source of strength Q, a distributed line sink of strength $k = Q/a$, and a uniform stream of velocity U along the x-axis, is

$$\psi = -\frac{Q}{4\pi}\cos\theta + \frac{Q}{4\pi a}(r - r_1) + \frac{1}{2}Ur^2\sin^2\theta. \tag{6.84}$$

A plot of the steady streamline pattern is shown in the bottom half of Figure 6.30, in which the top half shows instantaneous streamlines in a frame of reference at rest with the fluid at infinity.

Here we have assumed that the strength of the line sink is uniform along its length. Other interesting streamlines can be generated by assuming that the strength $k(\xi)$ is nonuniform.

21. Flow around an Arbitrary Body of Revolution

So far, in this chapter we have been *assuming* certain distributions of singularities, and determining what body shape results when the distribution is superposed on a uniform stream. The flow around a body of *given* shape can be simulated by superposing a uniform stream on a series of sources and sinks of unknown strength distributed on a line coinciding with the axis of the body. The strengths of the sources and sinks are then so adjusted that, when combined with a given uniform flow, a closed streamsurface coincides with the given body. The calculation is done numerically using a computer.

Let the body length L be divided into N equal segments of length $\Delta\xi$, and let k_n be the strength (m^2/s) of one of these line sources, which may be positive or negative (Figure 6.31). Then the streamfunction at any "body point" m due to the line source n is, using equation (6.83),

$$\psi_{mn} = -\frac{k_n}{4\pi}\left(r^m_{n-1} - r^m_n\right),$$

where the negative sign is introduced because equation (6.83) is for a sink. When combined with a uniform stream, the streamfunction at m due to all N line sources is

$$\psi_m = -\sum_{n=1}^{N}\frac{k_n}{4\pi}\left(r^m_{n-1} - r^m_n\right) + \tfrac{1}{2}UR^2_m.$$

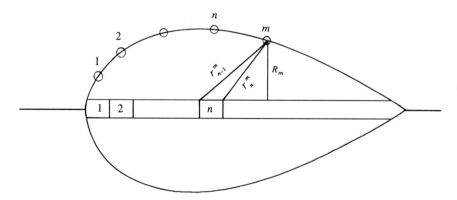

Figure 6.31   Flow around an arbitrary axisymmetric shape generated by superposition of a series of line sources.

Setting $\psi_m = 0$ for all N values of m, we obtain a set of N linear algebraic equations in N unknowns k_n ($n = 1, 2, \ldots, N$), which can be solved by the iteration technique described in Section 16 or some other matrix inversion routine.

22. Concluding Remarks

The theory of potential flow has reached a highly developed stage during the last 250 years because of the efforts of theoretical physicists such as Euler, Bernoulli, D'Alembert, Lagrange, Stokes, Helmholtz, Kirchhoff, and Kelvin. The special interest in the subject has resulted from the applicability of potential theory to other fields such as heat conduction, elasticity, and electromagnetism. When applied to fluid flows, however, the theory resulted in the prediction of zero drag on a body at variance with observations. Meanwhile, the theory of viscous flow was developed during the middle of the Nineteenth Century, after the Navier–Stokes equations were formulated. The viscous solutions generally applied either to very slow flows where the nonlinear advection terms in the equations of motion were negligible, or to flows in which the advective terms were identically zero (such as the viscous flow through a straight pipe). The viscous solutions were highly rotational, and it was not clear where the irrotational flow theory was applicable and why. This was left for Prandtl to explain, as will be shown in Chapter 10.

It is probably fair to say that the theory of irrotational flow does not occupy the center stage in fluid mechanics any longer, although it did so in the past. However, the subject is still quite useful in several fields, especially in aerodynamics. We shall see in Chapter 10 that the pressure distribution around streamlined bodies can still be predicted with a fair degree of accuracy from the irrotational flow theory. In Chapter 15 we shall see that the lift of an airfoil is due to the development of circulation around it, and the magnitude of the lift agrees with the Kutta–Zhukhovsky lift theorem. The technique of conformal mapping will also be essential in our study of flow around airfoil shapes.

Exercises

1. In Section 7, the doublet potential

$$w = \mu/z,$$

was derived by combining a source and a sink on the x-axis. Show that the same potential can also be obtained by superposing a clockwise vortex of circulation $-\Gamma$ on the y-axis at $y = \varepsilon$, and a counterclockwise vortex of circulation Γ at $y = -\varepsilon$, and letting $\varepsilon \to 0$.

2. By integrating pressure, show that the drag on a plane half-body (Section 8) is zero.

3. Graphically generate the streamline pattern for a plane half-body in the following manner. Take a source of strength $m = 200\,\text{m}^2/\text{s}$ and a uniform stream $U = 10\,\text{m/s}$. Draw radial streamlines from the source at equal intervals of $\Delta\theta = \pi/10$,

with the corresponding streamfunction interval

$$\Delta \psi_{\text{source}} = \frac{m}{2\pi} \Delta \theta = 10 \, \text{m}^2/\text{s}.$$

Now draw streamlines of the uniform flow with the same interval, that is,

$$\Delta \psi_{\text{stream}} = U \, \Delta y = 10 \, \text{m}^2/\text{s}.$$

This requires $\Delta y = 1$ m, which you can plot assuming a linear scale of 1 cm = 1 m. Now connect points of equal $\psi = \psi_{\text{source}} + \psi_{\text{stream}}$. (Most students enjoy doing this exercise!)

4. Take a plane source of strength m at point $(-a, 0)$, a plane sink of equal strength at $(a, 0)$, and superpose a uniform stream U directed along the x-axis. Show that there are two stagnation points located on the x-axis at points

$$\pm a \left(\frac{m}{\pi a U} + 1 \right)^{1/2}.$$

Show that the streamline passing through the stagnation points is given by $\psi = 0$. Verify that the line $\psi = 0$ represents a closed oval-shaped body, whose maximum width h is given by the solution of the equation

$$h = a \cot \left(\frac{\pi U h}{m} \right).$$

The body generated by the superposition of a uniform stream and a source–sink pair is called a *Rankine body*. It becomes a circular cylinder as the source–sink pair approach each other.

5. A two-dimensional potential vortex with clockwise circulation Γ is located at point $(0, a)$ above a flat plate. The plate coincides with the x-axis. A uniform stream U directed along the x-axis flows over the vortex. Sketch the flow pattern and show that it represents the flow over an oval-shaped body. [*Hint*: Introduce the image vortex and locate the two stagnation points on the x-axis.]

If the pressure at $x = \pm\infty$ is p_∞, and that *below* the plate is also p_∞, then show that the pressure at any point on the plate is given by

$$p_\infty - p = \frac{\rho \Gamma^2 a^2}{2\pi^2 (x^2 + a^2)^2} - \frac{\rho U \Gamma a}{\pi (x^2 + a^2)}.$$

Show that the total upward force on the plate is

$$F = \frac{\rho \Gamma^2}{4\pi a} - \rho U \Gamma.$$

6. Consider a point source of strength Q (m^3/s). Argue that the velocity components in spherical coordinates are $u_\theta = 0$ and $u_r = Q/4\pi r^2$ and that the velocity

potential and streamfunction must be of the form $\phi = \phi(r)$ and $\psi = \psi(\theta)$. Integrating the velocity, show that $\phi = -Q/4\pi r$ and $\psi = -Q\cos\theta/4\pi$.

7. Consider a point doublet obtained as the limiting combination of a point source and a point sink as the separation goes to zero. (See Section 7 for its two dimensional counterpart.) Show that the velocity potential and streamfunction in spherical coordinates are $\phi = m\cos\theta/r^2$ and $\psi = -m\sin^2\theta/r$, where m is the limiting value of $Q\,\delta s/4\pi$, with Q as the source strength and δs as the separation.

8. A solid hemisphere of radius a is lying on a flat plate. A uniform stream U is flowing over it. Assuming irrotational flow, show that the density of the material must be

$$\rho_h \geqslant \rho\left(1 + \frac{33}{64}\frac{U^2}{ag}\right),$$

to keep it on the plate.

9. Consider the plane flow around a circular cylinder. Use the Blasius theorem equation (6.45) to show that the drag is zero and the lift is $L = \rho U\Gamma$. (In Section 10, we derived these results by integrating the pressure.)

10. There is a point source of strength Q (m³/s) at the origin, and a uniform line sink of strength $k = Q/a$ extending from $x = 0$ to $x = a$. The two are combined with a uniform stream U parallel to the x-axis. Show that the combination represents the flow past a closed surface of revolution of airship shape, whose total length is the difference of the roots of

$$\frac{x^2}{a^2}\left(\frac{x}{a} \pm 1\right) = \frac{Q}{4\pi U a^2}.$$

11. Using a computer, determine the surface contour of an axisymmetric half-body formed by a line source of strength k (m²/s) distributed uniformly along the x-axis from $x = 0$ to $x = a$ and a uniform stream. Note that the nose is more pointed than that formed by the combination of a point source and a uniform stream. By a mass balance (see Section 8), show that the far downstream asymptotic radius of the half-body is $r = \sqrt{ak/\pi U}$.

12. For the flow described by equation (6.30) and sketched in Figure 6.8, show for $\mu > 0$ that $u < 0$ for $y < x$ and $u > 0$ for $y > x$. Also, show that $v < 0$ in the first quadrant and $v > 0$ in the second quadrant.

13. A hurricane is blowing over a long "Quonset hut," that is, a long half-circular cylindrical cross-section building, 6 m in diameter. If the velocity far upstream is $U_\infty = 40$ m/s and $p_\infty = 1.003 \times 10^5$ N/m, $\rho_\infty = 1.23$ kg/m³, find the force per unit depth on the building, assuming the pressure inside is p_∞.

14. In a two-dimensional constant density potential flow, a source of strength m is located a meters above an infinite plane. Find the velocity on the plane, the pressure on the plane, and the reaction force on the plane.

15. Consider a two-dimensional, constant density potential flow over a circular cylinder of radius $r = a$ with axis coincident with a right angle corner, as shown in the figure below. Solve for the streamfunction and velocity components.

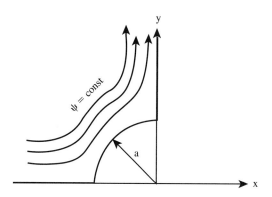

Literature Cited

Prandtl, L. (1952). *Essentials of Fluid Dynamics*, New York: Hafner Publishing.

Supplemental Reading

Batchelor, G. K. (1967). *An Introduction to Fluid Dynamics*, London: Cambridge University Press.
Milne-Thompson, L. M. (1962). *Theoretical Hydrodynamics*, London: Macmillan Press.
Shames, I. H. (1962). *Mechanics of Fluids*, New York: McGraw-Hill.
Vallentine, H. R. (1967). *Applied Hydrodynamics*, New York: Plenum Press.

Gravity Waves

©2010 Elsevier Inc. All rights reserved.
DOI: 10.1016/B978-0-12-381399-2.50007-1

1. Introduction

It is perhaps not an overstatement to say that wave motion is the most basic feature of all physical phenomena. Waves are the means by which information is transmitted between two points in space and time, without movement of the medium across the two points. The energy and phase of some disturbance travels during a wave motion, but motion of the matter is generally small. Waves are generated due to the existence of some kind of "restoring force" that tends to bring the system back to its undisturbed state, and of some kind of "inertia" that causes the system to overshoot after the system has returned to the undisturbed state. One type of wave motion is generated when the restoring forces are due to the compressibility or elasticity of the material medium, which can be a solid, liquid, or gas. The resulting wave motion, in which the particles move to and fro in the direction of wave propagation, is called a *compression wave, elastic wave, or pressure wave*. The small-amplitude variety of these is called a "sound wave." Another common wave motion, and the one we are most familiar with from everyday experience, is the one that occurs at the free surface of a liquid, with gravity playing the role of the restoring force. These are called *surface gravity waves*. Gravity waves, however, can also exist at the interface between two fluids of different density, in which case they are called *internal gravity waves*. The particle motion in gravity waves can have components both along and perpendicular to the direction of propagation, as we shall see.

In this chapter, we shall examine some basic features of wave motion and illustrate them with gravity waves because these are the easiest to comprehend physically. The wave frequency will be assumed much larger than the Coriolis frequency, in which case the wave motion is unaffected by the earth's rotation. Waves affected by planetary rotation will be considered in Chapter 14. Wave motion due to compressibility effects will be considered in Chapter 16. Unless specified otherwise, we shall assume that the waves have small amplitude, in which case the governing equation becomes linear.

2. The Wave Equation

Many simple "nondispersive" (to be defined later) wave motions of small amplitude obey the wave equation

$$\frac{\partial^2 \eta}{\partial t^2} = c^2 \nabla^2 \eta, \tag{7.1}$$

which is a linear partial differential equation of the hyperbolic type. Here η is any type of disturbance, for example the displacement of the free surface in a liquid, variation of density in a compressible medium, or displacement of a stretched string or membrane. The meaning of parameter c will become clear shortly. Waves traveling only in the x direction are described by

$$\frac{\partial^2 \eta}{\partial t^2} = c^2 \frac{\partial^2 \eta}{\partial x^2}, \tag{7.2}$$

which has a general solution of the form

$$\eta = f(x - ct) + g(x + ct), \tag{7.3}$$

where f and g are arbitrary functions. Equation (7.3), called *d'Alembert's solution*, signifies that any arbitrary function of the combination $(x \pm ct)$ is a solution of the wave equation; this can be verified by substitution of equation (7.3) into equation (7.2). It is easy to see that $f(x - ct)$ represents a wave propagating in the positive x direction with speed c, whereas $g(x + ct)$ propagates in the negative x direction at speed c. Figure 7.1 shows a plot of $f(x - ct)$ at $t = 0$. At a later time t, the distance x needs to be larger for the same value of $(x - ct)$. Consequently, $f(x - ct)$ has the same shape as $f(x)$, except displaced by an amount ct along the x-axis. Therefore, the speed of propagation of wave shape $f(x - ct)$ along the positive x-axis is c.

As an example of solution of the wave equation, assume initial conditions in the form

$$\eta(x, 0) = F(x) \quad \text{and} \quad \frac{\partial \eta}{\partial t}(x, 0) = G(x), \tag{7.4}$$

Then equation (7.3) requires that

$$f(x) + g(x) = F(x) \quad \text{and} \quad -\frac{df}{dx} + \frac{dg}{dx} = \frac{1}{c}G(x),$$

which gives the solution

$$f(x) = \frac{1}{2}\left[F(x) - \frac{1}{c}\int_{x_0}^{x} G(\xi)\, d\xi\right], \quad g(x) = \frac{1}{2}\left[F(x) + \frac{1}{c}\int_{x_0}^{x} G(\xi)\, d\xi\right], \tag{7.5}$$

The case of zero initial velocity $[G(x) = 0]$ is interesting. It corresponds to an initial displacement of the surface into an arbitrary profile $F(x)$, which is then left alone. In this case equation (7.5) reduces to $f(x) = g(x) = F(x)/2$, so that solution (7.5) becomes

$$\eta = \tfrac{1}{2}F(x - ct) + \tfrac{1}{2}F(x + ct), \tag{7.6}$$

The nature of this solution is illustrated in Figure 7.2. It is apparent that half the initial disturbance propagates to the right and the other half propagates to the left. Widths of the two components are equal to the width of the initial disturbance. Note that boundary conditions have not been considered in arriving at equation (7.6). Instead, the boundaries have been assumed to be so far away that the reflected waves do not return to the region of interest.

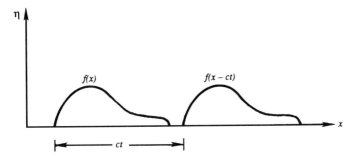

Figure 7.1 Profiles of $f(x - ct)$ at two times.

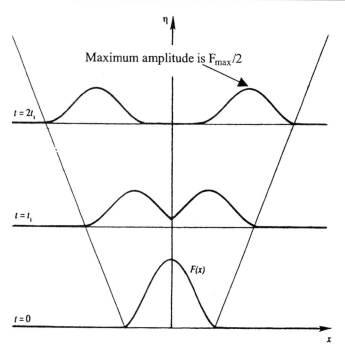

Figure 7.2   Wave profiles at three times. The initial profile is $F(x)$ and the initial velocity is assumed to be zero. Half the initial disturbance propagates to the right and the other half propagates to the left.

3.   Wave Parameters

According to Fourier's principle, any arbitrary disturbance can be decomposed into sinusoidal wave components of different wavelengths and amplitudes. Consequently, it is important to study sinusoidal waves of the form

$$\eta = a \sin\left[\frac{2\pi}{\lambda}(x - ct)\right]. \tag{7.7}$$

The argument $2\pi(x - ct)/\lambda$ is called the *phase* of the wave, and points of constant phase are those where the waveform has the same value, say a crest or trough. Since η varies between $\pm a$, a is called the *amplitude* of the wave. The parameter λ is called the *wavelength* because the value of η in equation (7.7) does not change if x is changed by $\pm\lambda$. Instead of using λ, it is more common to use the *wavenumber* defined as

$$k \equiv \frac{2\pi}{\lambda}, \tag{7.8}$$

which is the number of complete waves in a length 2π. It can be regarded as the "spatial frequency" (rad/m). The waveform equation (7.7) can then be written as

$$\eta = a \sin k(x - ct). \tag{7.9}$$

The *period T* of a wave is the time required for the condition at a point to repeat itself, and must equal the time required for the wave to travel one wavelength:

$$T = \frac{\lambda}{c}. \tag{7.10}$$

The number of oscillations at a point per unit time is the *frequency*, given by

$$\nu = \frac{1}{T}. \tag{7.11}$$

Clearly $c = \lambda \nu$. The quantity

$$\omega = 2\pi\nu = kc, \tag{7.12}$$

is called the *circular frequency*; it is also called the "radian frequency" because it is the rate of change of phase (in radians) per unit time. The speed of propagation of the waveform is related to k and ω by

$$\boxed{c = \frac{\omega}{k},} \tag{7.13}$$

which is called the *phase speed*, as it is the rate at which the "phase" of the wave (crests and troughs) propagates. We shall see that the phase speed may not be the speed at which the envelope of a *group* of waves propagates. In terms of ω and k, the waveform equation (7.7) is written as

$$\eta = a \, \sin(kx - \omega t). \tag{7.14}$$

So far we have been considering waves propagating in the x direction only. For three-dimensional waves of sinusoidal shape, equation (7.14) is generalized to

$$\eta = a \, \sin(kx + ly + mz - \omega t) = a \, \sin(\mathbf{K} \cdot \mathbf{x} - \omega t), \tag{7.15}$$

where $\mathbf{K} = (k, l, m)$ is a vector, called the *wavenumber vector*, whose magnitude is given by

$$K^2 = k^2 + l^2 + m^2. \tag{7.16}$$

It is easy to see that the wavelength of equation (7.15) is

$$\lambda = \frac{2\pi}{K}, \tag{7.17}$$

which is illustrated in Figure 7.3 in two dimensions. The magnitude of phase velocity is $c = \omega/K$, and the direction of propagation is that of $\mathbf{K}$. We can therefore write the phase velocity as the vector

$$\mathbf{c} = \frac{\omega}{K} \frac{\mathbf{K}}{K}, \tag{7.18}$$

where $\mathbf{K}/K$ represents the unit vector in the direction of $\mathbf{K}$.

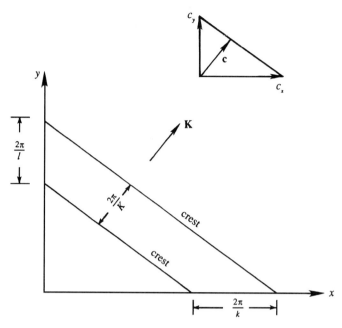

Figure 7.3 Wave propagating in the xy-plane. The inset shows how the components c_x and c_y are added to give the resultant $\mathbf{c}$.

From Figure 7.3, it is also clear that the phase speeds (that is, the speeds of propagation of lines of constant phase) in the three Cartesian directions are

$$c_x = \frac{\omega}{k} \qquad c_y = \frac{\omega}{l} \qquad c_z = \frac{\omega}{m}. \tag{7.19}$$

The preceding shows that the components c_x, c_y, and c_z are each larger than the resultant $c = \omega/K$. It is clear that the *components of the phase velocity vector* $\mathbf{c}$ *do not obey the rule of vector addition*. The method of obtaining $\mathbf{c}$ from the components c_x and c_y is illustrated at the top of Figure 7.3. The peculiarity of such an addition rule for the phase velocity vector merely reflects the fact that phase lines appear to propagate faster along directions not coinciding with the direction of propagation, say the x and y directions in Figure 7.3. In contrast, the components of the "group velocity" vector $\mathbf{c}_g$ do obey the usual vector addition rule, as we shall see later.

We have assumed that the waves exist without a mean flow. If the waves are superposed on a uniform mean flow $\mathbf{U}$, then the observed phase speed is

$$\mathbf{c}_0 = \mathbf{c} + \mathbf{U}.$$

A dot product of the forementioned with the wavenumber vector $\mathbf{K}$, and the use of equation (7.18), gives

$$\omega_0 = \omega + \mathbf{U} \cdot \mathbf{K}, \tag{7.20}$$

where ω_0 is the *observed frequency* at a fixed point, and ω is the *intrinsic frequency* measured by an observer moving with the mean flow. It is apparent that the frequency of a wave is *Doppler shifted* by an amount $\mathbf{U} \cdot \mathbf{K}$ due to the mean flow. Equation (7.20) is easy to understand by considering a situation in which the intrinsic frequency ω is zero and the flow pattern has a periodicity in the x direction of wavelength $2\pi/k$. If this sinusoidal pattern is translated in the x direction at speed U, then the observed frequency at a fixed point is $\omega_0 = Uk$.

The effects of mean flow on frequency will not be considered further in this chapter. Consequently, the involved frequencies should be interpreted as the intrinsic frequency.

4. Surface Gravity Waves

In this section we shall discuss gravity waves at the free surface of a sea of liquid of uniform depth H, which may be large or small compared to the wavelength λ. We shall assume that the amplitude a of oscillation of the free surface is small, in the sense that both a/λ and a/H are much smaller than one. The condition $a/\lambda \ll 1$ implies that the slope of the sea surface is small, and the condition $a/H \ll 1$ implies that the instantaneous depth does not differ significantly from the undisturbed depth. These conditions allow us to linearize the problem. The frequency of the waves is assumed large compared to the Coriolis frequency, so that the waves are unaffected by the earth's rotation. Here, we shall neglect surface tension; in water its effect is limited to wavelengths <7 cm, as discussed in Section 7. The fluid is assumed to have small viscosity, so that viscous effects are confined to boundary layers and do not affect the wave propagation significantly. The motion is assumed to be generated from rest, say, by wind action or by dropping a stone. According to Kelvin's circulation theorem, the resulting motion is *irrotational*, ignoring viscous effects, Coriolis forces, and stratification (density variation).

Formulation of the Problem

Consider a case where the waves propagate in the x direction only, and that the motion is two dimensional in the xz-plane (Figure 7.4). Let the vertical coordinate z

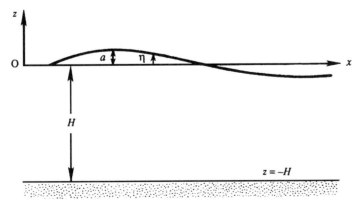

Figure 7.4 Wave nomenclature.

be measured upward from the undisturbed free surface. The free surface displacement is $\eta(x, t)$. Because the motion is irrotational, a velocity potential ϕ can be defined such that

$$u = \frac{\partial \phi}{\partial x} \qquad w = \frac{\partial \phi}{\partial z}. \tag{7.21}$$

Substitution into the continuity equation

$$\frac{\partial u}{\partial x} + \frac{\partial w}{\partial z} = 0, \tag{7.22}$$

gives the Laplace equation

$$\frac{\partial^2 \phi}{\partial x^2} + \frac{\partial^2 \phi}{\partial z^2} = 0. \tag{7.23}$$

Boundary conditions are to be satisfied at the free surface and at the bottom. The condition at the bottom is zero normal velocity, that is

$$w = \frac{\partial \phi}{\partial z} = 0 \quad \text{at} \quad z = -H. \tag{7.24}$$

At the free surface, a *kinematic boundary condition* is that the fluid particle never leaves the surface, that is

$$\frac{D\eta}{Dt} = w_\eta \quad \text{at} \quad z = \eta,$$

where $D/Dt = \partial/\partial t + u(\partial/\partial x)$, and w_η is the vertical component of fluid velocity at the free surface. This boundary condition is the specialization of that discussed in Chapter 4.19 to zero mass flow across the wave surface. The forementioned condition can be written as

$$\frac{\partial \eta}{\partial t} + u\frac{\partial \eta}{\partial x}\bigg|_{z=\eta} = \frac{\partial \phi}{\partial z}\bigg|_{z=\eta}. \tag{7.25}$$

For small-amplitude waves both u and $\partial \eta/\partial x$ are small, so that the quadratic term $u(\partial \eta/\partial x)$ is one order smaller than other terms in equation (7.25), which then simplifies to

$$\frac{\partial \eta}{\partial t} = \frac{\partial \phi}{\partial z}\bigg|_{z=\eta}, \tag{7.26}$$

We can simplify this condition still further by arguing that the right-hand side can be evaluated at $z = 0$ rather than at the free surface. To justify this, expand $\partial \phi/\partial z$ in a Taylor series around $z = 0$:

$$\frac{\partial \phi}{\partial z}\bigg|_{z=\eta} = \frac{\partial \phi}{\partial z}\bigg|_{z=0} + \eta\frac{\partial^2 \phi}{\partial z^2} + \bigg|_{z=0} \cdots \simeq \frac{\partial \phi}{\partial z}\bigg|_{z=0}.$$

Therefore, to the first order of accuracy desired here, $\partial \phi/\partial z$ in equation (7.26) can be evaluated at $z = 0$. We then have

$$\frac{\partial \eta}{\partial t} = \frac{\partial \phi}{\partial z} \quad \text{at} \quad z = 0. \tag{7.27}$$

The error involved in approximating equation (7.26) by (7.27) is explained again later in this section.

In addition to the kinematic condition at the surface, there is a *dynamic condition* that the pressure just below the free surface is always equal to the ambient pressure, with surface tension neglected. Taking the ambient pressure to be zero, the condition is

$$p = 0 \quad \text{at} \quad z = \eta. \tag{7.28}$$

Equation (7.28) follows from the boundary condition on $\boldsymbol{\tau} \cdot \mathbf{n}$, which is continuous across an interface as established in Chapter 4, Section 19. As before, we shall simplify this condition for small-amplitude waves. Since the motion is irrotational, Bernoulli's equation (see equation (4.81))

$$\frac{\partial \phi}{\partial t} + \frac{1}{2}(u^2 + w^2) + \frac{p}{\rho} + gz = F(t), \tag{7.29}$$

is applicable. Here, the function $F(t)$ can be absorbed in $\partial\phi/\partial t$ by redefining ϕ. Neglecting the nonlinear term $(u^2 + w^2)$ for small-amplitude waves, the linearized form of the unsteady Bernoulli equation is

$$\frac{\partial \phi}{\partial t} + \frac{p}{\rho} + gz = 0. \tag{7.30}$$

Substitution into the surface boundary condition (7.28) gives

$$\frac{\partial \phi}{\partial t} + g\eta = 0 \quad \text{at} \quad z = \eta. \tag{7.31}$$

As before, for small-amplitude waves, the term $\partial\phi/\partial t$ can be evaluated at $z = 0$ rather than at $z = \eta$ to give

$$\frac{\partial \phi}{\partial t} = -g\eta \quad \text{at} \quad z = 0. \tag{7.32}$$

Solution of the Problem

Recapitulating, we have to solve

$$\frac{\partial^2 \phi}{\partial x^2} + \frac{\partial^2 \phi}{\partial z^2} = 0. \tag{7.22}$$

subject to the conditions

$$\frac{\partial \phi}{\partial z} = 0 \qquad \text{at} \quad z = -H, \tag{7.24}$$

$$\frac{\partial \phi}{\partial z} = \frac{\partial \eta}{\partial t} \qquad \text{at} \quad z = 0, \tag{7.27}$$

$$\frac{\partial \phi}{\partial t} = -g\eta \quad \text{at} \quad z = 0. \tag{7.32}$$

In order to apply the boundary conditions, we need to assume a form for $\eta(x, t)$. The simplest case is that of a sinusoidal component with wavenumber k and frequency ω, for which

$$\eta = a \cos(kx - \omega t). \tag{7.33}$$

One motivation for studying sinusoidal waves is that small-amplitude waves on a water surface become roughly sinusoidal some time after their generation (unless the water depth is very shallow). This is due to the phenomenon of wave dispersion discussed in Section 10. A second, and stronger, motivation is that an arbitrary disturbance can be decomposed into various sinusoidal components by Fourier analysis, and the response of the system to an arbitrary small disturbance is the sum of the responses to the various sinusoidal components.

For a cosine dependence of η on $(kx - \omega t)$, conditions (7.27) and (7.32) show that ϕ must be a sine function of $(kx - \omega t)$. Consequently, we assume a separable solution of the Laplace equation in the form

$$\phi = f(z) \sin(kx - \omega t), \tag{7.34}$$

where $f(z)$ and $\omega(k)$ are to be determined. Substitution of equation (7.34) into the Laplace equation (7.22) gives

$$\frac{d^2 f}{dz^2} - k^2 f = 0,$$

whose general solution is

$$f(z) = Ae^{kz} + Be^{-kz}.$$

The velocity potential is then

$$\phi = (Ae^{kz} + Be^{-kz}) \sin(kx - \omega t). \tag{7.35}$$

The constants A and B are now determined from the boundary conditions (7.24) and (7.27). Condition (7.24) gives

$$B = Ae^{-2kH}. \tag{7.36}$$

Before applying condition (7.27) in the linearized form, let us explore what would happen if we applied it at $z = \eta$. From (7.35) we get

$$\left. \frac{\partial \phi}{\partial z} \right|_{z=\eta} = k(Ae^{k\eta} - Be^{-k\eta}) \sin(kx - \omega t),$$

Here we can set $e^{k\eta} \simeq e^{-k\eta} \simeq 1$ if $k\eta \ll 1$, valid for small slope of the free surface. This is effectively what we are doing by applying the surface boundary conditions equations (7.27) and (7.32) at $z = 0$ (instead of at $z = \eta$), which we justified previously by a Taylor series expansion.

Substitution of equations (7.33) and (7.35) into the surface velocity condition (7.27) gives

$$k(A - B) = a\omega. \tag{7.37}$$

The constants A and B can now be determined from equations (7.36) and (7.37) as

$$A = \frac{a\omega}{k(1 - e^{-2kH})} \qquad B = \frac{a\omega\, e^{-2kH}}{k(1 - e^{-2kH})}.$$

The velocity potential (7.35) then becomes

$$\phi = \frac{a\omega}{k} \frac{\cosh k(z + H)}{\sinh kH} \sin(kx - \omega t), \qquad (7.38)$$

from which the velocity components are found as

$$u = a\omega \frac{\cosh k(z + H)}{\sinh kH} \cos(kx - \omega t),$$
$$\qquad\qquad (7.39)$$
$$w = a\omega \frac{\sinh k(z + H)}{\sinh kH} \sin(kx - \omega t).$$

We have solved the Laplace equation using kinematic boundary conditions alone. This is typical of irrotational flows. In the last chapter we saw that the equation of motion, or its integral, the Bernoulli equation, is brought into play only to find the pressure distribution, after the problem has been solved from kinematic considerations alone. In the present case, we shall find that application of the dynamic free surface condition (7.32) gives a relation between k and ω.

Substitution of equations (7.33) and (7.38) into (7.32) gives the desired relation

$$\omega = \sqrt{gk \tanh kH}, \qquad (7.40)$$

The phase speed $c = \omega/k$ is related to the wave size by

$$\boxed{c = \sqrt{\frac{g}{k} \tanh kH} = \sqrt{\frac{g\lambda}{2\pi} \tanh \frac{2\pi H}{\lambda}},} \qquad (7.41)$$

This shows that the speed of propagation of a wave component depends on its wavenumber. Waves for which c is a function of k are called *dispersive* because waves of different lengths, propagating at different speeds, "disperse" or separate. (Dispersion is a word borrowed from optics, where it signifies separation of different colors due to the speed of light in a medium depending on the wavelength.) A relation such as equation (7.40), giving ω as a function of k, is called a *dispersion relation* because it expresses the nature of the dispersive process. Wave dispersion is a fundamental process in many physical phenomena; its implications in gravity waves are discussed in Sections 9 and 10.

5. *Some Features of Surface Gravity Waves*

Several features of surface gravity waves are discussed in this section. In particular, we shall examine the nature of pressure change, particle motion, and the energy flow due to a sinusoidal propagating wave. The water depth H is arbitrary; simplifications that result from assuming the depth to be shallow or deep are discussed in the next section.

Pressure Change Due to Wave Motion

It is sometimes possible to measure wave parameters by placing pressure sensors at the bottom or at some other suitable depth. One would therefore like to know how deep the pressure fluctuations penetrate into the water. Pressure is given by the linearized Bernoulli equation

$$\frac{\partial \phi}{\partial t} + \frac{p}{\rho} + gz = 0.$$

If we define

$$p' \equiv p + \rho gz, \tag{7.42}$$

as the *perturbation pressure*, that is, the pressure change from the undisturbed pressure of $-\rho gz$, then Bernoulli's equation gives

$$p' = -\rho \frac{\partial \phi}{\partial t}. \tag{7.43}$$

On substituting equation (7.38), we obtain

$$p' = \frac{\rho a \omega^2}{k} \frac{\cosh k(z + H)}{\sinh kH} \cos(kx - \omega t), \tag{7.44a}$$

which, on using the dispersion relation (7.40), becomes

$$p' = \rho ga \frac{\cosh k(z + H)}{\cosh kH} \cos(kx - \omega t). \tag{7.44b}$$

The perturbation pressure therefore decays into the water column, and whether it could be detected by a sensor depends on the magnitude of the water depth in relation to the wavelength. This is discussed further in Section 6.

Particle Path and Streamline

To examine particle orbits, we obviously need to use Lagrangian coordinates. (See Chapter 3, Section 2 for a discussion of the Lagrangian description.) Let $(x_0 + \xi, z_0 + \zeta)$ be the coordinates of a fluid particle whose rest position is (x_0, z_0), as shown in Figure 7.5. We can use (x_0, z_0) as a "tag" for particle identification, and write $\xi(x_0, z_0, t)$ and $\zeta(x_0, z_0, t)$ in the Lagrangian form. Then the velocity components are given by

$$
\begin{aligned}
u &= \frac{\partial \xi}{\partial t}, \\
w &= \frac{\partial \zeta}{\partial t},
\end{aligned}
\tag{7.45}
$$

where the partial derivative symbol is used because the particle identity (x_0, z_0) is kept fixed in the time derivatives. For small-amplitude waves, the particle excursion (ξ, ζ) is small, and the velocity of a particle along its path is nearly equal to the fluid velocity at the mean position (x_0, z_0) at that instant, given by equation (7.39).

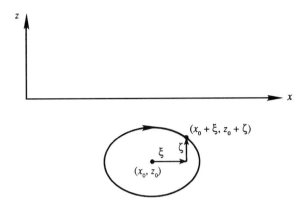

Figure 7.5   Orbit of a fluid particle whose mean position is (x_0, z_0).

Therefore, equation (7.45) gives

$$\frac{\partial \xi}{\partial t} = a\omega \frac{\cosh k(z_0 + H)}{\sinh kH} \cos(kx_0 - \omega t),$$

$$\frac{\partial \zeta}{\partial t} = a\omega \frac{\sinh k(z_0 + H)}{\sinh kH} \sin(kx_0 - \omega t).$$

Integrating in time, we obtain

$$\begin{aligned}
\xi &= -a \frac{\cosh k(z_0 + H)}{\sinh kH} \sin(kx_0 - \omega t), \\
\zeta &= a \frac{\sinh k(z_0 + H)}{\sinh kH} \cos(kx_0 - \omega t).
\end{aligned} \tag{7.46}$$

Elimination of $(kx_0 - \omega t)$ gives

$$\xi^2 \bigg/ \left[a \frac{\cosh k(z_0 + H)}{\sinh kH} \right]^2 + \zeta^2 \bigg/ \left[a \frac{\sinh k(z_0 + H)}{\sinh kH} \right]^2 = 1, \tag{7.47}$$

which represents ellipses. Both the semimajor axis, $a \cosh[k(z_0 + H)]/\sinh kH$ and the semiminor axis, $a \sinh[k(z_0 + H)]/\sinh kH$ decrease with depth, the minor axis vanishing at $z_0 = -H$ (Figure 7.6b). The distance between foci remains constant with depth. Equation (7.46) shows that the phase of the motion (that is, the argument of the sinusoidal term) is independent of z_0. Fluid particles in any vertical column are therefore in phase. That is, if one of them is at the top of its orbit, then all particles at the same x_0 are at the top of their orbits.

To find the streamline pattern, we need to determine the streamfunction ψ, related to the velocity components by

$$\frac{\partial \psi}{\partial z} = u = a\omega \frac{\cosh k(z + H)}{\sinh kH} \cos(kx - \omega t), \tag{7.48}$$

$$\frac{\partial \psi}{\partial x} = -w = -a\omega \frac{\sinh k(z + H)}{\sinh kH} \sin(kx - \omega t), \tag{7.49}$$

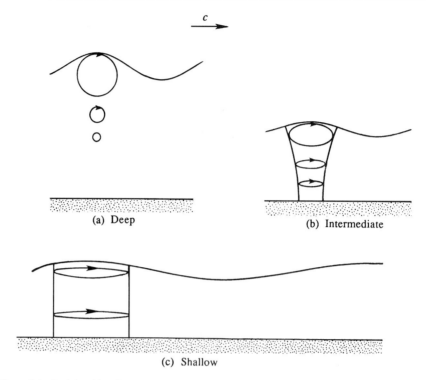

Figure 7.6   Particle orbits of wave motion in deep, intermediate and shallow seas.

where equation (7.39) has been introduced. Integrating equation (7.48) with respect to z, we obtain

$$\psi = \frac{a\omega}{k} \frac{\sinh k(z + H)}{\sinh kH} \cos(kx - \omega t) + F(x, t),$$

where $F(x, t)$ is an arbitrary function of integration. Similarly, integration of equation (7.49) with respect to x gives

$$\psi = \frac{a\omega}{k} \frac{\sinh k(z + H)}{\sinh kH} \cos(kx - \omega t) + G(z, t),$$

where $G(z, t)$ is another arbitrary function. Equating the two expressions for ψ we see that $F = G =$ function of time only; this can be set to zero if we regard ψ as due to wave motion only, so that $\psi = 0$ when $a = 0$. Therefore

$$\psi = \frac{a\omega}{k} \frac{\sinh k(z + H)}{\sinh kH} \cos(kx - \omega t). \tag{7.50}$$

Let us examine the streamline structure at a particular time, say, $t = 0$, when

$$\psi \propto \sinh k(z + H) \cos kx.$$

It is clear that $\psi = 0$ at $z = -H$, so that the bottom wall is a part of the $\psi = 0$ streamline. However, ψ is also zero at $kx = \pm\pi/2, \pm 3\pi/2, \ldots$ for any z. At these

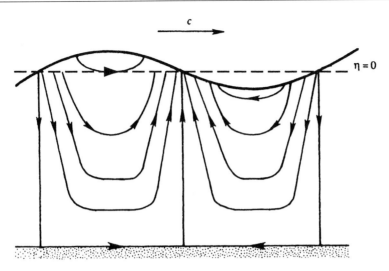

Figure 7.7   Instantaneous streamline pattern in a surface gravity wave propagating to the right.

values of kx, equation (7.33) shows that η vanishes. The resulting streamline pattern is shown in Figure 7.7. It is seen that the *velocity is in the direction of propagation (and horizontal) at all depths below the crests, and opposite to the direction of propagation at all depths below troughs.*

Energy Considerations

Surface gravity waves possess kinetic energy due to motion of the fluid and potential energy due to deformation of the free surface. Kinetic energy per unit horizontal area is found by integrating over the depth and averaging over a wavelength:

$$E_k = \frac{\rho}{2\lambda} \int_0^\lambda \int_{-H}^0 (u^2 + w^2)\, dz\, dx.$$

Here the z-integral is taken up to $z = 0$, because the integral up to $z = \eta$ gives a higher-order term. Substitution of the velocity components from equation (7.39) gives

$$E_k = \frac{\rho\omega^2}{2\sinh^2 kH} \left[\frac{1}{\lambda} \int_0^\lambda a^2 \cos^2(kx - \omega t)\, dx \int_{-H}^0 \cosh^2 k(z + H)\, dz \right.$$

$$\left. + \frac{1}{\lambda} \int_0^\lambda a^2 \sin^2(kx - \omega t)\, dx \int_{-H}^0 \sinh^2 k(z + H)\, dz \right]. \qquad (7.51)$$

In terms of free surface displacement η, the x-integrals in equation (7.51) can be written as

$$\frac{1}{\lambda} \int_0^\lambda a^2 \cos^2(kx - \omega t)\, dx = \frac{1}{\lambda} \int_0^\lambda a^2 \sin^2(kx - \omega t)\, dx$$

$$= \frac{1}{\lambda} \int_0^\lambda \eta^2\, dx = \overline{\eta^2},$$

where $\overline{\eta^2}$ is the mean square displacement. The z-integrals in equation (7.51) are easy to evaluate by expressing the hyperbolic functions in terms of exponentials. Using the dispersion relation (7.40), equation (7.51) finally becomes

$$E_k = \tfrac{1}{2}\rho g \overline{\eta^2}, \tag{7.52}$$

which is the kinetic energy of the wave motion per unit horizontal area.

Consider next the *potential energy* of the wave system, defined as the work done to deform a horizontal free surface into the disturbed state. It is therefore equal to the *difference* of potential energies of the system in the disturbed and undisturbed states. As the potential energy of an element in the fluid (per unit length in y) is $\rho g z \, dx \, dz$ (Figure 7.8), the potential energy of the wave system per unit horizontal area is

$$E_p = \frac{\rho g}{\lambda} \int_0^\lambda \int_{-H}^\eta z \, dz \, dx - \frac{\rho g}{\lambda} \int_0^\lambda \int_{-H}^0 z \, dz \, dx,$$

$$= \frac{\rho g}{\lambda} \int_0^\lambda \int_0^\eta z \, dz \, dx = \frac{\rho g}{2\lambda} \int_0^\lambda \eta^2 dx. \tag{7.53}$$

(An easier way to arrive at the expression for E_p is to note that the potential energy increase due to wave motion equals the work done in raising column A in Figure 7.8 to the location of column B, and integrating over *half the wavelength*. This is because an interchange of A and B over half a wavelength automatically forms a complete wavelength of the deformed surface. The mass of column A is $\rho \eta \, dx$, and the center of gravity is raised by η when A is taken to B. This agrees with the last form in equation (7.53).) Equation (7.53) can be written in terms of the mean square displacement as

$$E_p = \tfrac{1}{2}\rho g \overline{\eta^2}. \tag{7.54}$$

Comparison of equation (7.52) and equation (7.54) shows that the average kinetic and potential energies are equal. This is called the *principle of equipartition of energy*

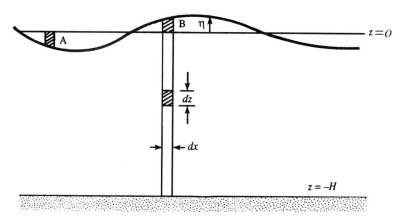

Figure 7.8   Calculation of potential energy of a fluid column.

and is valid in conservative dynamical systems undergoing small oscillations that are unaffected by planetary rotation. However, it is not valid when Coriolis forces are included, as will be seen in Chapter 14. The total wave energy in the water column per unit horizontal area is

$$E = E_p + E_k = \rho g \overline{\eta^2} = \tfrac{1}{2} \rho g a^2, \tag{7.55}$$

where the last form in terms of the amplitude a is valid if η is assumed sinusoidal, since the average of $\cos^2 x$ over a wavelength is 1/2.

Next, consider the rate of transmission of energy due to a single sinusoidal component of wavenumber k. The *energy flux* across the vertical plane $x = 0$ is the pressure work done by the fluid in the region $x < 0$ on the fluid in the region $x > 0$. Per unit length of crest, the time average energy flux is (writing p as the sum of a perturbation p' and a background pressure $-\rho g z$)

$$F = \left\langle \int_{-H}^{0} p u \, dz \right\rangle = \left\langle \int_{-H}^{0} p' u \, dz \right\rangle - \rho g \langle u \rangle \int_{-H}^{0} z \, dz$$

$$= \left\langle \int_{-H}^{0} p' u \, dz \right\rangle, \tag{7.56}$$

where $\langle \ \rangle$ denotes an average over a wave period; we have used the fact that $\langle u \rangle = 0$. Substituting for p' from equation (7.44a) and u from equation (7.39), equation (7.56) becomes

$$F = \langle \cos^2(kx - \omega t) \rangle \frac{\rho a^2 \omega^3}{k \sinh^2 kH} \int_{-H}^{0} \cosh^2 k(z + H) \, dz.$$

The time average of $\cos^2(kx - \omega t)$ is 1/2. The z-integral can be carried out by writing it in terms of exponentials. This finally gives

$$F = \left[\tfrac{1}{2} \rho g a^2 \right] \left[\frac{c}{2} \left(1 + \frac{2kH}{\sinh 2kH} \right) \right]. \tag{7.57}$$

The first factor is the wave energy given in equation (7.55). Therefore, the second factor must be the speed of propagation of wave energy of component k, called the *group speed*. This is discussed in Sections 9 and 10.

6. Approximations for Deep and Shallow Water

The analysis in the preceding section is applicable whatever the magnitude of λ is in relation to the water depth H. Interesting simplifications result for $H/\lambda \ll 1$ (shallow water) and $H/\lambda \gg 1$ (deep water). The expression for phase speed is given by equation (7.41), namely,

$$c = \sqrt{\frac{g\lambda}{2\pi} \tanh \frac{2\pi H}{\lambda}}. \tag{7.41}$$

Approximations are now derived under two limiting conditions in which equation (7.41) takes simple forms.

Deep-Water Approximation

We know that $\tanh x \to 1$ for $x \to \infty$ (Figure 7.9). However, x need not be very large for this approximation to be valid, because $\tanh x = 0.94138$ for $x = 1.75$. It follows that, with 3% accuracy, equation (7.41) can be approximated by

$$c = \sqrt{\frac{g\lambda}{2\pi}} = \sqrt{\frac{g}{k}}, \qquad (7.58)$$

for $H > 0.28\lambda$ (corresponding to $kH > 1.75$). Waves are therefore classified as *deep-water waves* if the depth is more than 28% of the wavelength. Equation (7.58) shows that longer waves in deep water propagate faster. This feature has interesting consequences and is discussed further in Sections 9 and 10.

A dominant period of wind-generated surface gravity waves in the ocean is ≈ 10 s, for which the dispersion relation (7.40) shows that the dominant wavelength is 150 m. The water depth on a typical continental shelf is ≈ 100 m, and in the open ocean it is about ≈ 4 km. It follows that the dominant wind waves in the ocean, even over the continental shelf, act as deep-water waves and do not feel the effects of the ocean

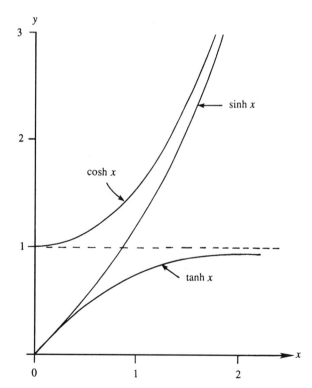

Figure 7.9 Behavior of hyperbolic functions.

bottom until they arrive near the beach. This is not true of gravity waves generated by tidal forces and earthquakes; these may have wavelengths of hundreds of kilometers.

In the preceding section we said that particle orbits in small-amplitude gravity waves describe ellipses given by equation (7.47). For $H > 0.28\lambda$, the semimajor and semiminor axes of these ellipses each become nearly equal to ae^{kz}. This follows from the approximation (valid for $kH > 1.75$)

$$\frac{\cosh k(z+H)}{\sinh kH} \simeq \frac{\sinh k(z+H)}{\sinh kH} \simeq e^{kz}.$$

(The various approximations for hyperbolic functions used in this section can easily be verified by writing them in terms of exponentials.) Therefore, for deep-water waves, particle orbits described by equation (7.46) simplify to

$$\xi = -a\,e^{kz_0}\sin(kx_0 - \omega t)$$

$$\zeta = a\,e^{kz_0}\cos(kx_0 - \omega t).$$

The orbits are therefore circles (Figure 7.6a), of which the radius at the surface equals a, the amplitude of the wave. The velocity components are

$$u = \frac{\partial \xi}{\partial t} = a\omega e^{kz}\cos(kx - \omega t)$$

$$w = \frac{\partial \zeta}{\partial t} = a\omega e^{kz}\sin(kx - \omega t),$$

where we have omitted the subscripts on (x_0, z_0). (For small amplitudes the difference in velocity at the present and mean positions of a particle is negligible. The distinction between mean particle positions and Eulerian coordinates is therefore not necessary, unless finite amplitude effects are considered, as we will see in Section 14.) The velocity vector therefore rotates clockwise (for a wave traveling in the positive x direction) at frequency ω, while its magnitude remains constant at $a\omega e^{kz_0}$.

For deep-water waves, the perturbation pressure given in equation (7.44b) simplifies to

$$p' = \rho g a e^{kz}\cos(kx - \omega t). \tag{7.59}$$

This shows that pressure change due to the presence of wave motion decays exponentially with depth, reaching 4% of its surface magnitude at a depth of $\lambda/2$. A sensor placed at the bottom cannot therefore detect gravity waves whose wavelengths are smaller than twice the water depth. Such a sensor acts like a "low-pass filter," retaining longer waves and rejecting shorter ones.

Shallow-Water Approximation

We know that $\tanh x \simeq x$ as $x \to 0$ (Figure 7.9). For $H/\lambda \ll 1$, we can therefore write

$$\tanh \frac{2\pi H}{\lambda} \simeq \frac{2\pi H}{\lambda},$$

in which case the phase speed equation (7.41) simplifies to

$$\boxed{c = \sqrt{gH}.}$$

(7.60)

The approximation gives a better than 3% accuracy if $H < 0.07\lambda$. Surface waves are therefore regarded as *shallow-water waves* if the water depth is <7% of the wavelength. (The water depth has to be really shallow for waves to behave as shallow-water waves. This is consistent with the comments made in what follows (equation (7.58)), that the water depth does not have to be really deep for water to behave as deep-water waves.) For these waves equation (7.60) shows that the wave speed is independent of wavelength and increases with water depth.

To determine the approximate form of particle orbits for shallow-water waves, we substitute the following approximations into equation (7.46):

$$\cosh k(z + H) \simeq 1$$

$$\sinh k(z + H) \simeq k(z + H)$$

$$\sinh kH \simeq kH.$$

The particle excursions given in equation (7.46) then become

$$\xi = -\frac{a}{kH} \sin(kx - \omega t)$$

$$\zeta = a \left(1 + \frac{z}{H}\right) \cos(kx - \omega t).$$

These represent thin ellipses (Figure 7.6c), with a depth-independent semimajor axis of a/kH and a semiminor axis of $a(1 + z/H)$, which linearly decreases to zero at the bottom wall. From equation (7.39), the velocity field is found as

$$
\begin{aligned}
u &= \frac{a\omega}{kH} \cos(kx - \omega t) \\
w &= a\omega\left(1 + \frac{z}{H}\right) \sin(kx - \omega t),
\end{aligned}
$$

(7.61)

which shows that the vertical component is much smaller than the horizontal component.

The pressure change from the undisturbed state is found from equation (7.44b) to be

$$p' = \rho g a \cos(kx - \omega t) = \rho g \eta,$$

(7.62)

where equation (7.33) has been used to express the pressure change in terms of η. This shows that the pressure change at any point is independent of depth, and equals the hydrostatic increase of pressure due to the surface elevation change η. *The pressure*

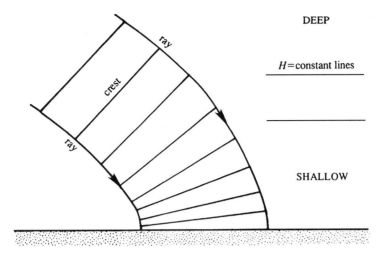

Figure 7.10   Refraction of a surface gravity wave approaching a sloping beach. Note that the crest lines tend to become parallel to the coast.

field is therefore completely hydrostatic in shallow-water waves. Vertical accelerations are negligible because of the small *w*-field. For this reason, shallow water waves are also called *hydrostatic waves.* It is apparent that a pressure sensor mounted at the bottom can sense these waves.

Wave Refraction in Shallow Water

We shall now qualitatively describe the commonly observed phenomenon of *refraction* of shallow-water waves. Consider a sloping beach, with depth contours parallel to the coastline (Figure 7.10). Assume that waves are propagating toward the coast from the deep ocean, with their crests at an angle to the coastline. Sufficiently near the coastline they begin to feel the effect of the bottom and finally become shallow-water waves. Their frequency does not change along the path (a fact that will be proved in Section 10), but the speed of propagation $c = \sqrt{gH}$ and the wavelength λ become smaller. Consequently, the crest lines, which are perpendicular to the local direction of c, tend to become parallel to the coast. *This is why we see that the waves coming toward the beach always seem to have their crests parallel to the coastline.*

An interesting example of wave refraction occurs when a deep-water wave with straight crests approaches an island (Figure 7.11). Assume that the water depth becomes shallower as the island is approached, and the constant depth contours are circles concentric with the island. Figure 7.11 shows that the waves always come in *toward* the island, even on the "shadow" side marked A!

The bending of wave paths in an inhomogeneous medium is called *wave refraction.* In this case the source of inhomogeneity is the spatial dependence of H. The analogous phenomenon in optics is the bending of light due to density changes in its path.

7. Influence of Surface Tension

It was explained in Section 1.5 that the interface between two immiscible fluids is in a state of tension. The tension acts as a restoring force, enabling the interface to support waves in a manner analogous to waves on a stretched membrane or string. Waves due to the presence of surface tension are called *capillary waves*. Although gravity is not needed to support these waves, the existence of surface tension alone without gravity is uncommon. We shall therefore examine the modification of the preceding results for pure gravity waves due to the inclusion of surface tension.

Let $PQ = ds$ be an element of arc on the free surface, whose local radius of curvature is r (Figure 7.12a). Suppose p_a is the pressure on the "atmospheric" side, and p is the pressure just inside the interface. The surface tension forces at P and Q,

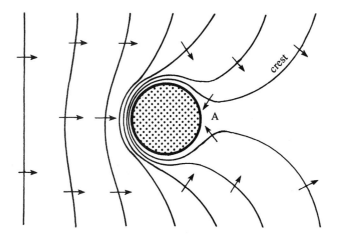

Figure 7.11 Refraction of a surface gravity wave approaching an island with sloping beach. Crest lines, perpendicular to the rays, are shown. Note that the crest lines come in toward the island, even on the shadow side A. *Reprinted with the permission of Mrs. Dorothy Kinsman Brown*: B. Kinsman, *Wind Waves*, Prentice-Hall Englewood Cliffs, NJ, 1965.

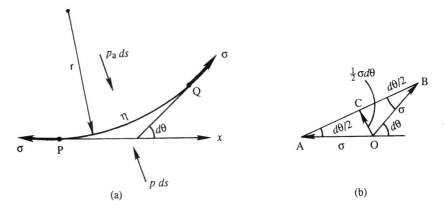

Figure 7.12 (a) Segment of a free surface under the action of surface tension; (b) net surface tension force on an element.

per unit length perpendicular to the plane of the paper, are each equal to σ and directed along the tangents at P and Q. Equilibrium of forces on the arc PQ is considered in Figure 7.12b. The force at P is represented by segment OA, and the force at Q is represented by segment OB. The resultant of OA and OB in a direction perpendicular to the arc PQ is represented by $2OC \simeq \sigma d\theta$. Therefore, the balance of forces in a direction perpendicular to the arc PQ requires

$$-p_a \, ds + p \, ds + \sigma d\theta = 0.$$

It follows that the pressure difference is related to the curvature by

$$p_a - p = \sigma \frac{d\theta}{ds} = \frac{\sigma}{r}.$$

The curvature $1/r$ of $\eta(x)$ is given by

$$\frac{1}{r} = \frac{\partial^2 \eta/\partial x^2}{[1 + (\partial \eta/\partial x)^2]^{3/2}} \simeq \frac{\partial^2 \eta}{\partial x^2},$$

where the approximate expression is for small slopes. Therefore,

$$p_a - p = \sigma \frac{\partial^2 \eta}{\partial x^2}.$$

Choosing the atmospheric pressure p_a to be zero, we obtain the condition

$$p = -\sigma \frac{\partial^2 \eta}{\partial x^2} \quad \text{at} \quad z = \eta. \tag{7.63}$$

Using the linearized Bernoulli equation

$$\frac{\partial \phi}{\partial t} + \frac{p}{\rho} + gz = 0,$$

condition (7.63) becomes

$$\frac{\partial \phi}{\partial t} = \frac{\sigma}{\rho} \frac{\partial^2 \eta}{\partial x^2} - g\eta \quad \text{at} \quad z = 0. \tag{7.64}$$

As before, for small-amplitude waves it is allowable to apply the surface boundary condition (7.64) at $z = 0$, instead at $z = \eta$.

Solution of the wave problem including surface tension is identical to the one for pure gravity waves presented in Section 4, except that the pressure boundary condition (7.32) is replaced by (7.64). This only changes the dispersion relation $\omega(k)$, which is found by substitution of (7.33) and (7.38) into (7.64), to give

$$\omega = \sqrt{k \left(g + \frac{\sigma k^2}{\rho} \right) \tanh kH}. \tag{7.65}$$

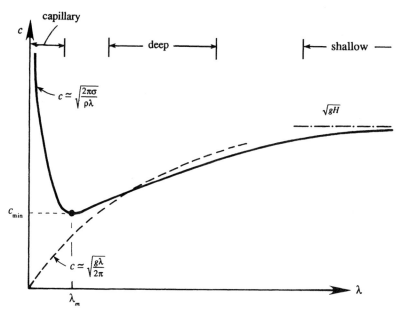

Figure 7.13   Sketch of phase velocity vs wavelength in a surface gravity wave.

The phase velocity is therefore

$$c = \sqrt{\left(\frac{g}{k} + \frac{\sigma k}{\rho}\right) \tanh kH} = \sqrt{\left(\frac{g\lambda}{2\pi} + \frac{2\pi\sigma}{\rho\lambda}\right) \tanh \frac{2\pi H}{\lambda}}. \qquad (7.66)$$

A plot of equation (7.66) is shown in Figure 7.13. It is apparent that the effect of surface tension is to increase c above its value for pure gravity waves at all wavelengths. This is because the free surface is now "tighter," and hence capable of generating more restoring forces. However, the effect of surface tension is only appreciable for very small wavelengths. A measure of these wavelengths is obtained by noting that there is a minimum phase speed at $\lambda = \lambda_m$, and surface tension dominates for $\lambda < \lambda_m$ (Figure 7.13). Setting $dc/d\lambda = 0$ in equation (7.66), and assuming the deep-water approximation $\tanh(2\pi H/\lambda) \simeq 1$ valid for $H > 0.28\lambda$, we obtain

$$c_{\min} = \left[\frac{4g\sigma}{\rho}\right]^{1/4} \quad \text{at} \quad \lambda_m = 2\pi\sqrt{\frac{\sigma}{\rho g}}. \qquad (7.67)$$

For an air–water interface at 20 °C, the surface tension is $\sigma = 0.074\,\text{N/m}$, giving

$$c_{\min} = 23.2\,\text{cm/s} \quad \text{at} \quad \lambda_m = 1.73\,\text{cm}. \qquad (7.68)$$

Only small waves (say, $\lambda < 7\,\text{cm}$ for an air–water interface), called *ripples*, are therefore affected by surface tension. Wavelengths <4 mm are dominated by surface tension and are rather unaffected by gravity. From equation (7.66), the phase speed of these *pure capillary waves* is

$$c = \sqrt{\frac{2\pi\sigma}{\rho\lambda}}, \tag{7.69}$$

where we have again assumed $\tanh(2\pi H/\lambda) \simeq 1$. The smallest of these, traveling at a relatively large speed, can be found leading the waves generated by dropping a stone into a pond.

8. Standing Waves

So far, we have been studying propagating waves. Nonpropagating waves can be generated by superposing two waves of the same amplitude and wavelength, but moving in opposite directions. The resulting surface displacement is

$$\eta = a \cos(kx - \omega t) + a \cos(kx + \omega t) = 2a \cos kx \cos \omega t.$$

It follows that $\eta = 0$ for $kx = \pm\pi/2, \pm3\pi/2\ldots$. Points of zero surface displacement are called *nodes*. The free surface therefore does not propagate, but simply oscillates up and down with frequency ω, keeping the nodal points fixed. Such waves are called *standing waves*. The corresponding streamfunction, using equation (7.50), is both for the $\cos(kx - \omega t)$ and $\cos(kx + \omega t)$ components, and for the sum. This gives

$$
\begin{aligned}
\psi &= \frac{a\omega}{k} \frac{\sinh k(z + H)}{\sinh kH} [\cos(kx - \omega t) - \cos(kx + \omega t)] \\
&= \frac{2a\omega}{k} \frac{\sinh k(z + H)}{\sinh kH} \sin kx \sin \omega t.
\end{aligned}
\tag{7.70}
$$

The instantaneous streamline pattern shown in Figure 7.14 should be compared with the streamline pattern for a propagating wave (Figure 7.7).

A limited body of water such as a lake forms standing waves by reflection from the walls. A standing oscillation in a lake is called a *seiche* (pronounced "saysh"),

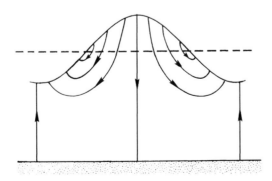

Figure 7.14 Instantaneous streamline pattern in a standing surface gravity wave. If this is mode $n = 0$, then two successive vertical streamlines are a distance L apart. If this is mode $n = 1$, then the first and third vertical streamlines are a distance L apart.

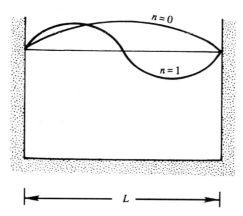

Figure 7.15   Normal modes in a lake, showing distributions of u for the first two modes. This is consistent with the streamline pattern of Figure 7.14.

in which only certain wavelengths and frequencies ω (eigenvalues) are allowed by the system. Let L be the length of the lake, and assume that the waves are invariant along y. The possible wavelengths are found by setting $u = 0$ at the two walls. Because $u = \partial\psi/\partial z$, equation (7.70) gives

$$u = 2a\omega \frac{\cosh k(z + H)}{\sinh kH} \sin kx \sin \omega t. \tag{7.71}$$

Taking the walls at $x = 0$ and L, the condition of no flow through the walls requires $\sin(kL) = 0$, that is,

$$kL = (n + 1)\pi \qquad n = 0, 1, 2, \ldots,$$

which gives the allowable wavelengths as

$$\lambda = \frac{2L}{n + 1}. \tag{7.72}$$

The largest wavelength is $2L$ and the next smaller is L (Figure 7.15). The allowed frequencies can be found from the dispersion relation (7.40), giving

$$\omega = \sqrt{\frac{\pi g(n + 1)}{L} \tanh\left[\frac{(n + 1)\pi H}{L}\right]}, \tag{7.73}$$

which are the natural frequencies of the lake.

9. *Group Velocity and Energy Flux*

An interesting set of phenomena takes place when the phase speed of a wave depends on its wavelength. The most common example is the deep water gravity wave, for which c is proportional to $\sqrt{\lambda}$. A wave phenomenon in which c depends on k is called *dispersive* because, as we shall see in the next section, the different wave components separate or "disperse" from each other.

In a dispersive system, the energy of a wave component does not propagate at the phase velocity $c = \omega/k$, but at the *group velocity* defined as $c_g = d\omega/dk$. To see this, consider the superposition of two sinusoidal components of equal amplitude but slightly different wavenumber (and consequently slightly different frequency because $\omega = \omega(k)$). Then the combination has a waveform

$$\eta = a\cos(k_1 x - \omega_1 t) + a\cos(k_2 x - \omega_2 t).$$

Applying the trigonometric identity for $\cos A + \cos B$, we obtain

$$\eta = 2a\cos\left[\tfrac{1}{2}(k_2 - k_1)x - \tfrac{1}{2}(\omega_2 - \omega_1)t\right]\cos\left[\tfrac{1}{2}(k_1 + k_2)x - \tfrac{1}{2}(\omega_1 + \omega_2)t\right].$$

Writing $k = (k_1 + k_2)/2$, $\omega = (\omega_1 + \omega_2)/2$, $dk = k_2 - k_1$, and $d\omega = \omega_2 - \omega_1$, we obtain

$$\eta = 2a\cos\left(\tfrac{1}{2}dk\,x - \tfrac{1}{2}d\omega\,t\right)\cos(kx - \omega t). \tag{7.74}$$

Here, $\cos(kx - \omega t)$ is a progressive wave with a phase speed of $c = \omega/k$. However, its amplitude $2a$ is modulated by a *slowly varying* function $\cos[dk\,x/2 - d\omega\,t/2]$, which has a large wavelength $4\pi/dk$, a large period $4\pi/d\omega$, and propagates at a speed (=wavelength/period) of

$$\boxed{c_g = \frac{d\omega}{dk}.} \tag{7.75}$$

Multiplication of a rapidly varying sinusoid and a slowly varying sinusoid, as in equation (7.74), generates repeating wave groups (Figure 7.16). The individual wave components propagate with the speed $c = \omega/k$, but the envelope of the wave groups travels with the speed c_g, which is therefore called the *group velocity*. If $c_g < c$, then the wave crests seem to appear from nowhere at a nodal point, proceed forward through the envelope, and disappear at the next nodal point. If, on the other hand, $c_g > c$, then the individual wave crests seem to emerge from a forward nodal point and vanish at a backward nodal point.

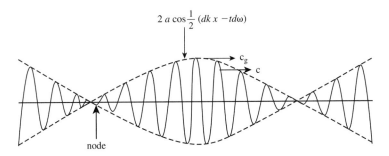

Figure 7.16 Linear combination of two sinusoids, forming repeated wave groups.

Equation (7.75) shows that the group speed of waves of a certain wavenumber k is given by the slope of the *tangent* to the dispersion curve $\omega(k)$. In contrast, the phase velocity is given by the slope of the radius vector (Figure 7.17).

A particularly illuminating example of the idea of group velocity is provided by the concept of a *wave packet*, formed by combining all wavenumbers in a certain narrow band δk around a central value k. In physical space, the wave appears nearly sinusoidal with wavelength $2\pi/k$, but the amplitude *dies away* in a length of order $1/\delta k$ (Figure 7.18). If the spectral width δk is narrow, then decay of the wave amplitude in physical space is slow. The concept of such a wave packet is more realistic than the one in Figure 7.16, which is rather unphysical because the wave groups repeat themselves. Suppose that, at some initial time, the wave group is represented by

$$\eta = a(x)\cos kx.$$

It can be shown (see, for example, Phillips (1977), p. 25) that for small times the subsequent evolution of the wave profile is approximately described by

$$\eta = a(x - c_g t)\cos(kx - \omega t), \tag{7.76}$$

where $c_g = d\omega/dk$. This shows that the *amplitude of a wave packet travels with the group speed*. It follows that c_g must equal the speed of propagation of *energy* of a certain wavelength. The fact that c_g is the speed of energy propagation is also evident in Figure 7.16 because the nodal points travel at c_g and no energy can cross the nodal points.

For surface gravity waves having the dispersion relation

$$\omega = \sqrt{gk \tanh kH}, \tag{7.40}$$

the group velocity is found to be

$$c_g = \frac{c}{2}\left[1 + \frac{2kH}{\sinh 2kH}\right]. \tag{7.77}$$

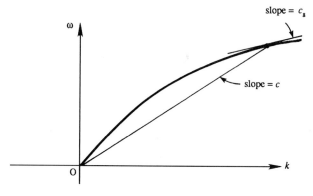

Figure 7.17   Finding c and c_g from dispersion relation $\omega(k)$.

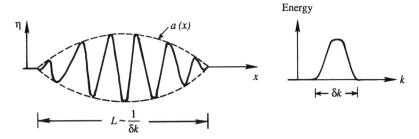

Figure 7.18 A wave packet composed of a narrow band of wavenumbers δk.

The two limiting cases are

$$c_{\text{g}} = \tfrac{1}{2}c \qquad \text{(deep water)},$$
$$c_{\text{g}} = c \qquad \text{(shallow water)}. \tag{7.78}$$

The group velocity of deep-water gravity waves is half the phase speed. Shallow-water waves, on the other hand, are *nondispersive*, for which $c = c_{\text{g}}$. For a linear nondispersive system, any waveform preserves its shape in time because all the wavelengths that make up the waveform travel at the same speed. For a pure capillary wave, the group velocity is $c_{\text{g}} = 3c/2$ (Exercise 3).

The rate of transmission of energy for gravity waves is given by equation (7.57), namely

$$F = E\frac{c}{2}\left[1 + \frac{2kH}{\sinh kH} \right],$$

where $E = \rho g a^2/2$ is the average energy in the water column per unit horizontal area. Using equation (7.77), we conclude that

$$\boxed{F = Ec_{\text{g}}.} \tag{7.79}$$

This signifies that the *rate of transmission of energy of a sinusoidal wave component is wave energy times the group velocity*. This reinforces our previous interpretation of the group velocity as the speed of propagation of energy.

We have discussed the concept of group velocity in one dimension only, taking ω to be a function of the wavenumber k in the direction of propagation. In three dimensions $\omega(k, l, m)$ is a function of the three components of the wavenumber vector $\mathbf{K} = (k, l, m)$ and, using Cartesian tensor notation, the group velocity vector is given by

$$c_{gi} = \frac{\partial \omega}{\partial K_i},$$

where K_i stands for any of the components of $\mathbf{K}$. The group velocity vector is then the gradient of ω in the wavenumber space.

10. Group Velocity and Wave Dispersion

Physical Motivation

We continue our discussion of group velocity in this section, focussing on how the different wavelength and frequency components are propagated. Consider waves in deep water, for which

$$c = \sqrt{\frac{g\lambda}{2\pi}} \qquad c_g = \frac{c}{2},$$

signifying that larger waves propagate faster. Suppose that a surface disturbance is generated by dropping a stone into a pool. The initial disturbance can be thought of as being composed of a great many wavelengths. A short time later, at $t = t_1$, the sea surface may have the rather irregular profile shown in Figure 7.19. The appearance of the surface at a later time t_2, however, is more regular, with the longer components (which have been traveling faster) out in front. The waves in front are the longest waves produced by the initial disturbance; we denote their length by λ_{max}, typically a few times larger than the stone. The leading edge of the wave system therefore propagates at the group speed corresponding to these wavelengths, that is, at the speed

$$c_{g\,max} = \frac{1}{2}\sqrt{\frac{g\lambda_{max}}{2\pi}}.$$

(Pure capillary waves can propagate faster than this speed, but they have small magnitude and get dissipated rather soon.) The region of initial disturbance becomes calm because there is a minimum group velocity of gravity waves due to the influence of surface tension, namely 17.8 cm/s (Exercise 4). The trailing edge of the wave system therefore travels at speed

$$c_{g\,min} = 17.8\,\text{cm/s}.$$

With $c_{g\,max} > 17.8$ cm/s for ordinary sizes of stones, the length of the disturbed region gets larger, as shown in Figure 7.19. The wave heights are correspondingly smaller because there is a fixed amount of energy in the wave system. (Wave dispersion, therefore, makes the linearity assumption more accurate.) The smoothening of the profile and the spreading of the region of disturbance continue until the amplitudes become imperceptible or the waves are damped by viscous dissipation. It is clear that the *initial superposition of various wavelengths, running for some time, will sort themselves out* in the sense that the different sinusoidal components, differing widely in their wavenumbers, become spatially *separated*, and are found in quite different places. This is a basic feature of the behavior of a dispersive system.

The wave group as a whole travels slower than the individual crests. Therefore, if we try to follow the last crest at the rear of the train, quite soon we find that it is the second one from the rear; a new crest has been born behind it. In fact, new crests are constantly "popping up from nowhere" at the rear of the train, propagating through the train, and finally disappearing in front of the train. This is because, by following a particular crest, we are traveling at twice the speed at which the energy of waves of a

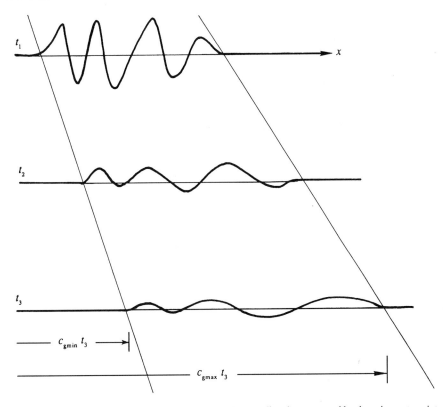

Figure 7.19   Surface profiles at three values of time due to a disturbance caused by dropping a stone into a pool.

particular length is traveling. Consequently, *we do not see a wave of fixed wavelength if we follow a particular crest.* In fact, an individual wave constantly becomes longer as it propagates through the train. When its length becomes equal to the longest wave generated initially, it cannot evolve any more and dies out. Clearly, the waves in front of the train are the longest Fourier components present in the initial disturbance.

Layer of Constant Depth

We shall now prove that an observer traveling at c_g would see no change in k if the layer depth H is uniform everywhere. Consider a wavetrain of "gradually varying wavelength," such as the one shown at later time values in Figure 7.19. By this we mean that the distance between successive crests varies slowly in space and time. Locally, we can describe the free surface displacement by

$$\eta = a(x, t) \cos[\theta(x, t)], \qquad (7.80)$$

where $a(x, t)$ is a slowly varying amplitude and $\theta(x, t)$ is the local phase. We know that the phase angle for a wavenumber k and frequency ω is $\theta = kx - \omega t$. For a gradually varying wavetrain, we can define a *local* wavenumber $k(x, t)$ and a *local* frequency $\omega(x, t)$ as the rate of change of phase in space and time, respectively. That is,

$$k = \frac{\partial \theta}{\partial x},$$

$$\omega = -\frac{\partial \theta}{\partial t}. \tag{7.81}$$

Cross differentiation gives

$$\frac{\partial k}{\partial t} + \frac{\partial \omega}{\partial x} = 0. \tag{7.82}$$

Now suppose we have a dispersion relation relating ω solely to k in the form $\omega = \omega(k)$. We can then write

$$\frac{\partial \omega}{\partial x} = \frac{d\omega}{dk} \frac{\partial k}{\partial x},$$

so that equation (7.82) becomes

$$\frac{\partial k}{\partial t} + c_g \frac{\partial k}{\partial x} = 0, \tag{7.83}$$

where $c_g = d\omega/dk$. The left-hand side of equation (7.83) is similar to the material derivative and gives the rate of change of k as seen by an observer traveling at speed c_g. Such an observer will always see the same wavelength. *Group velocity is therefore the speed at which wavenumbers are advected.* This is shown in the xt-diagram of Figure 7.20, where wave crests are followed along lines $dx/dt = c$ and wavelengths are preserved along the lines $dx/dt = c_g$. Note that the width of the disturbed region, bounded by the first and last thick lines in Figure 7.20, increases with time, and that the crests constantly appear at the back of the group and vanish at the front.

Layer of Variable Depth $H(x)$

The conclusion that an observer traveling at c_g sees only waves of the same length is true only for waves in a homogeneous medium, that is, a medium whose properties are uniform everywhere. In contrast, a sea of nonuniform depth $H(x)$ behaves like an inhomogeneous medium, provided the waves are shallow enough to feel the bottom. In such a case it is the *frequency* of the wave, and not its wavelength, that remains constant along the path of propagation of energy. To demonstrate this, consider a case where $H(x)$ is gradually varying (on the scale of a wavelength) so that we can still use the dispersion relation (7.40) with H replaced by $H(x)$:

$$\omega = \sqrt{gk \tanh[kH(x)]}.$$

Such a dispersion relation has a form

$$\omega = \omega(k, x). \tag{7.84}$$

In such a case we can find the group velocity at a point as

$$c_{\mathrm{g}}(k, x) = \frac{\partial \omega(k, x)}{\partial k},$$ (7.85)

which on multiplication by $\partial k / \partial t$ gives

$$c_{\mathrm{g}} \frac{\partial k}{\partial t} = \frac{\partial \omega}{\partial k} \frac{\partial k}{\partial t} = \frac{\partial \omega}{\partial t}.$$ (7.86)

Multiplying equation (7.82) by c_{g} and using equation (7.86) we obtain

$$\frac{\partial \omega}{\partial t} + c_{\mathrm{g}} \frac{\partial \omega}{\partial x} = 0.$$ (7.87)

In three dimensions, this is written as

$$\frac{\partial \omega}{\partial t} + \mathbf{c}_{\mathrm{g}} \cdot \nabla \omega = 0,$$

which shows that ω remains constant to an observer traveling with the group velocity in an inhomogeneous medium.

Summarizing, an observer traveling at c_{g} in a homogeneous medium sees constant values of k, $\omega(k)$, c, and $c_{\mathrm{g}}(k)$. Consequently, ray paths describing group velocity in the xt-plane are straight lines (Figure 7.20). In an inhomogeneous medium only ω

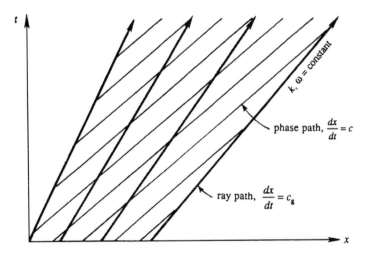

Figure 7.20 Propagation of a wave group in a homogeneous medium, represented on an xt-plot. Thin lines indicate paths taken by wave crests, and thick lines represent paths along which k and ω are constant. M. J. Lighthill, *Waves in Fluids*, 1978 and reprinted with the permission of Cambridge University Press, London.

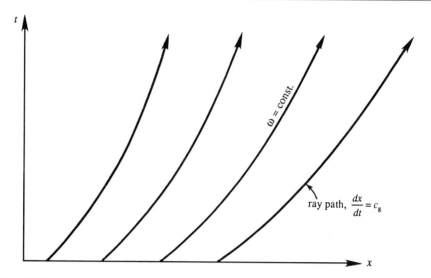

Figure 7.21 Propagation of a wave group in an inhomogeneous medium represented on an xt-plot. Only ray paths along which ω is constant are shown. M. J. Lighthill, *Waves in Fluids*, 1978 and reprinted with the permission of Cambridge University Press, London.

remains constant along the lines $dx/dt = c_g$, but k, c, and c_g can change. Consequently, ray paths are not straight in this case (Figure 7.21).

11. *Nonlinear Steepening in a Nondispersive Medium*

Until now we have assumed that the wave amplitude is small. This has enabled us to neglect the higher-order terms in the Bernoulli equation and to apply the boundary conditions at $z = 0$ instead of at the free surface $z = \eta$. One consequence of such linear analysis has been that waves of arbitrary shape propagate unchanged in form if the system is nondispersive, such as shallow water waves. The unchanging form is a result of the fact that all wavelengths, of which the initial waveform is composed, propagate at the same speed $c = \sqrt{gH}$, provided all the sinusoidal components satisfy the shallow-water approximation $Hk \ll 1$. We shall now see that the unchanging waveform result is no longer valid if *finite amplitude* effects are considered. Several other nonlinear effects will also be discussed in the following sections.

Finite amplitude effects can be formally treated by the *method of characteristics*; this is discussed, for example, in Liepmann and Roshko (1957) and Lighthill (1978). Instead, we shall adopt only a qualitative approach here. Consider a finite amplitude surface displacement consisting of an elevation and a depression, propagating in shallow-water of undisturbed depth H (Figure 7.22). Let a little wavelet be superposed on the elevation at point x, at which the water depth is $H'(x)$ and the fluid velocity due to the wave motion is $u(x)$. Relative to an observer moving with the fluid velocity u, the wavelet propagates at the local shallow-water speed $c' = \sqrt{gH'}$. The speed of the wavelet relative to a frame of reference fixed in the undisturbed fluid is therefore $c = c' + u$. It is apparent that the local wave speed c is no longer constant because

$c'(x)$ and $u(x)$ are variables. This is in contrast to the linearized theory in which u is negligible and c' is constant because $H' \simeq H$.

Let us now examine the effect of such a variable c on the wave profile. The value of c' is larger for points on the elevation than for points on the depression. From Figure 7.7 we also know that the fluid velocity u is positive (that is, in the direction of wave propagation) under an elevation and negative under a depression. It follows that wave speed c is larger for points on the hump than for points on the depression, so that the waveform undergoes a "shearing deformation" as it propagates, the region of elevation tending to overtake the region of depression (Figure 7.22).

We shall call the front face AB a "compression region" because the elevation here is rising with time. Figure 7.22 shows that the net effect of nonlinearity is a steepening of the compression region. For finite amplitude waves in a nondispersive medium like shallow water, therefore, there is an important distinction between compression and expansion regions. A compression region tends to steepen with time and form a jump, while an expansion region tends to flatten out. This eventually would lead to the shape shown at the top of Figure 7.22, implying the physically impossible situation of three values of surface elevation at a point. However, before this happens the wave slope becomes nearly infinite (profile at t_2 in Figure 7.22), so that dissipative

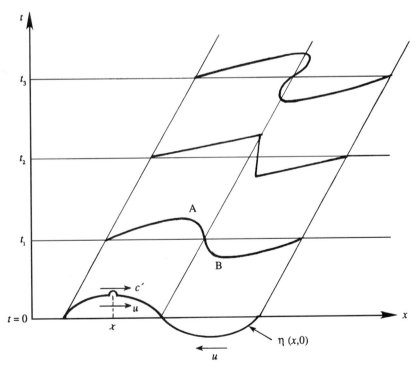

Figure 7.22 Wave profiles at four values of time. At t_2 the profile has formed a hydraulic jump. The profile at t_3 is impossible.

processes including wave breaking and foaming become important, and the previous inviscid arguments become inapplicable. Such a waveform has the form of a front and propagates into still fluid at constant speed that lies between $\sqrt{gH_1}$ and $\sqrt{gH_2}$, where H_1 and H_2 are the water depths on the two sides of the front. This is called the *hydraulic jump*, which is similar to the *shock wave* in a compressible flow. This is discussed further in the following section.

12. Hydraulic Jump

In the previous section we saw how steepening of the compression region of a surface wave in shallow water leads to the formation of a jump, which subsequently propagates into the undisturbed fluid at constant speed and without further change in form. In this section we shall discuss certain characteristics of flow across such a jump. Before we do so, we shall introduce certain definitions.

Consider the flow in a shallow canal of depth H. If the flow speed is u, we may define a nondimensional speed by

$$\mathrm{Fr} \equiv \frac{u}{\sqrt{gH}} = \frac{u}{c}.$$

This is called the *Froude number*, which is the ratio of the speed of flow to the speed of infinitesimal gravity waves. The flow is called *supercritical* if $\mathrm{Fr} > 1$, and *subcritical* if $\mathrm{Fr} < 1$. The Froude number is analogous to the *Mach number* in compressible flow, defined as the ratio of the speed of flow to the speed of sound in the medium.

It was seen in the preceding section that a hydraulic jump propagates into a still fluid at a speed (say, u_1) that lies between the long-wave speeds on the two sides, namely, $c_1 = \sqrt{gH_1}$ and $c_2 = \sqrt{gH_2}$ (Figure 7.23c). Now suppose a leftward propagating jump is made stationary by superposing a flow u_1 directed to the right. In this frame the fluid enters the jump at speed u_1 and exits at speed $u_2 < u_1$ (Figure 7.23b). Because $c_1 < u_1 < c_2$, it follows that $\mathrm{Fr}_1 > 1$ and $\mathrm{Fr}_2 < 1$. Just as a compressible flow suddenly changes from a supersonic to subsonic state by going through a shock wave (Section 16.6), a supercritical flow in a shallow canal can change into a subcritical state by going through a *hydraulic jump*. The depth of flow rises downstream of a hydraulic jump, just as the pressure rises downstream of a shock wave. To continue the analogy, mechanical energy is lost by dissipative processes both within the hydraulic jump and within the shock wave. A common example of a stationary hydraulic jump is found at the foot of a dam, where the flow almost always reaches a supercritical state because of the free fall (Figure 7.23a). A tidal bore propagating into a river mouth is an example of a propagating hydraulic jump.

Consider a control volume across a stationary hydraulic jump shown in Figure 7.23b. The depth rises from H_1 to H_2 and the velocity falls from u_1 to u_2. If Q is the volume rate of flow per unit width normal to the plane of the paper, then mass conservation requires

$$Q = u_1 H_1 = u_2 H_2.$$

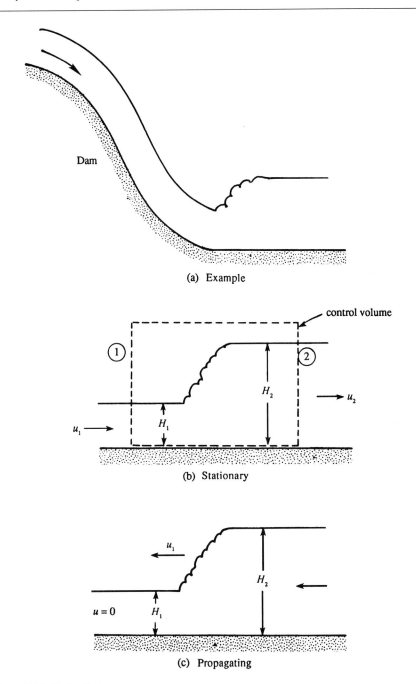

(a) Example

(b) Stationary

(c) Propagating

Figure 7.23 Hydraulic jump.

Now use the momentum principle (Section 4.8), which says that the sum of the forces on a control volume equals the momentum outflow rate at section 2 minus the momentum inflow rate at section 1. The force at section 1 is the average pressure $\rho g H_1/2$ times the area H_1; similarly, the force at section 2 is $\rho g H_2^2/2$. If the distance between sections 1 and 2 is small, then the force exerted by the bottom wall of the canal is negligible. Then the momentum theorem gives

$$\tfrac{1}{2}\rho g H_1^2 - \tfrac{1}{2}\rho g H_2^2 = \rho Q(u_2 - u_1).$$

Substituting $u_1 = Q/H_1$ and $u_2 = Q/H_2$ on the right-hand side, we obtain

$$\frac{g}{2}(H_1^2 - H_2^2) = Q\left(\frac{Q}{H_2} - \frac{Q}{H_1}\right). \tag{7.88}$$

Canceling the factor $(H_1 - H_2)$, we obtain

$$\left(\frac{H_2}{H_1}\right)^2 + \frac{H_2}{H_1} - 2\mathrm{Fr}_1^2 = 0,$$

where $\mathrm{Fr}_1^2 = Q^2/g H_1^3 = u_1^2/g H_1$. The solution is

$$\frac{H_2}{H_1} = \tfrac{1}{2}(-1 + \sqrt{1 + 8\mathrm{Fr}_1^2}). \tag{7.89}$$

For supercritical flows $\mathrm{Fr}_1 > 1$, for which equation (7.89) shows that $H_2 > H_1$. Therefore, depth of water increases downstream of the hydraulic jump.

Although the solution $H_2 < H_1$ for $\mathrm{Fr}_1 < 1$ is allowed by equation (7.89), such a solution violates the second law of thermodynamics, because it implies an increase of mechanical energy of the flow. To see this, consider the mechanical energy of a fluid particle at the surface, $E = u^2/2 + gH = Q^2/2H^2 + gH$. Eliminating Q by equation (7.88) we obtain, after some algebra,

$$E_2 - E_1 = -(H_2 - H_1)\frac{g(H_2 - H_1)^2}{4H_1 H_2}.$$

This shows that $H_2 < H_1$ implies $E_2 > E_1$, which violates the second law of thermodynamics. The mechanical energy, in fact, *decreases* in a hydraulic jump because of the eddying motion within the jump.

A hydraulic jump not only appears at the free surface, but also at density interfaces in a stratified fluid, in the laboratory as well as in the atmosphere and the ocean. (For example, see Turner (1973), Figure 3.11, for his photograph of an internal hydraulic jump on the lee side of a mountain.)

13. Finite Amplitude Waves of Unchanging Form in a Dispersive Medium

In Section 11 we considered a nondispersive medium, and found that nonlinear effects continually accumulate and add up until they become large changes. Such an accumulation is prevented in a dispersive medium because the different Fourier components propagate at different speeds and become separated from each other. In a dispersive system, then, nonlinear steepening could cancel out the dispersive spreading, resulting in finite amplitude waves of constant form. This is indeed the case. A brief description of the phenomenon is given here; further discussion can be found in Lighthill (1978), Whitham (1974), and LeBlond and Mysak (1978).

Note that if the amplitude is negligible, then in a dispersive system a wave of unchanging form can only be perfectly sinusoidal because the presence of any other Fourier component would cause the sinusoids to propagate at different speeds, resulting in a change in the wave shape.

Finite Amplitude Waves in Deep Water: The Stokes Wave

In 1847 Stokes showed that periodic waves of finite amplitude are possible in deep water. In terms of a power series in the amplitude a, he showed that the surface elevation of irrotational waves in deep water is given by

$$\eta = a \cos k(x - ct) + \tfrac{1}{2}ka^2 \cos 2k(x - ct) \\ + \tfrac{3}{8}k^2 a^3 \cos 3k(x - ct) + \cdots , \tag{7.90}$$

where the speed of propagation is

$$c = \sqrt{\frac{g}{k}(1 + k^2 a^2)}. \tag{7.91}$$

Equation (7.90) is the Fourier series for the waveform η. The addition of Fourier components of different wavelengths in equation (7.90) shows that the wave profile η is no longer exactly sinusoidal. The arguments in the cosine terms show that all the Fourier components propagate at the same speed c, so that the wave profile propagates unchanged in time. It has now been established that the existence of periodic wavetrains of unchanging form is a typical feature of nonlinear dispersive systems. Another important result, generally valid for nonlinear systems, is that the wave speed depends on the amplitude, as in equation (7.91).

Periodic finite-amplitude irrotational waves in deep water are frequently called *Stokes' waves*. They have a flattened trough and a peaked crest (Figure 7.24). The maximum possible amplitude is $a_{max} = 0.07\lambda$, at which point the crest becomes a sharp 120° angle. Attempts at generating waves of larger amplitude result in the appearance of foam (*white caps*) at these sharp crests. In finite amplitude waves, fluid particles no longer trace closed orbits, but undergo a slow drift in the direction of wave propagation; this is discussed in Section 14.

Figure 7.24 The Stokes wave. It is a finite amplitude periodic irrotational wave in deep water.

Finite Amplitude Waves in Fairly Shallow Water: Solitons

Next, consider nonlinear waves in a slightly dispersive system, such as "fairly long" waves with λ/H in the range between 10 and 20. In 1895 Korteweg and deVries showed that these waves approximately satisfy the nonlinear equation

$$\frac{\partial \eta}{\partial t} + c_0 \frac{\partial \eta}{\partial x} + \frac{3}{8}c_0 \frac{\eta}{H}\frac{\partial \eta}{\partial x} + \frac{1}{6}c_0 H^2 \frac{\partial^3 \eta}{\partial x^3} = 0, \qquad (7.92)$$

where $c_0 = \sqrt{gH}$. This is the *Korteweg–deVries equation*. The first two terms appear in the linear nondispersive limit. The third term is due to finite amplitude effects and the fourth term results from the weak dispersion due to the water depth being not shallow enough. (Neglecting the nonlinear term in equation (7.92), and substituting $\eta = a \exp(ikx - i\omega t)$, it is easy to show that the dispersion relation is $c = c_0(1 - (1/6)k^2 H^2)$. This agrees with the first two terms in the Taylor series expansion of the dispersion relation $c = \sqrt{(g/k)\tanh kH}$ for small kH, verifying that weak dispersive effects are indeed properly accounted for by the last term in equation (7.92).)

The ratio of nonlinear and dispersion terms in equation (7.92) is

$$\frac{\eta}{H}\frac{\partial \eta}{\partial x} \bigg/ H^2 \frac{\partial^3 \eta}{\partial x^3} \sim \frac{a\lambda^2}{H^3}.$$

When $a\lambda^2/H^3$ is larger than ≈ 16, nonlinear effects sharpen the forward face of the wave, leading to hydraulic jump, as discussed in Section 11. For lower values of $a\lambda^2/H^3$, a balance can be achieved between nonlinear steepening and dispersive spreading, and waves of unchanging form become possible. Analysis of the Korteweg–deVries equation shows that two types of solutions are then possible, a periodic solution and a solitary wave solution. The periodic solution is called *cnoidal*

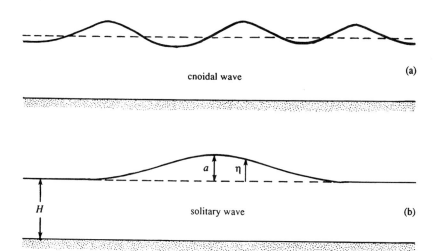

cnoidal wave   (a)

solitary wave   (b)

Figure 7.25 Cnoidal and solitary waves. Waves of unchanging form result because nonlinear steepening balances dispersive spreading.

wave, because it is expressed in terms of elliptic functions denoted by $cn(x)$. Its waveform is shown in Figure 7.25. The other possible solution of the Korteweg–deVries equation involves only a single hump and is called a *solitary wave* or *soliton*. Its profile is given by

$$\eta = a \operatorname{sech}^2 \left[\left(\frac{3a}{4H^3} \right)^{1/2} (x - ct) \right], \tag{7.93}$$

where the speed of propagation is

$$c = c_0 \left(1 + \frac{a}{2H} \right),$$

showing that the propagation velocity increases with the amplitude of the hump. The validity of equation (7.93) can be checked by substitution into equation (7.92). The waveform of the solitary wave is shown in Figure 7.25.

An isolated hump propagating at constant speed with unchanging form and in fairly shallow water was first observed experimentally by S. Russell in 1844. Solitons have been observed to exist not only as surface waves, but also as internal waves in stratified fluid, in the laboratory as well as in the ocean; (See Figure 3.3, Turner (1973)).

14. Stokes' Drift

Anyone who has observed the motion of a floating particle on the sea surface knows that the particle moves slowly in the direction of propagation of the waves. This is called *Stokes' drift*. It is a second-order or finite amplitude effect, due to which the particle orbit is not closed but has the shape shown in Figure 7.26. The mean velocity of a *fluid particle* (that is, the Lagrangian velocity) is therefore not zero, although the mean velocity *at a point* (the Eulerian velocity) must be zero if the process is periodic. The drift is essentially due to the fact that the particle moves forward faster (when it is at the top of its trajectory) than backward (when it is at the bottom of its orbit). Although it is a second-order effect, its magnitude is frequently significant.

To find an expression for Stokes' drift, we use Lagrangian specification, proceeding as in Section 5 but keeping a higher order of accuracy in the analysis. Our analysis is adapted from the presentation given in the work by Phillips (1977, p. 43). Let (x, z) be the instantaneous coordinates of a fluid particle whose position at $t = 0$ is (x_0, z_0). The initial coordinates (x_0, z_0) serve as a particle identification, and we can write its subsequent position as $x(x_0, z_0, t)$ and $z(x_0, z_0, t)$, using the Lagrangian form of specification. The velocity components of the "particle (x_0, z_0)" are $u_L(x_0, z_0, t)$ and $w_L(x_0, z_0, t)$. (Note that the subscript "L" was not introduced in Section 5, since to the lowest order we equated the velocity at time t of a particle with mean coordinates (x_0, z_0) to the Eulerian velocity at t at location (x_0, z_0). Here we are taking the analysis to a higher order of accuracy, and the use of a subscript "L" to denote Lagrangian velocity helps to avoid confusion.)

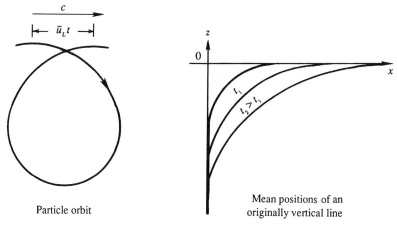

Figure 7.26   The Stokes drift.

The velocity components are

$$u_L = \frac{\partial x}{\partial t}$$
$$w_L = \frac{\partial z}{\partial t},$$

(7.94)

where the partial derivative signs mean that the initial position (serving as a particle tag) is kept fixed in the time derivative. The position of a particle is found by integrating equation (7.94):

$$x = x_0 + \int_0^t u_L(x_0, z_0, t')\, dt'$$
$$z = z_0 + \int_0^t w_L(x_0, z_0, t')\, dt'.$$

(7.95)

At time t the Eulerian velocity at (x, z) equals the Lagrangian velocity of particle (x_0, z_0) at the same time, if (x, z) and (x_0, z_0) are related by equation (7.95). (No approximation is involved here! The equality is merely a reflection of the fact that *particle* (x_0, z_0) occupies the *position* (x, z) at time t.) Denoting the Eulerian velocity components without subscript, we therefore have

$$u_L(x_0, z_0, t) = u(x, z, t).$$

Expanding the Eulerian velocity $u(x, z, t)$ in a Taylor series about (x_0, z_0), we obtain

$$u_L(x_0, z_0, t) = u(x_0, z_0, t) + (x - x_0)\left(\frac{\partial u}{\partial x}\right)_0 + (z - z_0)\left(\frac{\partial u}{\partial z}\right)_0 + \cdots, \quad (7.96)$$

and a similar expression for w_L. The Stokes drift is the time mean value of equation (7.96). As the time mean of the first term on the right-hand side of equation (7.96) is zero, the Stokes drift is given by the mean of the next two terms of equation (7.96). This was neglected in Section 5, and the result was closed orbits.

We shall now estimate the Stokes drift for gravity waves, using the deep water approximation for algebraic simplicity. The velocity components and particle displacements for this motion are given in Section 6 as

$$u(x_0, z_0, t) = a\omega e^{kz_0} \cos(kx_0 - \omega t),$$

$$x - x_0 = -ae^{kz_0} \sin(kx_0 - \omega t),$$

$$z - z_0 = ae^{kz_0} \cos(kx_0 - \omega t).$$

Substitution into the right-hand side of equation (7.96), taking time average, and using the fact that the time average of $\sin^2 t$ over a time period is $1/2$, we obtain

$$\bar{u}_L = a^2 \omega k e^{2kz_0}, \tag{7.97}$$

which is the *Stokes drift* in deep water. Its surface value is $a^2\omega k$, and the vertical decay rate is twice that for the Eulerian velocity components. It is therefore confined very close to the sea surface. For arbitrary water depth, it is easy to show that

$$\bar{u}_L = a^2 \omega k \frac{\cosh 2k(z_0 + H)}{2 \sinh^2 kH}. \tag{7.98}$$

The Stokes drift causes mass transport in the fluid, due to which it is also called the *mass transport velocity*. Vertical fluid lines marked, for example, by some dye gradually bend over (Figure 7.26). In spite of this mass transport, the mean Eulerian velocity anywhere *below the trough* is exactly zero (to any order of accuracy), if the flow is irrotational. This follows from the condition of irrotationality $\partial u/\partial z = \partial w/\partial x$, a vertical integral of which gives

$$u = u|_{z=-H} + \int_{-H}^{z} \frac{\partial w}{\partial x}\, dz,$$

showing that the mean of u is proportional to the mean of $\partial w/\partial x$ over a wavelength, which is zero for periodic flows.

15. Waves at a Density Interface between Infinitely Deep Fluids

To this point we have considered only waves at the free surface of a liquid. However, waves can also exist at the interface between two immiscible liquids of different densities. Such a sharp density gradient can, for example, be generated in the ocean by solar heating of the upper layer, or in an estuary (that is, a river mouth) or a fjord into which fresh (less saline) river water flows over oceanic water, which is more saline and consequently heavier. The situation can be idealized by considering a lighter fluid of density ρ_1 lying over a heavier fluid of density ρ_2 (Figure 7.27).

We assume that the fluids are infinitely deep, so that only those solutions that decay exponentially from the interface are allowed. In this section and in the rest of

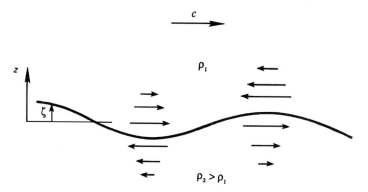

Figure 7.27   Internal wave at a density interface between two infinitely deep fluids.

the chapter, we shall make use of the convenience of *complex notation*. For example, we shall represent the interface displacement $\zeta = a\cos(kx - \omega t)$ by

$$\zeta = \mathcal{R}e\, a\, e^{i(kx-\omega t)},$$

where $\mathcal{R}e$ stands for "the real part of," and $i = \sqrt{-1}$. It is customary to omit the $\mathcal{R}e$ symbol and simply write

$$\zeta = a\, e^{i(kx-\omega t)}, \tag{7.99}$$

where it is implied that *only the real part of the equation is meant*. We are therefore carrying an extra imaginary part (which can be thought of as having no physical meaning) on the right-hand side of equation (7.99). The convenience of complex notation is that the algebra is simplified, essentially because differentiating exponentials is easier than differentiating trigonometric functions. If desired, the constant a in equation (7.99) can be considered to be a complex number. For example, the profile $\zeta = \sin(kx - \omega t)$ can be represented as the real part of $\zeta = -i\exp i(kx - \omega t)$.

We have to solve the Laplace equation for the velocity potential in both layers, subject to the continuity of p and w at the interface. The equations are, therefore,

$$\frac{\partial^2 \phi_1}{\partial x^2} + \frac{\partial^2 \phi_1}{\partial z^2} = 0$$

$$\frac{\partial^2 \phi_2}{\partial x^2} + \frac{\partial^2 \phi_2}{\partial z^2} = 0, \tag{7.100}$$

subject to

$$\phi_1 \to 0 \quad \text{as} \quad z \to \infty \tag{7.101}$$

$$\phi_2 \to 0 \quad \text{as} \quad z \to -\infty \tag{7.102}$$

$$\frac{\partial\phi_1}{\partial z} = \frac{\partial\phi_2}{\partial z} = \frac{\partial\zeta}{\partial t} \qquad \text{at} \quad z = 0 \qquad (7.103)$$

$$\rho_1\frac{\partial\phi_1}{\partial t} + \rho_1 g\zeta = \rho_2\frac{\partial\phi_2}{\partial t} + \rho_2 g\zeta \qquad \text{at} \quad z = 0. \qquad (7.104)$$

Equation (7.103) follows from equating the vertical velocity of the fluid on both sides of the interface to the rate of rise of the interface. Equation (7.104) follows from the continuity of pressure across the interface. As in the case of surface waves, the boundary conditions are linearized and applied at $z = 0$ instead of at $z = \zeta$. Conditions (7.101) and (7.102) require that the solutions of equation (7.100) must be of the form

$$\phi_1 = A\,e^{-kz}e^{i(kx-\omega t)}$$

$$\phi_2 = B\,e^{kz}e^{i(kx-\omega t)},$$

because a solution proportional to e^{kz} is not allowed in the upper fluid, and a solution proportional to e^{-kz} is not allowed in the lower fluid. Here A and B can be complex. As in Section 4, the constants are determined from the kinematic boundary conditions (7.103), giving

$$A = -B = i\omega a/k.$$

The dynamic boundary condition (7.104) then gives the dispersion relation

$$\omega = \sqrt{gk\left(\frac{\rho_2 - \rho_1}{\rho_2 + \rho_1}\right)} = \varepsilon\sqrt{gk}, \qquad (7.105)$$

where $\varepsilon^2 \equiv (\rho_2 - \rho_1)/(\rho_2 + \rho_1)$ is a small number if the density difference between the two liquids is small. The case of small density difference is relevant in geophysical situations; for example, a $10\,^\circ$C temperature change causes the density of the upper layer of the ocean to decrease by 0.3%. Equation (7.105) shows that waves at the interface between two liquids of infinite thickness travel like deep water surface waves, with ω proportional to $\sqrt{gk}$, but at a much reduced frequency. In general, therefore, *internal waves have a smaller frequency, and consequently a smaller phase speed, than surface waves.* As expected, equation (7.105) reduces to the expression for surface waves if $\rho_1 = 0$.

The kinetic energy of the field can be found by integrating $\rho(u^2 + w^2)/2$ over the range $z = \pm\infty$. This gives the average kinetic energy per unit horizontal area of (see Exercise 7):

$$E_k = \tfrac{1}{4}(\rho_2 - \rho_1)ga^2,$$

The potential energy can be calculated by finding the rate of work done in deforming a flat interface to the wave shape. In Figure 7.28, this involves a transfer of column A of density ρ_2 to location B, a simultaneous transfer of column B of density ρ_1 to location A, and integrating the work over *half the wavelength*, since the resulting

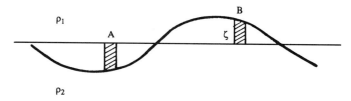

Figure 7.28 Calculation of potential energy of a two-layer fluid. The work done in transferring element A to B equals the weight of A times the vertical displacement of its center of gravity.

exchange forms a complete wavelength; see the previous discussion of Figure 7.8. The potential energy per unit horizontal area is therefore

$$E_p = \frac{1}{\lambda} \int_0^{\lambda/2} \rho_2 g \zeta^2 \, dx - \frac{1}{\lambda} \int_0^{\lambda/2} \rho_1 g \zeta^2 \, dx$$

$$= \frac{g(\rho_2 - \rho_1)}{2\lambda} \int_0^{\lambda/2} \zeta^2 \, dx = \frac{1}{4}(\rho_2 - \rho_1) g a^2.$$

The total wave energy per unit horizontal area is

$$E = E_k + E_p = \frac{1}{2}(\rho_2 - \rho_1) g a^2. \tag{7.106}$$

In a comparison with equation (7.55), it follows that the *amplitude of internal waves is usually much larger than those of surface waves if the same amount of energy is used to set off the motion.*

The horizontal velocity components in the two layers are

$$u_1 = \frac{\partial \phi_1}{\partial x} = -\omega a e^{-kz} e^{i(kx - \omega t)}$$

$$u_2 = \frac{\partial \phi_2}{\partial x} = \omega a e^{kz} e^{i(kx - \omega t)},$$

which show that the velocities in the two layers are oppositely directed (Figure 7.27). The interface is therefore a *vortex sheet*, which is a surface across which the tangential velocity is discontinuous. It can be expected that a continuously stratified medium, in which the density varies continuously as a function of z, will support internal waves whose vorticity is distributed throughout the flow. Consequently, *internal waves in a continuously stratified fluid are not irrotational and do not satisfy the Laplace equation.* This is discussed further in Section 16.

The existence of internal waves at a density discontinuity has explained an interesting phenomenon observed in Norwegian fjords (Gill, 1982). It was known for a long time that ships experienced unusually high drags on entering these fjords. The phenomenon was a mystery (and was attributed to "dead water"!) until Bjerknes, a Norwegian oceanographer, explained it as due to the internal waves at the interface generated by the motion of the ship (Figure 7.29). (Note that the product of the drag

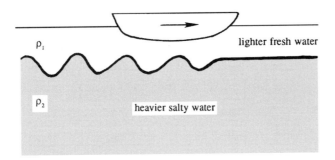

Figure 7.29   Phenomenon of "dead water" in Norwegian fjords.

times the speed of the ship gives the rate of generation of wave energy, with other
sources of resistance neglected.)

16.   *Waves in a Finite Layer Overlying an Infinitely Deep Fluid*

As a second example of an internal wave at a density discontinuity, consider the case
in which the upper layer is not infinitely thick but has a finite thickness; the lower
layer is initially assumed to be infinitely thick. The case of two infinitely deep liquids,
treated in the preceding section, is then a special case of the present situation. Whereas
only waves at the interface were allowed in the preceding section, the presence of the
free surface now allows an extra mode of surface waves. It is clear that the present
configuration will allow two modes of oscillation, one in which the free surface and
the interface are in phase and a second mode in which they are oppositely directed.

Let H be the thickness of the upper layer, and let the origin be placed at the mean
position of the free surface (Figure 7.30). The equations are

$$\frac{\partial^2 \phi_1}{\partial x^2} + \frac{\partial^2 \phi_1}{\partial z^2} = 0$$

$$\frac{\partial^2 \phi_2}{\partial x^2} + \frac{\partial^2 \phi_2}{\partial z^2} = 0,$$

subject to

$$\phi_2 \rightarrow 0 \quad \text{at} \quad z \rightarrow -\infty \tag{7.107}$$

$$\frac{\partial \phi_1}{\partial z} = \frac{\partial \eta}{\partial t} \quad \text{at} \quad z = 0 \tag{7.108}$$

$$\frac{\partial \phi_1}{\partial t} + g\eta = 0 \quad \text{at} \quad z = 0 \tag{7.109}$$

$$\frac{\partial \phi_1}{\partial z} = \frac{\partial \phi_2}{\partial z} = \frac{\partial \zeta}{\partial t} \quad \text{at} \quad z = -H \tag{7.110}$$

$$\rho_1 \frac{\partial \phi_1}{\partial t} + \rho_1 g\zeta = \rho_2 \frac{\partial \phi_2}{\partial t} + \rho_2 g\zeta \quad \text{at} \quad z = -H. \tag{7.111}$$

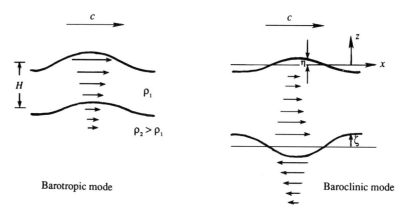

Figure 7.30   Two modes of motion of a layer of fluid overlying an infinitely deep fluid.

Assume a free surface displacement of the form

$$\eta = ae^{i(kx-\omega t)}, \tag{7.112}$$

and an interface displacement of the form

$$\zeta = be^{i(kx-\omega t)}. \tag{7.113}$$

As before, only the real part of the right-hand side is meant. Without losing generality, we can regard a as real, which means that we are considering a wave of the form $\eta = a\cos(kx - \omega t)$. The constant b should be left complex, because ζ and η may not be in phase. Solution of the problem determines such phase differences.

The velocity potentials in the layers must be of the form

$$\phi_1 = (A\,e^{kz} + B\,e^{-kz})\,e^{i(kx-\omega t)}, \tag{7.114}$$

$$\phi_2 = C\,e^{kz}\,e^{i(kx-\omega t)}. \tag{7.115}$$

The form (7.115) is chosen in order to satisfy equation (7.107). Conditions (7.108)–(7.110) give the constants in terms of the given amplitude a:

$$A = -\frac{ia}{2}\left(\frac{\omega}{k} + \frac{g}{\omega}\right), \tag{7.116}$$

$$B = \frac{ia}{2}\left(\frac{\omega}{k} - \frac{g}{\omega}\right), \tag{7.117}$$

$$C = -\frac{ia}{2}\left(\frac{\omega}{k} + \frac{g}{\omega}\right) - \frac{ia}{2}\left(\frac{\omega}{k} - \frac{g}{\omega}\right)e^{2kH}, \tag{7.118}$$

$$b = \frac{a}{2}\left(1 + \frac{gk}{\omega^2}\right)e^{-kH} + \frac{a}{2}\left(1 - \frac{gk}{\omega^2}\right)e^{kH}. \tag{7.119}$$

Substitution into equation (7.111) gives the required dispersion relation $\omega(k)$. After some algebraic manipulations, the result can be written as (Exercise 8)

$$\left(\frac{\omega^2}{gk} - 1\right)\left\{\frac{\omega^2}{gk}[\rho_1 \sinh kH + \rho_2 \cosh kH] - (\rho_2 - \rho_1)\sinh kH\right\} = 0. \quad (7.120)$$

The two possible roots of this equation are discussed in what follows.

Barotropic or Surface Mode

One possible root of equation (7.120) is

$$\omega^2 = gk, \quad (7.121)$$

which is the same as that for a deep water gravity wave. Equation (7.119) shows that in this case

$$b = ae^{-kH}, \quad (7.122)$$

implying that the amplitude at the interface is reduced from that at the surface by the factor e^{-kH}. Equation (7.122) also shows that the motions of the interface and the free surface are locked in phase; that is they go up or down simultaneously. This mode is similar to a gravity wave propagating on the free surface of the upper liquid, in which the motion decays as e^{-kz} from the free surface. It is called the *barotropic mode*, because the surfaces of constant pressure and density coincide in such a flow.

Baroclinic or Internal Mode

The other possible root of equation (7.120) is

$$\omega^2 = \frac{gk(\rho_2 - \rho_1)\sinh kH}{\rho_2 \cosh kH + \rho_1 \sinh kH}, \quad (7.123)$$

which reduces to equation (7.105) if $kH \to \infty$. Substitution of equation (7.123) into (7.119) shows that, after some straightforward algebra,

$$\eta = -\zeta\left(\frac{\rho_2 - \rho_1}{\rho_1}\right)e^{-kH}, \quad (7.124)$$

demonstrating that η and ζ have opposite signs and that the interface displacement is much larger than the surface displacement if the density difference is small. This mode of behavior is called the *baroclinic* or *internal mode* because the surfaces of constant pressure and density do not coincide. It can be shown that the horizontal velocity u changes sign across the interface. The existence of a density difference has therefore generated a motion that is quite different from the barotropic behavior. The case studied in the previous section, in which the fluids have infinite depth and no free surface, has only a baroclinic mode and no barotropic mode.

17. Shallow Layer Overlying an Infinitely Deep Fluid

A very common simplification, frequently made in geophysical situations in which large-scale motions are considered, involves assuming that the wavelengths are large compared to the upper layer depth. For example, the depth of the oceanic upper layer, below which there is a sharp density gradient, could be $\approx 50\,\text{m}$ thick, and we may be interested in interfacial waves that are much longer than this. The approximation $kH \ll 1$ is called the *shallow-water* or *long-wave approximation*. Using

$$\sinh kH \simeq kH,$$
$$\cosh kH \simeq 1,$$

the dispersion relation (7.123) corresponding to the baroclinic mode reduces to

$$\omega^2 = \frac{k^2 g H (\rho_2 - \rho_1)}{\rho_2}. \tag{7.125}$$

The phase velocity of waves at the interface is therefore

$$\boxed{c = \sqrt{g'H},} \tag{7.126}$$

where we have defined

$$\boxed{g' \equiv g\left(\frac{\rho_2 - \rho_1}{\rho_2}\right),} \tag{7.127}$$

which is called the *reduced gravity*. Equation (7.126) is similar to the corresponding expression for *surface* waves in a shallow homogeneous layer of thickness H, namely, $c = \sqrt{gH}$, except that its speed is reduced by the factor $\sqrt{(\rho_2 - \rho_1)/\rho_2}$. This agrees with our previous conclusion that internal waves generally propagate slower than surface waves. Under the shallow-water approximation, equation (7.124) reduces to

$$\eta = -\zeta\left(\frac{\rho_2 - \rho_1}{\rho_1}\right). \tag{7.128}$$

In Section 6 we noted that, for surface waves, the shallow-water approximation is equivalent to the hydrostatic approximation, and results in a depth-independent horizontal velocity. Such a conclusion also holds for interfacial waves. The fact that u_1 is independent of z follows from equation (7.114) on noting that $e^{kz} \simeq e^{-kz} \simeq 1$. To see that pressure is hydrostatic, the perturbation pressure in the upper layer determined from equation (7.114) is

$$p' = -\rho_1 \frac{\partial \phi_1}{\partial t} = i\rho_1 \omega (A + B)\, e^{i(kx - \omega t)} = \rho_1 g \eta, \tag{7.129}$$

where the constants given in equations (7.116) and (7.117) have been used. This shows that p' is independent of z and equals the hydrostatic pressure change due to the free surface displacement.

So far, the lower fluid has been assumed to be infinitely deep, resulting in an exponential decay of the flow field from the interface into the lower layer, with a decay scale of the order of the wavelength. If the lower layer is now considered thin compared to the wavelength, then the horizontal velocity will be depth independent, and the flow hydrostatic, in the lower layer. If *both* layers are considered thin compared to the wavelength, then the flow is hydrostatic (and the horizontal velocity field depth-independent) in *both* layers. This is the *shallow-water* or *long-wave approximation* for a two-layer fluid. In such a case the horizontal velocity field in the barotropic mode has a discontinuity at the interface, which vanishes in the Boussinesq limit $(\rho_2 - \rho_1)/\rho_1 \ll 1$. Under these conditions the two modes of a two-layer system have a simple structure (Figure 7.31): a barotropic mode in which the horizontal velocity is depth independent across the entire water column; and a baroclinic mode in which the horizontal velocity is directed in opposite directions in the two layers (but is depth independent in each layer).

We shall now summarize the results of interfacial waves presented in the preceding three sections. In the case of two infinitely deep fluids, only the baroclinic mode is possible, and it has a frequency of $\omega = \varepsilon\sqrt{gk}$. If the upper layer has finite thickness, then both baroclinic and barotropic modes are possible. In the barotropic mode, η and ζ are in phase, and the flow decreases exponentially away from the *free surface*. In the baroclinic mode, η and ζ are out of phase, the horizontal velocity changes direction across the interface, and the motion decreases exponentially away from the *interface*. If we also make the long-wave approximation for the upper layer, then the phase speed of interfacial waves in the baroclinic mode is $c = \sqrt{g'H}$, the fluid velocity in the upper layer is almost horizontal and depth independent, and the pressure in the upper layer is hydrostatic. If both layers are shallow, then the flow is depth independent and hydrostatic in both layers; the two modes in such a system have the simple structure shown in Figure 7.31.

18. Equations of Motion for a Continuously Stratified Fluid

We have considered surface gravity waves and internal gravity waves at a density discontinuity between two fluids. Internal waves also exist if the fluid is continuously

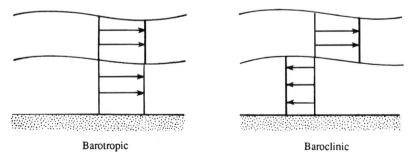

Barotropic   Baroclinic

Figure 7.31 Two modes of motion in a shallow-water, two-layer system in the Boussinesq limit.

stratified, in which the vertical density profile in a state of rest is a continuous function $\bar{\rho}(z)$. The equations of motion for internal waves in such a medium will be derived in this section, starting with the Boussinesq set (4.89) presented in Chapter 4. The Boussinesq approximation treats density as constant, except in the vertical momentum equation. We shall assume that the wave motion is inviscid. The amplitudes will be assumed to be small, in which case the nonlinear terms can be neglected. We shall also assume that the frequency of motion is much larger than the Coriolis frequency, which therefore does not affect the motion. Effects of the earth's rotation are considered in Chapter 14. The set (4.89) then simplifies to

$$\frac{\partial u}{\partial t} = -\frac{1}{\rho_0}\frac{\partial p}{\partial x}, \tag{7.130}$$

$$\frac{\partial v}{\partial t} = -\frac{1}{\rho_0}\frac{\partial p}{\partial y}, \tag{7.131}$$

$$\frac{\partial w}{\partial t} = -\frac{1}{\rho_0}\frac{\partial p}{\partial z} - \frac{\rho g}{\rho_0}, \tag{7.132}$$

$$\frac{D\rho}{Dt} = 0, \tag{7.133}$$

$$\frac{\partial u}{\partial x} + \frac{\partial v}{\partial y} + \frac{\partial w}{\partial z} = 0, \tag{7.134}$$

where ρ_0 is a constant reference density. As noted in Chapter 4, the equation $D\rho/Dt = 0$ is *not* an expression of conservation of mass, which is expressed by $\nabla \cdot \mathbf{u} = 0$ in the Boussinesq approximation. Rather, it expresses incompressibility of a fluid particle. If temperature is the only agency that changes the density, then $D\rho/Dt = 0$ follows from the heat equation in the nondiffusive form $DT/Dt = 0$ and an incompressible (that is, ρ is not a function of p) equation of state in the form $\delta\rho/\rho = -\alpha\,\delta T$, where α is the coefficient of thermal expansion. If the density changes are due to changes in the concentration S of a constituent, for example salinity in the ocean or water vapor in the atmosphere, then $D\rho/Dt = 0$ follows from $DS/Dt = 0$ (the nondiffusive form of conservation of the constituent) and an incompressible equation of state in the form of $\delta\rho/\rho = \beta\,\delta S$, where β is the coefficient describing how the density changes due to concentration of the constituent. In both cases, the principle underlying the equation $D\rho/Dt = 0$ is an incompressible equation of state. In terms of common usage, this equation is frequently called the "density equation," as opposed to the continuity equation $\nabla \cdot \mathbf{u} = 0$.

The equation set (7.130)–(7.134) consists of five equations in five unknowns (u, v, w, p, ρ). We first express the equations in terms of *changes* from a state of rest. That is, we assume that the flow is superimposed on a "background" state in which the density $\bar{\rho}(z)$ and pressure $\bar{p}(z)$ are in hydrostatic balance:

$$0 = -\frac{1}{\rho_0}\frac{d\bar{p}}{dz} - \frac{\bar{\rho}g}{\rho_0}. \tag{7.135}$$

When the motion develops, the pressure and density change to

$$p = \bar{p}(z) + p',$$

$$\rho = \bar{\rho}(z) + \rho'. \tag{7.136}$$

The density equation (7.133) then becomes

$$\frac{\partial}{\partial t}(\bar{\rho} + \rho') + u\frac{\partial}{\partial x}(\bar{\rho} + \rho') + v\frac{\partial}{\partial y}(\bar{\rho} + \rho') + w\frac{\partial}{\partial z}(\bar{\rho} + \rho') = 0. \tag{7.137}$$

Here, $\partial\bar{\rho}/\partial t = \partial\bar{\rho}/\partial x = \partial\bar{\rho}/\partial y = 0$. The nonlinear terms in the second, third, and fourth terms (namely, $u\,\partial\rho'/\partial x$, $v\,\partial\rho'/\partial y$, and $w\,\partial\rho'/\partial z$) are also negligible for small amplitude motions. The *linear* part of the fourth term, that is, $w\,d\bar{\rho}/dz$, represents a very important process and must be retained. Equation (7.137) then simplifies to

$$\frac{\partial\rho'}{\partial t} + w\frac{d\bar{\rho}}{dz} = 0, \tag{7.138}$$

which states that the density perturbation at a point is generated only by the vertical advection of the *background* density distribution. This is the linearized form of equation (7.133), with the vertical advection of density retained in a linearized form. We now introduce the definition

$$N^2 \equiv -\frac{g}{\rho_0}\frac{d\bar{\rho}}{dz}. \tag{7.139}$$

Here, $N(z)$ has the units of frequency (rad/s) and is called the *Brunt–Väisälä frequency* or *buoyancy frequency*. It plays a fundamental role in the study of stratified flows. We shall see in the next section that it has the significance of being the frequency of oscillation if a fluid particle is *vertically* displaced.

After substitution of equation (7.136), the equations of motion (7.130)–(7.134) become

$$\frac{\partial u}{\partial t} = -\frac{1}{\rho_0}\frac{\partial p'}{\partial x}, \tag{7.140}$$

$$\frac{\partial v}{\partial t} = -\frac{1}{\rho_0}\frac{\partial p'}{\partial y}, \tag{7.141}$$

$$\frac{\partial w}{\partial t} = -\frac{1}{\rho_0}\frac{\partial p'}{\partial z} - \frac{\rho'g}{\rho_0}, \tag{7.142}$$

$$\frac{\partial\rho'}{\partial t} - \frac{N^2\rho_0}{g}w = 0, \tag{7.143}$$

$$\frac{\partial u}{\partial x} + \frac{\partial v}{\partial y} + \frac{\partial w}{\partial z} = 0. \tag{7.144}$$

In deriving this set we have also used equation (7.135) and replaced the density equation by its linearized form (7.138). Comparing the sets (7.130)–(7.134) and (7.140)–(7.144), we see that the *equations satisfied by the perturbation density and pressure are identical to those satisfied by the total ρ and p.*

In deriving the equations for a stratified fluid, we have assumed that ρ is a function of temperature T and concentration S of a constituent, but not of pressure. At first this does not seem to be a good assumption. The compressibility effects in the atmosphere are certainly not negligible; even in the ocean the density changes due to the huge changes in the background pressure are as much as 4%, which is ≈ 10 times the density changes due to the variations of the salinity and temperature. The effects of compressibility, however, can be handled within the Boussinesq approximation if we regard $\bar{\rho}$ in the definition of N as the background *potential density*, that is the density distribution from which the adiabatic changes of density due to the changes of pressure have been subtracted out. The concept of potential density is explained in Chapter 1. Oceanographers account for compressibility effects by converting all their density measurements to the standard atmospheric pressure; thus, when they report variations in density (what they call "sigma tee") they are generally reporting variations due only to changes in temperature and salinity.

A useful equation for stratified flows is the one involving only w. The u and v can be eliminated by taking the time derivative of the continuity equation (7.144) and using the horizontal momentum equations (7.140) and (7.141). This gives

$$\frac{1}{\rho_0} \nabla_H^2 p' = \frac{\partial^2 w}{\partial z\, \partial t}, \tag{7.145}$$

where $\nabla_H^2 \equiv \partial^2/\partial x^2 + \partial^2/\partial y^2$ is the *horizontal* Laplacian operator. Elimination of ρ' from equations (7.142) and (7.143) gives

$$\frac{1}{\rho_0} \frac{\partial^2 p'}{\partial t\, \partial z} = -\frac{\partial^2 w}{\partial t^2} - N^2 w. \tag{7.146}$$

Finally, p' can be eliminated by taking ∇_H^2 of equation (7.146), and using equation (7.145). This gives

$$\frac{\partial^2}{\partial t\, \partial z}\left(\frac{\partial^2 w}{\partial t\, \partial z}\right) = -\nabla_H^2\left(\frac{\partial^2 w}{\partial t^2} + N^2 w\right),$$

which can be written as

$$\frac{\partial^2}{\partial t^2}\nabla^2 w + N^2 \nabla_H^2 w = 0, \tag{7.147}$$

where $\nabla^2 \equiv \partial^2/\partial x^2 + \partial^2/\partial y^2 + \partial^2/\partial z^2 = \nabla_H^2 + \partial^2/\partial z^2$ is the three-dimensional Laplacian operator. The w-equation will be used in the following section to derive the dispersion relation for internal gravity waves.

19. *Internal Waves in a Continuously Stratified Fluid*

In this chapter we have considered gravity waves at the surface or at a density discontinuity; these waves propagate only in the horizontal direction. Because every horizontal direction is alike, such waves are *isotropic*, in which only the magnitude of the wavenumber vector matters. By taking the x-axis along the direction of wave propagation, we obtained a dispersion relation $\omega(k)$ that depends only on the magnitude of the wavenumber. We found that phases and groups propagate in the same direction, although at different speeds. If, on the other hand, the fluid is *continuously* stratified, then the internal waves can propagate in any direction, at any angle to the vertical. In such a case the *direction* of the wavenumber vector becomes important. Consequently, we can no longer treat the wavenumber, phase velocity, and group velocity as scalars.

Any flow variable q can now be written as

$$q = q_0 \, e^{i(kx+ly+mz-\omega t)} = q_0 \, e^{i(\mathbf{K} \cdot \mathbf{x}-\omega t)},$$

where q_0 is the amplitude and $\mathbf{K} = (k, l, m)$ is the wavenumber vector with components k, l, and m in the three Cartesian directions. We expect that in this case the direction of wave propagation should matter because horizontal directions are basically different from the vertical direction, along which the all-important gravity acts. Internal waves in a continuously stratified fluid are therefore *anisotropic*, for which the frequency is a function of all three components of $\mathbf{K}$. This can be written in the following two ways:

$$\omega = \omega(k, l, m) = \omega(\mathbf{K}). \tag{7.148}$$

However, the waves are still *horizontally* isotropic because the dependence of the wave field on k and l is similar, although the dependence on k and m is dissimilar.

The propagation of internal waves is a baroclinic process, in which the surfaces of constant pressure do not coincide with the surfaces of constant density. It was shown in Section 5.4, in connection with the demonstration of Kelvin's circulation theorem, that baroclinic processes generate vorticity. *Internal waves in a continuously stratified fluid are therefore not irrotational. Waves at a density interface constitute a limiting case in which all the vorticity is concentrated in the form of a velocity discontinuity at the interface. The Laplace equation can therefore be used to describe the flow field within each layer. However, internal waves in a continuously stratified fluid cannot be described by the Laplace equation.*

The first task is to derive the dispersion relation. We shall simplify the analysis by assuming that N is depth independent, an assumption that may seem unrealistic at first. In the ocean, for example, N is large at a depth of ≈ 200 m and small elsewhere (see Figure 14.2). Figure 14.2 shows that $N < 0.01$ everywhere but N is largest between ≈ 200 m and 2 km. However, the results obtained by treating N as constant are *locally* valid if N varies slowly over the vertical wavelength $2\pi/m$ of the motion. The so-called *WKB approximation* of internal waves, in which such a slow variation of $N(z)$ is not neglected, is discussed in Chapter 14.

Consider a wave propagating in three dimensions, for which the vertical velocity is

$$w = w_0\, e^{i(kx+ly+mz-\omega t)}, \tag{7.149}$$

where w_0 is the amplitude of fluctuations. Substituting into the governing equation

$$\frac{\partial^2}{\partial t^2}\nabla^2 w + N^2 \nabla_H^2 w = 0, \tag{7.147}$$

gives the dispersion relation

$$\omega^2 = \frac{k^2 + l^2}{k^2 + l^2 + m^2}N^2. \tag{7.150}$$

For simplicity of discussion *we shall orient the xz-plane so as to contain the wavenumber vector* **K**. No generality is lost by doing this because the medium is horizontally isotropic. For this choice of reference axes we have $l = 0$; that is, the wave motion is two dimensional and invariant in the y-direction, and k represents the entire horizontal wavenumber. We can then write equation (7.150) as

$$\omega = \frac{kN}{\sqrt{k^2 + m^2}} = \frac{kN}{K}. \tag{7.151}$$

This is the dispersion relation for internal gravity waves and can also be written as

$$\boxed{\omega = N\cos\theta,} \tag{7.152}$$

where θ is the angle between the phase velocity vector **c** (and therefore **K**) and the horizontal direction (Figure 7.32). It follows that the frequency of an internal wave in a stratified fluid depends only on the *direction* of the wavenumber vector and not on the magnitude of the wavenumber. This is in sharp contrast with surface and interfacial gravity waves, for which frequency depends only on the magnitude. The frequency lies in the range $0 < \omega < N$, revealing one important significance of the buoyancy frequency: *N is the maximum possible frequency of internal waves in a stratified fluid.*

Before discussing the dispersion relation further, let us explore particle motion in an incompressible internal wave. The fluid motion can be written as

$$u = u_0\, e^{i(kx+ly+mz-\omega t)}, \tag{7.153}$$

plus two similar expressions for v and w. This gives

$$\frac{\partial u}{\partial x} = iku_0\, e^{i(kx+ly+mz-\omega t)} = iku.$$

The continuity equation then requires that $ku + lv + mw = 0$, that is,

$$\boxed{\mathbf{K}\cdot\mathbf{u} = 0,} \tag{7.154}$$

showing that *particle motion is perpendicular to the wavenumber vector* (Figure 7.32). Note that only two conditions have been used to derive this result, namely the incompressible continuity equation and a trigonometric behavior in *all* spatial directions. As such, the result is valid for many other wave systems that meet these two conditions. These waves are called *shear waves* (or transverse waves) because the fluid moves parallel to the constant phase lines. Surface or interfacial gravity waves do not have this property because the field varies *exponentially* in the vertical.

We can now interpret θ in the dispersion relation (7.152) as the angle between the particle motion and the *vertical* direction (Figure 7.32). The maximum frequency $\omega = N$ occurs when $\theta = 0$, that is, when the particles move up and down vertically. This case corresponds to $m = 0$ (see equation (7.151)), showing that the motion is independent of the z-coordinate. The resulting motion consists of a series of vertical columns, all oscillating at the buoyancy frequency N, the flow field varying in the horizontal direction only.

The $w = 0$ Limit

At the opposite extreme we have $\omega = 0$ when $\theta = \pi/2$, that is, when the particle motion is completely horizontal. In this limit our internal wave solution (7.151) would seem to require $k = 0$, that is, horizontal independence of the motion. However, such a conclusion is not valid; pure horizontal motion is not a limiting case of internal waves, and it is necessary to examine the basic equations to draw any conclusion for

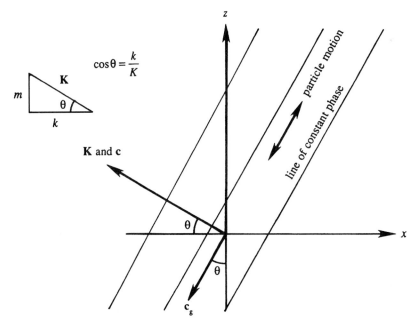

Figure 7.32 Basic parameters of internal waves. Note that **c** and **c**$_g$ are at right angles and have opposite vertical components.

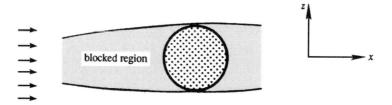

Figure 7.33 Blocking in strongly stratified flow. The circular region represents a two-dimensional body with its axis along the y direction.

this case. An examination of the governing set (7.140)–(7.144) shows that a possible steady solution is $w = p' = \rho' = 0$, with u and v *any* functions of x and y satisfying

$$\frac{\partial u}{\partial x} + \frac{\partial v}{\partial y} = 0. \tag{7.155}$$

The z-dependence of u and v is arbitrary. The motion is therefore two-dimensional in the horizontal plane, with the motion in the various horizontal planes decoupled from each other. This is why clouds in the upper atmosphere seem to move in flat horizontal sheets, as often observed in airplane flights (Gill, 1982). For a similar reason a cloud pattern pierced by a mountain peak sometimes shows *Karman vortex streets*, a two-dimensional feature; see the striking photograph in Figure 10.20. A restriction of strong stratification is necessary for such almost horizontal flows, for equation (7.143) suggests that the vertical motion is small if N is large.

The foregoing discussion leads to the interesting phenomenon of *blocking* in a strongly stratified fluid. Consider a two-dimensional body placed in such a fluid, with its axis horizontal (Figure 7.33). The two dimensionality of the body requires $\partial v/\partial y = 0$, so that the continuity equation (7.155) reduces to $\partial u/\partial x = 0$. A horizontal layer of fluid ahead of the body, bounded by tangents above and below it, is therefore blocked. (For photographic evidence see Figure 3.18 in the book by Turner (1973).) This happens because the strong stratification suppresses the w field and prevents the fluid from going around and over the body.

20. *Dispersion of Internal Waves in a Stratified Fluid*

In the case of isotropic gravity waves at a free surface and at a density discontinuity, we found that $\mathbf{c}$ and $\mathbf{c}_g$ are in the same direction, although their magnitudes can be different. This conclusion is no longer valid for the anisotropic internal waves in a continuously stratified fluid. In fact, as we shall see shortly, they are *perpendicular* to each other, violating all our intuitions acquired by observing surface gravity waves!

In three dimensions, the definition $c_g = d\omega/dk$ has to be generalized to

$$\mathbf{c}_g = \mathbf{i}_x \frac{\partial \omega}{\partial k} + \mathbf{i}_y \frac{\partial \omega}{\partial l} + \mathbf{i}_z \frac{\partial \omega}{\partial m}, \tag{7.156}$$

where $\mathbf{i}_x, \mathbf{i}_y, \mathbf{i}_z$ are the unit vectors in the three Cartesian directions. As in the preceding section, we orient the xz-plane so that the wavenumber vector $\mathbf{K}$ lies in this plane and $l = 0$. Substituting equation (7.151), this gives

$$\mathbf{c}_g = \frac{Nm}{K^3}(\mathbf{i}_x m - \mathbf{i}_z k).$$ (7.157)

The phase velocity is

$$\mathbf{c} = \frac{\omega}{K}\frac{\mathbf{K}}{K} = \frac{\omega}{K^2}(\mathbf{i}_x k + \mathbf{i}_z m),$$ (7.158)

where $\mathbf{K}/K$ represents the unit vector in the direction of $\mathbf{K}$. (Note that $\mathbf{c} \neq \mathbf{i}_x(\omega/k) + \mathbf{i}_z(\omega/m)$, as explained in Section 3.) It follows from equations (7.157) and (7.158) that

$$\boxed{\mathbf{c}_g \cdot \mathbf{c} = 0,}$$ (7.159)

showing that *phase and group velocity vectors are perpendicular.*

Equations (7.157) and (7.158) show that the horizontal components of $\mathbf{c}$ and $\mathbf{c}_g$ are in the same direction, while their vertical components are equal and opposite. In fact, $\mathbf{c}$ and $\mathbf{c}_g$ form two sides of a right-angled triangle whose hypotenuse is horizontal (Figure 7.34). Consequently, the phase velocity has an upward component when the group velocity has a downward component, and vice versa. Equations (7.154) and (7.159) are consistent because $\mathbf{c}$ and $\mathbf{K}$ are parallel and $\mathbf{c}_g$ and $\mathbf{u}$ are parallel. The fact that $\mathbf{c}$ and $\mathbf{c}_g$ are perpendicular, and have opposite vertical components, is illustrated in Figure 7.35. It shows that the phase lines are propagating toward the left and upward, whereas the wave groups are propagating to the left and downward. Wave crests are constantly appearing at one edge of the group, propagating through the group, and vanishing at the other edge.

The group velocity here has the usual significance of being the velocity of propagation of energy of a certain sinusoidal component. Suppose a source is oscillating at frequency ω. Then its energy will only be found radially outward along four beams oriented at an angle θ with the vertical, where $\cos\theta = \omega/N$. This has been verified in a laboratory experiment (Figure 7.36). The source in this case was a vertically oscillating cylinder with its axis perpendicular to the plane of paper. The frequency was $\omega < N$. The light and dark lines in the photograph are lines of constant density, made visible by an optical technique. The experiment showed that the energy radiated

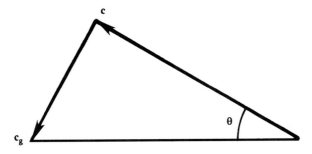

Figure 7.34   Orientation of phase and group velocity in internal waves.

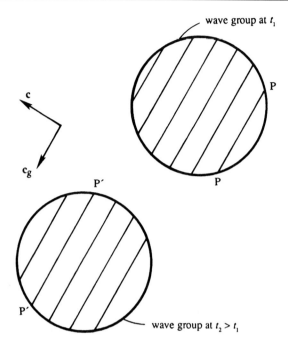

Figure 7.35   Illustration of phase and group propagation in internal waves. Positions of a wave group at two times are shown. The phase line PP at time t_1 propagates to P′P′ at t_2.

along four beams that became more vertical as the frequency was increased, which agrees with $\cos\theta = \omega/N$.

21. Energy Considerations of Internal Waves in a Stratified Fluid

In this section we shall derive the various commonly used expressions for potential energy of a continuously stratified fluid, and show that they are equivalent. We then show that the energy flux $\overline{p'u}$ is $\mathbf{c}_g$ times the wave energy.

A mechanical energy equation for internal waves can be derived from equations (7.140)–(7.142) by multiplying the first equation by $\rho_0 u$, the second by $\rho_0 v$, the third by $\rho_0 w$, and summing the results. This gives

$$\frac{\partial}{\partial t}\left[\frac{1}{2}\rho_0(u^2 + v^2 + w^2)\right] + g\rho'w + \nabla \cdot (p'\mathbf{u}) = 0. \qquad (7.160)$$

Here the continuity equation has been used to write $u\,\partial p'/\partial x + v\,\partial p'/\partial y + w\,\partial p'/\partial z = \nabla \cdot (p'\mathbf{u})$, which represents the net work done by pressure forces. Another interpretation is that $\nabla \cdot (p'\mathbf{u})$ is the divergence of the *energy flux* $p'\mathbf{u}$, which must change the wave energy at a point. As the first term in equation (7.160) is the rate of change of kinetic energy, we can anticipate that the second term $g\rho'w$ must be the rate of change of potential energy. This is consistent with the energy principle derived in

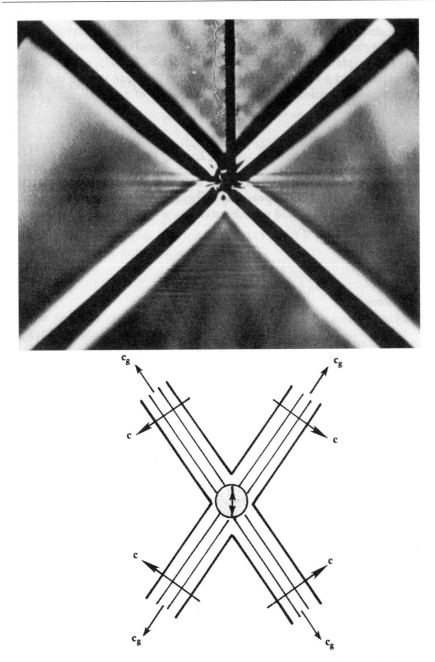

Figure 7.36 Waves generated in a stratified fluid of uniform buoyancy frequency $N = 1$ rad/s. The forcing agency is a horizontal cylinder, with its axis perpendicular to the plane of the paper, oscillating vertically at frequency $\omega = 0.71$ rad/s. With $\omega/N = 0.71 = \cos\theta$, this agrees with the observed angle of $\theta = 45°$ made by the beams with the horizontal. The vertical dark line in the upper half of the photograph is the cylinder support and should be ignored. The light and dark radial lines represent contours of constant ρ' and are therefore constant phase lines. The schematic diagram below the photograph shows the directions of **c** and **c**$_g$ for the four beams. Reprinted with the permission of Dr. T. Neil Stevenson, University of Manchester.

Chapter 4 (see equation (4.62)), except that ρ' and p' replace ρ and p because we have subtracted the mean state of rest here. Using the density equation (7.143), the rate of change of potential energy can be written as

$$\frac{\partial E_p}{\partial t} = g\rho'w = \frac{\partial}{\partial t}\left[\frac{g^2\rho'^2}{2\rho_0 N^2}\right], \tag{7.161}$$

which shows that the potential energy per unit volume must be the positive quantity $E_p = g^2\rho'^2/2\rho_0 N^2$. The potential energy can also be expressed in terms of the displacement ζ of a fluid particle, given by $w = \partial\zeta/\partial t$. Using the density equation (7.143), we can write

$$\frac{\partial\rho'}{\partial t} = \frac{N^2\rho_0}{g}\frac{\partial\zeta}{\partial t},$$

which requires that

$$\rho' = \frac{N^2\rho_0\zeta}{g}. \tag{7.162}$$

The potential energy *per unit volume* is therefore

$$E_p = \frac{g^2\rho'^2}{2\rho_0 N^2} = \frac{1}{2}N^2\rho_0\zeta^2. \tag{7.163}$$

This expression is consistent with our previous result from equation (7.106) for two infinitely deep fluids, for which the average potential energy of the entire water column *per unit horizontal area* was shown to be

$$\tfrac{1}{4}(\rho_2 - \rho_1)ga^2, \tag{7.164}$$

where the interface displacement is of the form $\zeta = a\cos(kx - \omega t)$ and $(\rho_2 - \rho_1)$ is the density discontinuity. To see the consistency, we shall symbolically represent the buoyancy frequency of a density discontinuity at $z = 0$ as

$$N^2 = -\frac{g}{\rho_0}\frac{d\bar{\rho}}{dz} = \frac{g}{\rho_0}(\rho_2 - \rho_1)\delta(z), \tag{7.165}$$

where $\delta(z)$ is the Dirac delta function. (As with other relations involving the delta function, equation (7.165) is valid in the *integral* sense, that is, the integral (across the origin) of the last two terms is equal because $\int \delta(z)\,dz = 1$.) Using equation (7.165), a vertical integral of equation (7.163), coupled with horizontal averaging over a wavelength, gives equation (7.164). Note that for surface or interfacial waves E_k and E_p represent kinetic and potential energies of the entire water column, per unit horizontal area. In a continuously stratified fluid, they represent energies per unit volume.

We shall now demonstrate that the average kinetic and potential energies are equal for internal wave motion. Substitute periodic solutions

$$[u, w, p', \rho'] = [\hat{u}, \hat{w}, \hat{p}, \hat{\rho}]\,e^{i(kx+mz-\omega t)}.$$

Then all variables can be expressed in terms of w:

$$p' = -\frac{\omega m \rho_0}{k^2} \hat{w} \, e^{i(kx+mz-\omega t)},$$

$$\rho' = \frac{i N^2 \rho_0}{\omega g} \hat{w} \, e^{i(kx+mz-\omega t)}, \qquad (7.166)$$

$$u = -\frac{m}{k} \hat{w} \, e^{i(kx+mz-\omega t)},$$

where p' is derived from equation (7.145), ρ' from equation (7.143), and u from equation (7.140). The average kinetic energy per unit volume is therefore

$$E_k = \frac{1}{2}\rho_0 \overline{(u^2 + w^2)} = \frac{1}{4}\rho_0 \left(\frac{m^2}{k^2} + 1\right) \hat{w}^2, \qquad (7.167)$$

where we have used the fact that the average of $\cos^2 x$ over a wavelength is $1/2$. The average potential energy per unit volume is

$$E_p = \frac{g^2 \overline{\rho'^2}}{2\rho_0 N^2} = \frac{N^2 \rho_0}{4\omega^2} \hat{w}^2, \qquad (7.168)$$

where we have used $\overline{\rho'^2} = \hat{w}^2 N^4 \rho_0^2 / 2\omega^2 g^2$, found from equation (7.166) after taking its real part. Use of the dispersion relation $\omega^2 = k^2 N^2 / (k^2 + m^2)$ shows that

$$E_k = E_p, \qquad (7.169)$$

which is a general result for small oscillations of a conservative system without Coriolis forces. The total wave energy is

$$E = E_k + E_p = \frac{1}{2}\rho_0 \left(\frac{m^2}{k^2} + 1\right) \hat{w}^2. \qquad (7.170)$$

Last, we shall show that $\mathbf{c}_g$ times the wave energy equals the energy flux. The average energy flux across a unit area can be found from equation (7.166):

$$\mathbf{F} = \overline{p'\mathbf{u}} = \mathbf{i}_x \overline{p'u} + \mathbf{i}_z \overline{p'w} = \frac{\rho_0 \omega m \hat{w}^2}{2k^2} \left(\mathbf{i}_x \frac{m}{k} - \mathbf{i}_z\right). \qquad (7.171)$$

Using equations (7.157) and (7.170), group velocity times wave energy is

$$\mathbf{c}_g E = \frac{Nm}{K^3}[\mathbf{i}_x m - \mathbf{i}_z k]\left[\frac{\rho_0}{2}\left(\frac{m^2}{k^2} + 1\right)\hat{w}^2\right],$$

which reduces to equation (7.171) on using the dispersion relation (7.151). It follows that

$$\boxed{\mathbf{F} = \mathbf{c}_g E.} \qquad (7.172)$$

This result also holds for surface or interfacial gravity waves. However, in that case **F** represents the flux per unit width perpendicular to the propagation direction (integrated over the entire depth), and E represents the energy per unit horizontal area. In equation (7.172), on the other hand, **F** is the flux per unit area, and E is the energy per unit volume.

Exercises

1. Consider stationary surface gravity waves in a rectangular container of length L and breadth b, containing water of undisturbed depth H. Show that the velocity potential

$$\phi = A \cos(m\pi x/L) \cos(n\pi y/b) \cosh k(z + H) e^{-i\omega t},$$

satisfies $\nabla^2 \phi = 0$ and the wall boundary conditions, if

$$(m\pi/L)^2 + (n\pi/b)^2 = k^2.$$

Here m and n are integers. To satisfy the free surface boundary condition, show that the allowable frequencies must be

$$\omega^2 = gk \tanh kH.$$

[*Hint*: combine the two boundary conditions (7.27) and (7.32) into a single equation $\partial^2 \phi / \partial t^2 = -g \, \partial \phi / \partial z$ at $z = 0$.]

2. This is a continuation of Exercise 1. A lake has the following dimensions

$$L = 30 \, \text{km} \quad b = 2 \, \text{km} \quad H = 100 \, \text{m}.$$

Suppose the relaxation of wind sets up the mode $m = 1$ and $n = 0$. Show that the period of the oscillation is 31.7 min.

3. Show that the group velocity of pure capillary waves in deep water, for which the gravitational effects are negligible, is

$$c_g = \tfrac{3}{2}c.$$

4. Plot the group velocity of surface gravity waves, including surface tension σ, as a function of λ. Assuming deep water, show that the group velocity is

$$c_g = \frac{1}{2}\sqrt{\frac{g}{k}} \frac{1 + 3\sigma k^2/\rho g}{\sqrt{1 + \sigma k^2/\rho g}}.$$

Show that this becomes minimum at a wavenumber given by

$$\frac{\sigma k^2}{\rho g} = \frac{2}{\sqrt{3}} - 1.$$

For water at 20 °C ($\rho = 1000 \, \text{kg/m}^3$ and $\sigma = 0.074 \, \text{N/m}$), verify that $c_{g \, \text{min}} = 17.8 \, \text{cm/s}$.

5. A *thermocline* is a thin layer in the upper ocean across which temperature and, consequently, density change rapidly. Suppose the thermocline in a very deep ocean is at a depth of 100 m from the ocean surface, and that the temperature drops across it from 30 to 20 °C. Show that the reduced gravity is $g' = 0.025 \, \text{m/s}^2$. Neglecting Coriolis effects, show that the speed of propagation of long gravity waves on such a thermocline is 1.58 m/s.

6. Consider internal waves in a continuously stratified fluid of buoyancy frequency $N = 0.02 \, \text{s}^{-1}$ and average density $800 \, \text{kg/m}^3$. What is the direction of ray paths if the frequency of oscillation is $\omega = 0.01 \, \text{s}^{-1}$? Find the energy flux per unit area if the amplitude of vertical velocity is $\hat{w} = 1$ cm/s and the horizontal wavelength is π meters.

7. Consider internal waves at a density interface between two infinitely deep fluids. Using the expressions given in Section 15, show that the average kinetic energy per unit horizontal area is $E_k = (\rho_2 - \rho_1)ga^2/4$. This result was quoted but not proved in Section 15.

8. Consider waves in a finite layer overlying an infinitely deep fluid, discussed in Section 16. Using the constants given in equations (7.116)–(7.119), prove the dispersion relation (7.120).

9. Solve the equation governing spherical waves $\partial^2 p/\partial t^2 = (c^2/r^2)(\partial/\partial r)$ $(r^2 \partial p/\partial r)$ subject to the initial conditions: $p(r, 0) = e^{-r}$, $(\partial p/\partial t)(r, 0) = 0$.

Literature Cited

Gill, A. (1982). *Atmosphere–Ocean Dynamics*, New York: Academic Press.
Kinsman, B. (1965). *Wind Waves*, Englewood Cliffs, New Jersey: Prentice-Hall.
LeBlond, P. H. and L. A. Mysak (1978). *Waves in the Ocean*, Amsterdam: Elsevier Scientific Publishing.
Liepmann, H. W. and A. Roshko (1957). *Elements of Gasdynamics*, New York: Wiley.
Lighthill, M. J. (1978). *Waves in Fluids*, London: Cambridge University Press.
Phillips, O. M. (1977). *The Dynamics of the Upper Ocean*, London: Cambridge University Press.
Turner, J. S. (1973). *Buoyancy Effects in Fluids*, London: Cambridge University Press.
Whitham, G. B. (1974). *Linear and Nonlinear Waves*, New York: Wiley.

Dynamic Similarity

1. Introduction

Two flows having different values of length scales, flow speeds, or fluid properties can apparently be different but still "dynamically similar". Exactly what is meant by dynamic similarity will be explained later in this chapter. At this point it is only necessary to know that in a class of dynamically similar flows we can predict flow properties if we have experimental data on one of them. In this chapter, we shall determine circumstances under which two flows can be dynamically similar to one another. We shall see that equality of certain relevant nondimensional parameters is a requirement for dynamic similarity. What these nondimensional parameters should be depends on the nature of the problem. For example, one nondimensional parameter must involve the fluid viscosity if the viscous effects are important in the problem.

The principle of dynamic similarity is at the heart of experimental fluid mechanics, in which the data should be unified and presented in terms of nondimensional parameters. The concept of similarity is also indispensable for designing models in which tests can be conducted for predicting flow properties of full-scale objects such as aircraft, submarines, and dams. An understanding of dynamic similarity is also important in theoretical fluid mechanics, especially when simplifications

©2010 Elsevier Inc. All rights reserved.
DOI: 10.1016/B978-0-12-381399-2.50008-3

are to be made. Under various limiting situations certain variables can be eliminated from our consideration, resulting in very useful relationships in which only the constants need to be determined from experiments. Such a procedure is used extensively in turbulence theory, and leads, for example, to the well-known $K^{-5/3}$ spectral law discussed in Chapter 13. Analogous arguments (applied to a different problem) are presented in Section 5 of the present chapter.

Nondimensional parameters for a problem can be determined in two ways. They can be deduced directly from the governing differential equations if these equations are known; this method is illustrated in the next section. If, on the other hand, the governing differential equations are unknown, then the nondimensional parameters can be determined by performing a simple dimensional analysis on the variables involved. This method is illustrated in Section 4.

The formulation of all problems in fluid mechanics is in terms of the conservation laws (mass, momentum, and energy), constitutive equations and equations of state to define the fluid, and boundary conditions to specify the problem. Most often, the conservation laws are written as partial differential equations and the conservation of momentum and energy may include the constitutive equations for stress and heat flux, respectively. Each term in the various equations has certain dimensions in terms of units of measurements. Of course, all of the terms in any given equation must have the same dimensions. Now, dimensions or units of measurement are human constructs for our convenience. No system of units has any inherent superiority over any other, despite the fact that in this text we exhibit a preference for the units ordained by Napoleon Bonaparte (of France) over those ordained by King Henry VIII (of England). The point here is that any physical problem must be expressible in completely dimensionless form. Moreover, the parameters used to render the dependent and independent variables dimensionless must appear in the equations or boundary conditions. One cannot define "reference" quantities that do not appear in the problem; spurious dimensionless parameters will be the result. If the procedure is done properly, there will be a reduction in the parametric dependence of the formulation, generally by the number of independent units. This is described in Sections 3 and 4 in this chapter. The parametric reduction is called a similitude. Similitudes greatly facilitate correlation of experimental data. In Chapter 9 we will encounter a situation in which there are no naturally occurring scales for length or time that can be used to render the formulation of a particular problem dimensionless. As the axiom that a dimensionless formulation is a physical necessity still holds, we must look for a dimensionless combination of the independent variables. This results in a contraction of the dimensionality of the space required for the solution, that is, a reduction by one in the number of independent varibles. Such a reduction is called a similarity and results in what is called a similarity solution.

2. *Nondimensional Parameters Determined from Differential Equations*

To illustrate the method of determining nondimensional parameters from the governing differential equations, consider a flow in which both viscosity and gravity are important. An example of such a flow is the motion of a ship, where the drag experienced is caused both by the generation of surface waves and by friction on the surface of the hull. All other effects such as surface tension and compressibility are neglected. The governing differential equation is the Navier–Stokes equation

$$\frac{\partial w}{\partial t} + u\frac{\partial w}{\partial x} + v\frac{\partial w}{\partial y} + w\frac{\partial w}{\partial z} = -\frac{1}{\rho}\frac{\partial p}{\partial z} - g + \frac{\mu}{\rho}\left(\frac{\partial^2 w}{\partial x^2} + \frac{\partial^2 w}{\partial y^2} + \frac{\partial^2 w}{\partial z^2}\right), \quad (8.1)$$

and two other equations for u and v. The equation can be nondimensionalized by defining a characteristic length scale l and a characteristic velocity scale U. In the present problem we can take l to be the length of the ship at the waterline and U to be the free-stream velocity at a large distance from the ship (Figure 8.1). The choice of these scales is dictated by their appearance in the boundary conditions; U is the boundary condition on the variable u and l occurs in the shape function of the ship hull. Dynamic similarity requires that the flows have *geometric similarity* of the boundaries, so that all characteristic lengths are proportional; for example, in Figure 8.1 we must have $d/l = d_1/l_1$. Dynamic similarity also requires that the flows should be *kinematically similar*, that is, they should have geometrically similar streamlines. The velocities at the same relative location are therefore proportional; if the velocity at point P in Figure 8.1a is $U/2$, then the velocity at the corresponding point P_1 in Figure 8.1b must be $U_1/2$. *All length and velocity scales are then proportional in a class of dynamically similar flows.* (Alternatively, we could take the characteristic length to be the depth d of the hull under water. Such a choice is, however, unconventional.) Moreover, a choice of l as the length of the ship makes the nondimensional distances of interest (that is, the magnitude of x/l in the region around the ship) of order one. Similarly, a choice of U as the free-stream velocity makes the maximum value of the nondimensional velocity u/U of order one. For reasons that will become more apparent in the later chapters, it is of value to have all dimensionless variables of finite order. Approximations may then be based on any extreme size of the dimensionless parameters that will preface some of the terms.

Accordingly, we introduce the following nondimensional variables, denoted by primes:

$$x' = \frac{x}{l} \qquad y' = \frac{y}{l} \qquad z' = \frac{z}{l} \qquad t' = \frac{tU}{l},$$

$$u' = \frac{u}{U} \qquad v' = \frac{v}{U} \qquad w' = \frac{w}{U} \qquad p' = \frac{p - p_\infty}{\rho U^2}. \qquad (8.2)$$

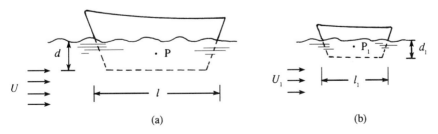

(a)   (b)

Figure 8.1 Two geometrically similar ships.

It is clear that the boundary conditions in terms of the nondimensional variables in equation (8.2) are independent of l and U. For example, consider the viscous flow over a circular cylinder of radius R. We choose the velocity scale U to be the free-stream velocity and the length scale to be the radius R. In terms of nondimensional velocity $u' = u/U$ and the nondimensional coordinate $r' = r/R$, the boundary condition at infinity is $u' \to 1$ as $r' \to \infty$, and the condition at the surface of the cylinder is $u' = 0$ at $r' = 1$. (Here, u is taken to be the r-component of velocity.)

There are instances where the shape function of a body may require two length scales, such as a length l and a thickness d. An additional dimensionless parameter, d/l would result to describe the slenderness of the body.

Normalization, that is, dimensionless representation of the pressure, depends on the dominant effect in the flow unless the flow is pressure-gradient driven. In the latter case for flow in ducts or tubes, the pressure should be made dimensionless by a characteristic pressure difference in the duct so that the dimensionless term is finite. In other cases, when the flow is not pressure-gradient driven, the pressure is a passive variable and should be normalized to balance the dominant effect in the flow. Because pressure enters only as a gradient, the pressure itself is not of consequence; only pressure differences are important. The conventional practice is to render $p - p_\infty$ dimensionless. Depending on the nature of the flow, this could be in terms of viscous stress $\mu U/l$, a hydrostatic pressure $\rho g l$, or as in the preceding, a dynamic pressure ρU^2.

Substitution of equation (8.2) into equation (8.1) gives

$$\frac{\partial w'}{\partial t'} + u'\frac{\partial w'}{\partial x'} + v'\frac{\partial w'}{\partial y'} + w'\frac{\partial w'}{\partial z'} = -\frac{\partial p'}{\partial z'} - \frac{gl}{U^2} + \frac{\nu}{Ul}\left(\frac{\partial^2 w'}{\partial x'^2} + \frac{\partial^2 w'}{\partial y'^2} + \frac{\partial^2 w'}{\partial z'^2}\right).$$

$$(8.3)$$

It is apparent that two flows (having different values of U, l, or ν), will obey the same nondimensional differential equation if the values of nondimensional groups gl/U^2 and ν/Ul are identical. Because the nondimensional boundary conditions are also identical in the two flows, it follows that *they will have the same nondimensional solutions*.

The nondimensional parameters Ul/ν and $U/\sqrt{gl}$ have been given special names:

$$\mathrm{Re} \equiv \frac{Ul}{\nu} = \text{Reynolds number},$$

$$\mathrm{Fr} \equiv \frac{U}{\sqrt{gl}} = \text{Froude number}.$$

$$(8.4)$$

Both Re and Fr have to be equal for dynamic similarity of two flows in which both viscous and gravitational effects are important. Note that the mere presence of gravity does not make the gravitational effects dynamically important. For flow around an object in a homogeneous fluid, gravity is important only if surface waves are generated. Otherwise, the effect of gravity is simply to add a hydrostatic pressure to the entire system, which can be eliminated by absorbing gravity into the pressure term.

Under dynamic similarity the nondimensional solutions are identical. Therefore, the local pressure at point $\mathbf{x} = (x, y, z)$ must be of the form

$$\frac{p\,(\mathbf{x}) - p_\infty}{\rho U^2} = f\left(\text{Fr}, \text{Re}; \frac{\mathbf{x}}{l}\right), \tag{8.5}$$

where $(p - p_\infty)/\rho U^2$ is called the *pressure coefficient*. Similar relations also hold for any other nondimensional flow variable such as velocity $\mathbf{u}/U$ and acceleration $\mathbf{a}l/U^2$. It follows that in dynamically similar flows the nondimensional local flow variables are identical at *corresponding points* (that is, for identical values of $\mathbf{x}/l$).

In the foregoing analysis we have assumed that the imposed boundary conditions are steady. However, we have retained the time derivative in equation (8.3) because the resulting flow can still be unsteady; for example, unstable waves can arise spontaneously under steady boundary conditions. Such unsteadiness must have a time scale proportional to l/U, as assumed in equation (8.2). Consider now a situation in which the imposed boundary conditions are unsteady. To be specific, consider an object having a characteristic length scale l oscillating with a frequency ω in a fluid at rest at infinity. This is a problem having an imposed length scale and an *imposed time scale* $1/\omega$. In such a case a velocity scale can be derived from ω and l to be $U = l\omega$. The preceding analysis then goes through, leading to the conclusion that $\text{Re} = Ul/\nu = \omega l^2/\nu$ and $\text{Fr} = U/\sqrt{gl} = \omega\sqrt{l/g}$ have to be duplicated for dynamic similarity of two flows in which viscous and gravitational effects are important.

All nondimensional quantities are identical for dynamically similar flows. For flow around an immersed body, we can define a nondimensional drag coefficient

$$C_{\mathrm{D}} \equiv \frac{D}{\rho U^2 l^2/2}, \tag{8.6}$$

where D is the drag experienced by the body; use of the factor of 1/2 in equation (8.6) is conventional but not necessary. Instead of writing C_{D} in terms of a length scale l, it is customary to define the drag coefficient more generally as

$$C_{\mathrm{D}} \equiv \frac{D}{\rho U^2 A/2},$$

where A is a characteristic area. For blunt bodies such as spheres and cylinders, A is taken to be a cross section perpendicular to the flow. Therefore, $A = \pi d^2/4$ for a sphere of diameter d, and $A = bd$ for a cylinder of diameter d and length b, with the axis of the cylinder perpendicular to the flow. For flow over a flat plate, on the other hand, A is taken to be the "wetted area", that is, $A = bl$; here, l is the length of the plate in the direction of flow and b is the width perpendicular to the flow.

The values of the drag coefficient C_{D} are identical for dynamically similar flows. In the present example in which the drag is caused both by gravitational and viscous effects, we must have a functional relation of the form

$$C_{\mathrm{D}} = f\,(\text{Fr}, \text{Re}). \tag{8.7}$$

For many flows the gravitational effects are unimportant. An example is the flow around the body, such as an airfoil, that does not generate gravity waves. In that case Fr is irrelevant, and

$$C_D = f\,(\mathrm{Re}).\tag{8.8}$$

We recall from the preceding discussion that speeds are low enough to ignore compressibility effects.

3. Dimensional Matrix

In many complicated flow problems the precise form of the differential equations may not be known. In this case the conditions for dynamic similarity can be determined by means of a dimensional analysis of the variables involved. A formal method of dimensional analysis is presented in the following section. Here we introduce certain ideas that are needed for performing a formal dimensional analysis.

The underlying principle in dimensional analysis is that of *dimensional homogeneity*, which states that all terms in an equation must have the same dimension. This is a basic check that we constantly apply when we derive an equation; if the terms do not have the same dimension, then the equation is not correct.

Fluid flow problems without electromagnetic forces and chemical reactions involve only mechanical variables (such as velocity and density) and thermal variables (such as temperature and specific heat). The dimensions of all these variables can be expressed in terms of four basic dimensions—mass M, length L, time T, and temperature θ. We shall denote the dimension of a variable q by $[q]$. For example, the dimension of velocity is $[u] = L/T$, that of pressure is $[p] = [\mathrm{force}]/[\mathrm{area}] = MLT^{-2}/L^2 = M/LT^2$, and that of specific heat is $[C] = [\mathrm{energy}]/[\mathrm{mass}][\mathrm{temperature}] = MLT^{-2}L/M\theta = L^2/\theta T^2$. When thermal effects are not considered, all variables can be expressed in terms of three fundamental dimensions, namely, M, L, and T. If temperature is considered only in combination with Boltzmann's constant $(k\theta)$ or a gas constant $(R\theta)$, then the units of the combination are simply L^2/T^2. Then only the three dimensions M, L, and T are required.

The method of dimensional analysis presented here uses the idea of a "dimensional matrix" and its rank. Consider the pressure drop Δp in a pipeline, which is expected to depend on the inside diameter d of the pipe, its length l, the average size e of the wall roughness elements, the average flow velocity U, the fluid density ρ, and the fluid viscosity μ. We can write the functional dependence as

$$f\,(\Delta p, d, l, e, U, \rho, \mu) = 0.\tag{8.9}$$

The dimensions of the variables can be arranged in the form of the following matrix:

$$
\begin{array}{c|ccccccc}
 & \Delta p & d & l & e & U & \rho & \mu \\
\hline
M & 1 & 0 & 0 & 0 & 0 & 1 & 1 \\
L & -1 & 1 & 1 & 1 & 1 & -3 & -1 \\
T & -2 & 0 & 0 & 0 & -1 & 0 & -1
\end{array}
\tag{8.10}
$$

Where we have written the variables $\Delta p, d, \ldots$ on the top and their dimensions in a vertical column underneath. For example, $[\Delta p] = ML^{-1}T^{-2}$. An array of dimensions such as equation (8.10) is called a *dimensional matrix*. The *rank* r of any matrix is defined to be the size of the largest square submatrix that has a nonzero determinant. Testing the determinant of the first three rows and columns, we obtain

$$\begin{vmatrix} 1 & 0 & 0 \\ -1 & 1 & 1 \\ -2 & 0 & 0 \end{vmatrix} = 0.$$

However, there does exist a nonzero third-order determinant, for example, the one formed by the last three columns:

$$\begin{vmatrix} 0 & 1 & 1 \\ 1 & -3 & -1 \\ -1 & 0 & -1 \end{vmatrix} = -1.$$

Thus, the rank of the dimensional matrix (8.10) is $r = 3$. If *all* possible third-order determinants were zero, we would have concluded that $r < 3$ and proceeded to test the second-order determinants.

It is clear that the rank is less than the number of rows only when one of the rows can be obtained by a linear combination of the other rows. For example, the matrix (not from equation (8.10)):

$$\begin{bmatrix} 0 & 1 & 0 & 1 \\ -1 & 2 & 1 & -2 \\ -1 & 4 & 1 & 0 \end{bmatrix}$$

has $r = 2$, as the last row can be obtained by adding the second row to twice the first row. A rank of less than 3 commonly occurs in problems of statics, in which the mass is really not relevant in the problem, although the dimensions of the variables (such as force) involve M. In most problems in fluid mechanics without thermal effects, $r = 3$.

4. Buckingham's Pi Theorem

Of the various formal methods of dimensional analysis, the one that we shall describe was proposed by Buckingham in 1914. Let $q_1, q_2, \ldots, q_n$ be n variables involved in a particular problem, so that there must exist a functional relationship of the form

$$f(q_1, q_2, \ldots, q_n) = 0. \tag{8.11}$$

Buckingham's theorem states that the *n variables can always be combined to form exactly $(n - r)$ independent nondimensional variables, where r is the rank of the dimensional matrix.* Each nondimensional parameter is called a "Π number," or more commonly a *nondimensional product*. (The symbol Π is used because the nondimensional parameter can be written as a *product* of the variables $q_1, \ldots, q_n$, raised to some power, as we shall see.) Thus, equation (8.11) can be written as a functional relationship

$$\phi(\Pi_1, \Pi_2, \ldots, \Pi_{n-r}) = 0. \tag{8.12}$$

It will be seen shortly that the nondimensional parameters are not unique. However, $(n - r)$ of them are *independent* and form a *complete set*.

The method of forming nondimensional parameters proposed by Buckingham is best illustrated by an example. Consider again the pipe flow problem expressed by

$$f(\Delta p, d, l, e, U, \rho, \mu) = 0, \tag{8.13}$$

whose dimensional matrix (8.10) has a rank of $r = 3$. Since there are $n = 7$ variables in the problem, the number of nondimensional parameters must be $n - r = 4$. We first select any $3 (= r)$ of the variables as "repeating variables", which we want to be repeated in all of our nondimensional parameters. These repeating variables must have different dimensions, and among them must contain all the fundamental dimensions M, L, and T. In many fluid flow problems we choose a characteristic velocity, a characteristic length, and a fluid property as the repeating variables. For the pipe flow problem, let us choose U, d, and ρ as the repeating variables. Although other choices would result in a different set of nondimensional products, we can always obtain other complete sets by combining the ones we have. Therefore, any choice of the repeating variables is satisfactory.

Each nondimensional product is formed by combining the three repeating variables with one of the remaining variables. For example, let the first dimensional product be taken as

$$\Pi_1 = U^a d^b \rho^c \Delta p.$$

The exponents a, b, and c are obtained from the requirement that Π_1 is dimensionless. This requires

$$M^0 L^0 T^0 = \left(LT^{-1}\right)^a (L)^b \left(ML^{-3}\right)^c \left(ML^{-1}T^{-2}\right) = M^{c+1} L^{a+b-3c-1} T^{-a-2}.$$

Equating indices, we obtain $a = -2, b = 0, c = -1$, so that

$$\Pi_1 = U^{-2} d^0 \rho^{-1} \Delta p = \frac{\Delta p}{\rho U^2}.$$

A similar procedure gives

$$\Pi_2 = U^a d^b \rho^c l = \frac{l}{d},$$

$$\Pi_3 = U^a d^b \rho^c e = \frac{e}{d},$$

$$\Pi_4 = U^a d^b \rho^c \mu = \frac{\mu}{\rho U d}.$$

Therefore, the nondimensional representation of the problem has the form

$$\frac{\Delta p}{\rho U^2} = \phi \left(\frac{l}{d}, \frac{e}{d}, \frac{\mu}{\rho U d} \right). \tag{8.14}$$

Other dimensionless products can be obtained by combining the four in the preceding. For example, a group $\Delta p d^2 \rho / \mu^2$ can be formed from Π_1 / Π_4^2. Also, different nondimensional groups would have been obtained had we taken variables other than $U, d,$ and ρ as the repeating variables. Whatever nondimensional groups we obtain, only four of these are independent for the pipe flow problem described by equation (8.13). However, the set in equation (8.14) contains the most commonly used nondimensional parameters, which have familiar physical interpretation and have been given special names. Several of the common dimensionless parameters will be discussed in Section 7.

The pi theorem is a formal method of forming dimensionless groups. With some experience, it becomes quite easy to form the dimensionless numbers by simple inspection. For example, since there are three length scales d, e, and l in equation (8.13), we can form two groups such as e/d and l/d. We can also form $\Delta p / \rho U^2$ as our dependent nondimensional variable; the Bernoulli equation tells us that ρU^2 has the same units as p. The nondimensional number that describes viscous effects is well known to be $\rho U d / \mu$. Therefore, with some experience, we can find all the nondimensional variables by inspection alone, thus no formal analysis is needed.

5. *Nondimensional Parameters and Dynamic Similarity*

Arranging the variables in terms of dimensionless products is especially useful in presenting experimental data. Consider the case of drag on a sphere of diameter d moving at a speed U through a fluid of density ρ and viscosity μ. The drag force can be written as

$$D = f (d, U, \rho, \mu). \tag{8.15}$$

If we do not form dimensionless groups, we would have to conduct an experiment to determine D vs d, keeping U, ρ, and μ fixed. We would then have to conduct an experiment to determine D as a function of U, keeping d, ρ, and μ fixed, and so on. However, such a duplication of effort is unnecessary if we write equation (8.15) in terms of dimensionless groups. A dimensional analysis of equation (8.15) gives

$$\frac{D}{\rho U^2 d^2} = f \left(\frac{\rho U d}{\mu} \right), \tag{8.16}$$

reducing the number of variables from five to two, and consequently a single experimental curve (Figure 8.2). Not only is the presentation of data united and simplified, the cost of experimentation is drastically reduced. It is clear that we need not vary the fluid viscosity or density at all; we could obtain all the data of Figure 8.2 in one wind tunnel experiment in which we determine D for various values of U. However, if we want to find the drag force for a fluid of different density or viscosity, we can

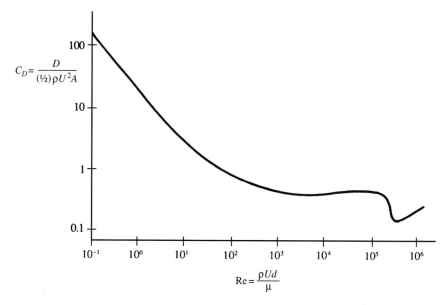

Figure 8.2   Drag coefficient for a sphere. The characteristic area is taken as $A = \pi d^2/4$. The reason for the sudden drop of C_D at Re $\sim 5 \times 10^5$ is the transition of the laminar boundary layer to a turbulent one, as explained in Chapter 10.

still use Figure 8.2. Note that the Reynolds number in equation (8.16) is written as the independent variable because it can be externally controlled in an experiment. In contrast, the drag coefficient is written as a dependent variable.

The idea of dimensionless products is intimately associated with the concept of similarity. In fact, a collapse of all the data on a single graph such as the one in Figure 8.2 is possible only because in this problem all flows having the same value of Re $= \rho U d/\mu$ are dynamically similar.

For flow around a sphere, the pressure at any point $\mathbf{x} = (x, y, z)$ can be written as

$$p\,(\mathbf{x}) - p_\infty = f\,(d, U, \rho, \mu; \mathbf{x})\,.$$

A dimensional analysis gives the local pressure coefficient:

$$\frac{p\,(\mathbf{x}) - p_\infty}{\rho U^2} = f\left(\frac{\rho U d}{\mu}; \frac{\mathbf{x}}{d}\right), \qquad (8.17)$$

requiring that nondimensional local flow variables be identical at corresponding points in dynamically similar flows. The difference between relations (8.16) and (8.17) should be noted. equation (8.16) is a relation between *overall* quantities (scales of motion), whereas (8.17) holds *locally* at a point.

Prediction of Flow Behavior from Dimensional Considerations

An interesting observation in Figure 8.2 is that $C_D \propto 1/\text{Re}$ at small Reynolds numbers. This can be justified solely on dimensional grounds as follows. At small values of Reynolds numbers we expect that the inertia forces in the equations of motion must become negligible. Then ρ drops out of equation (8.15), requiring

$$D = f(d, U, \mu).$$

The only dimensionless product that can be formed from the preceding is $D/\mu U d$. Because there is no other nondimensional parameter on which $D/\mu U d$ can depend, it can only be a constant:

$$D \propto \mu U d \qquad (\text{Re} \ll 1), \tag{8.18}$$

which is equivalent to $C_D \propto 1/\text{Re}$. It is seen that the *drag force in a low Reynolds number flow is linearly proportional to the speed U*; this is frequently called the *Stokes law of resistance*.

At the opposite extreme, Figure 8.2 shows that C_D becomes independent of Re for values of $\text{Re} > 10^3$. This is because the drag is now due mostly to the formation of a turbulent wake, in which the viscosity only has an indirect influence on the flow. (This will be clear in Chapter 13, where we shall see that the only effect of viscosity as $\text{Re} \to \infty$ is to dissipate the turbulent kinetic energy at increasingly smaller scales. The overall flow is controlled by inertia forces alone.) In this limit μ drops out of equation (8.15), giving

$$D = f(d, U, \rho).$$

The only nondimensional product is then $D/\rho U^2 d^2$, requiring

$$D \propto \rho U^2 d^2 \quad (\text{Re} \gg 1), \tag{8.19}$$

which is equivalent to $C_D = \text{const}$. It is seen that the *drag force is proportional to U^2 for high Reynolds number flows*. This rule is frequently applied to estimate various kinds of wind forces such as those on industrial structures, houses, automobiles, and the ocean surface. Consideration of surface tension effects may introduce additional dimensionless parameters depending on the nature of the problem. For example, if surface tension is to balance against a gravity body force, the Bond number $Bo = \rho g l^2/\sigma$ would be the appropriate dimensionless parameter to consider. If surface tension is in competition with a viscous stress, then it would be the capillary number, $Ca = \mu U/\sigma$. Similarly, the Weber number expresses the ratio of inertial forces to surface tension forces.

It is clear that very useful relationships can be established based on sound physical considerations coupled with a dimensional analysis. In the present case this procedure leads to $D \propto \mu U d$ for low Reynolds numbers, and $D \propto \rho U^2 d^2$ for high Reynolds numbers. Experiments can then be conducted to see if these relations do hold and to determine the unknown constants in these relations. Such arguments are constantly used in complicated fluid flow problems such as turbulence, where physical intuition plays a key role in research. A well-known example of this is the Kolmogorov $K^{-5/3}$ spectral law of isotropic turbulence presented in Chapter 13.

6. *Comments on Model Testing*

The concept of similarity is the basis of model testing, in which test data on one flow can be applied to other flows. The cost of experimentation with full-scale objects (which are frequently called *prototypes*) can be greatly reduced by experiments on a smaller geometrically similar model. Alternatively, experiments with a relatively inconvenient fluid such as air or helium can be substituted by an experiment with an easily workable fluid such as water. A model study is invariably undertaken when a new aircraft, ship, submarine, or harbor is designed.

In many flow situations both friction and gravity forces are important, which requires that both the Reynolds number and the Froude number be duplicated in a model testing. Since $Re = Ul/\nu$ and $Fr = U/\sqrt{gl}$, simultaneous satisfaction of both criteria would require $U \propto 1/l$ and $U \propto \sqrt{l}$ as the model length is varied. It follows that both the Reynolds and the Froude numbers cannot be duplicated simultaneously unless fluids of different viscosities are used in the model and the prototype flows. This becomes impractical, or even impossible, as the requirement sometimes needs viscosities that cannot be met by common fluids. It is then necessary to decide which of the two forces is more important in the flow, and a model is designed on the basis of the corresponding dimensionless number. Corrections can then be applied to account for the inequality of the remaining dimensionless group. This is illustrated in Example 8.1, which follows this section.

Although geometric similarity is a precondition to dynamic similarity, this is not always possible to attain. In a model study of a river basin, a geometrically similar model results in a stream so shallow that capillary and viscous effects become dominant. In such a case it is necessary to use a vertical scale larger than the horizontal scale. Such distorted models lack complete similitude, and their results are corrected before making predictions on the prototype.

Models of completely submerged objects are usually tested in a wind tunnel or in a towing tank where they are dragged through a pool of water. The *towing tank* is also used for testing models that are not completely submerged, for example, ship hulls; these are towed along the free surface of the liquid.

Example 8.1.   A ship 100 m long is expected to sail at 10 m/s. It has a submerged surface of 300 m². Find the model speed for a 1/25 scale model, neglecting frictional effects. The drag is measured to be 60 N when the model is tested in a towing tank at the model speed. Based on this information estimate the prototype drag after making corrections for frictional effects.

Solution: We first estimate the model speed neglecting frictional effects. Then the nondimensional drag force depends only on the Froude number:

$$D/\rho U^2 l^2 = f\left(U/\sqrt{gl}\right). \qquad (8.20)$$

Equating Froude numbers for the model (denoted by subscript "m") and prototype (denoted by subscript "p"), we get

$$U_m = U_p\sqrt{g_m l_m / g_p l_p} = 10\sqrt{1/25} = 2\,\text{m/s}.$$

The total drag on the model was measured to be 60 N at this model speed. Of the total measured drag, a part was due to frictional effects. The frictional drag can be estimated by treating the surface of the hull as a flat plate, for which the drag coefficient C_D is given in Figure 10.12 as a function of the Reynolds number. Using a value of $\nu = 10^{-6}\,\text{m}^2/\text{s}$ for water, we get

$$Ul/\nu \ (\text{model}) = [2\,(100/25)]/10^{-6} = 8 \times 10^6,$$

$$Ul/\nu \ (\text{prototype}) = 10\,(100)\,/10^{-6} = 10^9.$$

For these values of Reynolds numbers, Figure 10.12 gives the frictional drag coefficients of

$$C_D \ (\text{model}) = 0.003,$$
$$C_D \ (\text{prototype}) = 0.0015.$$

Using a value of $\rho = 1000\,\text{kg/m}^3$ for water, we estimate

$$\text{Frictional drag on model} = \tfrac{1}{2} C_D \rho U^2 A$$
$$= 0.5\,(0.003)\,(1000)\,(2)^2\left(300/25^2\right) = 2.88\,\text{N}$$

Out of the total model drag of 60 N, the wave drag is therefore $60 - 2.88 = 57.12\,\text{N}$.

Now the *wave drag* still obeys equation (8.20), which means that $D/\rho U^2 l^2$ for the two flows are identical, where D represents wave drag alone. Therefore

$$\text{Wave drag on prototype}$$
$$= (\text{Wave drag on model})\,\left(\rho_p/\rho_m\right)\left(l_p/l_m\right)^2\left(U_p/U_m\right)^2$$
$$= 57.12\,(1)\,(25)^2\,(10/2)^2 = 8.92 \times 10^5\,\text{N}$$

Having estimated the wave drag on the prototype, we proceed to determine its frictional drag. We obtain

$$\text{Frictional drag on prototype} = \tfrac{1}{2} C_D \rho U^2 A$$
$$= (0.5)\,(0.0015)\,(1000)\,(10)^2\,(300) = 0.225 \times 10^5\,\text{N}$$

Therefore, total drag on prototype $= (8.92 + 0.225) \times 10^5 = 9.14 \times 10^5\,\text{N}$.

If we did not correct for the frictional effects, and assumed that the measured model drag was all due to wave effects, then we would have found from equation (8.20) a prototype drag of

$$D_p = D_m\left(\rho_p/\rho_m\right)\left(l_p/l_m\right)^2\left(U_p/U_m\right)^2 = 60\,(1)\,(25)^2\,(10/2)^2 = 9.37 \times 10^5\,\text{N}.$$

7. Significance of Common Nondimensional Parameters

So far, we have encountered several nondimensional groups such as the pressure coefficient $(p - p_\infty)/\rho U^2$, the drag coefficient $2D/\rho U^2 l^2$, the Reynolds number $\mathrm{Re} = Ul/\nu$, and the Froude number $U/\sqrt{gl}$. Several independent nondimensional parameters that commonly enter fluid flow problems are listed and discussed briefly in this section. Other parameters will arise throughout the rest of the book.

Reynolds Number

The Reynolds number is the ratio of inertia force to viscous force:

$$\mathrm{Re} \equiv \frac{\text{Inertia force}}{\text{Viscous force}} \propto \frac{\rho u \, \partial u/\partial x}{\mu \partial^2 u/\partial x^2} \propto \frac{\rho U^2/l}{\mu U/l^2} = \frac{Ul}{\nu}.$$

Equality of Re is a requirement for the dynamic similarity of flows in which viscous forces are important.

Froude Number

The Froude number is defined as

$$\mathrm{Fr} = \left[\frac{\text{Inertia force}}{\text{Gravity force}} \right]^{1/2} \propto \left[\frac{\rho U^2/l}{\rho g} \right]^{1/2} = \frac{U}{\sqrt{gl}}.$$

Equality of Fr is a requirement for the dynamic similarity of flows with a free surface in which gravity forces are dynamically significant. Some examples of flows in which gravity plays a significant role are the motion of a ship, flow in an open channel, and the flow of a liquid over the spillway of a dam (Figure 8.3).

Internal Froude Number

In a density-stratified fluid the gravity force can play a significant role without the presence of a free surface. Then the effective gravity force in a two-layer situation is

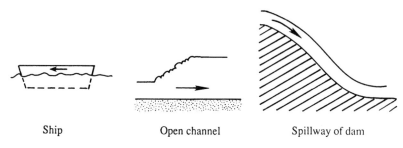

Ship   Open channel   Spillway of dam

Figure 8.3   Examples of flows in which gravity is important.

the "buoyancy" force $(\rho_2 - \rho_1)g$, as seen in the preceding chapter. In such a case we can define an internal Froude number as

$$\text{Fr}' \equiv \left[\frac{\text{Inertia force}}{\text{Buoyancy force}}\right]^{1/2} \propto \left[\frac{\rho_1 U^2/l}{(\rho_2 - \rho_1)g}\right]^{1/2} = \frac{U}{\sqrt{g'l}}, \qquad (8.21)$$

where $g' \equiv g(\rho_2 - \rho_1)/\rho_1$ is the "reduced gravity." For a continuously stratified fluid having a maximum buoyancy frequency N, we similarly define

$$\text{Fr}' \equiv \frac{U}{Nl},$$

which is analogous to equation (8.21) since $g' = g(\rho_2 - \rho_1)/\rho_1$ is similar to $-\rho_0^{-1}g(d\rho/dz)l = N^2l$.

Richardson Number

Instead of defining the internal Froude number, it is more common to define a nondimensional parameter that is equivalent to $1/\text{Fr}'^2$. This is called the Richardson number, and in a two-layer situation it is defined as

$$\text{Ri} \equiv \frac{g'l}{U^2}. \qquad (8.22)$$

In a continuously stratified flow, we can similarly define

$$\text{Ri} \equiv \frac{N^2 l^2}{U^2}. \qquad (8.23)$$

It is clear that the Richardson number has to be equal for the dynamic similarity of two density-stratified flows.

Equations (8.22) and (8.23) define overall or *bulk* Richardson numbers in terms of the *scales* l, N, and U. In addition, we can define a Richardson number involving the *local* values of velocity gradient and stratification at a certain depth z. This is called the *gradient Richardson number*, and it is defined as

$$\text{Ri}(z) \equiv \frac{N^2(z)}{(dU/dz)^2}.$$

Local Richardson numbers will be important in our studies of instability and turbulence in stratified fluids.

Mach Number

The Mach number is defined as

$$M \equiv \left[\frac{\text{Inertia force}}{\text{Compressibility force}}\right]^{1/2} \propto \left[\frac{\rho U^2/l}{\rho c^2/l}\right]^{1/2} = \frac{U}{c},$$

where c is the speed of sound. Equality of Mach numbers is a requirement for the dynamic similarity of compressible flows. For example, the drag experienced by a body in a flow with compressibility effects has the form

$$C_D = f\,(\mathrm{Re}, M)\,.$$

Flows in which $M < 1$ are called *subsonic*, whereas flows in which $M > 1$ are called *supersonic*. It will be shown in Chapter 16 that compressibility effects can be neglected if $M < 0.3$.

Prandtl Number

The Prandtl number enters as a nondimensional parameter in flows involving heat conduction. It is defined as

$$\mathrm{Pr} \equiv \frac{\text{Momentum diffusivity}}{\text{Heat diffusivity}} = \frac{\nu}{\kappa} = \frac{\mu/\rho}{k/\rho C_p} = \frac{C_p \mu}{k}\,.$$

It is therefore a fluid property and not a flow variable. For air at ordinary temperatures and pressures, $\mathrm{Pr} = 0.72$, which is close to the value of 0.67 predicted from a simplified kinetic theory model assuming hard spheres and monatomic molecules (Hirschfelder, Curtiss, and Bird (1954), pp. 9–16). For water at 20 °C, $\mathrm{Pr} = 7.1$. The dynamic similarity of flows involving thermal effects requires equality of Prandtl numbers.

Exercises

1. Suppose that the power to drive a propeller of an airplane depends on d (diameter of the propeller), U (free-stream velocity), ω (angular velocity of propeller), c (velocity of sound), ρ (density of fluid), and μ (viscosity). Find the dimensionless groups. In your opinion, which of these are the most important and should be duplicated in a model testing?

2. A 1/25 scale model of a submarine is being tested in a wind tunnel in which $p = 200\,\mathrm{kPa}$ and $T = 300\,\mathrm{K}$. If the prototype speed is 30 km/hr, what should be the free-stream velocity in the wind tunnel? What is the drag ratio? Assume that the submarine would not operate near the free surface of the ocean.

Literature Cited

Hirschfelder, J. O., C. F. Curtiss, and R. B. Bird (1954). *Molecular Theory of Gases and Liquids*, New York: John Wiley and Sons.

Supplemental Reading

Bridgeman, P. W. (1963). *Dimensional Analysis*, New Haven: Yale University Press.

Laminar Flow

1. Introduction

In Chapters 6 and 7 we studied inviscid flows in which the viscous terms in the Navier–Stokes equations were dropped. The underlying assumption was that the viscous forces were confined to thin boundary layers near solid surfaces, so that the bulk of the flow could be regarded as inviscid (Figure 6.1). We shall see in the next chapter that this is indeed valid if the Reynolds number is large. For low values of the Reynolds number, however, the *entire flow* may be dominated by viscosity, and the inviscid flow theory is of little use. The purpose of this chapter is to present certain solutions of the Navier–Stokes equations in some simple situations, retaining the viscous term $\mu \nabla^2 \mathbf{u}$ everywhere in the flow. While the inviscid flow theory allows the fluid to "slip" past a solid surface, real fluids will adhere to the surface because of

©2010 Elsevier Inc. All rights reserved.
DOI: 10.1016/B978-0-12-381399-2.50009-5

intermolecular interactions, that is, a real fluid satisfies the condition of zero relative velocity at a solid surface. This is the so-called *no-slip condition*.

Before presenting the solutions, we shall first discuss certain basic ideas about viscous flows. Flows in which the fluid viscosity is important can be of two types, namely, *laminar* and *turbulent*. The basic difference between the two flows was dramatically demonstrated in 1883 by Reynolds, who injected a thin stream of dye into the flow of water through a tube (Figure 9.1). At low rates of flow, the dye stream was observed to follow a well-defined straight path, indicating that the fluid moved in parallel layers (laminae) with no macroscopic mixing motion across the layers. This is called a *laminar flow*. As the flow rate was increased beyond a certain critical value, the dye streak broke up into an irregular motion and spread throughout the cross section of the tube, indicating the presence of macroscopic mixing motions perpendicular to the direction of flow. Such a chaotic fluid motion is called a *turbulent flow*. Reynolds demonstrated that the transition from laminar to turbulent flow always occurred at a fixed value of the ratio Re $= Vd/\nu \sim 3000$, where V is the velocity averaged over the cross section, d is the tube diameter, and ν is the kinematic viscosity.

Laminar flows in which viscous effects are important throughout the flow are the subject of the present chapter; laminar flows in which frictional effects are confined to boundary layers near solid surfaces are discussed in the next chapter. Chapter 12 considers the stability of laminar flows and their transition to turbulence; fully turbulent flows are discussed in Chapter 13. We shall assume here that the flow is incompressible, which is valid for Mach numbers less than 0.3. We shall also assume that the

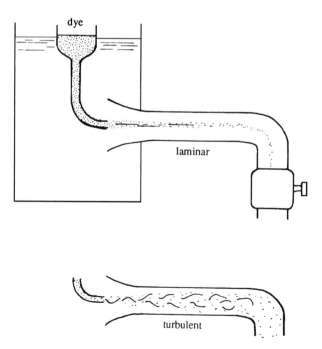

Figure 9.1   Reynolds's experiment to distinguish between laminar and turbulent flows.

flow is unstratified and observed in a nonrotating coordinate system. Some solutions of viscous flows in rotating coordinates, such as the Ekman layers, are presented in Chapter 14.

2. Analogy between Heat and Vorticity Diffusion

For two-dimensional flows that take place in the xy-plane, the vorticity equation is (see equation (5.13))

$$\frac{D\omega}{Dt} = \nu\nabla^2\omega,$$

where $\omega = \partial v/\partial x - \partial u/\partial y$. (For the sake of simplicity, we have avoided the vortex stretching term $\boldsymbol{\omega} \cdot \boldsymbol{\nabla u}$ by assuming two dimensionality.) This shows that the rate of change of vorticity $\partial\omega/\partial t$ at a point is due to advection $(-\mathbf{u} \cdot \boldsymbol{\nabla}\omega)$ and diffusion $(\nu\nabla^2\omega)$ of vorticity. The equation is similar to the heat equation

$$\frac{DT}{Dt} = \kappa\nabla^2 T,$$

where $\kappa = k/\rho C_p$ is the thermal diffusivity. The similarity of the equations suggests that vorticity diffuses in a manner analogous to the diffusion of heat. The similarity also brings out the fact that the diffusive effects are controlled by ν and κ, and not by μ and k. In fact, the momentum equation

$$\frac{D\mathbf{u}}{Dt} = \nu\nabla^2\mathbf{u} - \frac{1}{\rho}\boldsymbol{\nabla}p, \tag{9.1}$$

also shows that the acceleration due to viscous diffusion is proportional to ν. Thus, air ($\nu = 15 \times 10^{-6}\,\mathrm{m^2/s}$) is more diffusive than water ($\nu = 10^{-6}\,\mathrm{m^2/s}$), although μ for water is larger. Both ν and κ have the units of $\mathrm{m^2/s}$; the kinematic viscosity ν is therefore also called *momentum diffusivity*, in analogy with κ, which is called heat diffusivity. (However, velocity cannot be simply regarded as being diffused and advected in a flow because of the presence of the pressure gradient term in equation (9.1). The analogy between heat and vorticity is more appropriate.)

3. Pressure Change Due to Dynamic Effects

The equation of motion for the flow of a uniform density fluid is

$$\rho\frac{D\mathbf{u}}{Dt} = \rho\mathbf{g} - \boldsymbol{\nabla}p + \mu\nabla^2\mathbf{u}.$$

If the body of fluid is at rest, the pressure is hydrostatic:

$$0 = \rho\mathbf{g} - \boldsymbol{\nabla}p_s.$$

Subtracting, we obtain

$$\rho\frac{D\mathbf{u}}{Dt} = -\boldsymbol{\nabla}p_d + \mu\nabla^2\mathbf{u}, \tag{9.2}$$

where $p_d \equiv p - p_s$ is the pressure change due to dynamic effects. As there is no accepted terminology for p_d, we shall call it *dynamic pressure*, although the term is also used for $\rho q^2/2$, where q is the speed. Other common terms for p_d are "modified pressure" (Batchelor, 1967) and "excess pressure" (Lighthill, 1986).

For a fluid of uniform density, introduction of p_d eliminates gravity from the differential equation as in equation (9.2). However, the process may not eliminate gravity from the problem. Gravity reappears in the problem if the boundary conditions are given in terms of the total pressure p. An example is the case of surface gravity waves, where the total pressure is fixed at the free surface, and the mere introduction of p_d does not eliminate gravity from the problem. Without a free surface, however, gravity has no dynamic role. Its only effect is to add a hydrostatic contribution to the pressure field. In the applications that follow, we shall use equation (9.2), but the subscript on p will be omitted, as it is understood that p *stands for the dynamic pressure.*

4. *Steady Flow between Parallel Plates*

Because of the presence of the nonlinear advection term $\mathbf{u} \cdot \nabla \mathbf{u}$, very few exact solutions of the Navier–Stokes equations are known in closed form. In general, exact solutions are possible only when the nonlinear terms vanish identically. An example is the fully developed flow between infinite parallel plates. The term "fully developed" signifies that we are considering regions beyond the developing stage near the entrance (Figure 9.2), where the velocity profile changes in the direction of flow because of the development of boundary layers from the two walls. Within this "entrance length," which can be several times the distance between the walls, the velocity is uniform in the core increasing downstream and decreasing with x within the boundary layers. The derivative $\partial u/\partial x$ is therefore nonzero; the continuity equation $\partial u/\partial x + \partial v/\partial y = 0$

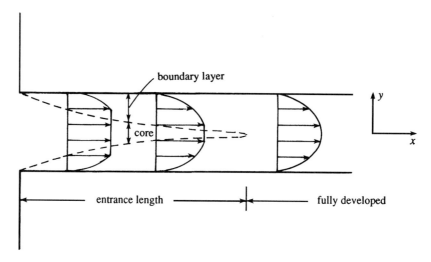

Figure 9.2 Developing and fully developed flows in a channel. The flow is fully developed after the boundary layers merge.

then requires that $v \neq 0$, so that the flow is *not* parallel to the walls within the entrance length.

Consider the fully developed stage of the steady flow between two infinite parallel plates. The flow is driven by a combination of an externally imposed pressure gradient (for example, maintained by a pump) and the motion of the upper plate at uniform speed U. Take the x-axis along the lower plate and in the direction of flow (Figure 9.3). Two dimensionality of the flow requires that $\partial/\partial z = 0$. Flow characteristics are also invariant in the x direction, so that continuity requires $\partial v/\partial y = 0$. Since $v = 0$ at $y = 0$, it follows that $v = 0$ everywhere, which reflects the fact that the flow is parallel to the walls. The x- and y-momentum equations are

$$0 = -\frac{1}{\rho}\frac{\partial p}{\partial x} + \nu\frac{d^2 u}{dy^2},$$

$$0 = -\frac{1}{\rho}\frac{\partial p}{\partial y}.$$

The y-momentum equation shows that p is not a function of y. In the x-momentum equation, then, the first term can only be a function of x, while the second term can only be a function of y. The only way this can be satisfied is for both terms to be constant. The *pressure gradient is therefore a constant*, which implies that the pressure varies linearly along the channel. Integrating the x-momentum equation twice, we obtain

$$0 = -\frac{y^2}{2}\frac{dp}{dx} + \mu u + Ay + B, \tag{9.3}$$

where we have written dp/dx because p is a function of x alone. The constants of integration A and B are determined as follows. The lower boundary condition $u = 0$ at $y = 0$ requires $B = 0$. The upper boundary condition $u = U$ at $y = 2b$ requires $A = b(dp/dx) - \mu U/2b$. The velocity profile equation (9.3) then becomes

$$u = \frac{yU}{2b} - \frac{y}{\mu}\frac{dp}{dx}\left(b - \frac{y}{2}\right). \tag{9.4}$$

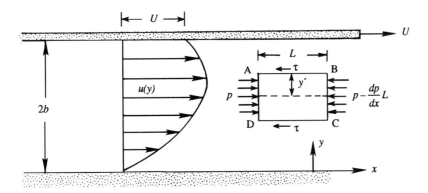

Figure 9.3   Flow between parallel plates.

The velocity profile is illustrated in Figure 9.4 for various cases.
The volume rate of flow per unit width of the channel is

$$Q = \int_0^{2b} u \, dy = Ub\left[1 - \frac{2b^2}{3\mu U}\frac{dp}{dx}\right],$$

so that the average velocity is

$$V \equiv \frac{Q}{2b} = \frac{U}{2}\left[1 - \frac{2b^2}{3\mu U}\frac{dp}{dx}\right].$$

Two cases of special interest are discussed in what follows.

Plane Couette Flow

The flow driven by the motion of the upper plate alone, without any externally imposed
pressure gradient, is called a plane Couette flow. In this case equation (9.4) reduces
to the linear profile (Figure 9.4c)

$$u = \frac{yU}{2b}. \tag{9.5}$$

The magnitude of shear stress is

$$\tau = \mu\frac{du}{dy} = \frac{\mu U}{2b},$$

which is uniform across the channel.

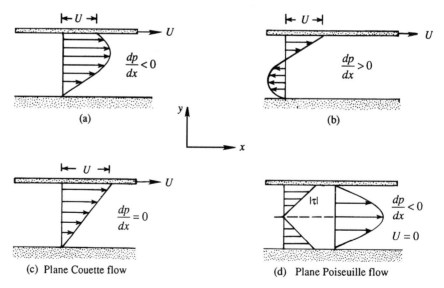

(a)

(b)

(c) Plane Couette flow

(d) Plane Poiseuille flow

Figure 9.4   Various cases of parallel flow in a channel.

Plane Poiseuille Flow

The flow driven by an externally imposed pressure gradient through two stationary flat walls is called a plane Poiseuille flow. In this case equation (9.4) reduces to the parabolic profile (Figure 9.4d)

$$u = -\frac{y}{\mu}\frac{dp}{dx}\left(b - \frac{y}{2}\right). \tag{9.6}$$

The magnitude of shear stress is

$$\tau = \mu\frac{du}{dy} = (b - y)\frac{dp}{dx},$$

which shows that the stress distribution is linear with a magnitude of $b(dp/dx)$ at the walls (Figure 9.4d).

It is important to note that the *constancy of the pressure gradient and the linearity of the shear stress distribution are general results for a fully developed channel flow and hold even if the flow is turbulent.* Consider a control volume ABCD shown in Figure 9.3, and apply the momentum principle (see equation (4.20)), which states that the net force on a control volume is equal to the net outflux of momentum through the surfaces. Because the momentum fluxes across surfaces AD and BC cancel each other, the forces on the control volume must be in balance; per unit width perpendicular to the plane of paper, the force balance gives

$$\left[p - \left(p - \frac{dp}{dx}L\right)\right]2y' = 2L\tau, \tag{9.7}$$

where y' is the distance measured from the center of the channel. In equation (9.7), $2y'$ is the area of surfaces AD and BC, and L is the area of surface AB or DC. Applying equation (9.7) at the wall, we obtain

$$\frac{dp}{dx}b = \tau_0, \tag{9.8}$$

which shows that the pressure gradient dp/dx is constant. Equations (9.7) and (9.8) give

$$\tau = \frac{y'}{b}\tau_0, \tag{9.9}$$

which shows that the magnitude of the shear stress increases linearly from the center of the channel (Figure 9.4d). Note that no assumption about the nature of the flow (laminar or turbulent) has been made in deriving equations (9.8) and (9.9).

Instead of applying the momentum principle, we could have reached the foregoing conclusions from the equation of motion in the form

$$\rho\frac{Du}{Dt} = -\frac{dp}{dx} + \frac{d\tau_{xy}}{dy},$$

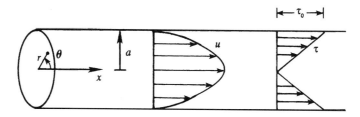

Figure 9.5   Laminar flow through a tube.

where we have introduced subscripts on τ and noted that the other stress components are zero. As the left-hand side of the equation is zero, it follows that dp/dx must be a constant and τ_{xy} must be linear in y.

5. Steady Flow in a Pipe

Consider the fully developed laminar motion through a tube of radius a. Flow through a tube is frequently called a *circular Poiseuille flow*. We employ cylindrical coordinates (r, θ, x), with the x-axis coinciding with the axis of the pipe (Figure 9.5). The only nonzero component of velocity is the axial velocity $u(r)$ (omitting the subscript "x" on u), and none of the flow variables depend on θ. The equations of motion in cylindrical coordinates are given in Appendix B. The radial equation of motion gives

$$0 = -\frac{\partial p}{\partial r},$$

showing that p is a function of x alone. The x-momentum equation gives

$$0 = -\frac{dp}{dx} + \frac{\mu}{r}\frac{d}{dr}\left(r\frac{du}{dr}\right).$$

As the first term can only be a function of x, and the second term can only be a function of r, it follows that both terms must be constant. The pressure therefore falls linearly along the length of pipe. Integrating twice, we obtain

$$u = \frac{r^2}{4\mu}\frac{dp}{dx} + A\ln r + B.$$

Because u must be bounded at $r = 0$, we must have $A = 0$. The wall condition $u = 0$ at $r = a$ gives $B = -(a^2/4\mu)(dp/dx)$. The velocity distribution therefore takes the parabolic shape

$$u = \frac{r^2 - a^2}{4\mu}\frac{dp}{dx}. \tag{9.10}$$

From Appendix B, the shear stress at any point is

$$\tau_{xr} = \mu \left[\frac{\partial u_r}{\partial x} + \frac{\partial u}{\partial r} \right].$$

In the present case the radial velocity u_r is zero. Dropping the subscript on τ, we obtain

$$\tau = \mu \frac{du}{dr} = \frac{r}{2} \frac{dp}{dx}, \tag{9.11}$$

which shows that the stress distribution is linear, having a maximum value at the wall of

$$\tau_0 = \frac{a}{2} \frac{dp}{dx}. \tag{9.12}$$

As in the previous section, equation (9.12) is also valid for turbulent flows.

The volume rate of flow is

$$Q = \int_0^a u 2 \pi r \, dr = -\frac{\pi a^4}{8\mu} \frac{dp}{dx},$$

where the negative sign offsets the negative value of dp/dx. The average velocity over the cross section is

$$V \equiv \frac{Q}{\pi a^2} = -\frac{a^2}{8\mu} \frac{dp}{dx}.$$

6. *Steady Flow between Concentric Cylinders*

Another example in which the nonlinear advection terms drop out of the equations of motion is the steady flow between two concentric, rotating cylinders. This is usually called the *circular Couette flow* to distinguish it from the plane Couette flow in which the walls are flat surfaces. Let the radius and angular velocity of the inner cylinder be R_1 and Ω_1 and those for the outer cylinder be R_2 and Ω_2 (Figure 9.6). Using cylindrical coordinates, the equations of motion in the radial and tangential directions are

$$-\frac{u_\theta^2}{r} = -\frac{1}{\rho} \frac{dp}{dr},$$

$$0 = \mu \frac{d}{dr} \left[\frac{1}{r} \frac{d}{dr} (r u_\theta) \right].$$

The r-momentum equation shows that the pressure increases radially outward due to the centrifugal force. The pressure distribution can therefore be determined once $u_\theta(r)$ has been found. Integrating the θ-momentum equation twice, we obtain

$$u_\theta = Ar + \frac{B}{r}. \tag{9.13}$$

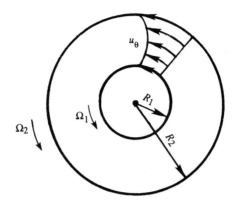

Figure 9.6   Circular Couette flow.

Using the boundary conditions $u_\theta = \Omega_1 R_1$ at $r = R_1$, and $u_\theta = \Omega_2 R_2$ at $r = R_2$, we obtain

$$A = \frac{\Omega_2 R_2^2 - \Omega_1 R_1^2}{R_2^2 - R_1^2},$$

$$B = \frac{(\Omega_1 - \Omega_2) R_1^2 R_2^2}{R_2^2 - R_1^2}.$$

Substitution into equation (9.13) gives the velocity distribution

$$u_\theta = \frac{1}{1 - (R_1/R_2)^2} \left\{ \left[\Omega_2 - \Omega_1 \left(\frac{R_1}{R_2} \right)^2 \right] r + \frac{R_1^2}{r} (\Omega_1 - \Omega_2) \right\}. \qquad (9.14)$$

Two limiting cases of the velocity distribution are considered in the following.

Flow Outside a Cylinder Rotating in an Infinite Fluid

Consider a long circular cylinder of radius R rotating with angular velocity Ω in an infinite body of viscous fluid (Figure 9.7). The velocity distribution for the present problem can be derived from equation (9.14) if we substitute $\Omega_2 = 0$, $R_2 = \infty$, $\Omega_1 = \Omega$, and $R_1 = R$. This gives

$$u_\theta = \frac{\Omega R^2}{r}, \qquad (9.15)$$

which shows that the velocity distribution is that of an irrotational vortex for which the tangential velocity is inversely proportional to r. As discussed in Chapter 5, Section 3, this is the only example in which the viscous solution is completely irrotational. Shear stresses do exist in this flow, but there is no *net* viscous force at a point. The shear stress at any point is given by

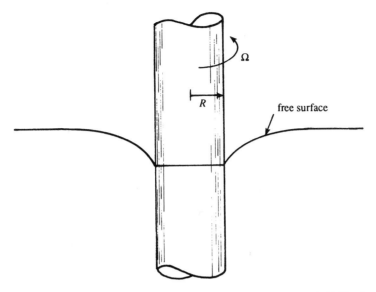

Figure 9.7 Rotation of a solid cylinder of radius R in an infinite body of viscous fluid. The shape of the free surface is also indicated. The flow field is viscous but irrotational.

$$\tau_{r\theta} = \mu \left[r \frac{\partial}{\partial r} \left(\frac{u_\theta}{r} \right) + \frac{1}{r} \frac{\partial u_r}{\partial \theta} \right],$$

which, for the present case, reduces to

$$\tau_{r\theta} = -\frac{2\mu\Omega R^2}{r^2}.$$

The forcing agent performs work on the fluid at the rate

$$2\pi R u_\theta \tau_{r\theta}.$$

It is easy to show that this rate of work equals the integral of the viscous dissipation over the flow field (Exercise 4).

Flow Inside a Rotating Cylinder

Consider the steady rotation of a cylindrical tank containing a viscous fluid. The radius of the cylinder is R, and the angular velocity of rotation is Ω (Figure 9.8). The flow would reach a steady state after the initial transients have decayed. The steady velocity distribution for this case can be found from equation (9.14) by substituting $\Omega_1 = 0$, $R_1 = 0$, $\Omega_2 = \Omega$, and $R_2 = R$. We get

$$u_\theta = \Omega r, \tag{9.16}$$

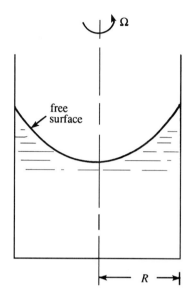

Figure 9.8   Steady rotation of a tank containing viscous fluid. The shape of the free surface is also indicated.

which shows that the tangential velocity is directly proportional to the radius, so that the fluid elements move as in a rigid solid. This flow was discussed in greater detail in Chapter 5, Section 3.

7. Impulsively Started Plate: Similarity Solutions

So far, we have considered steady flows with parallel streamlines, both straight and circular. The nonlinear terms dropped out and the velocity became a function of one spatial coordinate only. In the transient counterparts of these problems in which the flow is impulsively started from rest, the flow depends on a spatial coordinate and time. For these problems, exact solutions still exist because the nonlinear advection terms drop out again. One of these transient problems is given as Exercise 6. However, instead of considering the transient phase of all the problems already treated in the preceding sections, we shall consider several simpler and physically more revealing unsteady flow problems in this and the next three sections. First, consider the flow due to the impulsive motion of a flat plate parallel to itself, which is frequently called *Stokes' first problem*. (The problem is sometimes unfairly associated with the name of Rayleigh, who used Stokes' solution to predict the thickness of a developing boundary layer on a semi-infinite plate.)

Formulation of a Problem in Similarity Variables

Consider an infinite flat plate along $y = 0$, surrounded by fluid (with constant ρ and μ) for $y > 0$. The plate is impulsively given a velocity U at $t = 0$ (Figure 9.9). Since the resulting flow is invariant in the x direction, the continuity equation $\partial u/\partial x + \partial v/\partial y = 0$ requires $\partial v/\partial y = 0$. It follows that $v = 0$ everywhere because

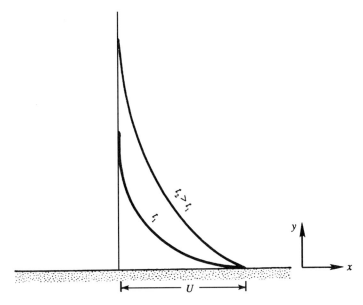

Figure 9.9 Laminar flow due to an impulsively started flat plate.

it is zero at $y = 0$. If the pressures at $x = \pm\infty$ are maintained at the same level, we can show that the pressure gradients are zero everywhere as follows. The x- and y-momentum equations are

$$\rho \frac{\partial u}{\partial t} = -\frac{\partial p}{\partial x} + \mu \frac{\partial^2 u}{\partial y^2},$$

$$0 = -\frac{\partial p}{\partial y}.$$

The y-momentum equation shows that p can only be a function of x and t. This can be consistent with the x-momentum equation, in which the first and the last terms can only be functions of y and t only if $\partial p/\partial x$ is independent of x. Maintenance of identical pressures at $x = \pm\infty$ therefore requires that $\partial p/\partial x = 0$. Alternatively, this can be established by observing that for an infinite plate the problem must be invariant under translation of coordinates by any finite constant in x.

The governing equation is therefore

$$\frac{\partial u}{\partial t} = \nu \frac{\partial^2 u}{\partial y^2}, \tag{9.17}$$

subject to

$$u(y, 0) = 0 \quad \text{[initial condition]}, \tag{9.18}$$
$$u(0, t) = U \quad \text{[surface condition]}, \tag{9.19}$$
$$u(\infty, t) = 0 \quad \text{[far field condition]}. \tag{9.20}$$

The problem is well posed, because equations (9.19) and (9.20) are conditions at two values of y, and equation (9.18) is a condition at one value of t; this is consistent with equation (9.17), which involves a first derivative in t and a second derivative in y.

The partial differential equation (9.17) can be transformed into an ordinary differential equation from dimensional considerations alone. Its real reason is the absence of scales for y and t as discussed on page 287. Let us write the solution as a functional relation

$$u = \phi(U, y, t, \nu). \qquad (9.21)$$

An examination of the equation set (9.17)–(9.20) shows that the parameter U appears only in the surface condition (9.19). This dependence on U can be eliminated from the problem by regarding u/U as the dependent variable, for then the equation set (9.17)–(9.20) can be written as

$$\frac{\partial u'}{\partial t} = \nu \frac{\partial^2 u'}{\partial y^2},$$
$$u'(y, 0) = 0,$$
$$u'(0, t) = 1,$$
$$u'(\infty, t) = 0,$$

where $u' \equiv u/U$. The preceding set is independent of U and must have a solution of the form

$$\frac{u}{U} = f(y, t, \nu). \qquad (9.22)$$

Because the left-hand side of equation (9.22) is dimensionless, the right-hand side can only be a dimensionless function of y, t, and ν. The only nondimensional variable formed from y, t, and ν is $y/\sqrt{\nu t}$, so that equation (9.22) must be of the form

$$\frac{u}{U} = F\left(\frac{y}{\sqrt{\nu t}}\right). \qquad (9.23)$$

Any function of $y/\sqrt{\nu t}$ would be dimensionless and could be used as the new independent variable. Why have we chosen to write it this way rather than $\nu t/y^2$ or some other equivalent form? We have done so because we want to solve for a velocity profile as a function of distance from the plate. By thinking of the solution to this problem in this way, our new dimensionless similarity variable will feature y in the numerator to the first power. We could have obtained equation (9.23) by applying Buckingham's pi theorem discussed in Chapter 8, Section 4. There are four variables in equation (9.22), and two basic dimensions are involved, namely, length and time. Two dimensionless variables can therefore be formed, and they are shown in equation (9.23).

We write equation (9.23) in the form

$$\frac{u}{U} = F(\eta), \qquad (9.24)$$

where η is the nondimensional distance given by

$$\eta \equiv \frac{y}{2\sqrt{vt}}.$$ (9.25)

We see that the absence of scales for length and time resulted in a reduction of the dimensionality of the space required for the solution (from 2 to 1). The factor of 2 has been introduced in the definition of η for eventual algebraic simplification. The equation set (9.17)–(9.20) can now be written in terms of η and $F(\eta)$. From equations (9.24) and (9.25), we obtain

$$\frac{\partial u}{\partial t} = U\frac{\partial F}{\partial t} = UF'\frac{\partial \eta}{\partial t} = -UF'\frac{y}{4\sqrt{v\,t^{3/2}}} = -\frac{UF'\eta}{2t},$$

$$\frac{\partial u}{\partial y} = U\frac{\partial F}{\partial y} = UF'\frac{\partial \eta}{\partial y} = UF'\frac{1}{2\sqrt{vt}},$$

$$\frac{\partial^2 u}{\partial y^2} = \frac{U}{2\sqrt{vt}}F''\frac{\partial \eta}{\partial y} = \frac{U}{4vt}F''.$$

Here, a prime on F denotes derivative with respect to η. With these substitutions, equation (9.17) reduces to the ordinary differential equation

$$-2\eta F' = F''.$$ (9.26)

The boundary conditions (9.18)–(9.20) reduce to

$$F(\infty) = 0,$$ (9.27)

$$F(0) = 1.$$ (9.28)

Note that *both* (9.18) and (9.20) reduce to the same condition $F(\infty) = 0$. This is expected because the original equation (9.17) was a partial differential equation and needed two conditions in y and one condition in t. In contrast, (9.26) is a second-order ordinary differential equation and needs only two boundary conditions.

Similarity Solution

Equation (9.26) can be integrated as follows:

$$\frac{dF'}{F'} = -2\eta\,d\eta.$$

Integrating once, we obtain

$$\ln F' = -\eta^2 + \text{const.}$$

which can be written as

$$\frac{dF}{d\eta} = A\,e^{-\eta^2},$$

where A is a constant of integration. Integrating again,

$$F(\eta) = A \int_0^\eta e^{-\eta^2} d\eta + B. \tag{9.29}$$

Condition (9.28) gives

$$F(0) = 1 = A \int_0^0 e^{-\eta^2} d\eta + B,$$

from which $B = 1$. Condition (9.27) gives

$$F(\infty) = 0 = A \int_0^\infty e^{-\eta^2} d\eta + 1 = \frac{A\sqrt{\pi}}{2} + 1,$$

(where we have used the result of a standard definite integral), from which $A = -2/\sqrt{\pi}$. Solution (9.29) then becomes

$$F = 1 - \frac{2}{\sqrt{\pi}} \int_0^\eta e^{-\eta^2} d\eta. \tag{9.30}$$

The function

$$\mathrm{erf}(\eta) \equiv \frac{2}{\sqrt{\pi}} \int_0^\eta e^{-\eta^2} d\eta,$$

is called the "error function" and is tabulated in mathematical handbooks. Solution (9.30) can then be written as

$$\frac{u}{U} = 1 - \mathrm{erf}\left[\frac{y}{2\sqrt{\nu t}}\right]. \tag{9.31}$$

It is apparent that *the solutions at different times all collapse into a single curve of u/U vs η, shown in Figure 9.10.*

The nature of the variation of u/U with y for various values of t is sketched in Figure 9.9. The solution clearly has a diffusive nature. At $t = 0$, a vortex sheet (that is, a velocity discontinuity) is created at the plate surface. The initial vorticity is in the form of a delta function, which is infinite at the plate surface and zero elsewhere. It can be shown that the integral $\int_0^\infty \omega \, dy$ is independent of time (see the following section for a demonstration), so that *no new vorticity is generated after the initial time.* The initial vorticity is simply diffused outward, resulting in an increase in the width of flow. The situation is analogous to a heat conduction problem in a semi-infinite solid extending from $y = 0$ to $y = \infty$. Initially, the solid has a uniform temperature, and at $t = 0$ the face $y = 0$ is suddenly brought to a different temperature. The temperature distribution for this problem is given by an equation similar to equation (9.31).

We may arbitrarily define the thickness of the diffusive layer as the distance at which u falls to 5% of U. From Figure 9.10, $u/U = 0.05$ corresponds to $\eta = 1.38$.

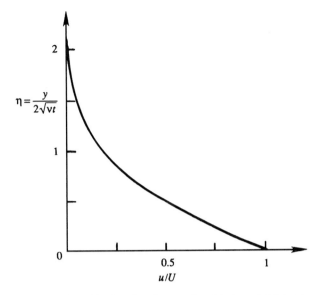

Figure 9.10   Similarity solution of laminar flow due to an impulsively started flat plate.

Therefore, in time t the diffusive effects propagate to a distance of order

$$\delta \sim 2.76\sqrt{\nu t} \qquad (9.32)$$

which increases as $\sqrt{t}$. Obviously, the factor of 2.76 in the preceding is somewhat arbitrary and can be changed by choosing a different ratio of u/U as the definition for the edge of the diffusive layer.

The present problem illustrates an important class of fluid mechanical problems that have *similarity solutions*. Because of the absence of suitable scales to render the independent variables dimensionless, the only possibility was a combination of variables that resulted in a reduction of independent variables (dimensionality of the space) required to describe the problem. In this case the reduction was from two (y, t) to one (η) so that the formulation reduced from a partial differential equation in y, t to an ordinary differential equation in η.

The solutions at different times are *self-similar* in the sense that they all collapse into a single curve if the velocity is scaled by U and y is scaled by the thickness of the layer taken to be $\delta(t) = 2\sqrt{\nu t}$. Similarity solutions exist in situations in which there is no natural scale in the direction of similarity. In the present problem, solutions at different t and y are similar because no length or time scales are imposed through the boundary conditions. Similarity would be violated if, for example, the boundary conditions are changed after a certain time t_1, which introduces a time scale into the problem. Likewise, if the flow was bounded above by a parallel plate at $y = b$, there could be no similarity solution.

An Alternative Method of Deducing the Form of η

Instead of arriving at the form of η from dimensional considerations, it could be derived by a different method as illustrated in the following. Denoting the thickness of the flow by $\delta(t)$, we assume similarity solutions in the form

$$\frac{u}{U} = F(\eta),$$

$$\eta = \frac{y}{\delta(t)}. \tag{9.33}$$

Then equation (9.17) becomes

$$UF'\frac{\partial \eta}{\partial t} = \nu U \frac{\partial^2 F}{\partial y^2}. \tag{9.34}$$

The derivatives in equation (9.34) are computed from equation (9.33):

$$\frac{\partial \eta}{\partial t} = -\frac{y}{\delta^2}\frac{d\delta}{dt} = -\frac{\eta}{\delta}\frac{d\delta}{dt},$$

$$\frac{\partial \eta}{\partial y} = \frac{1}{\delta},$$

$$\frac{\partial F}{\partial y} = F'\frac{\partial \eta}{\partial y} = \frac{F'}{\delta},$$

$$\frac{\partial^2 F}{\partial y^2} = \frac{1}{\delta}\frac{\partial F'}{\partial y} = \frac{F''}{\delta^2}.$$

Substitution into equation (9.34) and cancellation of factors give

$$-\left(\frac{\delta}{\nu}\frac{d\delta}{dt}\right)\eta F' = F''.$$

Since the right-hand side can only be an explicit function of η, the coefficient in parentheses on the left-hand side must be independent of t. This requires

$$\frac{\delta}{\nu}\frac{d\delta}{dt} = \text{const.} = 2, \quad \text{for example.}$$

Integration gives $\delta^2 = 4\nu t$, so that the flow thickness is $\delta = 2\sqrt{\nu t}$. Equation (9.33) then gives $\eta = y/(2\sqrt{\nu t})$, which agrees with our previous finding.

Method of Laplace Transform

Finally, we shall illustrate the method of Laplace transform for solving the problem. Let $\hat{u}(y, s)$ be the Laplace transform of $u(y, t)$. Taking the transform of equation (9.17), we obtain

$$s\hat{u} = \nu \frac{d^2\hat{u}}{dy^2}, \tag{9.35}$$

where the initial condition (9.18) of zero velocity has been used. The transform of the boundary conditions (9.19) and (9.20) are

$$\hat{u}(0, s) = \frac{U}{s},$$
(9.36)

$$\hat{u}(\infty, s) = 0.$$
(9.37)

Equation (9.35) has the general solution

$$\hat{u} = A \, e^{y\sqrt{s/\nu}} + B \, e^{-y\sqrt{s/\nu}},$$

where the constants $A(s)$ and $B(s)$ are to be determined from the boundary conditions. The condition (9.37) requires that $A = 0$, while equation (9.36) requires that $B = U/s$. We then have

$$\hat{u} = \frac{U}{s} e^{-y\sqrt{s/\nu}}.$$

The inverse transform of the preceding equation can be found in any mathematical handbook and is given by equation (9.31).

We have discussed this problem in detail because it illustrates the basic diffusive nature of viscous flows and also the mathematical techniques involved in finding similarity solutions. Several other problems of this kind are discussed in the following sections, but the discussions shall be somewhat more brief.

8. *Diffusion of a Vortex Sheet*

Consider the case in which the initial velocity field is in the form of a vortex sheet with $u = U$ for $y > 0$ and $u = -U$ for $y < 0$. We want to investigate how the vortex sheet decays by viscous diffusion. The governing equation is

$$\frac{\partial u}{\partial t} = \nu \frac{\partial^2 u}{\partial y^2},$$

subject to

$$u(y, 0) = U \, \text{sgn}(y),$$
$$u(\infty, t) = U,$$
$$u(-\infty, t) = -U,$$

where $\text{sgn}(y)$ is the "sign function," defined as 1 for positive y and -1 for negative y. As in the previous section, the parameter U can be eliminated from the governing set by regarding u/U as the dependent variable. Then u/U must be a function of (y, t, ν), and a dimensional analysis reveals that there must exist a similarity solution in the form

$$\frac{u}{U} = F(\eta),$$

$$\eta = \frac{y}{2\sqrt{vt}}.$$

The detailed arguments for the existence of a solution in this form are given in the preceding section. Substitution of the similarity form into the governing set transforms it into the ordinary differential equation

$$F'' = -2\eta F',$$
$$F(+\infty) = 1,$$
$$F(-\infty) = -1,$$

whose solution is

$$F(\eta) = \text{erf}(\eta).$$

The velocity distribution is therefore

$$u = U \text{ erf} \left[\frac{y}{2\sqrt{vt}} \right]. \tag{9.38}$$

A plot of the velocity distribution is shown in Figure 9.11. If we define the width of the transition layer as the distance between the points where $u = \pm 0.95U$, then the

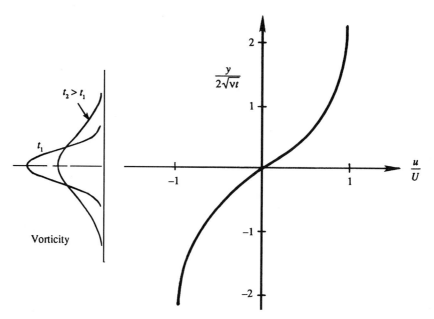

Figure 9.11 Viscous decay of a vortex sheet. The right panel shows the nondimensional solution and the left panel indicates the vorticity distribution at two times.

corresponding value of η is ± 1.38 and consequently the width of the transition layer is $5.52\sqrt{vt}$.

It is clear that the flow is essentially identical to that due to the impulsive start of a flat plate discussed in the preceding section. In fact, each half of Figure 9.11 is identical to Figure 9.10 (within an additive constant of ± 1). In both problems the initial delta-function-like vorticity is diffused away. In the present problem the magnitude of vorticity at any time is

$$\omega = \frac{\partial u}{\partial y} = \frac{U}{\sqrt{\pi vt}} e^{-y^2/4vt}. \tag{9.39}$$

This is a Gaussian distribution, whose width increases with time as $\sqrt{t}$, while the maximum value decreases as $1/\sqrt{t}$. The total amount of vorticity is

$$\int_{-\infty}^{\infty} \omega \, dy = 2\sqrt{vt} \int_{-\infty}^{\infty} \omega \, d\eta = \frac{2U}{\sqrt{\pi}} \int_{-\infty}^{\infty} e^{-\eta^2} \, d\eta = 2U,$$

which is independent of time, and equals the y-integral of the initial (delta-function-like) vorticity.

9. Decay of a Line Vortex

In Section 6 it was shown that when a solid cylinder of radius R is rotated at angular speed Ω in a viscous fluid, the resulting motion is irrotational with a velocity distribution $u_\theta = \Omega R^2/r$. The velocity distribution can be written as

$$u_\theta = \frac{\Gamma}{2\pi r},$$

where $\Gamma = 2\pi \Omega R^2$ is the circulation along any path surrounding the cylinder. Suppose the radius of the cylinder goes to zero while its angular velocity correspondingly increases in such a way that the product $\Gamma = 2\pi \Omega R^2$ is unchanged. In the limit we obtain a line vortex of circulation Γ, which has an infinite velocity discontinuity at the origin.

Now suppose that the limiting (infinitely thin and fast) cylinder suddenly stops rotating at $t = 0$, thereby reducing the velocity at the origin to zero impulsively. Then the fluid would gradually slow down from the initial distribution because of viscous diffusion from the region near the origin. The flow can therefore be regarded as that of the viscous decay of a line vortex, for which all the vorticity is initially concentrated at the origin. The problem is the circular analog of the decay of a plane vortex sheet discussed in the preceding section.

Employing cylindrical coordinates, the governing equation is

$$\frac{\partial u_\theta}{\partial t} = v \frac{\partial}{\partial r} \left[\frac{1}{r} \frac{\partial}{\partial r} (r u_\theta) \right], \tag{9.40}$$

subject to

$$u_\theta(r, 0) = \Gamma/2\pi r, \tag{9.41}$$

$$u_\theta(0, t) = 0, \tag{9.42}$$

$$u_\theta(r \to \infty, t) = \Gamma/2\pi r. \tag{9.43}$$

We expect similarity solutions here because there are no natural scales for r and t introduced from the boundary conditions. Conditions (9.41) and (9.43) show that the dependence of the solution on the parameter $\Gamma/2\pi r$ can be eliminated by defining a nondimensional velocity

$$u' \equiv \frac{u_\theta}{\Gamma/2\pi r}, \tag{9.44}$$

which must have a dependence of the form

$$u' = f(r, t, \nu).$$

As the left-hand side of the preceding equation is nondimensional, the right-hand side must be a nondimensional function of r, t, and ν. A dimensional analysis quickly shows that the only nondimensional group formed from these is $r/\sqrt{\nu t}$. Therefore, the problem must have a similarity solution of the form

$$\begin{aligned} u' &= F(\eta), \\ \eta &= \frac{r^2}{4\nu t}. \end{aligned} \tag{9.45}$$

(Note that we could have defined $\eta = r/2\sqrt{\nu t}$ as in the previous problems, but the algebra is slightly simpler if we define it as in equation (9.45).) Substitution of the similarity solution (9.45) into the governing set (9.40)–(9.43) gives

$$F'' + F' = 0,$$

subject to

$$F(\infty) = 1,$$
$$F(0) = 0.$$

The solution is

$$F = 1 - e^{-\eta}.$$

The dimensional velocity distribution is therefore

$$u_\theta = \frac{\Gamma}{2\pi r}[1 - e^{-r^2/4\nu t}]. \tag{9.46}$$

A sketch of the velocity distribution for various values of t is given in Figure 9.12. Near the center ($r \ll 2\sqrt{\nu t}$) the flow has the form of a rigid-body rotation, while in the outer region ($r \gg 2\sqrt{\nu t}$) the motion has the form of an irrotational vortex.

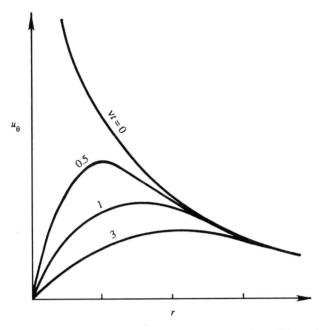

Figure 9.12   Viscous decay of a line vortex showing the tangential velocity at different times.

The foregoing discussion applies to the *decay* of a line vortex. Consider now the case where a line vortex is suddenly *introduced* into a fluid at rest. This can be visualized as the impulsive start of an infinitely thin and fast cylinder. It is easy to show that the velocity distribution is (Exercise 5)

$$u_\theta = \frac{\Gamma}{2\pi r} e^{-r^2/4\nu t}, \tag{9.47}$$

which should be compared to equation (9.46). The analogous problem in heat conduction is the sudden introduction of an infinitely thin and hot cylinder (containing a finite amount of heat) into a liquid having a different temperature.

10. *Flow Due to an Oscillating Plate*

The unsteady parallel flows discussed in the three preceding sections had similarity solutions, because there were no natural scales in space and time. We now discuss an unsteady parallel flow that does not have a similarity solution because of the existence of a natural time scale. Consider an infinite flat plate that executes sinusoidal oscillations parallel to itself. (This is sometimes called *Stokes' second problem.*) Only the steady periodic solution after the starting transients have died will be considered; thus there are no initial conditions to satisfy. The governing equation is

$$\frac{\partial u}{\partial t} = \nu \frac{\partial^2 u}{\partial y^2}, \tag{9.48}$$

subject to

$$u(0, t) = U \cos \omega t, \tag{9.49}$$

$$u(\infty, t) = \text{bounded.} \tag{9.50}$$

In the steady state, the flow variables must have a periodicity equal to the periodicity of the boundary motion. Consequently, we use a separable solution of the form

$$u = e^{i\omega t} f(y), \tag{9.51}$$

where what is meant is the real part of the right-hand side. (Such a complex form of representation is discussed in Chapter 7, Section 15.) Here, $f(y)$ is complex, thus $u(y, t)$ is allowed to have a phase difference with the wall velocity $U \cos \omega t$. Substitution of equation (9.51) into the governing equation (9.48) gives

$$i\omega f = \nu \frac{d^2 f}{dy^2}. \tag{9.52}$$

This is an equation with constant coefficients and must have exponential solutions. Substitution of a solution of the form $f = \exp(ky)$ gives $k = \sqrt{i\omega/\nu} = \pm(i+1)\sqrt{\omega/2\nu}$, where the two square roots of i have been used. Consequently, the solution of equation (9.52) is

$$f(y) = A\,e^{-(1+i)y\sqrt{\omega/2\nu}} + B\,e^{(1+i)y\sqrt{\omega/2\nu}}. \tag{9.53}$$

The condition (9.50), which requires that the solution must remain bounded at $y = \infty$, needs $B = 0$. The solution (9.51) then becomes

$$u = A\,e^{i\omega t}\,e^{-(1+i)y\sqrt{\omega/2\nu}}. \tag{9.54}$$

The surface boundary condition (9.49) now gives $A = U$. Taking the real part of equation (9.54), we finally obtain the velocity distribution for the problem:

$$u = U e^{-y\sqrt{\omega/2\nu}} \cos\left(\omega t - y\sqrt{\frac{\omega}{2\nu}}\right). \tag{9.55}$$

The cosine term in equation (9.55) represents a signal propagating in the direction of y, while the exponential term represents a decay in y. The flow therefore resembles a damped wave (Figure 9.13). However, this is a diffusion problem and *not* a wave-propagation problem because there are no restoring forces involved here. The apparent propagation is merely a result of the oscillating boundary condition. For $y = 4\sqrt{\nu/\omega}$, the amplitude of u is $U \exp(-4/\sqrt{2}) = 0.06U$, which means that the influence of the wall is confined within a distance of order

$$\delta \sim 4\sqrt{\nu/\omega}, \tag{9.56}$$

which decreases with frequency.

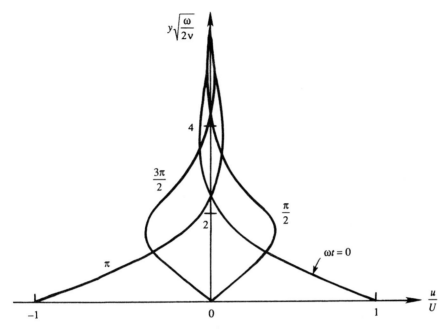

Figure 9.13 Velocity distribution in laminar flow near an oscillating plate. The distributions at $\omega t = 0$, $\pi/2$, π, and $3\pi/2$ are shown. The diffusive distance is of order $\delta = 4\sqrt{\nu/\omega}$.

Note that the solution (9.55) cannot be represented by a single curve in terms of the nondimensional variables. This is expected because the frequency of the boundary motion introduces a natural time scale $1/\omega$ into the problem, thereby violating the requirements of self-similarity. There are two parameters in the governing set (9.48)–(9.50), namely, U and ω. The parameter U can be eliminated by regarding u/U as the dependent variable. Thus the solution must have a form

$$\frac{u}{U} = f(y, t, \omega, \nu). \tag{9.57}$$

As there are five variables and two dimensions involved, it follows that there must be three dimensionless variables. A dimensional analysis of equation (9.57) gives u/U, ωt, and $y\sqrt{\omega/\nu}$ as the three nondimensional variables as in equation (9.55). Self-similar solutions exist only when there is an absence of such naturally occurring scales requiring a reduction in the dimensionality of the space.

An interesting point is that the oscillating plate has a constant diffusion distance $\delta = 4\sqrt{\nu/\omega}$ that is in contrast to the case of the impulsively started plate in which the diffusion distance increases with time. This can be understood from the governing equation (9.48). In the problem of sudden acceleration of a plate, $\partial^2 u/\partial y^2$ is positive for all y (see Figure 9.10), which results in a positive $\partial u/\partial t$ everywhere. The monotonic acceleration signifies that momentum is constantly diffused outward, which results in an ever-increasing width of flow. In contrast, in the case of an oscillating plate, $\partial^2 u/\partial y^2$ (and therefore $\partial u/\partial t$) constantly changes sign in y and t. Therefore, momentum cannot diffuse outward monotonically, which results in a constant width of flow.

The analogous problem in heat conduction is that of a semi-infinite solid, the surface of which is subjected to a periodic fluctuation of temperature. The resulting solution, analogous to equation (9.55), has been used to estimate the effective "eddy" diffusivity in the upper layer of the ocean from measurements of the phase difference (that is, the time lag between maxima) between the temperature fluctuations at two depths, generated by the diurnal cycle of solar heating.

11. High and Low Reynolds Number Flows

Many physical problems can be described by the behavior of a system when a certain parameter is either very small or very large. Consider the problem of steady flow around an object described by

$$\rho \mathbf{u} \cdot \nabla \mathbf{u} = -\nabla p + \mu \nabla^2 \mathbf{u}. \tag{9.58}$$

First, assume that the viscosity is small. Then the dominant balance in the flow is between the pressure and inertia forces, showing that pressure changes are of order ρU^2. Consequently, we nondimensionalize the governing equation (9.58) by scaling u by the free-stream velocity U, pressure by ρU^2, and distance by a representative length L of the body. Substituting the nondimensional variables (denoted by primes)

$$\mathbf{x}' = \frac{\mathbf{x}}{L} \qquad \mathbf{u}' = \frac{\mathbf{u}}{U} \qquad p' = \frac{p - p_\infty}{\rho U^2}, \tag{9.59}$$

the equation of motion (9.58) becomes

$$\mathbf{u}' \cdot \nabla \mathbf{u}' = -\nabla p' + \frac{1}{\mathrm{Re}} \nabla^2 \mathbf{u}', \tag{9.60}$$

where $\mathrm{Re} = Ul/\nu$ is the Reynolds number. For high Reynolds number flows, equation (9.60) is solved by treating $1/\mathrm{Re}$ as a small parameter. As a first approximation, we may set $1/\mathrm{Re}$ to zero everywhere in the flow, thus reducing equation (9.60) to the inviscid Euler equation. However, this omission of viscous terms cannot be valid near the body because the inviscid flow cannot satisfy the no-slip condition at the body surface. Viscous forces do become important near the body because of the high shear in a layer near the body surface. The scaling (9.59), which assumes that velocity gradients are proportional to U/L, is invalid in the boundary layer near the solid surface. We say that there is a region of *nonuniformity* near the body at which point a perturbation expansion in terms of the small parameter $1/\mathrm{Re}$ becomes *singular*. The proper scaling in the *boundary layer* and the procedure of solving high Reynolds number flows will be discussed in Chapter 10.

Now consider flows in the opposite limit of very low Reynolds numbers, that is, $\Re \to 0$. It is clear that low Reynolds number flows will have negligible inertia forces and therefore the viscous and pressure forces should be in approximate balance.

For the governing equations to display this fact, we should have a small parameter multiplying the *inertia forces* in this case. This can be accomplished if the variables are nondimensionalized properly to take into account the low Reynolds number nature of the flow. Obviously, the scaling (9.59), which leads to equation (9.60), is inappropriate in this case. For if equation (9.60) were multiplied by Re, then the small parameter Re would appear in front of not only the inertia force term but also the pressure force term, and the governing equation would reduce to $0 = \mu \nabla^2 \mathbf{u}$ as Re $\to 0$, which is *not* the balance for low Reynolds number flows. The source of the inadequacy of the nondimensionalization (9.59) for low Reynolds number flows is that the pressure is *not* of order ρU^2 in this case. As we noted in Chapter 8, for these external flows, pressure is a passive variable and it must be normalized by the dominant effect(s), which here are viscous forces. The purpose of scaling is to obtain nondimensional variables that are of order one, so that pressure should be scaled by ρU^2 only in high Reynolds number flows in which the pressure forces are of the order of the inertia forces. In contrast, in a low Reynolds number flow the pressure forces are of the order of the viscous forces. For ∇p to balance $\mu \nabla^2 \mathbf{u}$ in equation (9.58), the pressure changes must have a magnitude of the order

$$p \sim L\mu\nabla^2 u \sim \mu U/L.$$

Thus the proper nondimensionalization for low Reynolds number flows is

$$\mathbf{x}' = \frac{\mathbf{x}}{L} \qquad u' = \frac{\mathbf{u}}{U} \qquad p' = \frac{p - p_\infty}{\mu U/L}. \tag{9.61}$$

The variations of the nondimensional variables u' and p' in the flow field are now of order one. The pressure scaling also shows that p is proportional to μ in a low Reynolds number flow. A highly viscous oil is used in the bearing of a rotating shaft because the high pressure developed in the oil film of the bearing "lifts" the shaft and prevents metal-to-metal contact.

Substitution of equation (9.61) into (9.58) gives the nondimensional equation

$$\text{Re}\,\mathbf{u}' \cdot \nabla\mathbf{u}' = -\nabla p' + \nabla^2\mathbf{u}'. \tag{9.62}$$

In the limit Re $\to 0$, equation (9.62) becomes the linear equation

$$\nabla p = \mu\nabla^2\mathbf{u}, \tag{9.63}$$

where the variables have been converted back to their dimensional form.

Flows at Re $\ll 1$ are called *creeping motions*. They can be due to small velocity, large viscosity, or (most commonly) the small size of the body. Examples of such flows are the motion of a thin film of oil in the bearing of a shaft, settling of sediment particles near the ocean bottom, and the fall of moisture drops in the atmosphere. In the next section, we shall examine the creeping flow around a sphere.

Summary: The purpose of scaling is to generate nondimensional variables that are of order one in the flow field (except in singular regions or boundary layers).

The proper scales depend on the nature of the flow and are obtained by equating the terms that are most important in the flow field. For a high Reynolds number flow, the dominant terms are the inertia and pressure forces. This suggests the scaling (9.59), resulting in the nondimensional equation (9.60) in which the small parameter multiplies the subdominant term (except in boundary layers). In contrast, the dominant terms for a low Reynolds number flow are the pressure and viscous forces. This suggests the scaling (9.61), resulting in the nondimensional equation (9.62) in which the small parameter multiplies the subdominant term.

12. *Creeping Flow around a Sphere*

A solution for the creeping flow around a sphere was first given by Stokes in 1851. Consider the low Reynolds number flow around a sphere of radius a placed in a uniform stream U (Figure 9.14). The problem is axisymmetric, that is, the flow patterns are identical in all planes parallel to U and passing through the center of the sphere. Since Re $\rightarrow$ 0, as a first approximation we may neglect the inertia forces altogether and solve the equation

$$\nabla p = \mu \nabla^2 \mathbf{u}^*.$$

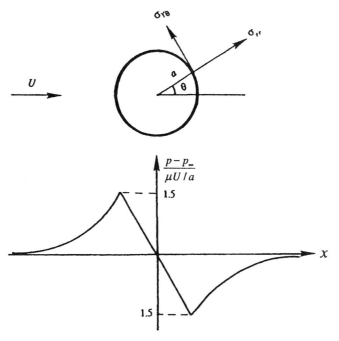

Figure 9.14 Creeping flow over a sphere. The upper panel shows the viscous stress components at the surface. The lower panel shows the pressure distribution in an axial ($\varphi = $ const.) plane.

We can form a vorticity equation by taking the curl of the preceding equation, obtaining

$$0 = \nabla^2 \boldsymbol{\omega}^*.$$

Here, we have used the fact that the curl of a gradient is zero, and that the order of the operators curl and ∇^2 can be interchanged. (The reader may verify this using indicial notation.) The only component of vorticity in this axisymmetric problem is ω_φ, the component perpendicular to $\varphi = $ const. planes in Figure 9.14, and is given by

$$\omega_\varphi = \frac{1}{r} \left[\frac{\partial (r u_\theta)}{\partial r} - \frac{\partial u_r}{\partial \theta} \right].$$

In axisymmetric flows we can define a streamfunction ψ given in Section 6.18. In spherical coordinates, it is defined as $\mathbf{u} = -\nabla \varphi \times \nabla \psi$, (6.74) so

$$u_r \equiv \frac{1}{r^2 \sin \theta} \frac{\partial \psi}{\partial \theta} \qquad u_\theta \equiv -\frac{1}{r \sin \theta} \frac{\partial \psi}{\partial r}.$$

In terms of the streamfunction, the vorticity becomes

$$\omega_\varphi = -\frac{1}{r} \left[\frac{1}{\sin \theta} \frac{\partial^2 \psi}{\partial r^2} + \frac{1}{r^2} \frac{\partial}{\partial \theta} \left(\frac{1}{\sin \theta} \frac{\partial \psi}{\partial \theta} \right) \right].$$

The governing equation is

$$\nabla^2 \omega_\varphi = 0^*.$$

Combining the last two equations, we obtain

$$\left[\frac{\partial^2}{\partial r^2} + \frac{\sin \theta}{r^2} \frac{\partial}{\partial \theta} \left(\frac{1}{\sin \theta} \frac{\partial}{\partial \theta} \right) \right]^2 \psi = 0. \tag{9.64}$$

The boundary conditions on the preceding equation are

$$\psi(a, \theta) = 0 \qquad\qquad [u_r = 0 \quad \text{at surface}], \tag{9.65}$$

$$\partial \psi / \partial r (a, \theta) = 0 \qquad\qquad [u_\theta = 0 \quad \text{at surface}], \tag{9.66}$$

$$\psi(\infty, \theta) = \tfrac{1}{2} U r^2 \sin^2 \theta \qquad [\text{uniform flow at } \infty]. \tag{9.67}$$

The last condition follows from the fact that the stream function for a uniform flow is $(1/2) U r^2 \sin^2 \theta$ in spherical coordinates (see equation (6.76)).

---

* In spherical polar coordinates, the operator in the footnoted equations is actually $-\nabla \times \nabla \times$ ($-$curl curl$_$), which is different from the Laplace operator defined in Appendix B. Eq. (9.64) is the square of the operator, and not the biharmonic.

The upstream condition (9.67) suggests a separable solution of the form

$$\psi = f(r)\sin^2\theta.$$

Substitution of this into the governing equation (9.64) gives

$$f^{iv} - \frac{4f''}{r^2} + \frac{8f'}{r^3} - \frac{8f}{r^4} = 0,$$

whose solution is

$$f = Ar^4 + Br^2 + Cr + \frac{D}{r}.$$

The upstream boundary condition (9.67) requires that $A = 0$ and $B = U/2$. The surface boundary conditions then give $C = -3\,Ua/4$ and $D = Ua^3/4$. The solution then reduces to

$$\psi = Ur^2\sin^2\theta\left[\frac{1}{2} - \frac{3a}{4r} + \frac{a^3}{4r^3}\right]. \tag{9.68}$$

The velocity components can then be found as

$$u_r = \frac{1}{r^2\sin\theta}\frac{\partial\psi}{\partial\theta} = U\cos\theta\left(1 - \frac{3a}{2r} + \frac{a^3}{2r^3}\right),$$

$$u_\theta = -\frac{1}{r\sin\theta}\frac{\partial\psi}{\partial r} = -U\sin\theta\left(1 - \frac{3a}{4r} - \frac{a^3}{4r^3}\right). \tag{9.69}$$

The pressure can be found by integrating the momentum equation $\nabla p = \mu\nabla^2\mathbf{u}$. The result is

$$p = -\frac{3a\mu U\cos\theta}{2r^2} + p_\infty \tag{9.70}$$

The pressure distribution is sketched in Figure 9.14. The pressure is maximum at the forward stagnation point where it equals $3\mu U/2a$, and it is minimum at the rear stagnation point where it equals $-3\mu U/2a$.

Let us determine the drag force D on the sphere. One way to do this is to apply the principle of mechanical energy balance over the entire flow field given in equation (4.63). This requires

$$DU = \int\phi\,dV,$$

which states that the work done by the sphere equals the viscous dissipation over the entire flow; here, ϕ is the viscous dissipation per unit volume. A more direct way to determine the drag is to integrate the stress over the surface of the sphere. The force per unit area normal to a surface, whose outward unit normal is $\mathbf{n}$ is

$$F_i = \tau_{ij}n_j = [-p\delta_{ij} + \sigma_{ij}]n_j = -pn_i + \sigma_{ij}n_j,$$

where τ_{ij} is the total stress tensor, and σ_{ij} is the viscous stress tensor. The component of the drag force per unit area in the direction of the uniform stream is therefore

$$[-p \cos \theta + \sigma_{rr} \cos \theta - \sigma_{r\theta} \sin \theta]_{r=a}, \tag{9.71}$$

which can be understood from Figure 9.14. The viscous stress components are

$$\sigma_{rr} = 2\mu \frac{\partial u_r}{\partial r} = 2\mu U \cos \theta \left[\frac{3a}{2r^2} - \frac{3a^3}{2r^4} \right],$$

$$\sigma_{r\theta} = \mu \left[r \frac{\partial}{\partial r} \left(\frac{u_\theta}{r} \right) + \frac{1}{r} \frac{\partial u_r}{\partial \theta} \right] = -\frac{3\mu U a^3}{2r^4} \sin \theta, \tag{9.72}$$

so that equation (9.71) becomes

$$\frac{3\mu U}{2a} \cos^2 \theta + 0 + \frac{3\mu U}{2a} \sin^2 \theta = \frac{3\mu U}{2a}.$$

The drag force is obtained by multiplying this by the surface area $4\pi a^2$ of the sphere, which gives

$$D = 6\pi \mu a U, \tag{9.73}$$

of which one-third is pressure drag and two-thirds is skin friction drag. It follows that the resistance in a creeping flow is proportional to the velocity; this is known as *Stokes' law of resistance*.

In a well-known experiment to measure the charge of an electron, Millikan used equation (9.73) to estimate the radius of an oil droplet falling through air. Suppose ρ' is the density of a spherical falling particle and ρ is the density of the surrounding fluid. Then the effective weight of the sphere is $4\pi a^3 g(\rho' - \rho)/3$, which is the weight of the sphere minus the weight of the displaced fluid. The falling body is said to reach the "terminal velocity" when it no longer accelerates, at which point the viscous drag equals the effective weight. Then

$$\tfrac{4}{3}\pi a^3 g(\rho' - \rho) = 6\pi \mu a U,$$

from which the radius a can be estimated.

Millikan was able to deduce the charge on an electron making use of Stokes' drag formula by the following experiment. Two horizontal parallel plates can be charged by a battery (see Fig. 9.15). Oil is sprayed through a very fine hole in the upper plate and develops static charge ($+$) by losing a few (n) electrons in passing through the small hole. If the plates are charged, then an electric force neE will act on each of the drops. Now n is not known but $E = -V_b/L$, where V_b is the battery voltage and L is the gap between the plates, provided that the charge density in the gap is very low. With the plates uncharged, measurement of the downward terminal velocity allowed the radius of a drop to be calculated assuming that the viscosity of the drop is much larger than the viscosity of the air. The switch is thrown to charge the upper

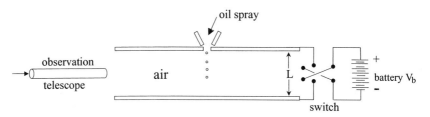

Figure 9.15   Millikan oil drop experiment.

plate negatively. The same droplet then reverses direction and is forced upwards. It quickly achieves its terminal velocity U_u by virtue of the balance of upward forces (electric + buoyancy) and downward forces (weight + drag). This gives

$$6\pi\mu U_u a + (4/3)\pi a^3 g(\rho' - \rho) = neE,$$

where U_u is measured by the observation telescope and the radius of the particle is now known. The data then allow for the calculation of ne. As n must be an integer, data from many droplets may be differenced to identify the minimum difference that must be e, the charge of a single electron.

The drag coefficient, defined as the drag force nondimensionalized by $\rho U^2/2$ and the projected area πa^2, is

$$C_D \equiv \frac{D}{\frac{1}{2}\rho U^2 \pi a^2} = \frac{24}{\text{Re}}, \tag{9.74}$$

where $\text{Re} = 2aU/\nu$ is the Reynolds number based on the diameter of the sphere. In Chapter 8, Section 5 it was shown that dimensional considerations alone require that C_D should be inversely proportional to Re for creeping motions. To repeat the argument, the drag force in a "massless" fluid (that is, $\text{Re} \ll 1$) can only have the dependence

$$D = f(\mu, U, a).$$

The preceding relation involves four variables and the three basic dimensions of mass, length, and time. Therefore, only one nondimensional parameter, namely, $D/\mu U a$, can be formed. As there is no second nondimensional parameter for it to depend on, $D/\mu U a$ must be a constant. This leads to $C_D \propto 1/\text{Re}$.

The flow pattern in a reference frame fixed to the fluid at infinity can be found by superposing a uniform velocity U to the left. This cancels out the first term in equation (9.68), giving

$$\psi = U r^2 \sin^2\theta \left[-\frac{3a}{4r} + \frac{a^3}{4r^3} \right],$$

which gives the streamline pattern as seen by an observer if the sphere is dragged in front of him from right to left (Figure 9.16). The pattern is symmetric between the upstream and the downstream directions, which is a result of the linearity of the

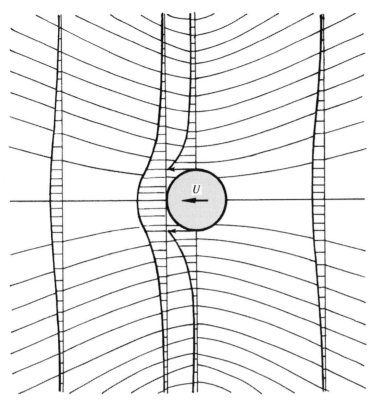

Figure 9.16 Streamlines and velocity distributions in Stokes' solution of creeping flow due to a moving sphere. Note the upstream and downstream symmetry, which is a result of complete neglect of nonlinearity.

governing equation (9.63); reversing the direction of the free-stream velocity merely changes **u** to −**u** and p to −p. The flow therefore does not have a "wake" behind the sphere.

13. Nonuniformity of Stokes' Solution and Oseen's Improvement

The Stokes solution for a sphere is not valid at large distances from the body because the advective terms are not negligible compared to the viscous terms at these distances. From equation (9.72), the largest viscous term is of the order

$$\text{viscous force/volume} = \text{stress gradient} \sim \frac{\mu U a}{r^3} \quad \text{as } r \to \infty,$$

while from equation (9.69) the largest inertia force is

$$\text{inertia force/volume} \sim \rho u_r \frac{\partial u_\theta}{\partial r} \sim \frac{\rho U^2 a}{r^2} \quad \text{as } r \to \infty.$$

Therefore,

$$\frac{\text{inertia force}}{\text{viscous force}} \sim \frac{\rho U a}{\mu} \frac{r}{a} = \Re \frac{r}{a} \quad \text{as } r \to \infty.$$

This shows that the inertia forces are not negligible for distances larger than $r/a \sim 1/\text{Re}$. At sufficiently large distances, no matter how small Re may be, the neglected terms become arbitrarily large.

Solutions of problems involving a small parameter can be developed in terms of the perturbation series in which the higher-order terms act as corrections on the lower-order terms. Perturbation expansions are discussed briefly in the following chapter. If we regard the Stokes solution as the first term of a series expansion in the small parameter Re, then the expansion is "nonuniform" because it breaks down at infinity. If we tried to calculate the next term (to order Re) of the perturbation series, we would find that the velocity corresponding to the higher-order term becomes unbounded at infinity.

The situation becomes worse for two-dimensional objects such as the circular cylinder. In this case, the Stokes balance $\nabla p = \mu \nabla^2 \mathbf{u}$ has *no solution at all* that can satisfy the uniform flow boundary condition at infinity. From this, Stokes concluded that steady, slow flows around cylinders cannot exist in nature. It has now been realized that the nonexistence of a first approximation of the Stokes flow around a cylinder is due to the *singular* nature of low Reynolds number flows in which there is a region of *nonuniformity* at infinity. The nonexistence of the second approximation for flow around a sphere is due to the same reason. In a different (and more familiar) class of singular perturbation problems, the region of nonuniformity is a thin layer (the "boundary layer") near the surface of an object. This is the class of flows with Re $\to \infty$, that will be discussed in the next chapter. For these high Reynolds number flows the small parameter $1/\text{Re}$ multiplies the *highest*-order derivative in the governing equations, so that the solution with $1/\text{Re}$ identically set to zero cannot satisfy all the boundary conditions. In low Reynolds number flows this classic symptom of the loss of the highest derivative is absent, but it is a singular perturbation problem nevertheless.

In 1910 Oseen provided an improvement to Stokes' solution by partly accounting for the inertia terms at large distances. He made the substitutions

$$u = U + u' \qquad v = v' \qquad w = w',$$

where (u', v', w') are the Cartesian components of the perturbation velocity, and are small at large distances. Substituting these, the advection term of the x-momentum equation becomes

$$u \frac{\partial u}{\partial x} + v \frac{\partial u}{\partial y} + w \frac{\partial u}{\partial z} = U \frac{\partial u'}{\partial x} + \left[u' \frac{\partial u'}{\partial x} + v' \frac{\partial u'}{\partial y} + w' \frac{\partial u'}{\partial z} \right].$$

Neglecting the quadratic terms, the equation of motion becomes

$$\rho U \frac{\partial u_i'}{\partial x} = -\frac{\partial p}{\partial x_i} + \mu \nabla^2 u_i',$$

where u_i' represents u', v', or w'. This is called *Oseen's equation*, and the approximation involved is called *Oseen's approximation*. In essence, the Oseen approximation linearizes the advective term $\mathbf{u} \cdot \nabla \mathbf{u}$ by $U(\partial \mathbf{u}/\partial x)$, whereas the Stokes approximation drops advection altogether. Near the body both approximations have the same order of accuracy. However, the Oseen approximation is better in the far field where the velocity is only slightly different than U. The Oseen equations provide a lowest-order solution that is uniformly valid everywhere in the flow field.

The boundary conditions for a moving sphere are

$$u' = v' = w' = 0 \quad \text{at infinity}$$
$$u' = -U, \quad v' = w' = 0 \quad \text{at surface.}$$

The solution found by Oseen is

$$\frac{\psi}{Ua^2} = \left[\frac{r^2}{2a^2} + \frac{a}{4r} \right] \sin^2 \theta - \frac{3}{\text{Re}}(1 + \cos \theta) \left\{ 1 - \exp \left[-\frac{\text{Re}}{4} \frac{r}{a}(1 - \cos \theta) \right] \right\},$$
(9.75)

where $\text{Re} = 2aU/\nu$ is the Reynolds number based on diameter. Near the surface $r/a \approx 1$, and a series expansion of the exponential term shows that Oseen's solution is identical to the Stokes solution (9.68) to the lowest order. The Oseen approximation predicts that the drag coefficient is

$$C_D = \frac{24}{\text{Re}} \left(1 + \frac{3}{16}\text{Re} \right),$$

which should be compared with the Stokes formula (9.74). Experimental results (see Figure 10.22 in the next chapter) show that the Oseen and the Stokes formulas for C_D are both fairly accurate for $\text{Re} < 5$.

The streamlines corresponding to the Oseen solution (9.75) are shown in Figure 9.17, where a uniform flow of U is added to the left so as to generate the pattern of flow due to a sphere moving in front of a stationary observer. It is seen that the flow is no longer symmetric, but has a wake where the streamlines are closer together than in the Stokes flow. The velocities in the wake are larger than in front of the sphere. Relative to the sphere, the flow is slower in the wake than in front of the sphere.

In 1957, Oseen's correction to Stokes' solution was rationalized independently by Kaplun and Proudman and Pearson in terms of matched asymptotic expansions. Here, we will obtain only the first-order correction. The full vorticity equation is

$$\nabla \times \nabla \times \boldsymbol{\omega} = \text{Re}\nabla \times (\mathbf{u} \times \boldsymbol{\omega}).$$
(9.76)

In terms of the Stokes streamfunction ψ, equation (9.64) is generalized to

$$D^4 \psi = \text{Re} \left[\frac{1}{r^2} \frac{\partial(\psi, D^2\psi)}{\partial(r, \mu)} + \frac{2}{r^2} D^2\psi L\psi \right],$$
(9.77)

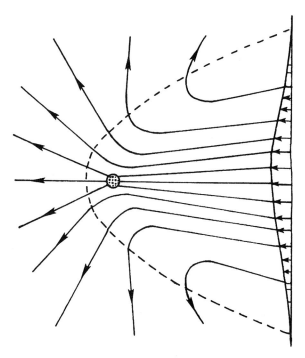

Figure 9.17 Streamlines and velocity distribution in Oseen's solution of creeping flow due to a moving sphere. Note the upstream and downstream asymmetry, which is a result of partial accounting for advection in the far field.

where $\partial(\psi, D^2\psi)/\partial(r, \mu)$ is shorthand notation for the Jacobian determinant with those four elements, $\mu = \cos\theta$, and the operators are

$$L = \frac{\mu}{1-\mu^2}\frac{\partial}{\partial r} + \frac{1}{r}\frac{\partial}{\partial \mu}, \qquad D^2 = \frac{\partial^2}{\partial r^2} + \frac{1-\mu^2}{r^2}\frac{\partial^2}{\partial \mu^2}.$$

We have seen that the right-hand side of equation (9.76) or (9.77) becomes of the same order as the left-hand side when Re $r/a \sim 1$ or $r/a \sim 1/$Re. We will define the "inner region" as $r/a \ll 1/$Re so that Stokes' solution holds approximately. To obtain a better approximation in the inner region, we will write

$$\psi(r, \mu; \text{Re}) = \psi_0(r, \mu) + \text{Re}\,\psi_1(r, \mu) + o(Re), \tag{9.78}$$

where the second correction "$o(\text{Re})$" means that it tends to zero faster than Re in the limit Re $\rightarrow$ 0. (See Chapter 10, Section 12. Here ψ is made dimensionless by Ua^2 and Re $= Ua/\nu$.) Substituting equation (9.78) into (9.77) and taking the limit Re $\rightarrow$ 0, we obtain $D^4\psi_0 = 0$ and recover Stokes' result

$$\psi_0 = -\frac{1}{2}\left(2r^2 - 3r + \frac{1}{r}\right)\frac{\mu^2 - 1}{2}. \tag{9.79}$$

Subtracting this, dividing by Re and taking the limit $Re \to 0$, we obtain

$$D^4 \psi_1 = \frac{1}{r^2} \frac{\partial(\psi_0, D^2 \psi_0)}{\partial(r, \mu)} + \frac{2}{r^2} D^2 \psi_0 L \psi_0,$$

which reduces to

$$D^4 \psi_1 = \frac{9}{4} \left(\frac{2}{r^2} - \frac{3}{r^3} + \frac{1}{r^5} \right) \mu(\mu^2 - 1), \qquad (9.80)$$

by using equation (9.79). This has the solution

$$\psi_1 = C_1 \left(2r^2 - 3r + \frac{1}{r} \right) \frac{\mu^2 - 1}{2} + \frac{3}{16} \left(2r^2 - 3r + 1 - \frac{1}{r} + \frac{1}{r^2} \right) \frac{\mu(\mu^2 - 1)}{2}, \tag{9.81}$$

where C_1 is a constant of integration for the solution to the homogeneous equation and is to be determined by matching with the outer region solution.

In the outer region $rRe = \rho$ is finite. The lowest-order outer solution must be uniform flow. Then we write the streamfuntion as

$$\Psi(\rho, \theta; Re) = \frac{1}{2} \frac{\rho^2}{Re^2} \sin^2 \theta + \frac{1}{Re} \Psi_1(\rho, \theta) + o \left(\frac{1}{Re} \right).$$

Substituting in equation (9.77) and taking the limit Re $\to 0$ yields

$$\left(\mathcal{D}^2 - \cos \theta \frac{\partial}{\partial \rho} + \frac{\sin \theta}{\rho} \frac{\partial}{\partial \theta} \right) \mathcal{D}^2 \Psi_1 = 0, \qquad (9.82)$$

where the operator

$$\mathcal{D}^2 = \partial^2/\partial \rho^2 + \frac{\sin \theta}{\rho^2} \left(\frac{\partial}{\partial \theta} \frac{1}{\sin \theta} \frac{\partial}{\partial \theta} \right).$$

The solution to equation (9.82) is found to be

$$\Psi_1(\rho, \theta) = -2C_2(1 + \cos \theta)[1 - e^{-\rho(1 - \cos \theta)/2}],$$

where the constant of integration C_2 is determined by matching in the overlap region between the inner and outer regions: $1 \ll r \ll 1/Re$, $Re \ll \rho \ll 1$.

The matching gives $C_2 = 3/4$ and $C_1 = -3/16$. Using this in equation (9.81) for the inner region solution, the $O(Re)$ correction to the stream function (equation (9.81)) has been obtained, from which the velocity components, shear stress, and pressure may be derived. Integrating over the surface of the sphere of radius $= a$, we obtain the final result for the drag force

$$D = 6\pi \mu U a [1 + 3Ua/(8\nu)],$$

which is consistent with Oseen's result. Higher-order corrections were obtained by Chester and Breach (1969).

14. Hele-Shaw Flow

Another low Reynolds number flow has seen wide application in flow visualization apparatus because of its peculiar and surprising property of reproducing the streamlines of potential flows (that is, infinite Reynolds number flows).

The Hele-Shaw flow is flow about a thin object filling a narrow gap between two parallel plates. Let the plates be located at $x = \pm b$ with Re $= U_0 b/\nu \ll 1$. Here, U_0 is the velocity upstream in the central plane (see Figure 9.18). Now place a circular cylinder of radius $= a$ and width $= 2b$ between the plates. We will require $b/a = \epsilon \ll 1$. The Hele-Shaw limit is Re $\ll \epsilon^2 \ll 1$. Imagine flow about a thin coin with parallel plates bounding the ends of the coin. We are interested in the streamlines of the flow around the cylinder. The origin of coordinates (R, θ, x) (Appendix B) is taken at the center of the cylinder.

Consider steady flow with constant density and viscosity in the absence of body forces. The dimensionless variables are, $x' = x/b$, $R' = r/a$, $\boldsymbol{v}' = \boldsymbol{v}/U_0$, $p' = (p - p_\infty)/(\mu U_0/b)$, Re $= U_0 b/\nu$, $\epsilon = b/a$. Conservation of mass and momentum then take the following form (primes suppressed):

$$\frac{\partial u_x}{\partial x} + \epsilon\left[\frac{1}{R}\frac{\partial}{\partial R}(R u_R) + \frac{1}{R}\frac{\partial u_\theta}{\partial \theta}\right] = 0.$$

$$\text{Re}\left[u_x\frac{\partial u_R}{\partial x} + \epsilon\left(u_R\frac{\partial u_R}{\partial R} + \frac{u_\theta}{R}\frac{\partial u_R}{\partial \theta} - \frac{u_\theta^2}{R}\right)\right]$$
$$= -\frac{\partial p}{\partial R} + \frac{\partial^2 u_R}{\partial x^2} + \epsilon^2\left(\frac{\partial^2 u_R}{\partial R^2} + \frac{1}{R}\frac{\partial u_R}{\partial R} + \frac{1}{R^2}\frac{\partial^2 u_R}{\partial \theta^2} - \frac{u_R}{R^2} - \frac{2}{R}\frac{\partial u_\theta}{\partial \theta}\right),$$

$$\text{Re}\left[u_x\frac{\partial u_\theta}{\partial x} + \epsilon\left(u_R\frac{\partial u_\theta}{\partial R} + \frac{u_\theta}{R}\frac{\partial u_\theta}{\partial \theta} - \frac{u_R u_\theta}{R}\right)\right]$$
$$= -\frac{1}{R}\frac{\partial p}{\partial \theta} + \frac{\partial^2 u_\theta}{\partial x^2} + \epsilon^2\left(\frac{\partial^2 u_\theta}{\partial R^2} + \frac{1}{R}\frac{\partial u_\theta}{\partial R} + \frac{1}{R^2}\frac{\partial^2 u_\theta}{\partial \theta^2} - \frac{2}{R^2}\frac{\partial u_R}{\partial \theta} - \frac{u_\theta}{R^2}\right),$$

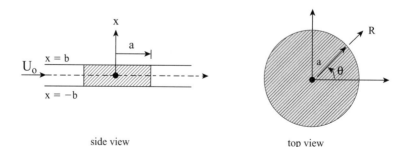

side view   top view

Figure 9.18   Hele-Shaw flow.

$$
\mathrm{Re}\left[u_x\frac{\partial u_x}{\partial x} + \epsilon\left(u_R\frac{\partial u_x}{\partial R} + \frac{u_\theta}{R}\frac{\partial u_x}{\partial \theta}\right)\right]
$$
$$
= -\frac{\partial p}{\partial x} + \epsilon^2\left(\frac{\partial^2 u_x}{\partial R^2} + \frac{1}{R}\frac{\partial u_x}{\partial R} + \frac{1}{R^2}\frac{\partial^2 u_x}{\partial \theta^2}\right).
$$

Because $\mathrm{Re} \ll \epsilon^2 \ll 1$, we take the limit $\mathrm{Re} \to 0$ first and drop the convective acceleration. Next, we take the limit $\epsilon \to 0$ to obtain the outer region flow:

$$
\frac{\partial u_x}{\partial x} = O(\epsilon) \to 0, \quad u_x(x = \pm 1) = 0, \quad \text{so } u_x = 0 \text{ throughout,}
$$
$$
\frac{\partial^2 u_R}{\partial x^2} = \frac{\partial p}{\partial R} + O(\epsilon^2),
$$
$$
\frac{\partial^2 u_\theta}{\partial x^2} = \frac{1}{R}\frac{\partial p}{\partial \theta} + O(\epsilon^2).
$$

With $u_x = O(\epsilon)$ at most, $\partial p/\partial x = O(\epsilon)$ at most so $p = p(R, \theta)$. Integrating the momentum equations with respect to x,

$$
u_R = -\frac{\partial}{\partial R}\left[\frac{1}{2}p(1 - x^2)\right], \qquad u_\theta = -\frac{1}{R}\frac{\partial}{\partial \theta}\left[\frac{1}{2}p(1 - x^2)\right],
$$

where no slip has been satisfied on $x = \pm 1$. Thus we can write $\mathbf{u} = \nabla\phi$ for the two-dimensional field u_R, u_θ. Here, $\phi = -\frac{1}{2}p(1 - x^2)$. Now we require that $u_x = o(\epsilon)$ so that the first term in the continuity equation is small compared with the others. Then

$$
\frac{1}{R}\frac{\partial}{\partial R}(Ru_R) + \frac{1}{R}\frac{\partial u_\theta}{\partial \theta} = o(1) \to 0 \quad \text{as} \quad \epsilon \to 0
$$

Substituting in terms of the velocity potential ϕ, we have $\nabla^2\phi = 0$ in R, θ subject to the boundary conditions:

$R = 1, \dfrac{\partial\phi}{\partial R} = 0$ (no mass flow normal to a solid boundary)

$R \to \infty, \phi \to R\cos\theta(1 - x^2)/2$ (uniform flow in each $x = $ constant plane)

The solution is just the potential flow over a circular cylinder (equation (6.35))

$$
\phi = R\cos\theta\left(1 + \frac{1}{R^2}\right)\frac{(1 - x^2)}{2},
$$

where x is just a parameter. Therefore, the streamlines corresponding to this velocity potential are identical to the potential flow streamlines of equation (6.35). This allows for the construction of an apparatus to visualize such potential flows by dye injection between two closely spaced glass plates. The velocity distribution of this flow is

$$u_\theta = -\sin\theta\left(1 + \frac{1}{R^2}\right)\left(\frac{1-x^2}{2}\right), \quad u_R = \cos\theta\left(1 - \frac{1}{R^2}\right)\left(\frac{1-x^2}{2}\right).$$

As $R \to 1$, $u_R \to 0$ but there is a slip velocity $u_\theta \to -2\sin\theta(1-x^2)/2$.

As this is a viscous flow, there must exist a thin region near $R = 1$ where the slip velocity u_θ decreases rapidly to zero to satisfy $u_\theta = 0$ on $R = 1$. This thin boundary layer is very close to the body surface $R = 1$. Thus, $u_R \approx 0$ and $\partial p/\partial R \approx 0$ throughout the layer. Now $p = -R\cos\theta(1 + 1/R^2)$ so for $R \approx 1$, $(1/R)\partial p/\partial\theta \approx 2\sin\theta$. In the θ momentum equation, R derivatives become very large so the dominant balance is

$$\frac{\partial^2 u_\theta}{\partial x^2} + \epsilon^2 \frac{\partial^2 u_\theta}{\partial R^2} = \frac{1}{R}\frac{\partial p}{\partial\theta} = 2\sin\theta.$$

It is clear from this balance that a stretching by $1/\epsilon$ is appropriate in the boundary layer: $\hat{R} = (R-1)/\epsilon$. In these terms

$$\frac{\partial^2 u_\theta}{\partial x^2} + \frac{\partial^2 u_\theta}{\partial\hat{R}^2} = 2\sin\theta,$$

subject to $u_\theta = 0$ on $\hat{R} = 0$ and $u_\theta \to -2\sin\theta(1-x^2)/2$ as $\hat{R} \to \infty$ (match with outer region). The solution to this problem is

$$u_\theta(\hat{R}, \theta, x) = -(1-x^2)\sin\theta + \sum_{n=0}^{\infty} A_n \cos(k_n x)e^{-k_n\hat{R}}\sin\theta, \quad k_n = \left(n + \frac{1}{2}\right)\pi,$$

$$A_n = \int_{-1}^{1}(1-x^2)\cos\left[\left(n + \frac{1}{2}\right)\pi x\right]dx.$$

We conclude that Hele-Shaw flow indeed simulates potential flow (inviscid) streamlines except for a very thin boundary layer of the order of the plate separation adjacent to the body surface.

15. Final Remarks

As in other fields, analytical methods in fluid flow problems are useful in understanding the physics and in making generalizations. However, it is probably fair to say that most of the analytically tractable problems in ordinary laminar flow have already been solved, and approximate methods are now necessary for further advancing our knowledge. One of these approximate techniques is the perturbation method, where the flow is assumed to deviate slightly from a basic linear state; perturbation methods are discussed in the following chapter. Another method that is playing an increasingly important role is that of solving the Navier–Stokes equations numerically using a computer. A proper application of such techniques requires considerable care and familiarity with various iterative techniques and their limitations. It is hoped that the reader will have the opportunity to learn numerical methods in a separate study. In Chapter 11, we will introduce several basic methods of computational fluid dynamics.

Exercises

1. Consider the laminar flow of a fluid layer falling down a plane inclined at an angle θ with the horizontal. If h is the thickness of the layer in the fully developed stage, show that the velocity distribution is

$$u = \frac{g \sin \theta}{2v}(h^2 - y^2),$$

where the x-axis points along the free surface, and the y-axis points toward the plane. Show that the volume flow rate per unit width is

$$Q = \frac{gh^3 \sin \theta}{3v},$$

and the frictional stress on the wall is

$$\tau_0 = \rho gh \sin \theta.$$

2. Consider the steady laminar flow through the annular space formed by two coaxial tubes. The flow is along the axis of the tubes and is maintained by a pressure gradient dp/dx, where the x direction is taken along the axis of the tubes. Show that the velocity at any radius r is

$$u(r) = \frac{1}{4\mu}\frac{dp}{dx}\left[r^2 - a^2 - \frac{b^2 - a^2}{\ln(b/a)}\ln\frac{r}{a}\right],$$

where a is the radius of the inner tube and b is the radius of the outer tube. Find the radius at which the maximum velocity is reached, the volume rate of flow, and the stress distribution.

3. A long vertical cylinder of radius b rotates with angular velocity Ω concentrically outside a smaller stationary cylinder of radius a. The annular space is filled with fluid of viscosity μ. Show that the steady velocity distribution is

$$u_\theta = \frac{r^2 - a^2}{b^2 - a^2}\frac{b^2\Omega}{r}.$$

Show that the torque exerted on either cylinder, per unit length, equals $4\pi\mu\Omega a^2 b^2/(b^2 - a^2)$.

4. Consider a solid cylinder of radius R, steadily rotating at angular speed Ω in an infinite viscous fluid. As shown in Section 6, the steady solution is irrotational:

$$u_\theta = \frac{\Omega R^2}{r}.$$

Show that the work done by the external agent in maintaining the flow (namely, the value of $2\pi R u_\theta \tau_{r\theta}$ at $r = R$) equals the total viscous dissipation rate in the flow field.

5. Suppose a line vortex of circulation Γ is suddenly introduced into a fluid at rest. Show that the solution is

$$u_\theta = \frac{\Gamma}{2\pi r} e^{-r^2/4\nu t}.$$

Sketch the velocity distribution at different times. Calculate and plot the vorticity, and observe how it diffuses outward.

6. Consider the development from rest of a plane Couette flow. The flow is bounded by two rigid boundaries at $y = 0$ and $y = h$, and the motion is started from rest by suddenly accelerating the lower plate to a steady velocity U. The upper plate is held stationary. Notice that similarity solutions cannot exist because of the appearance of the parameter h. Show that the velocity distribution is given by

$$u(y, t) = U \left(1 - \frac{y}{h}\right) - \frac{2U}{\pi} \sum_{n=1}^{\infty} \frac{1}{n} \exp\left(-n^2 \pi^2 \frac{\nu t}{h^2}\right) \sin \frac{n\pi y}{h}.$$

Sketch the flow pattern at various times, and observe how the velocity reaches the linear distribution for large times.

7. Planar Couette flow is generated by placing a viscous fluid between two infinite parallel plates and moving one plate (say, the upper one) at a velocity U with respect to the other one. The plates are a distance h apart. Two immiscible viscous liquids are placed between the plates as shown in the diagram. Solve for the velocity distributions in the two fluids.

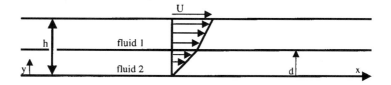

8. Calculate the drag on a spherical droplet of radius $r = a$, density ρ' and viscosity μ' moving with velocity U in an infinite fluid of density ρ and viscosity μ. Assume $\mathrm{Re} = \rho U a / \mu \ll 1$. Neglect surface tension.

9. Consider a very low Reynolds number flow over a circular cyclinder of radius $r = a$. For $r/a = O(1)$ in the $\mathrm{Re} = Ua/\nu \to 0$ limit, find the equation governing the streamfunction $\psi(r, \theta)$ and solve for ψ with the least singular behavior for large r. There will be one remaining constant of integration to be determined by asymptotic matching with the large r solution (which is not part of this problem). Find the domain of validity of your solution.

10. Consider a sphere of radius $r = a$ rotating with angular velocity ω about a diameter so that $\mathrm{Re} = \omega a^2 / \nu \ll 1$. Use the symmetries in the problem to solve the mass and momentum equations directly for the azimuthal velocity $v_\varphi(r, \theta)$. Then find the shear stress and torque on the sphere.

Literature Cited

Batchelor, G. K. (1967). *An Introduction to Fluid Dynamics*, London: Cambridge University Press.

Lighthill, M. J. (1986). *An Informal Introduction to Theoretical Fluid Mechanics*, Oxford, England: Clarendon Press.

Chester, W. and D. R. Breach (with I. Proudman) (1969). "On the flow past a sphere at low Reynolds number." *J. Fluid Mech.* **37**: 751–760.

Hele-Shaw, H. S. (1898). "Investigations of the Nature of Surface Resistance of Water and of Stream Line Motion Under Certain Experimental Conditions," *Trans. Roy. Inst. Naval Arch.* **40**: 21–46.

Kaplun, S. (1957). "Low Reynolds number flow past a circular cylinder." *J. Math. Mech.* **6**: 585–603.

Millikan, R. A. (1911). "The isolation of an ion, a precision measurement of its charge, and the correction of Stokes' law." *Phys. Rev.* **32**: 349–397.

Oseen, C. W. (1910). "Über die Stokes'sche Formel, und über eine verwandte Aufgabe in der Hydrodynamik." *Ark Math. Astrom. Fys.* **6**: No. 29.

Proudman, I. and J. R. A. Pearson (1957). "Expansions at small Reynolds numbers for the flow past a sphere and a circular cylinder." *J. Fluid Mech.* **2**: 237–262.

Supplemental Reading

Schlichting, H. (1979). *Boundary Layer Theory*, New York: McGraw-Hill.

Boundary Layers and Related Topics

©2010 Elsevier Inc. All rights reserved.
DOI: 10.1016/B978-0-12-381399-2.50010-1

1. Introduction

Until the beginning of the twentieth century, analytical solutions of steady fluid flows were generally known for two typical situations. One of these was that of parallel viscous flows and low Reynolds number flows, in which the nonlinear advective terms were zero and the balance of forces was that between the pressure and the viscous forces. The second type of solution was that of inviscid flows around bodies of various shapes, in which the balance of forces was that between the inertia and pressure forces. Although the equations of motion are nonlinear in this case, the velocity field can be determined by solving the linear Laplace equation. These irrotational solutions predicted pressure forces on a streamlined body that agreed surprisingly well with experimental data for flow of fluids of small viscosity. However, these solutions also predicted a zero drag force and a nonzero tangential velocity at the surface, features that did not agree with the experiments.

In 1905 Ludwig Prandtl, an engineer by profession and therefore motivated to find realistic fields near bodies of various shapes, first hypothesized that, for small viscosity, the viscous forces are negligible everywhere except close to the solid boundaries where the no-slip condition had to be satisfied. The thickness of these boundary layers approaches zero as the viscosity goes to zero. Prandtl's hypothesis reconciled two rather contradictory facts. On one hand he supported the intuitive idea that the effects of viscosity are indeed negligible in most of the flow field if ν is small. At the same time Prandtl was able to account for drag by insisting that the no-slip condition must be satisfied at the wall, no matter how small the viscosity. This reconciliation was Prandtl's aim, which he achieved brilliantly, and in such a simple way that it now seems strange that nobody before him thought of it. Prandtl also showed how the equations of motion within the boundary layer can be simplified. Since the time of Prandtl, the concept of the boundary layer has been generalized, and the mathematical techniques involved have been formalized, extended, and applied to various other branches of physical science. The concept of the boundary layer is considered one of the cornerstones in the history of fluid mechanics.

In this chapter we shall explore the boundary layer hypothesis and examine its consequences. We shall see that the equations of motion within the boundary layer can be simplified because of the layer's thinness, and solutions can be obtained in certain cases. We shall also explore approximate methods of solving the flow within a boundary layer. Some experimental data on the drag experienced by bodies of various shapes in high Reynolds number flows, including turbulent flows, will be examined. For those interested in sports, the mechanics of curving sports balls will be explored. Finally, the mathematical procedure of obtaining perturbation solutions in situations where there is a small parameter (such as $1/Re$ in boundary layer flows) will be briefly outlined.

2. Boundary Layer Approximation

In this section we shall see what simplifications of the equations of motion within the boundary layer are possible because of the layer's thinness. Across these layers, which exist only in high Reynolds number flows, the velocity varies rapidly enough for the

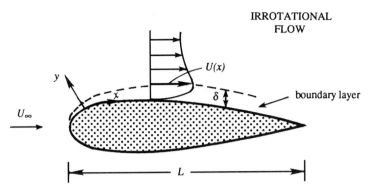

Figure 10.1 The boundary layer. Its thickness is greatly exaggerated in the figure. Here, U_∞ is the oncoming velocity and U is the velocity at the edge of the boundary layer.

viscous forces to be important. This is shown in Figure 10.1, where the boundary layer thickness is greatly exaggerated. (Around a typical airplane wing it is of order of a centimeter.) Thin viscous layers exist not only next to solid walls but also in the form of jets, wakes, and shear layers if the Reynolds number is sufficiently high. To be specific, we shall consider the case of a boundary layer next to a wall, adopting a curvilinear "boundary layer coordinate system" in which x is taken along the surface and y is taken normal to it. We shall refer to the solution of the irrotational flow outside the boundary layer as the "outer" problem and that of the boundary layer flow as the "inner" problem.

The thickness of the boundary layer varies with x; let $\bar{\delta}$ be the average thickness of the boundary layer over the length of the body. A measure of $\bar{\delta}$ can be obtained by considering the order of magnitude of the various terms in the equations of motion. The steady equation of motion for the longitudinal component of velocity is

$$u\frac{\partial u}{\partial x} + v\frac{\partial u}{\partial y} = -\frac{1}{\rho}\frac{\partial p}{\partial x} + \nu\left(\frac{\partial^2 u}{\partial x^2} + \frac{\partial^2 u}{\partial y^2}\right). \qquad (10.1)$$

The Cartesian form of the conservation laws is valid only when $\bar{\delta}/R \ll 1$, where R is the local radius of curvature of the body shape function. The more general curvilinear form for arbitrary $R(x)$ is given in Goldstein (1938) and Schlichting (1979). We generally expect $\bar{\delta}/R$ to be small for large Reynolds number flows over slender shapes. The first equation to be affected is the y-momentum equation where centrifugal acceleration will enter the normal component of the pressure gradient. In equation (10.1) we have also neglected body forces and any variations of ρ and μ. The essential features of viscous boundary layers can be more clearly illustrated without additional complications.

Let a characteristic magnitude of u in the flow field be U_∞, which can be identified with the upstream velocity at large distances from the body. Let L be the streamwise distance over which u changes appreciably. The longitudinal length of the body can serve as L, because u within the boundary layer does change by a large fraction of U_∞ in a distance L (Figure 10.2). A measure of $\partial u/\partial x$ is therefore U_∞/L, so that a

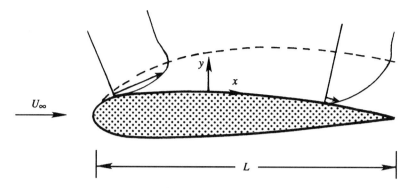

Figure 10.2   Velocity profiles at two positions within the boundary layer. The velocity arrows are drawn at the same distance y from the surface, showing that the variation of u with x is of the order of the free stream velocity U_∞. The boundary layer thickness is greatly exaggerated.

measure of the first advective term in equation (10.1) is

$$u\frac{\partial u}{\partial x} \sim \frac{U_\infty^2}{L}, \tag{10.2}$$

where $\sim$ is to be interpreted as "of order." We shall see shortly that the other advective term in equation (10.1) is of the same order. A measure of the viscous term in equation (10.1) is

$$\nu\frac{\partial^2 u}{\partial y^2} \sim \frac{\nu U_\infty}{\bar{\delta}^2}. \tag{10.3}$$

The magnitude of $\bar{\delta}$ can now be estimated by noting that the advective and viscous terms should be of the same order within the boundary layer, if viscous terms are to be important. Equating equations (10.2) and (10.3), we obtain

$$\bar{\delta} \sim \sqrt{\frac{\nu L}{U_\infty}} \quad \text{or} \quad \frac{\bar{\delta}}{L} \sim \frac{1}{\sqrt{\text{Re}}}.$$

This estimate of $\bar{\delta}$ can also be obtained by using results of unsteady parallel flows discussed in the preceding chapter, in which we saw that viscous effects diffuse to a distance of order $\sqrt{\nu t}$ in time t. As the time to flow along a body of length L is of order L/U_∞, the width of the diffusive layer at the end of the body is of order $\sqrt{\nu L/U_\infty}$.

A formal simplification of the equations of motion within the boundary layer can now be performed. The basic idea is that variations across the boundary layer are much faster than variations along the layer, that is

$$\frac{\partial}{\partial x} \ll \frac{\partial}{\partial y}, \qquad \frac{\partial^2}{\partial x^2} \ll \frac{\partial^2}{\partial y^2}.$$

The distances in the x-direction over which the velocity varies appreciably are of order L, but those in the y-direction are of order $\bar{\delta}$, which is much smaller than L.

Let us first determine a measure of the typical variation of v within the boundary layer. This can be done from an examination of the continuity equation $\partial u/\partial x + \partial v/\partial y = 0$. Because $u \gg v$ and $\partial/\partial x \ll \partial/\partial y$, we expect the two terms of the continuity equation to be of the same order. This requires $U_\infty/L \sim v/\bar{\delta}$, or that the variations of v are of order

$$v \sim \bar{\delta} U_\infty/L \sim U_\infty/\sqrt{\text{Re}}.$$

Next we estimate the magnitude of variation of pressure within the boundary layer. Experimental data on high Reynolds number flows show that the pressure distribution is nearly that in an irrotational flow around the body, implying that the pressure forces are of the order of the inertia forces. The requirement $\partial p/\partial x \sim \rho u(\partial u/\partial x)$ shows that the pressure variations within the flow field are of order

$$p - p_\infty \sim \rho U_\infty^2.$$

The proper nondimensional variables in the boundary layer are therefore

$$x' = \frac{x}{L}, \qquad y' = \frac{y}{\bar{\delta}} = \frac{y}{L}\sqrt{\text{Re}},$$

$$u' = \frac{u}{U_\infty}, \qquad v' = \frac{v}{\bar{\delta}U_\infty/L} = \frac{v}{U_\infty}\sqrt{\text{Re}}, \qquad p' = \frac{p - p_\infty}{\rho U^2}, \tag{10.4}$$

where $\bar{\delta} = \sqrt{vL/U_\infty}$. The important point to notice is that the distances across the boundary layer have been magnified or "stretched" by defining $y' = y/\bar{\delta} = (y/L)\sqrt{\text{Re}}$.

In terms of these nondimensional variables, the complete equations of motion for the boundary layer are

$$u'\frac{\partial u'}{\partial x'} + v'\frac{\partial u'}{\partial y'} = -\frac{\partial p'}{\partial x'} + \frac{1}{\text{Re}}\frac{\partial^2 u'}{\partial x'^2} + \frac{\partial^2 u'}{\partial y'^2}, \tag{10.5}$$

$$\frac{1}{\text{Re}}\left(u'\frac{\partial v'}{\partial x'} + v'\frac{\partial v'}{\partial y'}\right) = -\frac{\partial p'}{\partial y'} + \frac{1}{\text{Re}^2}\frac{\partial^2 v'}{\partial x'^2} + \frac{1}{\text{Re}}\frac{\partial^2 v'}{\partial y'^2}, \tag{10.6}$$

$$\frac{\partial u'}{\partial x'} + \frac{\partial v'}{\partial y'} = 0, \tag{10.7}$$

where we have defined $\text{Re} \equiv U_\infty L/v$ as an overall Reynolds number. In these equations, each of the nondimensional variables and their derivatives is of order one. For example, $\partial u'/\partial y' \sim 1$ in equation (10.5), essentially because the changes in u' and y' within the boundary layer are each of order one, a consequence of our normalization (10.4). It follows that the size of each term in the set (10.5) and (10.6) is determined by the presence of a multiplying factor involving the parameter Re. In particular, each term in equation (10.5) is of order one except the second term on the right-hand side, whose magnitude is of order $1/\text{Re}$. As $\text{Re} \to \infty$, these equations asymptotically become

$$u'\frac{\partial u'}{\partial x'} + v'\frac{\partial u'}{\partial y'} = -\frac{\partial p'}{\partial x'} + \frac{\partial^2 u'}{\partial y'^2},$$

$$0 = -\frac{\partial p'}{\partial y'},$$

$$\frac{\partial u'}{\partial x'} + \frac{\partial v'}{\partial y'} = 0.$$

The exercise of going through the nondimensionalization has been valuable, since it has shown what terms drop out under the boundary layer assumption. Transforming back to dimensional variables, the approximate equations of motion within the boundary layer are

$$u\frac{\partial u}{\partial x} + v\frac{\partial u}{\partial y} = -\frac{1}{\rho}\frac{\partial p}{\partial x} + \nu\frac{\partial^2 u}{\partial y^2}, \tag{10.8}$$

$$0 = -\frac{\partial p}{\partial y}, \tag{10.9}$$

$$\frac{\partial u}{\partial x} + \frac{\partial v}{\partial y} = 0. \tag{10.10}$$

Equation (10.9) says that the *pressure is approximately uniform across the boundary layer*, an important result. The pressure at the surface is therefore equal to that at the edge of the boundary layer, and so it can be found from a solution of the irrotational flow around the body. We say that the pressure is "imposed" on the boundary layer by the outer flow. *This justifies the experimental fact, pointed out in the preceding section, that the observed surface pressure is approximately equal to that calculated from the irrotational flow theory.* (A vanishing $\partial p/\partial y$, however, is not valid if the boundary layer separates from the wall or if the radius of curvature of the surface is not large compared with the boundary layer thickness. This will be discussed later in the chapter.) The pressure gradient at the edge of the boundary layer can be found from the inviscid Euler equation

$$-\frac{1}{\rho}\frac{dp}{dx} = U_e\frac{dU_e}{dx}, \tag{10.11}$$

or from its integral $p + \rho U_e^2/2 = \text{constant}$, which is the Bernoulli equation. This is because $v_e \sim 1/\sqrt{\text{Re}} \to 0$. Here $U_e(x)$ is the velocity at the *edge* of the boundary layer (Figure 10.1). This is the matching of the outer inviscid solution with the boundary layer solution in the overlap domain of common validity. However, instead of finding dp/dx at the edge of the boundary layer, as a first approximation we can apply equation (10.11) along the *surface* of the body, neglecting the existence of the boundary layer in the solution of the outer problem; the error goes to zero as the boundary layer becomes increasingly thin. In any event, the dp/dx term in equation (10.8) is to be regarded as known from an analysis of the outer problem, which must be solved before the boundary layer flow can be solved.

Equations (10.8) and (10.10) are then used to determine u and v in the boundary layer. The boundary conditions are

$$u(x, 0) = 0, \tag{10.12}$$

$$v(x, 0) = 0, \tag{10.13}$$

$$u(x, \infty) = U(x), \tag{10.14}$$

$$u(x_0, y) = u_{\text{in}}(y). \tag{10.15}$$

Condition (10.14) merely means that the boundary layer must join smoothly with the inviscid outer flow; points outside the boundary layer are represented by $y = \infty$, although we mean this strictly in terms of the nondimensional distance $y/\delta = (y/L)\sqrt{\text{Re}} \to \infty$. Condition (10.15) implies that an initial velocity profile $u_{\text{in}}(y)$ at some location x_0 is required for solving the problem. This is because the presence of the terms $u\,\partial u/\partial x$ and $v\,\partial^2 u/\partial y^2$ gives the boundary layer equations a parabolic character, with x playing the role of a timelike variable. Recall the Stokes problem of a suddenly accelerated plate, discussed in the preceding chapter, where the equation is $\partial u/\partial t = v\,\partial^2 u/\partial y^2$. In such problems governed by parabolic equations, the field at a certain time (or x in the problem here) depends only on its *past history*. Boundary layers therefore transfer effects only in the *downstream* direction. In contrast, the complete Navier–Stokes equations are of elliptic nature. Elliptic equations require specification on the bounding surface of the domain of solution. The Navier–Stokes equations are elliptic in velocity and thus require boundary conditions on the velocity (or its derivative normal to the boundary) upstream, downstream, and on the top and bottom boundaries, that is, all around. The upstream influence of the downstream boundary condition is always of concern in computations.

In summary, the simplifications achieved because of the thinness of the boundary layer are the following. First, diffusion in the x-direction is negligible compared to that in the y-direction. Second, the pressure field can be found from the irrotational flow theory, so that it is regarded as a known quantity in boundary layer analysis. Here, the boundary layer is so thin that the pressure does not change across it. Further, a crude estimate of the shear stress at the wall or skin friction is available from knowledge of the order of the boundary layer thickness $\tau_0 \sim \mu U/\bar{\delta} \sim (\mu U/L)\sqrt{\text{Re}}$. The skin friction coefficient is

$$\frac{\tau_0}{(1/2)\rho U^2} \sim \frac{2\mu U}{\rho L U^2}\sqrt{\text{Re}} \sim \frac{2}{\sqrt{\text{Re}}}.$$

As we shall see from the solutions to the problems in the following sections, this is indeed the correct order of magnitude. Only the finite numerical factor differs from problem to problem.

It is useful to compare equation (10.5) with equation (9.60), where we nondimensionalized both x- and y-directions by the same length scale. Notice that in equation (9.60) the Reynolds number multiplies *both* diffusion terms, whereas in equation (10.5) the diffusion term in the y-direction has been explicitly made order one by a normalization appropriate within the boundary layer.

3. Different Measures of Boundary Layer Thickness

As the velocity in the boundary layer smoothly joins that of the outer flow, we have to decide how to define the boundary layer thickness. The three common measures are described here.

The $u = 0.99U$ Thickness

One measure of the boundary thickness is the distance from the wall where the longitudinal velocity reaches 99% of the local free stream velocity, that is where $u = 0.99\,U$. We shall denote this as δ_{99}. This definition of the boundary layer thickness is however rather arbitrary, as we could very well have chosen the thickness as the point where $u = 0.95\,U$.

Displacement Thickness

A second measure of the boundary layer thickness, and one in which there is no arbitrariness, is the *displacement thickness* δ^*. This is defined as the distance by which the wall would have to be displaced outward in a hypothetical frictionless flow so as to maintain the same mass flux as in the actual flow. Let h be the distance from the wall to a point far outside the boundary layer (Figure 10.3). From the definition of δ^*, we obtain

$$\int_0^h u\,dy = U(h - \delta^*),$$

where the left-hand side is the actual mass flux below h and the right-hand side is the mass flux in the frictionless flow with the walls displaced by δ^*. Letting $h \to \infty$, the aforementioned gives

$$\boxed{\delta^* = \int_0^\infty \left(1 - \frac{u}{U}\right) dy.} \tag{10.16}$$

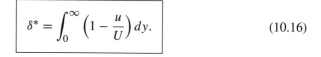

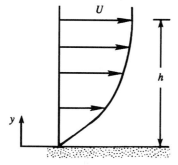 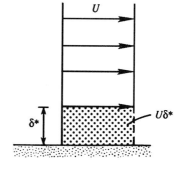

Figure 10.3   Displacement thickness.

The upper limit in equation (10.16) may be allowed to extend to infinity because, as we shall show in the following, $u/U \to 0$ exponentially fast in y as $y \to \infty$.

The concept of displacement thickness is used in the design of ducts, intakes of air-breathing engines, wind tunnels, etc. by first assuming a frictionless flow and then enlarging the passage walls by the displacement thickness so as to allow the same flow rate. Another use of δ^* is in finding dp/dx at the edge of the boundary layer, needed for solving the boundary layer equations. The first approximation is to neglect the existence of the boundary layer, and calculate the irrotational dp/dx over the body surface. A solution of the boundary layer equations gives the displacement thickness, using equation (10.16). The body surface is then displaced outward by this amount and a next approximation of dp/dx is found from a solution of the irrotational flow, and so on.

The displacement thickness can also be interpreted in an alternate and possibly more illuminating way. We shall now show that it is the distance by which the streamlines outside the boundary layer are displaced due to the presence of the boundary layer. Figure 10.4 shows the displacement of streamlines over a flat plate. Equating mass flux across two sections A and B, we obtain

$$Uh = \int_0^{h+\delta^*} u\,dy = \int_0^h u\,dy + U\delta^*,$$

which gives

$$U\delta^* = \int_0^h (U - u)\,dy.$$

Here h is any distance far from the boundary and can be replaced by ∞ without changing the integral, which then reduces to equation (10.16).

Momentum Thickness

A third measure of the boundary layer thickness is the momentum thickness θ, defined such that $\rho U^2 \theta$ is the momentum loss due to the presence of the boundary layer. Again choose a streamline such that its distance h is outside the boundary layer, and consider

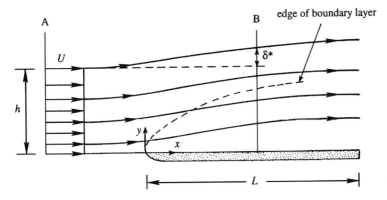

Figure 10.4 Displacement thickness and streamline displacement.

the momentum flux (=velocity times mass flow rate) below the streamline, per unit width. At section A the momentum flux is $\rho U^2 h$; that across section B is

$$\int_0^{h+\delta^*} \rho u^2 \, dy = \int_0^h \rho u^2 \, dy + \rho \, \delta^* U^2.$$

The loss of momentum due to the presence of the boundary layer is therefore the difference between the momentum fluxes across A and B, which is defined as $\rho U^2 \theta$:

$$\rho U^2 h - \int_0^h \rho u^2 \, dy - \rho \delta^* U^2 \equiv \rho U^2 \theta.$$

Substituting the expression for δ^* gives

$$\int_0^h (U^2 - u^2) \, dy - U^2 \int_0^h \left(1 - \frac{u}{U}\right) dy = U^2 \theta,$$

from which

$$\boxed{\theta = \int_0^\infty \frac{u}{U}\left(1 - \frac{u}{U}\right) dy,} \qquad (10.17)$$

where we have replaced h by ∞ because $u = U$ for $y > h$.

4. Boundary Layer on a Flat Plate with a Sink at the Leading Edge: Closed Form Solution

Although all other texts start their boundary layer discussion with the uniform flow over a semi-infinite flat plate, there is an even simpler related problem that can be solved in closed form in terms of elementary functions. We shall consider the large Reynolds number flow generated by a sink at the leading edge of a flat plate. The outer inviscid flow is represented by $\psi = m\theta/2\pi$, $m < 0$ so that $u_r = m/2\pi r$, $u_\theta = 0$ [Chapter 6, Section 5, equation (6.24) and Figure 6.6]. This represents radially inward flow towards the origin. A flat plate is now aligned with the x-axis so that its boundary is represented by $\theta = 0$. For large Re, the boundary layer is thin so $x = r \cos\theta \approx r$ because $\theta \ll 1$. For simplicity in what follows we shall absorb the 2π into the m by defining $m' = m/2\pi$ and then suppressing the prime. The velocity at the edge of the boundary layer is $U_e(x) = m/x$, $m < 0$ and the local Reynolds number is $U_e(x)x/\nu = m/\nu = \mathrm{Re}_x$. Boundary layer coordinates are used, as in Figure 10.1, with y normal to the plate and the origin at the leading edge.

The boundary layer equations (10.8)–(10.10) with equation (10.11) become

$$\frac{\partial u}{\partial x} + \frac{\partial v}{\partial y} = 0, \qquad u\frac{\partial u}{\partial x} + v\frac{\partial u}{\partial y} = -\frac{m^2}{x^3} + \nu\frac{\partial^2 u}{\partial y^2}$$

with the boundary conditions (10.12)–(10.15). We consider the limiting case $\mathrm{Re}_x = |m/\nu| \to \infty$. Because $m < 0$, the flow is from right (larger x) to left (smaller x), and the initial condition at $x = x_0$ is specified upstream, that is, at the largest x. The solution is then determined for all $x < x_0$, that is, downstream of the initial location. The natural way to make the variables dimensionless and finite in the boundary layer is to normalize x by x_0, y by $x_0/\sqrt{Re_x}$, u by m/x_0, v by $m/(x_0\sqrt{Re_x})$. The problem is fully two-dimensional and well posed for any reasonable initial condition (10.15). Now, suppress the initial condition. The length scale x_0, crucial to rendering the problem properly dimensionless, has disappeared. How is one to construct a dimensionless formulation? We have seen before that this situation results in a reduction in the dimensionality of the space required for the solution. The variable y can be made dimensionless only by x and must be stretched by $\sqrt{Re_x}$ to be finite in the boundary layer. The unique choice is then $(y/x)\sqrt{Re_x} = (y/x)\sqrt{|m/\nu|} = \eta$. This is consistent with the similarity variable for Stokes' first problem $\eta = y/\sqrt{\nu t}$ when t is taken to be x/U and $U = m/x$. Finite numerical factors are irrelevant here. Further, we note that we have found that $\delta \sim x_0/\sqrt{Re_{x_0}}$ so with the x_0 scale absent, $\delta \sim x/\sqrt{|m/\nu|}$ and $\eta = y/\delta$. Next we will reduce mass and momentum conservation to an ordinary differential equation for the xsimilarity streamfunction. To reverse the flow we will define the streamfunction ψ via $u = -\partial\psi/\partial y$, $v = \partial\psi/\partial x$ (note sign change). We now have:

$$\frac{\partial\psi}{\partial y}\frac{\partial^2\psi}{\partial y\,\partial x} - \frac{\partial\psi}{\partial x}\frac{\partial^2\psi}{\partial y^2} = -\frac{m^2}{x^3} - \nu\frac{\partial^3\psi}{\partial y^3},$$

$$y = 0: \quad \psi = \frac{\partial\psi}{\partial y} = 0,$$

$$y \to \text{overlap with inviscid flow:} \quad \frac{\partial\psi}{\partial y} \to \frac{m}{x}.$$

The streamfunction is made dimensionless by its order of magnitude and put in similarity form via

$$\psi(x, y) = U_e\delta(x)f(\eta) = U_e(x) \cdot \frac{x}{\sqrt{Re_x}}f(\eta)$$

$$= \sqrt{\nu U_e(x) \cdot x}\,f(\eta) = \sqrt{|\nu m|}\,f(\eta),$$

in this problem. The problem for f reduces to

$$f'''(\eta) - f'^2 = -1,$$

$$f(0) = 0, \qquad f'(0) = 0, \qquad f'(\infty) = 1.$$

This may be solved in closed form with the result

$$\frac{u}{U_e(x)} = f'(\eta) = 3\left[\frac{1 - \alpha e^{-\sqrt{2}\eta}}{1 + \alpha e^{-\sqrt{2}\eta}}\right]^2 - 2, \qquad \alpha = \frac{\sqrt{3} - \sqrt{2}}{\sqrt{3} + \sqrt{2}} = 0.101\ldots.$$

A result equivalent to this was first obtained by Pohlhausen (1921) in his solution for flow in a convergent channel. From this simple solution we can establish several properties characteristic of laminar boundary layers. First, as $\eta \to \infty$, the matching with the inviscid solution occurs exponentially fast, as $f'(\eta) \sim 1 - 12\alpha e^{-\sqrt{2}\eta} +$ smaller terms as $\eta \to \infty$.

Next v/U_e is of the correct small order,

$$\frac{v}{U_e} = \frac{y}{x} f'(\eta) = \frac{1}{\sqrt{\mathrm{Re}_x}} \eta f'(\eta) \sim \frac{1}{\sqrt{\mathrm{Re}_x}}.$$

The behavior of the displacement thickness is obtained from the definition

$$\delta^* = \int_0^\infty \left(1 - \frac{u}{U_e}\right) dy = \int_0^\infty [1 - f'(\eta)]\, d\eta \cdot \frac{x}{\sqrt{\mathrm{Re}_x}},$$

$$\frac{\delta^*}{x} = \frac{1}{\sqrt{\mathrm{Re}_x}} \int_0^\infty [1 - f'(\eta)]\, d\eta = \frac{12\alpha}{[(1+\alpha)\sqrt{2}\sqrt{\mathrm{Re}_x}]} = \frac{0.7785}{\sqrt{\mathrm{Re}_x}} \sim \frac{1}{\sqrt{\mathrm{Re}_x}}.$$

The shear stress at the wall is

$$\tau_0 = \mu \left.\frac{\partial u}{\partial y}\right|_0 = -\mu \frac{m}{x^2} \sqrt{\left|\frac{m}{\nu}\right|} f''(0), \qquad f''(0) = \frac{2}{\sqrt{3}}.$$

Then the skin friction coefficient is

$$C_f = \frac{\tau_0}{(1/2)\rho U_e^2} = \frac{-4/\sqrt{3}}{\sqrt{\mathrm{Re}_x}}, \qquad \mathrm{Re}_x = \left|\frac{m}{\nu}\right|.$$

Aside from numerical factors, which are obviously problem specific, the preceding results are universally valid for all similarity solutions of the laminar boundary layer equations. $U_e(x)$ is the velocity at the edge of the boundary layer and $\mathrm{Re}_x = U_e(x)x/\nu$. In these terms

$$\eta = \frac{y}{x}\sqrt{\mathrm{Re}_x}, \qquad \psi(x, y) = \sqrt{\nu U_e(x) \cdot x}\ f(\eta),$$

$f(\eta) = u/U_e(x) \to 1$ exponentially fast as $\eta \to \infty$. We find $v/U_e \sim 1/\sqrt{\mathrm{Re}_x}$, $\delta^*/x \sim 1/\sqrt{\mathrm{Re}_x}$, $C_f \sim 1/\sqrt{\mathrm{Re}_x}$.

Axisymmetric Problem

Now let us consider the axially symmetric version of the problem we just solved. This is the flow in the neighborhood of an infinite flat plate generated by a sink in the center of the plate. The inviscid outer flow is $u_r = -Q/r^2$ where r is the spherical radial coordinate centered on the sink. The boundary layer adjacent to the plate is best treated in cylindrical coordinates r, θ, z with $\partial/\partial\theta = 0$ (see Figure 10.5). Mass conservation for a constant density flow with symmetry about the z-axis is

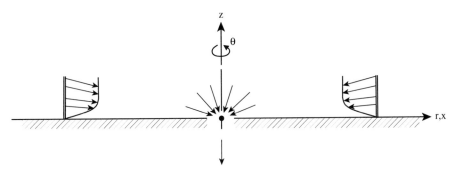

Figure 10.5 Axisymmetric flow into a sink at the center of an infinite plate.

$\partial/\partial r(ru_r) + \partial/\partial z(ru_z) = 0$. In the following, the streamwise coordinate r is replaced by x. Since $U_e = -Q/x^2$, the local Reynolds number can be written as $\text{Re}_x = U_e x/\nu = Q/x\nu$. Assuming this is sufficiently large, the full Navier–Stokes equations reduce to the boundary layer equations with an error that is small in powers of inverse Re_x. Thus we seek to solve

$$u\partial u/\partial x + w\partial u/\partial z = U_e dU_e/dx + \nu\partial^2 u/\partial z^2$$

subject to $u = w = 0$ on $z = 0$ and $u \rightarrow U_e$ as z leaves the boundary layer. A similarity solution can be obtained provided the requirement for an initial velocity distribution is not imposed. First, the streamwise momentum equation is put in terms of the axisymmetric streamfunction, $\mathbf{u} = -(\mathbf{1}_\theta/x) \times \nabla\psi$, so that $xu = -\partial\psi/\partial z$, $xw = \partial\psi/\partial x$. With the modification of the streamfunction due to axial symmetry, the universal dimensionless similarity form becomes

$$\psi(x, z) = x[\nu x U_e(x)]^{1/2} f(\eta) = (\nu Q x)^{1/2} f(\eta)$$

where $\eta = (z/x)(\text{Re}_x)^{1/2} = (Q/\nu)^{1/2} z/x^{3/2}$. The velocity components transform to $u = -x^{-1}\partial\psi/\partial z = U_e f'(\eta)$, $w = [(\nu Q)^{1/2}/(2x^{3/2})](f - 3\eta f') = \{U_e/[2(\text{Re}_x)^{1/2}]\}(f - 3\eta f')$.

The streamwise momentum equation transforms to

$$f''' - (1/2)ff'' + 2(1 - f'^2) = 0$$

subject to   (10.18)

$$f(0) = 0, \quad f'(0) = 0, \quad f'(\infty) = 1.$$

Rosenhead provides a tabulation of the solution to $f''' - ff'' + 4(1 - f'^2) = 0$, which is related to the equation above by the scaling $\eta/2^{1/2}$, and $f/2^{1/2}$. (We have tried not to add extraneous numerical factors to our universal dimensionless similarity scaling.) The solution to (10.18) is displayed in Figure 10.6.

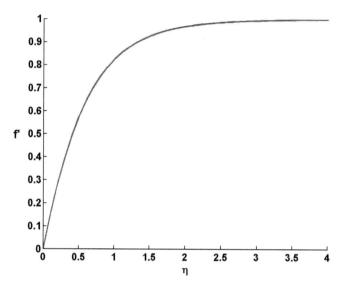

Figure 10.6   Dimensionless velocity profile for flow illustrated in Figure 10.5.

5. *Boundary Layer on a Flat Plate: Blasius Solution*

We shall next discuss the classic problem of the boundary layer on a semi-infinite flat plate. Equations (10.8)–(10.10) are a valid asymptotic representation of the full Navier–Stokes equations in the limit $Re_x \to \infty$. Thus with x measured from the leading edge, the initial station x_0 (see equation (10.15)) must be sufficiently far downstream that $U_e x_0/\nu \gg 1$. A major question in boundary layer theory is the extent of downstream memory of the initial state. If the external stream $U_e(x)$ admits a similarity solution, is the initial condition forgotten and how soon? Serrin (1967) and Peletier (1972) showed that for favorable pressure gradients ($U_e \, dU_e/dx$) of similarity form, the initial condition is forgotten and the larger the acceleration the sooner similarity is achieved. A decelerating flow will accentuate details of the initial state and similarity will never be found despite its mathematical admissability. This is consistent with the experimental findings of Gallo *et al.* (1970). A flat plate for which $U_e(x) = U = $ const. is the borderline case; similarity is eventually achieved here. In the previous problem, the sink creates a rapidly accelerating flow so that, if we could ever realize such a flow, similarity would be achieved quickly.

As the inviscid solution gives $u = U = $ const. everywhere, $\partial p/\partial x = 0$ and the equations become

$$u\frac{\partial u}{\partial x} + v\frac{\partial u}{\partial y} = \nu\frac{\partial^2 u}{\partial y^2},$$

$$\frac{\partial u}{\partial x} + \frac{\partial v}{\partial y} = 0,$$

(10.19)

subject to: $y = 0$: $u = v = 0$, $x > 0$

$$y \to \text{ overlap at edge of boundary layer: } \quad u \to U,$$

$$x = x_0: \quad u(y) \text{ given, } \text{Re}_{x_0} \gg 1. \tag{10.20}$$

For x large compared with x_0, we can argue that the initial condition is forgotten. With x_0 no longer available for rendering the independent variables dimensionless, a similarity solution will be obtained. Using our previous results,

$$\psi(x, y) = \sqrt{\nu U x}\, f(\eta), \quad \eta = \frac{y}{x}\sqrt{\text{Re}_x}, \quad \text{Re}_x = \frac{Ux}{\nu},$$

and $u = \partial\psi/\partial y$, $v = -\partial\psi/\partial x$. Now $u/U = f'(\eta)$ and

$$f''' + \tfrac{1}{2} f f'' = 0,$$

$$f(0) = f'(0) = 0, \quad f(\infty) = 1.$$

A different but equally correct method of obtaining the similarity form is described in what follows. The plate length L (Figure 10.4) has been taken very large so a solution independent of L has been sought. In addition, we limit our consideration to a domain far downstream of x_0 so the initial condition has been forgotten.

Similarity Solution—Alternative Procedure

We shall regard $\delta(x)$ as an unknown function in the following analysis; the form of $\delta(x)$ will follow from a requirement that a similarity solution must exist for this problem.

As there is no externally imposed length scale along x, the solutions at various downstream locations must be self similar. Blasius, a student of Prandtl, showed that a similarity solution can indeed be found for this problem. Clearly, the velocity distributions at various downstream points can collapse into a single curve only if the solution has the form

$$\frac{u}{U} = g(\eta), \tag{10.21}$$

where

$$\eta = \frac{y}{\delta(x)}. \tag{10.22}$$

At this point it is useful to pause a little and compare the situation with that of a suddenly accelerated plate (see Chapter 9, Section 7), for which similarity solutions exist. In that case we argued that the parameter U drops out of the equations and boundary conditions if we define u/U as the dependent variable, leading to $u/U = f(y, t, \nu)$. A dimensional analysis then immediately showed that the functional form must be $u/U = F[y/\delta(t)]$, where $\delta(t) \sim \sqrt{\nu t}$. In the current problem the downstream distance is timelike, but we cannot analogously write $u/U = f(y, x, \nu)$, because ν cannot be made nondimensional with the help of x or y. The dynamic reason for this is that U cannot be eliminated from the problem simply by regarding u/U as the dependent variable, because U still remains in the problem through the dependence of δ on U. The correct dimensional argument in this case is that we must have a solution of the form $u/U = g[y/\delta(x)]$, where $\delta(x)$ is a function of (U, x, ν) and therefore can only be of the form $\delta \sim \sqrt{\nu x/U}$.

We now resume our search for a similarity solution for the flat plate boundary layer. As the problem is two-dimensional, it is easier to work with the streamfunction defined by

$$u \equiv \frac{\partial \psi}{\partial y}, \quad v \equiv -\frac{\partial \psi}{\partial x}.$$

Using the similarity form (10.21), we obtain

$$\psi = \int_0^y u \, dy = \delta \int_0^\eta u \, d\eta = \delta \int_0^\eta U g(\eta) \, d\eta = U\delta f(\eta), \tag{10.23}$$

where we have defined

$$g(\eta) \equiv \frac{df}{d\eta}. \tag{10.24}$$

(Equation (10.23) shows that the similarity form for the stream function is $\psi/U\delta = f(\eta)$, signifying that the scale for the streamfunction is proportional to the local flow rate $U\delta$.)

In terms of the streamfunction, the governing sets (10.19) and (10.20) become

$$\frac{\partial \psi}{\partial y} \frac{\partial^2 \psi}{\partial x \, \partial y} - \frac{\partial \psi}{\partial x} \frac{\partial^2 \psi}{\partial y^2} = \nu \frac{\partial^3 \psi}{\partial y^3}, \tag{10.25}$$

subject to

$$\frac{\partial \psi}{\partial y} = \psi = 0 \quad \text{at } y = 0, \quad x > 0,$$

$$\frac{\partial \psi}{\partial y} \to U \qquad \text{as } \frac{y}{\delta} \to \infty. \tag{10.26}$$

To express sets (10.25) and (10.26) in terms of the similarity streamfunction $f(\eta)$, we find the following derivatives from equation (10.23):

$$\frac{\partial \psi}{\partial x} = U \left[f \frac{d\delta}{dx} + \delta \frac{\partial f}{\partial x} \right] = U \frac{d\delta}{dx} [f - f'\eta], \tag{10.27}$$

$$\frac{\partial^2 \psi}{\partial x \, \partial y} = U \frac{d\delta}{dx} \frac{\partial}{\partial y} [f - f'\eta] = -\frac{U\eta f''}{\delta} \frac{d\delta}{dx}, \tag{10.28}$$

$$\frac{\partial \psi}{\partial y} = Uf', \tag{10.29}$$

$$\frac{\partial^2 \psi}{\partial y^2} = \frac{U f''}{\delta},$$
(10.30)

$$\frac{\partial^3 \psi}{\partial y^3} = \frac{U f'''}{\delta^2},$$
(10.31)

where primes on f denote derivatives with respect to η. Substituting these derivatives in equation (10.25) and canceling terms, we obtain

$$-\left(\frac{U\delta}{\nu} \frac{d\delta}{dx}\right) ff'' = f'''.$$
(10.32)

In equation (10.32), f and its derivatives do not explicitly depend on x. The equation can be valid only if

$$\frac{U\delta}{\nu} \frac{d\delta}{dx} = \text{const.}$$

Choosing the constant to be $\frac{1}{2}$ for eventual algebraic simplicity, an integration gives

$$\delta = \sqrt{\frac{\nu x}{U}}.$$
(10.33)

Equation (10.32) then becomes

$$\frac{1}{2} ff'' + f''' = 0.$$
(10.34)

In terms of f, the boundary conditions (10.26) become

$$f'(\infty) = 1,$$
$$f(0) = f'(0) = 0.$$
(10.35)

A series solution of the nonlinear equation (10.34), subject to equation (10.35), was given by Blasius. It is much easier to solve the problem with a computer, using for example the Runge–Kutta technique. The resulting profile of $u/U = f'(\eta)$ is shown in Figure 10.7. The solution makes the profiles at various downstream distances collapse into a single curve of u/U vs $y\sqrt{U/\nu x}$, and is in excellent agreement with experimental data for laminar flows at high Reynolds numbers. The profile has a point of inflection (that is, zero curvature) at the wall, where $\partial^2 u/\partial y^2 = 0$. This is a result of the absence of pressure gradient in the flow and will be discussed in Section 7.

Matching with External Stream

We find in this case that the difference between f' and $1 \sim (1/\eta)e^{-\eta^2/4} \to 0$ exponentially fast as $\eta \to \infty$.

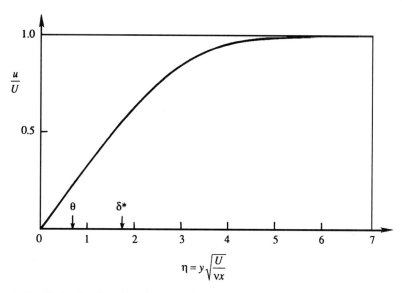

Figure 10.7   The Blasius similarity solution of velocity distribution in a laminar boundary layer on a flat plate. The momentum thickness θ and displacement δ^* are indicated by arrows on the horizontal axis.

Transverse Velocity

The lateral component of velocity is given by $v = -\partial\psi/\partial x$. From equation (10.27), this becomes

$$v = \frac{1}{2}\sqrt{\frac{\nu U}{x}}(\eta f' - f), \quad \frac{v}{U} = \frac{1}{2\sqrt{\text{Re}_x}}(\eta f' - f) \sim \frac{0.86}{\sqrt{\text{Re}_x}} \quad \text{as} \quad \eta \to \infty,$$

a plot of which is shown in Figure 10.8. The transverse velocity increases from zero at the wall to a maximum value at the edge of the boundary layer, a pattern that is in agreement with the streamline shapes sketched in Figure 10.4.

Boundary Layer Thickness

From Figure 10.7, the distance where $u = 0.99\,U$ is $\eta = 4.9$. Therefore

$$\boxed{\delta_{99} = 4.9\sqrt{\frac{\nu x}{U}}} \quad \text{or} \quad \frac{\delta_{99}}{x} = \frac{4.9}{\sqrt{\text{Re}_x}}, \tag{10.36}$$

where we have defined a *local* Reynolds number

$$\text{Re}_x \equiv \frac{Ux}{\nu}$$

The parabolic growth ($\delta \propto \sqrt{x}$) of the boundary layer thickness is in good agreement with experiments. For air at ordinary temperatures flowing at $U = 1\,\text{m/s}$, the Reynolds number at a distance of 1 m from the leading edge is $\text{Re}_x = 6 \times 10^4$, and equation (10.36) gives $\delta_{99} = 2\,\text{cm}$, showing that the boundary layer is indeed thin.

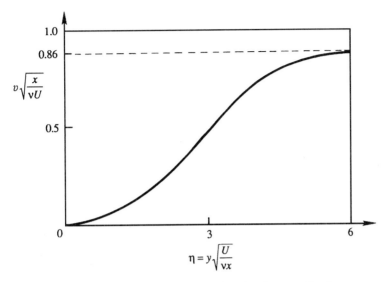

Figure 10.8   Transverse velocity component in a laminar boundary layer on a flat plate.

The displacement and momentum thicknesses, defined in equations (10.16) and (10.17), are found to be

$$\delta^* = 1.72\sqrt{\nu x/U},$$

$$\theta = 0.664\sqrt{\nu x/U}.$$

These thicknesses are indicated along the abscissa of Figure 10.7.

Skin Friction

The local wall shear stress is $\tau_0 = \mu(\partial u/\partial y)_0 = \mu(\partial^2 \psi/\partial y^2)_0$, where the subscript zero stands for $y = 0$. Using $\partial^2 \psi/\partial y^2 = Uf''/\delta$ given in equation (10.30), we obtain $\tau_0 = \mu U f''(0)/\delta$, and finally

$$\tau_0 = \frac{0.332\rho U^2}{\sqrt{\mathrm{Re}_x}}. \tag{10.37}$$

The wall shear stress therefore decreases as $x^{-1/2}$, a result of the thickening of the boundary layer and the associated decrease of the velocity gradient. Note that the wall shear stress at the leading edge is predicted to be infinite. Clearly the boundary layer theory breaks down near the leading edge where the assumption $\partial/\partial x \ll \partial/\partial y$ is invalid. The local Reynolds number Re_x in the neighborhood of the leading edge is of order 1, for which the boundary layer assumptions are not valid.

The wall shear stress is generally expressed in terms of the nondimensional *skin friction coefficient*

$$C_f \equiv \frac{\tau_0}{(1/2)\rho U^2} = \frac{0.664}{\sqrt{\mathrm{Re}_x}}. \tag{10.38}$$

The drag force per unit width on one side of a plate of length L is

$$D = \int_0^L \tau_0 \, dx = \frac{0.664\rho U^2 L}{\sqrt{\mathrm{Re}_L}},$$

where we have defined $\mathrm{Re}_L \equiv UL/\nu$ as the Reynolds number based on the plate length. This equation shows that the drag force is proportional to the $\frac{3}{2}$ power of velocity. This should be compared with small Reynolds number flows, where the drag is proportional to the first power of velocity. We shall see later in the chapter that the drag on a *blunt* body in a high Reynolds number flow is proportional to the *square* of velocity.

The overall *drag coefficient* defined in the usual manner is

$$C_D \equiv \frac{D}{(1/2)\rho U^2 L} = \frac{1.33}{\sqrt{\mathrm{Re}_L}}. \tag{10.39}$$

It is clear from equations (10.38) and (10.39) that

$$C_D = \frac{1}{L} \int_0^L C_f \, dx,$$

which says that the overall drag coefficient is the average of the local friction coefficient (Figure 10.9).

We must keep in mind that carrying out an integration from $x = 0$ is of questionable validity because the equations and solutions are valid only for very large Re_x.

Falkner–Skan Solution of the Laminar Boundary Layer Equations

No discussion of laminar boundary layer similarity solutions would be complete without mention of the work of V. W. Falkner and S. W. Skan (1931). They found

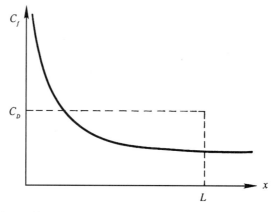

Figure 10.9   Friction coefficient and drag coefficient in a laminar boundary layer on a flat plate.

that $U_e(x) = ax^n$ admits a similarity solution, as follows. We assume that $\text{Re}_x = ax^{(n+1)}/\nu$ is sufficiently large so that the boundary layer equations are valid and any dependence on an initial condition has been forgotten. Then the initial station x_0 disappears from the problem and we may write

$$\psi(x, y) = \sqrt{\nu U_e(x) \cdot x}\; f(\eta) = \sqrt{\nu a}\; x^{(n+1)/2}\; f(\eta),$$

$$\eta = \frac{y}{x}\sqrt{\text{Re}_x} = \sqrt{\frac{a}{\nu}}\; yx^{(n-1)/2}.$$

Then $u/U_e = f'(\eta)$ and $U_e(dU_e/dx) = na^2x^{2n-1}$.

The x-momentum equation reduces to the similarity form

$$f''' + \frac{n+1}{2}\, ff'' - nf'^2 + n = 0, \tag{10.40}$$

$$f(0) = 0, \quad f'(0) = 0, \quad f'(\infty) = 1. \tag{10.41}$$

The Blasius equation (10.34) and (10.35) is a special case for $n = 0$, that is, $U_e(x) = U$. Although there are similarity solutions possible for $n < 0$, these are not likely to be seen in practice. For $n \geqslant 0$, all solutions of equations (10.40) and (10.41) have the proper behavior as detailed in the preceding. The numerical coefficients depend on n. Solutions to equations (10.40) and (10.41) are displayed in Figure 5.9.1 of Batchelor (1967) and reproduced here in Figure 10.10. They show a monotonically increasing shear stress $[f''(0)]$ as n increases. For $n = -0.0904$, $f''(0) = 0$ so $\tau_0 = 0$ and separation is imminent all along the surface. Solutions for $n < -0.0904$ do not represent boundary layers. In most real flows, similarity solutions are not available and the boundary layer equations with boundary and initial conditions as written in

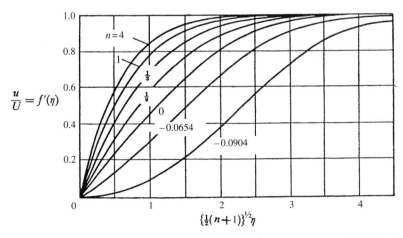

Figure 10.10 Velocity distribution in the boundary layer for external stream $U_e = ax^n$. G. K Batchelor, *An Introduction to Fluid Dynamics*, 1st ed. (1967), reprinted with the permission of Cambridge University Press.

equations (10.8)–(10.15) must be solved. A simple approximate procedure, the von Karman momentum integral, is discussed in the next section. More often the equations will be integrated numerically by procedures that are discussed in more detail in Chapter 11.

Breakdown of Laminar Solution

Agreement of the Blasius solution with experimental data breaks down at large downstream distances where the local Reynolds number Re_x is larger than some critical value, say Re_{cr}. At these Reynolds numbers the laminar flow becomes unstable and a transition to turbulence takes place. The critical Reynolds number varies greatly with the surface roughness, the intensity of existing fluctuations (that is, the degree of steadiness) within the outer irrotational flow, and the shape of the leading edge. For example, the critical Reynolds number becomes lower if either the roughness of the wall surface or the intensity of fluctuations in the free stream is increased. Within a factor of 5, the critical Reynolds number for a boundary layer over a flat plate is found to be

$$Re_{cr} \sim 10^6 \qquad \text{(flat plate)}.$$

Figure 10.11 schematically depicts the flow regimes on a semi-infinite flat plate. For finite $Re_x = Ux/\nu \sim 1$, the full Navier–Stokes equations are required to describe the leading edge region properly. As Re_x gets large at the downstream limit of the leading edge region, we can locate x_0 as the maximal upstream extent of the boundary layer

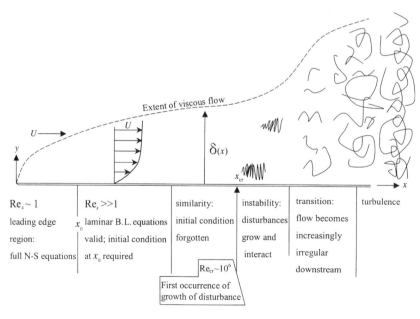

Figure 10.11   Schematic depiction of flow over a semiinfinite flat plate.

equations. For some distance $x > x_0$, the initial condition is remembered. Finally, the influence of the initial condition may be neglected and the solution becomes of similarity form. For somewhat larger Re_x, a bit farther downstream, the first instability appears. Then a band of waves becomes amplified and interacts nonlinearly through the advective acceleration. As Re_x increases, the flow becomes increasingly chaotic and irregular in the downstream direction. For lack of a better word, this is called transition. Eventually, the boundary layer becomes fully turbulent with a significant increase in shear stress at the plate τ_0.

After undergoing transition, the boundary layer thickness grows faster than $x^{1/2}$ (Figure 10.11), and the wall shear stress increases faster with U than in a laminar boundary layer; in contrast, the wall shear stress for a laminar boundary layer varies as $\tau_0 \propto U^{1.5}$. The increase in resistance is due to the greater macroscopic mixing in a turbulent flow.

Figure 10.12 sketches the nature of the observed variation of the drag coefficient in a flow over a flat plate, as a function of the Reynolds number. The lower curve applies if the boundary layer is laminar over the entire length of the plate, and the upper curve applies if the boundary layer is turbulent over the entire length. The curve joining the two applies if the boundary layer is laminar over the initial part and turbulent over the remaining part, as in Figure 10.11. The exact point at which the observed drag deviates from the wholly laminar behavior depends on experimental conditions and the transition shown in Figure 10.12 is at $\text{Re}_{cr} = 5 \times 10^5$.

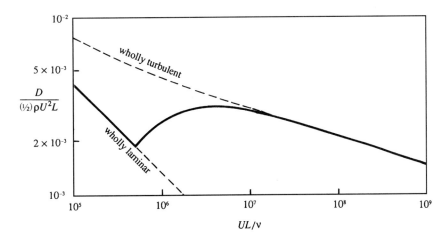

Figure 10.12 Measured drag coefficient for a boundary layer over a flat plate. The continuous line shows the drag coefficient for a plate on which the flow is partly laminar and partly turbulent, with the transition taking place at a position where the local Reynolds number is 5×10^5. The dashed lines show the behavior if the boundary layer was either completely laminar or completely turbulent over the entire length of the plate.

6. *von Karman Momentum Integral*

Exact solutions of the boundary layer equations are possible only in simple cases, such as that over a flat plate. In more complicated problems a frequently applied approximate method satisfies only an *integral* of the boundary layer equations across the layer thickness. The integral was derived by von Karman in 1921 and applied to several situations by Pohlhausen.

The point of an integral formulation is to obtain the information that is really required with minimum effort. The important results of boundary layer calculations are the wall shear stress, displacement thickness, and separation point. With the help of the von Karman momentum integral derived in what follows and additional correlations, these results can be obtained easily.

The equation is derived by integrating the boundary layer equation

$$u\frac{\partial u}{\partial x} + v\frac{\partial u}{\partial y} = U\frac{dU}{dx} + v\frac{\partial^2 u}{\partial y^2},$$

from $y = 0$ to $y = h$, where $h > \delta$ is any distance outside the boundary layer. Here the pressure gradient term has been expressed in terms of the velocity $U(x)$ at the edge of the boundary layer, where the inviscid Euler equation applies. Adding and subtracting $u(dU/dx)$, we obtain

$$(U - u)\frac{dU}{dx} + u\frac{\partial(U - u)}{\partial x} + v\frac{\partial(U - u)}{\partial y} = -v\frac{\partial^2 u}{\partial y^2}. \qquad (10.42)$$

Integrating from $y = 0$ to $y = h$, the various terms of this equation transform as follows.

The first term gives

$$\int_0^h (U - u)\frac{dU}{dx}\,dy = U\delta^*\frac{dU}{dx}.$$

Integrating by parts, the third term gives,

$$\int_0^h v\frac{\partial(U - u)}{\partial y}\,dy = \left[v(U - u)\right]_0^h - \int_0^h \frac{\partial v}{\partial y}(U - u)\,dy$$

$$= \int_0^h \frac{\partial u}{\partial x}(U - u)\,dy,$$

where we have used the continuity equation and the conditions that $v = 0$ at $y = 0$ and $u = U$ at $y = h$. The last term in equation (10.42) gives

$$-v\int_0^h \frac{\partial^2 u}{\partial y^2}\,dy = \frac{\tau_0}{\rho},$$

where τ_0 is the wall shear stress.

The integral of equation (10.42) is therefore

$$U\delta^*\frac{dU}{dx} + \int_0^h \left[u\frac{\partial(U-u)}{\partial x} + (U-u)\frac{\partial u}{\partial x} \right] dy = \frac{\tau_0}{\rho}. \tag{10.43}$$

The integral in equation (10.43) equals

$$\int_0^h \frac{\partial}{\partial x}[u(U-u)]\,dy = \frac{d}{dx}\int_0^h u(U-u)\,dy = \frac{d}{dx}(U^2\theta),$$

where θ is the momentum thickness defined by equation (10.17). Equation (10.43) then gives

$$\boxed{\frac{d}{dx}(U^2\theta) + \delta^*U\frac{dU}{dx} = \frac{\tau_0}{\rho},} \tag{10.44}$$

which is called the *Karman momentum integral equation*. In equation (10.44), θ, δ^*, and τ_0 are all unknown. Additional assumptions must be made or correlations provided to obtain a useful solution. It is valid for both laminar and turbulent boundary layers. In the latter case τ_0 cannot be equated to molecular viscosity times the velocity gradient and should be empirically specified. The procedure of applying the integral approach is to assume a reasonable velocity distribution, satisfying as many conditions as possible. Equation (10.44) then predicts the boundary layer thickness and other parameters.

The approximate method is only useful in situations where an exact solution does not exist. For illustrative purposes, however, we shall apply it to the boundary layer over a flat plate where $U(dU/dx) = 0$. Using definition (10.17) for θ, equation (10.44) reduces to

$$\frac{d}{dx}\int_0^\delta (U-u)u\,dy = \frac{\tau_0}{\rho}. \tag{10.45}$$

Assume a cubic profile

$$\frac{u}{U} = a + b\frac{y}{\delta} + c\frac{y^2}{\delta^2} + d\frac{y^3}{\delta^3}.$$

The four conditions that we can satisfy with this profile are chosen to be

$$u = 0, \qquad \frac{\partial^2 u}{\partial y^2} = 0 \quad \text{at} \quad y = 0,$$

$$u = U, \qquad \frac{\partial u}{\partial y} = 0 \quad \text{at} \quad y = \delta.$$

The condition that $\partial^2 u/\partial y^2 = 0$ at the wall is a requirement in a boundary layer over a flat plate, for which an application of the equation of motion (10.8) gives $v(\partial^2 u/\partial y^2)_0 = U(dU/dx) = 0$. Determination of the four constants reduces the assumed profile to

$$\frac{u}{U} = \frac{3}{2}\left(\frac{y}{\delta}\right) - \frac{1}{2}\left(\frac{y}{\delta}\right)^3.$$

The terms on the left- and right-hand sides of the momentum equation (10.45) are then

$$\int_0^\delta (U - u)u\,dy = \frac{39}{280}U^2\delta,$$

$$\frac{\tau_0}{\rho} = v\left(\frac{\partial u}{\partial y}\right)_0 = \frac{3}{2}\frac{U v}{\delta}.$$

Substitution into the momentum integral equation gives

$$\frac{39U^2}{280}\frac{d\delta}{dx} = \frac{3}{2}\frac{U v}{\delta}.$$

Integrating in x and using the condition $\delta = 0$ at $x = 0$, we obtain

$$\delta = 4.64\sqrt{vx/U},$$

which is remarkably close to the exact solution (10.36). The friction factor is

$$C_f = \frac{\tau_0}{(1/2)\rho U^2} = \frac{(3/2)U v/\delta}{(1/2)U^2} = \frac{0.646}{\sqrt{\mathrm{Re}_x}},$$

which is also very close to the exact solution of equation (10.38).

Pohlhausen found that a fourth-degree polynomial was necessary to exhibit sensitivity of the velocity profile to the pressure gradient. Adding another term below equation (10.45), $e(y/\delta)^4$ requires an additional boundary condition, $\partial^2 u/\partial y^2 = 0$ at $y = \delta$. With the assumption of a form for the velocity profile, equation (10.44) may be reduced to an equation with one unknown, $\delta(x)$ with $U(x)$, or the pressure gradient specified. This equation was solved approximately by Pohlhausen in 1921. This is described in Yih (1977, pp. 357–360). Subsequent improvements by Holstein and Bohlen (1940) are recounted in Schlichting (1979, pp. 206–217) and Rosenhead (1988, pp. 293–297). Sherman (1990, pp. 322–329) related the approximate solution due to Thwaites.

7. Effect of Pressure Gradient

So far we have considered the boundary layer on a flat plate, for which the pressure gradient of the external stream is zero. Now suppose that the surface of the body is curved (Figure 10.13). Upstream of the highest point the streamlines of the outer flow converge, resulting in an increase of the free stream velocity $U(x)$ and a consequent

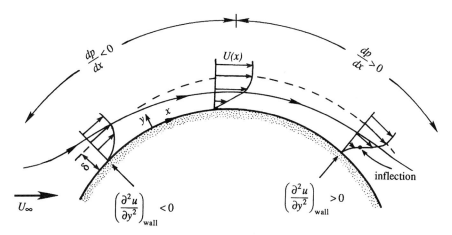

Figure 10.13   Velocity profiles across boundary layers with favorable and adverse pressure gradients.

fall of pressure with x. Downstream of the highest point the streamlines diverge, resulting in a decrease of $U(x)$ and a rise in pressure. In this section we shall investigate the effect of such a pressure gradient on the shape of the boundary layer profile $u(x, y)$. The boundary layer equation is

$$u\frac{\partial u}{\partial x} + v\frac{\partial u}{\partial y} = -\frac{1}{\rho}\frac{\partial p}{\partial x} + v\frac{\partial^2 u}{\partial y^2},$$

where the pressure gradient is found from the external velocity field as $dp/dx = -\rho U(dU/dx)$, with x taken along the surface of the body. At the wall, the boundary layer equation becomes

$$\mu\left(\frac{\partial^2 u}{\partial y^2}\right)_{\text{wall}} = \frac{\partial p}{\partial x}.$$

In an accelerating stream $dp/dx < 0$, and therefore

$$\left(\frac{\partial^2 u}{\partial y^2}\right)_{\text{wall}} < 0 \qquad \text{(accelerating)}. \tag{10.46}$$

As the velocity profile has to blend in smoothly with the external profile, the slope $\partial u/\partial y$ slightly below the edge of the boundary layer decreases with y from a positive value to zero; therefore, $\partial^2 u/\partial y^2$ slightly below the boundary layer edge is negative. Equation (10.46) then shows that $\partial^2 u/\partial y^2$ has the same sign at both the wall and the boundary layer edge, and presumably throughout the boundary layer. In contrast, for a decelerating external stream, the curvature of the velocity profile at the wall is

$$\left(\frac{\partial^2 u}{\partial y^2}\right)_{\text{wall}} > 0 \qquad \text{(decelerating)}. \tag{10.47}$$

so that the curvature changes sign somewhere within the boundary layer. In other words, the boundary layer profile in a decelerating flow has a *point of inflection* where $\partial^2 u/\partial y^2 = 0$. In the limiting case of a flat plate, the point of inflection is at the wall.

The shape of the velocity profiles in Figure 10.13 suggests that a decelerating pressure gradient tends to increase the thickness of the boundary layer. This can also be seen from the continuity equation

$$v(y) = -\int_0^y \frac{\partial u}{\partial x}\, dy.$$

Compared to a flat plate, a decelerating external stream causes a larger $-\partial u/\partial x$ within the boundary layer because the deceleration of the outer flow adds to the viscous deceleration within the boundary layer. It follows from the foregoing equation that the v-field, directed away from the surface, is larger for a decelerating flow. The boundary layer therefore thickens not only by viscous diffusion but also by advection away from the surface, resulting in a rapid increase in the boundary layer thickness with x.

If p falls along the direction of flow, $dp/dx < 0$ and we say that the pressure gradient is "favorable." If, on the other hand, the pressure rises along the direction of flow, $dp/dx > 0$ and we say that the pressure gradient is "adverse" or "uphill." The rapid growth of the boundary layer thickness in a decelerating stream, and the associated large v-field, causes the important phenomenon of *separation*, in which the external stream ceases to flow nearly parallel to the boundary surface. This is discussed in the next section.

8. Separation

We have seen in the last section that the boundary layer in a decelerating stream has a point of inflection and grows rapidly. The existence of the point of inflection implies a slowing down of the region next to the wall, a consequence of the uphill pressure gradient. Under a strong enough adverse pressure gradient, the flow next to the wall reverses direction, resulting in a region of backward flow (Figure 10.14). The reversed flow meets the forward flow at some point S at which the fluid near the surface is transported out into the mainstream. We say that the flow *separates* from the wall. The separation point S is defined as the boundary between the forward flow and backward flow of the fluid near the wall, where the stress vanishes:

$$\left(\frac{\partial u}{\partial y}\right)_{\text{wall}} = 0 \qquad \text{(separation)}.$$

It is apparent from the figure that one streamline intersects the wall at a definite angle at the point of separation.

At lower Reynolds numbers the reversed flow downstream of the point of separation forms part of a large steady vortex behind the surface (see Figure 10.17 in Section 9 for the range $4 < \text{Re} < 40$). At higher Reynolds numbers, when the flow

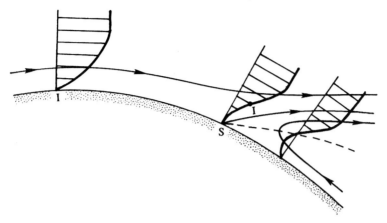

Figure 10.14 Streamlines and velocity profiles near a separation point S. Point of inflection is indicated by I. The dashed line represents $u = 0$.

has boundary layer characteristics, the flow downstream of separation is unsteady and frequently chaotic.

How strong an adverse pressure gradient the boundary layer can withstand without undergoing separation depends on the geometry of the flow, and whether the boundary layer is laminar or turbulent. A steep pressure gradient, such as that behind a blunt body, invariably leads to a quick separation. In contrast, the boundary layer on the trailing surface of a thin body can overcome the weak pressure gradients involved. Therefore, to avoid separation and large drag, the trailing section of a submerged body should be *gradually* reduced in size, giving it a so-called *streamlined* shape.

Evidence indicates that the point of separation is insensitive to the Reynolds number as long as the boundary layer is laminar. However, a *transition to turbulence delays boundary layer separation*; that is, a turbulent boundary layer is more capable of withstanding an adverse pressure gradient. This is because the velocity profile in a turbulent boundary layer is "fuller" (Figure 10.15) and has more energy. For example, the laminar boundary layer over a circular cylinder separates at 82° from the forward stagnation point, whereas a turbulent layer over the same body separates at 125° (shown later in Figure 10.17). Experiments show that the pressure remains fairly uniform downstream of separation and has a lower value than the pressures on the forward face of the body. The resulting drag due to pressure forces is called *form drag*, as it depends crucially on the shape of the body. For a blunt body the form drag is larger than the skin friction drag because of the occurrence of separation. (For a streamlined body, skin friction is generally larger than the form drag.) As long as the separation point is located at the same place on the body, the drag coefficient of a blunt body is nearly constant at high Reynolds numbers. However, the drag coefficient drops suddenly when the boundary layer undergoes transition to turbulence (see Figure 10.22 in Section 9). This is because the separation point then moves downstream, and the wake becomes narrower.

Separation takes place not only in external flows, but also in internal flows such as that in a highly divergent channel (Figure 10.16). Upstream of the throat the pressure

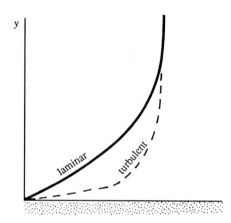

Figure 10.15   Comparison of laminar and turbulent velocity profiles in a boundary layer.

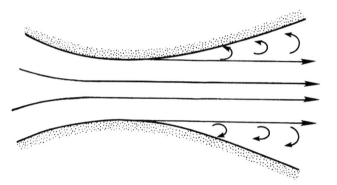

Figure 10.16   Separation of flow in a highly divergent channel.

gradient is favorable and the flow adheres to the wall. Downstream of the throat a large enough adverse pressure gradient can cause separation.

The boundary layer equations are valid only as far downstream as the point of separation. Beyond it the boundary layer becomes so thick that the basic underlying assumptions become invalid. Moreover, the parabolic character of the boundary layer equations requires that a numerical integration is possible only in the direction of advection (along which information is propagated), which is *upstream* within the reversed flow region. A forward (downstream) integration of the boundary layer equations therefore breaks down after the separation point. Last, we can no longer apply potential theory to find the pressure distribution in the separated region, as the effective boundary of the irrotational flow is no longer the solid surface but some unknown shape encompassing part of the body plus the separated region.

9. *Description of Flow past a Circular Cylinder*

In general, analytical solutions of viscous flows can be found (possibly in terms of perturbation series) only in two limiting cases, namely Re ≪ 1 and Re ≫ 1.

In the Re $\ll$ 1 limit the inertia forces are negligible over most of the flow field; the Stokes–Oseen solutions discussed in the preceding chapter are of this type. In the opposite limit of Re $\gg$ 1, the viscous forces are negligible everywhere except close to the surface, and a solution may be attempted by matching an irrotational outer flow with a boundary layer near the surface. In the intermediate range of Reynolds numbers, finding analytical solutions becomes almost an impossible task, and one has to depend on experimentation and numerical solutions. Some of these experimental flow patterns will be described in this section, taking the flow over a circular cylinder as an example. Instead of discussing only the intermediate Reynolds number range, we shall describe the experimental data for the entire range of small to very high Reynolds numbers.

Low Reynolds Numbers

Let us start with a consideration of the creeping flow around a circular cylinder, characterized by Re < 1. (Here we shall define Re $= U_\infty d/\nu$, based on the upstream velocity and the cylinder diameter.) Vorticity is generated close to the surface because of the no-slip boundary condition. In the Stokes approximation this vorticity is simply diffused, not advected, which results in a fore and aft symmetry. The Oseen approximation partially takes into account the advection of vorticity, and results in an asymmetric velocity distribution *far* from the body (which was shown in Figure 9.17). The vorticity distribution is qualitatively analogous to the dye distribution caused by a source of colored fluid at the position of the body. The color diffuses symmetrically in very slow flows, but at higher flow speeds the dye source is confined behind a parabolic boundary with the dye source at the focus.

As Re is increased beyond 1, the Oseen approximation breaks down, and the vorticity is increasingly confined behind the cylinder because of advection. For Re > 4, two small attached or "standing" eddies appear behind the cylinder. The wake is completely laminar and the vortices act like "fluidynamic rollers" over which the main stream flows (Figure 10.17). The eddies get longer as Re is increased.

von Karman Vortex Street

A very interesting sequence of events begins to develop when the Reynolds number is increased beyond 40, at which point the wake behind the cylinder becomes unstable. Photographs show that the wake develops a slow oscillation in which the velocity is periodic in time and downstream distance, with the amplitude of the oscillation increasing downstream. The oscillating wake rolls up into two staggered rows of vortices with opposite sense of rotation (Figure 10.18). von Karman investigated the phenomenon as a problem of superposition of irrotational vortices; he concluded that a nonstaggered row of vortices is unstable, and a staggered row is stable only if the ratio of lateral distance between the vortices to their longitudinal distance is 0.28. Because of the similarity of the wake with footprints in a street, the staggered row of vortices behind a blunt body is called a *von Karman vortex street*. The vortices move downstream at a speed smaller than the upstream velocity U_∞. This means that the vortex pattern slowly follows the cylinder if it is pulled through a stationary fluid.

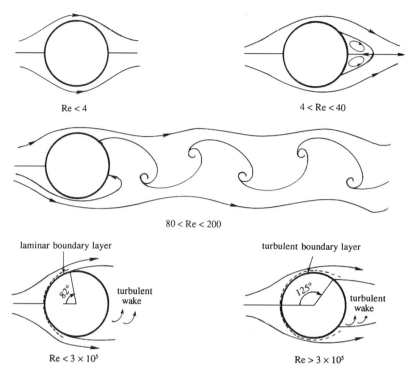

Figure 10.17   Some regimes of flow over a circular cylinder.

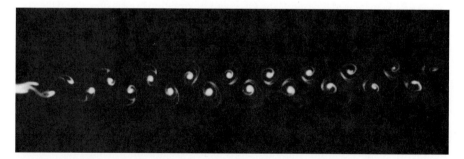

Figure 10.18   von Karman vortex street downstream of a circular cylinder at Re = 55. Flow visualized by condensed milk. S. Taneda, *Jour. Phys. Soc., Japan* **20**: 1714–1721, 1965, and reprinted with the permission of The Physical Society of Japan and Dr. Sadatoshi Taneda.

In the range $40 < \text{Re} < 80$, the vortex street does not interact with the pair of attached vortices. As Re is increased beyond 80 the vortex street forms closer to the cylinder, and the attached eddies (whose downstream length has now grown to be about twice the diameter of the cylinder) themselves begin to oscillate. Finally the attached eddies periodically break off alternately from the two sides of the cylinder. While an eddy on one side is shed, that on the other side forms, resulting in an unsteady flow near the cylinder. As vortices of opposite circulations are shed off alternately from the two sides, the circulation around the cylinder changes sign, resulting in an oscillating "lift" or lateral force. If the frequency of vortex shedding is close to the natural frequency of some mode of vibration of the cylinder body, then an appreciable lateral vibration has been observed to result. Engineering structures such as suspension bridges and oil drilling platforms are designed so as to break up a coherent shedding of vortices from cylindrical structures. This is done by including spiral blades protruding out of the cylinder surface, which break up the spanwise coherence of vortex shedding, forcing the vortices to detach at different times along the length of these structures (Figure 10.19).

The passage of regular vortices causes velocity measurements in the wake to have a dominant periodicity. The frequency n is expressed as a nondimensional parameter known as the *Strouhal number*, defined as

$$S \equiv \frac{nd}{U_\infty}.$$

Experiments show that for a circular cylinder the value of S remains close to 0.21 for a large range of Reynolds numbers. For small values of cylinder diameter and moderate values of U_∞, the resulting frequencies of the vortex shedding and oscillating lift lie in the acoustic range. For example, at $U_\infty = 10\,\text{m/s}$ and a wire diameter of $2\,\text{mm}$, the frequency corresponding to a Strouhal number of 0.21 is $n = 1050$ cycles per second. The "singing" of telephone and transmission lines has been attributed to this phenomenon.

Wen and Lin (2001) conducted very careful experiments that purported to be strictly two-dimensional by using both horizontal and vertical soap film water tunnels. They give a review of the recent literature on both the computational and experimental aspects of this problem. The asymptote cited here of $S = 0.21$ is for a flow including three-dimensional instabilities. Their experiments are in agreement with two-dimensional computations and the data are asymptotic to $S = 0.2417$.

Below Re = 200, the vortices in the wake are laminar and continue to be so for very large distances downstream. Above 200, the vortex street becomes unstable and

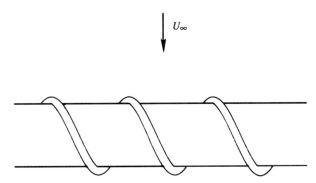

Figure 10.19 Spiral blades used for breaking up the spanwise coherence of vortex shedding from a cylindrical rod.

irregular, and the flow within the vortices themselves becomes chaotic. However, the flow in the wake continues to have a strong frequency component corresponding to a Strouhal number of $S = 0.21$. Above a very high Reynolds number, say 5000, the periodicity in the wake becomes imperceptible, and the wake may be described as completely turbulent.

Striking examples of vortex streets have also been observed in the atmosphere. Figure 10.20 shows a satellite photograph of the wake behind several isolated mountain peaks, through which the wind is blowing toward the southeast. The mountains pierce through the cloud level, and the flow pattern becomes visible by the cloud

Figure 10.20   A von Karman vortex street downstream of mountain peaks in a strongly stratified atmosphere. There are several mountain peaks along the linear, light-colored feature running diagonally in the upper left-hand corner of the photograph. North is upward, and the wind is blowing toward the southeast. R. E. Thomson and J. F. R. Gower, *Monthly Weather Review* **105**: 873–884, 1977, and reprinted with the permission of the American Meteorlogical Society.

pattern. The wakes behind at least two mountain peaks display the characteristics of a von Karman vortex street. The strong density stratification in this flow has prevented a vertical motion, giving the flow the two-dimensional character necessary for the formation of vortex streets.

High Reynolds Numbers

At high Reynolds numbers the frictional effects upstream of separation are confined near the surface of the cylinder, and the boundary layer approximation becomes valid as far downstream as the point of separation. For Re $< 3 \times 10^5$, the boundary layer remains laminar, although the wake may be completely turbulent. The laminar boundary layer separates at $\approx 82°$ from the forward stagnation point (Figure 10.17). The pressure in the wake downstream of the point of separation is nearly constant and lower than the upstream pressure (Figure 10.21). As the drag in this range is primarily due to the asymmetry in the pressure distribution caused by separation, and as the point of separation remains fairly stationary in this range, the drag coefficient also stays constant at $C_D \simeq 1.2$ (Figure 10.22).

Important changes take place beyond the critical Reynolds number of

$$\text{Re}_{\text{cr}} \sim 3 \times 10^5 \qquad \text{(circular cylinder)}.$$

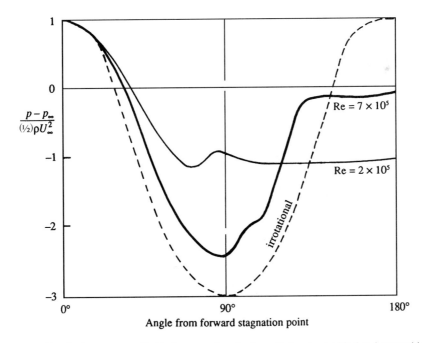

Figure 10.21 Surface pressure distribution around a circular cylinder at subcritical and supercritical Reynolds numbers. Note that the pressure is nearly constant within the wake and that the wake is narrower for flow at supercritical Re.

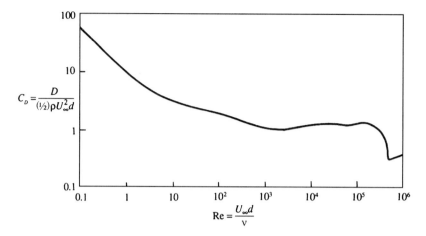

Figure 10.22 Measured drag coefficient of a circular cylinder. The sudden dip is due to the transition of the boundary layer to turbulence and the consequent downstream movement of the point of separation.

In the range $3 \times 10^5 < \text{Re} < 3 \times 10^6$, the laminar boundary layer becomes unstable and undergoes transition to turbulence. We have seen in the preceding section that because of its greater energy, a turbulent boundary layer, is able to overcome a larger adverse pressure gradient. In the case of a circular cylinder the turbulent boundary layer separates at 125° from the forward stagnation point, resulting in a thinner wake and a pressure distribution more similar to that of potential flow. Figure 10.21 compares the pressure distributions around the cylinder for two values of Re, one with a laminar and the other with a turbulent boundary layer. It is apparent that the pressures within the wake are higher when the boundary layer is turbulent, resulting in a sudden drop in the drag coefficient from 1.2 to 0.33 at the point of transition. For values of $\text{Re} > 3 \times 10^6$, the separation point slowly moves upstream as the Reynolds number is increased, resulting in an increase of the drag coefficient (Figure 10.22).

It should be noted that the critical Reynolds number at which the boundary layer undergoes transition is strongly affected by two factors, namely the intensity of fluctuations existing in the approaching stream and the roughness of the surface, an increase in either of which decreases Re_{cr}. The value of 3×10^5 is found to be valid for a smooth circular cylinder at low levels of fluctuation of the oncoming stream.

Before concluding this section we shall note an interesting anecdote about the von Karman vortex street. The pattern was investigated experimentally by the French physicist Henri Bénard, well-known for his observations of the instability of a layer of fluid heated from below. In 1954 von Karman wrote that Bénard became "jealous because the vortex street was connected with my name, and several times ... claimed priority for earlier observation of the phenomenon. In reply I once said 'I agree that what in Berlin and London is called *Karman Street* in Paris shall be called *Avenue de Henri Bénard*.' After this wisecrack we made peace and became good friends." von Karman also says that the phenomenon has been known for a long time and is even found in old paintings.

We close this section by noting that this flow illustrates three instances where the solution is counterintuitive. First, small causes can have large effects. If we solve for the flow of a fluid with zero viscosity around a circular cylinder, we obtain the results of Chapter 6, Section 9. The inviscid flow has fore-aft symmetry and the cylinder experiences zero drag. The bottom two panels of Figure 10.17 illustrate the flow for small viscosity. For viscosity as small as you choose, in the limit viscosity tends to zero, the flow must look like the last panel in which there is substantial fore-aft asymmetry, a significant wake, and significant drag. This is because of the necessity of a boundary layer and the satisfaction of the no-slip boundary condition on the surface so long as viscosity is not exactly zero. When viscosity is exactly zero, there is no boundary layer and there is slip at the surface. The resolution of d'Alembert's paradox is through the boundary layer, a singular perturbation of the Navier–Stokes equations in the direction normal to the boundary.

The second instance of counterintuitivity is that symmetric problems can have nonsymmetric solutions. This is evident in the intermediate Reynolds number middle panel of Figure 10.17. Beyond a Reynolds number of ≈ 40 the symmetric wake becomes unstable and a pattern of alternating vortices called a von Karman vortex street is established. Yet the equations and boundary conditions are symmetric about a central plane in the flow. If one were to solve only a half-problem, assuming symmetry, a solution would be obtained, but it would be unstable to infinitesimal disturbances and unlikely to be seen in the laboratory.

The third instance of counterintuitivity is that there is a range of Reynolds numbers where roughening the surface of the body can reduce its drag. This is true for all blunt bodies, such as a sphere (to be discussed in the next section). In this range of Reynolds numbers, the boundary layer on the surface of a blunt body is laminar, but sensitive to disturbances such as surface roughness, which would cause earlier transition of the boundary layer to turbulence than would occur on a smooth body. Although, as we shall see, the skin friction of a turbulent boundary layer is much larger than that of a laminar boundary layer, most of the drag is caused by incomplete pressure recovery on the downstream side of a blunt body as shown in Figure 10.21, rather than by skin friction. In fact, it is because the skin friction of a turbulent boundary layer is much larger, as a result of a larger velocity gradient at the surface, that a turbulent boundary layer can remain attached farther on the downstream side of a blunt body, leading to a narrower wake and more complete pressure recovery and thus reduced drag. The drag reduction attributed to the turbulent boundary layer is shown in Figure 10.22 for a circular cylinder and Figure 10.23 for a sphere.

10. Description of Flow past a Sphere

Several features of the description of flow over a circular cylinder qualitatively apply to flows over other two-dimensional blunt bodies. For example, a vortex street is observed in a flow perpendicular to a flat plate. The flow over a three-dimensional body, however, has one fundamental difference in that a regular vortex street is absent. For flow around a sphere at low Reynolds numbers, there is an attached eddy in the form of a doughnut-shaped ring; in fact, an axial section of the flow looks similar to

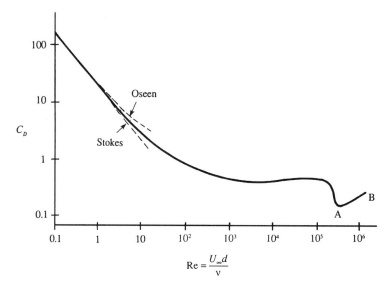

Figure 10.23   Measured drag coefficient of a smooth sphere. The Stokes solution is $C_D = 24/\text{Re}$, and the Oseen solution is $C_D = (24/\text{Re})(1 + 3\text{Re}/16)$; these two solutions are discussed in Chapter 9, Sections 12 and 13. The increase of drag coefficient in the range AB has relevance in explaining why the flight paths of sports balls bend in the air.

that shown in Figure 10.17 for the range $4 < \text{Re} < 40$. For $\text{Re} > 130$ the ring-eddy oscillates, and some of it breaks off periodically in the form of distorted vortex loops.

The behavior of the boundary layer around a sphere is similar to that around a circular cylinder. In particular it undergoes transition to turbulence at a critical Reynolds number of

$$\text{Re}_{\text{cr}} \sim 5 \times 10^5 \qquad \text{(sphere)},$$

which corresponds to a sudden dip of the drag coefficient (Figure 10.23). As in the case of a circular cylinder, the *separation point slowly moves upstream for postcritical Reynolds numbers*, accompanied by a rise in the drag coefficient. The behavior of the separation point for flow around a sphere at subcritical and supercritical Reynolds numbers is responsible for the bending in the flight paths of sports balls, as explained in the following section.

11. Dynamics of Sports Balls

The discussion of the preceding section could be used to explain why the trajectories of sports balls (such as those involved in tennis, cricket, and baseball games) bend in the air. The bending is commonly known as *swing, swerve, or curve*. The problem has been investigated by wind tunnel tests and by stroboscopic photographs of flight paths in field tests, a summary of which was given by Mehta (1985). Evidence indicates that the mechanics of bending is different for spinning and nonspinning balls. The following discussion gives a qualitative explanation of the mechanics of flight path bending. (Readers not interested in sports may omit this section!)

Cricket Ball Dynamics

The cricket ball has a prominent (1-mm high) seam, and tests show that the orientation of the seam is responsible for bending of the ball's flight path. It is known to bend when thrown at high speeds of around 30 m/s, which is equivalent to a Reynolds number of $Re = 10^5$. Here we shall define the Reynolds number as $Re = U_\infty d/\nu$, based on the translational speed U_∞ of the ball and its diameter d. The operating Reynolds number is somewhat less than the critical value of $Re_{cr} = 5 \times 10^5$ necessary for transition of the boundary layer on a smooth sphere into turbulence. However, the presence of the seam is able to trip the laminar boundary layer into turbulence on one side of the ball (the lower side in Figure 10.24), while the boundary layer on the other side remains laminar. We have seen in the preceding sections that because of greater energy a turbulent boundary layer separates later. Typically, the boundary layer on the laminar side separates at $\approx 85°$, whereas that on the turbulent side separates at $120°$. Compared to region B, the surface pressure near region A is therefore closer to that given by the potential flow theory (which predicts a suction pressure of $(p_{min} - p_\infty)/(\frac{1}{2}\rho U_\infty^2) = -1.25$; see equation (6.81)). In other words, the pressures are lower on side A, resulting in a downward force on the ball. (Note that Figure 10.24 is a view of the flow pattern looking downward on the ball, so that it corresponds to a ball that bends to the left in its flight. The flight of a cricket ball oriented as in Figure 10.24 is called an "outswinger" in cricket literature, in contrast to an "inswinger" for which the seam is oriented in the opposite direction so as to generate an upward force in Figure 10.24.)

Figure 10.25, a photograph of a cricket ball in a wind tunnel experiment, clearly shows the delayed separation on the seam side. Note that the wake has been deflected upward by the presence of the ball, implying that an upward force has been exerted

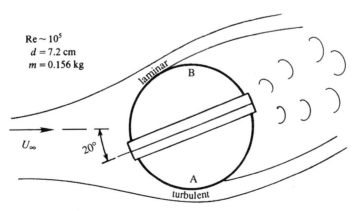

Figure 10.24 The swing of a cricket ball. The seam is oriented in such a way that the lateral force on the ball is downward in the figure.

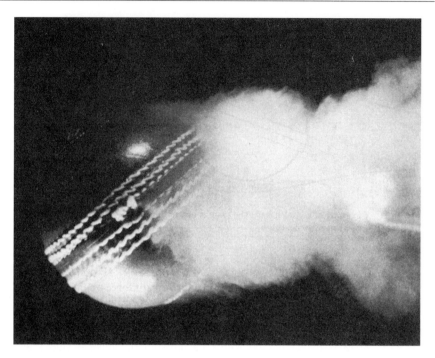

Figure 10.25 Smoke photograph of flow over a cricket ball. Flow is from left to right. Seam angle is 40°, flow speed is 17 m/s, Re $= 0.85 \times 10^5$. R. Mehta, *Ann. Rev Fluid Mech.* **17**: 151–189, 1985. Photograph reproduced with permission from the *Annual Review of Fluid Mechanics*, Vol. 17 © 1985 *Annual Reviews www.AnnualReviews.org*.

by the ball on the fluid. It follows that a downward force has been exerted by the fluid on the ball.

In practice some spin is invariably imparted to the ball. The ball is held along the seam and, because of the round arm action of the bowler, some backspin is always imparted *along* the seam. This has the important effect of stabilizing the orientation of the ball and preventing it from wobbling. A typical cricket ball can generate side forces amounting to almost 40% of its weight. A constant lateral force oriented in the same direction causes a deflection proportional to the square of time. The ball therefore travels in a parabolic path that can bend as much as 0.8 m by the time it reaches the batsman.

It is known that the trajectory of the cricket ball does not bend if the ball is thrown too slow or too fast. In the former case even the presence of the seam is not enough to trip the boundary layer into turbulence, and in the latter case the boundary layer on both sides could be turbulent; in both cases an asymmetric flow is prevented. It is also clear why only a new, shiny ball is able to swing, because the rough surface of an old ball causes the boundary layer to become turbulent on both sides. Fast bowlers in cricket maintain one hemisphere of the ball in a smooth state by constant polishing. It therefore seems that most of the known facts about the swing of a cricket ball have been adequately explained by scientific research. The feature that has not been explained is the universally observed fact that a cricket ball swings more in humid conditions. The changes in density and viscosity due to changes in humidity can change the Reynolds number by only 2%, which cannot explain this phenomenon.

Tennis Ball Dynamics

Unlike the cricket ball, the path of the tennis ball bends because of spin. A ball hit with topspin curves downward, whereas a ball hit with underspin travels in a much flatter trajectory. The direction of the lateral force is therefore in the same sense as that of the Magnus effect experienced by a circular *cylinder* in potential flow with circulation (see Chapter 6, Section 10). The mechanics, however, are different. The potential flow argument (involving the Bernoulli equation) offered to account for the lateral force around a circular cylinder cannot explain why a *negative* Magnus effect is universally observed at lower Reynolds numbers. (By a negative Magnus effect we mean a lateral force opposite to that experienced by a cylinder with a circulation of the same sense as the rotation of the sphere.) The correct argument seems to be the asymmetric boundary layer separation caused by the spin. In fact, the phenomenon was not properly explained until the boundary layer concepts were understood in the twentieth century. Some pioneering experimental work on the bending paths of spinning spheres was conducted by Robins about two hundred years ago; the deflection of rotating spheres is sometimes called the *Robins effect*.

Experimental data on nonrotating spheres (Figure 10.23) shows that the boundary layer on a sphere undergoes transition at a Reynolds number of $\approx \text{Re} = 5 \times 10^5$, indicated by a sudden drop in the drag coefficient. As discussed in the preceding section, this drop is due to the transition of the laminar boundary layer to turbulence. An important point for our discussion here is that for supercritical Reynolds numbers the separation point slowly moves upstream, as evidenced by the increase of the drag coefficient after the sudden drop shown in Figure 10.23.

With this background, we are now in a position to understand how a spinning ball generates a negative Magnus effect at $\text{Re} < \text{Re}_{cr}$ and a positive Magnus effect at $\text{Re} > \text{Re}_{cr}$. For a clockwise rotation of the ball, the fluid velocity *relative to the surface* is larger on the lower side (Figure 10.26). For the lower Reynolds number case (Figure 10.26a), this causes a transition of the boundary layer on the lower side, while the boundary layer on the upper side remains laminar. The result is a delayed separation and lower pressure on the bottom surface, and a consequent downward force on the ball. The force here is in a sense opposite to that of the Magnus effect.

The rough surface of a tennis ball lowers the critical Reynolds number, so that for a well-hit tennis ball the boundary layers on both sides of the ball have already undergone transition. Due to the higher relative velocity, the flow near the bottom has a higher Reynolds number, and is therefore farther along the Re-axis of Figure 10.23, in the range AB in which the separation point moves upstream with an increase of the Reynolds number. The separation therefore occurs *earlier* on the bottom side, resulting in a higher pressure there than on the top. This causes an upward lift force and a positive Magnus effect. Figure 10.26b shows that a tennis ball hit with underspin generates an upward force; this overcomes a large fraction of the weight of the ball, resulting in a much flatter trajectory than that of a tennis ball hit with topspin. A "slice serve," in which the ball is hit tangentially on the right-hand side, curves to

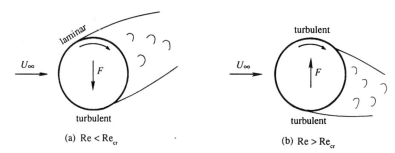

Figure 10.26 Bending of rotating spheres, in which F indicates the force exerted by the fluid: (a) negative Magnus effect; and (b) positive Magnus effect. A well-hit tennis ball is likely to display the positive Magnus effect.

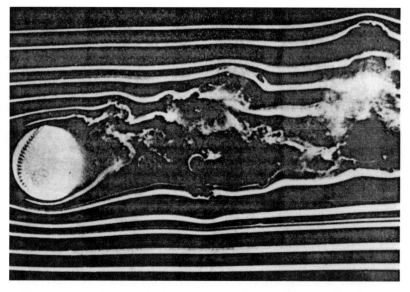

Figure 10.27 Smoke photograph of flow around a spinning baseball. Flow is from left to right, flow speed is 21 m/s, and the ball is spinning counterclockwise at 15 rev/s. [Photograph by F. N. M. Brown, University of Notre Dame.] Photograph reproduced with permission, from the *Annual Review of Fluid Mechanics*, Vol. 17 © 1985 by *Annual Reviews www.AnnualReviews.org.*

the left due to the same effect. (Presumably soccer balls curve in the air due to similar dynamics.)

Baseball Dynamics

A baseball pitcher uses different kinds of deliveries, a typical Reynolds number being 1.5×10^5. One type of delivery is called a "curveball," caused by sidespin imparted by the pitcher to bend away from the side of the throwing arm. A "screwball" has the opposite spin and curved trajectory. The dynamics of this is similar to that of a spinning tennis ball (Figure 10.26b). Figure 10.27 is a photograph of the flow over a spinning baseball, showing an asymmetric separation, a crowding together of streamlines at the bottom, and an upward deflection of the wake that corresponds to a downward force on the ball.

The knuckleball, on the other hand, is released without any spin. In this case the path of the ball bends due to an asymmetric separation caused by the orientation of the seam, much like the cricket ball. However, the cricket ball is released with spin along the seam, which stabilizes the orientation and results in a predictable bending. The knuckleball, on the other hand, tumbles in its flight because of a lack of stabilizing spin, resulting in an irregular orientation of the seam and a consequent irregular trajectory.

12. Two-Dimensional Jets

So far we have considered boundary layers over a solid surface. The concept of a boundary layer, however, is more general, and the approximations involved are applicable if the vorticity is confined in thin layers *without* the presence of a solid surface. Such a layer can be in the form of a jet of fluid ejected from an orifice, a wake (where the velocity is lower than the upstream velocity) behind a solid object, or a mixing layer (vortex sheet) between two streams of different speeds. As an illustration of the method of analysis of these "free shear flows," we shall consider the case of a laminar two-dimensional jet, which is an efflux of fluid from a long and narrow orifice. The surrounding is assumed to be made up of the same fluid as the jet itself, and some of this ambient fluid is carried along with the jet by the viscous drag at the outer edge of the jet (Figure 10.28). The process of drawing in the surrounding fluid from the sides of the jet by frictional forces is called *entrainment*.

The velocity distribution near the opening of the jet depends on the details of conditions upstream of the orifice exit. However, because of the absence of an externally imposed length scale in the downstream direction, the velocity profile in the jet approaches a self-similar shape not far from the exit, regardless of the velocity distribution at the orifice.

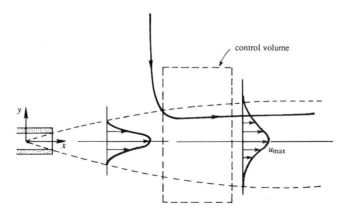

Figure 10.28 Laminar two-dimensional jet. A typical streamline showing entrainment of surrounding fluid is indicated.

For large Reynolds numbers, the jet is narrow and the boundary layer approximation can be applied. Consider a control volume with sides cutting across the jet axis at two sections (Figure 10.28); the other two sides of the control volume are taken at large distances from the jet axis. No external pressure gradient is maintained in the surrounding fluid, in which dp/dx is zero. According to the boundary layer approximation, the same zero pressure gradient is also impressed upon the jet. There is, therefore, no net force acting on the surfaces of the control volume, which requires that the rate of flow of x-momentum at the two sections across the jet are the same.

Let $u_o(x)$ be the streamwise velocity on the x-axis and assume $\mathrm{Re} = u_o x / v$ is sufficiently large for the boundary layer equations to be valid. The flow is steady, two-dimensional (x, y), without body forces, and with constant properties (ρ, μ). Then $\partial/\partial y \gg \partial/\partial x$, $v \ll u$, $\partial p/\partial y = 0$, so

$$\partial u/\partial x + \partial v/\partial y = 0, \tag{10.48}$$

$$u \partial u/\partial x + v \partial u/\partial y = v \partial^2 u/\partial y^2 \tag{10.49}$$

subject to the boundary conditions: $y \to \pm\infty : u = 0$; $y = 0 : v = 0$; $x = x_o$: $u = \tilde{u}(x_o, y)$. Form $u \cdot$ [equation (10.48)] + equation (10.49) and integrate over all y:

$$\int_{-\infty}^{\infty} 2u(\partial u/\partial x) dy + \int_{-\infty}^{\infty} (u \partial v/\partial y + v \partial u/\partial y) dy = v \partial u/\partial y |_{-\infty}^{\infty}$$

$$d/dx \int_{-\infty}^{\infty} u^2 dy + uv|_{-\infty}^{\infty} = v \partial u/\partial y |_{-\infty}^{\infty}.$$

Since $u(y = \pm\infty) = 0$, all derivatives of u with repect to y must also be zero at $y = \pm\infty$. Then the streamwise momentum flux must be preserved,

$$d/dx \int_{-\infty}^{\infty} \rho u^2 dy = 0 \tag{10.50}$$

Far enough downstream that (a) the boundary layer equations are valid, and (b) the initial distribution $\tilde{u}(x_o, y)$, specified at the upstream limit of validity of the boundary layer equations, is forgotten, a similarity solution is obtained. This similarity solution is of the universal dimensionless similarity form for the laminar boundary layer equations, that is,

$$\psi(x, y) = [x v u_o(x)]^{1/2} f(\eta), \, \eta = (y/x)[x u_o(x)/v]^{1/2}, \quad \mathrm{Re}_x = x u_o(x)/v \tag{10.51}$$

where ψ is the usual streamfunction, $\mathbf{u} = -\mathbf{k} \times \nabla \psi$, and f and η are dimensionless. We obtain the behavior of $u_o(x)$ by substitution of the similarity transformation (10.51) into the condition (10.50)

$$u = \partial\psi/\partial y = u_o(x)f'(\eta), \, dy = d\eta[\nu x/u_o(x)]^{1/2}$$

$$\rho d/dx\{u_o^2(x)[\nu x/u_o(x)]^{1/2} \int_{-\infty}^{\infty} f'^2(\eta)d\eta\} = 0.$$

Since the integral is a pure constant, we must have $u_o^{3/2}(x) \cdot x^{1/2} = C^{3/2}$ where C is a dimensional constant. Then $u_o = Cx^{-1/3}$. C is clearly related to the intensity or momentum flux in the jet. Now, (10.51) becomes

$$\psi(x, y) = (\nu C)^{1/2} \cdot x^{1/3} f(\eta), \, \eta = (C/\nu)^{1/2} \cdot y/x^{2/3}$$

In terms of the streamfunction, (10.49) may be written

$$\partial\psi/\partial y \cdot \partial^2\psi/\partial y\partial x - \partial\psi/\partial x \cdot \partial^2\psi/\partial y^2 = \nu\partial^3\psi/\partial y^3.$$

Evaluating the derivatives of the streamfunction and substituting into the x-momentum equation, we obtain

$$3f''' + ff'' + f'^2 = 0$$

subject to the boundary conditions

$$\eta = \pm\infty : f' = 0; \eta = 0 : f = 0.$$

Integrating once,

$$3f'' + ff' = C_1.$$

Evaluating at $\eta = \pm\infty$, $C_1 = 0$. Integrating again,

$$3f' + f^2/2 = 18C_2^2,$$

where the constant of integration is chosen to be "$18C_2^2$" for convenience in the next integration, as will be seen. Now consider the transformation $f/6 = g'/g$, so that $f'/6 = g''/g - g'^2/g^2$. This results in $g'' - C_2^2 g = 0$. The solution for g is

$$g = C_3 \exp(C_2\eta) + C_4 \exp(-C_2\eta).$$

Then

$$f = 6g'/g = 6C_2[C_3 \exp(C_2\eta) - C_4 \exp(-C_2\eta)]/[C_3 \exp(C_2\eta) \\ + C_4 \exp(-C_2\eta)].$$

Now, $f' = 6C_2^2 - f^2/6 = 6C_2^2\{1 - [(C_3e^{C_2\eta} - C_4e^{-C_2\eta})/(C_3e^{C_2\eta} + C_4e^{-C_2\eta})]^2\}$ must be even in η. Or, use the boundary condition $f(0) = 0$. This requires $C_3 = C_4$. Then

$$f'(\eta) = 6C_2^2[1 - \tanh^2(C_2\eta)] \text{ and } f(\eta) = 6C_2 \tanh(C_2\eta). \text{ Thus}$$
$$f'(\eta) = 6C_2^2 \text{ sech}^2(C_2\eta).$$

To obtain C_2, recall $u(x, y = 0) = u_o(x)f'(0) = Cx^{-1/3} \cdot 6C_2^2 = u_o(x)$ by our definition of $u_o(x)$.

Thus $6C_2^2 = 1$ and $C_2 = 1/\sqrt{6}$. Then $f'(\eta) = \text{sech}^2(\eta/\sqrt{6})$ and $u(x, y) = u_o(x)\,\text{sech}^2(\eta/\sqrt{6})$. The constant "$C$" in $u_o(x) = Cx^{-1/3}$ is related to the momentum flux in the jet via $F = \int\limits_{-\infty}^{\infty} \rho u^2\,dy = 2\rho C^{3/2}\nu^{1/2}\int\limits_{0}^{\infty}\text{sech}^4(\eta/\sqrt{6})d\eta =$ force per unit depth. Carrying out the integration, $F = (4\sqrt{6}/3)\rho C^{3/2}\nu^{1/2}$, so $C = [3F/(4\sqrt{6}\rho\nu^{1/2})]^{2/3}$, in terms of the jet force per unit depth or momentum flux. The mass flux in the jet is

$$\dot{m} = \int\limits_{-\infty}^{\infty} \rho u\,dy = \rho\int\limits_{-\infty}^{\infty} u_o(x)f'(\eta)d\eta \cdot [\nu x/u_o(x)]^{1/2} = (36\rho^2\nu F)^{1/3}x^{1/3}.$$

This grows downstream because of entrainment in the jet. The entrainment may be seen as inward flow (y component of velocity) from afar.

$$v = -\partial\psi/\partial x = -(\nu C)^{1/2}x^{-2/3}(f - 2\eta f')/3, \quad \text{so}$$

$$v/u_o = -(f - 2\eta f')/(3\sqrt{\text{Re}_x}), \quad \text{Re}_x = xu_o(x)/\nu.$$

As

$$\eta \to \infty, v/u_o \to -\sqrt{6}/(3\sqrt{\text{Re}_x}), \quad \text{downwards toward jet}$$

$$\eta \to -\infty, v/u_o \to +\sqrt{6}/(3\sqrt{\text{Re}_x}), \quad \text{upwards toward jet.}$$

Thus the entrainment is an inward flow of mass from above and below.

The jet spreads as it travels downstream. Now $f'(\eta) = \text{sech}^2(\eta/\sqrt{6})$. If $\eta = 5$ is taken as width of jet, $5/\sqrt{6} = 2.04$ and $f'(2.04) = .065$. Calling the transverse extent y of the jet, δ, we have $5 \approx (\delta/x)(Cx^{2/3}/\nu)^{1/2}$ so that $\delta \approx 5\sqrt{\nu/C}x^{2/3}$. The jet grows downstream $x^{2/3}$. We can express the Reynolds numbers in terms of the force or momentum flux in the jet, F

$$\text{Re}_x = Cx^{2/3}/\nu = [3Fx/(4\sqrt{6}\rho\nu^2)]^{2/3}, \quad \text{and}$$

$$\text{Re}_\delta = u_o\delta/\nu = 5 \cdot [3Fx/(4\sqrt{6}\rho\nu^2)]^{1/3}.$$

By drawing sketches of the profiles of u, u^2, and u^3, the reader can verify that, under similarity, the constraint

$$\frac{d}{dx}\int_{-\infty}^{\infty} u^2\,dy = 0,$$

must lead to

$$\frac{d}{dx}\int_{-\infty}^{\infty} u\,dy > 0,$$

and

$$\frac{d}{dx} \int_{-\infty}^{\infty} u^3 \, dy < 0.$$

The laminar jet solution given here is not readily observable because the flow easily breaks up into turbulence. The low critical Reynolds number for instability of a jet or wake is associated with the existence of a point of inflection in the velocity profile, as discussed in Chapter 12. Nevertheless, the laminar solution has revealed several significant ideas (namely constancy of momentum flux and increase of mass flux) that also apply to a turbulent jet. However, the rate of spreading of a turbulent jet is faster, being more like $\delta \propto x$ rather than $\delta \propto x^{2/3}$ (see Chapter 13).

The Wall Jet

An example of a two-dimensional jet that also shares some boundary layer characteristics is the "wall jet." The solution here is due to M. B. Glauert (1956). We consider a fluid exiting a narrow slot with its lower boundary being a planar wall taken along the x-axis (see Figure 10.29). Near the wall $y = 0$ and the flow behaves like a boundary layer, but far from the wall it behaves like a free jet. The boundary layer analysis shows that for large Re_x the jet is thin ($\delta/x \ll 1$) so $\partial p / \partial y \approx 0$ across it. The pressure is constant in the nearly stagnant outer fluid so $p \approx$ const. throughout the flow. The boundary layer equations are

$$\frac{\partial u}{\partial x} + \frac{\partial v}{\partial y} = 0, \tag{10.52}$$

$$u\frac{\partial u}{\partial x} + v\frac{\partial u}{\partial y} = \nu\frac{\partial^2 u}{\partial y^2}, \tag{10.53}$$

subject to the boundary conditions $y = 0$: $u = v = 0$; $y \to \infty$: $u \to 0$. With an initial velocity distribution forgotten sufficiently far downstream that $\mathrm{Re}_x \to \infty$, a similarity solution is available. However, unlike the free jet, the momentum flux is not constant; instead, it diminishes downstream because of the wall shear stress. To obtain the conserved property in the wall jet, we start by integrating equation (10.53) from y to ∞:

$$\int_y^{\infty} u\frac{\partial u}{\partial x} \, dy + \int_y^{\infty} v\frac{\partial u}{\partial y} \, dy = -\nu\frac{\partial u}{\partial y}.$$

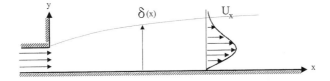

Figure 10.29 The planar wall jet.

Multiply this by u and integrate from 0 to ∞:

$$\int_0^\infty \left(u \frac{\partial}{\partial x} \int_y^\infty \frac{u^2}{2} \, dy \right) dy + \int_0^\infty \left(u \int_y^\infty v \frac{\partial u}{\partial y} \, dy \right) dy + \frac{\nu}{2} \int_0^\infty \frac{\partial}{\partial y} u^2 dy = 0.$$

The last term integrates to 0 because of the boundary conditions at both ends. Integrating the second term by parts and using equation (10.52) yields a term equal to the first term. Then we have

$$\int_0^\infty \left(u \frac{\partial}{\partial x} \int_y^\infty u^2 \, dy \right) dy - \int_0^\infty u^2 v \, dy = 0. \tag{10.54}$$

Now consider

$$\frac{d}{dx} \int_0^\infty \left(u \int_y^\infty u^2 \, dy \right) dy = \int_0^\infty \left(\frac{\partial u}{\partial x} \int_y^\infty u^2 \, dy \right) dy$$
$$+ \int_0^\infty \left(u \frac{\partial}{\partial x} \int_y^\infty u^2 \, dy \right) dy.$$

Using equation (10.52) in the first term on the right-hand side, integrating by parts, and using equation (10.54), we finally obtain

$$\frac{d}{dx} \int_0^\infty \left(u \int_y^\infty u^2 \, dy \right) dy = 0. \tag{10.55}$$

This says that the flux of exterior momentum flux is constant downstream and is used as the second condition to obtain the similarity exponents. Rewriting equation (10.53) in terms of the streamfunction $u = \partial\psi/\partial y$, $v = -\partial\psi/\partial x$, we obtain

$$\frac{\partial\psi}{\partial y} \frac{\partial^2\psi}{\partial y \partial x} - \frac{\partial\psi}{\partial x} \frac{\partial^2\psi}{\partial y^2} = \nu \frac{\partial^3\psi}{\partial y^3}, \tag{10.56}$$

subject to:

$$y = 0 : \psi = \frac{\partial\psi}{\partial y} = 0; \qquad y \to \infty : \frac{\partial\psi}{\partial y} \to 0. \tag{10.57}$$

Let $\bar{u}(x)$ be some average or characteristic speed of the wall jet. We will be able to relate this to the mass flow rate and width of the jet at the completion of this discussion. We can write the universal dimensionless similarity scaling for the laminar boundary layer equations in terms of $\bar{u}(x)$, via

$$\psi(x, y) = [\nu x \bar{u}(x)]^{1/2} \cdot f(\eta), \eta = (y/x)\sqrt{Re_x} = (y/x)[x\bar{u}(x)/\nu]^{1/2},$$

and expect this similarity to hold when $x \gg x_o$, where x_o is the location where the initial condition is specified, which we take to be the upstream extent of the validity of the boundary layer equations. Then $u(x, y) = \partial\psi/\partial y = \bar{u}(x)f'(\eta)$. Substituting this into the conserved flux [(10.55)], we obtain

$$d/dx\{\bar{u}(x)^3(\nu x/\bar{u}) \int_0^\infty f'[\int_y^\infty f'^2 d\eta] d\eta\} = 0,$$

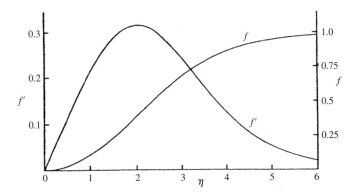

Figure 10.30 Variation of normalized mass flux (f) and normalized velocity (f') with similarly variable η. Reprinted with the permission of Cambridge University Press.

where we expect the integral to be independent of x. Then $(\bar{u})^2 x = C^2$, or $\bar{u}(x) = Cx^{-1/2}$. This gives us the similarity transformation

$$\psi(x, y) = \sqrt{\nu C} \cdot x^{1/4} f(\eta), \quad \text{where } \eta = (C/\nu)^{1/2} \cdot y/x^{3/4}.$$

Differentiating and substituting into (10.56), we obtain (after multiplication by $4x^2/C^2$),

$$4f''' + ff'' + 2f'^2 = 0$$

subject to the boundary conditions (10.57): $f(0) = 0$; $f'(0) = 0$; $f'(\infty) = 0$. This third order equation can be integrated once after multiplying by the integrating factor f, to yield $ff'' - f'^2/2 + f^2 f'/4 = 0$, where the constant of integration has been evaluated at $\eta = 0$. Dividing by the integrating factor $f^{3/2}$ gives an equation that can be integrated once more. The result is

$$f^{-1/2} f' + f^{3/2}/6 = C_1 \equiv f_\infty^{3/2}/6, \quad \text{where } f_\infty = f(\infty).$$

Since $f(0) = 0$, $f'(0) = 0$, a Tayor series for f starts with $f(\eta) = f''(0)\eta^2/2$. Then $f'^2(0)/f(0) = 2f''(0) = f_\infty^3/36$. Since f and η are dimensionless, f_∞ is a pure number. The final integration can be performed after one more transformation: $f/f_\infty = g^2(\bar{\eta})$, $\bar{\eta} = f_\infty \eta$. This results in the equation $dg/(1 - g^3) = d\bar{\eta}/12$. Now $1 - g^3 = (1 - g) \cdot (1 + g + g^2)$, so integration may be effected by partial fractions, with the result in implicit form,

$$-\ln(1-g) + \sqrt{3}\tan^{-1}[(2g+1)/\sqrt{3}] + \ln(1+g+g^2)^{1/2} = \bar{\eta}/4 + \sqrt{3}\tan^{-1}(1/\sqrt{3}),$$

where the boundary condition $g(0) = 0$ was used to evaluate the constant of integration. We can verify easily that $f' \to 0$ exponentially fast in η or $\bar{\eta}$ from our solution for $g(\bar{\eta})$. As $\bar{\eta} \to \infty$, $g \to 1$, so for large $\bar{\eta}$ the solution for g reduces to $-\ln(1 - g) + \sqrt{3}\tan^{-1}\sqrt{3} + (1/2)\ln 3 \cong \bar{\eta}/4 + \sqrt{3}\tan^{-1}(1/\sqrt{3})$. The first term on each side of the equation dominates, leaving $1 - g \approx e^{-(1/4)\bar{\eta}}$. Now

$f' = g(1 - g)(1 + g + g^2)/6 \approx (1/2)e^{-(1/4)\bar{\eta}}$. The mass flow rate in the jet is

$$\dot{m} = \int\limits_0^\infty \rho u \, dy = \rho \bar{u}(x) \int\limits_0^\infty f'(\eta) d\eta \sqrt{v/C} \cdot x^{3/4},$$

or since

$$\bar{u} = Cx^{-1/2}, \dot{m} = \rho\sqrt{vC} f_\infty x^{1/4},$$

indicating that entrainment increases the flow rate in the jet with $x^{1/4}$. If we define the edge of the jet as $\delta(x)$ and say it corresponds to $\bar{\eta} = 6$, for example, then $\delta = 6\sqrt{v/C} f_\infty^{-1} x^{3/4}$. If we define $\bar{u}$ by requiring $\dot{m} = \rho \bar{u}(x)\delta(x)$, the two forms for $\dot{m}$ are coincident if $f_\infty^2 = 6$. The entrainment is evident from the form of $v = -\partial\psi/\partial x = -\sqrt{vC}(f - 3\eta f')/(4x^{3/4}) \to -\sqrt{vC} f_\infty/(4x^{3/4})$ as $\eta \to \infty$, so the flow is downwards, toward the jet.

13. Secondary Flows

Large Reynolds number flows with curved streamlines tend to generate additional velocity components because of properties of the boundary layer. These components are called secondary flows and will be seen later in our discussion of instabilities (p. 454). An example of such a flow is made dramatically visible by putting finely crushed tea leaves, randomly dispersed, into a cup of water, and then stirring vigorously in a circular motion. When the motion has ceased, all of the particles have collected in a mound at the center of the bottom of the cup (see Figure 10.31). An explanation of this phenomenon is given in terms of thin boundary layers. The stirring motion imparts a primary velocity $u_\theta(R)$ (see Appendix B1 for coordinates) large enough for the Reynolds number to be large enough for the boundary layers on the sidewalls and bottom to be thin. The largest terms in the R-momentum equation are

$$\frac{\partial p}{\partial R} = \frac{\rho u_\theta^2}{R}.$$

Away from the walls, the flow is inviscid. As the boundary layer on the bottom is thin, boundary layer theory yields $\partial p/\partial x = 0$ from the x-momentum equation. Thus the pressure in the bottom boundary layer is the same as for the inviscid flow just outside the boundary layer. However, within the boundary layer, u_θ is less than the inviscid value at the edge. Thus $p(R)$ is everywhere larger in the boundary layer than that required for circular streamlines inside the boundary layer, pushing the streamlines inwards. That is, the pressure gradient within the boundary layer generates an inwardly directed u_R. This motion is fed by a downwardly directed flow in the sidewall boundary layer and an outwardly directed flow on the top surface. This secondary flow is closed by an upward flow along the center. The visualization is accomplished by crushed tea leaves which are slightly denser than water. They descend by gravity or are driven outwards by centrifugal acceleration. If they enter the sidewall boundary layer, they are transported downwards and thence to the center by the secondary flow. If the tea particles enter the bottom boundary layer from above, they are quickly swept

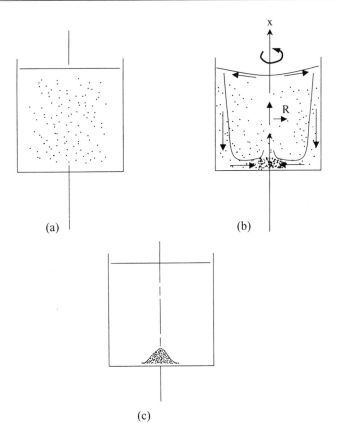

Figure 10.31 Secondary flow in a tea cup: (a) tea leaves randomly dispersed—initial state; (b) stirred vigorously—transient motion; and (c) final state.

to the center and dropped as the flow turns upwards. All the particles collect at the center of the bottom of the teacup. A practical application of this effect, illustrated in Exercise 9, relates to sand and silt transport by the Mississippi River.

14. Perturbation Techniques

The preceding sections, based on Prandtl's seminal idea, have revealed the physical basis of the boundary layer concept in a high Reynolds number flow. In recent years, the boundary layer method has become a powerful mathematical technique used to solve a variety of other physical problems. Some elementary ideas involved in these methods are discussed here. The interested reader should consult other specialized texts on the subject, such as van Dyke (1975), Bender and Orszag (1978), and Nayfeh (1981).

　　The essential idea is that the problem has a small parameter ε in either the governing equation or in the boundary conditions. In a flow at high Reynolds number the small parameter is $\varepsilon = 1/\text{Re}$, in a creeping flow $\varepsilon = \text{Re}$, and in flow around an airfoil ε is the ratio of thickness to chord length. The solutions to these problems can frequently be written in terms of a series involving the small parameter, the higher-order terms acting as a perturbation on the lower-order terms. These methods are called *perturbation techniques*. The perturbation expansions frequently break down in certain regions, where the field develops boundary layers. The boundary layers are treated differently than other regions by expressing the lateral coordinate y in terms of the boundary layer thickness δ and defining $\eta \equiv y/\delta$. The objective is to rescale variables so that they are all finite in the thin singular region.

Order Symbols and Gauge Functions

Frequently we have a complicated function $f(\varepsilon)$ and we want to determine the nature of variation of $f(\varepsilon)$ as $\varepsilon \to 0$. The three possibilities are

$$\left. \begin{array}{ll} f(\varepsilon) \to 0 & \text{(vanishing)} \\ f(\varepsilon) \to A & \text{(bounded)} \\ f(\varepsilon) \to \infty & \text{(unbounded)} \end{array} \right\} \quad \text{as } \varepsilon \to 0,$$

where A is finite. However, this behavior is rather vague because it does not say how fast $f(\varepsilon)$ goes to zero or infinity as $\varepsilon \to 0$. To describe this behavior, we compare the rate at which $f(\varepsilon)$ goes to zero or infinity with the rate at which certain familiar functions go to zero or infinity. The familiar functions used for comparison purposes are called *gauge functions*. The most common example of a sequence of gauge functions is $1, \varepsilon, \varepsilon^2, \varepsilon^3, \ldots$. As an example, suppose we want to find how $\sin \varepsilon$ goes to zero as $\varepsilon \to 0$. Using the Taylor series

$$\sin \varepsilon = \varepsilon - \frac{\varepsilon^3}{3!} + \frac{\varepsilon^5}{5!} - \cdots,$$

we find that

$$\lim_{\varepsilon \to 0} \frac{\sin \varepsilon}{\varepsilon} = \lim_{\varepsilon \to 0} \left(1 - \frac{\varepsilon^2}{3!} + \frac{\varepsilon^4}{5!} - \cdots \right) = 1,$$

which shows that $\sin \varepsilon$ tends to zero at the same rate at which ε tends to zero. Another way of expressing this is to say that $\sin \varepsilon$ is of order ε as $\varepsilon \to 0$, which we write as

$$\sin \varepsilon = O(\varepsilon) \quad \text{as } \varepsilon \to 0.$$

Other examples are that

$$\left. \begin{array}{l} \cos \varepsilon = O(1) \\ \cos \varepsilon - 1 = O(\varepsilon^2) \end{array} \right\} \quad \text{as } \varepsilon \to 0.$$

We can generalize the concept of "order" by the following statement. A function $f(\varepsilon)$ is considered to be of order of a gauge function $g(\varepsilon)$, and written

$$f(\varepsilon) = O[g(\varepsilon)] \quad \text{as } \varepsilon \to 0,$$

if

$$\lim_{\varepsilon \to 0} \frac{f(\varepsilon)}{g(\varepsilon)} = A,$$

where A is nonzero and finite. Note that the size of the constant A is immaterial as far as the mathematics is concerned. Thus, $\sin 7\varepsilon = O(\varepsilon)$ just as $\sin \varepsilon = O(\varepsilon)$, and likewise $1000 = O(1)$. Thus, the *mathematical order* considered here is different from the *physical order of magnitude*. However, if the physical problem has been properly nondimensionalized, with the relevant scales judiciously chosen, then the constant A will be of reasonable size. (Incidentally, we commonly regard a factor of 10 as a change of one physical order of magnitude, so when we say that the magnitude of u is of order $10 \, \text{cm/s}$, we mean that the magnitude of u is expected (or hoped!) to be between 30 and 3 cm/s.)

Sometimes a comparison in terms of a familiar gauge function is unavailable or inconvenient. We may say $f(\varepsilon) = o[g(\varepsilon)]$ in the limit $\varepsilon \to 0$ if

$$\lim_{\varepsilon \to 0} \frac{f(\varepsilon)}{g(\varepsilon)} = 0,$$

so that f is small compared with g as $\varepsilon \to 0$. For example, $|\ln \varepsilon| = o(1/\varepsilon)$ in the limit $\varepsilon \to 0$.

Asymptotic Expansion

An asymptotic expansion of a function, in terms of a given set of gauge functions, is essentially a series representation with a finite number of terms. Suppose the sequence of gauge functions is $g_n(\varepsilon)$, such that each one is smaller than the preceding one in the sense that

$$\lim_{\varepsilon \to 0} \frac{g_{n+1}}{g_n} = 0.$$

Then the *asymptotic expansion* of $f(\varepsilon)$ is of the form

$$f(\varepsilon) = a_0 + a_1 g_1(\varepsilon) + a_2 g_2(\varepsilon) + O[g_3(\varepsilon)], \tag{10.58}$$

where a_n are independent of ε. Note that the remainder, or the error, is of order of the first neglected term. We also write

$$f(\varepsilon) \sim a_0 + a_1 g_1(\varepsilon) + a_2 g_2(\varepsilon),$$

where $\sim$ means "asymptotically equal to." The asymptotic expansion of $f(\varepsilon)$ as $\varepsilon \to 0$ is not unique, because a different choice of the gauge functions $g_n(\varepsilon)$ would lead to a different expansion. A good choice leads to a good accuracy with only a few terms in the expansion. The most frequently used sequence of gauge functions is the power series ε^n. However, in many cases the series in integral powers of ε does not work, and other gauge functions must be used. There is a systematic way of arriving at the sequence of gauge functions, explained in van Dyke (1975), Bender and Orszag (1978), and Nayfeh (1981).

An asymptotic expansion is a finite sequence of limit statements of the type written in the preceding. For example, because $\lim_{\varepsilon \to 0}(\sin \varepsilon)/\varepsilon = 1$, $\sin \varepsilon = \varepsilon + o(\varepsilon)$. Following up using the powers of ε as gauge functions,

$$\lim_{\varepsilon \to 0}(\sin \varepsilon - \varepsilon)/\varepsilon^3 = -\frac{1}{3!}, \qquad \sin \varepsilon = \varepsilon - \frac{\varepsilon^3}{3!} + o(\epsilon^3).$$

By continuing this process we can establish that the term $o(\varepsilon^3)$ is better represented by $O(\varepsilon^5)$ and is in fact $\epsilon^5/5!$. The series terminates with the order symbol.

The interesting property of an asymptotic expansion is that the series (10.58) may not converge if extended indefinitely. Thus, for a fixed ε, the magnitude of a term may eventually increase as shown in Figure 10.32. Therefore, there is an optimum number of terms $N(\varepsilon)$ at which the series should be truncated. The number $N(\varepsilon)$ is difficult to guess, but that is of little consequence, because only one or two terms in the asymptotic expansion are calculated. *The accuracy of the asymptotic representation can be arbitrarily improved by keeping n fixed, and letting $\varepsilon \to 0$.*

We here emphasize the distinction between convergence and asymptoticity. In *convergence* we are concerned with terms far out in an infinite series, a_n. We must have $\lim_{n \to \infty} a_n = 0$ and, for example, $\lim_{n \to \infty}|a_{n+1}/a_n| < 1$ for convergence. *Asymptoticity* is a different limit: n is fixed at a finite number and the approximation is improved as ε (say) tends to its limit.

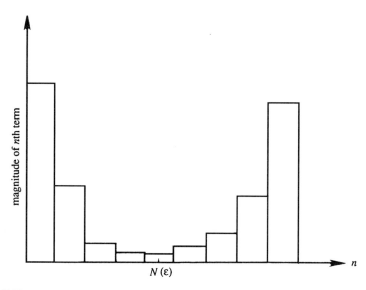

Figure 10.32 Terms in a divergent asymptotic series, in which $N(\varepsilon)$ indicates the optimum number of terms at which the series should be truncated. M. Van Dyke, *Perturbation Methods in Fluid Mechanics*, 1975 and reprinted with the permission of Prof. Milton Van Dyke for The Parabolic Press.

The value of an asymptotic expansion becomes clear if we compare the convergent series for a Bessel function $J_0(x)$, given by

$$J_0(x) = 1 - \frac{x^2}{2^2} + \frac{x^4}{2^2 4^2} - \frac{x^6}{2^2 4^2 6^2} + \cdots , \qquad (10.59)$$

with the first term of its asymptotic expansion

$$J_0(x) \sim \sqrt{\frac{2}{\pi x}} \cos \left(x - \frac{\pi}{4} \right) \quad \text{as } x \to \infty. \qquad (10.60)$$

The convergent series (10.59) is useful when x is small, but more than eight terms are needed for three-place accuracy when x exceeds 4. In contrast, the one-term asymptotic representation (10.60) gives three-place accuracy for $x > 4$. Moreover, the asymptotic expansion indicates the *shape* of the function, whereas the infinite series does not.

Nonuniform Expansion

In many situations we develop an asymptotic expansion for a function of two variables, say

$$f(x; \varepsilon) \sim \sum_n a_n(x) g_n(\varepsilon) \quad \text{as } \varepsilon \to 0. \qquad (10.61)$$

If the expansion holds for all values of x, it is called *uniformly valid* in x, and the problem is described as a *regular perturbation problem*. In this case any successive term is smaller than the preceding term *for all* x. In some interesting situations, however, the expansion may break down for certain values of x. For such values of x, $a_m(x)$ increases faster with m than $g_m(\varepsilon)$ decreases with m, so that the term $a_m(x)g_m(\varepsilon)$ is not smaller than the preceding term. When the asymptotic expansion (10.61) breaks down for certain values of x, it is called a *nonuniform* expansion, and the problem is called a singular perturbation problem. For example, the series

$$\frac{1}{1 + \varepsilon x} = 1 - \varepsilon x + \varepsilon^2 x^2 - \varepsilon^3 x^3 + \cdots , \qquad (10.62)$$

is nonuniformly valid, because it breaks down when $\varepsilon x = O(1)$. No matter how small we make ε, the second term is not a correction of the first term for $x > 1/\varepsilon$. We say that the singularity of the perturbation expansion (10.62) is at large x or at infinity. On the other hand, the expansion

$$\sqrt{x + \varepsilon} = \sqrt{x} \left(1 + \frac{\varepsilon}{x} \right)^{1/2} = \sqrt{x} \left(1 + \frac{\varepsilon}{2x} - \frac{\varepsilon^2}{8x^2} + \cdots \right), \qquad (10.63)$$

is nonuniform because it breaks down when $\varepsilon/x = O(1)$. The singularity of this expansion is at $x = 0$, because it is not valid for $x < \varepsilon$. *The regions of nonuniformity are called boundary layers*; for equation (10.62) it is $x > 1/\varepsilon$, and for equation (10.63) it is $x < \varepsilon$. To obtain expansions that are valid within these singular regions, we need to write the solution in terms of a variable η which is of order 1 within the region of nonuniformity. It is evident that $\eta = \varepsilon x$ for equation (10.62), and $\eta = x/\varepsilon$ for equation (10.63).

In many cases, singular perturbation problems are associated with the small parameter ε multiplying the highest order derivative (as in viscous boundary layer problems), so that the differential equation drops by one order as $\varepsilon \to 0$, resulting in an inability to satisfy all the boundary conditions. In several other singular perturbation problems the small parameter does not multiply the highest-order derivative. An example is low Reynolds number flows, for which the nondimensional governing equation is

$$\varepsilon \mathbf{u} \cdot \nabla \mathbf{u} = -\nabla p + \nabla^2 \mathbf{u},$$

where $\varepsilon = \mathrm{Re} \ll 1$. In this case the singularity or nonuniformity is at infinity. This is discussed in Section 9.13.

15. An Example of a Regular Perturbation Problem

As a simple example of a perturbation expansion that is uniformly valid everywhere, consider a plane Couette flow with a uniform suction across the flow (Figure 10.33). The upper plate is moving parallel to itself at speed U and the lower plate is stationary. The distance between the plates is d and there is a uniform downward suction velocity v'_s, with the fluid coming in through the upper plate and going out through the bottom. For notational simplicity, we shall denote dimensional variables by a prime and nondimensional variables without primes:

$$y = \frac{y'}{d}, \qquad u = \frac{u'}{U}, \qquad v = \frac{v'}{U}.$$

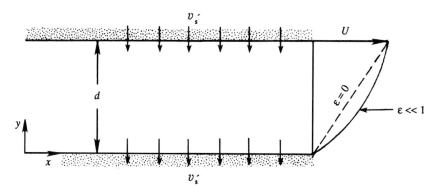

Figure 10.33   Uniform suction in a Couette flow, showing the velocity profile $u(y)$ for $\varepsilon = 0$ and $\varepsilon \ll 1$.

As $\partial/\partial x = 0$ for all variables, the nondimensional equations are

$$\frac{\partial v}{\partial y} = 0 \qquad \text{(continuity)}, \tag{10.64}$$

$$v\frac{du}{dy} = \frac{1}{\text{Re}}\frac{d^2 u}{dy^2} \qquad \text{(x-momentum)}, \tag{10.65}$$

subject to

$$v(0) = v(1) = -v_s, \tag{10.66}$$
$$u(0) = 0, \tag{10.67}$$
$$u(1) = 1, \tag{10.68}$$

where $\text{Re} = U\,d/v$, and $v_s = v_s'/U$.

The continuity equation shows that the lateral flow is independent of y and therefore must be

$$v(y) = -v_s,$$

to satisfy the boundary conditions on v. The x-momentum equation then becomes

$$\frac{d^2 u}{dy^2} + \varepsilon\frac{du}{dy} = 0, \tag{10.69}$$

where $\varepsilon = v_s\text{Re} = v_s'd/v$. We assume that the suction velocity is small, so that $\varepsilon \ll 1$. The problem is to solve equation (10.69), subject to equations (10.67) and (10.68). An exact solution can easily be found for this problem, and will be presented at the end of this section. However, an exact solution may not exist in more complicated problems, and we shall illustrate the perturbation approach. We try a perturbation solution in integral powers of ε, of the form,

$$u(y) = u_0(y) + \varepsilon u_1(y) + \varepsilon^2 u_2(y) + \text{O}(\varepsilon^3). \tag{10.70}$$

(A power series in ε may not always be possible, as remarked upon in the preceding section.) Our task is to determine $u_0(y)$, $u_1(y)$, etc.

Substituting equation (10.70) into equations (10.69), (10.67), and (10.68), we obtain

$$\frac{d^2 u_0}{dy^2} + \varepsilon\left[\frac{du_0}{dy} + \frac{d^2 u_1}{dy^2}\right] + \varepsilon^2\left[\frac{du_1}{dy} + \frac{d^2 u_2}{dy^2}\right] + \text{O}(\varepsilon^3) = 0, \tag{10.71}$$

subject to

$$u_0(0) + \varepsilon u_1(0) + \varepsilon^2 u_2(0) + \text{O}(\varepsilon^3) = 0, \tag{10.72}$$
$$u_0(1) + \varepsilon u_1(1) + \varepsilon^2 u_2(1) + \text{O}(\varepsilon^3) = 1. \tag{10.73}$$

Equations for the various orders are obtained by taking the limits of equations (10.71)–(10.73) as $\varepsilon \to 0$, then dividing by ε and taking the limit $\varepsilon \to 0$ again, and so on. This is equivalent to equating terms with like powers of ε. Up to order ε, this gives the following sets:

Order ε^0:

$$\frac{d^2 u_0}{dy^2} = 0,$$
$$u_0(0) = 0, \qquad u_0(1) = 1. \tag{10.74}$$

Order ε^1:

$$\frac{d^2 u_1}{dy^2} = -\frac{du_0}{dy},$$
$$u_1(0) = 0, \qquad u_1(1) = 0. \tag{10.75}$$

The solution of the zero-order problem (10.74) is

$$u_0 = y. \tag{10.76}$$

Substituting this into the first-order problem (10.75), we obtain the solution

$$u_1 = \frac{y}{2}(1 - y).$$

The complete solution up to order ε is then

$$u(y) = y + \frac{\varepsilon}{2}[y(1 - y)] + O(\varepsilon^2). \tag{10.77}$$

In this expansion the second term is less than the first term for all values of y as $\varepsilon \to 0$. The expansion is therefore uniformly valid for all y and the perturbation problem is regular. A sketch of the velocity profile (10.77) is shown in Figure 10.33.

It is of interest to compare the perturbation solution (10.77) with the exact solution. The exact solution of (10.69), subject to equations (10.67) and (10.68), is easily found to be

$$u(y) = \frac{1 - e^{-\varepsilon y}}{1 - e^{-\varepsilon}}. \tag{10.78}$$

For $\varepsilon \ll 1$, Equation (10.78) can be expanded in a power series of ε, where the first few terms are identical to those in equation (10.77).

16. *An Example of a Singular Perturbation Problem*

Consider again the problem of uniform suction across a plane Couette flow, discussed in the preceding section. For the case of weak suction, namely $\varepsilon = v_s' d/\nu \ll 1$, we saw that the perturbation problem is regular and the series is uniformly valid for all values of y. A more interesting case is that of *strong suction*, defined as $\varepsilon \gg 1$, for which we shall now see that the perturbation expansion breaks down near one of the walls. As before, the v-field is uniform everywhere:

$$v(y) = -v_s.$$

The governing equation is (10.69), which we shall now write as

$$\delta \frac{d^2 u}{dy^2} + \frac{du}{dy} = 0, \tag{10.79}$$

subject to

$$u(0) = 0, \tag{10.80}$$

$$u(1) = 1, \tag{10.81}$$

where we have defined

$$\delta \equiv \frac{1}{\varepsilon} = \frac{v}{v_s' d} \ll 1,$$

as the small parameter. We try an expansion in powers of δ:

$$u(y) = u_0(y) + \delta u_1(y) + \delta^2 u_2(y) + O(\delta^3). \tag{10.82}$$

Substitution into equation (10.79) leads to

$$\frac{du_0}{dy} = 0. \tag{10.83}$$

The solution of this equation is $u_0 = $ const., which cannot satisfy conditions at *both* $y = 0$ and $y = 1$. This is expected, because as $\delta \to 0$ the *highest order derivative drops out of the governing equation* (10.79), and the approximate solution cannot satisfy all the boundary conditions. This happens no matter how many terms are included in the perturbation series. A boundary layer is therefore expected near one of the walls, where the solution varies so rapidly that the two terms in equation (10.79) are of the same order.

The expansion (10.82), valid outside the boundary layers, is the "outer" expansion, the first term of which is governed by equation (10.83). If the outer expansion satisfies the boundary condition (10.80), then the first term in the expansion is $u_0 = 0$; if on the other hand the outer expansion satisfies the condition (10.81), then $u_0 = 1$. The outer expansion should be smoothly matched to an "inner" expansion valid within the boundary layer. The two possibilities are sketched in Figure 10.34, where it is evident that a boundary layer occurs at the top plate if $u_0 = 0$, and it occurs at the bottom plate if $u_0 = 1$. Physical reasons suggest that a strong suction would tend to keep the profile of the longitudinal velocity uniform near the wall through which the fluid *enters*, so that a boundary layer at the lower wall seems more reasonable. Moreover, the $\varepsilon \gg 1$ case is then a continuation of the $\varepsilon \ll 1$ behavior (Figure 10.33).

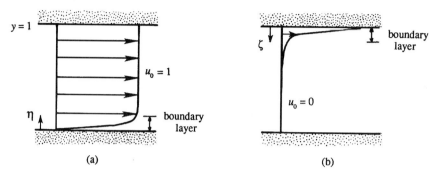

Figure 10.34   Couette flow with strong suction, showing two possible locations of the boundary layer. The one shown in (a) is the correct one.

We shall therefore proceed with this assumption and verify later in the section that it is not mathematically possible to have a boundary layer at $y = 1$.

The location of the boundary layer is determined by the sign of the ratio of the dominant terms in the boundary layer. This is the case because the boundary layer must always decay into the domain and the decay is generally exponential. The inward decay is required so as to match with the outer region solution. Thus a ratio of signs that is positive (when both terms are on the same side of the equation) requires the boundary layer to be at the left or bottom, that is, the boundary with the smaller coordinate.

The first task is to determine the natural distance within the boundary layer, where both terms in equation (10.79) must be of the same order. If y is a typical distance within the boundary layer, this requires that $\delta/y^2 = O(1/y)$, that is

$$y = O(\delta),$$

showing that the natural scale for measuring distances within the boundary layer is δ. We therefore define a boundary layer coordinate

$$\eta = \frac{y}{\delta},$$

which transforms the governing equation (10.79) to

$$-\frac{du}{d\eta} = \frac{d^2u}{d\eta^2}. \tag{10.84}$$

As in the Blasius solution, $\eta = O(1)$ within the boundary layer and $\eta \to \infty$ far outside of it.

The solution of equation (10.84) as $\eta \to \infty$ is to be matched to the solution of equation (10.79) as $y \to 0$. Another way to solve the problem is to write a *composite expansion* consisting of *both* the outer and the inner solutions:

$$u(y) = [u_0(y) + \delta u_1(y) + \cdots] + \{\hat{u}_0(\eta) + \delta \hat{u}_1(\eta) + \cdots\}, \tag{10.85}$$

where the term within { } is regarded as the *correction* to the outer solution within the boundary layer. All terms in the boundary layer correction { } go to zero as $\eta \to \infty$. Substituting equation (10.85) into equation (10.79), we obtain

$$
\frac{du_0}{dy} + \delta \left[\frac{du_1}{dy} + \frac{d^2 u_0}{dy^2} \right] + \delta^2 \left[\right] + O(\delta^3)
$$
$$
+ \delta^{-1} \left[\frac{d\hat{u}_0}{d\eta} + \frac{d^2 \hat{u}_0}{d\eta^2} \right] + \left[\frac{d\hat{u}_1}{d\eta} + \frac{d^2 \hat{u}_1}{d\eta^2} \right] + O(\delta) = 0. \tag{10.86}
$$

A systematic procedure is to multiply equation (10.86) by powers of δ and take limits as $\delta \to 0$, with first y held fixed and then η held fixed. When y is held fixed (which we write as $y = O(1)$) and $\delta \to 0$, the boundary layer becomes progressively thinner and we move outside and into the outer region. When η is held fixed (i.e, $\eta = O(1)$) and $\delta \to 0$, we obtain the behavior within the boundary layer.

Multiplying equation (10.86) by δ and taking the limit as $\delta \to 0$, with $\eta = O(1)$, we obtain

$$
\frac{d\hat{u}_0}{d\eta} + \frac{d^2 \hat{u}_0}{d\eta^2} = 0, \tag{10.87}
$$

which governs the first term of the boundary layer correction. Next, the limit of equation (10.86) as $\delta \to 0$, with $y = O(1)$, gives

$$
\frac{du_0}{dy} = 0, \tag{10.88}
$$

which governs the first term of the outer solution. (Note that in this limit $\eta \to \infty$, and consequently we move outside the boundary layer where all correction terms go to zero, that is $d\hat{u}_1/d\eta \to 0$ and $d^2 \hat{u}_1/d\eta^2 \to 0$.) The next largest term in equation (10.86) is obtained by considering the limit $\delta \to 0$ with $\eta = O(1)$, giving

$$
\frac{d\hat{u}_1}{d\eta} + \frac{d^2 \hat{u}_1}{d\eta^2} = 0,
$$

and so on. It is clear that our formal limiting procedure is equivalent to setting the coefficients of like powers of δ in equation (10.86) to zero, with the boundary layer terms treated *separately*.

As the composite expansion holds everywhere, all boundary conditions can be applied on it. With the assumed solution of equation (10.85), the boundary condition equations (10.80) and (10.81) give

$$
u_0(0) + \hat{u}_0(0) + \delta[u_1(0) + \hat{u}_1(0)] + \cdots = 0, \tag{10.89}
$$
$$
u_0(1) + 0 + \delta[u_1(1) + 0] + \cdots = 1. \tag{10.90}
$$

Equating like powers of δ, we obtain the following conditions

$$u_0(0) + \hat{u}_0(0) = 0, \qquad u_1(0) + \hat{u}_1(0) = 0, \tag{10.91}$$
$$u_0(1) = 1, \qquad u_1(1) = 0. \tag{10.92}$$

We can now solve equation (10.88) along with the first condition in equation (10.92), obtaining

$$u_0(y) = 1. \tag{10.93}$$

Next, we can solve equation (10.87), along with the first condition in equation (10.91), namely

$$\hat{u}_0(0) = -u_0(0) = -1,$$

and the condition $\hat{u}_0(\infty) = 0$. The solution is

$$\hat{u}_0(\eta) = -e^{-\eta}.$$

To the lowest order, the composite expansion is, therefore,

$$u(y) = 1 - e^{-\eta} = 1 - e^{-y/\delta}, \tag{10.94}$$

which we have written in terms of both the inner variable η and the outer variable y, because the composite expansion is valid everywhere. The first term is the lowest-order outer solution, and the second term is the lowest-order correction in the boundary layer.

Comparison with Exact Solution

The exact solution of the problem is (see equation (10.78)):

$$u(y) = \frac{1 - e^{-y/\delta}}{1 - e^{-1/\delta}}. \tag{10.95}$$

We want to write the exact solution in powers of δ and compare with our perturbation solution. An important result to remember is that $\exp(-1/\delta)$ decays faster than *any* power of δ as $\delta \to 0$, which follows from the fact that

$$\lim_{\delta \to 0} \frac{e^{-1/\delta}}{\delta^n} = \lim_{\varepsilon \to \infty} \frac{\varepsilon^n}{e^\varepsilon} = 0, \qquad e^{-1/\delta} = o(\delta^n), \quad n > 0,$$

for any n, as can be verified by applying the l'Hôpital rule n times. Thus, the denominator in equation (10.95) exponentially approaches 1, with *no* contribution in powers of δ. It follows that the expansion of the exact solution in terms of a power series in δ is

$$u(y) \simeq 1 - e^{-y/\delta}, \tag{10.96}$$

which agrees with our composite expansion (10.94). Note that no terms in powers of δ enter in equation (10.96). Although in equation (10.94) we did not try to continue our series to terms of order δ and higher, the form of equation (10.96) shows that these extra terms would have turned out to be zero if we had calculated them. However, the nonexistence of terms proportional to δ and higher is special to the current problem, and not a frequent event.

Why There Cannot Be a Boundary Layer at $y = 1$

So far we have assumed that the boundary layer could occur only at $y = 0$. Let us now investigate what would happen if we assumed that the boundary layer happened to be at $y = 1$. In this case we define a boundary layer coordinate

$$\zeta \equiv \frac{1 - y}{\delta}, \tag{10.97}$$

which increases into the fluid from the upper wall (Figure 10.34b). Then the lowest-order terms in the boundary conditions (10.91) and (10.92) are replaced by

$$u_0(0) = 0,$$
$$u_0(1) + \hat{u}_0(0) = 1,$$

where $\hat{u}_0(0)$ represents the value of $\hat{u}_0$ at the *upper* wall where $\zeta = 0$. The first condition gives the lowest-order outer solution $u_0(y) = 0$. To find the lowest-order boundary layer correction $\hat{u}_0(\zeta)$, note that the equation governing it (obtained by substituting equation (10.97) into equation (10.87)) is

$$\frac{d\hat{u}_0}{d\zeta} - \frac{d^2\hat{u}_0}{d\zeta^2} = 0, \tag{10.98}$$

subject to

$$\hat{u}_0(0) = 1 - u_0(1) = 1,$$
$$\hat{u}_0(\infty) = 0.$$

A substitution of the form $\hat{u}_0(\zeta) = \exp(a\zeta)$ into equation (10.103) shows that $a = +1$, so that the solution to equation (10.98) is exponentially increasing in ζ and cannot satisfy the condition at $\zeta = \infty$.

17. Decay of a Laminar Shear Layer

It is shown in Chapter 12 (pp. 515–516) that flows exhibiting an inflection point in the streamwise velocity profile are highly unstable. These results, based on the Orr-Sommerfeld analysis for parallel flows, have been re-examined recently and it has been found that the instability is not manifested until sufficiently far downstream for similarity to develop. This is discussed in more detail in Chapter 12. Here we note that a detailed treatment of the decay of a laminar shear layer illustrates some interesting points. The problem of the downstream smoothing of an initial velocity discontinuity has not been completely solved even now, although considerable literature might suggest otherwise. Thus it is appropriate to close this chapter with a problem that remains to be put to rest. See Figure 10.35 for a general sketch of the problem. The basic parameter is $\mathrm{Re}_x = U_1 x/\nu$. In these terms the problem splits into distinct regions as illustrated in Figure 10.11. This shown in the paper by Alston and Cohen (1992), which also contains a brief historical summary. In the region for which Re_x is finite, the full Navier–Stokes equations are required for a solution. As Re_x becomes large, $\delta \ll x$, $v \ll u$ and the Navier–Stokes equations asymptotically decay to the boundary layer equations. The boundary layer equations require an initial condition, which is provided by the downstream limit of the solution in the finite Reynolds number region. Here we see that, because they are of elliptic form, the full Navier–Stokes equations require downstream boundary conditions on u and v (which would have to be provided by an asymptotic matching). Paradoxically it seems, the downstream limit of the Navier–Stokes equations, represented by the boundary layer equations, cannot accept a downstream boundary condition because they are of parabolic form. The boundary layer equations govern the downstream evolution from a specified initial station of the streamwise velocity profile. In this problem there must be a matching between the downstream limit of the initial finite Reynolds number region and the initial condition for the boundary layer equations. Although the boundary layer equations are a subset of the full Navier–Stokes equations and are generally appreciated to be the resolution of d'Alembert's paradox via a singular perturbation in the normal (say y) direction, they are also a singular perturbation in the streamwise (say x) direction. That is, the highest x derivative is dropped in the boundary layer approximation and the boundary condition that must be dropped is the one downstream. This becomes an issue in numerical solutions of the full Navier–Stokes equations. It arises downstream in this problem as well.

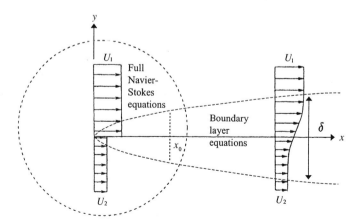

Figure 10.35   Decay of a laminar shear layer.

If in Figure 10.35 the pressure in the top and bottom flow is the same, the boundary layer formulation valid for $x > x_0$, $Re_{x_0} \gg 1$ is

$$\frac{\partial u}{\partial x} + \frac{\partial v}{\partial y} = 0, \quad u\frac{\partial u}{\partial x} + v\frac{\partial u}{\partial y} = \nu\frac{\partial^2 u}{\partial y^2},$$

$$y \to +\infty : u \to U_1, \quad y \to -\infty : u \to U_2,$$

$x = x_0 : U(x_0, y)$ specified (initial condition). One boundary condition on v is required.

We can look for a solution sufficiently far downstream that the initial condition has been forgotten so that the similarity form has been achieved. Then,

$$\eta = \frac{y}{x}\sqrt{\frac{U_1 x}{\nu}} \quad \text{and} \quad \psi(x, y) = \sqrt{\nu U_1 x} f(\eta).$$

In these terms $u/U_1 = f'(\eta)$ and

$$f''' + \tfrac{1}{2} f f'' = 0, \quad f'(\infty) = 1, \quad f'(-\infty) = U_2/U_1.$$

Of course a third boundary condition is required for a unique solution. This represents the need to specify one boundary condition on v. Let us see how far we can go towards a solution and what the missing boundary condition actually pins down. Consider the transformation $f'(\eta) = F(f) = u/U_1$. Then

$$\frac{d^2 f}{d\eta^2} = F\frac{dF}{df}$$

and

$$\frac{d^3 f}{d\eta^3} = \left[F\frac{d^2 F}{df^2} + \left(\frac{dF}{df}\right)^2 \right] F.$$

The Blasius equation transforms to

$$F\frac{d^2 F}{df^2} + \left(\frac{dF}{df}\right)^2 + \frac{1}{2} f\frac{dF}{df} = 0, \tag{10.99}$$

$$F(f = \infty) = 1, \quad F(f = -\infty) = U_2/U_1. \tag{10.100}$$

This has a unique solution for the streamwise velocity $u/U_1 = F$ in terms of the similarity streamfunction $f(\eta)$ with the expected properties, which are shown in Figure 10.36(a) and (b). The exact solution varies more steeply than the linearized solution for small velocity difference, with the greatest difference between solutions at the region of maximum curvature at the low velocity end. This difference is shown more clearly in the magnified insets of each frame. The difference increases as the normalized velocity difference, $(U_1 - U_2)/U_1$, increases. We can see from the (Blasius)

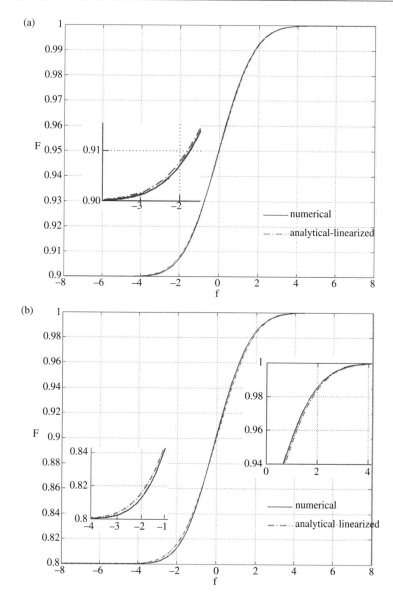

Figure 10.36 Solution for $F(f)$ from equation (10.99) subject to boundary conditions (10.100) when (a) $U_2/U_1 = 0.9$, and (b) $U_2/U_1 = 0.8$. The "analytical—linearized" approximation is the asymptotic solution for $(U_1 - U_2)/U_1 \ll 1 : F = 1 - [(U_1 - U_2)/(2U_1)]erfc(f/2)$. Magnified insets show the difference between the two curves.

equation in η-space that the maximum of the shear stress occurs where $f = 0$. This is the inflection point in the velocity profile in η or y. However, the inflection point in the $F(f)$ curve is located where $f = -2\, dF/df < 0$. This is below the dividing streamline $f = 0$. To put this back in physical space (x, y), the transformation must be inverted, $\int d\eta = \int df/F(f)$.

The integral on the right-hand side can be calculated exactly but the correspondence between any integration limit on the right-hand side and that on the left-hand side is ambiguous. This solution admits a translation of η by any constant. The ambiguity in the location in y (or η) of the calculated profile was known to Prandtl. In the literature, five different third boundary conditions have been used. They are as follows:

(a) $f(\eta = 0) = 0$ ($v = 0$ on y or $\eta = 0$);
(b) $f'(\eta = 0) = (1 + U_2/U_1)/2$ (average velocity on the axis);
(c) $\eta f' - f \to 0$ as $\eta \to \infty$ ($v \to 0$ as $\eta \to \infty$);
(d) $\eta f' - f \to 0$ as $\eta \to -\infty$ ($v \to 0$ as $\eta \to -\infty$); and
(e) $uv]_\infty + uv]_{-\infty} = 0$ or $f'(\eta f' - f)]_\infty + f'(\eta f' - f)]_{-\infty} = 0$ (von Karman; zero net transverse force).

Alston and Cohen (1992) considered the limit of small velocity difference $(U_2 - U_1)/U_1 \ll 1$ and showed that none of these third boundary conditions are correct. As the normalized velocity difference increases, we expect the error in using any of the incorrect boundary conditions to increase. Of all of them, the last (e) is closest to the correct result. D.-C. Hwang, in his doctoral dissertation (2005) has shown, that as the normalized velocity difference $(U_1 - U_2)/U_1$ increases, the trends seen by Alston continue. Figure 10.37 shows that the streamwise velocity on the dividing streamline ($f = 0$) is larger than the average velocity of the two streams, when the upper stream is the faster one. What is not determined from the solution to (10.99)

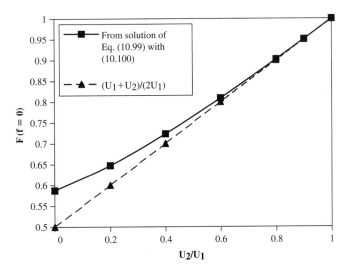

Figure 10.37 Comparison of streamwise velocity at dividing streamline ($f = 0$) with average velocity (dashed line).

subject to (10.100) is the location of the dividing streamline, $f = 0$, because that depends on the inverse transformation, which requires one more boundary condition for a unique specification. When $U_1 > U_2$, the dividing streamline $\psi = 0$, which starts at the origin, bends slowly downwards and its path can be tracked only by starting the solution at the origin and following the evolution of the equations downstream. Thus, no simple statement of a third boundary condition is possible to complete the similarity formulation. Following Klemp and Acrivos (1972), the deflection of the dividing streamline from the x-axis was written by Hwang (2005) as $\varphi(x)$. Then a modified similarity variable $\eta = (U_1/v)^{1/2}(y - \varphi)/\sqrt{2x}$ was defined. In these terms, the dividing streamline is $\eta = 0$ so that the boundary condition on the dividing streamline is $f(\eta = 0) = 0$. However, the behavior of $\varphi(x)$ had to be computed by solving the Navier-Stokes equations from the origin. The downstream asymptote for $\varphi(x)$, which can be used as the third boundary condition for the similarity solution, is shown in Figure 10.38 as a function of velocity ratio. The downstream distance to achieve similarity is found to be given by $\text{Re} = U_1 x/v \approx 10^4$. Although the results of Alston and Cohen are difficult to distinguish on the graph from the von Karman condition (e), the numerical calculation shows a clear distinction. Deviations become larger as U_2/U_1 diminishes. Beyond the point shown, computations became excessively tedious.

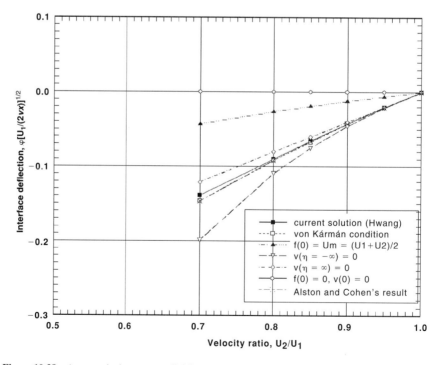

Figure 10.38   Asymptotic downstream dividing streamline deflection as a function of velocity ratio.

Exercises

1. Solve the Blasius sets (10.34) and (10.35) with a computer, using the Runge–Kutta scheme of numerical integration.

2. A flat plate 4 m wide and 1 m long (in the direction of flow) is immersed in kerosene at 20 °C ($\nu = 2.29 \times 10^{-6}$ m^2/s, $\rho = 800$ kg/m^3) flowing with an undisturbed velocity of 0.5 m/s. Verify that the Reynolds number is less than critical everywhere, so that the flow is laminar. Show that the thickness of the boundary layer and the shear stress at the center of the plate are $\delta = 0.74$ cm and $\tau_0 = 0.2$ N/m^2, and those at the trailing edge are $\delta = 1.05$ cm and $\tau_0 = 0.14$ N/m^2. Show also that the total frictional drag on one side of the plate is 1.14 N. Assume that the similarity solution holds for the entire plate.

3. Air at 20 °C and 100 kPa ($\rho = 1.167$ kg/m^3, $\nu = 1.5 \times 10^{-5}$ m^2/s) flows over a thin plate with a free-stream velocity of 6 m/s. At a point 15 cm from the leading edge, determine the value of y at which $u/U = 0.456$. Also calculate v and $\partial u/\partial y$ at this point. [*Answer:* $y = 0.857$ mm, $v = 0.39$ cm/s, $\partial u/\partial y = 3020$ s^{-1}. You may not be able to get this much accuracy, because your answer will probably use certain figures in the chapter.]

4. Assume that the velocity in the laminar boundary layer on a flat plate has the profile

$$\frac{u}{U} = \sin\frac{\pi y}{2\delta}.$$

Using the von Karman momentum integral equation, show that

$$\frac{\delta}{x} = \frac{4.795}{\sqrt{\mathrm{Re}_x}}, \qquad C_f = \frac{0.655}{\sqrt{\mathrm{Re}_x}}.$$

Notice that these are very similar to the Blasius solution.

5. Water flows over a flat plate 30 m long and 17 m wide with a free-stream velocity of 1 m/s. Verify that the Reynolds number at the end of the plate is larger than the critical value for transition to turbulence. Using the drag coefficient in Figure 10.12, estimate the drag on the plate.

6. Find the diameter of a parachute required to provide a fall velocity no larger than that caused by jumping from a 2.5 m height, if the total load is 80 kg. Assume that the properties of air are $\rho = 1.167$ kg/m^3, $\nu = 1.5 \times 10^{-5}$ m^2/s, and treat the parachute as a hemispherical shell with $C_D = 2.3$. [*Answer:* 3.9 m]

7. Consider the roots of the algebraic equation

$$x^2 - (3 + 2\varepsilon)x + 2 + \varepsilon = 0,$$

for $\varepsilon \ll 1$. By a perturbation expansion, show that the roots are

$$x = \begin{cases} 1 - \varepsilon + 3\varepsilon^2 + \cdots, \\ 2 + 3\varepsilon - 3\varepsilon^2 + \cdots. \end{cases}$$

(From Nayfeh, 1981, p. 28 and reprinted by permission of John Wiley & Sons, Inc.)

8. Consider the solution of the equation

$$\varepsilon \frac{d^2 y}{dx^2} - (2x + 1)\frac{dy}{dx} + 2y = 0, \quad \varepsilon \ll 1,$$

with the boundary conditions

$$y(0) = \alpha, \qquad y(1) = \beta.$$

Convince yourself that a boundary layer at the left end does not generate "matchable" expansions, and that a boundary layer at $x = 1$ is necessary. Show that the composite expansion is

$$y = \alpha(2x + 1) + (\beta - 3\alpha)e^{-3(1-x)/\varepsilon} + \cdots .$$

For the two values $\varepsilon = 0.1$ and 0.01, sketch the solution if $\alpha = 1$ and $\beta = 0$. (From Nayfeh, 1981, p. 284 and reprinted by permission of John Wiley & Sons, Inc.)

9. Consider incompressible, slightly viscous flow over a semi-infinite flat plate with constant suction. The suction velocity $v(x, y = 0) = v_0 < 0$ is ordered by $O(\text{Re}^{-1/2}) < v_0/U < O(1)$ where $\text{Re} = Ux/\nu \to \infty$. The flow upstream is parallel to the plate with speed U. Solve for u, v in the boundary layer.

10. Mississippi River boatmen know that when rounding a bend in the river, they must stay close to the outer bank or else they will run aground. Explain in fluid mechanical terms the reason for the cross-sectional shape of the river at the bend:

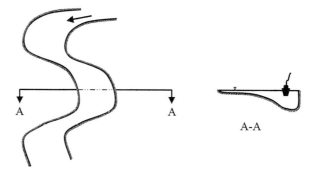

11. Solve to leading order in ε in the limit $\varepsilon \to 0$

$$\varepsilon[x^{-2} + \cos(\ln x)]\frac{d^2 f}{dx^2} + \cos x \frac{df}{dx} + \sin x f = 0,$$
$$1 \leqslant x \leqslant 2, \quad f(1) = 0, \quad f(2) = \cos 2.$$

12. A laminar shear layer develops immediately downstream of a velocity discontinuity. Imagine parallel flow upstream of the origin with a velocity discontinuity

at $x = 0$ so that $u = U_1$ for $y > 0$ and $u = U_2$ for $y < 0$. The density may be assumed constant and the appropriate Reynolds number is sufficiently large that the shear layer is thin (in comparison to distance from the origin). Assume the static pressures are the same in both halves of the flow at $x = 0$. Describe any ambiguities or nonuniquenesses in a similarity formulation and how they may be resolved. In the special case of small velocity difference, solve explicitly to first order in the smallness parameter (velocity difference normalized by average velocity, say) and show where the nonuniqueness enters.

13. Solve equation (10.99) subject to equation (10.100) asymptotically for small velocity difference and obtain the result in the caption to Figure 10.36.

Literature Cited

Alston, T. M. and I. M. Cohen (1992). "Decay of a laminar shear layer." *Phys. Fluids* A**4**: 2690–2699.

Bender, C. M. and S. A. Orszag (1978). *Advanced Mathematical Methods for Scientists and Engineers*. New York: McGraw-Hill.

Falkner, V. W. and S. W. Skan (1931). "Solutions of the boundary layer equations." *Phil. Mag. (Ser. 7)* **12**: 865–896.

Gallo, W. F., J. G. Marvin, and A. V. Gnos (1970). "Nonsimilar nature of the laminar boundary layer." *AIAA J.* **8**: 75–81.

Glauert, M. B. (1956). "The Wall Jet." *J. Fluid Mech.* **1**: 625–643.

Goldstein, S. (ed.). (1938). *Modern Developments in Fluid Dynamics*, London: Oxford University Press; Reprinted by Dover, New York (1965).

Holstein, H. and T. Bohlen (1940). "Ein einfaches Verfahren zur Berechnung laminarer Reibungsschichten die dem Näherungsverfahren von K. Pohlhausen genügen." *Lilienthal-Bericht*. S. 10: 5–16.

Hwang, Din-Chih (2005). "Evolution of a laminar mixing layer." Ph.D. dissertation, University of Pennsylvania. Submitted for publication.

Klemp, J. B. and A. Acrivos (1972). "A note on the laminar mixing of two uniform parallel semi-infinite streams." *Journal of Fluid Mechanics* **55**: 25–30.

Mehta, R. (1985). "Aerodynamics of sports balls." *Annual Review of Fluid Mechanics* **17**, 151–189.

Nayfeh, A. H. (1981). *Introduction to Perturbation Techniques*, New York: Wiley.

Peletier, L. A. (1972). "On the asymptotic behavior of velocity profiles in laminar boundary layers." *Arch. for Rat. Mech. and Anal.* **45**: 110–119.

Pohlhausen, K. (1921). "Zur näherungsweisen Integration der Differentialgleichung der laminaren Grenzschicht." *Z. Angew. Math. Mech.* **1**: 252–268.

Rosenhead, L. (ed.). (1988). *Laminar Boundary Layers*, New York: Dover.

Schlichting, H. (1979). *Boundary Layer Theory*, 7th ed., New York: McGraw-Hill.

Serrin, J. (1967). "Asymptotic behaviour of velocity profiles in the Prandtl boundary layer theory." *Proc. Roy. Soc.* A**299**: 491–507.

Sherman, F. S. (1990). *Viscous Flow*, New York: McGraw-Hill.

Taneda, S. (1965). "Experimental investigation of vortex streets." *J. Phys. Soc. Japan* **20**: 1714–1721.

Thomson, R. E. and J. F. R. Gower (1977). "Vortex streets in the wake of the Aleutian Islands." *Monthly Weather Review* **105**: 873–884.

Thwaites, B. (1949). "Approximate calculation of the laminar boundary layer." *Aero. Quart.* **1**: 245–280.

van Dyke, M. (1975). *Perturbation Methods in Fluid Mechanics*, Stanford, CA: The Parabolic Press.

von Karman, T. (1921). "Über laminare und turbulente Reibung." *Z. Angew. Math. Mech.* **1**: 233–252.

Wen, C.-Y. and C.-Y. Lin (2001). "Two-dimensional vortex shedding of a circular cylinder." *Phys. Fluids* **13**: 557–560.

Yih, C. S. (1977). *Fluid Mechanics: A Concise Introduction to the Theory*, Ann Arbor, MI: West River Press.

Supplemental Reading

Batchelor, G. K. (1967). *An Introduction to Fluid Dynamics*, London: Cambridge University Press.

Friedrichs, K. O. (1955). "Asymptotic phenomena in mathematical physics." *Bull. Am. Math. Soc.* **61**: 485–504.

Lagerstrom, P. A. and R. G. Casten (1972). "Basic concepts underlying singular perturbation techniques." *SIAM Review* **14**: 63–120.

Meksyn, D. (1961). *New Methods in Laminar Boundary Layer Theory*, New York: Pergamon Press.

Panton, R. L. (1984). *Incompressible Flow*, New York: Wiley.

Chapter 11

Computational Fluid Dynamics

by Howard H. Hu
University of Pennsylvania
Philadelphia, PA, USA

1. Introduction

Computational Fluid Dynamics (CFD) is a science that, with the help of digital computers, produces quantitative predictions of fluid-flow phenomena based on those conservation laws (conservation of mass, momentum, and energy) governing fluid motion. These predictions normally occur under those conditions defined in terms of flow geometry, the physical properties of a fluid, and the boundary and initial conditions of a flow field. The prediction generally concerns sets of values of the flow variables, for example, velocity, pressure, or temperature at selected locations in the domain and for selected times. It may also evaluate the overall behavior of the flow, such as the flow rate or the hydrodynamic force acting on an object in the flow.

©2010 Elsevier Inc. All rights reserved.
DOI: 10.1016/B978-0-12-381399-2.50011-3

During the past four decades different types of numerical methods have been developed to simulate fluid flows involving a wide range of applications. These methods include finite difference, finite element, finite volume, and spectral methods. Some of them will be discussed in this chapter.

The CFD predictions are never completely exact. Because many sources of error are involved in the predictions, one has to be very careful in interpreting the results produced by CFD techniques. The most common sources of error are:

- *Discretization error.* This is intrinsic to all numerical methods. This error is incurred whenever a continuous system is approximated by a discrete one where a finite number of locations in space (grids) or instants of time may have been used to resolve the flow field. Different numerical schemes may have different orders of magnitude of the discretization error. Even with the same method, the discretization error will be different depending upon the distribution of the grids used in a simulation. In most applications, one needs to properly select a numerical method and choose a grid to control this error to an acceptable level.

- *Input data error.* This is due to the fact that both flow geometry and fluid properties may be known only in an approximate way.

- *Initial and boundary condition error.* It is common that the initial and boundary conditions of a flow field may represent the real situation too crudely. For example, flow information is needed at locations where fluid enters and leaves the flow geometry. Flow properties generally are not known exactly and are thus only approximated.

- *Modeling error.* More complicated flows may involve physical phenomena that are not perfectly described by current scientific theories. Models used to solve these problems certainly contain errors, for example, turbulence modeling, atmospheric modeling, problems in multiphase flows, and so on.

As a research and design tool, CFD normally complements experimental and theoretical fluid dynamics. However, CFD has a number of distinct advantages:

- It can be produced inexpensively and quickly. Although the price of most items is increasing, computing costs are falling. According to Moore's law based on the observation of the data for the last 40 years, the CPU power will double every 18 months into the foreseeable future.

- It generates complete information. CFD produces detailed and comprehensive information of all relevant variables throughout the domain of interest. This information can also be easily accessed.

- It allows easy change of the parameters. CFD permits input parameters to be varied easily over wide ranges, thereby facilitating design optimization.

- It has the ability to simulate realistic conditions. CFD can simulate flows directly under practical conditions, unlike experiments, where a small- or a large-scale model may be needed.

- It has the ability to simulate ideal conditions. CFD provides the convenience of switching off certain terms in the governing equations, which allows one to focus attention on a few essential parameters and eliminate all irrelevant features.

- It permits exploration of unnatural events. CFD allows events to be studied that every attempt is made to prevent, for example, conflagrations, explosions, or nuclear power plant failures.

2. Finite Difference Method

The key to various numerical methods is to convert the partial different equations that govern a physical phenomenon into a system of algebraic equations. Different techniques are available for this conversion. The finite difference method is one of the most commonly used.

Approximation to Derivatives

Consider the one-dimensional transport equation,

$$\frac{\partial T}{\partial t} + u\frac{\partial T}{\partial x} = D\frac{\partial^2 T}{\partial x^2} \quad \text{for} \quad 0 \leqslant x \leqslant L. \tag{11.1}$$

This is the classic convection-diffusion problem for $T(x, t)$, where u is a convective velocity and D is a diffusion coefficient. For simplicity, let us assume that u and D are two constants. This equation is written in nondimensional form. The boundary conditions for this problem are

$$T(0, t) = g \quad \text{and} \quad \frac{\partial T}{\partial x}(L, t) = q, \tag{11.2}$$

where g and q are two constants. The initial condition is

$$T(x, 0) = T_0(x) \quad \text{for} \quad 0 \leqslant x \leqslant L, \tag{11.3}$$

where $T_0(x)$ is a given function that satisfies the boundary conditions (11.2).

Let us first discretize the transport equation (11.1) on a uniform grid with a grid spacing Δx, as shown in Figure 11.1. Equation (11.1) is evaluated at spatial location $x = x_i$ and time $t = t_n$. Define $T(x_i, t_n)$ as the exact value of T at the location $x = x_i$ and time $t = t_n$, and let T_i^n be its approximation. Using the Taylor series expansion, we have

$$T_{i+1}^n = T_i^n + \Delta x \left[\frac{\partial T}{\partial x}\right]_i^n + \frac{\Delta x^2}{2}\left[\frac{\partial^2 T}{\partial x^2}\right]_i^n + \frac{\Delta x^3}{6}\left[\frac{\partial^3 T}{\partial x^3}\right]_i^n + \frac{\Delta x^4}{24}\left[\frac{\partial^4 T}{\partial x^4}\right]_i^n + O\left(\Delta x^5\right), \tag{11.4}$$

$$T_{i-1}^n = T_i^n - \Delta x \left[\frac{\partial T}{\partial x}\right]_i^n + \frac{\Delta x^2}{2}\left[\frac{\partial^2 T}{\partial x^2}\right]_i^n - \frac{\Delta x^3}{6}\left[\frac{\partial^3 T}{\partial x^3}\right]_i^n + \frac{\Delta x^4}{24}\left[\frac{\partial^4 T}{\partial x^4}\right]_i^n + O\left(\Delta x^5\right), \tag{11.5}$$

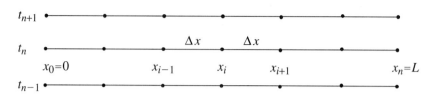

Figure 11.1 Uniform grid in space and time.

where $O\left(\Delta x^5\right)$ means terms of the order of Δx^5. Therefore, the first spatial derivative may be approximated as

$$
\left[\frac{\partial T}{\partial x}\right]_i^n = \frac{T_{i+1}^n - T_i^n}{\Delta x} + O\left(\Delta x\right) \qquad \text{(forward difference)}
$$

$$
= \frac{T_i^n - T_{i-1}^n}{\Delta x} + O\left(\Delta x\right) \qquad \text{(backward difference)} \qquad (11.6)
$$

$$
= \frac{T_{i+1}^n - T_{i-1}^n}{2\Delta x} + O\left(\Delta x^2\right) \quad \text{(centered difference)}
$$

and the second order derivative may be approximated as

$$
\left[\frac{\partial^2 T}{\partial x^2}\right]_i^n = \frac{T_{i+1}^n - 2T_i^n + T_{i-1}^n}{\Delta x^2} + O\left(\Delta x^2\right). \qquad (11.7)
$$

The orders of accuracy of the approximations (truncation errors) are also indicated in the expressions of (11.6) and (11.7). More accurate approximations generally require more values of the variable on the neighboring grid points. Similar expressions can be derived for nonuniform grids.

In the same fashion, the time derivative can be discretized as

$$
\left[\frac{\partial T}{\partial t}\right]_i^n = \frac{T_i^{n+1} - T_i^n}{\Delta t} + O\left(\Delta t\right)
$$

$$
= \frac{T_i^n - T_i^{n-1}}{\Delta t} + O\left(\Delta t\right) \qquad (11.8)
$$

$$
= \frac{T_i^{n+1} - T_i^{n-1}}{2\Delta t} + O\left(\Delta t^2\right)
$$

where $\Delta t = t_{n+1} - t_n = t_n - t_{n-1}$ is the constant time step.

Discretization and Its Accuracy

A discretization of the transport equation (11.1) is obtained by evaluating the equation at fixed spatial and temporal grid points and using the approximations for the individual derivative terms listed in the preceding section. When the first expression in (11.8) is used, together with (11.7) and the centered difference in (11.6), (11.1) may be discretized by

$$\frac{T_i^{n+1} - T_i^n}{\Delta t} + u \frac{T_{i+1}^n - T_{i-1}^n}{2\Delta x} = D \frac{T_{i+1}^n - 2T_i^n + T_{i-1}^n}{\Delta x^2} + O\left(\Delta t, \Delta x^2\right), \quad (11.9)$$

or

$$T_i^{n+1} \approx T_i^n - u\Delta t \frac{T_{i+1}^n - T_{i-1}^n}{2\Delta x} + D\Delta t \frac{T_{i+1}^n - 2T_i^n + T_{i-1}^n}{\Delta x^2} \qquad (11.10)$$

$$= T_i^n - \alpha \left(T_{i+1}^n - T_{i-1}^n\right) + \beta \left(T_{i+1}^n - 2T_i^n + T_{i-1}^n\right),$$

where

$$\alpha = u \frac{\Delta t}{2\Delta x}, \quad \beta = D \frac{\Delta t}{\Delta x^2}. \qquad (11.11)$$

Once the values of T_i^n are known, starting with the initial condition (11.3), the expression (11.10) simply updates the variable for the next time step $t = t_{n+1}$. This scheme is known as an explicit algorithm. The discretization (11.10) is first order accurate in time and second order accurate in space.

As another example, when the backward difference expression in (11.8) is used, we will have

$$\frac{T_i^n - T_i^{n-1}}{\Delta t} + u \frac{T_{i+1}^n - T_{i-1}^n}{2\Delta x} = D \frac{T_{i+1}^n - 2T_i^n + T_{i-1}^n}{\Delta x^2} + O\left(\Delta t, \Delta x^2\right), \quad (11.12)$$

or

$$T_i^n + \alpha \left(T_{i+1}^n - T_{i-1}^n\right) - \beta \left(T_{i+1}^n - 2T_i^n + T_{i-1}^n\right) \approx T_i^{n-1}. \qquad (11.13)$$

At each time step $t = t_n$, here a system of algebraic equations needs to be solved to advance the solution. This scheme is known as an implicit algorithm. Obviously, for the same accuracy, the explicit scheme (11.10) is much simpler than the implicit one (11.13). However, the explicit scheme has limitations.

Convergence, Consistency, and Stability

The result from the solution of the explicit scheme (11.10) or the implicit scheme (11.13) represents an approximate numerical solution to the original partial differential equation (11.1). One certainly hopes that the approximate solution will be close to the exact one. Thus we introduce the concepts of *convergence*, *consistency*, and *stability* of the numerical solution.

The approximate solution is said to be *convergent* if it approaches the exact solution as the grid spacings Δx and Δt tend to zero. We may define the solution error as the difference between the approximate solution and the exact solution,

$$e_i^n = T_i^n - T\left(x_i, t_n\right). \qquad (11.14)$$

Thus the approximate solution converges when $e_i^n \to 0$ as $\Delta x, \Delta t \to 0$. For a convergent solution, some measure of the solution error can be estimated as

$$\left\| e_i^n \right\| \leqslant K \Delta x^a \Delta t^b, \tag{11.15}$$

where the measure may be the root mean square (rms) of the solution error on all the grid points; K is a constant independent of the grid spacing Δx and the time step Δt; the indices a and b represent the convergence rates at which the solution error approaches zero.

One may reverse the discretization process, and examine the limit of the discretized equations (11.10) and (11.13), as the grid spacing tends to zero. The discretized equation is said to be *consistent* if it recovers the original partial differential equation (11.1) in the limit of zero grid spacing.

Let us consider the explicit scheme (11.10). Substitution of the Taylor series expansions (11.4) and (11.5) into this scheme (11.10) produces,

$$\left[\frac{\partial T}{\partial t} \right]_i^n + u \left[\frac{\partial T}{\partial x} \right]_i^n - D \left[\frac{\partial^2 T}{\partial x^2} \right]_i^n + E_i^n = 0, \tag{11.16}$$

where

$$E_i^n = \frac{\Delta t}{2} \left[\frac{\partial^2 T}{\partial t^2} \right]_i^n + u \frac{\Delta x^2}{6} \left[\frac{\partial^3 T}{\partial x^3} \right]_i^n - D \frac{\Delta x^2}{12} \left[\frac{\partial^4 T}{\partial x^4} \right]_i^n + O\left(\Delta t^2, \Delta x^4 \right), \tag{11.17}$$

is the truncation error. Obviously, as the grid spacing $\Delta x, \Delta t \to 0$, this truncation error is of the order of $O\left(\Delta t, \Delta x^2 \right)$ and tends to zero. Therefore, the explicit scheme (11.10) or expression (11.16) recovers the original partial differential equation (11.1), or it is consistent. It is said to be first-order accurate in time and second-order accurate in space, according to the order of magnitude of the truncation error.

In addition to the truncation error introduced in the discretization process, other sources of error may be present in the approximate solution. Spontaneous disturbances (such as the round-off error) may be introduced during either the evaluation or the numerical solution process. A numerical approximation is said to be *stable* if these disturbances decay and do not affect the solution.

The stability of the explicit scheme (11.10) may be examined using the von Neumann method. Let us consider the error at a grid point,

$$\xi_i^n = T_i^n - \overline{T}_i^n, \tag{11.18}$$

where T_i^n is the exact solution of the discretized system (11.10) and $\overline{T}_i^n$ is the approximate numerical solution of the same system. This error could be introduced due to the round-off error at each step of the computation. We need to monitor its decay/growth with time. It can be shown that the evolution of this error satisfies the same homogeneous algebraic system (11.10) or

$$\xi_i^{n+1} = (\alpha + \beta) \, \xi_{i-1}^n + (1 - 2\beta) \, \xi_i^n + (\beta - \alpha) \, \xi_{i+1}^n. \tag{11.19}$$

The error distributed along the grid line can always be decomposed in Fourier space as

$$\xi_i^n = \sum_{k=-\infty}^{\infty} g^n(k)\, e^{i\pi k x_i} \tag{11.20}$$

where $i = \sqrt{-1}$, k is the wavenumber in Fourier space, and g^n represents the function g at time $t = t_n$. As the system is linear, we can examine one component of (11.20) at a time,

$$\xi_i^n = g^n(k) e^{i\pi k x_i}. \tag{11.21}$$

The component at the next time level has a similar form

$$\xi_i^{n+1} = g^{n+1}(k) e^{i\pi k x_i}. \tag{11.22}$$

Substituting the preceding two equations (11.21) and (11.22) into error equation (11.19), we obtain,

$$g^{n+1} e^{i\pi k x_i} = g^n [(\alpha + \beta) e^{i\pi k x_{i-1}} + (1 - 2\beta) e^{i\pi k x_i} + (\beta - \alpha) e^{i\pi k x_{i+1}}] \tag{11.23}$$

or

$$\frac{g^{n+1}}{g^n} = [(\alpha + \beta) e^{-i\pi k \Delta x} + (1 - 2\beta) + (\beta - \alpha) e^{i\pi k \Delta x}]. \tag{11.24}$$

This ratio g^{n+1}/g^n is called the amplification factor. The condition for stability is that the magnitude of the error should decay with time, or

$$\left| \frac{g^{n+1}}{g^n} \right| \leqslant 1, \tag{11.25}$$

for any value of the wavenumber k. For this explicit scheme, the condition for stability (11.25) can be expressed as

$$\left(1 - 4\beta \sin^2\left(\frac{\theta}{2}\right)\right)^2 + (2\alpha \sin \theta)^2 \leqslant 1, \tag{11.26}$$

where $\theta = k\pi \Delta x$. The stability condition (11.26) also can be expressed as (Noye, 1983),

$$0 \leqslant 4\alpha^2 \leqslant 2\beta \leqslant 1. \tag{11.27}$$

For the pure diffusion problem ($u = 0$), the stability condition (11.27) for this explicit scheme requires that

$$0 \leqslant \beta \leqslant \frac{1}{2} \quad or \quad \Delta t \leqslant \frac{1}{2}\frac{\Delta x^2}{D}, \tag{11.28}$$

which limits the size of the time step. For the pure convection problem ($D = 0$), condition (11.27) will never be satisfied, which indicates that the scheme is always unstable and it means that any error introduced during the computation will explode with time. Thus, this explicit scheme is useless for pure convection problems. To improve the stability of the explicit scheme for the convection problem, one may use an upwind scheme to approximate the convective term,

$$T_i^{n+1} = T_i^n - 2\alpha \left(T_i^n - T_{i-1}^n\right), \tag{11.29}$$

where the stability condition requires that

$$u\frac{\Delta t}{\Delta x} \leqslant 1. \tag{11.30}$$

The condition (11.30) is known as the Courant-Friedrichs-Lewy (CFL) condition. This condition indicates that a fluid particle should not travel more than one spatial grid in one time step.

It can easily be shown that the implicit scheme (11.13) is also consistent and unconditionally stable.

It is normally difficult to show the convergence of an approximate solution theoretically. However, the *Lax Equivalence Theorem* (Richtmyer and Morton, 1967) states that: *for an approximation to a well-posed linear initial value problem, which satisfies the consistency condition, stability is a necessary and sufficient condition for the convergence of the solution.*

For convection-diffusion problems, the exact solution may change significantly in a narrow boundary layer. If the computational grid is not sufficiently fine to resolve the rapid variation of the solution in the boundary layer, the numerical solution may present unphysical oscillations adjacent to or in the boundary layer. To prevent the oscillatory solution, a condition on the cell Peclét number (or Reynolds number) is normally required (see Section 4),

$$R_{cell} = u\frac{\Delta x}{D} \leqslant 2. \tag{11.31}$$

3. Finite Element Method

The finite element method was developed initially as an engineering procedure for stress and displacement calculations in structural analysis. The method was subsequently placed on a sound mathematical foundation with a variational interpretation of the potential energy of the system. For most fluid dynamics problems, finite element applications have used the Galerkin finite element formulation on which we will focus in this section.

Weak or Variational Form of Partial Differential Equations

Let us consider again the one-dimensional transport problem (11.1). The form of (11.1) with the boundary condition (11.2) and the initial conditions (11.3) is called the strong (or classical) form of the problem.

We first define a collection of trial solutions, which consists of all functions that have square-integrable first derivatives (H^1 functions, i.e. $\int_0^L (T_{,x})^2 dx < \infty$ if

$T \in H^1$) and satisfy the Dirichlet type of boundary condition (where the value of the variable is specified) at $x = 0$. This is expressed as the trial functional space,

$$S = \left\{ T \ \middle| \ T \in H^1, \ T(0) = g \right\}. \tag{11.32}$$

The variational space of the trial solution is defined as

$$V = \left\{ w \ \middle| \ w \in H^1, \ w(0) = 0 \right\}, \tag{11.33}$$

which requires a corresponding homogeneous boundary condition.

We next multiply the transport equation (11.1) by a function in the variational space ($w \in V$), and integrate the product over the domain where the problem is defined,

$$\int_0^L \left(\frac{\partial T}{\partial t} w \right) dx + u \int_0^L \left(\frac{\partial T}{\partial x} w \right) dx = D \int_0^L \left(\frac{\partial^2 T}{\partial x^2} w \right) dx. \tag{11.34}$$

Integrating the right-hand-side of (11.34) by parts, we have

$$\int_0^L \left(\frac{\partial T}{\partial t} w \right) dx + u \int_0^L \left(\frac{\partial T}{\partial x} w \right) dx + D \int_0^L \left(\frac{\partial T}{\partial x} \frac{\partial w}{\partial x} \right) dx = D \left[\frac{\partial T}{\partial x} w \right]_0^L$$
$$= Dqw(L), \tag{11.35}$$

where the boundary conditions $\partial T/\partial x = q$ and $w(0) = 0$ are applied. The integral equation (11.35) is called the weak form of this problem. Therefore, the weak form can be stated as: Find $T \in S$ such that for all $w \in V$,

$$\int_0^L \left(\frac{\partial T}{\partial t} w \right) dx + u \int_0^L \left(\frac{\partial T}{\partial x} w \right) dx + D \int_0^L \left(\frac{\partial T}{\partial x} \frac{\partial w}{\partial x} \right) dx = Dqw(L). \tag{11.36}$$

It can be formally shown that the solution of the weak problem is identical to that of the strong problem, or that the strong and weak forms of the problem are *equivalent*. Obviously, if T is a solution of the strong problem (11.1) and (11.2), it must also be a solution of the weak problem (11.36) using the procedure for derivation of the weak formulation. However, let us assume that T is a solution of the weak problem (11.36). By reversing the order in deriving the weak formulation, we have

$$\int_0^L \left(\frac{\partial T}{\partial t} + u \frac{\partial T}{\partial x} - D \frac{\partial^2 T}{\partial x^2} \right) w dx + D \left[\frac{\partial T}{\partial x}(L) - q \right] w(L) = 0. \tag{11.37}$$

Satisfying (11.37) for all possible functions of $w \in V$ requires that

$$\frac{\partial T}{\partial t} + u \frac{\partial T}{\partial x} - D \frac{\partial^2 T}{\partial x^2} = 0 \text{ for } x \in (0, L), \quad \text{and} \quad \frac{\partial T}{\partial x}(L) - q = 0, \tag{11.38}$$

which means that this solution T will be also a solution of the strong problem. It should be noted that the Dirichlet type of boundary condition (where the value of the variable is specified) is built into the trial functional space S, and is thus called an essential boundary condition. However, the Neumann type of boundary condition (where the derivative of the variable is imposed) is implied by the weak formulation as indicated in (11.38) and is referred to as a natural boundary condition.

Galerkin's Approximation and Finite Element Interpolations

As we have shown, the strong and weak forms of the problem are equivalent, and there is no approximation involved between these two formulations. Finite element methods start with the weak formulation of the problem. Let us construct finite-dimensional approximations of S and V, which are denoted by S^h and V^h, respectively. The superscript refers to a discretization with a characteristic grid size h. The weak formulation (11.36) can be rewritten using these new spaces, as: Find $T^h \in S^h$ such that for all $w^h \in V^h$,

$$\int_0^L \left(\frac{\partial T^h}{\partial t} w^h\right) dx + u \int_0^L \left(\frac{\partial T^h}{\partial x} w^h\right) dx + D \int_0^L \left(\frac{\partial T^h}{\partial x}\frac{\partial w^h}{\partial x}\right) dx = Dq w^h(L).$$

$$(11.39)$$

Normally, S^h and V^h will be subsets of S and V, respectively. This means that if a function $\phi \in S^h$ then $\phi \in S$, and if another function $\psi \in V^h$ then $\psi \in V$. Therefore, (11.39) defines an approximate solution T^h to the exact weak form of the problem (11.36).

It should be noted that, up to the boundary condition $T(0) = g$, the function spaces S^h and V^h are composed of identical collections of functions. We may take out this boundary condition by defining a new function

$$v^h(x, t) = T^h(x, t) - g^h(x),$$
$$(11.40)$$

where g^h is a specific function that satisfies the boundary condition $g^h(0) = g$. Thus, the functions v^h and w^h belong to the same space V^h. Equation (11.39) can be rewritten in terms of the new function v^h: Find $T^h = v^h + g^h$, where $v^h \in V^h$, such that for all $w^h \in V^h$,

$$\int_0^L \left(\frac{\partial v^h}{\partial t} w^h\right) dx + a\left(w^h, v^h\right) = Dq w^h(L) - a\left(w^h, g^h\right).$$
$$(11.41)$$

The operator $a(\cdot, \cdot)$ is defined as

$$a(w, v) = u \int_0^L \left(\frac{\partial v}{\partial x} w\right) dx + D \int_0^L \left(\frac{\partial v}{\partial x}\frac{\partial w}{\partial x}\right) dx.$$
$$(11.42)$$

The formulation (11.41) is called a Galerkin formulation, because the solution and the variational functions are in the same space. Again, the Galerkin formulation of the problem is an approximation to the weak formulation (11.36). Other classes of approximation methods, called Petrov-Galerkin methods, are those in which the solution function may be contained in a collection of functions other than V^h.

Next we need to explicitly construct the finite-dimensional variational space V^h. Let us assume that the dimension of the space is n and that the basis (shape or interpolation) functions for the space are

$$N_A(x), \quad A = 1, 2, ..., n. \tag{11.43}$$

Each shape function has to satisfy the boundary condition at $x = 0$,

$$N_A(0) = 0, \quad A = 1, 2, ..., n, \tag{11.44}$$

which is required by the space V^h. The form of the shape functions will be discussed later. Any function $w^h \in V^h$ can be expressed as a linear combination of these shape functions,

$$w^h = \sum_{A=1}^{n} c_A N_A(x), \tag{11.45}$$

where the coefficients c_A are independent of x and uniquely define this function. We may introduce one additional function N_0 to specify the function g^h in (11.40) related to the essential boundary condition. This shape function has the property

$$N_0(0) = 1. \tag{11.46}$$

Therefore, the function g^h can be expressed as

$$g^h(x) = g N_0(x), \quad \text{and} \quad g^h(0) = g. \tag{11.47}$$

With these definitions, the approximate solution can be written as

$$v^h(x, t) = \sum_{A=1}^{n} d_A(t) N_A(x), \tag{11.48}$$

and

$$T^h(x, t) = \sum_{A=1}^{n} d_A(t) N_A(x) + g N_0(x), \tag{11.49}$$

where d_A's are functions of time only for time dependent problems.

Matrix Equations, Comparison with Finite Difference Method

With the construction of the finite-dimensional space V^h, the Galerkin formulation of the problem (11.41) leads to a coupled system of ordinary differential equations.

Substitution of the expressions for the variational function (11.45) and for the approximate solution (11.48) into the Galerkin formulation (11.41) yields

$$
\int_0^L \left(\sum_{B=1}^n \dot{d}_B N_B \sum_{A=1}^n c_A N_A \right) dx + a \left(\sum_{A=1}^n c_A N_A, \sum_{B=1}^n d_B N_B \right)
$$

$$
= Dq \sum_{A=1}^n c_A N_A (L) - a \left(\sum_{A=1}^n c_A N_A, g N_0 \right) \tag{11.50}
$$

where $\dot{d}_B = d(d_B)/dt$. Rearranging the terms, (11.50) reduces to

$$
\sum_{A=1}^n c_A G_A = 0, \tag{11.51}
$$

where

$$
G_A = \sum_{B=1}^n \dot{d}_B \int_0^L (N_A N_B) \, dx + \sum_{B=1}^n d_B a \, (N_A, N_B) - Dq N_A (L) + ga \, (N_A, N_0) . \tag{11.52}
$$

As the Galerkin formulation (11.41) should hold for all possible functions of $w^h \in V^h$, the coefficients, c_A, should be arbitrary. The necessary requirement for (11.51) to hold is that each G_A must be zero, that is,

$$
\sum_{B=1}^n \dot{d}_B \int_0^L (N_B N_A) \, dx + \sum_{B=1}^n d_B a \, (N_A, N_B) = Dq N_A (L) - ga \, (N_A, N_0) \tag{11.53}
$$

for $A = 1, 2, \ldots, n$. System of equations (11.53) constitutes a system of n first-order ordinary differential equations (ODEs) for the d_Bs. It can be put into a more concise matrix form. Let us define,

$$
\mathbf{M} = [M_{AB}], \quad \mathbf{K} = [K_{AB}], \quad \mathbf{F} = \{F_A\}, \quad \mathbf{d} = \{d_B\}, \tag{11.54}
$$

where

$$
M_{AB} = \int_0^L (N_A N_B) \, dx, \tag{11.55}
$$

$$
K_{AB} = u \int_0^L \left(N_{B,x} N_A \right) dx + D \int_0^L \left(N_{B,x} N_{A,x} \right) dx, \tag{11.56}
$$

$$
F_A = Dq N_A (L) - gu \int_0^L \left(N_{0,x} N_A \right) dx - gD \int_0^L \left(N_{0,x} N_{A,x} \right) dx. \tag{11.57}
$$

Equation (11.53) can then be written as

$$
\mathbf{M\dot{d}} + \mathbf{Kd} = \mathbf{F}. \tag{11.58}
$$

The system of equations (11.58) is also termed the matrix form of the problem. Usually, **M** is called the mass matrix, **K** is the stiffness matrix, **F** is the force vector, and **d** is the displacement vector. This system of ODE's can be integrated by numerical methods, for example, Runge-Kutta methods, or discretized (in time) by finite difference schemes as described in the previous section. The initial condition (11.3) will be used for integration. An alternative approach is to use a finite difference approximation to the time derivative term in the transport equation (11.1) at the beginning of the process, for example, by replacing $\partial T/\partial t$ with $\left(T^{n+1} - T^{n}\right)/\Delta t$, and then using the finite element method to discretize the resulting equation.

Now let us consider the actual construction of the shape functions for the finite dimensional variational space. The simplest example is to use piecewise-linear finite element space. We first partition the domain $[0, L]$ into n nonoverlapping subintervals (elements). A typical one is denoted as $\left[x_A, x_{A+1}\right]$. The shape functions associated with the interior nodes, $A = 1, 2, \ldots, n - 1$, are defined as

$$N_A(x) = \begin{cases} \dfrac{x - x_{A-1}}{x_A - x_{A-1}}, & x_{A-1} \leqslant x < x_A, \\ \dfrac{x_{A+1} - x}{x_{A+1} - x_A}, & x_A \leqslant x \leqslant x_{A+1}, \\ 0, & \text{elsewhere.} \end{cases} \tag{11.59}$$

Further, for the boundary nodes, the shape functions are defined as

$$N_n(x) = \frac{x - x_{n-1}}{x_n - x_{n-1}}, \quad x_{n-1} \leqslant x \leqslant x_n, \tag{11.60}$$

and

$$N_0(x) = \frac{x_1 - x}{x_1 - x_0}, \quad x_0 \leqslant x \leqslant x_1. \tag{11.61}$$

These shape functions are graphically plotted in Figure 11.2. It should be noted that these shape functions have very compact (local) support and satisfy $N_A(x_B) = \delta_{AB}$, where δ_{AB} is the Kronecker delta (i.e. $\delta_{AB} = 1$ if $A = B$, whereas $\delta_{AB} = 0$ if $A \neq B$).

With the construction of the shape functions, the coefficients, d_As, in the expression for the approximate solution (11.49) represent the values of T^h at the nodes $x = x_A$ $(A = 1, 2, \ldots, n)$, or

$$d_A = T^h(x_A) = T_A. \tag{11.62}$$

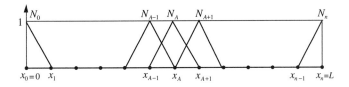

Figure 11.2 Piecewise linear finite element space.

To compare the discretized equations generated from the finite element method with those from finite difference methods, we substitute (11.59) into (11.53) and evaluate the integrals. For an interior node x_A ($A = 1, 2, \ldots, n-1$), we have

$$\frac{d}{dt}\left(\frac{T_{A-1}}{6} + \frac{2T_A}{3} + \frac{T_{A+1}}{6}\right) + \frac{u}{2h}\left(T_{A+1} - T_{A-1}\right) - \frac{D}{h^2}\left(T_{A-1} - 2T_A + T_{A+1}\right) = 0,$$

$$(11.63)$$

where h is the uniform mesh size. The convective and diffusive terms in expression (11.63) have the same forms as those discretized using the standard second-order finite difference method (centered difference) in (11.12). However, in the finite element scheme, the time derivative term is presented with a three-point spatial average of the variable T, which differs from the finite difference method. In general, the Galerkin finite element formulation is equivalent to a finite difference method. The advantage of the finite element method lies in its flexibility to handle complex geometries.

Element Point of View of the Finite Element Method

So far we have been using a global view of the finite element method. The shape functions are defined on the global domain, as shown in Figure 11.2. However, it is also convenient to present the finite element method using a local (or element) point of view. This viewpoint is useful for the evaluation of the integrals in (11.55) to (11.57) and the actual computer implementation of the finite element method.

Figure 11.3 depicts the global and local descriptions of the eth element. The global description of the element e is just the "local" view of the full domain shown in Figure 11.2. Only two shape functions are nonzero within this element, N_{A-1} and N_A. Using the local coordinate in the standard element (parent domain) as shown on the right of Figure 11.3, we can write the standard shape functions as

$$N_1(\xi) = \frac{1}{2}(1 - \xi) \quad \text{and} \quad N_2(\xi) = \frac{1}{2}(1 + \xi).$$

$$(11.64)$$

Clearly, the standard shape function N_1 (or N_2) corresponds to the global shape function N_{A-1} (or N_A). The mapping between the domains of the global and local descriptions can easily be generated with the help of these shape functions,

$$x(\xi) = N_1(\xi)x_1^e + N_2(\xi)x_2^e = \frac{1}{2}\left[(x_A - x_{A-1})\xi + x_A + x_{A-1}\right],$$

$$(11.65)$$

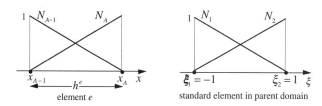

Figure 11.3   Global and local descriptions of an element.

with the notation that $x_1^e = x_{A-1}$ and $x_2^e = x_A$. One can also solve (11.65) for the inverse map

$$\xi(x) = \frac{2x - x_A - x_{A-1}}{x_A - x_{A-1}}. \tag{11.66}$$

Within the element e, the derivative of the shape functions can be evaluated using the mapping equation (11.66),

$$\frac{dN_A}{dx} = \frac{dN_A}{d\xi}\frac{d\xi}{dx} = \frac{2}{x_A - x_{A-1}}\frac{dN_1}{d\xi} = \frac{-1}{x_A - x_{A-1}} \tag{11.67}$$

and

$$\frac{dN_{A+1}}{dx} = \frac{dN_{A+1}}{d\xi}\frac{d\xi}{dx} = \frac{2}{x_A - x_{A-1}}\frac{dN_2}{d\xi} = \frac{1}{x_A - x_{A-1}}. \tag{11.68}$$

The global mass matrix (11.55), the global stiffness matrix (11.56), and the global force vector (11.57) have been defined as the integrals over the global domain $[0, L]$. These integrals may be written as the summation of integrals over each element's domain. Thus

$$\mathbf{M} = \sum_{e=1}^{n_{el}} \mathbf{M}^e, \quad \mathbf{K} = \sum_{e=1}^{n_{el}} \mathbf{K}^e, \quad \mathbf{F} = \sum_{e=1}^{n_{el}} \mathbf{F}^e, \tag{11.69}$$

$$\mathbf{M}^e = \left[M_{AB}^e \right], \quad \mathbf{K}^e = \left[K_{AB}^e \right], \quad \mathbf{F}^e = \left\{ F_A^e \right\} \tag{11.70}$$

where n_{el} is the total number of finite elements (in this case $n_{el} = n$), and

$$M_{AB}^e = \int_{\Omega^e} (N_A N_B)\, dx, \tag{11.71}$$

$$K_{AB}^e = u \int_{\Omega^e} (N_{B,x} N_A)\, dx + D \int_{\Omega^e} (N_{B,x} N_{A,x})\, dx, \tag{11.72}$$

$$F_A^e = Dq\delta_{en_{el}}\delta_{An} - gu \int_{\Omega^e} (N_{0,x} N_A)\, dx - gD \int_{\Omega^e} (N_{0,x} N_{A,x})\, dx \tag{11.73}$$

and $\Omega^e = \left[x_1^e, x_2^e \right] = \left[x_{A-1}, x_A \right]$ is the domain of the eth element; and the first term on right-hand-side of (11.73) is nonzero only for $e = n_{el}$ and $A = n$.

Given the construction of the shape functions, most of the element matrices and force vectors in (11.71) to (11.73) will be zero. The non-zero ones require that $A = e$ or $e + 1$ and $B = e$ or $e + 1$. We may collect these nonzero terms and arrange them into the element mass matrix, stiffness matrix, and force vector as follows:

$$\mathbf{m}^e = \left[m_{ab}^e \right], \quad \mathbf{k}^e = \left[k_{ab}^e \right], \quad \mathbf{f}^e = \left\{ f_a^e \right\}, \quad a, b = 1, 2 \tag{11.74}$$

where

$$m_{ab}^e = \int_{\Omega^e} (N_a N_b)\, dx, \tag{11.75}$$

$$k_{ab}^e = u \int_{\Omega^e} \left(N_{b,x} N_a\right) dx + D \int_{\Omega^e} \left(N_{b,x} N_{a,x}\right) dx, \qquad (11.76)$$

$$f_a^e = \begin{cases} -g k_{a1}^e & e = 1, \\ 0 & e = 2, 3, \ldots, n_{el} - 1, \\ Dq\delta_{a2} & e = n_{el}. \end{cases} \qquad (11.77)$$

Here, $\mathbf{m}^e$, $\mathbf{k}^e$ and $\mathbf{f}^e$ are defined with the local (element) ordering, and represent the nonzero terms in the corresponding $\mathbf{M}^e$, $\mathbf{K}^e$ and $\mathbf{F}^e$ with the global ordering. The terms in the local ordering need to be mapped back into the global ordering. For this example, the mapping is defined as

$$A = \begin{cases} e - 1 & \text{if } a = 1 \\ e & \text{if } a = 2 \end{cases} \qquad (11.78)$$

for element e.

Therefore, in the element viewpoint, the global matrices and the global vector can be constructed by summing the contributions of the element matrices and the element vector, respectively. The evaluation of both the element matrices and the element vector can be performed on a standard element using the mapping between the global and local descriptions.

The finite element methods for two- or three-dimensional problems will follow the same basic steps introduced in this section. However, the data structure and the forms of the elements or the shape functions will be more complicated. Refer to Hughes (1987) for a detailed discussion. In Section 5, we will present an example of a two dimensional flow over a circular cylinder.

4. Incompressible Viscous Fluid Flow

In this section, we will discuss numerical schemes for solving incompressible viscous fluid flows. We will focus on techniques using the primitive variables (velocity and pressure). Other formulations using streamfunction and vorticity are available in the literature (see Fletcher 1988, Vol. II) and will not be discussed here since their extensions to three-dimensional flows are not straightforward. The schemes to be discussed normally apply to laminar flows. However, by incorporating additional appropriate turbulence models, these schemes will also be effective for turbulent flows.

For an incompressible Newtonian fluid, the fluid motion satisfies the Navier-Stokes equation,

$$\rho \left(\frac{\partial \mathbf{u}}{\partial t} + (\mathbf{u} \cdot \nabla)\mathbf{u} \right) = \rho \mathbf{g} - \nabla p + \mu \nabla^2 \mathbf{u}, \qquad (11.79)$$

and the continuity equation,

$$\nabla \cdot \mathbf{u} = 0, \qquad (11.80)$$

where $\mathbf{u}$ is the velocity vector, $\mathbf{g}$ is the body force per unit mass, which could be the gravitational acceleration, p is the pressure, and ρ, μ are the density and viscosity of the fluid, respectively. With the proper scaling, (11.79) can be written in the dimensionless form,

$$\frac{\partial \mathbf{u}}{\partial t} + (\mathbf{u} \cdot \nabla)\mathbf{u} = \mathbf{g} - \nabla p + \frac{1}{\mathrm{Re}}\nabla^2 \mathbf{u} \tag{11.81}$$

where Re is the Reynolds number of the flow. In some approaches, the convective term is rewritten in conservative form,

$$(\mathbf{u} \cdot \nabla)\,\mathbf{u} = \nabla \cdot (\mathbf{u}\mathbf{u})\,, \tag{11.82}$$

because $\mathbf{u}$ is solenoidal.

In order to guarantee that a flow problem is well-posed, appropriate initial and boundary conditions for the problem must be specified. For time-dependent flow problems, the initial condition for the velocity,

$$\mathbf{u}\,(\mathbf{x}, t = 0) = \mathbf{u}_0\,(\mathbf{x})\,, \tag{11.83}$$

is required. The initial velocity field has to satisfy the continuity equation $\nabla \cdot \mathbf{u}_0 = 0$. At a solid surface, the fluid velocity should equal the surface velocity (no-slip condition). No boundary condition for the pressure is required at a solid surface. If the computational domain contains a section where the fluid enters the domain, the fluid velocity (and the pressure) at this inflow boundary should be specified. If the computational domain contains a section where the fluid leaves the domain (outflow section), appropriate outflow boundary conditions include zero tangential velocity and zero normal stress, or zero velocity derivatives, as further discussed in Gresho (1991). Because the conditions at the outflow boundary are artificial, it should be checked that the numerical results are not sensitive to the location of this boundary. In order to solve the Navier-Stokes equations, it is also appropriate to specify the value of the pressure at one reference point in the domain, because the pressure appears only as a gradient and can be determined up to a constant.

There are two major difficulties in solving the Navier-Stokes equations numerically. One is related to the unphysical oscillatory solution often found in a convection-dominated problem. The other is the treatment of the continuity equation that is a constraint on the flow to determine the pressure.

Convection-Dominated Problems

As mentioned in Section 2, the exact solution may change significantly in a narrow boundary layer for convection dominated transport problems. If the computational grid is not sufficiently fine to resolve the rapid variation of the solution in the boundary layer, the numerical solution may present unphysical oscillations adjacent the boundary. Let us examine the steady transport problem in one dimension,

$$u\frac{\partial T}{\partial x} = D\frac{\partial^2 T}{\partial x^2} \quad \text{for} \quad 0 \leqslant x \leqslant L, \tag{11.84}$$

with two boundary conditions

$$T\,(0) = 0 \quad \text{and} \quad T\,(L) = 1. \tag{11.85}$$

The exact solution for this problem is

$$T = \frac{e^{Rx/L} - 1}{e^R - 1} \tag{11.86}$$

where

$$R = uL/D \tag{11.87}$$

is the global Peclét number. For large values of R, the solution (11.86) behaves as

$$T = e^{-R(1-x/L)}. \tag{11.88}$$

The essential feature of this solution is the existence of a boundary layer at $x = L$, and its thickness δ is of the order of,

$$\frac{\delta}{L} = O\left(\frac{1}{|R|}\right). \tag{11.89}$$

At $1 - x/L = 1/R$, T is about 37% of the boundary value; while at $1 - x/L = 2/R$, T is about 13.5% of the boundary value.

If centered differences are used to discretize the steady transport equation (11.84) using the grid shown in Figure 11.1, the resulting finite difference scheme is,

$$\frac{u\Delta x}{2D}\left(T_{j+1} - T_{j-1}\right) = \left(T_{j+1} - 2T_j + T_{j-1}\right), \tag{11.90}$$

or

$$0.5R_{cell}\left(T_{j+1} - T_{j-1}\right) = \left(T_{j+1} - 2T_j + T_{j-1}\right), \tag{11.91}$$

where the grid spacing $\Delta x = L/n$ and the cell Peclét number $R_{cell} = u\Delta x/D = R/n$. From the scaling of the boundary thickness (11.89) we know that it is of the order,

$$\delta = O\left(\frac{L}{nR_{cell}}\right) = O\left(\frac{\Delta x}{R_{cell}}\right). \tag{11.92}$$

Physically, if T represents the temperature in the transport problem (11.84), the convective term brings the heat toward the boundary $x = L$, whereas the diffusive term conducts the heat away through the boundary. These two terms have to be balanced. The discretized equation (11.91) has the same physical meaning. Let us examine this balance for a node next to the boundary, $j = n - 1$. When the cell Peclét number $R_{cell} > 2$, according to (11.92) the thickness of the boundary layer is less than half the grid spacing, and the exact solution (11.86) indicates that the temperatures T_j and T_{j-1} are already outside the boundary layer and are essentially zero. Thus, the two sides of the discretized equation (11.91) cannot balance, or the conduction term is not strong enough to remove the heat convected to the boundary, assuming the solution is smooth. In order to force the heat balance, an unphysical oscillatory solution with $T_j < 0$ is generated to enhance the conduction term in the discretized problem (11.91). To prevent the oscillatory solution, the cell Peclét number is normally required to be less than two, which can be achieved by refining the grid to resolve the flow inside the boundary layer. In some respect, an oscillatory solution may be a virtue since it provides a warning that a physically important feature is not being properly resolved. To reduce the overall computational cost, non-uniform grids with local fine grid spacing inside the boundary layer will frequently be used to resolve the variables there.

Another common method to avoid the oscillatory solution is to use a first-order upwind scheme,

$$R_{cell} \left(T_j - T_{j-1} \right) = \left(T_{j+1} - 2T_j + T_{j-1} \right),\tag{11.93}$$

where a forward difference scheme is used to discretize the convective term. It is easy to see that this scheme reduces the heat convected to the boundary and thus prevents the oscillatory solution. However, the upwind scheme is not very accurate (only first-order accurate). It can be easily shown that the upwind scheme (11.93) does not recover the original transport equation (11.84). Instead it is consistent with a slightly different transport equation (when the cell Peclét number is kept finite during the process),

$$u \frac{\partial T}{\partial x} = D \left(1 + 0.5 R_{cell} \right) \frac{\partial^2 T}{\partial x^2}.\tag{11.94}$$

Thus, another way to view the effect of the first-order upwind scheme (11.93) is that it introduces a numerical diffusivity of the value of $0.5 R_{cell} D$, which enhances the conduction of heat through the boundary. For an accurate solution, one normally requires that $0.5 R_{cell} << 1$, which is very restrictive and does not offer any advantage over the centered difference scheme (11.91).

Higher-order upwind schemes may be introduced to obtain more accurate non-oscillatory solutions without excessive grid refinement. However, those schemes may be less robust. Refer to Fletcher (1988, vol.I, chapter 9) for discussions.

Similarly, there are upwind schemes for finite element methods to solve convection-dominated problems. Most of those are based on Petrov-Galerkin approach that permit an effective upwind treatment of the convective term along local streamlines (Brooks and Hughes, 1982). More recently, stabilized finite element methods have been developed where a least-square term is added to the momentum balance equation to provide the necessary stability for convection-dominated flows (see Franca *et al.*, 1992).

Incompressibility Condition

In solving the Navier-Stokes equations using the primitive variables (velocity and pressure), another numerical difficulty lies in the continuity equation: The continuity equation can be regarded either as a constraint on the flow field to determine the pressure or the pressure plays the role of the Lagrange multiplier to satisfy the continuity equation.

In a flow field, the information (or disturbance) travels with both the flow and the speed of sound in the fluid. Since the speed of sound is infinite in an incompressible fluid, part of the information (pressure disturbance) is propagated instantaneously throughout the domain. In many numerical schemes the pressure is often obtained by solving a Poisson equation. The Poisson equation may occur in either continuous form or discrete form. Some of these schemes will be described here. In some of them, solving the pressure Poisson equation is the most costly step.

Another common technique to surmount the difficulty of the incompressible limit is to introduce an artificial compressibility (Chorin, 1967). This formulation is normally used for steady problems with a pseudo-transient formulation. In the formulation, the continuity equation is replaced by,

$$\frac{\partial p}{\partial t} + c^2 \nabla \cdot \mathbf{u} = 0, \tag{11.95}$$

where c is an arbitrary constant and could be the artificial speed of sound in a corresponding compressible fluid with the equation of state $p = c^2 \rho$. The formulation is called pseudo-transient because (11.95) does not have any physical meaning before the steady state is reached. However, when c is large, (11.95) can be considered as an approximation to the unsteady solution of the incompressible Navier-Stokes problem.

Explicit MacCormack Scheme

Instead of using the artificial compressibility in (11.95), one may start with the exact compressible Navier-Stokes equations. In Cartesian coordinates, the component form of the continuity equation (4.8) and compressible Navier-Stokes equation (4.44) in two dimensions can be explicitly written as

$$\frac{\partial \rho}{\partial t} + \frac{\partial (\rho u)}{\partial x} + \frac{\partial (\rho v)}{\partial y} = 0, \tag{11.96}$$

$$\frac{\partial}{\partial t} (\rho u) + \frac{\partial}{\partial x} \left(\rho u^2 \right) + \frac{\partial}{\partial y} (\rho v u) = \rho g_x - \frac{\partial p}{\partial x} + \mu \nabla^2 u + \frac{\mu}{3} \frac{\partial}{\partial x} \left(\frac{\partial u}{\partial x} + \frac{\partial v}{\partial y} \right), \tag{11.97}$$

$$\frac{\partial}{\partial t} (\rho v) + \frac{\partial}{\partial x} (\rho u v) + \frac{\partial}{\partial y} \left(\rho v^2 \right) = \rho g_y - \frac{\partial p}{\partial y} + \mu \nabla^2 v + \frac{\mu}{3} \frac{\partial}{\partial y} \left(\frac{\partial u}{\partial x} + \frac{\partial v}{\partial y} \right), \tag{11.98}$$

with the equation of state, $p = c^2 \rho$ \hfill (11.99)

where c is speed of sound in the medium. As long as the flows are limited to low Mach numbers and the conditions are almost isothermal, the solution to this set of equations should approximate the incompressible limit.

The explicit MacCormack scheme, after R.W. MacCormack (1969), is essentially a predictor-corrector scheme, similar to a second-order Runge-Kutta method commonly used to solve ordinary differential equations. For a system of equations of the form,

$$\frac{\partial \mathbf{U}}{\partial t} + \frac{\partial \mathbf{E}(\mathbf{U})}{\partial x} + \frac{\partial \mathbf{F}(\mathbf{U})}{\partial y} = 0, \tag{11.100}$$

the explicit MacCormack scheme consists of two steps,

$$\text{predictor}: \mathbf{U}_{i,j}^* = \mathbf{U}_{i,j}^n - \frac{\Delta t}{\Delta x}\left(\mathbf{E}_{i+1,j}^n - \mathbf{E}_{i,j}^n\right) - \frac{\Delta t}{\Delta y}\left(\mathbf{F}_{i,j+1}^n - \mathbf{F}_{i,j}^n\right), \tag{11.101}$$

$$\text{corrector}: \mathbf{U}_{i,j}^{n+1} = \frac{1}{2}\left[\mathbf{U}_{i,j}^n + \mathbf{U}_{i,j}^* - \frac{\Delta t}{\Delta x}\left(\mathbf{E}_{i,j}^* - \mathbf{E}_{i-1,j}^*\right) - \frac{\Delta t}{\Delta y}\left(\mathbf{F}_{i,j}^* - \mathbf{F}_{i,j-1}^*\right)\right] \tag{11.102}$$

Notice that the spatial derivatives in (11.100) are discretized with opposite one-sided finite differences in the predictor and corrector stages. The star variables are all evaluated at time level t_{n+1}. This scheme is second-order accurate in both time and space.

Applying the MacCormack scheme to the compressible Navier-Stokes equations (11.96) to (11.98) and replacing the pressure with (11.99), we have the predictor step,

$$\rho_{i,j}^* = \rho_{i,j}^n - c_1\left[(\rho u)_{i+1,j}^n - (\rho u)_{i,j}^n\right] - c_2\left[(\rho v)_{i,j+1}^n - (\rho v)_{i,j}^n\right] \tag{11.103}$$

$$
\begin{aligned}
(\rho u)_{i,j}^* = (\rho u)_{i,j}^n &- c_1\left[\left(\rho u^2 + c^2\rho\right)_{i+1,j}^n - \left(\rho u^2 + c^2\rho\right)_{i,j}^n\right] \\
&- c_2\left[(\rho uv)_{i,j+1}^n - (\rho uv)_{i,j}^n\right] + \frac{4}{3}c_3\left(u_{i+1,j}^n - 2u_{i,j}^n + u_{i-1,j}^n\right) \\
&+ c_4\left(u_{i,j+1}^n - 2u_{i,j}^n + u_{i,j-1}^n\right) \\
&+ c_5\left(v_{i+1,j+1}^n + v_{i-1,j-1}^n - v_{i+1,j-1}^n - v_{i-1,j+1}^n\right) \tag{11.104}
\end{aligned}
$$

$$
\begin{aligned}
(\rho v)_{i,j}^* = (\rho v)_{i,j}^n &- c_1\left[(\rho uv)_{i+1,j}^n - (\rho uv)_{i,j}^n\right] \\
&- c_2\left[\left(\rho v^2 + c^2\rho\right)_{i,j+1}^n - \left(\rho v^2 + c^2\rho\right)_{i,j}^n\right] \\
&+ c_3\left(v_{i+1,j}^n - 2v_{i,j}^n + v_{i-1,j}^n\right) + \frac{4}{3}c_4\left(v_{i,j+1}^n - 2v_{i,j}^n + v_{i,j-1}^n\right) \\
&+ c_5\left(u_{i+1,j+1}^n + u_{i-1,j-1}^n - u_{i+1,j-1}^n - u_{i-1,j+1}^n\right) \tag{11.105}
\end{aligned}
$$

Similarly, the corrector step is given by

$$2\rho_{i,j}^{n+1} = \rho_{i,j}^n + \rho_{i,j}^* - c_1\left[(\rho u)_{i,j}^* - (\rho u)_{i-1,j}^*\right] - c_2\left[(\rho v)_{i,j}^* - (\rho v)_{i,j-1}^*\right] \tag{11.106}$$

$$2 (\rho u)_{i,j}^{n+1} = (\rho u)_{i,j}^n + (\rho u)_{i,j}^* - c_1 \left[\left(\rho u^2 + c^2 \rho \right)_{i,j}^* - \left(\rho u^2 + c^2 \rho \right)_{i-1,j}^* \right]$$

$$- c_2 \left[(\rho u v)_{i,j}^* - (\rho u v)_{i,j-1}^* \right] + \frac{4}{3} c_3 \left(u_{i+1,j}^* - 2 u_{i,j}^* + u_{i-1,j}^* \right)$$

$$+ c_4 \left(u_{i,j+1}^* - 2 u_{i,j}^* + u_{i,j-1}^* \right)$$

$$+ c_5 \left(v_{i+1,j+1}^* + v_{i-1,j-1}^* - v_{i+1,j-1}^* - v_{i-1,j+1}^* \right) \qquad (11.107)$$

$$2 (\rho v)_{i,j}^{n+1} = (\rho v)_{i,j}^n + (\rho v)_{i,j}^* - c_1 \left[(\rho u v)_{i,j}^* - (\rho u v)_{i-1,j}^* \right]$$

$$- c_2 \left[\left(\rho v^2 + c^2 \rho \right)_{i,j}^* - \left(\rho v^2 + c^2 \rho \right)_{i,j-1}^* \right]$$

$$+ c_3 \left(v_{i+1,j}^* - 2 v_{i,j}^* + v_{i-1,j}^* \right) + \frac{4}{3} c_4 \left(v_{i,j+1}^* - 2 v_{i,j}^* + v_{i,j-1}^* \right)$$

$$+ c_5 \left(u_{i+1,j+1}^* + u_{i-1,j-1}^* - u_{i+1,j-1}^* - u_{i-1,j+1}^* \right) \qquad (11.108)$$

The coefficients are defined as,

$$c_1 = \frac{\Delta t}{\Delta x}, \quad c_2 = \frac{\Delta t}{\Delta y}, \quad c_3 = \frac{\mu \Delta t}{(\Delta x)^2}, \quad c_4 = \frac{\mu \Delta t}{(\Delta y)^2}, \quad c_5 = \frac{\mu \Delta t}{12 \Delta x \Delta y}. \qquad (11.109)$$

In both the predictor and corrector steps the viscous terms (the second-order derivative terms) are all discretized with centered-differences to maintain second-order accuracy. For brevity, body force terms in the momentum equations are neglected here.

During the predictor and corrector stages of the explicit MacCormack scheme (11.103) to (11.108), one-sided differences are arranged in the *FF* and *BB* fashion, respectively. Here, in the notation *FF*, the first *F* denotes the forward difference in the *x*-direction and the second *F* denotes the forward difference in the *y*-direction. Similarly, *BB* stands for backward differences in both *x* and *y* directions. We denote this arrangement as *FF/BB*. Similary, one may get *BB/FF*, *FB/BF*, *BF/FB* arrangements. It is noted that some balanced cyclings of these arrangements generate better results than others.

Tannehill, Anderson and Pletcher (1997) give the following semi-empirical stability criterion for the explicit MacCormack scheme,

$$\Delta t \leqslant \frac{\sigma}{\left(1 + 2 / Re_\Delta \right)} \left[\frac{|u|}{\Delta x} + \frac{|v|}{\Delta y} + c \sqrt{\frac{1}{\Delta x^2} + \frac{1}{\Delta y^2}} \right]^{-1}, \qquad (11.110)$$

where σ is a safety factor (≈ 0.9), $Re_\Delta = \min \left(\rho |u| \Delta x / \mu, \; \rho |v| \Delta y / \mu \right)$ is the minimum mesh Reynolds number. This condition is quite conservative for flows with small mesh Reynolds numbers.

One key issue for the explicit MacCormack scheme to work properly is the boundary conditions for density (thus pressure). We leave this issue to the next section where its implementation in two sample problems will be demonstrated.

MAC Scheme

Most of numerical schemes developed for computational fluid dynamics problems can be characterized as operator splitting algorithms. The operator splitting algorithms divide each time step into several substeps. Each substep solves one part of the operator and thus decouples the numerical difficulties associated with each part of the operator. For example, consider a system,

$$\frac{d\phi}{dt} + A(\phi) = f, \tag{11.111}$$

with initial condition $\phi(0) = \phi_0$, where the operator A may be split into two operators

$$A(\phi) = A_1(\phi) + A_2(\phi). \tag{11.112}$$

Using a simple first-order accurate Marchuk-Yanenko fractional-step scheme (Yanenko, 1971, and Marchuk, 1975), the solution of the system at each time step $\phi^{n+1} = \phi((n+1)\Delta t)$ $(n = 0,1, \ldots)$ is approximated by solving the following two successive problems:

$$\frac{\phi^{n+1/2} - \phi^n}{\Delta t} + A_1\left(\phi^{n+1/2}\right) = f_1^{n+1}, \tag{11.113}$$

$$\frac{\phi^{n+1} - \phi^{n+1/2}}{\Delta t} + A_2\left(\phi^{n+1}\right) = f_2^{n+1}, \tag{11.114}$$

where $\phi^0 = \phi_0$, $\Delta t = t_{n+1} - t_n$, and $f_1^{n+1} + f_2^{n+1} = f^{n+1} = f((n+1)\Delta t)$. The time discretizations in (11.113) and (11.114) are implicit. Some schemes to be discussed in what follows actually use explicit discretizations. However, the stability conditions for those explicit schemes must be satisfied.

The MAC (marker-and-cell) method was first proposed by Harlow and Welsh (1965) to solve flow problems with free surfaces. There are many variations of this method. It basically uses a finite difference discretization for the Navier-Stokes equations and splits the equations into two operators

$$\mathbf{A}_1(\mathbf{u}, p) = \begin{pmatrix} (\mathbf{u} \cdot \nabla)\mathbf{u} - \frac{1}{\mathrm{Re}}\nabla^2\mathbf{u} \\ 0 \end{pmatrix}, \quad \text{and} \quad \mathbf{A}_2(\mathbf{u}, p) = \begin{pmatrix} \nabla p \\ \nabla \cdot \mathbf{u} \end{pmatrix}. \tag{11.115}$$

Each time step is divided into two substeps as discussed in the Marchuk-Yanenko fractional-step scheme (11.113) and (11.114). The first step solves a convection and diffusion problem, which is discretized explicitly,

$$\frac{\mathbf{u}^{n+1/2} - \mathbf{u}^n}{\Delta t} + (\mathbf{u}^n \cdot \nabla)\mathbf{u}^n - \frac{1}{\mathrm{Re}}\nabla^2\mathbf{u}^n = \mathbf{g}^{n+1}. \tag{11.116}$$

In the second step, the pressure gradient operator is added (implicitly) and, at the same time, the incompressible condition is enforced,

$$\frac{\mathbf{u}^{n+1} - \mathbf{u}^{n+1/2}}{\Delta t} + \nabla p^{n+1} = \mathbf{0}, \tag{11.117}$$

and
$$\nabla \cdot \mathbf{u}^{n+1} = 0. \tag{11.118}$$

This step is also called a projection step to satisfy the incompressibility condition.

Normally, the MAC scheme is presented in a discretized form. A preferred feature of the MAC method is the use of the staggered grid. An example of a staggered grid in two dimensions is shown in Figure 11.4. On this staggered grid, pressure variables are defined at the centers of the cells and velocity components are defined at the cell faces, as shown in Figure 11.4.

Using the staggered grid, two components of the transport equation (11.116) can be written as,

$$u_{i+1/2,j}^{n+1/2} = u_{i+1/2,j}^{n} - \Delta t \left(u\frac{\partial u}{\partial x} + v\frac{\partial u}{\partial y} - \frac{1}{Re}\nabla^2 u \right)_{i+1/2,j}^{n} + \Delta t \; f_{i+1/2,j}^{n+1}, \tag{11.119}$$

$$v_{i,j+1/2}^{n+1/2} = v_{i,j+1/2}^{n} - \Delta t \left(u\frac{\partial v}{\partial x} + v\frac{\partial v}{\partial y} - \frac{1}{Re}\nabla^2 v \right)_{i,j+1/2}^{n} + \Delta t \; g_{i,j+1/2}^{n+1}, \tag{11.120}$$

where $\mathbf{u} = (u, v)$, $\mathbf{g} = (f, g)$, $\left(u\dfrac{\partial u}{\partial x} + v\dfrac{\partial u}{\partial y} - \dfrac{1}{Re}\nabla^2 u \right)_{i+1/2,j}^{n}$ and $\left(u\dfrac{\partial v}{\partial x} + v\dfrac{\partial v}{\partial y} \right.$

$\left. -\dfrac{1}{Re}\nabla^2 v \right)_{i,j+1/2}^{n}$ are the functions interpolated at the grid locations for the

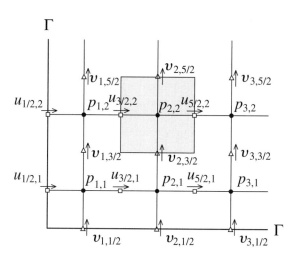

Figure 11.4   Staggered grid and a typical cell around $p_{2,2}$.

x-component of the velocity at $(i + 1/2, j)$ and for the y-component of the velocity at $(i, j + 1/2)$, respectively, and at the previous time $t = t_n$. The discretized form of (11.117) is

$$u_{i+1/2,j}^{n+1} = u_{i+1/2,j}^{n+1/2} - \frac{\Delta t}{\Delta x} \left(p_{i+1,j}^{n+1} - p_{i,j}^{n+1} \right), \qquad (11.121)$$

$$v_{i,j+1/2}^{n+1} = v_{i,j+1/2}^{n+1/2} - \frac{\Delta t}{\Delta y} \left(p_{i,j+1}^{n+1} - p_{i,j}^{n+1} \right), \qquad (11.122)$$

where $\Delta x = x_{i+1} - x_i$ and $\Delta y = y_{j+1} - y_j$ are the uniform grid spacing in the x and y directions, respectively. The discretized continuity equation (11.118) can be written as,

$$\frac{u_{i+1/2,j}^{n+1} - u_{i-1/2,j}^{n+1}}{\Delta x} + \frac{v_{i,j+1/2}^{n+1} - v_{i,j-1/2}^{n+1}}{\Delta y} = 0. \qquad (11.123)$$

Substitution of the two velocity components from (11.121) and (11.122) into the discretized continuity equation (11.123) generates a discrete Poisson equation for the pressure,

$$\nabla_d^2 p_{i,j}^{n+1} \equiv \frac{1}{\Delta x^2} \left(p_{i+1,j}^{n+1} - 2p_{i,j}^{n+1} + p_{i-1,j}^{n+1} \right) + \frac{1}{\Delta y^2} \left(p_{i,j+1}^{n+1} - 2p_{i,j}^{n+1} + p_{i,j-1}^{n+1} \right)$$

$$= \frac{1}{\Delta t} \left(\frac{u_{i+1/2,j}^{n+1/2} - u_{i-1/2,j}^{n+1/2}}{\Delta x} + \frac{v_{i,j+1/2}^{n+1/2} - v_{i,j-1/2}^{n+1/2}}{\Delta y} \right). \qquad (11.124)$$

The major advantage of the staggered grid is that it prevents the appearance of oscillatory solutions. On a normal grid, the pressure gradient would have to be approximated using two alternate grid points (not the adjacent ones) when a central difference scheme is used, that is

$$\left(\frac{\partial p}{\partial x} \right)_{i,j} = \frac{p_{i+1,j} - p_{i-1,j}}{2\Delta x} \quad \text{and} \quad \left(\frac{\partial p}{\partial y} \right)_{i,j} = \frac{p_{i,j+1} - p_{i,j-1}}{2\Delta y}. \qquad (11.125)$$

Thus a wavy pressure field (in a zigzag pattern) would be felt like a uniform one by the momentum equation. However, on a staggered grid, the pressure gradient is approximated by the difference of the pressures between two adjacent grid points. Consequently, a pressure field with a zigzag pattern would no longer be felt as a uniform pressure field and could not arise as a possible solution. It is also seen that the discretized continuity equation (11.123) contains the differences of the adjacent velocity components, which would prevent a wavy velocity field from satisfying the continuity equation.

Another advantage of the staggered grid is its accuracy. For example, the truncation error for (11.123) is $O\left(\Delta x^2, \Delta y^2\right)$ even though only four grid points are involved. The pressure gradient evaluated at the cell faces,

$$\left(\frac{\partial p}{\partial x}\right)_{i+1/2,j} = \frac{p_{i+1,j} - p_{i,j}}{\Delta x}, \quad \text{and} \quad \left(\frac{\partial p}{\partial y}\right)_{i,j+1/2} = \frac{p_{i,j+1} - p_{i,j}}{\Delta y}, \quad (11.126)$$

are all second-order accurate.

On the staggered grid, the MAC method does not require boundary conditions for the pressure equation (11.124). Let us examine a pressure node next to the boundary, for example $p_{1,2}$ as shown in Figure 11.4. When the normal velocity is specified at the boundary, $u_{1/2,2}^{n+1}$ is known. In evaluating the discrete continuity equation (11.123) at the pressure node $(1, 2)$, the velocity $u_{1/2,2}^{n+1}$ should not be expressed in terms of $u_{1/2,2}^{n+1/2}$ using (11.121). Therefore $p_{0,2}$ will not appear in equation (11.120), and no boundary condition for the pressure is needed. It should also be noted that (11.119) and (11.120) only update the velocity components for the interior grid points, and their values at the boundary grid points are not needed in the MAC scheme. Peyret and Taylor (1983, chapter 6) also noticed that the numerical solution in the MAC method is independent of the boundary values of $u^{n+1/2}$ and $v^{n+1/2}$, and a zero normal pressure gradient on the boundary would give satisfactory results. However, their explanation was more cumbersome.

In summary, for each time step in the MAC scheme, the intermediate velocity components, $u_{i+1/2,j}^{n+1/2}$ and $v_{i,j+1/2}^{n+1/2}$, in the interior of the domain are first evaluated using (11.119) and (11.120), respectively. Next, the discrete pressure Poisson equation (11.124) is solved. Finally, the velocity components at the new time step are obtained from (11.121) and (11.122). In the MAC scheme, the most costly step is the solution of the Poisson equation for the pressure (11.124).

Chorin (1968) and Temam (1969) independently presented a numerical scheme for the incompressible Navier-Stokes equations, termed the projection method. The projection method was initially proposed using the standard grid. However, when it is applied in an explicit fashion on the MAC staggered grid, it is identical to the MAC method as long as the boundary conditions are not considered, as shown in Peyret and Taylor (1983, chapter 6).

A physical interpretation of the MAC scheme or the projection method is that the explicit update of the velocity field does not generate a divergence free velocity field in the first step. Thus an irrotational correction field, in the form of a velocity potential which is proportional to the pressure, is added to the nondivergence-free velocity field in the second step in order to enforce the incompressibility condition.

As the MAC method uses an explicit scheme in the convection-diffusion step, the stability conditions for this method are (Peyret and Taylor, 1983, chapter 6),

$$\frac{1}{2}\left(u^2 + v^2\right)\Delta t \mathrm{Re} \leqslant 1, \quad (11.127)$$

and
$$\frac{4\Delta t}{\mathrm{Re}\Delta x^2} \leqslant 1, \tag{11.128}$$

when $\Delta x = \Delta y$. The stability conditions (11.127) and (11.128) are quite restrictive on the size of the time step. These restrictions can be removed by using implicit schemes for the convection-diffusion step.

Θ-Scheme

The MAC algorithm described in the preceding section is only first-order accurate in time. In order to have a second-order accurate scheme for the Navier-Stokes equations, the Θ-scheme of Glowinski (1991) may be used. The Θ-scheme splits each time step symmetrically into three substeps, which are described here.

- Step 1:

$$\frac{\mathbf{u}^{n+\theta} - \mathbf{u}^n}{\theta \Delta t} - \frac{\alpha}{\mathrm{Re}}\nabla^2\mathbf{u}^{n+\theta} + \nabla p^{n+\theta} = \mathbf{g}^{n+\theta} + \frac{\beta}{\mathrm{Re}}\nabla^2\mathbf{u}^n - (\mathbf{u}^n \cdot \nabla)\mathbf{u}^n, \tag{11.129}$$

$$\nabla \cdot \mathbf{u}^{n+\theta} = 0. \tag{11.130}$$

- Step 2:

$$\frac{\mathbf{u}^{n+1-\theta} - \mathbf{u}^{n+\theta}}{(1 - 2\theta)\Delta t} - \frac{\beta}{\mathrm{Re}}\nabla^2\mathbf{u}^{n+1-\theta} + (\mathbf{u}^* \cdot \nabla)\mathbf{u}^{n+1-\theta}$$
$$= \mathbf{g}^{n+1-\theta} + \frac{\alpha}{\mathrm{Re}}\nabla^2\mathbf{u}^{n+\theta} - \nabla p^{n+\theta}. \tag{11.131}$$

- Step 3:

$$\frac{\mathbf{u}^{n+1} - \mathbf{u}^{n+1-\theta}}{\theta \Delta t} - \frac{\alpha}{\mathrm{Re}}\nabla^2\mathbf{u}^{n+1} + \nabla p^{n+1} = \mathbf{g}^{n+1} + \frac{\beta}{\mathrm{Re}}\nabla^2\mathbf{u}^{n+1-\theta}$$
$$- (\mathbf{u}^{n+1-\theta} \cdot \nabla)\mathbf{u}^{n+1-\theta}, \tag{11.132}$$

$$\nabla \cdot \mathbf{u}^{n+1} = 0. \tag{11.133}$$

It was shown that when $\theta = 1 - 1/\sqrt{2} = 0.29289\ldots, \alpha + \beta = 1$ and $\beta = \theta/(1 - \theta)$, the scheme is second-order accurate. The first and third steps of the Θ-scheme are identical and are the Stokes flow problems. The second step, (11.131), represents a nonlinear convection-diffusion problem if $\mathbf{u}^* = \mathbf{u}^{n+1-\theta}$. However, it was concluded that there is practically no loss in accuracy and stability if $\mathbf{u}^* = \mathbf{u}^{n+\theta}$ is used. Numerical techniques for solving these substeps are discussed in Glowinski (1991).

Mixed Finite Element Formulation

The weak formulation described in Section 3 can be directly applied to the Navier-Stokes equations (11.81) and (11.80), and it gives

$$\int_\Omega \left(\frac{\partial \mathbf{u}}{\partial t} + \mathbf{u} \cdot \nabla \mathbf{u} - \mathbf{g} \right) \cdot \tilde{\mathbf{u}} d\Omega + \frac{2}{\mathrm{Re}} \int_\Omega \mathbf{D}\left[\mathbf{u}\right] : \mathbf{D}\left[\tilde{\mathbf{u}}\right] d\Omega - \int_\Omega p\left(\nabla \cdot \tilde{\mathbf{u}}\right) d\Omega = 0,$$
(11.134)

$$\int_\Omega \tilde{p} \nabla \cdot \mathbf{u} d\Omega = 0,$$
(11.135)

where $\tilde{\mathbf{u}}$ and $\tilde{p}$ are the variations of the velocity and pressure, respectively. The rate of strain tensor is given by

$$\mathbf{D}\left[\mathbf{u}\right] = \frac{1}{2}\left[\nabla \mathbf{u} + (\nabla \mathbf{u})^T\right].$$
(11.136)

The Galerkin finite element formulation for the problem is identical to (11.134) and (11.135), except that all the functions are chosen from finite dimensional subspaces and represented in the form of basis or interpolation functions.

The main difficulty with this finite element formulation is the choice of the interpolation functions (or the type of the elements) for velocity and pressure. The finite element approximations that use the same interpolation functions for velocity and pressure suffer from a highly oscillatory pressure field. As described in the previous section, a similar behavior in the finite difference scheme is prevented by introducing the staggered grid. There are a number of options to overcome this problem with spurious pressure. One of them is the mixed finite element formulation that uses different interpolation functions (or finite elements) for velocity and pressure. The requirement for the mixed finite element approach is related to the so-called Babuska-Brezzi (or LBB) stability condition, or *inf-sup* condition. The detailed discussions for this condition can be found in Oden and Carey (1984). A common practice in the mixed finite element formulation is to use a pressure interpolation function that is one order lower than a velocity interpolation function. As an example in two dimensions, a triangular element is shown in Figure 11.5a. On this mixed element, quadratic interpolation functions are used for the velocity components and are defined on all six nodes, while linear interpolation functions are used for the pressure and are defined

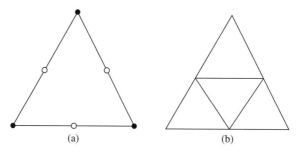

(a)   (b)

Figure 11.5   Mixed finite elements.

on three vertices only. A slightly different approach is to use a pressure grid that is twice coarser than the velocity one, and then use the same interpolation functions on both grids (Glowinski, 1991). For example, a piecewise-linear pressure is defined on the outside (coarser) triangle; while a piecewise-linear velocity is defined on all four subtriangles, as shown in Figure 11.5b.

Another option to prevent a spurious pressure field is to use the stabilized finite element formulation while keeping the equal order interpolations for velocity and pressure. A general formulation in this approach is the Galerkin/least-squares (GLS) stabilization (Tezduyar, 1992). In the GLS stabilization, the stabilizing terms are obtained by minimizing the squared residual of the momentum equation integrated over each element domain. The choice of the stabilization parameter is discussed in Franca *et al.* (1992) and Franca and Frey (1992).

Comparing the mixed and the stabilized finite element formulations, the mixed finite element method is parameter free, as pointed out in Glowinski (1991). There is no need to adjust the stabilization parameters, which could be a delicate problem. More importantly, for a given flow problem the desired finite element mesh size is generally determined based on the velocity behavior (e.g., it is defined by the boundary or shear layer thickness). Therefore, equal order interpolation will be more costly from the pressure point of view but without further gains in accuracy. However, the GLS-stabilized finite element formulation has the additional benefit of preventing oscillatory solutions produced in the Galerkin finite element method due to the large convective term in high Reynolds number flows.

Once the interpolation functions for the velocity and pressure in the mixed finite element approximations are determined, the matrix form of equations (11.134) and (11.135) can be written as

$$\begin{pmatrix} \mathbf{M\dot{u}} \\ \mathbf{0} \end{pmatrix} + \begin{pmatrix} \mathbf{A} & \mathbf{B} \\ \mathbf{B}^T & \mathbf{0} \end{pmatrix} \begin{pmatrix} \mathbf{u} \\ \mathbf{p} \end{pmatrix} = \begin{pmatrix} \mathbf{f}_u \\ \mathbf{f}_p \end{pmatrix}, \tag{11.137}$$

where $\mathbf{u}$ and $\mathbf{p}$ are the vectors containing all unknown values of the velocity components and pressure defined on the finite element mesh, respectively. $\mathbf{\dot{u}}$ is the first time derivative of $\mathbf{u}$. $\mathbf{M}$ is the mass matrix corresponding to the time derivative term in equation (11.134). Matrix $\mathbf{A}$ depends on the value of $\mathbf{u}$ due to the nonlinear convective term in the momentum equation. The symmetry in the pressure terms in (11.134) and (11.135) results in the symmetric arrangement of $\mathbf{B}$ and $\mathbf{B}^T$ in the algebraic system (11.137). Vectors $\mathbf{f}_u$ and $\mathbf{f}_p$ come from the body force term in the momentum equation and from the application of the boundary conditions.

The ordinary differential equation (11.137) can be further discretized in time with finite difference methods. The resulting nonlinear system of equations is typically solved iteratively using Newton's method. At each stage of the nonlinear iteration, the sparse linear algebraic equations are normally solved either by using a direct solver such as the Gauss elimination procedure for small system sizes or by using an iterative solver such as the generalized minimum residual method (GMRES) for large systems. Other iterative solution methods for sparse nonsymmetric systems can be found in Saad (1996). An application of the mixed finite element method is discussed as one of the examples in the next section.

5. Three Examples

In this section, we will solve three sample problems. The first one is the classic driven cavity flow problem. The second is flow around a square block confined between two parallel plates. These two problems will be solved by using the explicit MacCormack scheme, with details in Perrin and Hu (2006). The contribution by Andrew Perrin in preparing results for these two problems is greatly appreciated. The last problem is flow around a circular cylinder confined between two parallel plates. It will be solved by using a mixed finite element formulation.

Explicit MacCormack Scheme for Driven Cavity Flow Problem

The driven cavity flow problem, in which a fluid-filled square box ("cavity") is swirled by a uniformly translating lid as shown in Figure 11.6, is a classic problem in CFD. This problem is unambiguous with easily applied boundary conditions and has a wealth of documented analytical and computational results, for example Ghia *et al.* (1982). We will solve this flow using the explicit MacCormack scheme discussed in the previous section.

We may nondimensionalize the problem with the following scaling: lengths with D, velocity with U, time with D/U, density with a reference density ρ_0, and pressure with $\rho_0 U^2$. Using this scaling, the equation of state (11.99) becomes $p = \rho/M^2$, where $M = U/c$ is the Mach number. The Reynolds number is defined as Re $= \rho_0 U D/\mu$.

The boundary conditions for this problem are relatively simple. The velocity components on all four sides of the cavity are well defined. There are two singularities of velocity gradient at the two top corners where velocity u drops from U to 0 directly underneath the sliding lid. However, these singularities will be smoothed out on a given grid, since the change of the velocity occurs linearly between two grid points. The boundary conditions for the density (hence the pressure) are more involved. Since the density is not specified on a solid surface, we need to generate an update scheme for

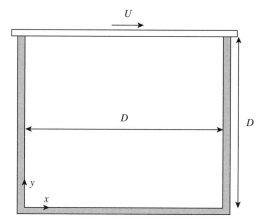

Figure 11.6 Driven cavity flow problem. The cavity is filled with a fluid with the top lid sliding at a constant velocity U.

values of density on all boundary points. A natural option is to derive that using the continuity equation.

Consider the boundary on the left (at $x = 0$). Since $v = 0$ along the surface, the continuity equation (11.96) reduces to

$$\frac{\partial \rho}{\partial t} + \frac{\partial \rho u}{\partial x} = 0. \tag{11.138}$$

We may use a predictor-corrector scheme to update density on this surface with a one-sided second-order accurate discretization for the spatial derivative,

$$\left(\frac{\partial f}{\partial x}\right)_i = \frac{1}{2\Delta x} (-f_{i+2} + 4 f_{i+1} - 3 f_i) + O\left(\Delta x^2\right)$$

$$\text{or} \quad \left(\frac{\partial f}{\partial x}\right)_i = \frac{-1}{2\Delta x} (-f_{i-2} + 4 f_{i-1} - 3 f_i) + O\left(\Delta x^2\right).$$

Therefore, on the surface of $x = 0$ (for $i = 0$ including two corner points on the left), we have the following update scheme for density,

$$\text{predictor} \quad \rho_{i,j}^* = \rho_{i,j}^n - \frac{\Delta t}{2\Delta x} \left[-(\rho u)_{i+2,j}^n + 4 (\rho u)_{i+1,j}^n - 3 (\rho u)_{i,j}^n \right],$$

$$\tag{11.139}$$

$$\text{corrector} \quad 2\rho_{i,j}^{n+1} = \rho_{i,j}^n + \rho_{i,j}^* - \frac{\Delta t}{2\Delta x} \left[-(\rho u)_{i+2,j}^* + 4 (\rho u)_{i+1,j}^* - 3 (\rho u)_{i,j}^* \right].$$

$$\tag{11.140}$$

Similarly, on the right side of the cavity $x = D$ (for $i = n_x - 1$, where n_x is the number of grid points in the x-direction, including two corner points on the right), we have

$$\text{predictor} \quad \rho_{i,j}^* = \rho_{i,j}^n + \frac{\Delta t}{2\Delta x} \left[-(\rho u)_{i-2,j}^n + 4 (\rho u)_{i-1,j}^n - 3 (\rho u)_{i,j}^n \right],$$

$$\tag{11.141}$$

$$\text{corrector} \quad 2\rho_{i,j}^{n+1} = \rho_{i,j}^n + \rho_{i,j}^* + \frac{\Delta t}{2\Delta x} \left[-(\rho u)_{i-2,j}^* + 4 (\rho u)_{i-1,j}^* - 3 (\rho u)_{i,j}^* \right].$$

$$\tag{11.142}$$

On the bottom of the cavity $y = 0$ ($j = 0$),

$$\text{predictor} \quad \rho_{i,j}^* = \rho_{i,j}^n - \frac{\Delta t}{2\Delta y} \left[-(\rho v)_{i,j+2}^n + 4 (\rho v)_{i,j+1}^n - 3 (\rho v)_{i,j}^n \right], \tag{11.143}$$

$$\text{corrector} \quad 2\rho_{i,j}^{n+1} = \rho_{i,j}^n + \rho_{i,j}^* - \frac{\Delta t}{2\Delta y} \left[-(\rho v)_{i,j+2}^* + 4 (\rho v)_{i,j+1}^* - 3 (\rho v)_{i,j}^* \right].$$

$$\tag{11.144}$$

Finally, on the top of the cavity $y = D$ ($j = n_y - 1$ where n_y is the number of grid points in the y-direction), the density needs to be updated from slightly different expressions since $\partial \rho u / \partial x = U \partial \rho / \partial x$ is not zero there, predictor

$$\rho^*_{i,j} = \rho^n_{i,j} - \frac{\Delta t \, U}{2\Delta x}\left[\rho^n_{i+1,j} - \rho^n_{i-1,j}\right] + \frac{\Delta t}{2\Delta y}\left[-(\rho v)^n_{i,j-2} + 4\,(\rho v)^n_{i,j-1} - 3\,(\rho v)^n_{i,j}\right],$$

(11.145)

corrector

$$2\rho^{n+1}_{i,j} = \rho^n_{i,j} + \rho^*_{i,j} - \frac{\Delta t \, U}{2\Delta x}\left[\rho^*_{i+1,j} - \rho^*_{i-1,j}\right] + \frac{\Delta t}{2\Delta y}\left[-(\rho v)^*_{i,j-2} + 4\,(\rho v)^*_{i,j-1}\right.$$
$$\left. - 3\,(\rho v)^*_{i,j}\right].$$

(11.146)

In summary, we may organize the explicit MacCormack scheme at each time step (11.103) to (11.108) into the following six substeps.

Step 1:   For $0 \leqslant i < n_x$ and $0 \leqslant j < n_y$ (all nodes):

$$u_{i,j} = (\rho u)^n_{i,j}\Big/\rho^n_{i,j}, \quad v_{i,j} = (\rho v)^n_{i,j}\Big/\rho^n_{i,j}.$$

Step 2:   For $1 \leqslant i < n_x - 1$ and $1 \leqslant j < n_y - 1$ (all interior nodes):

$$\rho^*_{i,j} = \rho^n_{i,j} - a_1\left[(\rho u)^n_{i+1,j} - (\rho u)^n_{i,j}\right] - a_2\left[(\rho v)^n_{i,j+1} - (\rho v)^n_{i,j}\right],$$

$$(\rho u)^*_{i,j} = (\rho u)^n_{i,j} - a_3\left(\rho^n_{i+1,j} - \rho^n_{i,j}\right) - a_1\left[\left(\rho u^2\right)^n_{i+1,j} - \left(\rho u^2\right)^n_{i,j}\right]$$

$$- a_2\left[(\rho u v)^n_{i,j+1} - (\rho u v)^n_{i,j}\right] - a_{10}u_{i,j} + a_5\left(u_{i+1,j} + u_{i-1,j}\right)$$

$$+ a_6\left(u_{i,j+1} + u_{i,j-1}\right) + a_9\left(v_{i+1,j+1} + v_{i-1,j-1} - v_{i+1,j-1} - v_{i-1,j+1}\right),$$

$$(\rho v)^*_{i,j} = (\rho v)^n_{i,j} - a_4\left(\rho^n_{i,j+1} - \rho^n_{i,j}\right) - a_1\left[(\rho u v)^n_{i+1,j} - (\rho u v)^n_{i,j}\right]$$

$$- a_2\left[\left(\rho v^2\right)^n_{i,j+1} - \left(\rho v^2\right)^n_{i,j}\right] - a_{11}v_{i,j} + a_7\left(v_{i+1,j} + v_{i-1,j}\right)$$

$$+ a_8\left(v_{i,j+1} + v_{i,j-1}\right) + a_9\left(u_{i+1,j+1} + u_{i-1,j-1} - u_{i+1,j-1} - u_{i-1,j+1}\right).$$

Step 3: Impose boundary conditions (at time t_{n+1}) for $\rho^*_{i,j}$, $(\rho u)^*_{i,j}$ and $(\rho v)^*_{i,j}$.
Step 4: For $0 \leqslant i < n_x$ and $0 \leqslant j < n_y$ (all nodes):

$$u^*_{i,j} = (\rho u)^*_{i,j}\Big/\rho^*_{i,j}, \quad v^*_{i,j} = (\rho v)^*_{i,j}\Big/\rho^*_{i,j}.$$

Step 5: For $1 \leqslant i < n_x - 1$ and $1 \leqslant j < n_y - 1$ (all interior nodes):

$$2\rho_{i,j}^{n+1} = \left(\rho_{i,j}^n + \rho_{i,j}^*\right) - a_1 \left[(\rho u)_{i,j}^* - (\rho u)_{i-1,j}^*\right] - a_2 \left[(\rho v)_{i,j}^* - (\rho v)_{i,j-1}^*\right],$$

$$2(\rho u)_{i,j}^{n+1} = (\rho u)_{i,j}^n + (\rho u)_{i,j}^* - a_3 \left(\rho_{i,j}^* - \rho_{i-1,j}^*\right) - a_1 \left[\left(\rho u^2\right)_{i,j}^* - \left(\rho u^2\right)_{i-1,j}^*\right]$$

$$- a_2 \left[(\rho u v)_{i,j}^* - (\rho u v)_{i,j-1}^*\right] - a_{10} u_{i,j}^* + a_5 \left(u_{i+1,j}^* + u_{i-1,j}^*\right)$$

$$+ a_6 \left(u_{i,j+1}^* + u_{i,j-1}^*\right) + a_9 \left(v_{i+1,j+1}^* + v_{i-1,j-1}^* - v_{i+1,j-1}^* - v_{i-1,j+1}^*\right),$$

$$2(\rho v)_{i,j}^{n+1} = (\rho v)_{i,j}^n + (\rho v)_{i,j}^* - a_4 \left(\rho_{i,j}^* - \rho_{i,j-1}^*\right) - a_1 \left[(\rho u v)_{i,j}^* - (\rho u v)_{i-1,j}^*\right]$$

$$- a_2 \left[\left(\rho v^2\right)_{i,j}^* - \left(\rho v^2\right)_{i,j-1}^*\right] - a_{11} v_{i,j}^* + a_7 \left(v_{i+1,j}^* + v_{i-1,j}^*\right)$$

$$+ a_8 \left(v_{i,j+1}^* + v_{i,j-1}^*\right) + a_9 \left(u_{i+1,j+1}^* + u_{i-1,j-1}^* - u_{i+1,j-1}^* - u_{i-1,j+1}^*\right).$$

Step 6: Impose boundary conditions for $\rho_{i,j}^{n+1}$, $(\rho u)_{i,j}^{n+1}$ and $(\rho v)_{i,j}^{n+1}$.
The coefficients are defined as,

$$a_1 = \frac{\Delta t}{\Delta x}, \quad a_2 = \frac{\Delta t}{\Delta y}, \quad a_3 = \frac{\Delta t}{\Delta x M^2}, \quad a_4 = \frac{\Delta t}{\Delta y M^2}, \quad a_5 = \frac{4\Delta t}{3\mathrm{Re}\,(\Delta x)^2},$$

$$a_6 = \frac{\Delta t}{\mathrm{Re}\,(\Delta y)^2}, \quad a_7 = \frac{\Delta t}{\mathrm{Re}\,(\Delta x)^2}, \quad a_8 = \frac{4\Delta t}{3\mathrm{Re}\,(\Delta y)^2}, \quad a_9 = \frac{\Delta t}{12\mathrm{Re}\,\Delta x \Delta y},$$

$$a_{10} = 2\,(a_5 + a_6), \quad a_{11} = 2\,(a_7 + a_8).$$

For coding purposes, the variables $u_{i,j}$ ($v_{i,j}$) and $u_{i,j}^*$ ($v_{i,j}^*$) can take the same storage space. At the end of each time step, the starting values of $\rho_{i,j}^n$, $(\rho u)_{i,j}^n$ and $(\rho v)_{i,j}^n$ will be replaced with the corresponding new values of $\rho_{i,j}^{n+1}$, $(\rho u)_{i,j}^{n+1}$ and $(\rho v)_{i,j}^{n+1}$.

Next we present some of the results and compare them with those in the paper by Hou *et al.* (1995) obtained by a lattice Boltzmann method. To keep the flow almost incompresible, the Mach number is chosen as $M = 0.1$. Flows with two Reynolds numbers, $\mathrm{Re} = \rho_0 U D/\mu = 100$ and 400 are simulated. At these Reynolds numbers, the flow will eventually be steady. Thus calculations need to be run long enough to get to the steady state. A uniform grid of 256 by 256 was used for this example.

Figure 11.7 shows comparisons of the velocity field calculated by the explicit MacCormack scheme with the streamlines from Hou (1995) at $\mathrm{Re} = 100$ and 400. The agreement seems reasonable. It was also observed that the location of the center of the primary eddy agrees even better. When $\mathrm{Re} = 100$, the center of the primary eddy is found at $(0.62 \pm 0.02, 0.74 \pm 0.02)$ from the MacCormack scheme in comparison with $(0.6196, 0.7373)$ from Hou. When $\mathrm{Re} = 400$, the center of the primary eddy is found at $(0.57 \pm 0.02, 0.61 \pm 0.02)$ from the MacCormack scheme in comparison with $(0.5608, 0.6078)$ from Hou.

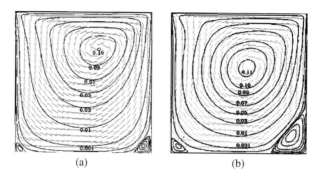

Figure 11.7 Comparisons of results from the explicit MacCormack scheme (light gray, velocity vector field) and those from Hou, et al. (1995) (dark solid streamlines) calculated using a Lattice Boltzmann Method. (a) Re = 100, (b) at Re = 400.

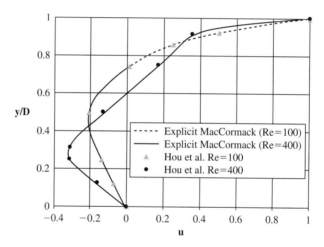

Figure 11.8 Comparison of velocity profiles along a line cut through the center of the cavity ($x = 0.5\,D$) at Re = 100 and 400.

For a more quantitative comparison, Figure 11.8 plots the velocity profile along a vertical line cut through the center of the cavity ($x = 0.5D$). The velocity profiles for two Reynolds numbers, Re = 100 and 400, are compared. The results from the explicit MacCormack scheme are shown in solid and dashed lines. The data points in symbols were directly converted from Hou's paper. The agreement is excellent.

Explicit MacCormack Scheme for Flow Over a Square Block

For the second example, we consider flow around a square block confined between two parallel plates. Fluid comes in from the left with a uniform velocity profile U, and the plates are sliding with the same velocity, as indicated in Figure 11.9. This flow corresponds to the block moving left with velocity U along channel's center line. In the calculation we set the channel width $H = 3D$, the channel length $L = 35D$ with

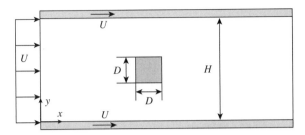

Figure 11.9 Flow around a square block between two parallel plates.

$15D$ ahead of the block and $19D$ behind. The Mach number is set at $M = 0.05$ to approximate the incompressible limit.

The velocity boundary conditions in this problem are specified as shown in Figure 11.9, except that at the outflow section, conditions $\partial \rho u / \partial x = 0$ and $\partial \rho v / \partial x = 0$ are used. The density (or pressure) boundary conditions are much more complicated, especially on the block surface. On all four sides of the outer boundary (top and bottom plates, inflow and outflow), the continuity equation is used to update density as in the previous example. However, on the block surface, it was found that the conditions derived from the momentum equations give better results. Let us consider the front section of the block, and evaluate the x-component of the momentum equation (11.97) with $u = v = 0$,

$$
\begin{aligned}
\frac{\partial \rho}{\partial x} &= M^2 \left[\frac{1}{\text{Re}} \left(\frac{4}{3} \frac{\partial^2 u}{\partial x^2} + \frac{1}{3} \frac{\partial^2 v}{\partial x \partial y} + \frac{\partial^2 u}{\partial y^2} \right) \right. \\
&\quad \left. - \frac{\partial}{\partial x} \left(\rho u^2 \right) - \frac{\partial}{\partial y} \left(\rho v u \right) - \frac{\partial}{\partial t} \left(\rho u \right) \right]_{\text{front suface}} \\
&= \frac{M^2}{\text{Re}} \left(\frac{4}{3} \frac{\partial^2 u}{\partial x^2} + \frac{1}{3} \frac{\partial^2 v}{\partial x \partial y} \right) .
\end{aligned}
\tag{11.147}
$$

In (11.147), the variables are non-dimensionalized with the same scaling as the driven cavity flow problem except that the block size D is used for length. Furthermore, the density gradient may be approximated with a second order backward finite difference scheme,

$$
\left(\frac{\partial \rho}{\partial x} \right)_{i,j} = \frac{-1}{2 \Delta x} \left(-\rho_{i-2,j} + 4\rho_{i-1,j} - 3\rho_{i,j} \right) + O(\Delta x^2).
\tag{11.148}
$$

And the second order derivatives for the velocities are expressed as,

$$
\left(\frac{\partial^2 u}{\partial x^2} \right)_{i,j} = \frac{1}{\Delta x^2} \left(2u_{i,j} - 5u_{i-1,j} + 4u_{i-2,j} - u_{i-3,j} \right) + O(\Delta x^2)
\tag{11.149}
$$

and

$$\left(\frac{\partial^2 v}{\partial x \partial y}\right)_{i,j} = \frac{-1}{4\Delta x \Delta y}\left[-\left(v_{i-2,j+1}-v_{i-2,j-1}\right)+4\left(v_{i-1,j+1}-v_{i-1,j-1}\right)\right.$$

$$\left. -3\left(v_{i,j+1}-v_{i,j-1}\right)\right]+O\left(\Delta x^2,\ \Delta x \Delta y,\ \Delta y^2\right).\qquad(11.150)$$

Substituting (11.148) to (11.150) into (11.147), we have an expression for density at the front of the block,

$$\rho_{i,j}\big|_{front} = \frac{1}{3}\left(4\rho_{i-1,j}-\rho_{i-2,j}\right)+\frac{8}{9\Delta x}\frac{M^2}{Re}\left(-5u_{i-1,j}+4u_{i-2,j}-u_{i-3,j}\right)$$

$$-\frac{1}{18\Delta y}\frac{M^2}{Re}\left[-\left(v_{i-2,j+1}-v_{i-2,j-1}\right)+4\left(v_{i-1,j+1}-v_{i-1,j-1}\right)\right.$$

$$\left. -3\left(v_{i,j+1}-v_{i,j-1}\right)\right].\qquad(11.151)$$

Similarly at the back of the block,

$$\rho_{i,j}\big|_{back} = \frac{1}{3}\left(4\rho_{i+1,j}-\rho_{i+2,j}\right)-\frac{8}{9\Delta x}\frac{M^2}{Re}\left(-5u_{i+1,j}+4u_{i+2,j}-u_{i+3,j}\right)$$

$$-\frac{1}{18\Delta y}\frac{M^2}{Re}\left[-\left(v_{i+2,j+1}-v_{i+2,j-1}\right)+4\left(v_{i+1,j+1}-v_{i+1,j-1}\right)\right.$$

$$\left. -3\left(v_{i,j+1}-v_{i,j-1}\right)\right].\qquad(11.152)$$

At the top of the block, the y-component of the momentum equation should be used, and it is easy to find that

$$\rho_{i,j}\big|_{top} = \frac{1}{3}\left(4\rho_{i,j+1}-\rho_{i,j+2}\right)-\frac{8}{9\Delta y}\frac{M^2}{Re}\left(-5v_{i,j+1}+4v_{i,j+2}-v_{i,j+3}\right)$$

$$-\frac{1}{18\Delta x}\frac{M^2}{Re}\left[-\left(u_{i+1,j+2}-u_{i-1,j+2}\right)+4\left(u_{i+1,j+1}-u_{i-1,j+1}\right)\right.$$

$$\left. -3\left(u_{i+1,j}-u_{i-1,j}\right)\right],\qquad(11.153)$$

and finally at the bottom of the block,

$$\rho_{i,j}\big|_{bottom} = \frac{1}{3}\left(4\rho_{i,j-1}-\rho_{i,j-2}\right)+\frac{8}{9\Delta y}\frac{M^2}{Re}\left(-5v_{i,j-1}+4v_{i,j-2}-v_{i,j-3}\right)$$

$$-\frac{1}{18\Delta x}\frac{M^2}{Re}\left[-\left(u_{i+1,j-2}-u_{i-1,j-2}\right)+4\left(u_{i+1,j-1}-u_{i-1,j-1}\right)\right.$$

$$\left. -3\left(u_{i+1,j}-u_{i-1,j}\right)\right].\qquad(11.154)$$

At the four corners of the block, the average values from the two corresponding sides may be used.

In computation, double precision numbers should be used: otherwise cumulative round-off error may corrupt the simulation, especially for long runs. It is also helpful to introduce a new variable for density, $\rho' = \rho - 1$, such that only the density variation is computed. For this example, we may extend the *FF/BB* form of the explicit MacCormack scheme to have a *FB/BF* arrangement for one time step and a *BF/FB* arrangement for the subsequent time step. This cycling seems to generate better results.

We first plot the drag coefficient, $C_D = Drag/(\frac{1}{2}\rho_0 U^2 D)$, and the lift coefficient, $C_L = Lift/(\frac{1}{2}\rho_0 U^2 D)$, as functions of time for flows at two Reynolds numbers, Re = 20 and 100, in Figure 11.10. For Re = 20, after the initial messy transient (corresponding to sound waves bouncing around the block and reflecting at the outflow) the flow eventually settles into a steady state. The drag coefficient stabilizes at a constant value around $C_D = 6.94$ (obtained on a grid of 701x61). Calculation on a finer grid (1401x121) yields $C_D = 7.003$. This is in excellent agreement with the value of $C_D = 7.005$ obtained from an implicit finite element calculation for incompressible flows (similar to the one used in the next example in this section) on a similar mesh to 1401×121. There is a small lift ($C_L = 0.014$) due to asymmetries in the numerical scheme. The lift reduces to $C_L = 0.003$ on the finer grid of 1401x121. For Re = 100, periodic vortex shedding occurs. Drag and lift coefficients are shown in Figure 11.10(b). The mean value of the drag coefficient and the amplitude of the lift coefficient are $C_D = 3.35$ and $C_L = 0.77$, respectively. The finite element results are $C_D = 3.32$ and $C_L = 0.72$ under similar conditions.

The flow field around the block at Re = 20 is shown in Figure 11.11. A steady wake is attached behind the block, and the circulation within the wake is clearly visible. Figure 11.12 displays a sequence of the flow field around the block during one cycle of vortex shedding at Re = 100.

Figure 11.13 shows the convergence of the drag coefficient as the grid spacing is reduced. Tests for two Reynolds numbers, Re = 20 and 100, are plotted. It seems that the solution with 20 grid points across the block ($\Delta x = \Delta y = 0.05$) reasonably resolves the drag coefficient and the singularity at the block corners does not affect this convergence very much.

The explicit MacCormack scheme can be quite efficient to compute flows at high Reynolds numbers where small time steps are naturally needed to resolve high frequencies in the flow and the stability condition for the time step is no longer too restrictive. Since with $\Delta x = \Delta y$ and large (grid) Reynolds numbers, the stability condition (11.110) becomes approximately,

$$\Delta t \leqslant \frac{\sigma}{\sqrt{2}} M \Delta x. \tag{11.155}$$

As a more complicated example, the flow around a circular cylinder confined between two parallel plates (the same geometry as the fourth example later in this section) is calculated at Re = 1000 using the explicit MacCormack scheme. For flow visualization, a

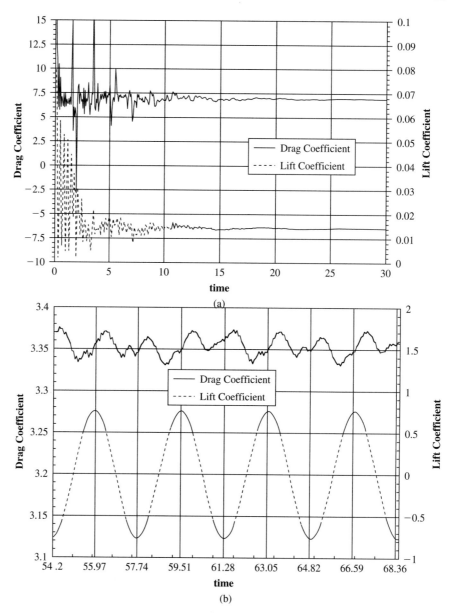

Figure 11.10 Drag and lift coefficients as functions of time for flow over a block. (a) Re = 20, on a grid of 701 × 61, (b) Re = 100, on a grid of 1401 × 121.

smoke line is introduced at the inlet. Numerically, an additional convection-diffusion equation for smoke concentration is solved similarly, with an explicit scheme at each time step coupled with the computed flow field. Two snap shots of the flow field are displayed in Figure 11.14. In this calculation, the flow Mach number is set at

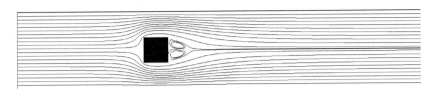

Figure 11.11   Streamlines for flow around a block at Re = 20.

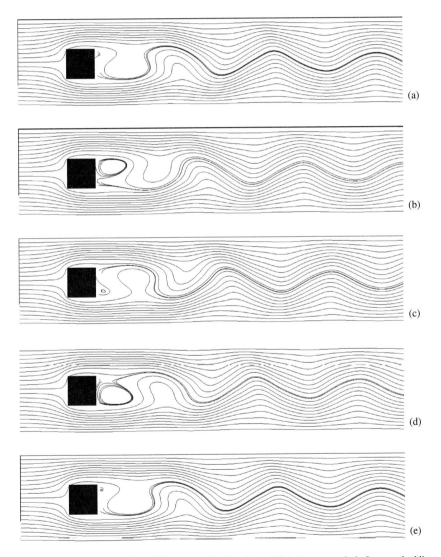

Figure 11.12   A sequence of flow fields around a block at Re = 100 during one period of vortex shedding.
(a) $t = 40.53$, (b) $t = 41.50$, (c) $t = 42.48$, (d) $t = 43.45$, (e) $t = 44.17$.

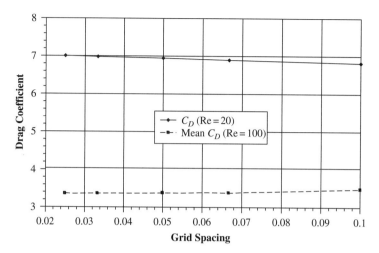

Figure 11.13   Convergence tests for the drag coefficient as the grid spacing decreases. The grid spacing is equal in both directions $\Delta x = \Delta y$, and time step Δt is determined by the stability condition.

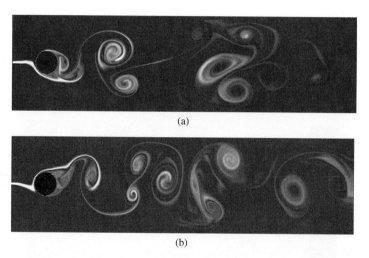

Figure 11.14   Smoke lines in flow around a circular cylinder between two parallel plates at Re = 1000. The flow geometry is the same as in the fourth example later in this section.

$M = 0.3$, and a uniform fine grid with 100 grid points across the cylinder diameter is used.

Finite Element Formulation for Flow Over a Cylinder Confined in a Channel

We next consider the flow over a circular cylinder moving along the center of a channel. In the computation, we fix the cylinder, and use the flow geometry as shown in Figure 11.15. The flow comes from the left with a uniform velocity U. Both plates of the channel are sliding to the right with the same velocity U. The diameter of the cylinder is d and the width of the channel is $W = 4d$. The boundary sections for the

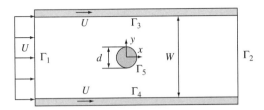

Figure 11.15 Flow geometry of flow around a cylinder in a channel.

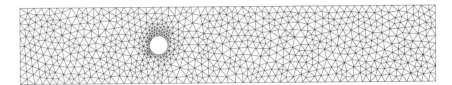

Figure 11.16 A finite element mesh around a cylinder.

computational domain are indicated in the figure. The location of the inflow boundary Γ_1 is selected to be at $x_{\min} = -7.5d$, and the location of the outflow boundary section Γ_2 is at $x_{\max} = 15d$. They are both far away from the cylinder so as to minimize their influence on the flow field near the cylinder. In order to compute the flow at higher Reynolds numbers, we relax the assumptions that the flow is symmetric and steady. We will compute unsteady flow (with vortex shedding) in the full geometry and using the Cartesian coordinates shown in Figure 11.15.

The first step in the finite element method is to discretize (mesh) the computational domain described in Figure 11.15. We cover the domain with triangular elements. A typical mesh is presented in Figure 11.16. The mesh size is distributed in a way that finer elements are used next to the cylinder surface to better resolve the local flow field. For this example, the mixed finite element method will be used, such that each triangular element will have six nodes as shown Figure 11.5a. This element allows for curved sides that better capture the surface of the circular cylinder. The mesh in Figure 11.16 has 3320 elements, 6868 velocity nodes, and 1774 pressure nodes.

The weak formulation of the Navier-Stokes equations is given in (11.134) and (11.135). For this example the body force term is zero, $\mathbf{g} = 0$. In Cartesian coordinates, the weak form of the momentum equation (11.134) can be written explicitly as

$$\int_\Omega \left(\frac{\partial \mathbf{u}}{\partial t} + u\frac{\partial \mathbf{u}}{\partial x} + v\frac{\partial \mathbf{u}}{\partial y}\right) \cdot \tilde{\mathbf{u}} d\Omega + \frac{2}{\text{Re}} \int_\Omega \left[\frac{\partial u}{\partial x}\frac{\partial \tilde{u}}{\partial x} + \frac{1}{2}\left(\frac{\partial u}{\partial y} + \frac{\partial v}{\partial x}\right)\left(\frac{\partial \tilde{u}}{\partial y} + \frac{\partial \tilde{v}}{\partial x}\right)\right.$$

$$+ \left.\frac{\partial v}{\partial y}\frac{\partial \tilde{v}}{\partial y}\right] d\Omega - \int_\Omega p\left(\frac{\partial \tilde{u}}{\partial x} + \frac{\partial \tilde{v}}{\partial y}\right) d\Omega = 0, \tag{11.156}$$

where Ω is the computational domain and $\tilde{\mathbf{u}} = (\tilde{u}, \tilde{v})$. Since the variational functions $\tilde{u}$ and $\tilde{v}$ are independent, the weak formulation (11.156) can be separated into two equations,

$$\int_{\Omega} \left(\frac{\partial u}{\partial t} + u \frac{\partial u}{\partial x} + v \frac{\partial u}{\partial y} \right) \tilde{u} d\Omega - \int_{\Omega} p \frac{\partial \tilde{u}}{\partial x} d\Omega$$

$$+ \frac{1}{Re} \int_{\Omega} \left[2 \frac{\partial u}{\partial x} \frac{\partial \tilde{u}}{\partial x} + \left(\frac{\partial u}{\partial y} + \frac{\partial v}{\partial x} \right) \frac{\partial \tilde{u}}{\partial y} \right] d\Omega = 0, \tag{11.157}$$

$$\int_{\Omega} \left(\frac{\partial v}{\partial t} + u \frac{\partial v}{\partial x} + v \frac{\partial v}{\partial y} \right) \tilde{v} d\Omega - \int_{\Omega} p \frac{\partial \tilde{v}}{\partial y} d\Omega$$

$$+ \frac{1}{Re} \int_{\Omega} \left[\left(\frac{\partial u}{\partial y} + \frac{\partial v}{\partial x} \right) \frac{\partial \tilde{v}}{\partial x} + 2 \frac{\partial v}{\partial y} \frac{\partial \tilde{v}}{\partial y} \right] d\Omega = 0. \tag{11.158}$$

The weak form of the continuity equation (11.135) is expressed as

$$- \int_{\Omega} \left(\frac{\partial u}{\partial x} + \frac{\partial v}{\partial y} \right) \tilde{p} \, d\Omega = 0. \tag{11.159}$$

Given a triangulation of the computational domain, for example, the mesh shown in Figure 11.16, the weak formulation of (11.157) to (11.159) can be approximated by the Galerkin finite element formulation based on the finite-dimensional discretization of the flow variables. The Galerkin formulation can be written as,

$$\int_{\Omega^h} \left(\frac{\partial u^h}{\partial t} + u^h \frac{\partial u^h}{\partial x} + v^h \frac{\partial u^h}{\partial y} \right) \tilde{u}^h d\Omega - \int_{\Omega^h} p^h \frac{\partial \tilde{u}^h}{\partial x} d\Omega$$

$$+ \frac{1}{Re} \int_{\Omega^h} \left[2 \frac{\partial u^h}{\partial x} \frac{\partial \tilde{u}^h}{\partial x} + \left(\frac{\partial u^h}{\partial y} + \frac{\partial v^h}{\partial x} \right) \frac{\partial \tilde{u}^h}{\partial y} \right] d\Omega = 0, \tag{11.160}$$

$$\int_{\Omega^h} \left(\frac{\partial v^h}{\partial t} + u^h \frac{\partial v^h}{\partial x} + v^h \frac{\partial v^h}{\partial y} \right) \tilde{v}^h d\Omega - \int_{\Omega^h} p^h \frac{\partial \tilde{v}^h}{\partial y} d\Omega$$

$$+ \frac{1}{Re} \int_{\Omega^h} \left[\left(\frac{\partial u^h}{\partial y} + \frac{\partial v^h}{\partial x} \right) \frac{\partial \tilde{v}^h}{\partial x} + 2 \frac{\partial v^h}{\partial y} \frac{\partial \tilde{v}^h}{\partial y} \right] d\Omega = 0, \tag{11.161}$$

and

$$- \int_{\Omega^h} \left(\frac{\partial u^h}{\partial x} + \frac{\partial v^h}{\partial y} \right) \tilde{p}^h \, d\Omega = 0, \tag{11.162}$$

where h indicates a given triangulation of the computational domain.

The time derivatives in (11.160) and (11.161) can be discretized by finite difference methods. We first evaluate all the terms in (11.160) to (11.162) at a given time instant $t = t_{n+1}$ (fully implicit discretization). Then the time derivative in (11.160) and (11.161) can be approximated as

$$\frac{\partial \mathbf{u}}{\partial t} (\mathbf{x}, t_{n+1}) \approx \alpha \frac{\mathbf{u} (\mathbf{x}, t_{n+1}) - \mathbf{u} (\mathbf{x}, t_n)}{\Delta t} - \beta \frac{\partial \mathbf{u}}{\partial t} (\mathbf{x}, t_n), \tag{11.163}$$

where $\Delta t = t_{n+1} - t_n$ is the time step. The approximation in (11.163) is first-order accurate in time when $\alpha = 1$ and $\beta = 0$. It can be improved to second-order accurate by selecting $\alpha = 2$ and $\beta = 1$ which is a variation of the well-known Crank-Nicolson scheme.

As (11.160) and (11.161) are nonlinear, iterative methods are often used for the solution. In Newton's method, the flow variables at the current time $t = t_{n+1}$ are often expressed as

$$\mathbf{u}^h(\mathbf{x}, t_{n+1}) = \mathbf{u}^*(\mathbf{x}, t_{n+1}) + \mathbf{u}'(\mathbf{x}, t_{n+1}),$$
$$p^h(\mathbf{x}, t_{n+1}) = p^*(\mathbf{x}, t_{n+1}) + p'(\mathbf{x}, t_{n+1}), \tag{11.164}$$

where $\mathbf{u}^*$ and p^* are the guesstimated values of velocity and pressure during the iteration. $\mathbf{u}'$ and p' are the corrections sought at each iteration.

Substituting (11.163) and (11.164) into Galerkin formulation (11.160) to (11.162), and linearizing the equations with respect to the correction variables, we have

$$\int_{\Omega^h} \left(\frac{\alpha}{\Delta t} u' + u^* \frac{\partial u'}{\partial x} + v^* \frac{\partial u'}{\partial y} + \frac{\partial u^*}{\partial x} u' + \frac{\partial u^*}{\partial y} v' \right) \tilde{u}^h \, d\Omega - \int_{\Omega^h} p' \frac{\partial \tilde{u}^h}{\partial x} \, d\Omega$$

$$+ \frac{1}{Re} \int_{\Omega^h} \left[2 \frac{\partial u'}{\partial x} \frac{\partial \tilde{u}^h}{\partial x} + \left(\frac{\partial u'}{\partial y} + \frac{\partial v'}{\partial x} \right) \frac{\partial \tilde{u}^h}{\partial y} \right] d\Omega$$

$$= - \int_{\Omega^h} \left[\frac{\alpha}{\Delta t} (u^* - u(t_n)) - \beta \frac{\partial u}{\partial t}(t_n) + u^* \frac{\partial u^*}{\partial x} + v^* \frac{\partial u^*}{\partial y} \right] \tilde{u}^h \, d\Omega$$

$$+ \int_{\Omega^h} p^* \frac{\partial \tilde{u}^h}{\partial x} \, d\Omega - \frac{1}{Re} \int_{\Omega^h} \left[2 \frac{\partial u^*}{\partial x} \frac{\partial \tilde{u}^h}{\partial x} + \left(\frac{\partial u^*}{\partial y} + \frac{\partial v^*}{\partial x} \right) \frac{\partial \tilde{u}^h}{\partial y} \right] d\Omega, \tag{11.165}$$

$$\int_{\Omega^h} \left(\frac{\alpha}{\Delta t} v' + u^* \frac{\partial v'}{\partial x} + v^* \frac{\partial v'}{\partial y} + \frac{\partial v^*}{\partial x} u' + \frac{\partial v^*}{\partial y} v' \right) \tilde{v}^h \, d\Omega - \int_{\Omega^h} p' \frac{\partial \tilde{v}^h}{\partial y} \, d\Omega$$

$$+ \frac{1}{Re} \int_{\Omega^h} \left[\left(\frac{\partial u'}{\partial y} + \frac{\partial v'}{\partial x} \right) \frac{\partial \tilde{v}^h}{\partial x} + 2 \frac{\partial v'}{\partial y} \frac{\partial \tilde{v}^h}{\partial y} \right] d\Omega$$

$$= - \int_{\Omega^h} \left[\frac{\alpha}{\Delta t} (v^* - v(t_n)) - \beta \frac{\partial v^*}{\partial t}(t_n) + u^* \frac{\partial v^*}{\partial x} + v^* \frac{\partial v^*}{\partial y} \right] \tilde{v}^h \, d\Omega$$

$$+ \int_{\Omega^h} p^* \frac{\partial \tilde{v}^h}{\partial y} \, d\Omega - \frac{1}{Re} \int_{\Omega^h} \left[\left(\frac{\partial u^*}{\partial y} + \frac{\partial v^*}{\partial x} \right) \frac{\partial \tilde{v}^h}{\partial x} + 2 \frac{\partial v^*}{\partial y} \frac{\partial \tilde{v}^h}{\partial y} \right] d\Omega, \tag{11.166}$$

and

$$- \int_{\Omega^h} \left(\frac{\partial u'}{\partial x} + \frac{\partial v'}{\partial y} \right) \tilde{p}^h \, d\Omega = \int_{\Omega^h} \left(\frac{\partial u^*}{\partial x} + \frac{\partial v^*}{\partial y} \right) \tilde{p}^h \, d\Omega. \tag{11.167}$$

As the functions in the integrals, unless specified otherwise, are all evaluated at the current time instant t_{n+1}, the temporal discretization in (11.165) and (11.166) is fully implicit and unconditionally stable. The terms on the right-hand-side of (11.165) to (11.167) represent the residuals of the corresponding equations and can be used to monitor the convergence of the nonlinear iteration.

Similar to the one-dimensional case in Section 3, the finite-dimensional discretization of the flow variables can be constructed using shape (or interpolation) functions,

$$u' = \sum_A u_A N_A^u (x, y), \quad v' = \sum_A v_A N_A^u (x, y), \quad p' = \sum_B p_B N_B^p (x, y),$$

(11.168)

where $N_A^u (x, y)$ and $N_B^p (x, y)$ are the shape functions for the velocity and the pressure, respectively. They are not necessarily the same. In order to satisfy the LBB stability condition, the shape function $N_A^u (x, y)$ in the mixed finite element formulation should be one order higher than $N_B^p (x, y)$, as discussed in Section 4. The summation over A is through all the velocity nodes, while the summation over B runs through all the pressure nodes. The variational functions may be expressed in terms of the same shape functions,

$$\tilde{u}^h = \sum_A \tilde{u}_A N_A^u (x, y), \quad \tilde{v}^h = \sum_A \tilde{v}_A N_A^u (x, y), \quad \tilde{p}^h = \sum_B \tilde{p}_B N_B^p (x, y).$$

(11.169)

Since the Galerkin formulation (11.165) to (11.167) is valid for all possible choices of the variational functions, the coefficients in (11.169) should be arbitrary. In this way, the Galerkin formulation (11.165) to (11.167) reduces to a system of algebraic equations,

$$\sum_{A'} u_{A'} \int_{\Omega^h} \left[\left(\frac{\alpha}{\Delta t} N_{A'}^u + u^* \frac{\partial N_{A'}^u}{\partial x} + v^* \frac{\partial N_{A'}^u}{\partial y} + \frac{\partial u^*}{\partial x} N_{A'}^u \right) N_A^u \right.$$

$$\left. + \frac{1}{Re} \left(2 \frac{\partial N_{A'}^u}{\partial x} \frac{\partial N_A^u}{\partial x} + \frac{\partial N_{A'}^u}{\partial y} \frac{\partial N_A^u}{\partial y} \right) \right] d\Omega$$

$$+ \sum_{A'} v_{A'} \int_{\Omega^h} \left(\frac{\partial u^*}{\partial y} N_{A'}^u N_A^u + \frac{1}{Re} \frac{\partial N_{A'}^u}{\partial x} \frac{\partial N_A^u}{\partial y} \right) d\Omega - \sum_{B'} p_{B'} \int_{\Omega^h} N_{B'}^p \frac{\partial N_A^u}{\partial x} d\Omega$$

$$= - \int_{\Omega^h} \left[\frac{\alpha}{\Delta t} \left(u^* - u(t_n) \right) - \beta \frac{\partial u}{\partial t} (t_n) + u^* \frac{\partial u^*}{\partial x} + v^* \frac{\partial u^*}{\partial y} \right] N_A^u d\Omega$$

$$+ \int_{\Omega^h} p^* \frac{\partial N_A^u}{\partial x} d\Omega - \frac{1}{Re} \int_{\Omega^h} \left[2 \frac{\partial u^*}{\partial x} \frac{\partial N_A^u}{\partial x} + \left(\frac{\partial u^*}{\partial y} + \frac{\partial v^*}{\partial x} \right) \frac{\partial N_A^u}{\partial y} \right] d\Omega,$$

(11.170)

$$\sum_{A'} v_{A'} \int_{\Omega^h} \left[\left(\frac{\alpha}{\Delta t} N_{A'}^u + u^* \frac{\partial N_{A'}^u}{\partial x} + v^* \frac{\partial N_{A'}^u}{\partial y} + \frac{\partial u^*}{\partial y} N_{A'}^u \right) N_A^u \right.$$

$$+ \frac{1}{Re} \left(\frac{\partial N_{A'}^u}{\partial x} \frac{\partial N_A^u}{\partial x} + 2 \frac{\partial N_{A'}^u}{\partial y} \frac{\partial N_A^u}{\partial y} \right) \bigg] d\Omega$$

$$+ \sum_{A'} u_{A'} \int_{\Omega^h} \left(\frac{\partial v^*}{\partial x} N_{A'}^u N_A^u + \frac{1}{Re} \frac{\partial N_{A'}^u}{\partial y} \frac{\partial N_A^u}{\partial x} \right) d\Omega - \sum_{B'} p_{B'} \int_{\Omega^h} N_{B'}^p \frac{\partial N_A^u}{\partial y} d\Omega$$

$$= - \int_{\Omega^h} \left[\frac{\alpha}{\Delta t} \left(v^* - v \left(t_n \right) \right) - \beta \frac{\partial v^*}{\partial t} \left(t_n \right) + u^* \frac{\partial v^*}{\partial x} + v^* \frac{\partial v^*}{\partial y} \right] N_A^u d\Omega$$

$$+ \int_{\Omega^h} p^* \frac{\partial N_A^u}{\partial y} d\Omega - \frac{1}{Re} \int_{\Omega^h} \left[\left(\frac{\partial u^*}{\partial y} + \frac{\partial v^*}{\partial x} \right) \frac{\partial N_A^u}{\partial x} + 2 \frac{\partial v^*}{\partial y} \frac{\partial N_A^u}{\partial y} \right] d\Omega, \tag{11.171}$$

and

$$- \sum_{A'} u_{A'} \int_{\Omega^h} \left(\frac{\partial N_{A'}^u}{\partial x} N_B^p \right) d\Omega - \sum_{A'} v_{A'} \int_{\Omega^h} \left(\frac{\partial N_{A'}^u}{\partial y} N_B^p \right) d\Omega$$

$$= \int_{\Omega^h} \left(\frac{\partial u^*}{\partial x} + \frac{\partial v^*}{\partial y} \right) N_B^p \, d\Omega, \tag{11.172}$$

for all the velocity nodes A and pressure nodes B. Equations (11.170) to (11.172) can be organized into a matrix form,

$$\begin{pmatrix} \mathbf{A}_{uu} & \mathbf{A}_{uv} & \mathbf{B}_{up} \\ \mathbf{A}_{vu} & \mathbf{A}_{vv} & \mathbf{B}_{vp} \\ \mathbf{B}_{up}^T & \mathbf{B}_{vp}^T & 0 \end{pmatrix} \begin{pmatrix} \mathbf{u} \\ \mathbf{v} \\ \mathbf{p} \end{pmatrix} = \begin{pmatrix} \mathbf{f}_u \\ \mathbf{f}_v \\ \mathbf{f}_p \end{pmatrix}, \tag{11.173}$$

where

$$\mathbf{A}_{uu} = \left[A_{AA'}^{uu} \right], \quad \mathbf{A}_{uv} = \left[A_{AA'}^{uv} \right], \quad \mathbf{B}_{up} = \left[B_{AB'}^{up} \right],$$
$$\mathbf{A}_{vu} = \left[A_{AA'}^{vu} \right], \quad \mathbf{A}_{vv} = \left[A_{AA'}^{vv} \right], \quad \mathbf{B}_{vp} = \left[B_{AB'}^{vp} \right], \tag{11.174}$$
$$\mathbf{u} = \{ u_{A'} \}, \quad \mathbf{v} = \{ v_{A'} \}, \quad \mathbf{p} = \{ p_{B'} \},$$
$$\mathbf{f}_u = \left\{ f_A^u \right\}, \quad \mathbf{f}_v = \left\{ f_A^v \right\}, \quad \mathbf{f}_p = \left\{ f_B^p \right\},$$

and

$$A_{AA'}^{uu} = \int_{\Omega^h} \left[\left(\frac{\alpha}{\Delta t} N_{A'}^u + u^* \frac{\partial N_{A'}^u}{\partial x} + v^* \frac{\partial N_{A'}^u}{\partial y} + \frac{\partial u^*}{\partial x} N_{A'}^u \right) N_A^u \right.$$

$$+ \frac{1}{Re} \left(2 \frac{\partial N_{A'}^u}{\partial x} \frac{\partial N_A^u}{\partial x} + \frac{\partial N_{A'}^u}{\partial y} \frac{\partial N_A^u}{\partial y} \right) \bigg] d\Omega, \tag{11.175}$$

$$A_{AA'}^{uv} = \int_{\Omega^h} \left(\frac{\partial u^*}{\partial y} N_{A'}^u N_A^u + \frac{1}{Re} \frac{\partial N_{A'}^u}{\partial x} \frac{\partial N_A^u}{\partial y} \right) d\Omega, \tag{11.176}$$

$$A_{AA'}^{vu} = \int_{\Omega^h} \left(\frac{\partial v^*}{\partial x} N_{A'}^u N_A^u + \frac{1}{Re} \frac{\partial N_{A'}^u}{\partial y} \frac{\partial N_A^u}{\partial x} \right) d\Omega, \tag{11.177}$$

$$A_{AA'}^{vv} = \int_{\Omega^h} \left[\left(\frac{\alpha}{\Delta t} N_{A'}^u + u^* \frac{\partial N_{A'}^u}{\partial x} + v^* \frac{\partial N_{A'}^u}{\partial y} + \frac{\partial v^*}{\partial y} N_{A'}^u \right) N_A^u \right.$$
$$\left. + \frac{1}{Re} \left(\frac{\partial N_{A'}^u}{\partial x} \frac{\partial N_A^u}{\partial x} + 2 \frac{\partial N_{A'}^u}{\partial y} \frac{\partial N_A^u}{\partial y} \right) \right] d\Omega, \tag{11.178}$$

$$B_{AB'}^{up} = -\int_{\Omega^h} N_{B'}^p \frac{\partial N_A^u}{\partial x} d\Omega, \tag{11.179}$$

$$B_{AB'}^{vp} = -\int_{\Omega^h} N_{B'}^p \frac{\partial N_A^u}{\partial y} d\Omega, \tag{11.180}$$

$$f_A^u = -\int_{\Omega^h} \left[\frac{\alpha}{\Delta t} \left(u^* - u\left(t_n\right) \right) - \beta \frac{\partial u}{\partial t}\left(t_n\right) + u^* \frac{\partial u^*}{\partial x} + v^* \frac{\partial u^*}{\partial y} \right] N_A^u d\Omega$$
$$+ \int_{\Omega^h} p^* \frac{\partial N_A^u}{\partial x} d\Omega - \frac{1}{Re} \int_{\Omega^h} \left[2 \frac{\partial u^*}{\partial x} \frac{\partial N_A^u}{\partial x} + \left(\frac{\partial u^*}{\partial y} + \frac{\partial v^*}{\partial x} \right) \frac{\partial N_A^u}{\partial y} \right] d\Omega, \tag{11.181}$$

$$f_A^v = -\int_{\Omega^h} \left[\frac{\alpha}{\Delta t} \left(v^* - v\left(t_n\right) \right) - \beta \frac{\partial v^*}{\partial t}\left(t_n\right) + u^* \frac{\partial v^*}{\partial x} + v^* \frac{\partial v^*}{\partial y} \right] N_A^u d\Omega$$
$$+ \int_{\Omega^h} p^* \frac{\partial N_A^u}{\partial y} d\Omega - \frac{1}{Re} \int_{\Omega^h} \left[\left(\frac{\partial u^*}{\partial y} + \frac{\partial v^*}{\partial x} \right) \frac{\partial N_A^u}{\partial x} + 2 \frac{\partial v^*}{\partial y} \frac{\partial N_A^u}{\partial y} \right] d\Omega, \tag{11.182}$$

$$f_B^p = \int_{\Omega^h} \left(\frac{\partial u^*}{\partial x} + \frac{\partial v^*}{\partial y} \right) N_B^p \, d\Omega. \tag{11.183}$$

The practical evaluation of the integrals in (11.175) to (11.183) is done element-wise. We need to construct the shape functions locally and transform these global integrals into local integrals over each element.

In the finite element method, the global shape functions have very compact support. They are zero everywhere except in the neighborhood of the corresponding grid point in the mesh. It is convenient to cast the global formulation using the element point of view (Section 3). In this element view, the local shape functions are defined inside each element. The global shape functions are the assembly of the relevant local ones. For example, the global shape function corresponding to the grid point A in the finite element mesh consists of the local shape functions of all the elements that share this grid point. An element in the physical space can be mapped into a standard element, as shown in Figure 11.17 and the local shape functions can be defined on this standard element. The mapping is given by

$$x(\xi, \eta) = \sum_{a=1}^{6} x_a^e \phi_a(\xi, \eta) \quad \text{and} \quad y(\xi, \eta) = \sum_{a=1}^{6} y_a^e \phi_a(\xi, \eta), \quad (11.184)$$

where $\left(x_a^e, y_a^e\right)$ are the coordinates of the nodes in the element e. The local shape functions are ϕ_a. For a quadratic triangular element they are defined as

$$\phi_1 = \zeta\,(2\zeta - 1), \ \phi_2 = \xi\,(2\xi - 1), \ \phi_3 = \eta\,(2\eta - 1), \ \phi_4 = 4\xi\zeta, \ \phi_5 = 4\xi\eta,$$
$$\phi_6 = 4\eta\zeta, \quad (11.185)$$

where $\zeta = 1 - \xi - \eta$. As shown in Figure 11.17 the mapping (11.184) is able to handle curved triangles. The variation of the flow variables within this element can also be expressed in terms of their values at the nodes of the element and the local shape functions,

$$u' = \sum_{a=1}^{6} u_a^e \phi_a\,(\xi, \eta), \quad v' = \sum_{a=1}^{6} v_a^e \phi_a\,(\xi, \eta), \quad p' = \sum_{b=1}^{3} p_b^e \psi_b\,(\xi, \eta). \quad (11.186)$$

Here the shape functions for velocities are quadratic and the same as the coordinates. The shape functions for the pressure are chosen to be linear, thus one order less than those for the velocities. They are given by,

$$\psi_1 = \zeta, \ \psi_2 = \xi, \ \psi_3 = \eta. \quad (11.187)$$

Furthermore, the integration over the global computational domain can be written as the summation of the integrations over all the elements in the domain. As most of

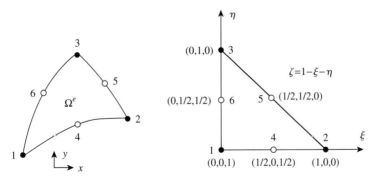

Figure 11.17 A quadratic triangular finite element mapping into the standard element.

these integrations will be zero, the non-zero ones are grouped as element matrices and vectors,

$$
\begin{aligned}
\mathbf{A}_{uu}^e &= \left[A_{aa'}^{euu} \right], \quad \mathbf{A}_{uv}^e = \left[A_{aa'}^{euv} \right], \quad \mathbf{B}_{up}^e = \left[B_{ab'}^{eup} \right], \\
\mathbf{A}_{vu}^e &= \left[A_{aa'}^{evu} \right], \quad \mathbf{A}_{vv}^e = \left[A_{aa'}^{evv} \right], \quad \mathbf{B}_{vp}^e = \left[B_{ab'}^{evp} \right], \\
\mathbf{f}_u^e &= \left\{ f_a^{eu} \right\}, \quad \mathbf{f}_v^e = \left\{ f_a^{ev} \right\}, \quad \mathbf{f}_p^e = \left\{ f_b^{ep} \right\},
\end{aligned}
\tag{11.188}
$$

where

$$
\begin{aligned}
A_{aa'}^{euu} = \int_{\Omega^e} &\left[\left(\frac{\alpha}{\Delta t} \phi_{a'} + u^* \frac{\partial \phi_{a'}}{\partial x} + v^* \frac{\partial \phi_{a'}}{\partial y} + \frac{\partial u^*}{\partial x} \phi_{a'} \right) \phi_a \right. \\
&\left. + \frac{1}{\text{Re}} \left(2 \frac{\partial \phi_{a'}}{\partial x} \frac{\partial \phi_a}{\partial x} + \frac{\partial \phi_{a'}}{\partial y} \frac{\partial \phi_a}{\partial y} \right) \right] d\Omega,
\end{aligned}
\tag{11.189}
$$

$$
A_{aa'}^{euv} = \int_{\Omega^e} \left(\frac{\partial u^*}{\partial y} \phi_{a'} \phi_a + \frac{1}{\text{Re}} \frac{\partial \phi_{a'}}{\partial x} \frac{\partial \phi_a}{\partial y} \right) d\Omega,
\tag{11.190}
$$

$$
A_{aa'}^{evu} = \int_{\Omega^e} \left(\frac{\partial v^*}{\partial x} \phi_{a'} \phi_a + \frac{1}{\text{Re}} \frac{\partial \phi_{a'}}{\partial y} \frac{\partial \phi_a}{\partial x} \right) d\Omega,
\tag{11.191}
$$

$$
\begin{aligned}
A_{aa'}^{evv} = \int_{\Omega^e} &\left[\left(\frac{\alpha}{\Delta t} \phi_{a'} + u^* \frac{\partial \phi_{a'}}{\partial x} + v^* \frac{\partial \phi_{a'}}{\partial y} + \frac{\partial v^*}{\partial y} \phi_{a'} \right) \phi_a \right. \\
&\left. + \frac{1}{Re} \left(\frac{\partial \phi_{a'}}{\partial x} \frac{\partial \phi_a}{\partial x} + 2 \frac{\partial \phi_{a'}}{\partial y} \frac{\partial \phi_a}{\partial y} \right) \right] d\Omega,
\end{aligned}
\tag{11.192}
$$

$$
B_{ab'}^{eup} = - \int_{\Omega^e} \psi_{b'} \frac{\partial \phi_a}{\partial x} d\Omega,
\tag{11.193}
$$

$$
B_{ab'}^{evp} = - \int_{\Omega^e} \psi_{b'} \frac{\partial \phi_a}{\partial y} d\Omega,
\tag{11.194}
$$

$$
\begin{aligned}
f_a^{eu} = &- \int_{\Omega^e} \left[\frac{\alpha}{\Delta t} \left(u^* - u\left(t_n\right)\right) - \beta \frac{\partial u}{\partial t} \left(t_n\right) + u^* \frac{\partial u^*}{\partial x} + v^* \frac{\partial u^*}{\partial y} \right] \phi_a d\Omega \\
&+ \int_{\Omega^e} p^* \frac{\partial \phi_a}{\partial x} d\Omega - \frac{1}{\text{Re}} \int_{\Omega^e} \left[2 \frac{\partial u^*}{\partial x} \frac{\partial \phi_a}{\partial x} + \left(\frac{\partial u^*}{\partial y} + \frac{\partial v^*}{\partial x} \right) \frac{\partial \phi_a}{\partial y} \right] d\Omega,
\end{aligned}
\tag{11.195}
$$

$$
\begin{aligned}
f_a^{ev} = &- \int_{\Omega^e} \left[\frac{\alpha}{\Delta t} \left(v^* - v\left(t_n\right)\right) - \beta \frac{\partial v^*}{\partial t} \left(t_n\right) + u^* \frac{\partial v^*}{\partial x} + v^* \frac{\partial v^*}{\partial y} \right] \phi_a d\Omega \\
&+ \int_{\Omega^e} p^* \frac{\partial \phi_a}{\partial y} d\Omega - \frac{1}{\text{Re}} \int_{\Omega^e} \left[\left(\frac{\partial u^*}{\partial y} + \frac{\partial v^*}{\partial x} \right) \frac{\partial \phi_a}{\partial x} + 2 \frac{\partial v^*}{\partial y} \frac{\partial \phi_a}{\partial y} \right] d\Omega,
\end{aligned}
\tag{11.196}
$$

$$f_b^{ep} = \int_{\Omega^e} \left(\frac{\partial u^*}{\partial x} + \frac{\partial v^*}{\partial y} \right) \psi_b \, d\Omega. \tag{11.197}$$

The indices a and a' run from 1 to 6, and b and b' run from 1 to 3.

The integrals in the above expressions can be evaluated by numerical integration rules,

$$\int_{\Omega^e} f(x, y) \, d\Omega = \int_0^1 \int_0^{1-\eta} f(\xi, \eta) J(\xi, \eta) \, d\xi \, d\eta = \frac{1}{2} \sum_{l=1}^{N_{int}} f(\xi_l, \eta_l) J(\xi_l, \eta_l) W_l, \tag{11.198}$$

where the Jacobian of the mapping (11.184) is given by $J = x_\xi y_\eta - x_\eta y_\xi$. Here N_{int} is the number of numerical integration points and W_l is the weight of the lth integration point. For this example, a seven-point integration formula with degree of precision of 5 (see Hughes, 1987) was used.

The global matrices and vectors in (11.173) are the summations of the element matrices and vectors in (11.188) over all the elements. In the process of summation (assembly), a mapping of the local nodes in each element to the global node numbers is needed. This information is commonly available for any finite element mesh.

Once the matrix equation (11.173) is generated, we may impose the essential boundary conditions for the velocities. One simple method is to use the equation of the boundary condition to replace the corresponding equation in the original matrix or one can multiply a large constant by the equation of the boundary condition and add this equation to the original system of equations in order to preserve the structure of the matrix. The resulting matrix equation may be solved using common direct or iterative solvers for a linear algebraic system of equations.

Figures 11.18 and 11.19 display the streamlines and vorticity lines around the cylinder at three Reynolds numbers $Re = 1, 10$, and 40. For these Reynolds numbers,

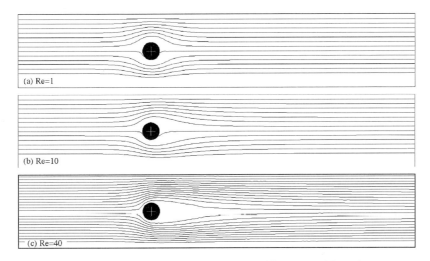

(a) Re=1

(b) Re=10

(c) Re=40

Figure 11.18 Streamlines for flow around a cylinder at three different Reynolds numbers.

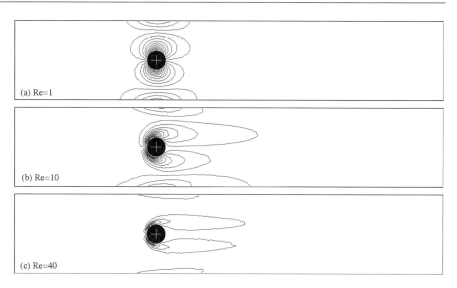

Figure 11.19   Vorticity lines for flow around a cylinder at three different Reynolds numbers.

the flow is steady and should be symmetric above and below the cylinder. However, due to the imperfection in the mesh used for the calculation and as shown in Figure 11.16, the calculated flow field is not perfectly symmetric. From Figure 11.18 we observe the increase in the size of the wake behind the cylinder as the Reynolds number increases. In Figure 11.19, we see the effects of the Reynolds number in the vorticity build up in front of the cylinder, and in the convection of the vorticity by the flow.

We next compute the case with Reynolds number Re $= 100$. In this case, the flow is expected to be unsteady. Periodic vortex shedding occurs. In order to capture the details of the flow, we used a finer mesh than the one shown in Figure 11.16. The finer mesh has 9222 elements, 18816 velocity nodes and 4797 pressure nodes. In this calculation, the flow starts from rest. Initially, the flow is symmetric, and the wake behind the cylinder grows bigger and stronger. Then, the wake becomes unstable, undergoes a supercritical Hopf bifurcation, and sheds periodically away from the cylinder. The periodic vortex shedding forms the well-known von Karman vortex street. The vorticity lines are presented in Figure 11.20 for a complete cycle of vortex shedding.

For this case with Re $= 100$, we plot in Figure 11.21 the history of the forces and torque acting on the cylinder. The oscillations shown in the lift and torque plots are typical for the supercritical Hopf bifurcation. The nonzero mean value of the torque shown in Figure 11.21c is due to the asymmetry in the finite element mesh. It is clear that the flow becomes fully periodic at the times shown in Figure 11.20. The period of the oscillation is measured as $\tau = 0.0475s$ or $\bar{\tau} = 4.75$ in the non-dimensional units. This period corresponds to a nondimensional Strouhal number $S = nd/U = 0.21$, where n is the frequency of the shedding. In the literature, the value of the Strouhal number for an unbounded uniform flow around a cylinder is found to be around 0.167

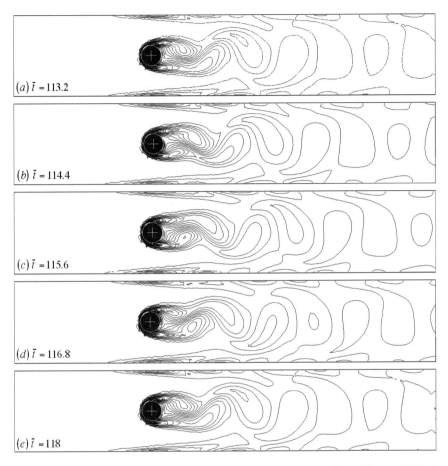

$(a)\ \bar{t} = 113.2$

$(b)\ \bar{t} = 114.4$

$(c)\ \bar{t} = 115.6$

$(d)\ \bar{t} = 116.8$

$(e)\ \bar{t} = 118$

Figure 11.20 Vorticity lines for flow around a cylinder at Reynolds number Re = 100. $\bar{t} = tU/d$ is the dimensionless time.

at Re = 100 (e.g., see Wen and Lin, 2001). The difference could be caused by the geometry in which the cylinder is confined in a channel.

6. Concluding Remarks

It should be strongly emphasized that CFD is merely a tool for analyzing fluid-flow problems. If it is used correctly, it would provide useful information cheaply and quickly. However, it could easily be misused or even abused. In today's computer age, people have a tendency to trust the output from a computer, especially when they do not understand what is behind the computer. One certainly should be aware of the assumptions used in producing the results from a CFD model.

As we have previously discussed, CFD is never exact. There are uncertainties involved in CFD predictions. However, one is able to gain more confidence in CFD predictions by following a few steps. Tests on some benchmark problems with known solutions are often encouraged. A mesh refinement test is normally a must in order

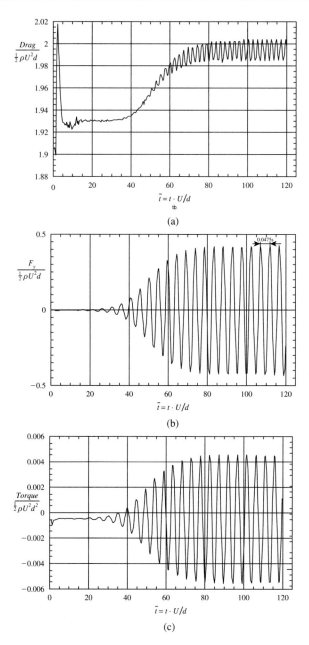

Figure 11.21   History of forces and torque acting on the cylinder at Re $= 100$: (a) drag coefficient; (b) lift coefficient; and (c) coefficient for the torque.

to be sure that the numerical solution converges to something meaningful. A similar test with the time step for unsteady flow problems is often desired. If the boundary locations and conditions are in doubt, their effects on the CFD predictions should be minimized. Furthermore, the sensitivity of the CFD predictions to some key parameters in the problem should be investigated for practical design problems.

In this chapter, we have discussed the basics of the finite difference and finite element methods and their applications in CFD. There are other kinds of numerical methods, for example, the spectral method and the spectral element method, which are often used in CFD. They share the common approach that discretizes the Navier-Stokes equations into a system of algebraic equations. However, a class of new numerical techniques including lattice gas cellular automata, lattice Boltzmann method, and dissipative particle dynamics do not start from the continuum Navier-Stokes equations. Unlike the conventional methods discussed in this chapter, they are based on simplified kinetic models that incorporate the essential physics of the microscopic or mesoscopic processes so that the macroscopic-averaged properties obey the desired macroscopic Navier-Stokes equations.

Exercises

1. Show that the stability condition for the explicit scheme (11.10) is the condition (11.26).

2. For the heat conduction equation $\partial T/\partial t - D\left(\partial^2 T/\partial x^2\right) = 0$, one of the discretized forms is

$$-sT_{j+1}^{n+1} + (1+2s)\,T_j^{n+1} - sT_{j-1}^{n+1} = T_j^n$$

where $s = D\Delta t/\Delta x^2$. Show that this implicit algorithm is always stable.

3. An insulated rod initially has a temperature of $T(x,0) = 0°\,C$ $(0 \leqslant x \leqslant 1)$. At $t = 0$ hot reservoirs $(T = 100°C)$ are brought into contact with the two ends, $A\,(x = 0)$ and $B\,(x = 1)$: $T(0,t) = T(1,t) = 100°C$. Numerically find the temperature $T(x,t)$ of any point in the rod. The governing equation of the problem is the heat conduction equation $\partial T/\partial t - D\left(\partial^2 T/\partial x^2\right) = 0$. The exact solution to this problem is

$$T^*\left(x_j, t_n\right) = 100 - \sum_{m=1}^{M} \frac{400}{(2m-1)\,\pi} \sin\left[(2m-1)\,\pi x_j\right] \exp\left[-D\,(2m-1)^2\,\pi^2 t_n\right]$$

(11.199)

where M is the number of terms used in the approximation.

(a). Try to solve the problem with the explicit forward time, centered space (FTCS) scheme. Use the parameter $s = D\Delta t/\Delta x^2 = 0.5$ and 0.6 to test the stability of the scheme.

(b). Solve the problem with a stable explicit or implicit scheme. Test the rate of convergence numerically using the error at $x = 0.5$.

4. Derive the weak form, Galerkin form, and the matrix form of the following strong problem:

Given functions $D(x)$, $f(x)$, and constants g, h, find $u(x)$ such that
$$[D(x)u_{,x}]_{,x} + f(x) = 0 \quad \text{on } \Omega = (0, 1),$$
with $u(0) = g$ and $-u_{,x}(1) = h$.

Write a computer program solving this problem using piecewise-linear shape functions. You may set $D = 1, g = 1, h = 1$ and $h = 1$. Check your numerical result with the exact solution.

5. Solve numerically the steady convective transport equation,

$$u \frac{\partial T}{\partial x} = D \frac{\partial^2 T}{\partial x^2}, \quad \text{for } 0 \leqslant x \leqslant 1,$$

with two boundary conditions $T(0) = 0$ and $T(1) = 1$, where u and D are two constants,
(a) using the centered finite difference scheme in equation (11.91), and compare with the exact solution; and
(b) using the upwind scheme (11.93), and compare with the exact solution.

6. Code the explicit MacCormack scheme with the FF/BB arrangement for the driven cavity flow problem as described in Section 5. Compute the flow field at Re = 100 and 400, and explore effects of Mach number and the stability condition (11.110).

Literature Cited

Brooks, A. N. and T. J. R. Hughes (1982). "Streamline-upwinding/Petrov-Galerkin formulation for convection dominated flows with particular emphasis on incompressible Navier-Stokes equation." *Comput. Methods Appl. Mech. Engrg.* **30**: 199–259.

Chorin, A. J. (1967). "A numerical method for solving incompressible viscous flow problems." *J. Comput. Phys.* **2**: 12–26.

Chorin, A. J. (1968). "Numerical solution of the Navier-Stokes equations." *Math. Comput.* **22**: 745–762.

Dennis, S. C. R. and G. Z. Chang (1970). "Numerical solutions for steady flow past a circular cylinder at Reynolds numbers up to 100." *J. Fluid Mech.* **42**: 471–489.

Fletcher, C. A. J. (1988). *Computational Techniques for Fluid Dynamics, I–Fundamental and General Techniques, and II–Special Techniques for Different Flow Categories*, New York: Springer-Verlag.

Franca, L. P., S. L. Frey and T. J. R. Hughes (1992). "Stabilized finite element methods: I. Application to the advective-diffusive model." *Comput. Methods Appl. Mech. Engrg.* **95**: 253–276.

Franca, L. P. and S. L. Frey (1992). "Stabilized finite element methods: II. The incompressible Navier-Stokes equations,"*Comput. Methods Appl. Mech. Engrg.* **99**: 209–233.

Ghia, U., K. N. Ghia, C. T. Shin (1982) îHigh-Re solutions for incompressible flow using the Navier-Stokes equations and a multigrid method.î *J. Comput. Phys.* **48**: 387–411.

Glowinski, R. (1991). "Finite element methods for the numerical simulation of incompressible viscous flow, introduction to the control of the Navier-Stokes equations," in *Lectures in Applied Mathematics* Vol.**28**: 219–301. Providence, R.I.: American Mathematical Society.

Gresho, P. M. (1991). "Incompressible fluid dynamics: Some fundamental formulation issues," *Annu. Rev. Fluid Mech.* **23**: 413–453.

Harlow, F. H. and J. E. Welch (1965). "Numerical calculation of time-dependent viscous incompressible flow of fluid with free surface." *Phys. Fluids* **8**: 2182–2189.

Hou, S., Q. Zou, S. Chen, G. D. Doolen and A. C. Cogley (1995). ìSimulation of cavity flow by the lattice Boltzmann method.î *J. Comp. Phys.* **118**: 329–347.

Hughes, T. J. R. (1987). *The Finite Element Method, Linear Static and Dynamic Finite Element Analysis*, Englewood Cliffs, NJ: Prentice-Hall.

MacCormack, R. W. (1969). "The effect of viscosity in hypervelocity impact cratering." *AIAA Paper* 69–354, Cincinnati, Ohio.

Marchuk, G. I. (1975). *Methods of Numerical Mathematics,* New York: Springer-Verlag.

Noye, J (1983). Chapter 2 in *Numerical Solution of Differential Equations,* J. Noye, ed., Amsterdam: North-Holland.

Oden, J. T. and G. F. Carey (1984). *Finite Elements: Mathematical Aspects, Vol. IV*, Englewood Cliffs, N.J.: Prentice-Hall.

Perrin, A. and H.H. Hu (2006). "An explicit finite-difference scheme for simulation of moving particles," *J. Comput. Phys.* **212**: 166–187.

Peyret, R. and T. D. Taylor (1983). *Computational Methods for Fluid Flow*, New York: Springer-Verlag.

Richtmyer, R. D. and K. W. Morton (1967). *Difference Methods for Initial-Value Problems*, New York: Interscience.

Saad, Y. (1996). *Iterative Methods for Sparse Linear Systems*, Boston: PWS Publishing Company.

Sucker, D. and H. Brauer (1975). "Fluiddynamik bei der angeströmten Zylindern." Wärme-Stoffübertrag. **8**: 149.

Takami, H. and H. B. Keller (1969). "Steady two-dimensional viscous flow of an incompressible fluid past a circular cylinder." *Phys. Fluids* **12**: Suppl.II, II-51-II-56.

Tannehill, J. C., D. A. Anderson, R. H. Pletcher (1997), *Computational Fluid Mechanics and Heat Transfer*, Washington, DC, Taylor & Francis.

Temam, R. (1969). "Sur l'approximation des équations de Navier-Stokes par la méthode de pas fraction-aires." *Archiv. Ration. Mech. Anal.* **33**: 377–385.

Tezduyar, T. E. (1992). "Stabilized Finite Element Formulations for Incompressible Flow Computations," *in Advances in Applied Mechanics* , J.W. Hutchinson and T.Y. Wu eds., Vol. **28**: 1–44. New York: Academic Press.

Yanenko, N. N. (1971). *The Method of Fractional Steps,* New York: Springer-Verlag.

Wen, C. Y. and C. Y. Lin (2001). "Two dimensional vortex shedding of a circular cylinder," *Phys. Fluids* **13**: 557–560.

Chapter 12

Instability

1. Introduction

A phenomenon that may satisfy all conservation laws of nature exactly, may still be unobservable. For the phenomenon to occur in nature, it has to satisfy one more condition, namely, it must be stable to small disturbances. In other words, infinitesimal disturbances, which are invariably present in any real system, must not amplify

©2010 Elsevier Inc. All rights reserved.
DOI: 10.1016/B978-0-12-381399-2.50012-5

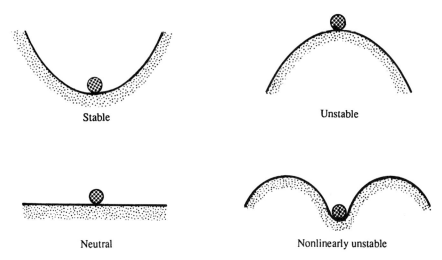

Stable   Unstable

Neutral   Nonlinearly unstable

Figure 12.1 Stable and unstable systems.

spontaneously. A perfectly vertical rod satisfies all equations of motion, but it does
not occur in nature. A smooth ball resting on the surface of a hemisphere is stable (and
therefore observable) if the surface is concave upwards, but unstable to small displacements if the surface is convex upwards (Figure 12.1). In fluid flows, smooth laminar
flows are stable to small disturbances only when certain conditions are satisfied. For
example, in flows of homogeneous viscous fluids in a channel, the Reynolds number
must be less than some critical value, and in a stratified shear flow, the Richardson
number must be larger than a critical value. When these conditions are not satisfied,
infinitesimal disturbances grow spontaneously. Sometimes the disturbances can grow
to a finite amplitude and reach equilibrium, resulting in a new steady state. The new
state may then become unstable to other types of disturbances, and may grow to yet
another steady state, and so on. Finally, the flow becomes a superposition of various
large disturbances of random phases, and reaches a chaotic condition that is commonly described as "turbulent." Finite amplitude effects, including the development
of chaotic solutions, will be examined briefly later in the chapter.

The primary objective of this chapter, however, is the examination of stability
of certain fluid flows with respect to infinitesimal disturbances. We shall introduce
perturbations on a particular flow, and determine whether the equations of motion
demand that the perturbations should grow or decay with time. In this analysis the
problem is linearized by neglecting terms quadratic in the perturbation variables
and their derivatives. This linear method of analysis, therefore, only examines the
initial behavior of the disturbances. The loss of stability does not in itself constitute
a transition to turbulence, and the linear theory can at best describe only the very
beginning of the process of transition to turbulence. Moreover, a real flow may be
stable to infinitesimal disturbances (linearly stable), but still can be unstable to sufficiently large disturbances (nonlinearly unstable); this is schematically represented in
Figure 12.1. These limitations of the linear stability analysis should be kept in mind.

Nevertheless, the successes of the linear stability theory have been considerable. For example, there is almost an exact agreement between experiments and theoretical prediction of the onset of thermal convection in a layer of fluid, and of the onset of the Tollmien–Schlichting waves in a viscous boundary layer. Taylor's experimental verification of his own theoretical prediction of the onset of secondary flow in a rotating Couette flow is so striking that it has led people to suggest that Taylor's work is the first rigorous confirmation of Navier–Stokes equations, on which the calculations are based.

For our discussion we shall choose problems that are of importance in geophysical as well as engineering applications. None of the problems discussed in this chapter, however, contains Coriolis forces; the problem of "baroclinic instability," which does contain the Coriolis frequency, is discussed in Chapter 14. Some examples will also be chosen to illustrate the basic physics rather than any potential application. Further details of these and other problems can be found in the books by Chandrasekhar (1961, 1981) and Drazin and Reid (1981). The review article by Bayly, Orszag, and Herbert (1988) is recommended for its insightful discussions after the reader has read this chapter.

2. *Method of Normal Modes*

The method of linear stability analysis consists of introducing sinusoidal disturbances on a *basic state* (also called background or initial state), which is the flow whose stability is being investigated. For example, the velocity field of a basic state involving a flow parallel to the x-axis, and varying along the y-axis, is $\mathbf{U} = [U(y), 0, 0]$. On this background flow we superpose a disturbance of the form

$$u(\mathbf{x}, t) = \hat{u}(y)\, e^{ikx+imz+\sigma t}, \tag{12.1}$$

where $\hat{u}(y)$ is a complex amplitude; it is understood that the real part of the right-hand side is taken to obtain physical quantities. (The complex form of notation is explained in Chapter 7, Section 15.) The reason solutions exponential in (x, z, t) are allowed in equation (12.1) is that, as we shall see, the coefficients of the differential equation governing the perturbation in this flow are independent of (x, z, t). The flow field is assumed to be unbounded in the x and z directions, hence the wavenumber components k and m can only be real in order that the dependent variables remain bounded as x, $z \to \infty$; $\sigma = \sigma_r + i\sigma_i$ is regarded as complex.

The behavior of the system for *all* possible $\mathbf{K} = [k, 0, m]$ is examined in the analysis. If σ_r is positive for *any* value of the wavenumber, the system is unstable to disturbances of this wavenumber. If no such unstable state can be found, the system is stable. We say that

$$\sigma_r < 0: \quad \text{stable,}$$

$$\sigma_r > 0: \quad \text{unstable,}$$

$$\sigma_r = 0: \quad \text{neutrally stable.}$$

The method of analysis involving the examination of Fourier components such as equation (12.1) is called the *normal mode method*. An arbitrary disturbance can be decomposed into a complete set of normal modes. In this method the stability of each of the modes is examined separately, as the linearity of the problem implies that the various modes do not interact. The method leads to an eigenvalue problem, as we shall see.

The boundary between stability and instability is called the *marginal state*, for which $\sigma_r = 0$. There can be two types of marginal states, depending on whether σ_i is also zero or nonzero in this state. If $\sigma_i = 0$ in the marginal state, then equation (12.1) shows that the marginal state is characterized by a *stationary pattern* of motion; we shall see later that the instability here appears in the form of *cellular convection* or *secondary flow* (see Figure 12.12 later). For such marginal states one commonly says that the *principle of exchange of stabilities* is valid. (This expression was introduced by Poincaré and Jeffreys, but its significance or usefulness is not entirely clear.)

If, on the other hand, $\sigma_i \neq 0$ in the marginal state, then the instability sets in as oscillations of growing amplitude. Following Eddington, such a mode of instability is frequently called "overstability" because the restoring forces are so strong that the system overshoots its corresponding position on the other side of equilibrium. We prefer to avoid this term and call it the *oscillatory mode* of instability.

The difference between the *neutral state* and the *marginal state* should be noted as both have $\sigma_r = 0$. However, the marginal state has the additional constraint that it lies at the *borderline* between stable and unstable solutions. That is, a slight change of parameters (such as the Reynolds number) from the marginal state can take the system into an unstable regime where $\sigma_r > 0$. In many cases we shall find the stability criterion by simply setting $\sigma_r = 0$, without formally demonstrating that it is indeed at the borderline of unstable and stable states.

3. Thermal Instability: The Bénard Problem

A layer of fluid heated from below is "top heavy," but does not necessarily undergo a convective motion. This is because the viscosity and thermal diffusivity of the fluid try to prevent the appearance of convective motion, and only for large enough temperature gradients is the layer unstable. In this section we shall determine the condition necessary for the onset of thermal instability in a layer of fluid.

The first intensive experiments on instability caused by heating a layer of fluid were conducted by Bénard in 1900. Bénard experimented on only very thin layers (a millimeter or less) that had a free surface and observed beautiful hexagonal cells when the convection developed. Stimulated by these experiments, Rayleigh in 1916 derived the theoretical requirement for the development of convective motion in a layer of fluid with two free surfaces. He showed that the instability would occur when the adverse temperature gradient was large enough to make the ratio

$$\boxed{\mathrm{Ra} = \frac{g\alpha\Gamma d^4}{\kappa\nu},}$$

(12.2)

exceed a certain critical value. Here, g is the acceleration due to gravity, α is the coefficient of thermal expansion, $\Gamma = -d\bar{T}/dz$ is the vertical temperature gradient of the background state, d is the depth of the layer, κ is the thermal diffusivity, and ν is the kinematic viscosity. The parameter Ra is called the *Rayleigh number*, and we shall see shortly that it represents the ratio of the destabilizing effect of buoyancy force to the stabilizing effect of viscous force. It has been recognized only recently that most of the *motions observed by Bénard were instabilities driven by the variation of surface tension with temperature and not the thermal instability due to a top-heavy density gradient* (Drazin and Reid 1981, p. 34). The importance of instabilities driven by surface tension decreases as the layer becomes thicker. Later experiments on thermal convection in thicker layers (with or without a free surface) have obtained convective cells of many forms, not just hexagonal. Nevertheless, the phenomenon of thermal convection in a layer of fluid is still commonly called the *Bénard convection*.

Rayleigh's solution of the thermal convection problem is considered a major triumph of the linear stability theory. The concept of critical Rayleigh number finds application in such geophysical problems as solar convection, cloud formation in the atmosphere, and the motion of the earth's core.

Formulation of the Problem

Consider a layer confined between two isothermal walls, in which the lower wall is maintained at a higher temperature. We start with the Boussinesq set

$$\frac{\partial \tilde{u}_i}{\partial t} + \tilde{u}_j \frac{\partial \tilde{u}_i}{\partial x_j} = -\frac{1}{\rho_0} \frac{\partial \tilde{p}}{\partial x_i} - g[1 - \alpha(\tilde{T} - T_0)]\delta_{i3} + \nu\nabla^2 \tilde{u}_i,$$

$$\frac{\partial \tilde{T}}{\partial t} + \tilde{u}_j \frac{\partial \tilde{T}}{\partial x_j} = \kappa\nabla^2 \tilde{T},$$

(12.3)

along with the continuity equation $\partial \tilde{u}_i / \partial x_i = 0$. Here, the density is given by the equation of state $\tilde{\rho} = \rho_0[1 - \alpha(\tilde{T} - T_0)]$, with ρ_0 representing the reference density at the reference temperature T_0. The total flow variables (background plus perturbation) are represented by a tilde ($\tilde{\ }$), a convention that will also be used in the following chapter. We decompose the motion into a background state of no motion, plus perturbations:

$$\tilde{u}_i = 0 + u_i(\mathbf{x}, t),$$

$$\tilde{T} = \bar{T}(z) + T'(\mathbf{x}, t),$$

(12.4)

$$\tilde{p} = P(z) + p(\mathbf{x}, t),$$

where the z-axis is taken vertically upward. The variables in the basic state are represented by uppercase letters except for the temperature, for which the symbol is $\bar{T}$. The basic state satisfies

$$0 = -\frac{1}{\rho_0} \frac{\partial P}{\partial x_i} - g[1 - \alpha(\bar{T} - T_0)]\delta_{i3},$$

$$0 = \kappa \frac{d^2\bar{T}}{dz^2}.$$

(12.5)

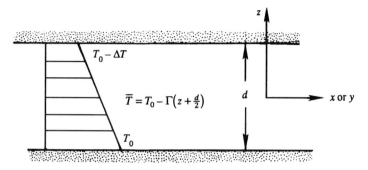

Figure 12.2 Definition sketch for the Bénard problem.

The preceding heat equation gives the linear vertical temperature distribution

$$\bar{T} = T_0 - \Gamma(z + d/2), \tag{12.6}$$

where $\Gamma \equiv \Delta T/d$ is the magnitude of the vertical temperature gradient, and T_0 is the temperature of the lower wall (Figure 12.2). Substituting equation (12.4) into equation (12.3), we obtain

$$\frac{\partial u_i}{\partial t} + u_j \frac{\partial u_i}{\partial x_j} = -\frac{1}{\rho_0} \frac{\partial}{\partial x_i}(P + p)$$
$$- g[1 - \alpha(\bar{T} + T' - T_0)]\delta_{i3} + \nu\nabla^2 u_i, \tag{12.7}$$

$$\frac{\partial T'}{\partial t} + u_j \frac{\partial}{\partial x_j}(\bar{T} + T') = \kappa\nabla^2(\bar{T} + T').$$

Subtracting the mean state equation (12.5) from the perturbed state equation (12.7), and neglecting squares of perturbations, we have

$$\frac{\partial u_i}{\partial t} = -\frac{1}{\rho_0} \frac{\partial p}{\partial x_i} + g\alpha T'\delta_{i3} + \nu\nabla^2 u_i, \tag{12.8}$$

$$\frac{\partial T'}{\partial t} - \Gamma w = \kappa\nabla^2 T', \tag{12.9}$$

where w is the vertical component of velocity. The advection term in equation (12.9) results from $u_j(\partial\bar{T}/\partial x_j) = w(d\bar{T}/dz) = -w\Gamma$. Equations (12.8) and (12.9) govern the behavior of perturbations on the system.

At this point it is useful to pause and show that the Rayleigh number defined by equation (12.2) is the ratio of buoyancy force to viscous force. From equation (12.9), the velocity scale is found by equating the advective and diffusion terms, giving

$$w \sim \frac{\kappa T'/d^2}{\Gamma} \sim \frac{\kappa\Gamma/d}{\Gamma} = \frac{\kappa}{d}.$$

An examination of the last two terms in equation (12.8) shows that

$$\frac{\text{Buoyancy force}}{\text{Viscous force}} \sim \frac{g\alpha T'}{\nu w/d^2} \sim \frac{g\alpha\Gamma d}{\nu w/d^2} = \frac{g\alpha\Gamma d^4}{\nu\kappa},$$

which is the Rayleigh number.

We now write the perturbation equations in terms of w and T' only. Taking the Laplacian of the $i = 3$ component of equation (12.8), we obtain

$$\frac{\partial}{\partial t}(\nabla^2 w) = -\frac{1}{\rho_0}\nabla^2\frac{\partial p}{\partial z} + g\alpha\nabla^2 T' + \nu\nabla^4 w. \tag{12.10}$$

The pressure term in equation (12.10) can be eliminated by taking the divergence of equation (12.8) and using the continuity equation $\partial u_i/\partial x_i = 0$. This gives

$$0 = -\frac{1}{\rho_0}\frac{\partial^2 p}{\partial x_i\,\partial x_i} + g\alpha\frac{\partial T'}{\partial x_i}\delta_{i3} + 0.$$

Differentiating with respect to z, we obtain

$$0 = -\frac{1}{\rho_0}\nabla^2\frac{\partial p}{\partial z} + g\alpha\frac{\partial^2 T'}{\partial z^2},$$

so that equation (12.10) becomes

$$\frac{\partial}{\partial t}(\nabla^2 w) = g\alpha\nabla_{\mathrm{H}}^2 T' + \nu\nabla^4 w, \tag{12.11}$$

where $\nabla_{\mathrm{H}}^2 \equiv \partial^2/\partial x^2 + \partial^2/\partial y^2$ is the horizontal Laplacian operator.

Equations (12.9) and (12.11) govern the development of perturbations on the system. The boundary conditions on the upper and lower rigid surfaces are that the no-slip condition is satisfied and that the walls are maintained at constant temperatures. These conditions require $u = v = w = T' = 0$ at $z = \pm d/2$. Because the conditions on u and v hold for all x and y, it follows from the continuity equation that $\partial w/\partial z = 0$ at the walls. The boundary conditions therefore can be written as

$$w = \frac{\partial w}{\partial z} = T' = 0 \quad \text{at } z = \pm\frac{d}{2}. \tag{12.12}$$

We shall use dimensionless independent variables in the rest of the analysis. For this, we make the transformation

$$t \to \frac{d^2}{\kappa}t,$$
$$(x, y, z) \to (xd, yd, zd),$$

where the old variables are on the left-hand side and the new variables are on the right-hand side; note that we are avoiding the introduction of new symbols for the nondimensional variables. Equations (12.9), (12.11), and (12.12) then become

$$\left(\frac{\partial}{\partial t} - \nabla^2\right) T' = \frac{\Gamma d^2}{\kappa} w, \tag{12.13}$$

$$\left(\frac{1}{\mathrm{Pr}}\frac{\partial}{\partial t} - \nabla^2\right)\nabla^2 w = \frac{g\alpha d^2}{\nu}\nabla_\mathrm{H}^2 T', \tag{12.14}$$

$$w = \frac{\partial w}{\partial z} = T' = 0 \quad \text{at } z = \pm\frac{1}{2} \tag{12.15}$$

where $\mathrm{Pr} \equiv \nu/\kappa$ is the Prandtl number.

The method of normal modes is now introduced. Because the coefficients of the governing set (12.13) and (12.14) are independent of x, y, and t, solutions exponential in these variables are allowed. We therefore assume normal modes of the form

$$w = \hat{w}(z)\,e^{ikx+ily+\sigma t},$$

$$T' = \hat{T}(z)\,e^{ikx+ily+\sigma t}.$$

The requirement that solutions remain bounded as x, $y \to \infty$ implies that the wavenumbers k and l must be real. In other words, the normal modes must be periodic in the directions of unboundedness. The growth rate $\sigma = \sigma_r + i\sigma_i$ is allowed to be complex. With this dependence, the operators in equations (12.13) and (12.14) transform as follows:

$$\frac{\partial}{\partial t} \to \sigma,$$

$$\nabla_\mathrm{H}^2 \to -K^2,$$

$$\nabla^2 \to \frac{d^2}{dz^2} - K^2,$$

where $K = \sqrt{k^2 + l^2}$ is the magnitude of the (nondimensional) horizontal wavenumber. Equations (12.13) and (12.14) then become

$$[\sigma - (D^2 - K^2)]\hat{T} = \frac{\Gamma d^2}{\kappa}\hat{w}, \tag{12.16}$$

$$\left[\frac{\sigma}{\mathrm{Pr}} - (D^2 - K^2)\right](D^2 - K^2)\hat{w} = -\frac{g\alpha d^2 K^2}{\nu}\hat{T}, \tag{12.17}$$

where $D \equiv d/dz$. Making the substitution

$$\frac{\Gamma d^2}{\kappa}\hat{w} \equiv W.$$

Equations (12.16) and (12.17) become

$$[\sigma - (D^2 - K^2)]\hat{T} = W, \tag{12.18}$$

$$\left[\frac{\sigma}{\mathrm{Pr}} - (D^2 - K^2)\right](D^2 - K^2)W = -\mathrm{Ra}\,K^2\hat{T}, \tag{12.19}$$

where

$$\mathrm{Ra} \equiv \frac{g\alpha\Gamma d^4}{\kappa\nu},$$

is the Rayleigh number. The boundary conditions (12.15) become

$$W = DW = \hat{T} = 0 \quad \text{at } z = \pm\frac{1}{2}. \tag{12.20}$$

Before we can proceed further, we need to show that σ in this problem can only be real.

Proof That σ Is Real for $\mathrm{Ra} > 0$

The sign of the real part of σ ($= \sigma_r + i\sigma_i$) determines whether the flow is stable or unstable. We shall now show that for the Bénard problem σ is real, and the *marginal state* that separates stability from instability is governed by $\sigma = 0$. To show this, multiply equation (12.18) by $\hat{T}^*$ (the complex conjugate of $\hat{T}$), and integrate between $\pm\frac{1}{2}$, by parts if necessary, using the boundary conditions (12.20). The various terms transform as follows:

$$\sigma \int \hat{T}^* \hat{T} \, dz = \sigma \int |\hat{T}|^2 \, dz,$$

$$\int \hat{T}^* D^2 \hat{T} \, dz = [\hat{T}^* D\hat{T}]_{-1/2}^{1/2} - \int D\hat{T}^* D\hat{T} \, dz = -\int |D\hat{T}|^2 \, dz,$$

where the limits on the integrals have not been explicitly written. Equation (12.18) then becomes

$$\sigma \int |\hat{T}|^2 \, dz + \int |D\hat{T}|^2 \, dz + K^2 \int |\hat{T}|^2 \, dz = \int \hat{T}^* W \, dz,$$

which can be written as

$$\sigma I_1 + I_2 = \int \hat{T}^* W \, dz, \tag{12.21}$$

where

$$I_1 \equiv \int |\hat{T}|^2 \, dz,$$

$$I_2 \equiv \int [|D\hat{T}|^2 + K^2 |\hat{T}|^2] \, dz.$$

Similarly, multiply equation (12.19) by W^* and integrate by parts. The first term in equation (12.19) gives

$$\frac{\sigma}{\text{Pr}} \int W^*(D^2 - K^2)W \, dz = \frac{\sigma}{\text{Pr}} \int W^* D^2 W \, dz - \frac{\sigma K^2}{\text{Pr}} \int W^* W \, dz$$

$$= -\frac{\sigma}{\text{Pr}} \int [|DW|^2 + K^2 |W|^2] \, dz. \qquad (12.22)$$

The second term in (12.19) gives

$$\int W^*(D^2 - K^2)(D^2 - K^2)W \, dz$$

$$= \int W^*(D^4 + K^4 - 2K^2 D^2)W \, dz$$

$$= \int W^* D^4 W \, dz + K^4 \int W^* W \, dz - 2K^2 \int W^* D^2 W \, dz$$

$$= [W^* D^3 W]_{-1/2}^{1/2} - \int DW^* D^3 W \, dz + K^4 \int |W|^2 \, dz$$

$$\quad - 2K^2 [W^* DW]_{-1/2}^{1/2} + 2K^2 \int DW^* DW \, dz$$

$$= \int [|D^2 W|^2 + 2K^2 |DW|^2 + K^4 |W|^2] \, dz. \qquad (12.23)$$

Using equations (12.22) and (12.23), the integral of equation (12.19) becomes

$$\frac{\sigma}{\text{Pr}} J_1 + J_2 = \text{Ra} \, K^2 \int W^* \hat{T} \, dz, \qquad (12.24)$$

where

$$J_1 \equiv \int [|DW|^2 + K^2 |W|^2] \, dz,$$

$$J_2 \equiv \int [|D^2 W|^2 + 2K^2 |DW|^2 + K^4 |W|^2] \, dz.$$

Note that the four integrals I_1, I_2, J_1, and J_2 are all positive. Also, the right-hand side of equation (12.24) is $\text{Ra} \, K^2$ times the complex conjugate of the right-hand side of equation (12.21). We can therefore eliminate the integral on the right-hand side of these equations by taking the complex conjugate of equation (12.21) and substituting into equation (12.24). This gives

$$\frac{\sigma}{\text{Pr}} J_1 + J_2 = \text{Ra} \, K^2 (\sigma^* I_1 + I_2).$$

Equating imaginary parts

$$\sigma_i \left[\frac{J_1}{\text{Pr}} + \text{Ra} \, K^2 I_1 \right] = 0.$$

We consider only the top-heavy case, for which Ra > 0. The quantity within [   ] is then positive, and the preceding equation requires that $\sigma_i = 0$.

The Bénard problem is one of two well-known problems in which σ is real. (The other one is the Taylor problem of Couette flow between rotating cylinders, discussed in the following section.) In most other problems σ is complex, and the marginal state ($\sigma_r = 0$) contains propagating waves. In the Bénard and Taylor problems, however, the marginal state corresponds to $\sigma = 0$, and is therefore *stationary* and does not contain propagating waves. In these the onset of instability is marked by a transition from the background state to another *steady* state. In such a case we commonly say that the *principle of exchange of stabilities* is valid, and the instability sets in as a *cellular convection*, which will be explained shortly.

Solution of the Eigenvalue Problem with Two Rigid Plates

First, we give the solution for the case that is easiest to realize in a laboratory experiment, namely, a layer of fluid confined between two rigid plates where no-slip conditions are satisfied. The solution to this problem was first given by Jeffreys in 1928. A much simpler solution exists for a layer of fluid with two stress-free surfaces. This will be discussed later.

For the marginal state $\sigma = 0$, and the set (12.18) and (12.19) becomes

$$
\begin{aligned}
(D^2 - K^2)\hat{T} &= -W, \\
(D^2 - K^2)^2 W &= \text{Ra}\, K^2 \hat{T}.
\end{aligned}
\tag{12.25}
$$

Eliminating $\hat{T}$, we obtain

$$(D^2 - K^2)^3 W = -\text{Ra}\, K^2 W. \tag{12.26}$$

The boundary condition (12.20) becomes

$$W = DW = (D^2 - K^2)^2 W = 0 \quad \text{at } z = \pm \frac{1}{2}. \tag{12.27}$$

We have a sixth-order homogeneous differential equation with six homogeneous boundary conditions. Nonzero solutions for such a system can only exist for a particular value of Ra (for a given K). It is therefore an eigenvalue problem. Note that the Prandtl number has dropped out of the marginal state.

The point to observe is that the problem is symmetric with respect to the two boundaries, thus the eigenfunctions fall into two distinct classes—those with the vertical velocity symmetric about the midplane $z = 0$, and those with the vertical velocity antisymmetric about the midplane (Figure 12.3). The gravest even mode therefore has one row of cells, and the gravest odd mode has two rows of cells. It can be shown that the smallest critical Rayleigh number is obtained by assuming disturbances in the form of the gravest even mode, which also agrees with experimental findings of a single row of cells.

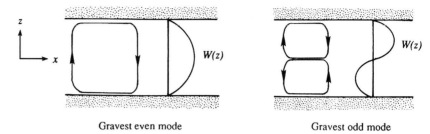

Gravest even mode   Gravest odd mode

Figure 12.3   Flow pattern and eigenfunction structure of the gravest even mode and the gravest odd mode in the Bénard problem.

Because the coefficients of the governing equations (12.26) are independent of z, the general solution can be expressed as a superposition of solutions of the form

$$W = e^{qz},$$

where the six roots of q are given by

$$(q^2 - K^2)^3 = -\text{Ra} \, K^2.$$

The three roots of this equation are

$$q^2 = -K^2\left[\left(\frac{\text{Ra}}{K^4}\right)^{1/3} - 1\right],$$

$$q^2 = K^2\left[1 + \frac{1}{2}\left(\frac{\text{Ra}}{K^4}\right)^{1/3}(1 \pm i\sqrt{3})\right]. \tag{12.28}$$

Taking square roots, the six roots finally become

$$\pm iq_0, \qquad \pm q, \qquad \text{and} \qquad \pm q^*,$$

where

$$q_0 = K\left[\left(\frac{\text{Ra}}{K^4}\right)^{1/3} - 1\right]^{1/2},$$

and q and its conjugate q^* are given by the two roots of equation (12.28).

The even solution of equation (12.26) is therefore

$$W = A \cos q_0 z + B \cosh qz + C \cosh q^* z.$$

To apply the boundary conditions on this solution, we find the following derivatives:

$$DW = -Aq_0 \sin q_0 z + Bq \sinh qz + Cq^* \sinh q^* z,$$

$$(D^2 - K^2)^2 W = A(q_0^2 + K^2)^2 \cos q_0 z + B(q^2 - K^2)^2 \cosh qz$$

$$+ C(q^{*2} - K^2)^2 \cosh q^* z.$$

The boundary conditions (12.27) then require

$$
\begin{bmatrix}
\cos\dfrac{q_0}{2} & \cosh\dfrac{q}{2} & \cosh\dfrac{q^*}{2} \\[2mm]
-q_0\sin\dfrac{q_0}{2} & q\sinh\dfrac{q}{2} & q^*\sinh\dfrac{q^*}{2} \\[2mm]
(q_0^2+K^2)^2\cos\dfrac{q_0}{2} & (q^2-K^2)^2\cosh\dfrac{q}{2} & (q^{*2}-K^2)^2\cosh\dfrac{q^*}{2}
\end{bmatrix}
\begin{bmatrix} A \\[2mm] B \\[2mm] C \end{bmatrix} = 0.
$$

Here, A, B, and C cannot all be zero if we want to have a nonzero solution, which requires that the determinant of the matrix must vanish. This gives a relation between Ra and the corresponding eigenvalue K (Figure 12.4). Points on the curve $K(\text{Ra})$ represent marginally stable states, which separate regions of stability and instability. The lowest value of Ra is found to be $\text{Ra}_{cr} = 1708$, attained at $K_{cr} = 3.12$. As *all* values of K are allowed by the system, the flow first becomes unstable when the Rayleigh number reaches a value of

$$
\boxed{\text{Ra}_{cr} = 1708.}
$$

The wavelength at the onset of instability is

$$
\lambda_{cr} = \frac{2\pi d}{K_{cr}} \simeq 2d.
$$

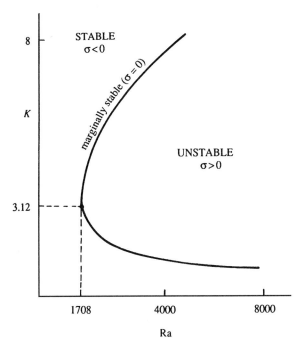

Figure 12.4   Stable and unstable regions for Bénard convection.

Laboratory experiments agree remarkably well with these predictions, and the solution of the Bénard problem is considered one of the major successes of the linear stability theory.

Solution with Stress-Free Surfaces

We now give the solution for a layer of fluid with stress-free surfaces. This case can be approximately realized in a laboratory experiment if a layer of liquid is floating on top of a somewhat heavier liquid. The main interest in the problem, however, is that it allows a simple solution, which was first given by Rayleigh. In this case the boundary conditions are $w = T' = \mu(\partial u/\partial z + \partial w/\partial x) = \mu(\partial v/\partial z + \partial w/\partial y) = 0$ at the surfaces, the latter two conditions resulting from zero stress. Because w vanishes (for all x and y) on the boundaries, it follows that the vanishing stress conditions require $\partial u/\partial z = \partial v/\partial z = 0$ at the boundaries. On differentiating the continuity equation with respect to z, it follows that $\partial^2 w/\partial z^2 = 0$ on the free surfaces. In terms of the complex amplitudes, the eigenvalue problem is therefore

$$(D^2 - K^2)^3 W = -\text{Ra}\, K^2 W, \tag{12.29}$$

with $W = (D^2 - K^2)^2 W = D^2 W = 0$ at the surfaces. By expanding $(D^2 - K^2)^2$, the boundary conditions can be written as

$$W = D^2 W = D^4 W = 0 \quad \text{at } z = \pm\frac{1}{2},$$

which should be compared with the conditions (12.27) for rigid boundaries.

   Successive differentiation of equation (12.29) shows that *all* even derivatives of W vanish on the boundaries. The eigenfunctions must therefore be

$$W = A \sin n\pi z,$$

where A is any constant and n is an integer. Substitution into equation (12.29) leads to the eigenvalue relation

$$\text{Ra} = (n^2\pi^2 + K^2)^3/K^2, \tag{12.30}$$

which gives the Rayleigh number in the marginal state. For a given K^2, the lowest value of Ra occurs when $n = 1$, which is the gravest mode. The critical Rayleigh number is obtained by finding the minimum value of Ra as K^2 is varied, that is, by setting $d\,\text{Ra}/dK^2 = 0$. This gives

$$\frac{d\,\text{Ra}}{dK^2} = \frac{3(\pi^2 + K^2)^2}{K^2} - \frac{(\pi^2 + K^2)^3}{K^4} = 0,$$

which requires $K_{\text{cr}}^2 = \pi^2/2$. The corresponding value of Ra is

$$\text{Ra}_{\text{cr}} = \tfrac{27}{4}\pi^4 = 657.$$

   For a layer with a free upper surface (where the stress is zero) and a rigid bottom wall, the solution of the eigenvalue problem gives $\text{Ra}_{\text{cr}} = 1101$ and $K_{\text{cr}} = 2.68$.

This case is of interest in laboratory experiments having the most visual effects, as originally conducted by Bénard.

Cell Patterns

The linear theory specifies the horizontal wavelength at the onset of instability, but not the horizontal pattern of the convective cells. This is because a given wavenumber vector $\mathbf{K}$ can be decomposed into two orthogonal components in an infinite number of ways. If we assume that the experimental conditions are horizontally isotropic, with no preferred directions, then regular polygons in the form of equilateral triangles, squares, and regular hexagons are all possible structures. Bénard's original experiments showed only hexagonal patterns, but we now know that he was observing a different phenomenon. The observations summarized in Drazin and Reid (1981) indicate that hexagons frequently predominate initially. As Ra is increased, the cells tend to merge and form rolls, on the walls of which the fluid rises or sinks (Figure 12.5). The cell structure becomes more chaotic as Ra is increased further, and the flow becomes turbulent when Ra $> 5 \times 10^4$.

The magnitude or direction of flow in the cells cannot be predicted by linear theory. After a short time of exponential growth, the flow becomes large enough for the nonlinear terms to be important and reaches a nonlinear equilibrium stage. The flow pattern for a hexagonal cell is sketched in Figure 12.6. Particles in the middle of the cell usually rise in a liquid and fall in a gas. This has been attributed to the

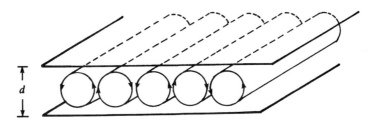

Figure 12.5   Convection rolls in a Bénard problem.

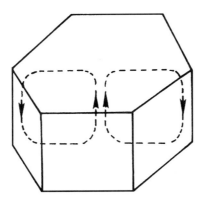

Figure 12.6   Flow pattern in a hexagonal Bénard cell.

property that the viscosity of a liquid decreases with temperature, whereas that of a gas increases with temperature. The rising fluid loses heat by thermal conduction at the top wall, travels horizontally, and then sinks. For a steady cellular pattern, the continuous generation of kinetic energy is balanced by viscous dissipation. The generation of kinetic energy is maintained by continuous release of potential energy due to heating at the bottom and cooling at the top.

4. Double-Diffusive Instability

An interesting instability results when the density of the fluid depends on two opposing gradients. The possibility of this phenomenon was first suggested by Stommel *et al.* (1956), but the dynamics of the process was first explained by Stern (1960). Turner (1973), and review articles by Huppert and Turner (1981), and Turner (1985) discuss the dynamics of this phenomenon and its applications to various fields such as astrophysics, engineering, and geology. Historically, the phenomenon was first suggested with oceanic application in mind, and this is how we shall present it. For sea water the density depends on the temperature $\tilde{T}$ and salt content $\tilde{s}$ (kilograms of salt per kilograms of water), so that the density is given by

$$\tilde{\rho} = \rho_0[1 - \alpha(\tilde{T} - T_0) + \beta(\tilde{s} - s_0)],$$

where the value of α determines how fast the density decreases with temperature, and the value of β determines how fast the density increases with salinity. As defined here, both α and β are positive. The key factor in this instability is that the diffusivity κ_s of salt in water is only 1% of the thermal diffusivity κ. *Such a system can be unstable even when the density decreases upwards.* By means of the instability, the flow releases the potential energy of the *component* that is "heavy at the top." Therefore, the effect of diffusion in such a system can be to *destabilize* a stable density gradient. This is in contrast to a medium containing a single diffusing component, for which the analysis of the preceding section shows that the effect of diffusion is to *stabilize* the system even when it is heavy at the top.

Finger Instability

Consider the two situations of Figure 12.7, both of which can be unstable although each is stably stratified in density ($d\bar{\rho}/dz < 0$). Consider first the case of hot and salty water lying over cold and fresh water (Figure 12.7a), that is, when the system is top heavy in salt. In this case both $d\bar{T}/dz$ and dS/dz are positive, and we can arrange the composition of water such that the density decreases upward. Because $\kappa_s \ll \kappa$, a displaced particle would be near thermal equilibrium with the surroundings, but would exchange negligible salt. A rising particle therefore would be constantly lighter than the surroundings because of the salinity deficit, and would continue to rise. A parcel displaced downward would similarly continue to plunge downward. The basic state shown in Figure 12.7a is therefore unstable. Laboratory observations show that the instability in this case appears in the form of a forest of long narrow convective cells, called *salt fingers* (Figure 12.8). Shadowgraph images in the deep ocean have confirmed their existence in nature.

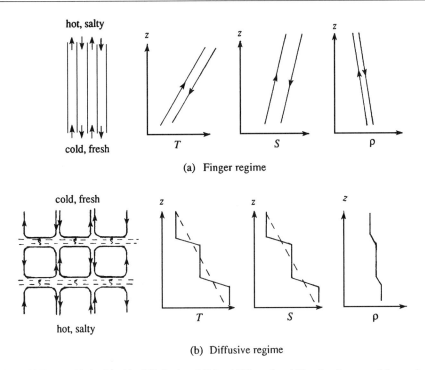

(a) Finger regime

(b) Diffusive regime

Figure 12.7 Two kinds of double-diffusive instabilities. (a) Finger instability, showing up- and downgoing salt fingers and their temperature, salinity, and density. Arrows indicate direction of motion. (b) Oscillating instability, finally resulting in a series of convecting layers separated by "diffusive" interfaces. Across these interfaces T and S vary sharply, but heat is transported much faster than salt.

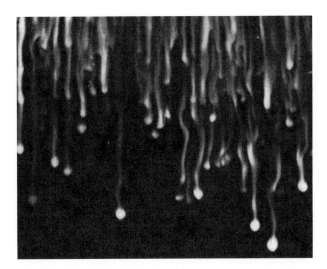

Figure 12.8 Salt fingers, produced by pouring salt solution on top of a stable temperature gradient. Flow visualization by fluorescent dye and a horizontal beam of light. J. Turner, *Naturwissenschaften* **72**: 70–75, 1985 and reprinted with the permission of Springer-Verlag GmbH & Co.

We can derive a criterion for instability by generalizing our analysis of the Bénard convection so as to include salt diffusion. Assume a layer of depth d confined between stress-free boundaries maintained at constant temperature and constant salinity. If we repeat the derivation of the perturbation equations for the normal modes of the system, the equations that replace equation (12.25) are found to be

$$(D^2 - K^2)\hat{T} = -W,$$
$$\frac{\kappa_s}{\kappa}(D^2 - K^2)\hat{s} = -W, \qquad (12.31)$$
$$(D^2 - K^2)^2 W = -\text{Ra}\, K^2 \hat{T} + \text{Rs}'\, K^2 \hat{s},$$

where $\hat{s}(z)$ is the complex amplitude of the salinity perturbation, and we have defined

$$\text{Ra} \equiv \frac{g\alpha d^4 (d\bar{T}/dz)}{\nu\kappa},$$

and

$$\text{Rs}' \equiv \frac{g\beta d^4 (dS/dz)}{\nu\kappa}.$$

Note that κ (and not κ_s) appears in the definition of Rs'. In contrast to equation (12.31), a positive sign appeared in equation (12.25) in front of Ra because in the preceding section Ra was defined to be positive for a top-heavy situation.

It is seen from the first two of equations (12.31) that the equations for $\hat{T}$ and $\hat{s}\kappa_s/\kappa$ are the same. The boundary conditions are also the same for these variables:

$$\hat{T} = \frac{\kappa_s \hat{s}}{\kappa} = 0 \quad \text{at } z = \pm\frac{1}{2}.$$

It follows that we must have $\hat{T} = \hat{s}\kappa_s/\kappa$ everywhere. Equations (12.31) therefore become

$$(D^2 - K^2)\hat{T} = -W,$$
$$(D^2 - K^2)^2 W = (\text{Rs} - \text{Ra})K^2 \hat{T},$$

where

$$\text{Rs} \equiv \frac{\text{Rs}' \kappa}{\kappa_s} = \frac{g\beta d^4 (dS/dz)}{\nu\kappa_s}.$$

The preceding set is now identical to the set (12.25) for the Bénard convection, with $(\text{Rs} - \text{Ra})$ replacing Ra. For stress-free boundaries, solution of the preceding section shows that the critical value is

$$\text{Rs} - \text{Ra} = \tfrac{27}{4}\pi^4 = 657,$$

which can be written as

$$\frac{gd^4}{\nu}\left[\frac{\beta}{\kappa_s}\frac{dS}{dz} - \frac{\alpha}{\kappa}\frac{d\bar{T}}{dz}\right] = 657. \tag{12.32}$$

Even if $\alpha(d\bar{T}/dz) - \beta(dS/dz) > 0$ (i.e., $\bar{\rho}$ decreases upward), the condition (12.32) can be quite easily satisfied because κ_s is much smaller than κ. The flow can therefore be made unstable simply by ensuring that the factor within [ ] is positive and making d large enough.

The analysis predicts that the lateral width of the cell is of the order of d, but such wide cells are not observed at supercritical stages when (Rs − Ra) far exceeds 657. Instead, long thin salt fingers are observed, as shown in Figure 12.8. If the salinity gradient is large, then experiments as well as calculations show that a deep layer of salt fingers becomes unstable and breaks down into a series of convective layers, with fingers confined to the interfaces. Oceanographic observations frequently show a series of staircase-shaped vertical distributions of salinity and temperature, with a positive overall dS/dz and $d\bar{T}/dz$; this can indicate salt finger activity.

Oscillating Instability

Consider next the case of cold and fresh water lying over hot and salty water (Figure 12.7b). In this case both $d\bar{T}/dz$ and dS/dz are negative, and we can choose their values such that the density decreases upwards. Again the system is unstable, but the dynamics are different. A particle displaced upward loses heat but no salt. Thus it becomes heavier than the surroundings and buoyancy forces it back toward its initial position, resulting in an oscillation. However, a stability calculation shows that a less than perfect heat conduction results in a growing oscillation, although some energy is dissipated. In this case the growth rate σ is complex, in contrast to the situation of Figure 12.7a where it is real.

Laboratory experiments show that the initial oscillatory instability does not last long, and eventually results in the formation of a number of horizontal *convecting layers*, as sketched in Figure 12.7b. Consider the situation when a stable salinity gradient in an isothermal fluid is heated from below (Figure 12.9). The initial instability starts as a growing oscillation near the bottom. As the heating is continued beyond the initial appearance of the instability, a well-mixed layer develops, capped by a salinity step, a temperature step, and no density step. The heat flux through this step forms a thermal boundary layer, as shown in Figure 12.9. As the well-mixed layer grows, the temperature step across the thermal boundary layer becomes larger. Eventually, the Rayleigh number across the thermal boundary layer becomes critical, and a second convecting layer forms on top of the first. The second layer is maintained by heat flux (and negligible salt flux) across a sharp laminar interface on top of the first layer. This process continues until a stack of horizontal layers forms one upon another. From comparison with the Bénard convection, it is clear that inclusion of a stable salinity gradient has prevented a complete overturning from top to bottom.

The two examples in this section show that in a double-component system in which the diffusivities for the two components are different, the effect of diffusion

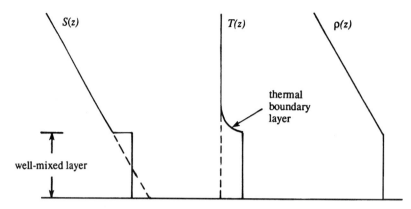

Figure 12.9 Distributions of salinity, temperature, and density, generated by heating a linear salinity gradient from below.

can be destabilizing, even if the system is judged hydrostatically stable. In contrast, diffusion is stabilizing in a single-component system, such as the Bénard system. The two requirements for the double-diffusive instability are that the diffusivities of the components be different, and that the components make opposite contributions to the vertical density gradient.

5. Centrifugal Instability: Taylor Problem

In this section we shall consider the instability of a Couette flow between concentric rotating cylinders, a problem first solved by Taylor in 1923. In many ways the problem is similar to the Bénard problem, in which there is a potentially unstable arrangement of an "adverse" temperature gradient. In the Couette flow problem the source of the instability is the adverse gradient of angular momentum. Whereas convection in a heated layer is brought about by buoyant forces becoming large enough to overcome the viscous resistance, the convection in a Couette flow is generated by the centrifugal forces being able to overcome the viscous forces. We shall first present Rayleigh's discovery of an inviscid stability criterion for the problem and then outline Taylor's solution of the viscous case. Experiments indicate that the instability initially appears in the form of axisymmetric disturbances, for which $\partial/\partial\theta = 0$. Accordingly, we shall limit ourselves only to the axisymmetric case.

Rayleigh's Inviscid Criterion

The problem was first considered by Rayleigh in 1888. Neglecting viscous effects, he discovered the source of instability for this problem and demonstrated a necessary and sufficient condition for instability. Let $U_\theta(r)$ be the velocity at any radial distance. For inviscid flows $U_\theta(r)$ can be *any* function, but only certain distributions can be stable. Imagine that two fluid rings of equal masses at radial distances r_1 and r_2 ($>r_1$) are interchanged. As the motion is inviscid, Kelvin's theorem requires that the circulation $\Gamma = 2\pi r U_\theta$ (proportional to the angular momentum $r U_\theta$) should remain constant during the interchange. That is, after the interchange, the fluid at r_2 will have the circulation (namely, Γ_1) that it had at r_1 before the interchange. Similarly, the fluid at r_1 will have the circulation (namely, Γ_2) that it had at r_2 before the interchange. The conservation of circulation requires that the kinetic energy E must change during the interchange. Because $E = U_\theta^2/2 = \Gamma^2/8\pi^2 r^2$, we have

$$E_{\text{final}} = \frac{1}{8\pi^2} \left[\frac{\Gamma_2^2}{r_1^2} + \frac{\Gamma_1^2}{r_2^2} \right],$$

$$E_{\text{initial}} = \frac{1}{8\pi^2} \left[\frac{\Gamma_1^2}{r_1^2} + \frac{\Gamma_2^2}{r_2^2} \right],$$

so that the kinetic energy change per unit mass is

$$\Delta E = E_{\text{final}} - E_{\text{initial}} = \frac{1}{8\pi^2} (\Gamma_2^2 - \Gamma_1^2) \left(\frac{1}{r_1^2} - \frac{1}{r_2^2} \right).$$

Because $r_2 > r_1$, a velocity distribution for which $\Gamma_2^2 > \Gamma_1^2$ would make ΔE positive, which implies that an external source of energy would be necessary to perform the interchange of the fluid rings. Under this condition a *spontaneous* interchange of the rings is not possible, and the flow is stable. On the other hand, if Γ^2 decreases with r, then an interchange of rings will result in a release of energy; such a flow is unstable. It can be shown that in this situation the centrifugal force in the new location of an outwardly displaced ring is larger than the prevailing (radially inward) pressure gradient force.

Rayleigh's criterion can therefore be stated as follows: *An inviscid Couette flow is unstable if*

$$\frac{d\Gamma^2}{dr} < 0 \qquad \text{(unstable)}.$$

The criterion is analogous to the inviscid requirement for static instability in a density stratified fluid:

$$\frac{d\bar{\rho}}{dz} > 0 \qquad \text{(unstable)}.$$

Therefore, the "stratification" of angular momentum in a Couette flow is unstable if it decreases radially outwards. Consider a situation in which the outer cylinder is held stationary and the inner cylinder is rotated. Then $d\Gamma^2/dr < 0$, and Rayleigh's criterion implies that the flow is inviscidly unstable. As in the Bénard problem, however, merely having a potentially unstable arrangement does not cause instability in a viscous medium. The inviscid Rayleigh criterion is modified by Taylor's solution of the viscous problem, outlined in what follows.

Formulation of the Problem

Using cylindrical polar coordinates (r, θ, z) and assuming axial symmetry, the equations of motion are

$$
\begin{aligned}
\frac{D\tilde{u}_r}{Dt} - \frac{\tilde{u}_\theta^2}{r} &= -\frac{1}{\rho}\frac{\partial \tilde{p}}{\partial r} + \nu\left(\nabla^2 \tilde{u}_r - \frac{\tilde{u}_r}{r^2}\right), \\
\frac{D\tilde{u}_\theta}{Dt} + \frac{\tilde{u}_r \tilde{u}_\theta}{r} &= \nu\left(\nabla^2 \tilde{u}_\theta - \frac{\tilde{u}_\theta}{r^2}\right), \\
\frac{D\tilde{u}_z}{Dt} &= -\frac{1}{\rho}\frac{\partial \tilde{p}}{\partial z} + \nu\nabla^2 \tilde{u}_z, \\
\frac{\partial \tilde{u}_r}{\partial r} + \frac{\tilde{u}_r}{r} + \frac{\partial \tilde{u}_z}{\partial z} &= 0,
\end{aligned}
\tag{12.33}
$$

where

$$
\frac{D}{Dt} \equiv \frac{\partial}{\partial t} + \tilde{u}_r \frac{\partial}{\partial r} + \tilde{u}_z \frac{\partial}{\partial z},
$$

and

$$
\nabla^2 \equiv \frac{\partial^2}{\partial r^2} + \frac{1}{r}\frac{\partial}{\partial r} + \frac{\partial^2}{\partial z^2}.
$$

We decompose the motion into a background state plus perturbation:

$$
\begin{aligned}
\tilde{\mathbf{u}} &= \mathbf{U} + \mathbf{u}, \\
\tilde{p} &= P + p.
\end{aligned}
\tag{12.34}
$$

The background state is given by (see Chapter 9, Section 6)

$$
U_r = U_z = 0, \qquad U_\theta = V(r), \qquad \frac{1}{\rho}\frac{dP}{dr} = \frac{V^2}{r},
\tag{12.35}
$$

where

$$
V = Ar + B/r,
\tag{12.36}
$$

with constants defined as

$$
A \equiv \frac{\Omega_2 R_2^2 - \Omega_1 R_1^2}{R_2^2 - R_1^2}, \qquad B \equiv \frac{(\Omega_1 - \Omega_2)R_1^2 R_2^2}{R_2^2 - R_1^2}.
$$

Here, Ω_1 and Ω_2 are the angular speeds of the inner and outer cylinders, respectively, and R_1 and R_2 are their radii (Figure 12.10).

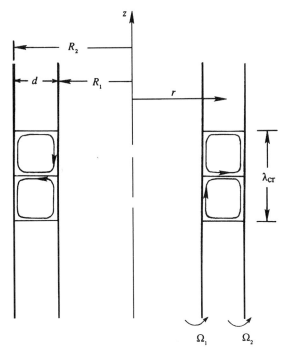

Figure 12.10 Definition sketch of instability in rotating Couette flow.

Substituting equation (12.34) into the equations of motion (12.33), neglecting nonlinear terms, and subtracting the background state (12.35), we obtain the perturbation equations

$$\frac{\partial u_r}{\partial t} - \frac{2V}{r} u_\theta = -\frac{1}{\rho}\frac{\partial p}{\partial r} + \nu\left(\nabla^2 u_r - \frac{u_r}{r^2}\right),$$

$$\frac{\partial u_\theta}{\partial t} + \left(\frac{dV}{dr} + \frac{V}{r}\right) u_r = \nu\left(\nabla^2 u_\theta - \frac{u_\theta^2}{r}\right),$$

$$\frac{\partial u_z}{\partial t} = -\frac{1}{\rho}\frac{\partial p}{\partial z} + \nu\nabla^2 u_z, \qquad (12.37)$$

$$\frac{\partial u_r}{\partial r} + \frac{u_r}{r} + \frac{\partial u_z}{\partial z} = 0.$$

As the coefficients in these equations depend only on r, the equations admit solutions that depend on z and t exponentially. We therefore consider normal mode solutions of the form

$$(u_r, u_\theta, u_z, p) = (\hat{u}_r, \hat{u}_\theta, \hat{u}_z, \hat{p})\, e^{\sigma t + ikz}.$$

The requirement that the solutions remain bounded as $z \to \pm\infty$ implies that the axial wavenumber k must be real. After substituting the normal modes into (12.37) and eliminating $\hat{u}_z$ and $\hat{p}$, we get a coupled system of equations in $\hat{u}_r$ and $\hat{u}_\theta$. Under the *narrow-gap approximation*, for which $d = R_2 - R_1$ is much smaller than $(R_1 + R_2)/2$, these equations finally become (see Chandrasekhar (1961) for details)

$$(D^2 - k^2 - \sigma)(D^2 - k^2)\hat{u}_r = (1 + \alpha x)\hat{u}_\theta,$$
$$(D^2 - k^2 - \sigma)\hat{u}_\theta = -\mathrm{Ta}\, k^2 \hat{u}_r, \tag{12.38}$$

where

$$\alpha \equiv \frac{\Omega_2}{\Omega_1} - 1,$$
$$x \equiv \frac{r - R_1}{d},$$
$$d \equiv R_2 - R_1,$$
$$D \equiv \frac{d}{dr}.$$

We have also defined the *Taylor number*

$$\mathrm{Ta} \equiv 4 \left(\frac{\Omega_1 R_1^2 - \Omega_2 R_2^2}{R_2^2 - R_1^2} \right) \frac{\Omega_1 d^4}{\nu^2}. \tag{12.39}$$

It is the ratio of the centrifugal force to viscous force, and equals $2(V_1 d/\nu)^2 (d/R_1)$ when only the inner cylinder is rotating and the gap is narrow.

The boundary conditions are

$$\hat{u}_r = D\hat{u}_r = \hat{u}_\theta = 0 \qquad \text{at } x = 0, 1. \tag{12.40}$$

The eigenvalues k at the marginal state are found by setting the real part of σ to zero. On the basis of experimental evidence, Taylor assumed that the principle of exchange of stabilities must be valid for this problem, and the marginal states are given by $\sigma = 0$. This was later proven to be true for cylinders rotating in the same directions, but a general demonstration for all conditions is still lacking.

Discussion of Taylor's Solution

A solution of the eigenvalue problem (12.38), subject to equation (12.40), was obtained by Taylor. Figure 12.11 shows the results of his calculations and his own experimental verification of the analysis. The vertical axis represents the angular velocity of the inner cylinder (taken positive), and the horizontal axis represents the angular velocity of the outer cylinder. Cylinders rotating in opposite directions are represented by a negative Ω_2. Taylor's solution of the marginal state is indicated, with the region above the curve corresponding to instability. Rayleigh's inviscid criterion is also indicated by the straight dashed line. It is apparent that the presence of viscosity can stabilize a flow. Taylor's viscous solution indicates that the flow remains stable until a critical Taylor number of

$$\mathrm{Ta}_{\mathrm{cr}} = \frac{1708}{(1/2)\,(1 + \Omega_2/\Omega_1)}, \tag{12.41}$$

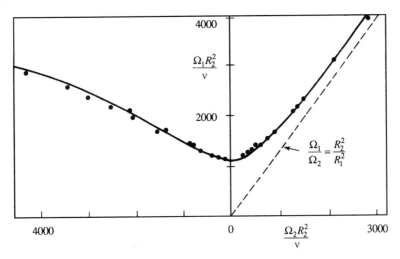

Figure 12.11 Taylor's observation and narrow-gap calculation of marginal stability in rotating Couette flow of water. The ratio of radii is $R_2/R_1 = 1.14$. The region above the curve is unstable. The dashed line represents Rayleigh's inviscid criterion, with the region to the left of the line representing instability.

is attained. The nondimensional axial wavenumber at the onset of instability is found to be $k_{cr} = 3.12$, which implies that the wavelength at onset is $\lambda_{cr} = 2\pi d/k_{cr} \simeq 2d$. The height of one cell is therefore nearly equal to d, so that the cross-section of a cell is nearly a square. In the limit $\Omega_2/\Omega_1 \to 1$, the critical Taylor number is identical to the critical Rayleigh number for thermal convection discussed in the preceding section, for which the solution was given by Jeffreys five years later. The agreement is expected, because in this limit $\alpha = 0$, and the eigenvalue problem (12.38) reduces to that of the Bénard problem (12.25). For cylinders rotating in opposite directions the Rayleigh criterion predicts instability, but the viscous solution can be stable.

Taylor's analysis of the problem was enormously satisfying, both experimentally and theoretically. He measured the wavelength at the onset of instability by injecting dye and obtained an almost exact agreement with his calculations. The observed onset of instability in the $\Omega_1\Omega_2$-plane (Figure 12.11) was also in remarkable agreement. This has prompted remarks such as "the closeness of the agreement between his theoretical and experimental results was without precedent in the history of fluid mechanics" (Drazin and Reid 1981, p. 105). It even led some people to suggest happily that the agreement can be regarded as a verification of the underlying Navier–Stokes equations, which make a host of assumptions including a linearity between stress and strain rate.

The instability appears in the form of counter-rotating toroidal (or doughnut-shaped) vortices (Figure 12.12a) called *Taylor vortices*. The streamlines are in the form of helixes, with axes wrapping around the annulus, somewhat like the stripes on a barber's pole. These vortices themselves become unstable at higher values of Ta, when they give rise to wavy vortices for which $\partial/\partial\theta \neq 0$ (Figure 12.12b). In effect, the flow has now attained the next higher mode. The number of waves around the annulus depends on the Taylor number, and the wave pattern travels around the

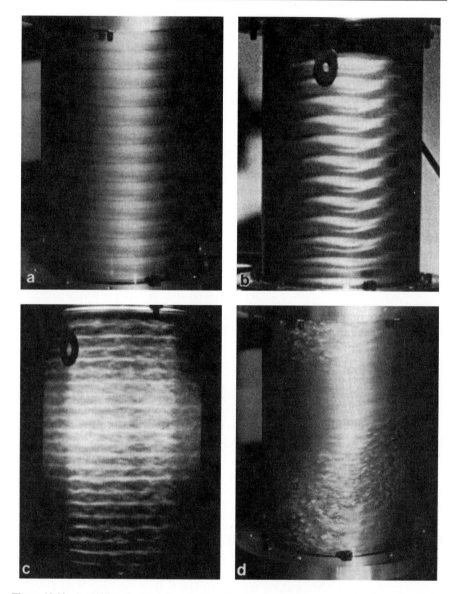

Figure 12.12   Instability of rotating Couette flow. Panels a, b, c, and d correspond to increasing Taylor number. D. Coles, *Journal of Fluid Mechanics* **21**: 385–425, 1965 and reprinted with the permission of Cambridge University Press.

annulus. More complicated patterns of vortices result at a higher rates of rotation, finally resulting in the occasional appearance of turbulent patches (Figure 12.12d), and then a fully turbulent flow.

Phenomena analogous to the Taylor vortices are called *secondary flows* because they are superposed on a primary flow (such as the Couette flow in the present case).

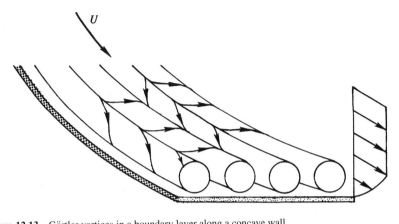

Figure 12.13 Görtler vortices in a boundary layer along a concave wall.

There are two other situations where a combination of curved streamlines (which give rise to centrifugal forces) and viscosity result in instability and steady secondary flows in the form of vortices. One is the flow through a curved channel, driven by a pressure gradient. The other is the appearance of *Görtler vortices* in a boundary layer flow along a concave wall (Figure 12.13). The possibility of secondary flows signifies that the *solutions of the Navier–Stokes equations are nonunique* in the sense that more than one steady solution is allowed under the same boundary conditions. We can derive the form of the primary flow only if we exclude the secondary flow by appropriate assumptions. For example, we can derive the expression (12.36) for Couette flow by *assuming* that $U_r = 0$ and $U_z = 0$, which rule out the secondary flow.

6. *Kelvin–Helmholtz Instability*

Instability at the interface between two horizontal parallel streams of different velocities and densities, with the heavier fluid at the bottom, is called the *Kelvin–Helmholtz instability*. The name is also commonly used to describe the instability of the more general case where the variations of velocity and density are continuous and occur over a finite thickness. The more general case is discussed in the following section.

Assume that the layers have infinite depth and that the interface has zero thickness. Let U_1 and ρ_1 be the velocity and density of the basic state in the upper layer and U_2 and ρ_2 be those in the bottom layer (Figure 12.14). By Kelvin's circulation theorem, the perturbed flow must be irrotational in each layer because the motion develops from an irrotational basic flow of uniform velocity in each layer. The flow can therefore be described by a velocity potential that satisfies the Laplace equation. Let the variables in the perturbed state be denoted by a tilde ($\tilde{\ }$). Then

$$\nabla^2 \tilde{\phi}_1 = 0, \qquad \nabla^2 \tilde{\phi}_2 = 0. \tag{12.42}$$

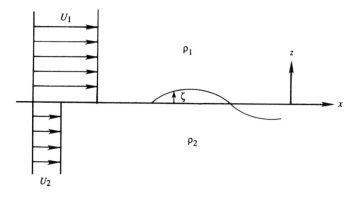

Figure 12.14 Discontinuous shear across a density interface.

The flow is decomposed into a basic state plus perturbations:

$$\tilde{\phi}_1 = U_1 x + \phi_1,$$
$$\tilde{\phi}_2 = U_2 x + \phi_2,$$
(12.43)

where the first terms on the right-hand side represent the basic flow of uniform streams. Substitution into equation (12.42) gives the perturbation equations

$$\nabla^2 \phi_1 = 0, \qquad \nabla^2 \phi_2 = 0,$$
(12.44)

subject to

$$\phi_1 \to 0 \qquad \text{as } z \to \infty,$$
$$\phi_2 \to 0 \qquad \text{as } z \to -\infty.$$
(12.45)

As discussed in Chapter 7, there are kinematic and dynamic conditions to be satisfied at the interface. The kinematic boundary condition is that the fluid particles at the interface must move with the interface. Considering particles just above the interface, this requires

$$\frac{\partial \phi_1}{\partial z} = \frac{D\zeta}{Dt} = \frac{\partial \zeta}{\partial t} + (U_1 + u_1)\frac{\partial \zeta}{\partial x} + v_1\frac{\partial \zeta}{\partial y} \qquad \text{at } z = \zeta.$$

This condition can be linearized by applying it at $z = 0$ instead of at $z = \zeta$ and by neglecting quadratic terms. Writing a similar equation for the lower layer, the kinematic boundary conditions are

$$\frac{\partial \phi_1}{\partial z} = \frac{\partial \zeta}{\partial t} + U_1\frac{\partial \zeta}{\partial x} \qquad \text{at } z = 0,$$
(12.46)

$$\frac{\partial \phi_2}{\partial z} = \frac{\partial \zeta}{\partial t} + U_2\frac{\partial \zeta}{\partial x} \qquad \text{at } z = 0.$$
(12.47)

The dynamic boundary condition at the interface is that the pressure must be continuous across the interface (if surface tension is neglected), requiring $p_1 = p_2$

at $z = \zeta$. The unsteady Bernoulli equations are

$$\frac{\partial \tilde{\phi}_1}{\partial t} + \frac{1}{2}(\nabla \tilde{\phi}_1)^2 + \frac{\tilde{p}_1}{\rho_1} + gz = C_1,$$

$$\frac{\partial \tilde{\phi}_2}{\partial t} + \frac{1}{2}(\nabla \tilde{\phi}_2)^2 + \frac{\tilde{p}_2}{\rho_2} + gz = C_2.$$

(12.48)

In order that the pressure be continuous in the *undisturbed state* ($P_1 = P_2$ at $z = 0$), the Bernoulli equation requires

$$\rho_1 \left(\frac{1}{2}U_1^2 - C_1 \right) = \rho_2 \left(\frac{1}{2}U_2^2 - C_2 \right).$$

(12.49)

Introducing the decomposition (12.43) into the Bernoulli equations (12.48), and requiring $\tilde{p}_1 = \tilde{p}_2$ at $z = \zeta$, we obtain the following condition at the interface:

$$\rho_1 C_1 - \rho_1 \frac{\partial \phi_1}{\partial t} - \frac{\rho_1}{2}[(U_1 + u_1)^2 + v_1^2 + w_1^2] - \rho_1 g \zeta$$

$$= \rho_2 C_2 - \rho_2 \frac{\partial \phi_2}{\partial t} - \frac{\rho_2}{2}[(U_2 + u_2)^2 + v_2^2 + w_2^2] - \rho_2 g \zeta.$$

Subtracting the basic state condition (12.49) and neglecting nonlinear terms, we obtain

$$\rho_1 \left[\frac{\partial \phi_1}{\partial t} + U_1 \frac{\partial \phi_1}{\partial x} + g \zeta \right]_{z=0} = \rho_2 \left[\frac{\partial \phi_2}{\partial t} + U_2 \frac{\partial \phi_2}{\partial x} + g \zeta \right]_{z=0}.$$

(12.50)

The perturbations therefore satisfy equation (12.44), and conditions (12.45), (12.46), (12.47), and (12.50). Assume normal modes of the form

$$(\zeta, \phi_1, \phi_2) = (\hat{\zeta}, \hat{\phi}_1, \hat{\phi}_2) \, e^{ik(x-ct)},$$

where k is real (and can be taken positive without loss of generality), but $c = c_r + ic_i$ is complex. The flow is unstable if there exists a positive c_i. (Note that in the preceding sections we assumed a time dependence of the form $\exp(\sigma t)$, which is more convenient when the instability appears in the form of convective cells.) Substitution of the normal modes into the Laplace equations (12.44) requires solutions of the form

$$\hat{\phi}_1 = A \, e^{-kz},$$

$$\hat{\phi}_2 = B \, e^{kz},$$

where solutions exponentially increasing from the interface are ignored because of equation (12.45).

Now equations (12.46), (12.47), and (12.50) give three homogeneous linear algebraic equations for determining the three unknowns $\hat{\zeta}$, A, and B; solutions can therefore exist only for certain values of $c(k)$. The kinematic conditions (12.46) and (12.47) give

$$A = -i(U_1 - c)\hat{\zeta},$$
$$B = i(U_2 - c)\hat{\zeta}.$$

The Bernoulli equation (12.50) gives

$$\rho_1[ik(U_1 - c)A + g\hat{\zeta}] = \rho_2[ik(U_2 - c)B + g\hat{\zeta}].$$

Substituting for A and B, this gives the eigenvalue relation for $c(k)$:

$$k\rho_2(U_2 - c)^2 + k\rho_1(U_1 - c)^2 = g(\rho_2 - \rho_1),$$

for which the solutions are

$$c = \frac{\rho_2 U_2 + \rho_1 U_1}{\rho_2 + \rho_1} \pm \left[\frac{g}{k}\frac{\rho_2 - \rho_1}{\rho_2 + \rho_1} - \rho_1\rho_2\left(\frac{U_1 - U_2}{\rho_2 + \rho_1}\right)^2\right]^{1/2}. \qquad (12.51)$$

It is seen that both solutions are neutrally stable (c real) as long as the second term within the square root is smaller than the first; this gives the stable waves of the system. However, there is a growing solution ($c_i > 0$) if

$$g(\rho_2^2 - \rho_1^2) < k\rho_1\rho_2(U_1 - U_2)^2.$$

Equation (12.51) shows that for each growing solution there is a corresponding decaying solution. As explained more fully in the following section, this happens because the coefficients of the differential equation and the boundary conditions are all real. Note also that the dispersion relation of free waves in an initial static medium, given by Equation (7.105), is obtained from equation (12.51) by setting $U_1 = U_2 = 0$.

If $U_1 \neq U_2$, then one can always find a large enough k that satisfies the requirement for instability. Because all wavelengths must be allowed in an instability analysis, we can say that the *flow is always unstable (to short waves) if $U_1 \neq U_2$*.

Consider now the flow of a homogeneous fluid ($\rho_1 = \rho_2$) with a velocity discontinuity, which we can call a *vortex sheet*. Equation (12.51) gives

$$c = \frac{1}{2}(U_1 + U_2) \pm \frac{i}{2}(U_1 - U_2).$$

The vortex sheet is therefore always unstable to all wavelengths. It is also seen that the unstable wave moves with a phase velocity equal to the average velocity of the basic flow. This must be true from symmetry considerations. In a frame of reference moving with the average velocity, the basic flow is symmetric and the wave therefore should have no preference between the positive and negative x directions (Figure 12.15).

The Kelvin–Helmholtz instability is caused by the destabilizing effect of shear, which overcomes the stabilizing effect of stratification. This kind of instability is easy

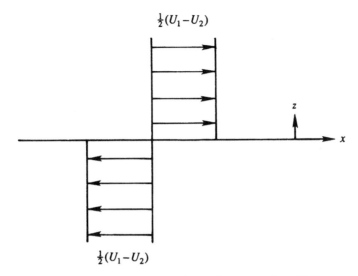

Figure 12.15 Background velocity field as seen by an observer moving with the average velocity $(U_1 + U_2)/2$ of two layers.

Figure 12.16 Kelvin–Helmholtz instability generated by tilting a horizontal channel containing two liquids of different densities. The lower layer is dyed. Mean flow in the lower layer is down the plane and that in the upper layer is up the plane. S. A. Thorpe, *Journal of Fluid Mechanics* **46**: 299–319, 1971 and reprinted with the permission of Cambridge University Press.

to generate in the laboratory by filling a horizontal glass tube (of rectangular cross section) containing two liquids of slightly different densities (one colored) and gently tilting it. This starts a current in the lower layer down the plane and a current in the upper layer up the plane. An example of instability generated in this manner is shown in Figure 12.16.

Shear instability of stratified fluids is ubiquitous in the atmosphere and the ocean and believed to be a major source of internal waves in them. Figure 12.17 is a striking photograph of a cloud pattern, which is clearly due to the existence of high shear across a sharp density gradient. Similar photographs of injected dye have been recorded in oceanic thermoclines (Woods, 1969).

Figure 12.17   Billow cloud near Denver, Colorado. P. G. Drazin and W. H. Reid, *Hydrodynamic Stability*, 1981 and reprinted with the permission of Cambridge University Press.

Figures 12.16 and 12.17 show the advanced nonlinear stage of the instability in which the interface is a rolled-up layer of vorticity. Such an observed evolution of the interface is in agreement with results of numerical calculations in which the nonlinear terms are retained (Figure 12.18).

The source of energy for generating the Kelvin–Helmholtz instability is derived from the kinetic energy of the shear flow. The disturbances essentially smear out the gradients until they cannot grow any longer. Figure 12.19 shows a typical behavior, in which the unstable waves at the interface have transformed the sharp density profile ACDF to ABEF and the sharp velocity profile MOPR to MNQR. The high-density fluid in the depth range DE has been raised upward (and mixed with the lower-density fluid in the depth range BC), which means that the potential energy of the system has increased after the instability. The required energy has been drawn from the kinetic energy of the basic field. It is easy to show that the kinetic energy of the initial profile MOPR is larger than that of the final profile MNQR. To see this, assume that the initial velocity of the lower layer is zero and that of the upper layer is U_1. Then the linear velocity profile after mixing is given by

$$U(z) = U_1 \left(\frac{1}{2} + \frac{z}{2h} \right) \quad -h \leqslant z \leqslant h.$$

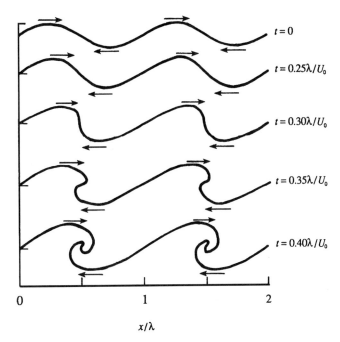

Figure 12.18 Nonlinear numerical calculation of the evolution of a vortex sheet that has been given a small sinusoidal displacement of wavelength λ. The density difference across the interface is zero, and U_0 is the velocity difference across the sheet. J. S. Turner, *Buoyancy Effects in Fluids*, 1973 and reprinted with the permission of Cambridge University Press.

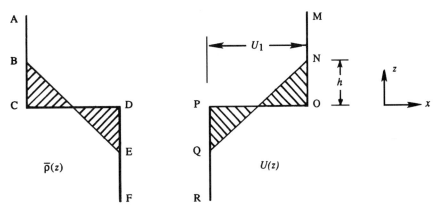

Figure 12.19 Smearing out of sharp density and velocity profiles, resulting in an increase of potential energy and a decrease of kinetic energy.

Consider the change in kinetic energy only in the depth range $-h < z < h$, as the energy outside this range does not change. Then the initial and final kinetic energies per unit width are

$$E_{\text{initial}} = \frac{\rho}{2} U_1^2 h,$$

$$E_{\text{final}} = \frac{\rho}{2} \int_{-h}^{h} U^2(z)\, dz = \frac{\rho}{3} U_1^2 h.$$

The kinetic energy of the flow has therefore decreased, although the total momentum $(= \int U\, dz)$ is unchanged. This is a general result: If the integral of $U(z)$ does not change, then the integral of $U^2(z)$ decreases if the gradients decrease.

In this section we have considered the case of a discontinuous variation across an infinitely thin interface and shown that the flow is always unstable. The case of continuous variation is considered in the following section. We shall see that a certain condition must be satisfied in order for the flow to be unstable.

7. Instability of Continuously Stratified Parallel Flows

An instability of great geophysical importance is that of an inviscid stratified fluid in horizontal parallel flow. If the density and velocity vary discontinuously across an interface, the analysis in the preceding section shows that the flow is unconditionally unstable. Although only the discontinuous case was studied by Kelvin and Helmholtz, the more general case of continuous distribution is also commonly called the *Kelvin–Helmholtz instability*.

The problem has a long history. In 1915, Taylor, on the basis of his calculations with assumed distributions of velocity and density, *conjectured* that a gradient Richardson number (to be defined shortly) must be less than $\frac{1}{4}$ for instability. Other values of the critical Richardson number (ranging from 2 to $\frac{1}{4}$) were suggested by Prandtl, Goldstein, Richardson, Synge, and Chandrasekhar. Finally, Miles (1961) was able to prove Taylor's conjecture, and Howard (1961) immediately and elegantly generalized Miles' proof. A short record of the history is given in Miles (1986). In this section we shall prove the Richardson number criterion in the manner given by Howard.

Taylor–Goldstein Equation

Consider a horizontal parallel flow $U(z)$ directed along the x-axis. The z-axis is taken vertically upwards. The basic flow is in equilibrium with the undisturbed density field $\bar{\rho}(z)$ and the basic pressure field $P(z)$. We shall only consider two-dimensional disturbances on this basic state, assuming that they are more unstable than three-dimensional disturbances; this is called *Squires' theorem* and is demonstrated in Section 8 in another context. The disturbed state has velocity, pressure, and density fields of

$$[U + u, 0, w], \qquad P + p, \qquad \bar{\rho} + \rho.$$

The continuity equation reduces to

$$\frac{\partial u}{\partial x} + \frac{\partial w}{\partial z} = 0.$$

The disturbed velocity field is assumed to satisfy the Boussinesq equation

$$\frac{\partial}{\partial t}(U_i + u_i) + (U_j + u_j)\frac{\partial}{\partial x_j}(U_i + u_i) = -\frac{g}{\rho_0}(\bar{\rho} + \rho)\delta_{i3} - \frac{1}{\rho_0}\frac{\partial}{\partial x_i}(P + p),$$

where the density variations are neglected except in the vertical equation of motion. Here, ρ_0 is a reference density. The basic flow satisfies

$$0 = -\frac{g\bar{\rho}}{\rho_0}\delta_{i3} - \frac{1}{\rho_0}\frac{\partial P}{\partial x_i}.$$

Subtracting the last two equations and dropping nonlinear terms, we obtain the perturbation equation of motion

$$\frac{\partial u_i}{\partial t} + u_j\frac{\partial U_i}{\partial x_j} + U_j\frac{\partial u_i}{\partial x_j} = -\frac{g\rho}{\rho_0}\delta_{i3} - \frac{1}{\rho_0}\frac{\partial p}{\partial x_i}.$$

The $i = 1$ and $i = 3$ components of the preceding equation are

$$\frac{\partial u}{\partial t} + w\frac{\partial U}{\partial z} + U\frac{\partial u}{\partial x} = -\frac{1}{\rho_0}\frac{\partial p}{\partial x},$$
$$\frac{\partial w}{\partial t} + U\frac{\partial w}{\partial x} = -\frac{g\rho}{\rho_0} - \frac{1}{\rho_0}\frac{\partial p}{\partial z}. \tag{12.52}$$

In the absence of diffusion the density is conserved along the motion, which requires that $D(\text{density})/Dt = 0$, or that

$$\frac{\partial}{\partial t}(\bar{\rho} + \rho) + (U + u)\frac{\partial}{\partial x}(\bar{\rho} + \rho) + w\frac{\partial}{\partial z}(\bar{\rho} + \rho) = 0.$$

Keeping only the linear terms, and using the fact that $\bar{\rho}$ is a function of z only, we obtain

$$\frac{\partial \rho}{\partial t} + U\frac{\partial \rho}{\partial x} + w\frac{d\bar{\rho}}{dz} = 0,$$

which can be written as

$$\frac{\partial \rho}{\partial t} + U\frac{\partial \rho}{\partial x} - \frac{\rho_0 N^2 w}{g} = 0, \tag{12.53}$$

where we have defined

$$N^2 \equiv -\frac{g}{\rho_0}\frac{d\bar{\rho}}{dz}.$$

N is the buoyancy frequency. The last term in equation (12.53) represents the density change at a point due to the vertical advection of the basic density field across the point.

The continuity equation can be satisfied by defining a streamfunction through

$$u = \frac{\partial \psi}{\partial z}, \qquad w = -\frac{\partial \psi}{\partial x}.$$

Equations (12.52) and (12.53) then become

$$\psi_{zt} - \psi_x U_z + \psi_{xz} U = -\frac{1}{\rho_0} p_x,$$

$$-\psi_{xt} - \psi_{xx} U = -\frac{g\rho}{\rho_0} - \frac{1}{\rho_0} p_z, \qquad (12.54)$$

$$\rho_t + U\rho_x + \frac{\rho_0 N^2}{g} \psi_x = 0,$$

where *subscripts denote partial derivatives*.

As the coefficients of equation (12.54) are independent of x and t, exponential variations in these variables are allowed. Consequently, we assume normal mode solutions of the form

$$[\rho, p, \psi] = [\hat{\rho}(z), \hat{p}(z), \hat{\psi}(z)]\, e^{ik(x-ct)},$$

where quantities denoted by ($\hat{}$) are complex amplitudes. Because the flow is unbounded in x, the wavenumber k must be real. The eigenvalue $c = c_r + ic_i$ can be complex, and the solution is unstable if there exists a $c_i > 0$. Substituting the normal modes, equation (12.54) becomes

$$(U - c)\hat{\psi}_z - U_z\hat{\psi} = -\frac{1}{\rho_0} \hat{p}, \qquad (12.55)$$

$$k^2(U - c)\hat{\psi} = -\frac{g\hat{\rho}}{\rho_0} - \frac{1}{\rho_0} \hat{p}_z, \qquad (12.56)$$

$$(U - c)\hat{\rho} + \frac{\rho_0 N^2}{g} \hat{\psi} = 0. \qquad (12.57)$$

We want to obtain a single equation in $\hat{\psi}$. The pressure can be eliminated by taking the z-derivative of equation (12.55) and subtracting equation (12.56). The density can be eliminated by equation (12.57). This gives

$$(U - c)\left(\frac{d^2}{dz^2} - k^2\right)\hat{\psi} - U_{zz}\hat{\psi} + \frac{N^2}{U - c}\hat{\psi} = 0. \qquad (12.58)$$

This is the *Taylor–Goldstein equation*, which governs the behavior of perturbations in a stratified parallel flow. Note that the complex conjugate of the equation is also a valid equation because we can take the imaginary part of the equation, change the sign, and add to the real part of the equation. Now because the Taylor–Goldstein equation does not involve any i, a complex conjugate of the equation shows that if $\hat{\psi}$ is an eigenfunction with eigenvalue c for some k, then $\hat{\psi}^*$ is a possible eigenfunction with eigenvalue c^* for the same k. Therefore, to each eigenvalue with a positive c_i there is a corresponding eigenvalue with a negative c_i. In other words, *to each growing mode there is a corresponding decaying mode.* A nonzero c_i therefore ensures instability.

The boundary conditions are that $w = 0$ on rigid boundaries at $z = 0, d$. This requires $\psi_x = ik\hat{\psi} \exp(ikx - ikct) = 0$ at the walls, which is possible only if

$$\hat{\psi}(0) = \hat{\psi}(d) = 0. \qquad (12.59)$$

Richardson Number Criterion

A necessary condition for linear instability of inviscid stratified parallel flows can be derived by defining a new variable ϕ by

$$\phi \equiv \frac{\hat{\psi}}{\sqrt{U - c}} \quad \text{or} \quad \hat{\psi} = (U - c)^{1/2}\phi.$$

Then we obtain the derivatives

$$\hat{\psi}_z = (U - c)^{1/2}\phi_z + \frac{\phi U_z}{2(U - c)^{1/2}},$$

$$\hat{\psi}_{zz} = (U - c)^{1/2}\phi_{zz} + \frac{U_z\phi_z + (1/2)\phi U_{zz}}{(U - c)^{1/2}} - \frac{1}{4}\frac{\phi U_z^2}{(U - c)^{3/2}}.$$

The Taylor–Goldstein equation then becomes, after some rearrangement,

$$\frac{d}{dz}\{(U - c)\phi_z\} - \left\{ k^2(U - c) + \frac{1}{2}U_{zz} + \frac{(1/4)U_z^2 - N^2}{U - c} \right\}\phi = 0. \qquad (12.60)$$

Now multiply equation (12.60) by ϕ^* (the complex conjugate of ϕ), integrate from $z = 0$ to $z = d$, and use the boundary conditions $\phi(0) = \phi(d) = 0$. The first term gives

$$\int \frac{d}{dz}\{(U - c)\phi_z\}\phi^*\, dz = \int \left[\frac{d}{dz}\{(U - c)\phi_z\phi^*\} - (U - c)\phi_z\phi_z^* \right] dz$$

$$= -\int (U - c)|\phi_z|^2\, dz,$$

where we have used $\phi = 0$ at the boundaries. Integrals of the other terms in equation (12.60) are also simple to manipulate. We finally obtain

$$\int \frac{N^2 - (1/4)U_z^2}{U - c}|\phi|^2\, dz = \int (U - c)\{|\phi_z|^2 + k^2|\phi|^2\}\, dz$$

$$+ \frac{1}{2}\int U_{zz}|\phi|^2\, dz. \qquad (12.61)$$

The last term in the preceding is real. The imaginary part of the first term can be found by noting that

$$\frac{1}{U - c} = \frac{U - c^*}{|U - c|^2} = \frac{U - c_r + ic_i}{|U - c|^2}.$$

Then the imaginary part of equation (12.61) gives

$$c_i \int \frac{N^2 - (1/4)U_z^2}{|U - c|^2} |\phi|^2 \, dz = -c_i \int \{|\phi_z|^2 + k^2 |\phi|^2\} \, dz.$$

The integral on the right-hand side is positive. If the flow is such that $N^2 > U_z^2/4$ everywhere, then the preceding equation states that c_i times a positive quantity equals c_i times a negative quantity; this is impossible and requires that $c_i = 0$ for such a case. Defining the *gradient Richardson number*

$$\mathrm{Ri}(z) \equiv \frac{N^2}{U_z^2}, \tag{12.62}$$

we can say that *linear stability is guaranteed if the inequality*

$$\boxed{\mathrm{Ri} > \tfrac{1}{4}} \quad \text{(stable)}, \tag{12.63}$$

is satisfied everywhere in the flow.

Note that the criterion does not state that the flow is necessarily unstable if $\mathrm{Ri} < \frac{1}{4}$ somewhere, or even everywhere, in the flow. Thus $\mathrm{Ri} < \frac{1}{4}$ is a *necessary* but not sufficient condition for instability. For example, in a jetlike velocity profile $u \propto \mathrm{sech}^2 z$ and an exponential density profile, the flow does not become unstable until the Richardson number falls below 0.214. A critical Richardson number lower than $\frac{1}{4}$ is also found in the presence of boundaries, which stabilize the flow. In fact, there is no unique critical Richardson number that applies to all distributions of $U(z)$ and $N(z)$. However, several calculations show that in many shear layers (having linear, tanh, or error function profiles for velocity and density) the flow does become unstable to disturbances of certain wavelengths if the minimum value of Ri in the flow (which is generally at the center of the shear layer) is less than $\frac{1}{4}$. The "most unstable" wave, defined as the first to become unstable as Ri is reduced below $\frac{1}{4}$, is found to have a wavelength $\lambda \simeq 7h$, where h is the thickness of the shear layer. Laboratory (Scotti and Corcos, 1972) as well as geophysical observations (Eriksen, 1978) show that the requirement

$$\mathrm{Ri}_{\min} < \tfrac{1}{4},$$

is a useful guide for the prediction of instability of a stratified shear layer.

Howard's Semicircle Theorem

A useful result concerning the behavior of the complex phase speed c in an inviscid parallel shear flow, valid both with and without stratification, was derived by Howard (1961). To derive this, first substitute

$$F \equiv \frac{\hat{\psi}}{U - c},$$

in the Taylor–Goldstein equation (12.58). With the derivatives

$$\hat{\psi}_z = (U - c)F_z + U_z F,$$

$$\hat{\psi}_{zz} = (U - c)F_{zz} + 2U_z F_z + U_{zz} F,$$

Equation (12.58) gives

$$(U - c)[(U - c)F_{zz} + 2U_z F_z - k^2(U - c)F] + N^2 F = 0,$$

where terms involving U_{zz} have canceled out. This can be rearranged in the form

$$\frac{d}{dz}[(U - c)^2 F_z] - k^2(U - c)^2 F + N^2 F = 0.$$

Multiplying by F^*, integrating (by parts if necessary) over the depth of flow, and using the boundary conditions, we obtain

$$-\int (U - c)^2 F_z F_z^* \, dz - k^2 \int (U - c)^2 |F|^2 \, dz + \int N^2 |F|^2 \, dz = 0,$$

which can be written as

$$\int (U - c)^2 Q \, dz = \int N^2 |F|^2 \, dz,$$

where

$$Q \equiv |F_z|^2 + k^2 |F|^2,$$

is positive. Equating real and imaginary parts, we obtain

$$\int [(U - c_r)^2 - c_i^2] Q \, dz = \int N^2 |F|^2 \, dz, \qquad (12.64)$$

$$c_i \int (U - c_r) Q \, dz = 0. \qquad (12.65)$$

For instability $c_i \neq 0$, for which equation (12.65) shows that $(U - c_r)$ must change sign somewhere in the flow, that is,

$$\boxed{U_{min} < c_r < U_{max},} \qquad (12.66)$$

which states that c_r lies in the range of U. Recall that we have assumed solutions of the form

$$e^{ik(x-ct)} = e^{ik(x-c_r t)} e^{kc_i t},$$

which means that c_r is the phase velocity in the positive x direction, and kc_i is the growth rate. Equation (12.66) shows that c_r is positive if U is everywhere positive, and is negative if U is everywhere negative. In these cases we can say that unstable waves propagate in the direction of the background flow.

Limits on the maximum growth rate can also be predicted. Equation (12.64) gives

$$\int [U^2 + c_r^2 - 2Uc_r - c_i^2]Q \, dz > 0,$$

which, on using equation (12.65), becomes

$$\int (U^2 - c_r^2 - c_i^2)Q \, dz > 0. \tag{12.67}$$

Now because $(U_{min} - U) < 0$ and $(U_{max} - U) > 0$, it is always true that

$$\int (U_{min} - U)(U_{max} - U)Q \, dz \leqslant 0,$$

which can be recast as

$$\int [U_{max}U_{min} + U^2 - U(U_{max} + U_{min})]Q \, dz \leqslant 0.$$

Using equation (12.67), this gives

$$\int [U_{max}U_{min} + c_r^2 + c_i^2 - U(U_{max} + U_{min})]Q \, dz \leqslant 0.$$

On using equation (12.65), this becomes

$$\int [U_{max}U_{min} + c_r^2 + c_i^2 - c_r(U_{max} + U_{min})]Q \, dz \leqslant 0.$$

Because the quantity within [ ] is independent of z, and $\int Q \, dz > 0$, we must have [ ] $\leqslant 0$. With some rearrangement, this condition can be written as

$$\left[c_r - \frac{1}{2}(U_{max} + U_{min})\right]^2 + c_i^2 \leqslant \left[\frac{1}{2}(U_{max} - U_{min})\right]^2.$$

This shows that the *complex wave velocity c of any unstable mode of a disturbance in parallel flows of an inviscid fluid must lie inside the semicircle in the upper half of the c-plane, which has the range of U as the diameter* (Figure 12.20). This is called the *Howard semicircle theorem.* It states that the maximum growth rate is limited by

$$kc_i < \frac{k}{2}(U_{max} - U_{min}).$$

The theorem is very useful in searching for eigenvalues $c(k)$ in numerical solution of instability problems.

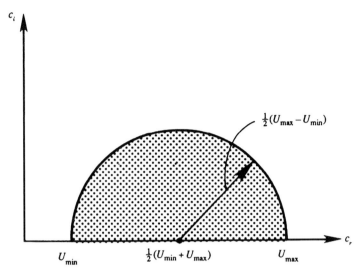

$\frac{1}{2}(U_{max} - U_{min})$

U_{min}

$\frac{1}{2}(U_{min} + U_{max})$

U_{max}

Figure 12.20 The Howard semicircle theorem. In several inviscid parallel flows the complex eigenvalue c must lie within the semicircle shown.

8. Squire's Theorem and Orr–Sommerfeld Equation

In our studies of the Bénard and Taylor problems, we encountered two flows in which viscosity has a stabilizing effect. Curiously, viscous effects can also be *destabilizing*, as indicated by several calculations of wall-bounded parallel flows. In this section we shall derive the equation governing the stability of parallel flows of a homogeneous viscous fluid. Let the primary flow be directed along the x direction and vary in the y direction so that $\mathbf{U} = [U(y), 0, 0]$. We decompose the total flow as the sum of the basic flow plus the perturbation:

$$\tilde{\mathbf{u}} = [U + u, v, w],$$
$$\tilde{p} = P + p.$$

Both the background and the perturbed flows satisfy the Navier–Stokes equations. The perturbed flow satisfies the x-momentum equation

$$\frac{\partial u}{\partial t} + (U + u)\frac{\partial}{\partial x}(U + u) + v\frac{\partial}{\partial y}(U + u)$$
$$= -\frac{\partial}{\partial x}(P + p) + \frac{1}{\mathrm{Re}}\nabla^2(U + u), \qquad (12.68)$$

where the variables have been nondimensionalized by a characteristic length scale L (say, the width of flow), and a characteristic velocity U_0 (say, the maximum velocity of the basic flow); time is scaled by L/U_0 and the pressure is scaled by ρU_0^2. The Reynolds number is defined as $\mathrm{Re} = U_0 L/\nu$.

The background flow satisfies

$$0 = -\frac{\partial P}{\partial x} + \frac{1}{\text{Re}}\nabla^2 U.$$

Subtracting from equation (12.68) and neglecting terms nonlinear in the perturbations, we obtain the x-momentum equation for the perturbations:

$$\frac{\partial u}{\partial t} + U\frac{\partial u}{\partial x} + v\frac{\partial U}{\partial y} = -\frac{\partial p}{\partial x} + \frac{1}{\text{Re}}\nabla^2 u. \tag{12.69}$$

Similarly the y-momentum, z-momentum, and continuity equations for the perturbations are

$$\frac{\partial v}{\partial t} + U\frac{\partial v}{\partial x} = -\frac{\partial p}{\partial y} + \frac{1}{\text{Re}}\nabla^2 v,$$

$$\frac{\partial w}{\partial t} + U\frac{\partial w}{\partial x} = -\frac{\partial p}{\partial z} + \frac{1}{\text{Re}}\nabla^2 w, \tag{12.70}$$

$$\frac{\partial u}{\partial x} + \frac{\partial v}{\partial y} + \frac{\partial w}{\partial z} = 0.$$

The coefficients in the perturbation equations (12.69) and (12.70) depend only on y, so that the equations admit solutions exponential in x, z, and t. Accordingly, we assume normal modes of the form

$$[\mathbf{u}, p] = [\hat{\mathbf{u}}(y),\ \hat{p}(y)]\, e^{i(kx+mz-kct)}. \tag{12.71}$$

As the flow is unbounded in x and z, the wavenumber components k and m must be real. The wave speed $c = c_r + ic_i$ may be complex. Without loss of generality, we can consider only positive values for k and m; the sense of propagation is then left open by keeping the sign of c_r unspecified. The normal modes represent waves that travel obliquely to the basic flow with a wavenumber of magnitude $\sqrt{k^2 + m^2}$ and have an amplitude that varies in time as $\exp(kc_i t)$. Solutions are therefore stable if $c_i < 0$ and unstable if $c_i > 0$.

On substitution of the normal modes, the perturbation equations (12.69) and (12.70) become

$$ik(U - c)\hat{u} + \hat{v}U_y = -ik\hat{p} + \frac{1}{\text{Re}}[\hat{u}_{yy} - (k^2 + m^2)\hat{u}],$$

$$ik(U - c)\hat{v} = -\hat{p}_y + \frac{1}{\text{Re}}[\hat{v}_{yy} - (k^2 + m^2)\hat{v}],$$

$$ik(U - c)\hat{w} = -im\hat{p} + \frac{1}{\text{Re}}[\hat{w}_{yy} - (k^2 + m^2)\hat{w}], \tag{12.72}$$

$$ik\hat{u} + \hat{v}_y + im\hat{w} = 0,$$

where subscripts denote derivatives with respect to y. These are the normal mode equations for three-dimensional disturbances. Before proceeding further, we shall first show that only two-dimensional disturbances need to be considered.

Squire's Theorem

A very useful simplification of the normal mode equations was achieved by Squire in 1933, showing that *to each unstable three-dimensional disturbance there corresponds a more unstable two-dimensional one*. To prove this theorem, consider the *Squire transformation*

$$\begin{aligned}
\bar{k} &= (k^2 + m^2)^{1/2}, & \bar{c} &= c, \\
\bar{k}\bar{u} &= k\hat{u} + m\hat{w}, & \bar{v} &= \hat{v}, \\
\frac{\bar{p}}{\bar{k}} &= \frac{\hat{p}}{k}, & \bar{k}\,\overline{\text{Re}} &= k\,\text{Re}.
\end{aligned} \tag{12.73}$$

In substituting these transformations into equation (12.72), the first and third of equation (12.72) are added; the rest are simply transformed. The result is

$$i\bar{k}(U - c)\bar{u} + \bar{v}U_y = -i\bar{k}\bar{p} + \frac{1}{\overline{\text{Re}}}[\bar{u}_{yy} - \bar{k}^2\bar{u}],$$

$$i\bar{k}(U - c)\bar{v} = -\bar{p}_y + \frac{1}{\overline{\text{Re}}}[\bar{v}_{yy} - \bar{k}^2\bar{v}],$$

$$i\bar{k}\bar{u} + \bar{v}_y = 0.$$

These equations are exactly the same as equation (12.72), but with $m = \hat{w} = 0$. Thus, to each three-dimensional problem corresponds an equivalent two-dimensional one. Moreover, Squire's transformation (12.73) shows that the equivalent two-dimensional problem is associated with a *lower* Reynolds number as $\bar{k} > k$. It follows that the critical Reynolds number at which the instability starts is lower for two-dimensional disturbances. Therefore, we only need to consider a two-dimensional disturbance if we want to determine the minimum Reynolds number for the onset of instability.

The three-dimensional disturbance (12.71) is a wave propagating obliquely to the basic flow. If we orient the coordinate system with the new x-axis in this direction, the equations of motion are such that only the component of basic flow in this direction affects the disturbance. Thus, the effective Reynolds number is reduced.

An argument without using the Reynolds number is now given because Squire's theorem also holds for several other problems that do not involve the Reynolds number. Equation (12.73) shows that the growth rate for a two-dimensional disturbance is $\exp(\bar{k}\bar{c}_i t)$, whereas equation (12.71) shows that the growth rate of a three-dimensional disturbance is $\exp(kc_i t)$. The two-dimensional growth rate is therefore larger because Squire's transformation requires $\bar{k} > k$ and $\bar{c} = c$. We can therefore say that the two-dimensional disturbances are more unstable.

Orr–Sommerfeld Equation

Because of Squire's theorem, we only need to consider the set (12.72) with $m = \hat{w} = 0$. The two-dimensionality allows the definition of a streamfunction $\psi(x, y, t)$ for the perturbation field by

$$u = \frac{\partial \psi}{\partial y}, \qquad v = -\frac{\partial \psi}{\partial x}.$$

We assume normal modes of the form

$$[u, v, \psi] = [\hat{u}, \hat{v}, \phi]\, e^{ik(x-ct)}.$$

(To be consistent, we should denote the complex amplitude of ψ by $\hat{\psi}$; we are using ϕ instead to follow the standard notation for this variable in the literature.) Then we must have

$$\hat{u} = \phi_y, \qquad \hat{v} = -ik\phi.$$

A single equation in terms of ϕ can now be found by eliminating the pressure from the set (12.72). This gives

$$(U - c)(\phi_{yy} - k^2\phi) - U_{yy}\phi = \frac{1}{ik\,\mathrm{Re}}[\phi_{yyyy} - 2k^2\phi_{yy} + k^4\phi], \qquad (12.74)$$

where subscripts denote derivatives with respect to y. It is a fourth-order ordinary differential equation. The boundary conditions at the walls are the no-slip conditions $u = v = 0$, which require

$$\phi = \phi_y = 0 \quad \text{at } y = y_1 \text{ and } y_2. \qquad (12.75)$$

Equation (12.74) is the well-known *Orr–Sommerfeld equation*, which governs the stability of nearly parallel viscous flows such as those in a straight channel or in a boundary layer. It is essentially a vorticity equation because the pressure has been eliminated. Solutions of the Orr–Sommerfeld equations are difficult to obtain, and only the results of some simple flows will be discussed in the later sections. However, we shall first discuss certain results obtained by ignoring the viscous term in this equation.

9. *Inviscid Stability of Parallel Flows*

Useful insights into the viscous stability of parallel flows can be obtained by first assuming that the disturbances obey inviscid dynamics. The governing equation can be found by letting $\mathrm{Re} \to \infty$ in the Orr–Sommerfeld equation, giving

$$(U - c)[\phi_{yy} - k^2\phi] - U_{yy}\phi = 0, \qquad (12.76)$$

which is called the *Rayleigh equation*. If the flow is bounded by walls at y_1 and y_2 where $v = 0$, then the boundary conditions are

$$\phi = 0 \quad \text{at } y = y_1 \text{ and } y_2. \qquad (12.77)$$

The set (12.76) and (12.77) defines an eigenvalue problem, with $c(k)$ as the eigenvalue and ϕ as the eigenfunction. As the equations do not involve i, taking the complex conjugate shows that if ϕ is an eigenfunction with eigenvalue c for some k, then ϕ^* is also an eigenfunction with eigenvalue c^* for the same k. Therefore, to each eigenvalue with a positive c_i there is a corresponding eigenvalue with a negative c_i.

In other words, *to each growing mode there is a corresponding decaying mode*. Stable solutions therefore can have only a real c. Note that this is true of inviscid flows only. The viscous term in the full Orr–Sommerfeld equation (12.74) involves an i, and the foregoing conclusion is no longer valid.

We shall now show that certain velocity distributions $U(y)$ are potentially unstable according to the inviscid Rayleigh equation (12.76). In this discussion it should be noted that we are only assuming that the *disturbances* obey inviscid dynamics; the background flow $U(y)$ may be chosen to be chosen to be any profile, for example, that of viscous flows such as Poiseuille flow or Blasius flow.

Rayleigh's Inflection Point Criterion

Rayleigh proved that *a necessary (but not sufficient) criterion for instability of an inviscid parallel flow is that the basic velocity profile $U(y)$ has a point of inflection.*

To prove the theorem, rewrite the Rayleigh equation (12.76) in the form

$$\phi_{yy} - k^2\phi - \frac{U_{yy}}{U - c}\phi = 0,$$

and consider the unstable mode for which $c_i > 0$, and therefore $U - c \neq 0$. Multiply this equation by ϕ^*, integrate from y_1 to y_2, by parts if necessary, and apply the boundary condition $\phi = 0$ at the boundaries. The first term transforms as follows:

$$\int \phi^* \phi_{yy}\, dy = [\phi^* \phi_y]_{y_1}^{y_2} - \int \phi_y^* \phi_y\, dy = -\int |\phi_y|^2\, dy,$$

where the limits on the integrals have not been explicitly written. The Rayleigh equation then gives

$$\int [|\phi_y|^2 + k^2|\phi|^2]\, dy + \int \frac{U_{yy}}{U - c}|\phi|^2\, dy = 0. \tag{12.78}$$

The first term is real. The imaginary part of the second term can be found by multiplying the numerator and denominator by $(U - c^*)$. The imaginary part of equation (12.78) then gives

$$c_i \int \frac{U_{yy}|\phi|^2}{|U - c|^2}\, dy = 0. \tag{12.79}$$

For the unstable case, for which $c_i \neq 0$, equation (12.79) can be satisfied only if U_{yy} changes sign at least once in the *open* interval $y_1 < y < y_2$. In other words, for instability the background velocity distribution must have at least one point of inflection (where $U_{yy} = 0$) within the flow. Clearly, the existence of a point of inflection does not guarantee a nonzero c_i. The inflection point is therefore a necessary but not sufficient condition for inviscid instability.

Fjortoft's Theorem

Some seventy years after Rayleigh's discovery, the Swedish meteorologist Fjortoft in 1950 discovered a stronger necessary condition for the instability of inviscid parallel flows. He showed that *a necessary condition for instability of inviscid parallel flows is that $U_{yy}(U - U_I) < 0$ somewhere in the flow*, where U_I is the value of U at the point of inflection. To prove the theorem, take the real part of equation (12.78):

$$\int \frac{U_{yy}(U - c_r)}{|U - c|^2} |\phi|^2 \, dy = - \int [|\phi_y|^2 + k^2 |\phi|^2] \, dy < 0. \tag{12.80}$$

Suppose that the flow is unstable, so that $c_i \neq 0$, and a point of inflection does exist according to the Rayleigh criterion. Then it follows from equation (12.79) that

$$(c_r - U_I) \int \frac{U_{yy}|\phi|^2}{|U - c|^2} \, dy = 0. \tag{12.81}$$

Adding equations (12.80) and (12.81), we obtain

$$\int \frac{U_{yy}(U - U_I)}{|U - c|^2} |\phi|^2 \, dy < 0,$$

so that $U_{yy}(U - U_I)$ must be negative somewhere in the flow.

Some common velocity profiles are shown in Figure 12.21. Only the two flows shown in the bottom row can possibly be unstable, for only they satisfy Fjortoft's theorem. Flows (a), (b), and (c) do not have any inflection point: flow (d) does satisfy Rayleigh's condition but not Fjortoft's because $U_{yy}(U - U_I)$ is positive. Note that an alternate way of stating Fjortoft's theorem is that *the magnitude of vorticity of the basic flow must have a maximum within the region of flow*, not at the boundary. In flow (d), the maximum magnitude of vorticity occurs at the walls.

The criteria of Rayleigh and Fjortoft essentially point to the importance of having a point of inflection in the velocity profile. They show that flows in jets, wakes, shear layers, and boundary layers with adverse pressure gradients, all of which have a point of inflection and satisfy Fjortoft's theorem, are potentially unstable. On the other hand, plane Couette flow, Poiseuille flow, and a boundary layer flow with zero or favorable pressure gradient have no point of inflection in the velocity profile, and are stable in the inviscid limit.

However, neither of the two conditions is sufficient for instability. An example is the sinusoidal profile $U = \sin y$, with boundaries at $y = \pm b$. It has been shown that the flow is stable if the width is restricted to $2b < \pi$, although it has an inflection point at $y = 0$.

Critical Layers

Inviscid parallel flows satisfy Howard's semicircle theorem, which was proved in Section 7 for the more general case of a stratified shear flow. The theorem states that the phase speed c_r has a value that lies between the minimum and the maximum values of $U(y)$ in the flow field. Now growing and decaying modes are characterized by a nonzero c_i, whereas neutral modes can have only a real $c = c_r$. It follows that neutral modes must have $U = c$ somewhere in the flow field. The neighborhood y around y_c at which $U = c = c_r$ is called a *critical layer*. The point y_c is a critical

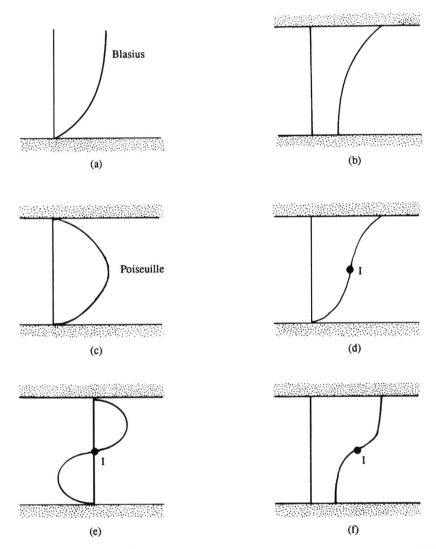

Figure 12.21 Examples of parallel flows. Points of inflection are denoted by I. Only (e) and (f) satisfy Fjortoft's criterion of inviscid instability.

point of the inviscid governing equation (12.76), because the highest derivative drops out at this value of y. The solution of the eigenfunction is discontinuous across this layer. The full Orr–Sommerfeld equation (12.74) has no such critical layer because the highest-order derivative does not drop out when $U = c$. It is apparent that in a real flow a viscous boundary layer must form at the location where $U = c$, and the layer becomes thinner as Re $\rightarrow \infty$.

The streamline pattern in the neighborhood of the critical layer where $U = c$ was given by Kelvin in 1888; our discussion here is adapted from Drazin and Reid (1981).

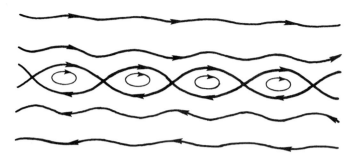

Figure 12.22 The Kelvin cat's eye pattern near a critical layer, showing streamlines as seen by an observer moving with the wave.

Consider a flow viewed by an observer moving with the phase velocity $c = c_r$. Then the basic velocity field seen by this observer is $(U - c)$, so that the streamfunction due to the basic flow is

$$\Psi = \int (U - c) \, dy.$$

The total streamfunction is obtained by adding the perturbation:

$$\tilde{\psi} = \int (U - c) \, dy + A\phi(y) \, e^{ikx}, \tag{12.82}$$

where A is an arbitrary constant, and we have omitted the time factor on the second term because we are considering only neutral disturbances. Near the critical layer $y = y_c$, a Taylor series expansion shows that equation (12.82) is approximately

$$\tilde{\psi} = \frac{1}{2} U_{yc} (y - y_c)^2 + A\phi(y_c) \cos kx,$$

where U_{yc} is the value of U_y at y_c; we have taken the real part of the right-hand side, and taken $\phi(y_c)$ to be real. The streamline pattern corresponding to the preceding equation is sketched in Figure 12.22, showing the so-called *Kelvin cat's eye pattern*.

10. Some Results of Parallel Viscous Flows

Our intuitive expectation is that viscous effects are stabilizing. The thermal and centrifugal convections discussed earlier in this chapter have confirmed this intuitive expectation. However, the conclusion that the effect of viscosity is stabilizing is not always true. Consider the Poiseuille flow and the Blasius boundary layer profiles in Figure 12.21, which do not have any inflection point and are therefore inviscidly stable. These flows are known to undergo transition to turbulence at some Reynolds number, which suggests that inclusion of viscous effects may in fact be *destabilizing* in these flows. Fluid viscosity may thus have a dual effect in the sense that it can be stabilizing as well as destabilizing. This is indeed true as shown by stability calculations of parallel viscous flows.

The analytical solution of the Orr–Sommerfeld equation is notoriously complicated and will not be presented here. The viscous term in (12.74) contains the highest-order derivative, and therefore the eigenfunction may contain regions of rapid variation in which the viscous effects become important. Sophisticated asymptotic techniques are therefore needed to treat these boundary layers. Alternatively, solutions can be obtained numerically. For our purposes, we shall discuss only certain features of these calculations. Additional information can be found in Drazin and Reid (1981), and in the review article by Bayly, Orszag, and Herbert (1988).

Mixing Layer

Consider a mixing layer with the velocity profile

$$U = U_0 \tanh \frac{y}{L}.$$

A stability diagram for solution of the Orr–Sommerfeld equation for this velocity distribution is sketched in Figure 12.23. It is seen that at all Reynolds numbers the flow is unstable to waves having low wavenumbers in the range $0 < k < k_u$, where the upper limit k_u depends on the Reynolds number $\mathrm{Re} = U_0 L / \nu$. For high values of Re, the range of unstable wavenumbers increases to $0 < k < 1/L$, which corresponds to a wavelength range of $\infty > \lambda > 2\pi L$. It is therefore essentially a long wavelength instability.

Figure 12.23 implies that the critical Reynolds number in a mixing layer is zero. In fact, viscous calculations for all flows with "inflectional profiles" show a small critical Reynolds number; for example, for a jet of the form $u = U \operatorname{sech}^2(y/L)$, it is $\mathrm{Re}_{\mathrm{cr}} = 4$. These wall-free shear flows therefore become unstable very quickly, and the inviscid criterion that these flows are always unstable is a fairly good description. The reason the inviscid analysis works well in describing the stability characteristics of free shear

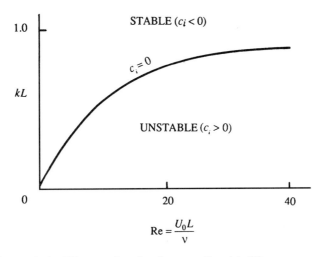

Figure 12.23 Marginal stability curve for a shear layer $u = U_0 \tanh(y/L)$.

flows can be explained as follows. For flows with inflection points the eigenfunction of the inviscid solution is smooth. On this zero-order approximation, the viscous term acts as a *regular* perturbation, and the resulting correction to the eigenfunction and eigenvalues can be computed as a perturbation expansion in powers of the small parameter 1/Re. This is true even though the viscous term in the Orr–Sommerfeld equation contains the highest-order derivative.

The instability in flows with inflection points is observed to form rolled-up blobs of vorticity, much like in the calculations of Figure 12.18 or in the photograph of Figure 12.16. This behavior is robust and insensitive to the detailed experimental conditions. They are therefore easily observed. In contrast, the unstable waves in a wall-bounded shear flow are extremely difficult to observe, as discussed in the next section.

Plane Poiseuille Flow

The flow in a channel with parabolic velocity distribution has no point of inflection and is inviscidly stable. However, linear viscous calculations show that the flow becomes unstable at a critical Reynolds number of 5780. Nonlinear calculations, which consider the distortion of the basic profile by the finite amplitude of the perturbations, give a critical number of 2510, which agrees better with the observed transition. In any case, the interesting point is that viscosity is *destabilizing* for this flow. The solution of the Orr–Sommerfeld equation for the Poiseuille flow and other parallel flows with rigid boundaries, which do not have an inflection point, is complicated. In contrast to flows with inflection points, the viscosity here acts as a *singular* perturbation, and the eigenfunction has viscous boundary layers on the channel walls and around critical layers where $U = c_r$. The waves that cause instability in these flows are called *Tollmien–Schlichting* waves, and their experimental detection is discussed in the next section. In his text, C. S. Yih gives a thorough discussion of the solution of the Orr-Sommerfeld equation using asymptotic expansions in the limit sequence $Re \rightarrow \infty$, then $k \rightarrow 0$ (but $k Re \gg 1$). He follows closely the analysis of W. Heisenberg (1924). Yih presents C. C. Lin's improvements on Heisenberg's analysis with S. F. Shen's calculations of the stability curves.

Plane Couette Flow

This is the flow confined between two parallel plates; it is driven by the motion of one of the plates parallel to itself. The basic velocity profile is linear, with $U = \Gamma y$. Contrary to the experimentally observed fact that the flow does become turbulent at high values of Re, all linear analyses have shown that the flow is stable to small disturbances. It is now believed that the instability is caused by disturbances of finite magnitude.

Pipe Flow

The absence of an inflection point in the velocity profile signifies that the flow is inviscidly stable. All linear stability calculations of the *viscous* problem have also shown that the flow is stable to small disturbances. In contrast, most experiments show that the transition to turbulence takes place at a Reynolds number of about $\text{Re} = U_{\max} d/\nu \sim 3000$. However, careful experiments, some of them performed by Reynolds in his classic investigation of the onset of turbulence, have been able to maintain laminar flow until $\text{Re} = 50{,}000$. Beyond this the observed flow is invariably turbulent. The observed transition has been attributed to one of the following effects: (1) It could be a finite amplitude effect; (2) the turbulence may be initiated at the entrance of the tube by boundary layer instability (Figure 9.2); and (3) the instability could be caused by a slow rotation of the inlet flow which, when added to the Poiseuille distribution, has been shown to result in instability. This is still under investigation. New insights into the instability and transition of pipe flow were described by Eckhardt et al. (2007) by analysis via dynamical systems theory and comparison with recent very carefully crafted experiments by them and others. They characterized the turbulent state as a "chaotic saddle in state space." The boundary between laminar and turbulent flow was found to be exquisitely sensitive to initial conditions. Because pipe flow is linearly stable, finite amplitude disturbances are necessary to cause transition, but as Reynolds number increases, the amplitude of the critical disturbance diminishes. The boundary between laminar and turbulent states appears to be characterized by a pair of vortices closer to the walls which give the strongest amplification of the initial disturbance.

Boundary Layers with Pressure Gradients

Recall from Chapter 10, Section 7 that a pressure falling in the direction of flow is said to have a "favorable" gradient, and a pressure rising in the direction of flow is said to have an "adverse" gradient. It was shown there that boundary layers with an adverse pressure gradient have a point of inflection in the velocity profile. This has a dramatic effect on the stability characteristics. A schematic plot of the marginal stability curve for a boundary layer with favorable and adverse gradients of pressure is shown in Figure 12.24. The ordinate in the plot represents the longitudinal wavenumber, and the abscissa represents the Reynolds number based on the free-stream velocity and the displacement thickness δ^* of the boundary layer. The marginal stability curve divides stable and unstable regions, with the region within the "loop" representing instability. Because the boundary layer thickness grows along the direction of flow, Re_δ increases with x, and points at various downstream distances are represented by larger values of Re_δ.

The following features can be noted in the figure. The flow is stable for low Reynolds numbers, although it is unstable at higher Reynolds numbers. The effect of increasing viscosity is therefore stabilizing in this range. For boundary layers with a zero pressure gradient (Blasius flow) or a favorable pressure gradient, the instability loop shrinks to zero as $\text{Re}_\delta \to \infty$. This is consistent with the fact that these flows do not have a point of inflection in the velocity profile and are therefore inviscidly stable. In contrast, for boundary layers with an adverse pressure gradient, the instability loop does not shrink to zero; the upper branch of the marginal stability curve now becomes flat with a limiting value of k_∞ as $\text{Re}_\delta \to \infty$. The flow is then unstable to disturbances of wavelengths in the range $0 < k < k_\infty$. This is consistent with the existence of a point of inflection in the velocity profile, and the results of the mixing

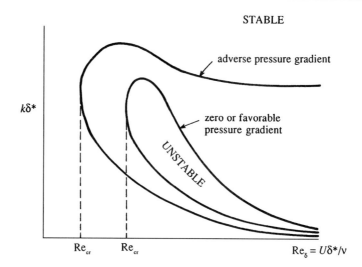

Figure 12.24   Sketch of marginal stability curves for a boundary layer with favorable and adverse pressure gradients.

TABLE 12.1   Linear Stability Results of Common Viscous Parallel Flows

Flow	$U(y)/U_0$	Re_{cr}	Remarks
Jet	$sech^2(y/L)$	4	
Shear layer	$tanh(y/L)$	0	Always unstable
Blasius		520	Re based on δ^*
Plane Poiseuille	$1 - (y/L)^2$	5780	$L =$ half-width
Pipe flow	$1 - (r/R)^2$	∞	Always stable
Plane Couette	y/L	∞	Always stable

layer calculation (Figure 12.23). Note also that the critical Reynolds number is lower for flows with adverse pressure gradients.

Table 12.1 summarizes the results of the linear stability analyses of some common parallel viscous flows.

The first two flows in the table have points of inflection in the velocity profile and are inviscidly unstable; the viscous solution shows either a zero or a small critical Reynolds number. The remaining flows are stable in the inviscid limit. Of these, the Blasius boundary layer and the plane Poiseuille flow are unstable in the presence of viscosity, but have high critical Reynolds numbers.

In Section 10.17 we discussed the decay of a laminar shear layer. Mass conservation requires that a transverse velocity be generated so the flow cannot be parallel. Although the idealized *tanh* profile for a shear layer, assuming straight and parallel streamlines is immediately unstable, recent work by Bhattacharya et al. (2006), which allowed for the basic flow to be two-dimensional, has yielded a finite critical Reynolds number, modifying somewhat Table 12.1 (above).

How can Viscosity Destabilize a Flow?

Let us examine how viscous effects can be destabilizing. For this we derive an integral form of the kinetic energy equation in a viscous flow. The Navier–Stokes equation for the disturbed flow is

$$\frac{\partial}{\partial t}(U_i + u_i) + (U_j + u_j)\frac{\partial}{\partial x_j}(U_i + u_i)$$

$$= -\frac{1}{\rho}\frac{\partial}{\partial x_i}(P + p) + \nu\frac{\partial^2}{\partial x_j\,\partial x_j}(U_i + u_i).$$

Subtracting the equation of motion for the basic state, we obtain

$$\frac{\partial u_i}{\partial t} + u_j\frac{\partial u_i}{\partial x_j} + U_j\frac{\partial u_i}{\partial x_j} + u_j\frac{\partial U_i}{\partial x_j} = -\frac{1}{\rho}\frac{\partial p}{\partial x_i} + \nu\frac{\partial^2 u_i}{\partial x_j^2},$$

which is the equation of motion of the disturbance. The integrated mechanical energy equation for the disturbance motion is obtained by multiplying this equation by u_i and integrating over the region of flow. The control volume is chosen to coincide with the walls where no-slip conditions are satisfied, and the length of the control volume in the direction of periodicity is chosen to be an integral number of wavelengths (Figure 12.25). The various terms of the energy equation then simplify as follows:

$$\int u_i\frac{\partial u_i}{\partial t}\,dV = \frac{d}{dt}\int\frac{u_i^2}{2}\,dV,$$

$$\int u_i u_j\frac{\partial u_i}{\partial x_j}\,dV = \frac{1}{2}\int\frac{\partial}{\partial x_j}(u_i^2 u_j)\,dV = \frac{1}{2}\int u_i^2 u_j\,dA_j = 0,$$

$$\int u_i U_j\frac{\partial u_i}{\partial x_j}\,dV = \frac{1}{2}\int\frac{\partial}{\partial x_j}(u_i^2 U_j)\,dV = \frac{1}{2}\int u_i^2 U_j\,dA_j = 0,$$

$$\int u_i\frac{\partial p}{\partial x_i}\,dV = \int\frac{\partial}{\partial x_i}(pu_i)\,dV = \int pu_i\,dA_i = 0,$$

$$\int u_i\frac{\partial^2 u_i}{\partial x_j^2}\,dV = \int\frac{\partial}{\partial x_j}\left(u_i\frac{\partial u_i}{\partial x_j}\right)dV - \int\frac{\partial u_i}{\partial x_j}\frac{\partial u_i}{\partial x_j}\,dV$$

$$= -\int\frac{\partial u_i}{\partial x_j}\frac{\partial u_i}{\partial x_j}\,dV.$$

Here, $d\mathbf{A}$ is an element of surface area of the control volume, and dV is an element of volume. In these the continuity equation $\partial u_i/\partial x_i = 0$, Gauss' theorem, and the no-slip and periodic boundary conditions have been used to show that the divergence terms drop out in an integrated energy balance. We finally obtain

$$\frac{d}{dt}\int\frac{1}{2}u_i^2\,dV = -\int u_i u_j\frac{\partial U_i}{\partial x_j}\,dV - \phi,$$

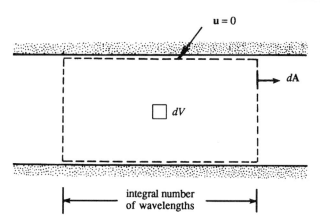

Figure 12.25   A control volume with zero net flux across boundaries.

where $\phi = \nu \int (\partial u_i / \partial x_i)^2 \, dV$ is the viscous dissipation. For two-dimensional disturbances in a shear flow defined by $\mathbf{U} = [U(y), 0, 0]$, the energy equation becomes

$$\frac{d}{dt} \int \frac{1}{2} (u^2 + v^2) \, dV = - \int uv \frac{\partial U}{\partial y} \, dV - \phi. \qquad (12.83)$$

This equation has a simple interpretation. The first term is the rate of change of kinetic energy of the disturbance, and the second term is the rate of production of disturbance energy by the interaction of the "Reynolds stress" uv and the mean shear $\partial U / \partial y$. The concept of Reynolds stress will be explained in the following chapter. The point to note here is that the value of the product uv averaged over a period is zero if the velocity components u and v are out of phase of $90°$; for example, the mean value of uv is zero if $u = \sin t$ and $v = \cos t$.

In inviscid parallel flows without a point of inflection in the velocity profile, the u and v components are such that the disturbance field cannot extract energy from the basic shear flow, thus resulting in stability. The presence of viscosity, however, changes the phase relationship between u and v, which causes Reynolds stresses such that the mean value of $-uv(\partial U / \partial y)$ over the flow field is positive and larger than the viscous dissipation. This is how viscous effects can cause instability.

11. Experimental Verification of Boundary Layer Instability

In this section we shall present the results of stability calculations of the Blasius boundary layer profile and compare them with experiments. Because of the nearly parallel nature of the Blasius flow, most stability calculations are based on an analysis of the Orr–Sommerfeld equation, which assumes a parallel flow. The first calculations were performed by Tollmien in 1929 and Schlichting in 1933. Instead of assuming exactly the Blasius profile (which can be specified only numerically), they used the profile

$$
\frac{U}{U_\infty} = \begin{cases} 1.7\,(y/\delta) & 0 \leqslant y/\delta \leqslant 0.1724, \\ 1 - 1.03\,[1 - (y/\delta)^2] & 0.1724 \leqslant y/\delta \leqslant 1, \\ 1 & y/\delta \geqslant 1, \end{cases}
$$

which, like the Blasius profile, has a zero curvature at the wall. The calculations of Tollmien and Schlichting showed that unstable waves appear when the Reynolds number is high enough; the unstable waves in a viscous boundary layer are called *Tollmien–Schlichting waves*. Until 1947 these waves remained undetected, and the experimentalists of the period believed that the transition in a real boundary layer was probably a finite amplitude effect. The speculation was that large disturbances cause locally adverse pressure gradients, which resulted in a local separation and consequent transition. The theoretical view, in contrast, was that small disturbances of the right frequency or wavelength can amplify if the Reynolds number is large enough.

Verification of the theory was finally provided by some clever experiments conducted by Schubauer and Skramstad in 1947. The experiments were conducted in a "low turbulence" wind tunnel, specially designed such that the intensity of fluctuations of the free stream was small. The experimental technique used was novel. Instead of depending on natural disturbances, they introduced periodic disturbances of known frequency by means of a vibrating metallic ribbon stretched across the flow close to the wall. The ribbon was vibrated by passing an alternating current through it in the field of a magnet. The subsequent development of the disturbance was followed downstream by hot wire anemometers. Such techniques have now become standard.

The experimental data are shown in Figure 12.26, which also shows the calculations of Schlichting and the more accurate calculations of Shen. Instead of the wavenumber, the ordinate represents the frequency of the wave, which is easier to measure. It is apparent that the agreement between Shen's calculations and the experimental data is very good.

The detection of the Tollmien–Schlichting waves is regarded as a major accomplishment of the linear stability theory. The ideal conditions for their existence require two dimensionality and consequently a negligible intensity of fluctuations of the free stream. These waves have been found to be very sensitive to small deviations from the ideal conditions. That is why they can be observed only under very carefully controlled experimental conditions and require artificial excitation. People who care about historical fairness have suggested that the waves should only be referred to as TS waves, to honor Tollmien, Schlichting, Schubauer, and Skramstad. The TS waves have also been observed in natural flow (Bayly *et al.*, 1988).

Nayfeh and Saric (1975) treated Falkner-Skan flows in a study of nonparallel stability and found that generally there is a decrease in the critical Reynolds number. The decrease is least for favorable pressure gradients, about 10% for zero pressure gradient, and grows rapidly as the pressure gradient becomes more adverse. Grabowski (1980) applied linear stability theory to the boundary layer near a stagnation point on a body of revolution. His stability predictions were found to be close to those of parallel flow stability theory obtained from solutions of the Orr–Sommerfeld equation.

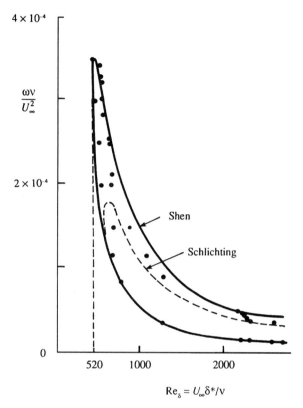

$$\frac{\omega v}{U_\infty^2}$$

$$\mathrm{Re}_\delta = U_\infty \delta^*/v$$

Figure 12.26 Marginal stability curve for a Blasius boundary layer. Theoretical solutions of Shen and Schlichting are compared with experimental data of Schubauer and Skramstad.

Reshotko (2001) provides a review of temporally and spatially transient growth as a path from subcritical (Tollmien–Schlichting) disturbances to transition. Growth or decay is studied from the Orr–Sommerfeld and Squire equations. Growth may occur because eigenfunctions of these equations are not orthogonal as the operators are not self-adjoint. Results for Poiseuille pipe flow and compressible blunt body flows are given.

Fransson and Alfredsson (2003) have shown that the asymptotic suction profile [solved in Exercise 9, Chapter 10] significantly delays transition stimulated by free stream turbulence or by Tollmien-Schlichting waves. Specifically, the value of $\mathrm{Re}_{cr} = 520$ based on δ_* in Table 12.1 is increased for suction velocity ratio to $v_o/U_\infty = -.00288$ over 54000. The very large stabilizing effect is a result of the change in the shape of the streamwise velocity from the Blasius profile to an exponential. The normal suction velocity has a very small effect on stability.

12. Comments on Nonlinear Effects

To this point we have discussed only linear stability theory, which considers infinitesimal perturbations and predicts exponential growth when the relevant parameter exceeds a critical value. The effect of the perturbations on the basic field is neglected in the linear theory. An examination of equation (12.83) shows that the perturbation field must be such that the mean Reynolds stress $\overline{uv}$ (the "mean" being over a wavelength) be nonzero for the perturbations to extract energy from the basic shear; similarly, the heat flux $\overline{\mathbf{u}T'}$ must be nonzero in a thermal convection problem. These rectified fluxes of momentum and heat change the *basic* velocity and temperature fields. The linear instability theory neglects these changes of the basic state. A consequence of the constancy of the basic state is that the growth rate of the perturbations is also constant, leading to an exponential growth. Within a short time of such initial growth the perturbations become so large that the rectified fluxes of momentum and heat significantly change the basic state, which in turn alters the growth of the perturbations.

A frequent effect of nonlinearity is to change the basic state in such a way as to stop the growth of the disturbances after they have reached significant amplitude through the initial exponential growth. (Note, however, that the effect of nonlinearity can sometimes be destabilizing; for example, the instability in a pipe flow may be a finite amplitude effect because the flow is stable to infinitesimal disturbances.) Consider the thermal convection in the annular space between two vertical cylinders rotating at the same speed. The outer wall of the annulus is heated and the inner wall is cooled. For small heating rates the flow is steady. For large heating rates a system of regularly spaced waves develop and progress azimuthally at a uniform speed without changing their shape. (This is the equilibrated form of baroclinic instability, discussed in Chapter 14, Section 17.) At still larger heating rates an irregular, aperiodic, or chaotic flow develops. The chaotic response to constant forcing (in this case the heating rate) is an interesting nonlinear effect and is discussed further in Section 14. Meanwhile, a brief description of the transition from laminar to turbulent flow is given in the next section.

13. Transition

The process by which a laminar flow changes to a turbulent one is called *transition*. Instability of a laminar flow does not immediately lead to turbulence, which is a severely nonlinear and chaotic stage characterized by macroscopic "mixing" of fluid particles. After the initial breakdown of laminar flow because of amplification of small disturbances, the flow goes through a complex sequence of changes, finally resulting in the chaotic state we call turbulence. The process of transition is greatly affected by such experimental conditions as intensity of fluctuations of the free stream, roughness of the walls, and shape of the inlet. The sequence of events that lead to turbulence is also greatly dependent on boundary geometry. For example, the scenario of transition in a wall-bounded shear flow is different from that in free shear flows such as jets and wakes.

Early stages of the transition consist of a succession of instabilities on increasingly complex basic flows, an idea first suggested by Landau in 1944. The basic state of wall-bounded parallel shear flows becomes unstable to two-dimensional TS waves, which grow and eventually reach equilibrium at some finite amplitude. This steady state can be considered a new background state, and calculations show that it is generally unstable to *three-dimensional* waves of short wavelength, which vary in the "spanwise" direction. (If x is the direction of flow and y is the directed normal to the boundary, then the z-axis is spanwise.) We shall call this the *secondary instability*. Interestingly, the secondary instability does not reach equilibrium at finite amplitude but directly evolves to a fully turbulent flow. Recent calculations of the secondary instability have been quite successful in reproducing critical Reynolds numbers for various wall-bounded flows, as well as predicting three-dimensional structures observed in experiments.

A key experiment on the three-dimensional nature of the transition process in a boundary layer was performed by Klebanoff, Tidstrom, and Sargent (1962). They conducted a series of controlled experiments by which they introduced three-dimensional disturbances on a field of TS waves in a boundary layer. The TS waves were as usual artificially generated by an electromagnetically vibrated ribbon, and the three dimensionality of a particular spanwise wavelength was introduced by placing spacers (small pieces of transparent tape) at equal intervals underneath the vibrating ribbon (Figure 12.27). When the amplitude of the TS waves became roughly 1% of the free-stream velocity, the three-dimensional perturbations grew rapidly and resulted

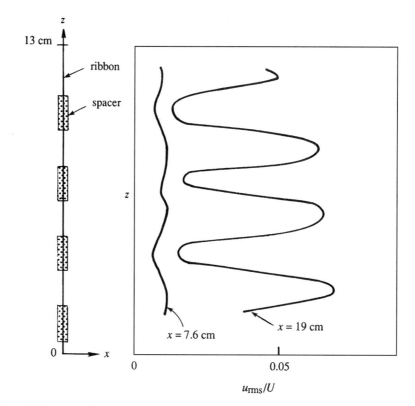

Figure 12.27 Three-dimensional unstable waves initiated by vibrating ribbon. Measured distributions of intensity of the u-fluctuation at two distances from the ribbon are shown. P. S. Klebanoff *et al.*, *Journal of Fluid Mechanics* **12**: 1–34, 1962 and reprinted with the permission of Cambridge University Press.

in a spanwise irregularity of the streamwise velocity displaying peaks and valleys in the amplitude of u. The three-dimensional disturbances continued to grow until the boundary layer became fully turbulent. The chaotic flow seems to result from the nonlinear evolution of the secondary instability, and recent numerical calculations have accurately reproduced several characteristic features of real flows (see Figures 7 and 8 in Bayly *et al.*, 1988).

It is interesting to compare the chaos observed in turbulent shear flows with that in controlled low-order dynamical systems such as the Bérnard convection or Taylor vortex flow. In these low-order flows only a very small number of modes participate in the dynamics because of the strong constraint of the boundary conditions. All but a few low modes are identically zero, and the chaos develops in an orderly way. As the constraints are relaxed (we can think of this as increasing the number of allowed Fourier modes), the evolution of chaos becomes less orderly.

Transition in a free shear layer, such as a jet or a wake, occurs in a different manner. Because of the inflectional velocity profiles involved, these flows are unstable at a very low Reynolds numbers, that is, of order 10 compared to about 10^3 for a wall-bounded flow. The breakdown of the laminar flow therefore occurs quite readily and close to the origin of such a flow. Transition in a free shear layer is characterized by the appearance of a rolled-up row of vortices, whose wavelength corresponds to the one with the largest growth rate. Frequently, these vortices group themselves in the form of pairs and result in a dominant wavelength twice that of the original wavelength. Small-scale turbulence develops within these larger scale vortices, finally leading to turbulence.

14. Deterministic Chaos

The discussion in the previous section has shown that dissipative nonlinear systems such as fluid flows reach a random or chaotic state when the parameter measuring nonlinearity (say, the Reynolds number or the Rayleigh number) is large. The change to the chaotic stage generally takes place through a sequence of transitions, with the exact route depending on the system. It has been realized that chaotic behavior not only occurs in continuous systems having an infinite number of degrees of freedom, but also in discrete nonlinear systems having only a small number of degrees of freedom, governed by ordinary nonlinear differential equations. In this context, a *chaotic system* is defined as one in which the solution is *extremely sensitive to initial conditions*. That is, solutions with arbitrarily close initial conditions evolve into quite different states. Other symptoms of a chaotic system are that the solutions are *aperiodic*, and that the spectrum is broadband instead of being composed of a few discrete lines.

Numerical integrations (to be shown later in this section) have recently demonstrated that nonlinear systems governed by a finite set of deterministic ordinary differential equations allow chaotic solutions in response to a steady forcing. This fact is interesting because in a dissipative *linear* system a constant forcing ultimately (after the decay of the transients) leads to constant response, a periodic forcing leads to periodic response, and a random forcing leads to random response. In the presence of nonlinearity, however, a constant forcing can lead to a variable response, both periodic and aperiodic. Consider again the experiment mentioned in Section 12, namely, the thermal convection in the annular space between two vertical cylinders rotating at the same speed. The outer wall of the annulus is heated and the inner wall is cooled. For small heating rates the flow is steady. For large heating rates a system of regularly spaced waves develops and progresses azimuthally at a uniform speed, without the waves changing shape. At still larger heating rates an irregular, aperiodic, or chaotic flow develops. This experiment shows that both periodic and aperiodic flow can result in a nonlinear system even when the forcing (in this case the heating rate) is constant. Another example is the periodic oscillation in the flow behind a blunt body at Re $\sim$ 40 (associated with the initial appearance of the von Karman vortex street) and the breakdown of the oscillation into turbulent flow at larger values of the Reynolds number.

It has been found that transition to chaos in the solution of ordinary nonlinear differential equations displays a certain *universal* behavior and proceeds in one of a few different ways. At the moment it is unclear whether the transition in fluid flows is closely related to the development of chaos in the solutions of these simple systems; this is under intense study. In this section we shall discuss some of the elementary ideas involved, starting with certain definitions. An introduction to the subject of chaos is given by Bergé, Pomeau, and Vidal (1984); a useful review is given in Lanford (1982). The subject has far-reaching cosmic consequences in physics and evolutionary biology, as discussed by Davies (1988).

Phase Space

Very few nonlinear equations have analytical solutions. For nonlinear systems, a typical procedure is to find a numerical solution and display its properties in a space whose axes are the *dependent* variables. Consider the equation governing the motion of a simple pendulum of length l:

$$\ddot{X} + \frac{g}{l} \sin X = 0,$$

where X is the *angular* displacement and $\ddot{X}$ $(= d^2X/dt^2)$ is the angular acceleration. (The component of gravity parallel to the trajectory is $-g \sin X$, which is balanced by the linear acceleration $l\ddot{X}$.) The equation is nonlinear because of the $\sin X$ term. The second-order equation can be split into two coupled first-order equations

$$\dot{X} = Y,$$
$$\dot{Y} = -\frac{g}{l} \sin X. \tag{12.84}$$

Starting with some initial conditions on X and Y, one can integrate set (12.84). The behavior of the system can be studied by describing how the variables Y $(=\dot{X})$ and X vary as a function of time. For the pendulum problem, the space whose axes are $\dot{X}$ and X is called a *phase space*, and the evolution of the system is described by a *trajectory* in this space. The dimension of the phase space is called the *degree of freedom* of the system; it equals the number of independent initial conditions necessary to specify the system. For example, the degree of freedom for the set (12.84) is two.

Attractor

Dissipative systems are characterized by the existence of *attractors*, which are structures in the phase space toward which neighboring trajectories approach as $t \to \infty$. An attractor can be a *fixed point* representing a stable steady flow or a closed curve (called a *limit cycle*) representing a stable oscillation (Figure 12.28a, b). The nature of the attractor depends on the value of the nonlinearity parameter, which

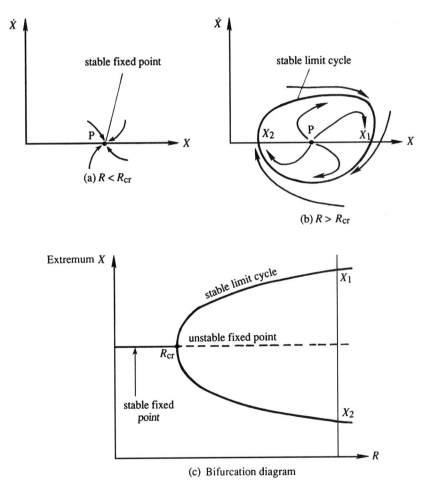

(a) $R < R_{cr}$

(b) $R > R_{cr}$

(c) Bifurcation diagram

Figure 12.28 Attractors in a phase plane. In (a), point P is an attractor. For a larger value of R, panel (b) shows that P becomes an unstable fixed point (a "repeller"), and the trajectories are attracted to a limit cycle. Panel (c) is the bifurcation diagram.

will be denoted by R in this section. As R is increased, the fixed point representing a steady solution may change from being an attractor to a repeller with spirally outgoing trajectories, signifying that the steady flow has become unstable to infinitesimal perturbations. Frequently, the trajectories are then attracted by a limit cycle, which means that the unstable steady solution gives way to a steady oscillation (Figure 12.28b). For example, the steady flow behind a blunt body becomes oscillatory as Re is increased, resulting in the periodic von Karman vortex street (Figure 10.18).

The branching of a solution at a critical value R_{cr} of the nonlinearity parameter is called a *bifurcation*. Thus, we say that the stable steady solution of Figure 12.28a bifurcates to a stable limit cycle as R increases through R_{cr}. This can be represented on the graph of a dependent variable (say, X) vs R (Figure 12.28c). At $R = R_{cr}$, the solution curve branches into two paths; the two values of X on these branches (say, X_1 and X_2) correspond to the maximum and minimum values of X in Figure 12.28b. It is seen that the size of the limit cycle grows larger as $(R - R_{cr})$ becomes larger. Limit cycles, representing oscillatory response with amplitude independent of initial conditions, are characteristic features of nonlinear systems. Linear stability theory predicts an exponential growth of the perturbations if $R > R_{cr}$, but a nonlinear theory frequently shows that the perturbations eventually equilibrate to a steady oscillation whose amplitude increases with $(R - R_{cr})$.

The Lorenz Model of Thermal Convection

Taking the example of thermal convection in a layer heated from below (the Bénard problem), Lorenz (1963) demonstrated that the development of chaos is associated with the attractor acquiring certain strange properties. He considered a layer with stress-free boundaries. Assuming nonlinear disturbances in the form of rolls invariant in the y direction, and defining a streamfunction in the xz-plane by $u = -\partial \psi / \partial z$ and $w = \partial \psi / \partial x$, he substituted solutions of the form

$$\psi \propto X(t) \cos \pi z \sin kx,$$
$$T' \propto Y(t) \cos \pi z \cos kx + Z(t) \sin 2\pi z, \qquad (12.85)$$

into the equations of motion (12.7). Here, T' is the departure of temperature from the state of no convection, k is the wavenumber of the perturbation, and the boundaries are at $z = \pm \frac{1}{2}$. It is clear that X is proportional to the intensity of convective motion, Y is proportional to the temperature difference between the ascending and descending currents, and Z is proportional to the distortion of the average vertical profile of temperature from linearity. (Note in equation (12.85) that the x-average of the term multiplied by $Y(t)$ is zero, so that this term does not cause distortion of the basic temperature profile.) As discussed in Section 3, Rayleigh's linear analysis showed that solutions of the form (12.85), with X and Y constants and $Z = 0$, would develop if Ra slightly exceeds the critical value $\text{Ra}_{cr} = 27 \pi^4 / 4$. Equations (12.85) are expected to give realistic results when Ra is slightly supercritical but not when strong convection occurs because only the lowest terms in a "Galerkin expansion" are retained.

On substitution of equation (12.85) into the equations of motion, Lorenz finally obtained

$$\dot{X} = \Pr(Y - X),$$
$$\dot{Y} = -XZ + rX - Y, \qquad\qquad (12.86)$$
$$\dot{Z} = XY - bZ,$$

where Pr is the Prandtl number, $r = \mathrm{Ra}/\mathrm{Ra_{cr}}$, and $b = 4\pi^2/(\pi^2 + k^2)$. Equations (12.86) represent a set of nonlinear equations with three degrees of freedom, which means that the phase space is three-dimensional.

Equations (12.86) allow the steady solution $X = Y = Z = 0$, representing the state of no convection. For $r > 1$ the system possesses two additional steady-state solutions, which we shall denote by $\bar{X} = \bar{Y} = \pm\sqrt{b(r - 1)}, \bar{Z} = r - 1$; the two signs correspond to the two possible senses of rotation of the rolls. (The fact that these steady solutions satisfy equation (12.86) can easily be checked by substitution and setting $\dot{X} = \dot{Y} = \dot{Z} = 0$.) Lorenz showed that the steady-state convection becomes unstable if r is large. Choosing $\Pr = 10$, $b = 8/3$, and $r = 28$, he numerically integrated the set and found that the solution never repeats itself; it is aperiodic and wanders about in a chaotic manner. Figure 12.29 shows the variation of $X(t)$, starting with some initial conditions. (The variables $Y(t)$ and $Z(t)$ also behave in a similar way.) It is seen that the amplitude of the convecting motion initially oscillates around one of the steady values $\bar{X} = \pm\sqrt{b(r - 1)}$, with the oscillations growing in magnitude. When it is large enough, the amplitude suddenly goes through zero to start oscillations of opposite sign about the other value of $\bar{X}$. That is, the motion switches in a chaotic

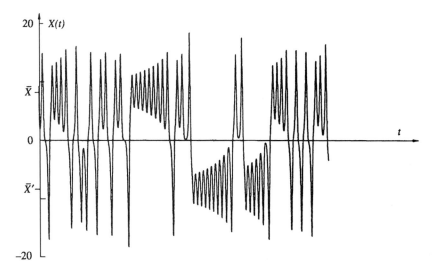

Figure 12.29 Variation of $X(t)$ in the Lorenz model. Note that the solution oscillates erratically around the two steady values $\bar{X}$ and $\bar{X}'$. P. Bergé, Y. Pomeau, and C. Vidal, *Order Within Chaos*, 1984 and reprinting permitted by Heinemann Educational, a division of Reed Educational & Professional Publishing Ltd.

manner between two oscillatory limit cycles, with the number of oscillations between transitions seemingly random. Calculations show that the variables X, Y, and Z have continuous spectra and that the solution is extremely sensitive to initial conditions.

Strange Attractors

The trajectories in the phase plane in the Lorenz model of thermal convection are shown in Figure 12.30. The centers of the two loops represent the two steady convections $\bar{X} = \bar{Y} = \pm\sqrt{b(r-1)}$, $\bar{Z} = r - 1$. The structure resembles two rather flat loops of ribbon, one lying slightly in front of the other along a central band with the two joined together at the bottom of that band. The trajectories go clockwise around the left loop and counterclockwise around the right loop; two trajectories never intersect. The structure shown in Figure 12.30 is an attractor because orbits starting with initial conditions *outside of the attractor* merge on it and then follow it. The attraction is a result of dissipation in the system. The aperiodic attractor, however, is unlike the normal attractor in the form of a fixed point (representing steady motion) or a closed curve (representing a limit cycle). This is because two trajectories *on the aperiodic attractor*, with infinitesimally different initial conditions, follow each other closely only for a while, eventually diverging to very different final states. This is the basic reason for sensitivity to initial conditions.

For these reasons the aperiodic attractor is called a *strange attractor*. The idea of a strange attractor is quite nonintuitive because it has the dual property of attraction and divergence. Trajectories are attracted from the neighboring region of phase space, but once on the attractor the trajectories eventually diverge and result in chaos. An ordinary attractor "forgets" slightly different initial conditions, whereas the strange

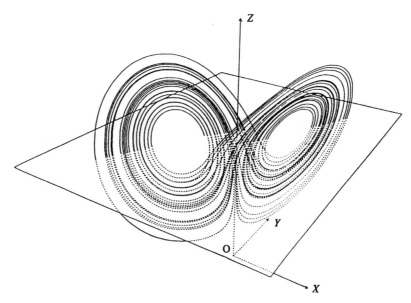

Figure 12.30 The Lorenz attractor. Centers of the two loops represent the two steady solutions $(\bar{X}, \bar{Y}, \bar{Z})$.

attractor ultimately accentuates them. The idea of the strange attractor was first conceived by Lorenz, and since then attractors of other chaotic systems have also been studied. They all have the common property of aperiodicity, continuous spectra, and sensitivity to initial conditions.

Scenarios for Transition to Chaos

Thus far we have studied discrete dynamical systems having only a small number of degrees of freedom and seen that aperiodic or chaotic solutions result when the nonlinearity parameter is large. Several routes or scenarios of transition to chaos in such systems have been identified. Two of these are described briefly here.

(1) *Transition through subharmonic cascade*: As R is increased, a typical nonlinear system develops a limit cycle of a certain frequency ω. With further increase of R, several systems are found to generate additional frequencies $\omega/2$, $\omega/4$, $\omega/8$, ... The addition of frequencies in the form of *subharmonics* does not change the periodic nature of the solution, but the period doubles each time a lower harmonic is added. The period doubling takes place more and more rapidly as R is increased, until an *accumulation point* (Figure 12.31) is reached, beyond which the solution wanders about in a chaotic manner. At this point the peaks disappear from the spectrum, which becomes continuous. Many systems approach chaotic behavior through period doubling. Feigenbaum (1980) proved the important result that this kind of transition develops in a *universal* way, independent of the particular nonlinear systems studied. If R_n represents the value for development of a new subharmonic, then R_n converges in a geometric series with

$$\frac{R_n - R_{n-1}}{R_{n+1} - R_n} \to 4.6692 \qquad \text{as } n \to \infty.$$

That is, the horizontal gap between two bifurcation points is about a fifth of the previous gap. The vertical gap between the branches of the bifurcation diagram also decreases, with each gap about two-fifths of the previous gap. In other words, the bifurcation diagram (Figure 12.31) becomes "self similar" as the accumulation point is approached. (Note that Figure 12.31 has not been drawn to scale, for illustrative purposes.) Experiments in low Prandtl number fluids (such as liquid metals) indicate that Bénard convection in the form of rolls develops oscillatory motion of a certain frequency ω at Ra = 2Ra$_{cr}$. As Ra is further increased, additional frequencies $\omega/2$, $\omega/4$, $\omega/8$, $\omega/16$, and $\omega/32$ have been observed. The convergence ratio has been measured to be 4.4, close to the value of 4.669 predicted by Feigenbaum's theory. The experimental evidence is discussed further in Bergé, Pomeau, and Vidal (1984).

(2) *Transition through quasi-periodic regime*: Ruelle and Takens (1971) have mathematically proved that certain systems need only a *small number* of bifurcations to produce chaotic solutions. As the nonlinearity parameter is increased, the steady solution loses stability and bifurcates to an oscillatory

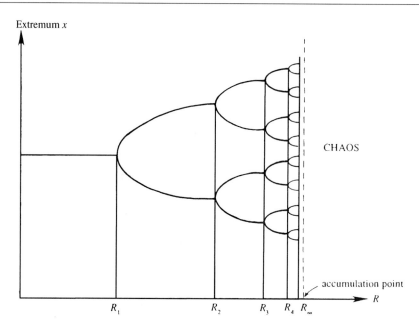

Figure 12.31 Bifurcation diagram during period doubling. The period doubles at each value R_n of the nonlinearity parameter. For large n the "bifurcation tree" becomes self similar. Chaos sets in beyond the accumulation point R_∞.

limit cycle with frequency ω_1. As R is increased, two more frequencies (ω_2 and ω_3) appear through additional bifurcations. In this scenario the ratios of the three frequencies (such as ω_1/ω_2) are *irrational* numbers, so that the motion consisting of the three frequencies is not exactly periodic. (When the ratios are rational numbers, the motion is exactly periodic. To see this, think of the Fourier series of a periodic function in which the various terms represent sinusoids of the fundamental frequency ω and its harmonics 2ω, 3ω, ... Some of the Fourier coefficients could be zero.) The spectrum for these systems suddenly develops broadband characteristics of chaotic motion as soon as the third frequency ω_3 appears. The exact point at which chaos sets in is not easy to detect in a measurement; in fact the third frequency may not be identifiable in the spectrum before it becomes broadband. The Ruelle–Takens theory is fundamentally different from that of Landau, who conjectured that turbulence develops due to an *infinite* number of bifurcations, each generating a new higher frequency, so that the spectrum becomes saturated with peaks and resembles a continuous one. According to Bergé, Pomeau, and Vidal (1984), the Bénard convection experiments in *water* seem to suggest that turbulence in this case probably sets in according to the Ruelle–Takens scenario.

The development of chaos in the Lorenz attractor is more complicated and does not follow either of the two routes mentioned in the preceding.

Closure

Perhaps the most intriguing characteristic of a chaotic system is the extreme *sensitivity to initial conditions*. That is, solutions with arbitrarily close initial conditions evolve into two quite different states. Most nonlinear systems are susceptible to chaotic behavior. The extreme sensitivity to initial conditions implies that nonlinear phenomena (including the weather, in which Lorenz was primarily interested when he studied the convection problem) are essentially unpredictable, no matter how well we know the governing equations or the initial conditions. Although the subject of chaos has become a scientific revolution recently, the central idea was conceived by Henri Poincaré in 1908. He did not, of course, have the computing facilities to demonstrate it through numerical integration.

It is important to realize that the behavior of chaotic systems is not *intrinsically* indeterministic; as such the implication of deterministic chaos is different from that of the uncertainty principle of quantum mechanics. In any case, the extreme sensitivity to initial conditions implies that the **future is essentially unknowable** because it is never possible to know the initial conditions *exactly*. As discussed by Davies (1988), this fact has interesting philosophical implications regarding the evolution of the universe, including that of living species.

We have examined certain elementary ideas about how chaotic behavior may result in simple nonlinear systems having only a small number of degrees of freedom. Turbulence in a continuous fluid medium is capable of displaying an infinite number of degrees of freedom, and it is unclear whether the study of chaos can throw a great deal of light on more complicated transitions such as those in pipe or boundary layer flow. However, the fact that nonlinear systems can have chaotic solutions for a large value of the nonlinearity parameter (see Figure 12.29 again) is an important result by itself.

Exercises

1. Consider the thermal instability of a fluid confined between two rigid plates, as discussed in Section 3. It was stated there without proof that the minimum critical Rayleigh number of $Ra_{cr} = 1708$ is obtained for the gravest *even* mode. To verify this, consider the gravest *odd* mode for which

$$W = A \sin q_0 \, z + B \sinh q \, z + C \sinh q^* z.$$

(Compare this with the gravest even mode structure: $W = A \cos q_0 z + B \cosh q \, z + C \cosh q^* z$.) Following Chandrasekhar (1961, p. 39), show that the minimum Rayleigh number is now 17,610, reached at the wavenumber $K_{cr} = 5.365$.

2. Consider the centrifugal instability problem of Section 5. Making the narrow-gap approximation, work out the algebra of going from equation (12.37) to equation (12.38).

3. Consider the centrifugal instability problem of Section 5. From equations (12.38) and (12.40), the eigenvalue problem for determining the marginal

state ($\sigma = 0$) is

$$(D^2 - k^2)^2 \hat{u}_r = (1 + \alpha x)\hat{u}_\theta, \tag{12.87}$$

$$(D^2 - k^2)^2 \hat{u}_\theta = -\text{Ta}\, k^2 \hat{u}_r, \tag{12.88}$$

with $\hat{u}_r = D\hat{u}_r = \hat{u}_\theta = 0$ at $x = 0$ and 1. Conditions on $\hat{u}_\theta$ are satisfied by assuming solutions of the form

$$\hat{u}_\theta = \sum_{m=1}^{\infty} C_m \sin m\pi x. \tag{12.89}$$

Inserting this in equation (12.87), obtain an equation for $\hat{u}_r$, and arrange so that the solution satisfies the four remaining conditions on $\hat{u}_r$. With $\hat{u}_r$ determined in this manner and $\hat{u}_\theta$ given by equation (12.89), equation (12.88) leads to an eigenvalue problem for Ta(k). Following Chandrasekhar (1961, p. 300), show that the minimum Taylor number is given by equation (12.41) and is reached at $k_{\text{cr}} = 3.12$.

4. Consider an infinitely deep fluid of density ρ_1 lying over an infinitely deep fluid of density $\rho_2 > \rho_1$. By setting $U_1 = U_2 = 0$, equation (12.51) shows that

$$c = \sqrt{\frac{g}{k}\frac{\rho_2 - \rho_1}{\rho_2 + \rho_1}}. \tag{12.90}$$

Argue that if the whole system is given an upward vertical acceleration a, then g in equation (12.90) is replaced by $g' = g + a$. It follows that there is instability if $g' < 0$, that is, the system is given a downward acceleration of magnitude larger than g. This is called the *Rayleigh–Taylor instability*, which can be observed simply by rapidly accelerating a beaker of water downward.

5. Consider the inviscid instability of parallel flows given by the Rayleigh equation

$$(U - c)(\hat{v}_{yy} - k^2\hat{v}) - U_{yy}\hat{v} = 0, \tag{12.91}$$

where the y-component of the perturbation velocity is $v = \hat{v}\exp(i\,kx - i\,kct)$.

(i)   Note that this equation is identical to the Rayleigh equation (12.76) written in terms of the stream function amplitude ϕ, as it must because $\hat{v} = -ik\phi$. For a flow bounded by walls at y_1 and y_2, note that the boundary conditions are identical in terms of ϕ and $\hat{v}$.

(ii) Show that if c is an eigenvalue of equation (12.91), then so is its conjugate $c^* = c_r - i\,c_i$. What aspect of equation (12.91) allows the result to be valid?

(iii) Let $U(y)$ be an *antisymmetric* jet, so that $U(y) = -U(-y)$. Demonstrate that if $c(k)$ is an eigenvalue, then $-c(k)$ is also an eigenvalue. Explain the result physically in terms of the possible directions of propagation of perturbations in an antisymmetric flow.

(iv) Let $U(y)$ be a *symmetric* jet. Show that in this case $\hat{v}$ is either symmetric or antisymmetric about $y = 0$.

[*Hint*: Letting $y \to -y$, show that the solution $\hat{v}(-y)$ satisfies equation (12.91) with the same eigenvalue c. Form a symmetric solution $S(y) = \hat{v}(y) + \hat{v}(-y) = S(-y)$, and an antisymmetric solution $A(y) = \hat{v}(y) - \hat{v}(-y) = -A(-y)$. Then write $A[S\text{-eqn}] - S[A\text{-eqn}] = 0$, where S-eqn indicates the differential equation (12.91) in terms of S. Canceling terms this reduces to $(SA' - AS')' = 0$, where the prime $(')$ indicates y-derivative. Integration gives $SA' - AS' = 0$, where the constant of integration is zero because of the boundary condition. Another integration gives $S = bA$, where b is a constant of integration. Because the symmetric and antisymmetric functions cannot be proportional, it follows that one of them must be zero.]

Comments: If v is symmetric, then the cross-stream velocity has the same sign across the entire jet, although the sign alternates every half of a wavelength along the jet. This mode is consequently called *sinuous*. On the other hand, if v is antisymmetric, then the shape of the jet expands and contracts along the length. This mode is now generally called the *sausage* instability because it resembles a line of linked sausages.

6. For a Kelvin–Helmholtz instability in a continuously stratified ocean, obtain a globally integrated energy equation in the form

$$\frac{1}{2}\frac{d}{dt}\int (u^2 + w^2 + g^2\rho^2/\rho_0^2 N^2)\, dV = -\int uwU_z\, dV.$$

(As in Figure 12.25, the integration in x takes place over an integral number of wavelengths.) Discuss the physical meaning of each term and the mechanism of instability.

Literature Cited

Bayly, B. J., S. A. Orszag, and T. Herbert (1988). "Instability mechanisms in shear-flow transition." *Annual Review of Fluid Mechanics* **20**: 359–391.

Bergé, P., Y. Pomeau, and C. Vidal (1984). *Order Within Chaos,* New York: Wiley.

Bhattacharya, P., M. P. Manoharan, R. Govindarajan, and R. Narasimha (2006). "The critical Reynolds number of a laminar incompressible mixing layer from minimal composite theory." *Journal of Fluid Mechanics* **565**: 105–114.

Chandrasekhar, S. (1961). *Hydrodynamic and Hydromagnetic Stability,* London: Oxford University Press; New York: Dover reprint, 1981.

Coles, D. (1965). "Transition in circular Couette flow." *Journal of Fluid Mechanics* **21**: 385–425.

Davies, P. (1988). *Cosmic Blueprint,* New York: Simon and Schuster.

Drazin, P. G. and W. H. Reid (1981). *Hydrodynamic Stability,* London: Cambridge University Press.

Eckhardt, B., T. M. Schneider, B. Hof, and J. Westerweel (2007). "Turbulence transition in pipe flow." *Annual Review of Fluid Mechanics* **39**: 447–468.

Eriksen, C. C. (1978). "Measurements and models of fine structure, internal gravity waves, and wave breaking in the deep ocean." *Journal of Geophysical Research* **83**: 2989–3009.

Feigenbaum, M. J. (1978). "Quantitative universality for a class of nonlinear transformations." *Journal of Statistical Physics* **19**: 25–52.

Fransson, J. H. M. and P. H. Alfredsson (2003). "On the disturbance growth in an asymptotic suction boundary layer." *Journal of Fluid Mechanics* **482**: 51–90.

Grabowski, W. J. (1980). "Nonparallel stability analysis of axisymmetric stagnation point flow." *Physics of Fluids* **23**: 1954–1960.

Heisenberg, W. (1924). "Über Stabilität und Turbulenz von Flüssigkeitsströmen." *Annalen der Physik (Leipzig)* (4) **74**: 577–627.

Howard, L. N. (1961). "Note on a paper of John W. Miles." *Journal of Fluid Mechanics* **13**: 158–160.

Huppert, H. E. and J. S. Turner (1981). "Double-diffusive convection." *Journal of Fluid Mechanics* **106**: 299–329.

Klebanoff, P. S., K. D. Tidstrom, and L. H. Sargent (1962). "The three-dimensional nature of boundary layer instability". *Journal of Fluid Mechanics* **12**: 1–34.

Lanford, O. E. (1982). "The strange attractor theory of turbulence." *Annual Review of Fluid Mechanics* **14**: 347–364.

Lin, C. C. (1955). *The Theory of Hydrodynamic Stability*, London: Cambridge University Press, Chapter 8.

Lorenz, E. (1963). "Deterministic nonperiodic flows." *Journal of Atmospheric Sciences* **20**: 130–141.

Miles, J. W. (1961). "On the stability of heterogeneous shear flows." *Journal of Fluid Mechanics* **10**: 496–508.

Miles, J. W. (1986). "Richardson's criterion for the stability of stratified flow." *Physics of Fluids* **29**: 3470–3471.

Nayfeh, A. H. and W. S. Saric (1975). "Nonparallel stability of boundary layer flows." *Physics of Fluids* **18**: 945–950.

Reshotko, E. (2001). "Transient growth: A factor in bypass transition." *Physics of Fluids* **13**: 1067–1075.

Ruelle, D. and F. Takens (1971). "On the nature of turbulence." *Communications in Mathematical Physics* **20**: 167–192.

Scotti, R. S. and G. M. Corcos (1972). "An experiment on the stability of small disturbances in a stratified free shear layer." *Journal of Fluid Mechanics* **52**: 499–528.

Shen, S. F. (1954). "Calculated amplified oscillations in plane Poiseuille and Blasius Flows." *Journal of the Aeronautical Sciences* **21**: 62–64.

Stern, M. E. (1960). "The salt fountain and thermohaline convection." *Tellus* **12**: 172–175.

Stommel, H., A. B. Arons, and D. Blanchard (1956). "An oceanographic curiosity: The perpetual salt fountain." *Deep-Sea Research* **3**: 152–153.

Thorpe, S. A. (1971). "Experiments on the instability of stratified shear flows: Miscible fluids." *Journal of Fluid Mechanics* **46**: 299–319.

Turner, J. S. (1973). *Buoyancy Effects in Fluids,* London: Cambridge University Press.

Turner, J. S. (1985). "Convection in multicomponent systems." *Naturwissenschaften* **72**: 70–75.

Woods, J. D. (1969). "On Richardson's number as a criterion for turbulent–laminar transition in the atmosphere and ocean." *Radio Science* **4**: 1289–1298.

Yih, C. S. (1979). *Fluid Mechanics: A Concise Introduction to the Theory*, Ann Arbor, MI: West River Press, pp. 469–496.

Turbulence

1. Introduction

Most flows encountered in engineering practice and in nature are turbulent. The boundary layer on an aircraft wing is likely to be turbulent, the atmospheric boundary layer over the earth's surface is turbulent, and the major oceanic currents are turbulent. In this chapter we shall discuss certain elementary ideas about the dynamics of

©2010 Elsevier Inc. All rights reserved.
DOI: 10.1016/B978-0-12-381399-2.50013-7

turbulent flows. We shall see that such flows do not allow a strict analytical study, and one depends heavily on physical intuition and dimensional arguments. In spite of our everyday experience with it, turbulence is not easy to define precisely. In fact, there is a tendency to confuse turbulent flows with "random flows." With some humor, Lesieur (1987) said that "turbulence is a dangerous topic which is at the origin of serious fights in scientific meetings since it represents extremely different points of view, all of which have in common their complexity, as well as an inability to solve the problem. It is even difficult to agree on what exactly is the problem to be solved."

Some characteristics of turbulent flows are the following:

(1) *Randomness*: Turbulent flows seem irregular, chaotic, and unpredictable.
(2) *Nonlinearity*: Turbulent flows are highly nonlinear. The nonlinearity serves two purposes. First, it causes the relevant nonlinearity parameter, say the Reynolds number Re, the Rayleigh number Ra, or the inverse Richardson number Ri^{-1}, to exceed a critical value. In unstable flows small perturbations grow spontaneously and frequently equilibrate as finite amplitude disturbances. On further exceeding the stability criteria, the new state can become unstable to more complicated disturbances, and the flow eventually reaches a chaotic state. Second, the nonlinearity of a turbulent flow results in vortex stretching, a key process by which three-dimensional turbulent flows maintain their vorticity.
(3) *Diffusivity*: Due to the macroscopic mixing of fluid particles, turbulent flows are characterized by a rapid rate of diffusion of momentum and heat.
(4) *Vorticity*: Turbulence is characterized by high levels of fluctuating vorticity. The identifiable structures in a turbulent flow are vaguely called *eddies*. Flow visualization of turbulent flows shows various structures—coalescing, dividing, stretching, and above all spinning. A characteristic feature of turbulence is the existence of an enormous range of eddy sizes. The large eddies have a size of order of the width of the region of turbulent flow; in a boundary layer this is the thickness of the layer (Figure 13.1). The large eddies contain most of the

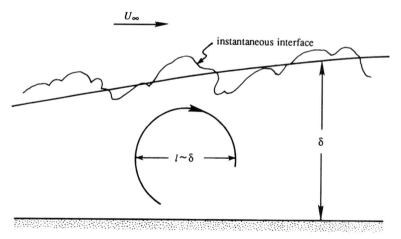

Figure 13.1 Turbulent flow in a boundary layer, showing that a large eddy has a size of the order of boundary layer thickness.

energy. The energy is handed down from large to small eddies by nonlinear interactions, until it is dissipated by viscous diffusion in the smallest eddies, whose size is of the order of millimeters.

(5) *Dissipation*: The vortex stretching mechanism transfers energy and vorticity to increasingly smaller scales, until the gradients become so large that they are smeared out (i.e., dissipated) by viscosity. Turbulent flows therefore require a continuous supply of energy to make up for the viscous losses.

These features of turbulence suggest that many flows that seem "random," such as gravity waves in the ocean or the atmosphere, are not turbulent because they are not dissipative, vortical, and nonlinear.

Incompressible turbulent flows in systems not large enough to be influenced by the Coriolis force will be studied in this chapter. These flows are three-dimensional. In large-scale geophysical systems, on the other hand, the existence of stratification and Coriolis force severely restricts vertical motion and leads to a chaotic flow that is nearly "geostropic" and two-dimensional. *Geostrophic turbulence* is briefly commented on in Chapter 14.

2. *Historical Notes*

Turbulence research is currently at the forefront of modern fluid dynamics, and some of the well-known physicists of this century have worked in this area. Among them are G. I. Taylor, Kolmogorov, Reynolds, Prandtl, von Karman, Heisenberg, Landau, Millikan, and Onsagar. A brief historical outline is given in what follows; further interesting details can be found in Monin and Yaglom (1971). The reader is expected to fully appreciate these historical remarks only after reading the chapter.

The first systematic work on turbulence was carried out by Osborne Reynolds in 1883. His experiments in pipe flows, discussed in Section 9.1, showed that the flow becomes turbulent or irregular when the nondimensional ratio $\mathrm{Re} = UL/\nu$, later named the Reynolds number by Sommerfeld, exceeds a certain critical value. (Here ν is the kinematic viscosity, U is the velocity scale, and L is the length scale.) This nondimensional number subsequently proved to be the parameter that determines the dynamic similarity of viscous flows. Reynolds also separated turbulent variables as the sum of a mean and a fluctuation and arrived at the concept of turbulent stress. The discovery of the significance of Reynolds number and turbulent stress has proved to be of fundamental importance in our present knowledge of turbulence.

In 1921 the British physicist G. I. Taylor, in a simple and elegant study of turbulent diffusion, introduced the idea of a correlation function. He showed that the rms distance of a particle from its source point initially increases with time as $\propto t$, and subsequently as $\propto \sqrt{t}$, as in a random walk. Taylor continued his outstanding work in a series of papers during 1935–1936 in which he laid down the foundation of the statistical theory of turbulence. Among the concepts he introduced were those of homogeneous and isotropic turbulence and of turbulence spectrum. Although real turbulent flows are not isotropic (the turbulent stresses, in fact, vanish for isotropic flows), the mathematical techniques involved have proved valuable for describing the *small scales* of turbulence, which are isotropic. In 1915 Taylor also introduced the idea of mixing length, although it is generally credited to Prandtl for making full use of the idea.

During the 1920s Prandtl and his student von Karman, working in Göttingen, Germany, developed the semiempirical theories of turbulence. The most successful of these was the mixing length theory, which is based on an analogy with the concept of mean free path in the kinetic theory of gases. By guessing at the correct form for the mixing length, Prandtl was able to deduce that the velocity profile near a solid wall is logarithmic, one of the most reliable results of turbulent flows. It is for this reason that subsequent textbooks on fluid mechanics have for a long time glorified the mixing length theory. Recently, however, it has become clear that the mixing length theory is not helpful since there is really no rational way of predicting the form of the mixing length. In fact, the logarithmic law can be justified from dimensional considerations alone.

Some very important work was done by the British meteorologist Lewis Richardson. In 1922 he wrote the very first book on numerical weather prediction. In this book he proposed that the turbulent kinetic energy is transferred from large to small eddies, until it is destroyed by viscous dissipation. This idea of a spectral energy cascade is at the heart of our present understanding of turbulent flows. However, Richardson's work was largely ignored at the time, and it was not until some 20 years later that the idea of a spectral cascade took a quantitative shape in the hands of Kolmogorov and Obukhov in Russia. Richardson also did another important piece of work that displayed his amazing physical intuition. On the basis of experimental data on the movement of balloons in the atmosphere, he proposed that the effective diffusion coefficient of a patch of turbulence is proportional to $l^{4/3}$, where l is the scale of the patch. This is called Richardson's four-third law, which has been subsequently found to be in agreement with Kolmogorov's five-third law of spectrum.

The Russian mathematician Kolmogorov, generally regarded as the greatest probabilist of the twentieth century, followed up on Richardson's idea of a spectral energy cascade. He hypothesized that the statistics of small scales are isotropic and depend on only two parameters, namely viscosity ν and the rate of dissipation ε. On dimensional grounds, he derived that the smallest scales must be of size $\eta = (\nu^3/\varepsilon)^{1/4}$. His second hypothesis was that, at scales much smaller than l and much larger than η, there must exist an inertial subrange in which ν plays no role; in this range the statistics depend only on a single parameter ε. Using this idea, in 1941 Kolmogorov and Obukhov independently derived that the spectrum in the inertial subrange must be proportional to $\varepsilon^{2/3} K^{-5/3}$, where K is the wavenumber. The five-third law is one of the most important results of turbulence theory and is in agreement with observations.

There has been much progress in recent years in both theory and observations. Among these may be mentioned the experimental work on the coherent structures near a solid wall. Observations in the ocean and the atmosphere (which von Karman called "a giant laboratory for turbulence research"), in which the Reynolds numbers are very large, are shedding new light on the structure of stratified turbulence.

3. Averages

The variables in a turbulent flow are not deterministic in details and have to be treated as *stochastic* or *random* variables. In this section we shall introduce certain definitions and nomenclature used in the theory of random variables.

Let $u(t)$ be any measured variable in a turbulent flow. Consider first the case when the "average characteristics" of $u(t)$ do not vary with time (Figure 13.2a). In such a case we can define the average variable as the time mean

$$\bar{u} \equiv \lim_{t_0 \to \infty} \frac{1}{t_0} \int_0^{t_0} u(t)\, dt. \tag{13.1}$$

Now consider a situation in which the average characteristics do vary with time. An example is the decaying series shown in Figure 13.2b, which could represent the velocity of a jet as the pressure in the supply tank falls. In this case the average is a function of time and cannot be formally defined by using equation (13.1), because we cannot specify how large the averaging interval t_0 should be made in evaluating the integral (13.1). If we take t_0 to be very large then we may not get a "local" average, and if we take t_0 to be very small then we may not get a reliable average. In such a case, however, one can still define an average by performing a large number of experiments, conducted under identical conditions. To define this average precisely, we first need to introduce certain terminology.

A *collection* of experiments, performed under an identical set of experimental conditions, is called an *ensemble*, and an average over the collection is called an *ensemble average*, or *expected value*. Figure 13.3 shows an example of several records of a random variable, for example, the velocity in the atmospheric boundary layer measured from 8 AM to 10 AM in the morning. Each record is measured at the same place, supposedly under identical conditions, on different days. The ith record is denoted by $u^i(t)$. (Here the superscript does not stand for power.) All records in the figure show that for some dynamic reason the velocity is decaying with time. In other words, the expected velocity at 8 AM is larger than that at 10 AM. It is clear that the average velocity at 9 AM can be found by adding together the velocity at 9 AM for

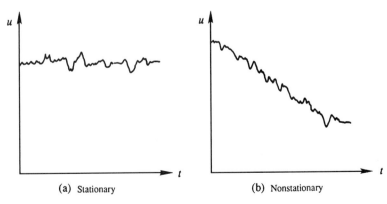

(a) Stationary   (b) Nonstationary

Figure 13.2   Stationary and nonstationary time series.

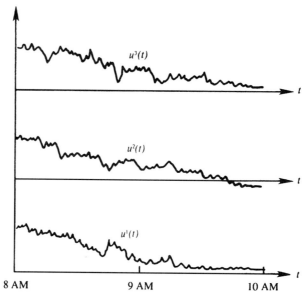

Figure 13.3 An ensemble of functions $u(t)$.

each record and dividing the sum by the number of records. We therefore define the *ensemble average* of u at time t to be

$$\bar{u}(t) \equiv \frac{1}{N} \sum_{i=1}^{N} u^i(t), \tag{13.2}$$

where N is a large number. From this it follows that the average derivative at a certain time is

$$\overline{\frac{\partial u}{\partial t}} = \frac{1}{N} \left[\frac{\partial u^1(t)}{\partial t} + \frac{\partial u^2(t)}{\partial t} + \frac{\partial u^3(t)}{\partial t} + \cdots \right]$$
$$= \frac{\partial}{\partial t} \left[\frac{1}{N} \{ u^1(t) + u^2(t) + \cdots \} \right] = \frac{\partial \bar{u}}{\partial t}.$$

This shows that the operation of differentiation commutes with the operation of ensemble averaging, so that their orders can be interchanged. In a similar manner we can show that the operation of integration also commutes with ensemble averaging. We therefore have the rules

$$\overline{\frac{\partial u}{\partial t}} = \frac{\partial \bar{u}}{\partial t}, \tag{13.3}$$

$$\overline{\int_a^b u \, dt} = \int_a^b \bar{u} \, dt. \tag{13.4}$$

Similar rules also hold when the variable is a function of space:

$$\overline{\frac{\partial u}{\partial x_i}} = \frac{\partial \bar{u}}{\partial x_i}, \tag{13.5}$$

$$\overline{\int u \, d\mathbf{x}} = \int \bar{u} \, d\mathbf{x}. \tag{13.6}$$

The rules of commutation (13.3)–(13.6) will be constantly used in the algebraic manipulations throughout the chapter.

As there is no way of controlling natural phenomena in the atmosphere and the ocean, it is very difficult to obtain observations under identical conditions. Consequently, in a nonstationary process such as the one shown in Figure 13.2b, the average value of u at a certain time is sometimes determined by using equation (13.1) and choosing an appropriate averaging time t_0, small compared to the time during which the average properties change appreciably. In any case, for theoretical discussions, all averages defined by overbars in this chapter are to be regarded as ensemble averages. If the process also happens to be stationary, then the overbar can be taken to mean the time average.

The various averages of a random variable, such as its mean and rms value, are collectively called the *statistics* of the variable. When the statistics of a random variable are independent of time, we say that the underlying process is *stationary*. Examples of stationary and nonstationary processes are shown in Figure 13.2. *For a stationary process the time average* (i.e., the average over a single record, defined by equation (13.1)) *can be shown to equal the ensemble average*, resulting in considerable simplification. Similarly, we define a *homogeneous* process as one whose statistics are independent of space, for which the ensemble average equals the spatial average.

The mean square value of a variable is called the *variance*. The square root of variance is called the *root-mean-square* (rms) value:

$$\text{variance} \equiv \overline{u^2},$$

$$u_{\text{rms}} \equiv (\overline{u^2})^{1/2}.$$

The time series $[u(t) - \bar{u}]$, obtained after subtracting the mean $\bar{u}$ of the series, represents the fluctuation of the variable about its mean. The rms value of the fluctuation is called the *standard deviation*, defined as

$$u_{\text{SD}} \equiv [\overline{(u - \bar{u})^2}]^{1/2}.$$

4. Correlations and Spectra

The autocorrelation of a single variable $u(t)$ at two times t_1 and t_2 is defined as

$$R(t_1, t_2) \equiv \overline{u(t_1)u(t_2)}. \tag{13.7}$$

In the general case when the series is not stationary, the overbar is to be regarded as an ensemble average. Then the correlation can be computed as follows: Obtain a number of records of $u(t)$, and on each record read off the values of u at t_1 and t_2. Then multiply the two values of u in each record and calculate the average value of the product over the ensemble.

The magnitude of this average product is small when a positive value of $u(t_1)$ is associated with both positive and negative values of $u(t_2)$. In such a case the magnitude of $R(t_1, t_2)$ is small, and we say that the values of u at t_1 and t_2 are "weakly correlated." If, on the other hand, a positive value of $u(t_1)$ is mostly associated with a positive value of $u(t_2)$, and a negative value of $u(t_1)$ is mostly associated with a negative value of $u(t_2)$, then the magnitude of $R(t_1, t_2)$ is large and positive; in such a case we say that the values of $u(t_1)$ and $u(t_2)$ are "strongly correlated." We may also have a case with $R(t_1, t_2)$ large and negative, in which one sign of $u(t_1)$ is mostly associated with the opposite sign of $u(t_2)$.

For a stationary process the statistics (i.e., the various kinds of averages) are independent of the origin of time, so that we can shift the origin of time by any amount. Shifting the origin by t_1, the autocorrelation (13.7) becomes $\overline{u(0)u(t_2 - t_1)}$ $= \overline{u(0)u(\tau)}$, where $\tau = t_2 - t_1$ is the *time lag*. It is clear that we can also write this correlation as $\overline{u(t)u(t + \tau)}$, which is a function of τ only, t being an arbitrary origin of measurement. We can therefore define an *autocorrelation function* of a stationary process by

$$R(\tau) = \overline{u(t)u(t + \tau)}.$$

As we have assumed stationarity, the overbar in the aforementioned expression can also be regarded as a time average. In such a case the method of estimating the correlation is to align the series $u(t)$ with $u(t + \tau)$ and multiply them vertically (Figure 13.4).

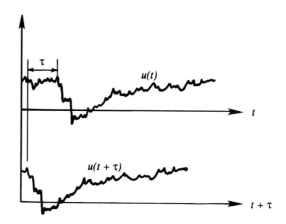

Figure 13.4 Method of calculating autocorrelation $\overline{u(t)u(t + \tau)}$.

We can also define a *normalized autocorrelation function*

$$r(\tau) \equiv \frac{\overline{u(t)u(t+\tau)}}{\overline{u^2}}, \tag{13.8}$$

where $\overline{u^2}$ is the mean square value. For any function $u(t)$, it can be proved that

$$\overline{u(t_1)u(t_2)} \leqslant [\overline{u^2}(t_1)]^{1/2}[\overline{u^2}(t_2)]^{1/2}, \tag{13.9}$$

which is called the *Schwartz inequality*. It is analogous to the rule that the inner product of two vectors cannot be larger than the product of their magnitudes. For a stationary process the mean square value is independent of time, so that the right-hand side of equation (13.9) equals $\overline{u^2}$. Using equation (13.9), it follows from equation (13.8) that

$$r \leqslant 1.$$

Obviously, $r(0) = 1$. For a stationary process the autocorrelation is a symmetric function, because then

$$R(\tau) = \overline{u(t)u(t+\tau)} = \overline{u(t-\tau)u(t)} = \overline{u(t)u(t-\tau)} = R(-\tau).$$

A typical autocorrelation plot is shown in Figure 13.5. Under normal conditions r goes to 0 as $\tau \to \infty$, because a process becomes uncorrelated with itself after a long time. A measure of the width of the correlation function can be obtained by replacing the measured autocorrelation distribution by a rectangle of height 1 and width $\mathcal{T}$ (Figure 13.5), which is therefore given by

$$\mathcal{T} \equiv \int_0^\infty r(\tau)\, d\tau. \tag{13.10}$$

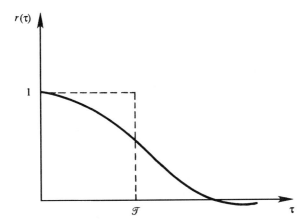

Figure 13.5 Autocorrelation function and the integral time scale.

This is called the *integral time scale*, which is a measure of the time over which $u(t)$ is highly correlated with itself. In other words, $\mathcal{T}$ is a measure of the "memory" of the process.

Let $S(\omega)$ denote the Fourier transform of the autocorrelation function $R(\tau)$. By definition, this means that

$$S(\omega) \equiv \frac{1}{2\pi} \int_{-\infty}^{\infty} e^{-i\omega\tau} R(\tau) \, d\tau. \tag{13.11}$$

It can be shown that, for equation (13.11) to be true, $R(\tau)$ must be given in terms of $S(\omega)$ by

$$R(\tau) \equiv \int_{-\infty}^{\infty} e^{i\omega\tau} S(\omega) \, d\omega. \tag{13.12}$$

We say that equations (13.11) and (13.12) define a "Fourier transform pair." The relationships (13.11) and (13.12) are not special for the autocorrelation function, but hold for any function for which a Fourier transform can be defined. Roughly speaking, a Fourier transform can be defined if the function decays to zero fast enough at infinity.

It is easy to show that $S(\omega)$ is real and symmetric if $R(\tau)$ is real and symmetric (Exercise 1). Substitution of $\tau = 0$ in equation (13.12) gives

$$\overline{u^2} = \int_{-\infty}^{\infty} S(\omega) \, d\omega. \tag{13.13}$$

This shows that $S(\omega)\, d\omega$ is the energy (more precisely, variance) in a frequency band $d\omega$ centered at ω. Therefore, the function $S(\omega)$ represents the way energy is distributed as a function of frequency ω. We say that $S(\omega)$ is the *energy spectrum*, and equation (13.11) shows that it is simply the Fourier transform of the autocorrelation function. From equation (13.11) it also follows that

$$S(0) = \frac{1}{2\pi} \int_{-\infty}^{\infty} R(\tau) \, d\tau = \frac{\overline{u^2}}{\pi} \int_{0}^{\infty} r(\tau) \, d\tau = \frac{\overline{u^2}\mathcal{T}}{\pi},$$

which shows that the value of the spectrum at zero frequency is proportional to the integral time scale.

So far we have considered u as a function of time and have defined its autocorrelation $R(\tau)$. In a similar manner we can define an autocorrelation as a function of the spatial separation between measurements of the same variable at two points. Let $u(\mathbf{x}_0, t)$ and $u(\mathbf{x}_0 + \mathbf{x}, t)$ denote the measurements of u at points $\mathbf{x}_0$ and $\mathbf{x}_0 + \mathbf{x}$. Then the spatial correlation is defined as $\overline{u(\mathbf{x}_0, t)u(\mathbf{x}_0 + \mathbf{x}, t)}$. If the field is spatially homogeneous, then the statistics are independent of the location $\mathbf{x}_0$, so that the correlation depends only on the separation $\mathbf{x}$:

$$R(\mathbf{x}) = \overline{u(\mathbf{x}_0, t)u(\mathbf{x}_0 + \mathbf{x}, t)}.$$

We can now define an energy spectrum $S(\mathbf{K})$ as a function of the wavenumber vector $\mathbf{K}$ by the Fourier transform

$$S(\mathbf{K}) = \frac{1}{(2\pi)^{1/3}} \int_{-\infty}^{\infty} e^{-i\mathbf{K}\cdot\mathbf{x}} R(\mathbf{x}) \, d\mathbf{x}, \tag{13.14}$$

where

$$R(\mathbf{x}) = \int_{-\infty}^{\infty} e^{i\mathbf{K}\cdot\mathbf{x}} S(\mathbf{K}) \, d\mathbf{K}. \tag{13.15}$$

The pair (13.14) and (13.15) is analogous to equations (13.11) and (13.12). In the integral (13.14), $d\mathbf{x}$ is the shorthand notation for the volume element $dx \, dy \, dz$. Similarly, in the integral (13.15), $d\mathbf{K} = dk \, dl \, dm$ is the volume element in the wavenumber space (k, l, m).

It is clear that we need an instantaneous measurement $u(\mathbf{x})$ as a function of position to calculate the spatial correlations $R(\mathbf{x})$. This is a difficult task and so we frequently determine this value approximately by rapidly moving a probe in a desired direction. If the speed U_0 of traversing of the probe is rapid enough, we can assume that the turbulence field is "frozen" and does not change during the measurement. Although the probe actually records a time series $u(t)$, we may then transform it to a spatial series $u(x)$ by replacing t by x/U_0. The assumption that the turbulent fluctuations at a point are caused by the advection of a frozen field past the point is called *Taylor's hypothesis*.

So far we have defined autocorrelations involving measurements of the same variable u. We can also define a *cross-correlation function* between two stationary variables $u(t)$ and $v(t)$ as

$$C(\tau) \equiv \overline{u(t)v(t + \tau)}.$$

Unlike the autocorrelation function, the cross-correlation function is *not* a symmetric function of the time lag τ, because $C(-\tau) = \overline{u(t)v(t - \tau)} \neq C(\tau)$. The value of the cross-correlation function at zero lag, that is $\overline{u(t)v(t)}$, is simply written as $\overline{uv}$ and called the "correlation" of u and v.

5. *Averaged Equations of Motion*

A turbulent flow instantaneously satisfies the Navier–Stokes equations. However, it is virtually impossible to predict the flow in detail, as there is an enormous range of scales to be resolved, the smallest spatial scales being less than millimeters and the smallest time scales being milliseconds. Even the most powerful of today's computers would take an enormous amount of computing time to predict the details of an ordinary turbulent flow, resolving all the fine scales involved. Fortunately, we are generally interested in finding only the gross characteristics in such a flow, such as the distributions of mean velocity and temperature. In this section we shall derive the equations of motion for the mean state in a turbulent flow and examine what effect the turbulent fluctuations may have on the mean flow.

We assume that the density variations are caused by temperature fluctuations alone. The density variations due to other sources such as the concentration of a solute can be handled within the present framework by defining an equivalent temperature. Under the Boussinesq approximation, the equations of motion for the instantaneous variables are

$$\frac{\partial \tilde{u}_i}{\partial t} + \tilde{u}_j \frac{\partial \tilde{u}_i}{\partial x_j} = -\frac{1}{\rho_0} \frac{\partial \tilde{p}}{\partial x_i} - g[1 - \alpha(\tilde{T} - T_0)]\delta_{i3} + \nu \frac{\partial^2 \tilde{u}_i}{\partial x_j \, \partial x_j}, \qquad (13.16)$$

$$\frac{\partial \tilde{u}_i}{\partial x_i} = 0, \qquad (13.17)$$

$$\frac{\partial \tilde{T}}{\partial t} + \tilde{u}_j \frac{\partial \tilde{T}}{\partial x_j} = \kappa \frac{\partial^2 \tilde{T}}{\partial x_j \, \partial x_j}. \qquad (13.18)$$

As in the preceding chapter, we are denoting the instantaneous quantities by a tilde ($\tilde{\ }$). Let the variables be decomposed into their mean part and a deviation from the mean:

$$\begin{aligned}
\tilde{u}_i &= U_i + u_i, \\
\tilde{p} &= P + p, \\
\tilde{T} &= \bar{T} + T'.
\end{aligned} \qquad (13.19)$$

(The corresponding density is $\tilde{\rho} = \bar{\rho} + \rho'$.) This is called the *Reynolds decomposition*. As in the preceding chapter, *the mean velocity and the mean pressure are denoted by uppercase letters, and their turbulent fluctuations are denoted by lowercase letters. This convention is impossible to use for temperature and density, for which we use an overbar for the mean state and a prime for the turbulent part.* The mean quantities $(U, P, \bar{T})$ are to be regarded as ensemble averages; for stationary flows they can also be regarded as time averages. Taking the average of both sides of equation (13.19), we obtain

$$\bar{u}_i = \bar{p} = \overline{T'} = 0,$$

showing that the fluctuations have zero mean.

The equations satisfied by the mean flow are obtained by substituting the Reynolds decomposition (13.19) into the instantaneous Navier–Stokes equations (13.16)–(13.18) and taking the average of the equations. The three equations transform as follows.

Mean Continuity Equation

Averaging the continuity equation (13.17), we obtain

$$\overline{\frac{\partial}{\partial x_i}(U_i + u_i)} = \frac{\partial U_i}{\partial x_i} + \overline{\frac{\partial u_i}{\partial x_i}} = \frac{\partial U_i}{\partial x_i} + \frac{\partial \bar{u}_i}{\partial x_i} = 0,$$

where we have used the commutation rule (13.5). Using $\bar{u}_i = 0$, we obtain

$$\frac{\partial U_i}{\partial x_i} = 0, \tag{13.20}$$

which is the continuity equation for the mean flow. Subtracting this from the continuity equation (13.17) for the total flow, we obtain

$$\frac{\partial u_i}{\partial x_i} = 0, \tag{13.21}$$

which is the continuity equation for the turbulent fluctuation field. It is therefore seen that the instantaneous, the mean, and the turbulent parts of the velocity field are all nondivergent.

Mean Momentum Equation

The momentum equation (13.16) gives

$$\frac{\partial}{\partial t}(U_i + u_i) + (U_j + u_j)\frac{\partial}{\partial x_j}(U_i + u_i)$$

$$= -\frac{1}{\rho_0}\frac{\partial}{\partial x_i}(P + p) - g[1 - \alpha(\bar{T} + T' - T_0)]\delta_{i3} + \nu\frac{\partial^2}{\partial x_j^2}(U_i + u_i). \tag{13.22}$$

We shall take the average of each term of this equation. The average of the time derivative term is

$$\overline{\frac{\partial}{\partial t}(U_i + u_i)} = \frac{\partial U_i}{\partial t} + \frac{\overline{\partial u_i}}{\partial t} = \frac{\partial U_i}{\partial t} + \frac{\partial \bar{u}_i}{\partial t} = \frac{\partial U_i}{\partial t},$$

where we have used the commutation rule (13.3), and $\bar{u}_i = 0$. The average of the advective term is

$$\overline{(U_j + u_j)\frac{\partial}{\partial x_j}(U_i + u_i)} = U_j\frac{\partial U_i}{\partial x_j} + U_j\frac{\partial \bar{u}_i}{\partial x_j} + \bar{u}_j\frac{\partial U_i}{\partial x_j} + \overline{u_j\frac{\partial u_i}{\partial x_j}}$$

$$= U_j\frac{\partial U_i}{\partial x_j} + \frac{\partial}{\partial x_j}(\overline{u_i u_j}),$$

where we have used the commutation rule (13.5) and $\bar{u}_i = 0$; the continuity equation $\partial u_j/\partial x_j = 0$ has also been used in obtaining the last term.

The average of the pressure gradient term is

$$\overline{\frac{\partial}{\partial x_i}(P + p)} = \frac{\partial P}{\partial x_i} + \frac{\partial \bar{p}}{\partial x_i} = \frac{\partial P}{\partial x_i}.$$

The average of the gravity term is

$$\overline{g[1 - \alpha(\bar{T} + T' - T_0)]} = g[1 - \alpha(\bar{T} - T_0)],$$

where we have used $\bar{T}' = 0$. The average of the viscous term is

$$\nu \overline{\frac{\partial^2}{\partial x_j \partial x_j}(U_i + u_i)} = \nu \frac{\partial^2 U_i}{\partial x_j \partial x_j}.$$

Collecting terms, the mean of the momentum equation (13.22) takes the form

$$\frac{\partial U_i}{\partial t} + U_j \frac{\partial U_i}{\partial x_j} + \frac{\partial}{\partial x_j}(\overline{u_i u_j}) = -\frac{1}{\rho_0} \frac{\partial P}{\partial x_i} - g[1 - \alpha(\bar{T} - T_0)]\,\delta_{i3} + \nu \frac{\partial^2 U_i}{\partial x_j \partial x_j}.$$
(13.23)

The correlation $\overline{u_i u_j}$ in equation (13.23) is generally nonzero, although $\bar{u}_i = 0$. This is discussed further in what follows.

Reynolds Stress

Writing the term $\overline{u_i u_j}$ on the right-hand side, the mean momentum equation (13.23) becomes

$$\frac{DU_i}{Dt} = -\frac{1}{\rho_0} \frac{\partial P}{\partial x_i} - g[1 - \alpha(\bar{T} - T_0)]\,\delta_{i3} + \frac{\partial}{\partial x_j}\left[\nu \frac{\partial U_i}{\partial x_j} - \overline{u_i u_j}\right], \qquad (13.24)$$

which can be written as

$$\boxed{\frac{DU_i}{Dt} = \frac{1}{\rho_0} \frac{\partial \bar{\tau}_{ij}}{\partial x_j} - g[1 - \alpha(\bar{T} - T_0)]\,\delta_{i3},} \qquad (13.25)$$

where

$$\boxed{\bar{\tau}_{ij} = -P\delta_{ij} + \mu \left(\frac{\partial U_i}{\partial x_j} + \frac{\partial U_j}{\partial x_i}\right) - \rho_0 \overline{u_i u_j}.} \qquad (13.26)$$

Compare equations (13.25) and (13.26) with the corresponding equations for the instantaneous flow, given by (see equations (4.13) and (4.36))

$$\frac{D\tilde{u}_i}{Dt} = \frac{1}{\rho_0} \frac{\partial \tilde{\tau}_{ij}}{\partial x_j} - g[1 - \alpha(\tilde{T} - T_0)]\,\delta_{i3},$$

$$\tilde{\tau}_{ij} = -\tilde{p}\delta_{ij} + \mu \left(\frac{\partial \tilde{u}_i}{\partial x_j} + \frac{\partial \tilde{u}_j}{\partial x_i}\right).$$

It is seen from equation (13.25) that there is an *additional* stress $-\rho_0 \overline{u_i u_j}$ acting in a mean turbulent flow. In fact, these extra stresses on the mean field of a turbulent flow are much larger than the viscous contribution $\mu(\partial U_i/\partial x_j + \partial U_i/\partial x_j)$, except very close to a solid surface where the fluctuations are small and mean flow gradients are large.

The tensor $-\rho_0\overline{u_i u_j}$ is called the *Reynolds stress tensor* and has the nine Cartesian components

$$
\begin{bmatrix}
-\rho_0\overline{u^2} & -\rho_0\overline{uv} & -\rho_0\overline{uw} \\
-\rho_0\overline{uv} & -\rho_0\overline{v^2} & -\rho_0\overline{vw} \\
-\rho_0\overline{uw} & -\rho_0\overline{vw} & -\rho_0\overline{w^2}
\end{bmatrix}.
$$

This is a symmetric tensor; its diagonal components are normal stresses, and the off-diagonal components are shear stresses. If the turbulent fluctuations are completely isotropic, that is, if they do not have any directional preference, then the off-diagonal components of $\overline{u_i u_j}$ vanish, and $\overline{u^2} = \overline{v^2} = \overline{w^2}$. This is shown in Figure 13.6, which shows a cloud of data points (sometimes called a "scatter plot") on a uv-plane. The dots represent the instantaneous values of the uv-pair at different times. In the isotropic case there is no directional preference, and the dots form a spherically symmetric pattern. In this case a positive u is equally likely to be associated with both a positive and a negative v. Consequently, *the average value of the product uv is zero if the turbulence is isotropic*. In contrast, the scatter plot in an anisotropic turbulent field has a polarity. The figure shows a case where a positive u is mostly associated with a negative v, giving $\overline{uv} < 0$.

It is easy to see why the average product of the velocity fluctuations in a turbulent flow is not expected to be zero. Consider a shear flow where the mean shear dU/dy is positive (Figure 13.7). Assume that a particle at level y is instantaneously traveling upward ($v > 0$). On the average the particle retains its original velocity during the migration, and when it arrives at level $y + dy$ it finds itself in a region where a larger velocity prevails. Thus the particle tends to slow down the neighboring fluid particles

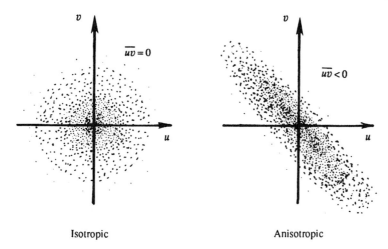

Isotropic   Anisotropic

Figure 13.6 Isotropic and anisotropic turbulent fields. Each dot represents a uv-pair at a certain time.

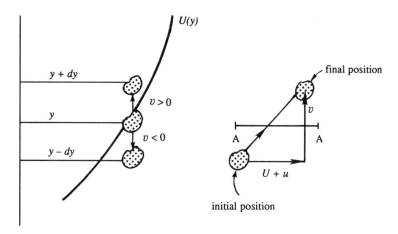

Figure 13.7   Movement of a particle in a turbulent shear flow.

after it has reached the level $y + dy$, and causes a negative u. Conversely, the particles that travel downward ($v < 0$) tend to cause a positive u in the new level $y - dy$. On the average, therefore, a positive v is mostly associated with a negative u, and a negative v is mostly associated with a positive u. The correlation $\overline{uv}$ is therefore negative for the velocity field shown in Figure 13.7, where $dU/dy > 0$. This makes sense, since in this case the x-momentum should tend to flow in the negative y-direction as the turbulence tends to diffuse the gradients and decrease dU/dy.

The procedure of deriving equation (13.26) shows that the Reynolds stress arises out of the nonlinear term $\tilde{u}_j (\partial \tilde{u}_i / \partial x_j)$ of the equation of motion. It is a stress exerted by the turbulent fluctuations on the mean flow. Another way to interpret the Reynolds stress is that it is the rate of mean momentum transfer by turbulent fluctuations. Consider again the shear flow $U(y)$ shown in Figure 13.7, where the instantaneous velocity is $(U + u, v, w)$. The fluctuating velocity components constantly transport fluid particles, and associated momentum, across a plane AA normal to the y-direction. The instantaneous rate of mass transfer across a unit area is $\rho_0 v$, and consequently the instantaneous rate of x-momentum transfer is $\rho_0 (U + u)v$. Per unit area, the average rate of flow of x-momentum in the y-direction is therefore

$$\rho_0 \overline{(U + u)v} = \rho_0 U \bar{v} + \rho_0 \overline{uv} = \rho_0 \overline{uv}.$$

Generalizing, $\rho_0 \overline{u_i u_j}$ *is the average flux of j-momentum along the i-direction, which also equals the average flux of i-momentum along the j-direction.*

The sign convention for the Reynolds stress is the same as that explained in Chapter 2, Section 4: On a surface whose outward normal points in the positive x-direction, a positive τ_{xy} points along the y-direction. According to this convention, the Reynolds stresses $-\rho_0 \overline{uv}$ on a rectangular element are directed as in Figure 13.8, if they are positive. The discussion in the preceding paragraph shows that such a Reynolds stress causes a mean flow of x-momentum along the negative y-direction.

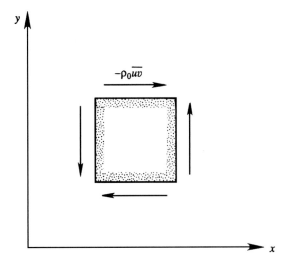

Figure 13.8   Positive directions of Reynolds stresses on a rectangular element.

Mean Heat Equation

The heat equation (13.18) is

$$\frac{\partial}{\partial t}(\bar{T} + T') + (U_j + u_j)\frac{\partial}{\partial x_j}(\bar{T} + T') = \kappa\frac{\partial^2}{\partial x_j^2}(\bar{T} + T').$$

The average of the time derivative term is

$$\overline{\frac{\partial}{\partial t}(\bar{T} + T')} = \frac{\partial \bar{T}}{\partial t} + \frac{\partial \bar{T}'}{\partial t} = \frac{\partial \bar{T}}{\partial t}.$$

The average of the advective term is

$$\overline{(U_j + u_j)\frac{\partial}{\partial x_j}(\bar{T} + T')} = U_j\frac{\partial \bar{T}}{\partial x_j} + U_j\frac{\partial \bar{T}'}{\partial x_j} + \bar{u}_j\frac{\partial \bar{T}}{\partial x_j} + \overline{u_j\frac{\partial T'}{\partial x_j}}$$

$$= U_j\frac{\partial \bar{T}}{\partial x_j} + \frac{\partial}{\partial x_j}(\overline{u_j T'}).$$

The average of the diffusion term is

$$\overline{\frac{\partial^2}{\partial x_j^2}(\bar{T} + T')} = \frac{\partial^2 \bar{T}}{\partial x_j^2} + \frac{\partial^2 \bar{T}'}{\partial x_j^2} = \frac{\partial^2 \bar{T}}{\partial x_j^2}.$$

Collecting terms, the mean heat equation takes the form

$$\frac{\partial \bar{T}}{\partial t} + U_j\frac{\partial \bar{T}}{\partial x_j} + \frac{\partial}{\partial x_j}(\overline{u_j T'}) = \kappa\frac{\partial^2 \bar{T}}{\partial x_j^2},$$

which can be written as

$$\frac{D\bar{T}}{Dt} = \frac{\partial}{\partial x_j}\left(\kappa\frac{\partial\bar{T}}{\partial x_j} - \overline{u_j T'}\right).$$ (13.27)

Multiplying by $\rho_0 C_p$, we obtain

$$\rho_0 C_p \frac{D\bar{T}}{Dt} = -\frac{\partial Q_j}{\partial x_j},$$ (13.28)

where the heat flux is given by

$$Q_j = -k\frac{\partial\bar{T}}{\partial x_j} + \rho_0 C_p \overline{u_j T'},$$ (13.29)

and $k = \rho_0 C_p \kappa$ is the thermal conductivity. Equation (13.29) shows that the fluctuations cause an additional mean *turbulent heat flux of* $\rho_0 C_p \overline{\mathbf{u}T'}$, in addition to the molecular heat flux of $-k\nabla\bar{T}$. For example, the surface of the earth becomes hot during the day, resulting in a decrease of the mean temperature with height, and an associated turbulent convective motion. An upward fluctuating motion is then mostly associated with a positive temperature fluctuation, giving rise to an upward heat flux $\rho_0 C_p \overline{wT'} > 0$.

6. Kinetic Energy Budget of Mean Flow

In this section we shall examine the sources and sinks of mean kinetic energy of a turbulent flow. As shown in Chapter 4, Section 13, a kinetic energy equation can be obtained by multiplying the equation for DU/Dt by $\mathbf{U}$. The equation of motion for the mean flow is, from equations (13.25) and (13.26),

$$\frac{\partial U_i}{\partial t} + U_j\frac{\partial U_i}{\partial x_j} = \frac{1}{\rho_0}\frac{\partial\bar{\tau}_{ij}}{\partial x_j} - \frac{g}{\rho_0}\bar{\rho}\delta_{i3},$$ (13.30)

where the stress is given by

$$\bar{\tau}_{ij} = -P\delta_{ij} + 2\mu E_{ij} - \rho_0\overline{u_i u_j}.$$ (13.31)

Here we have introduced the mean strain rate

$$E_{ij} \equiv \frac{1}{2}\left(\frac{\partial U_i}{\partial x_j} + \frac{\partial U_j}{\partial x_i}\right).$$

Multiplying equation (13.30) by U_i (and, of course, summing over i), we obtain

$$\frac{\partial}{\partial t}\left(\frac{1}{2}U_i^2\right) + U_j\frac{\partial}{\partial x_j}\left(\frac{1}{2}U_i^2\right) = \frac{1}{\rho_0}\frac{\partial}{\partial x_j}(U_i\bar{\tau}_{ij}) - \frac{1}{\rho_0}\bar{\tau}_{ij}\frac{\partial U_i}{\partial x_j} - \frac{g}{\rho_0}\bar{\rho}U_i\delta_{i3}.$$

On introducing expression (13.31) for $\bar{\tau}_{ij}$, we obtain

$$\frac{D}{Dt}\left(\frac{1}{2}U_i^2\right) = \frac{\partial}{\partial x_j}\left(-\frac{1}{\rho_0}U_i P\delta_{ij} + 2\nu U_i E_{ij} - \overline{u_i u_j}U_i\right)$$
$$+ \frac{1}{\rho_0}P\delta_{ij}\frac{\partial U_i}{\partial x_j} - 2\nu E_{ij}\frac{\partial U_i}{\partial x_j} + \overline{u_i u_j}\frac{\partial U_i}{\partial x_j} - \frac{g}{\rho_0}\bar{\rho}U_3.$$

The fourth term on the right-hand side is proportional to $\delta_{ij}(\partial U_i/\partial x_j) = \partial U_i/\partial x_i = 0$ by continuity. The mean kinetic energy balance then becomes

$$\frac{D}{Dt}\left(\frac{1}{2}U_i^2\right) = \underbrace{\frac{\partial}{\partial x_j}\left(-\frac{PU_j}{\rho_0} + 2\nu U_i E_{ij} - \overline{u_i u_j}U_i\right)}_{\text{transport}}$$

$$\underbrace{- 2\nu E_{ij}E_{ij}}_{\substack{\text{viscous} \\ \text{dissipation}}} + \underbrace{\overline{u_i u_j}\frac{\partial U_i}{\partial x_j}}_{\substack{\text{loss to} \\ \text{turbulence}}} - \underbrace{\frac{g}{\rho_0}\bar{\rho}U_3}_{\substack{\text{loss to} \\ \text{potential} \\ \text{energy}}}. \tag{13.32}$$

The left-hand side represents the rate of change of mean kinetic energy, and the right-hand side represents the various mechanisms that bring about this change. The first three terms are in the "flux divergence" form. If equation (13.32) is integrated over all space to obtain the rate of change of the total (or global) kinetic energy, then the divergence terms can be transformed into a surface integral by Gauss' theorem. These terms then would not contribute if the flow is confined to a limited region in space, with $\mathbf{U} = 0$ at sufficient distance. It follows that the first three terms can only *transport* or redistribute energy from one region to another, but cannot generate or dissipate it. The first term represents the transport of mean kinetic energy by the mean pressure, the second by the mean viscous stresses $2\nu E_{ij}$, and the third by Reynolds stresses.

The fourth term is the product of the mean strain rate E_{ij} and the mean viscous stress $2\nu E_{ij}$. It is a loss at every point in the flow and represents the *direct viscous dissipation* of mean kinetic energy. The energy is lost to the *agency* that generates the viscous stress, and so reappears as the kinetic energy of molecular motion (heat).

The fifth term is analogous to the fourth term. It can be written as $\overline{u_i u_j}(\partial U_i/\partial x_j) = \overline{u_i u_j}E_{ij}$, so that it is a product of the turbulent stress and the mean strain rate field. (Note that the doubly contracted product of a symmetric tensor $\overline{u_i u_j}$ and any tensor $\partial U_i/\partial x_j$ is equal to the product of $\overline{u_i u_j}$ and *symmetric* part of $\partial U_i/\partial x_j$, namely, E_{ij}; this is proved in Chapter 2, Section 11.) If the mean flow is given by $U(y)$, then $\overline{u_i u_j}(\partial U_i/\partial x_j) = \overline{uv}(dU/dy)$. We saw in the preceding section that $\overline{uv}$ is likely to be negative if dU/dy is positive. The fifth term $\overline{u_i u_j}(\partial U_i/\partial x_j)$ is therefore likely to be negative in shear flows. By analogy with the fourth term, it must represent an energy loss to the agency that generates turbulent stress, namely the fluctuating field. Indeed, we shall see in the following section that this term appears on the right-hand side of an equation for the rate of change of *turbulent* kinetic energy, but *with the sign reversed*. Therefore, this term generally results in a loss of mean kinetic energy and a gain of turbulent kinetic energy. We shall call this term the *shear production* of turbulence by the interaction of Reynolds stresses and the mean shear.

The sixth term represents the work done by gravity on the mean vertical motion. For example, an upward mean motion results in a loss of mean kinetic energy, which is accompanied by an increase in the potential energy of the mean field.

The two viscous terms in equation (13.32), namely, the viscous transport $2\nu \partial(U_i E_{ij})/\partial x_j$ and the viscous dissipation $-2\nu E_{ij} E_{ij}$, are small in a fully turbulent flow at high Reynolds numbers. Compare, for example, the viscous dissipation and the shear production terms:

$$\frac{2\nu E_{ij}^2}{\overline{u_i u_j}(\partial U_i/\partial x_j)} \sim \frac{\nu(U/L)^2}{u_{\text{rms}}^2 U/L} \sim \frac{\nu}{UL} \ll 1,$$

where U is the scale for mean velocity, L is a length scale (for example, the width of the boundary layer), and u_{rms} is the rms value of the turbulent fluctuation; we have also assumed that u_{rms} and U are of the same order, since experiments show that u_{rms} is a substantial fraction of U. The direct influence of viscous terms is therefore negligible on the *mean* kinetic energy budget. We shall see in the following section that this is *not* true for the *turbulent* kinetic energy budget, in which the viscous terms play a major role. What happens is the following: The mean flow loses energy to the turbulent field by means of the shear production; the *turbulent* kinetic energy so generated is then dissipated by viscosity.

7. Kinetic Energy Budget of Turbulent Flow

An equation for the turbulent kinetic energy is obtained by first finding an equation for $\partial \mathbf{u}/\partial t$ and taking the scalar product with $\mathbf{u}$. The algebra becomes compact if we use the "comma notation," introduced in Chapter 2, Section 15, namely, that a comma denotes a spatial derivative:

$$A_{,i} \equiv \frac{\partial A}{\partial x_i},$$

where A is any variable. (This notation is very simple and handy, but it may take a little practice to get used to it. It is used in this book only if the algebra would become cumbersome otherwise. There is only one other place in the book where this notation has been applied, namely Section 5.7. With a little initial patience, the reader will quickly see the convenience of this notation.)

Equations of motion for the total and mean flows are, respectively,

$$\frac{\partial}{\partial t}(U_i + u_i) + (U_j + u_j)(U_i + u_i)_{,j}$$

$$= -\frac{1}{\rho_0}(P + p)_{,i} - g[1 - \alpha(\bar{T} + T' - T_0)]\delta_{i3} + \nu(U_i + u_i)_{,jj},$$

$$\frac{\partial U_i}{\partial t} + U_j U_{i,j} = -\frac{1}{\rho_0}P_{,i} - g[1 - \alpha(\bar{T} - T_0)]\delta_{i3} + \nu U_{i,jj} - (\overline{u_i u_j})_{,j}.$$

Subtracting, we obtain the equation of motion for the turbulent velocity u_i:

$$\frac{\partial u_i}{\partial t} + U_j u_{i,j} + u_j U_{i,j} + u_j u_{i,j} - \overline{(u_i u_j)}_{,j} = -\frac{1}{\rho_0} p_{,i} + g\alpha T' \delta_{i3} + \nu u_{i,jj}.$$

(13.33)

The equation for the turbulent kinetic energy is obtained by multiplying this equation by u_i and averaging.

The first two terms on the left-hand side of equation (13.33) give

$$\overline{u_i \frac{\partial u_i}{\partial t}} = \frac{\partial}{\partial t} \left(\frac{1}{2} \overline{u_i^2} \right),$$

$$\overline{u_i U_j u_{i,j}} = U_j \left(\frac{1}{2} \overline{u_i^2} \right)_{,j}.$$

The third, fourth and fifth terms on the left-hand side of equation (13.33) give

$$\overline{u_i u_j U_{i,j}} = \overline{u_i u_j} U_{i,j},$$

$$\overline{u_i u_j u_{i,j}} = (\tfrac{1}{2} \overline{u_i^2 u_j})_{,j} - \tfrac{1}{2} \overline{u_i^2 u_{j,j}} = \tfrac{1}{2} \overline{(u_i^2 u_j)}_{,j},$$

$$\overline{-u_i (\overline{u_i u_j})}_{,j} = -\bar{u}_i (\overline{u_i u_j})_{,j} = 0,$$

where we have used the continuity equation $u_{i,i} = 0$ and $\bar{u}_i = 0$.

The first and second terms on the right-hand side of equation (13.33) give

$$\overline{-u_i \frac{1}{\rho_0} p_{,i}} = -\frac{1}{\rho_0} \overline{(u_i p)}_{,i},$$

$$\overline{u_i g\alpha T' \delta_{i3}} = g\alpha \overline{wT'}.$$

The last term on the right-hand side of equation (13.33) gives

$$\nu \overline{u_i u_{i,jj}} = \nu \{ \overline{u_i u_{i,jj}} + \tfrac{1}{2} \overline{(u_{i,j} + u_{j,i})(u_{i,j} - u_{j,i})} \},$$

where we have added the doubly contracted product of a symmetric tensor $(u_{i,j} + u_{j,i})$ and an antisymmetric tensor $(u_{i,j} - u_{j,i})$, such a product being zero. In the first term on the right-hand side, we can write $u_{i,jj} = (u_{i,j} + u_{j,i})_{,j}$ because of the continuity equation. Then we can write

$$\nu \overline{u_i u_{i,jj}} = \nu \{ \overline{u_i (u_{i,j} + u_{j,i})_{,j}} + \overline{(u_{i,j} + u_{j,i})(u_{i,j} - \tfrac{1}{2} u_{i,j} - \tfrac{1}{2} u_{j,i})} \}$$

$$= \nu \{ \overline{[u_i (u_{i,j} + u_{j,i})]}_{,j} - \tfrac{1}{2} \overline{(u_{i,j} + u_{j,i})^2} \}.$$

Defining the fluctuating strain rate by

$$e_{ij} \equiv \tfrac{1}{2} (u_{i,j} + u_{j,i}),$$

we finally obtain

$$\nu \overline{u_i u_{i,jj}} = 2\nu[\overline{u_i e_{ij}}]_{,j} - 2\nu \overline{e_{ij} e_{ij}}.$$

Collecting terms, the turbulent energy equation becomes

$$\frac{D}{Dt}\overline{\left(\frac{1}{2}u_i^2\right)} = -\frac{\partial}{\partial x_j}\underbrace{\left(\frac{1}{\rho_0}\overline{p u_j} + \frac{1}{2}\overline{u_i^2 u_j} - 2\nu\overline{u_i e_{ij}}\right)}_{\text{transport}}$$

$$\underbrace{-\,\overline{u_i u_j}U_{i,j}}_{\text{shear prod}} + \underbrace{g\alpha\overline{wT'}}_{\text{buoyant prod}} - \underbrace{2\nu\overline{e_{ij} e_{ij}}}_{\text{viscous diss}}. \qquad (13.34)$$

The first three terms on the right-hand side are in the flux divergence form and consequently represent the spatial transport of turbulent kinetic energy. The first two terms represent the transport by turbulence itself, whereas the third term is viscous transport.

The fourth term $\overline{u_i u_j}U_{i,j}$ also appears in the kinetic energy budget of the mean flow with its sign reversed, as seen by comparing equation (13.32) and equation (13.34). As argued in the preceding section, $-\overline{u_i u_j}U_{i,j}$ is usually positive, so that this term represents a loss of mean kinetic energy and a gain of turbulent kinetic energy. It must then represent the rate of generation of turbulent kinetic energy by the interaction of the Reynolds stress with the mean shear $U_{i,j}$. Therefore,

$$\boxed{\text{Shear production} = -\overline{u_i u_j}\frac{\partial U_i}{\partial x_j}.} \qquad (13.35)$$

The fifth term $g\alpha\overline{wT'}$ can have either sign, depending on the nature of the background temperature distribution $\bar{T}(z)$. In a stable situation in which the background temperature increases upward (as found, e.g., in the atmospheric boundary layer at night), rising fluid elements are likely to be associated with a negative temperature fluctuation, resulting in $\overline{wT'} < 0$, which means a downward turbulent heat flux. In such a stable situation $g\alpha\overline{wT'}$ represents the rate of turbulent energy loss by working against the stable background density gradient. In the opposite case, when the background density profile is unstable, the turbulent heat flux $\overline{wT'}$ is upward, and convective motions cause an increase of turbulent kinetic energy (Figure 13.9). We shall call $g\alpha\overline{wT'}$ the *buoyant production* of turbulent kinetic energy, keeping in mind that it can also be a buoyant "destruction" if the turbulent heat flux is downward. Therefore,

$$\boxed{\text{Buoyant production} = g\alpha\overline{wT'}.} \qquad (13.36)$$

The buoyant generation of turbulent kinetic energy lowers the potential energy of the mean field. This can be understood from Figure 13.9, where it is seen that the heavier fluid has moved downward in the final state as a result of the heat flux. This

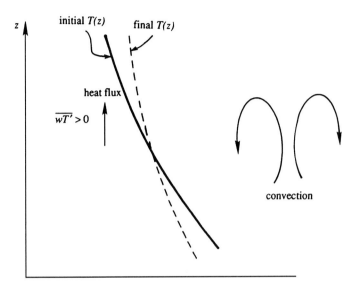

Figure 13.9 Heat flux in an unstable environment, generating turbulent kinetic energy and lowering the mean potential energy.

can also be demonstrated by deriving an equation for the mean potential energy, in which the term $g\alpha\overline{wT'}$ appears with a *negative* sign on the right-hand side. Therefore, the *buoyant generation* of turbulent kinetic energy by the upward heat flux occurs at the expense of the mean *potential* energy. This is in contrast to the *shear production* of turbulent kinetic energy, which occurs at the expense of the mean *kinetic* energy.

The sixth term $2\nu\overline{e_{ij}e_{ij}}$ is the *viscous dissipation of turbulent kinetic energy*, and is usually denoted by ε:

$$\boxed{\varepsilon = \text{Viscous dissipation} = 2\nu\overline{e_{ij}e_{ij}}.} \tag{13.37}$$

This term is *not* negligible in the turbulent kinetic energy equation, although an analogous term (namely $2\nu E_{ij}^2$) is negligible in the *mean* kinetic energy equation, as discussed in the preceding section. In fact, the viscous dissipation ε is of the order of the turbulence production terms ($\overline{u_i u_j}U_{i,j}$ or $g\alpha\overline{wT'}$) in most locations.

8. Turbulence Production and Cascade

Evidence suggests that the large eddies in a turbulent flow are anisotropic, in the sense that they are "aware" of the direction of mean shear or of background density gradient. In a completely isotropic field the off-diagonal components of the Reynolds stress $\overline{u_i u_j}$ are zero (see Section 5 here), as is the upward heat flux $\overline{wT'}$ because there is no preference between the upward and downward directions. In such an isotropic case no turbulent energy can be extracted from the mean field. Therefore, turbulence must develop anisotropy if it has to sustain itself against viscous dissipation.

A possible mechanism of generating anisotropy in a turbulent shear flow is discussed by Tennekes and Lumley (1972, p. 41). Consider a parallel shear flow $U(y)$

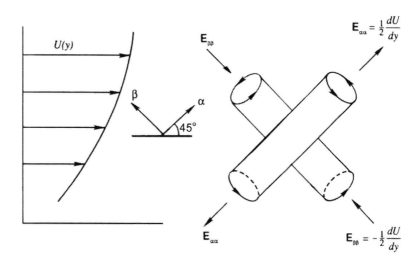

Figure 13.10   Large eddies oriented along the principal directions of a parallel shear flow. Note that the vortex aligned with the α-axis has a positive v when u is negative and a negative v when u is positive, resulting in $\overline{uv} < 0$.

shown in Figure 13.10, in which the fluid elements translate, rotate, and undergo shearing deformation. The nature of deformation of an element depends on the orientation of the element. An element oriented parallel to the xy-axes undergoes only a shear strain rate $E_{xy} = \frac{1}{2}\,dU/dy$, but no linear strain rate ($E_{xx} = E_{yy} = 0$). The strain rate tensor in the xy-coordinate system is therefore

$$\mathbf{E} = \begin{bmatrix} 0 & \frac{1}{2}\,dU/dy \\ \frac{1}{2}\,dU/dy & 0 \end{bmatrix}.$$

As shown in Chapter 3, Section 10, such a symmetric tensor can be diagonalized by rotating the coordinate system by $45°$. Along these principal axes (denoted by α and β in Figure 13.10), the strain rate tensor is

$$\mathbf{E} = \begin{bmatrix} \frac{1}{2}\,dU/dy & 0 \\ 0 & -\frac{1}{2}\,dU/dy \end{bmatrix},$$

so that there is a linear extension rate of $E_{\alpha\alpha} = \frac{1}{2}\,dU/dy$, a linear compression rate of $E_{\beta\beta} = -\frac{1}{2}\,dU/dy$, and no shear ($E_{\alpha\beta} = 0$). The kinematics of stretching and compression along the principal directions in a parallel shear flow is discussed further in Chapter 3, Section 10.

The large eddies with vorticity oriented along the α-axis intensify in strength due to the vortex stretching, and the ones with vorticity oriented along the β-axis decay in strength. The net effect of the mean shear on the turbulent field is therefore to cause a predominance of eddies with vorticity oriented along the α-axis. As is evident in Figure 13.10, these eddies are associated with a positive u when v is negative, and with a negative u when v is positive, resulting in a positive value for the shear production $-\overline{uv}(dU/dy)$.

The largest eddies are of order of the width of the shear flow, for example the diameter of a pipe or the width of a boundary layer along a wall or along the upper surface of the ocean. These eddies extract kinetic energy from the mean field. The eddies that are somewhat smaller than these are strained by the velocity field of the largest eddies, and extract energy from the larger eddies by the same mechanism of vortex stretching. The much smaller eddies are essentially advected in the velocity field of the large eddies, as the scales of the strain rate field of the large eddies are much larger than the size of a small eddy. Therefore, the small eddies do not interact with either the large eddies or the mean field. *The kinetic energy is therefore cascaded down from large to small eddies in a series of small steps. This process of energy cascade is essentially inviscid, as the vortex stretching mechanism arises from the nonlinear terms of the equations of motion.*

In a fully turbulent shear flow (i.e., for large Reynolds numbers), therefore, the viscosity of the fluid does not affect the shear production, if all other variables are held constant. The viscosity does, however, determine the *scales* at which turbulent energy is dissipated into heat. From the expression $\varepsilon = 2\nu\overline{e_{ij}e_{ij}}$, it is clear that the energy dissipation is effective only at very small scales, which have high fluctuating strain rates. The continuous stretching and cascade generate long and thin filaments, somewhat like "spaghetti." When these filaments become thin enough, molecular diffusive effects are able to smear out their velocity gradients. These are the smallest scales in a turbulent flow and are responsible for the dissipation of the turbulent kinetic energy. Figure 13.11 illustrates the deformation of a fluid particle in a turbulent motion, suggesting that molecular effects can act on thin filaments generated by continuous stretching. The large mixing rates in a turbulent flow, therefore, are essentially a result of the turbulent fluctuations generating the large *surfaces* on which the molecular diffusion finally acts.

It is clear that ε does not depend on ν, but is determined by the *inviscid* properties of the large eddies, which supply the energy to the dissipating scales. Suppose l is a typical length scale of the large eddies (which may be taken equal to the integral length scale defined from a spatial correlation function, analogous to the integral time scale defined by equation (13.10)), and u' is a typical scale of the fluctuating velocity

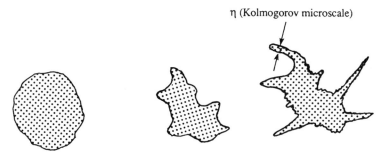

Figure 13.11 Successive deformations of a marked fluid element. Diffusive effects cause smearing when the scale becomes of the order of the Kolmogorov microscale.

(which may be taken equal to the rms fluctuating speed). Then the time scale of large eddies is of order l/u'. Observations show that the large eddies lose much of their energy during the time they turn over one or two times, so that the rate of energy transferred from large eddies is proportional to u'^2 times their frequency u'/l. The dissipation rate must then be of order

$$\boxed{\varepsilon \sim \frac{u'^3}{l},}$$

$$(13.38)$$

signifying that the viscous dissipation is determined by the inviscid large-scale dynamics of the turbulent field.

Kolmogorov suggested in 1941 that the size of the dissipating eddies depends on those parameters that are relevant to the smallest eddies. These parameters are the rate ε at which energy has to be dissipated by the eddies and the diffusivity ν that does the smearing out of the velocity gradients. As the unit of ε is m^2/s^3, dimensional reasoning shows that the length scale formed from ε and ν is

$$\boxed{\eta = \left(\frac{\nu^3}{\varepsilon}\right)^{1/4},}$$

$$(13.39)$$

which is called the *Kolmogorov microscale. A decrease of ν merely decreases the scale at which viscous dissipation takes place, and not the rate of dissipation ε.* Estimates show that η is of the order of millimeters in the ocean and the atmosphere. In laboratory flows the Kolmogorov microscale is much smaller because of the larger rate of viscous dissipation. Landahl and Mollo-Christensen (1986) give a nice illustration of this. Suppose we are using a 100-W household mixer in 1 kg of water. As all the power is used to generate the turbulence, the rate of dissipation is $\varepsilon = 100\,W/kg = 100\,m^2/s^3$. Using $\nu = 10^{-6}\,m^2/s$ for water, we obtain $\eta = 10^{-2}\,mm$.

9. Spectrum of Turbulence in Inertial Subrange

In Section 4 we defined the wavenumber spectrum $S(\mathbf{K})$, representing turbulent kinetic energy as a function of the wavenumber vector $\mathbf{K}$. If the turbulence is isotropic, then the spectrum becomes independent of the orientation of the wavenumber vector and depends on its magnitude K only. In that case we can write

$$\overline{u^2} = \int_0^\infty S(K)\,dK.$$

In this section we shall derive the form of $S(K)$ in a certain range of wavenumbers in which the turbulence is nearly isotropic.

Somewhat vaguely, we shall associate a wavenumber K with an eddy of size K^{-1}. Small eddies are therefore represented by large wavenumbers. Suppose l is the scale of the large eddies, which may be the width of the boundary layer. At the relatively small scales represented by wavenumbers $K \gg l^{-1}$, there is no direct interaction between the turbulence and the motion of the large, energy-containing eddies. This is because the small scales have been generated by a long series of small steps, losing information at each step. *The spectrum in this range of large wavenumbers is nearly isotropic*, as only the large eddies are aware of the directions of mean gradients. The spectrum here does not depend on how much energy is present at large scales (where most of the energy is contained), or the scales at which most of the energy is present. The spectrum in this range depends only on the parameters that determine the nature of the small-scale flow, so that we can write

$$S = S(K, \varepsilon, \nu) \qquad K \gg l^{-1}.$$

The range of wavenumbers $K \gg l^{-1}$ is usually called the *equilibrium range*. The dissipating wavenumbers with $K \sim \eta^{-1}$, beyond which the spectrum falls off very rapidly, form the high end of the equilibrium range (Figure 13.12). The lower end of this range, for which $l^{-1} \ll K \ll \eta^{-1}$, is called the *inertial subrange*, as only the transfer of energy by inertial forces (vortex stretching) takes place in this range. Both production and dissipation are small in the inertial subrange. The production of energy by large eddies causes a peak of S at a certain $K \simeq l^{-1}$, and the dissipation of energy causes a sharp drop of S for $K > \eta^{-1}$. The question is, how does S vary with K between the two limits in the inertial subrange?

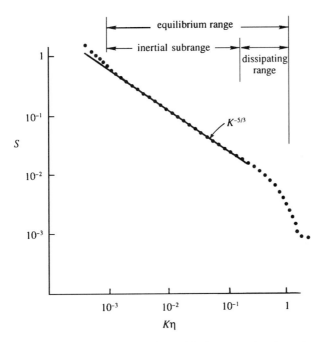

Figure 13.12 A typical wavenumber spectrum observed in the ocean, plotted on a log–log scale. The unit of S is arbitrary, and the dots represent hypothetical data.

Kolmogorov argued that, in the inertial subrange part of the equilibrium range, S is independent of ν also, so that

$$S = S(K, \varepsilon) \qquad l^{-1} \ll K \ll \eta^{-1}.$$

Although little dissipation takes place in the inertial subrange, the spectrum here does depend on ε. This is because the energy that is dissipated must be transferred across the inertial subrange, from low to high wavenumbers. As the unit of S is m^3/s^2 and that of ε is m^2/s^3, dimensional reasoning gives

$$\boxed{S = A\varepsilon^{2/3} K^{-5/3} \quad l^{-1} \ll K \ll \eta^{-1},} \qquad (13.40)$$

where $A \simeq 1.5$ has been found to be a universal constant, valid for all turbulent flows. Equation (13.40) is usually called *Kolmogorov's $K^{-5/3}$ law*. If the Reynolds number of the flow is large, then the dissipating eddies are much smaller than the energy-containing eddies, and the inertial subrange is quite broad.

Because very large Reynolds numbers are difficult to generate in the laboratory, the Kolmogorov spectral law was not verified for many years. In fact, doubts were being raised about its theoretical validity. The first confirmation of the Kolmogorov law came from the oceanic observations of Grant *et al.* (1962), who obtained a velocity spectrum in a tidal flow through a narrow passage between two islands near the west coast of Canada. The velocity fluctuations were measured by hanging a hot film anemometer from the bottom of a ship. Based on the depth of water and the average flow velocity, the Reynolds number was of order 10^8. Such large Reynolds numbers are typical of geophysical flows, since the length scales are very large. The $K^{-5/3}$ law has since been verified in the ocean over a wide range of wavenumbers, a typical behavior being sketched in Figure 13.12. Note that the spectrum drops sharply at $K\eta \sim 1$, where viscosity begins to affect the spectral shape. The figure also shows that the spectrum departs from the $K^{-5/3}$ law for small values of the wavenumber, where the turbulence production by large eddies begins to affect the spectral shape. Laboratory experiments are also in agreement with the Kolmogorov spectral law, although in a narrower range of wavenumbers because the Reynolds number is not as large as in geophysical flows. The $K^{-5/3}$ law has become one of the most important results of turbulence theory.

10. *Wall-Free Shear Flow*

Nearly parallel shear flows are divided into two classes—wall-free shear flows and wall-bounded shear flows. In this section we shall examine some aspects of turbulent flows that are free of solid boundaries. Common examples of such flows are jets, wakes, and shear layers (Figure 13.13). For simplicity we shall consider only plane two-dimensional flows. Axisymmetric flows are discussed in Townsend (1976) and Tennekes and Lumley (1972).

Intermittency

Consider a turbulent flow confined to a limited region. To be specific we shall consider the example of a wake (Figure 13.13b), but our discussion also applies to a jet, a shear layer, or the outer part of a boundary layer on a wall. The fluid outside the turbulent

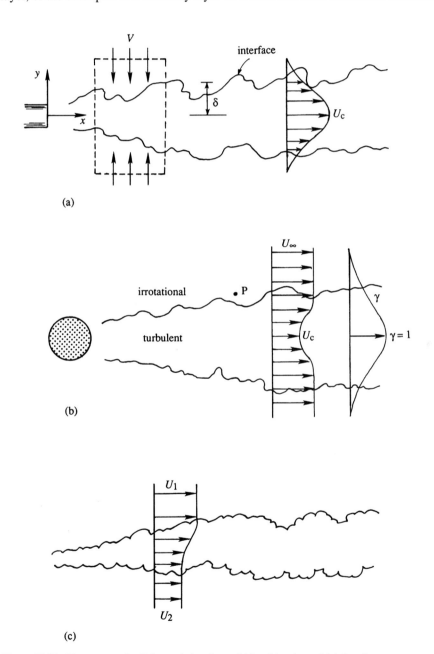

(a)

(b)

(c)

Figure 13.13 Three types of wall-free turbulent flows: (a) jet; (b) wake; and (c) shear layer.

region is either in irrotational motion (as in the case of a wake or a boundary layer), or nearly static (as in the case of a jet). Observations show that the instantaneous interface between the turbulent and nonturbulent fluid is very sharp. In fact, the thickness of the interface must equal the size of the smallest scales in the flow, namely the Kolmogorov microscale. The interface is highly contorted due to the presence of eddies of various sizes. However, a photograph exposed for a long time does not show such an irregular and sharp interface but rather a gradual and smooth transition region.

Measurements at a fixed point in the outer part of the turbulent region (say at point P in Figure 13.13b) show periods of high-frequency fluctuations as the point P moves into the turbulent flow and quiet periods as the point moves out of the turbulent region. Intermittency γ is defined as the fraction of time the flow at a point is turbulent. The variation of γ across a wake is sketched in Figure 13.13b, showing that $\gamma = 1$ near the center where the flow is always turbulent, and $\gamma = 0$ at the outer edge of the flow.

Entrainment

A flow can slowly pull the surrounding irrotational fluid inward by "frictional" effects; the process is called *entrainment*. The source of this "friction" is viscous in laminar flow and inertial in turbulent flow. The entrainment of a laminar jet was discussed in Chapter 10, Section 12. The entrainment in a turbulent flow is similar, but the rate is much larger. After the irrotational fluid is drawn inside a turbulent region, the new fluid must be made turbulent. This is initiated by small eddies (which are dominated by viscosity) acting at the sharp interface between the turbulent and the nonturbulent fluid (Figure 13.14).

The foregoing discussion of intermittency and entrainment applies not only to wall-free shear flows but also to the outer edge of boundary layers.

Self-Preservation

Far downstream, experiments show that the mean field in a wall-free shear flow becomes approximately self-similar at various downstream distances. As the mean field is affected by the Reynolds stress through the equations of motion, this means that the various turbulent quantities (such as Reynolds stress) also must reach self-similar

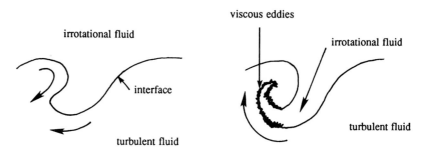

Figure 13.14 Entrainment of a nonturbulent fluid and its assimilation into turbulent fluid by viscous action at the interface.

states. This is indeed found to be approximately true (Townsend, 1976). The flow is then in a state of "moving equilibrium," in which both the mean and the turbulent fields are determined solely by the *local* scales of length and velocity. This is called *self-preservation*. In the self-similar state, the mean velocity at various downstream distances is given by

$$\frac{U}{U_c} = f\left(\frac{y}{\delta}\right) \qquad \text{(jet)},$$

$$\frac{U_\infty - U}{U_\infty - U_c} = f\left(\frac{y}{\delta}\right) \qquad \text{(wake)}, \qquad (13.41)$$

$$\frac{U - U_1}{U_2 - U_1} = f\left(\frac{y}{\delta}\right) \qquad \text{(shear layer)}.$$

Here $\delta(x)$ is the width of flow, $U_c(x)$ is the centerline velocity for the jet and the wake, and U_1 and U_2 are the velocities of the two streams in a shear layer (Figure 13.13).

Consequence of Self-Preservation in a Plane Jet

We shall now derive how the centerline velocity and width in a plane jet must vary if we assume that the mean velocity profiles at various downstream distances are self similar. This can be done by examining the equations of motion in differential form. An alternate way is to examine an integral form of the equation of motion, derived in Chapter 10, Section 12. It was shown there that the momentum flux $M = \rho \int U^2 \, dy$ across the jet is independent of x, while the mass flux $\rho \int U \, dy$ increases downstream due to entrainment. Exactly the same constraint applies to a turbulent jet. For the sake of readers who find cross references annoying, the integral constraint for a two-dimensional jet is rederived here.

Consider a control volume shown by the dotted line in Figure 13.13a, in which the horizontal surfaces of the control volume are assumed to be at a large distance from the jet axis. At these large distances, there is a mean V field toward the jet axis due to entrainment, but no U field. Therefore, the flow of x-momentum through the horizontal surfaces of the control volume is zero. The pressure is uniform throughout the flow, and the viscous forces are negligible. The net force on the surface of the control volume is therefore zero. The momentum principle for a control volume (see Chapter 4, Section 8) states that the net x-directed force on the boundary equals the *net* rate of outflow of x-momentum through the control surfaces. As the net force here is zero, the influx of x-momentum must equal the outflow of x-momentum. That is

$$M = \rho \int_{-\infty}^{\infty} U^2 \, dy = \text{independent of } x, \qquad (13.42)$$

where M is the momentum flux of the jet (= integral of mass flux $\rho U \, dy$ times velocity U). The momentum flux is the basic externally controlled parameter for a jet and is known from an evaluation of equation (13.42) at the orifice opening. The mass flux $\rho \int U \, dy$ across the jet must increase because of entrainment of the surrounding fluid.

The assumption of self similarity can now be used to predict how δ and U_c in a jet should vary with x. Substitution of the self-similarity assumption (13.41) into the integral constraint (13.42) gives

$$M = \rho U_c^2 \delta \int_{-\infty}^{\infty} f^2 \, d\left(\frac{y}{\delta}\right).$$

The preceding integral is a constant because it is completely expressed in terms of the nondimensional function $f(y/\delta)$. As M is also a constant, we must have

$$U_c^2 \delta = \text{const.} \tag{13.43}$$

At this point we make another important assumption. We assume that the Reynolds number is large, so that the gross characteristics of the flow are independent of the Reynolds number. This is called *Reynolds number similarity*. The assumption is expected to be valid in a wall-free shear flow, as viscosity does not directly affect the motion; a decrease of ν, for example, merely decreases the scale of the dissipating eddies, as discussed in Section 8. (The principle is not valid near a smooth wall, and as a consequence the drag coefficient for a smooth flat plate does not become independent of the Reynolds number as Re $\to \infty$; see Figure 10.12.) For large Re, then, U_c is independent of viscosity and can only depend on x, ρ, and M:

$$U_c = U_c(x, \rho, M).$$

A dimensional analysis shows that

$$U_c \propto \sqrt{\frac{M}{\rho x}} \qquad (\text{jet}), \tag{13.44}$$

so that equation (13.43) requires

$$\delta \propto x \qquad (\text{jet}). \tag{13.45}$$

This should be compared with the $\delta \propto x^{2/3}$ behavior of a *laminar* jet, derived in Chapter 10, Section 12. Experiments show that the width of a turbulent jet does grow linearly, with a spreading angle of 4°.

For two-dimensional wakes and shear layers, it can be shown (Townsend, 1976; Tennekes and Lumley, 1972) that the assumption of self similarity requires

$$U_\infty - U_c \propto x^{-1/2}, \quad \delta \propto \sqrt{x} \qquad (\text{wake}),$$
$$U_1 - U_2 = \text{const.}, \quad \delta \propto x \qquad (\text{shear layer}).$$

Turbulent Kinetic Energy Budget in a Jet

The turbulent kinetic energy equation derived in Section 7 will now be applied to a two-dimensional jet. The energy budget calculation uses the experimentally measured distributions of turbulence intensity and Reynolds stress across the jet. Therefore, we present the distributions of these variables first. Measurements show that

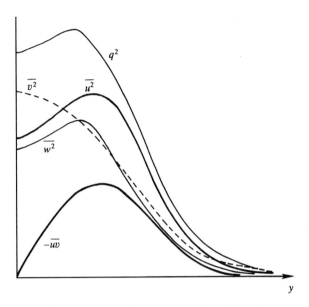

Figure 13.15   Sketch of observed variation of turbulent intensity and Reynolds stress across a jet.

the turbulent intensities and Reynolds stress are distributed as in Figure 13.15. Here $\overline{u^2}$ is the intensity of fluctuation in the downstream direction x, $\overline{v^2}$ is the intensity along the cross-stream direction y, and $\overline{w^2}$ is the intensity in the z-direction; $q^2 \equiv (\overline{u^2} + \overline{v^2} + \overline{w^2})/2$ is the turbulent kinetic energy per unit mass. The Reynolds stress is zero at the center of the jet by symmetry, since there is no reason for v at the center to be mostly of one sign if u is either positive or negative. The Reynolds stress reaches a maximum magnitude roughly where $\partial U/\partial y$ is maximum. This is also close to the region where the turbulent kinetic energy reaches a maximum.

Consider now the kinetic energy budget. For a two-dimensional jet under the boundary layer assumption $\partial/\partial x \ll \partial/\partial y$, equation (13.34) becomes

$$0 = -U\frac{\partial q^2}{\partial x} - V\frac{\partial q^2}{\partial y} - \overline{uv}\frac{\partial U}{\partial y} - \frac{\partial}{\partial y}\left[\overline{q^2 v} + \overline{pv}/\rho\right] - \varepsilon, \qquad (13.46)$$

where the left-hand side represents $\partial q^2/\partial t = 0$. Here the viscous transport and a term $(\overline{v^2} - \overline{u^2})(\partial U/\partial x)$ arising out of the shear production have been neglected on the right-hand side because they are small. The balance of terms is analyzed in Townsend (1976), and the results are shown in Figure 13.16, where T denotes turbulent transport represented by the fourth term on the right-hand side of (13.46). The shear production is zero at the center where both $\partial U/\partial y$ and $\overline{uv}$ are zero, and reaches a maximum close to the position of the maximum Reynolds stress. Near the center, the dissipation is primarily balanced by the downstream advection $-U(\partial q^2/\partial x)$, which is positive because the turbulent intensity q^2 decays downstream. Away from the center, but not too close to the outer edge of the jet, the production and dissipation terms balance. In the outer parts of the jet, the transport term balances the cross-stream

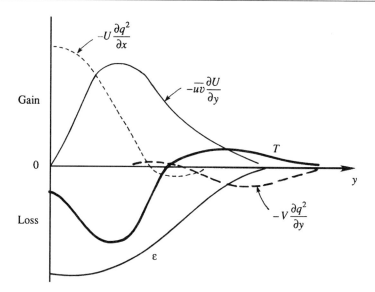

Figure 13.16   Sketch of observed kinetic energy budget in a turbulent jet. Turbulent transport is indicated by T.

advection. In this region V is negative (i.e., toward the center) due to entrainment of the surrounding fluid, and also q^2 decreases with y. Therefore the cross-stream advection $-V(\partial q^2/\partial y)$ is negative, signifying that the entrainment velocity V tends to decrease the turbulent kinetic energy at the outer edge of the jet. The stationary state is therefore maintained by the transport term T carrying turbulent kinetic energy away from the center (where $T < 0$) into the outer parts of the jet (where $T > 0$).

11.   Wall-Bounded Shear Flow

The gross characteristics of free shear flows, discussed in the preceding section, are independent of viscosity. This is not true of a turbulent flow bounded by a solid wall, in which the presence of viscosity affects the motion near the wall. The effect of viscosity is reflected in the fact that the drag coefficient of a smooth flat plate depends on the Reynolds number even for Re $\to \infty$, as seen in Figure 10.12. Therefore, the concept of Reynolds number similarity, which says that the gross characteristics are independent of Re when Re $\to \infty$, no longer applies. In this section we shall examine how the properties of a turbulent flow near a wall are affected by viscosity. Before doing this, we shall examine how the Reynolds stress should vary with distance from the wall.

Consider first a fully developed turbulent flow in a channel. By "fully developed" we mean that the flow is no longer changing in x (see Figure 9.2). Then the mean equation of motion is

$$0 = -\frac{\partial P}{\partial x} + \frac{\partial \bar{\tau}}{\partial y},$$

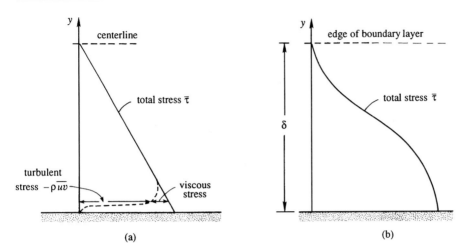

Figure 13.17 Variation of shear stress across a channel and a boundary layer: (a) channel; and (b) boundary layer.

where $\bar{\tau} = \mu(dU/dy) - \rho\overline{uv}$ is the total stress. Because $\partial P/\partial x$ is a function of x alone and $\partial\bar{\tau}/\partial y$ is a function of y alone, both of them must be constants. The stress distribution is then linear (Figure 13.17a). Away from the wall $\bar{\tau}$ is due mostly to the Reynolds stress, but close to the wall the viscous contribution dominates. In fact, at the wall the velocity fluctuations and consequently the Reynolds stresses vanish, so that the stress is entirely viscous.

In a boundary layer on a flat plate there is no pressure gradient and the mean flow equation is

$$\rho U\frac{\partial U}{\partial x} + \rho V\frac{\partial U}{\partial y} = \frac{\partial\bar{\tau}}{\partial y},$$

where $\bar{\tau}$ is a function of x and y. The variation of the stress across a boundary layer is sketched in Figure 13.17b.

Inner Layer: Law of the Wall

Consider the flow near the wall of a channel, pipe, or boundary layer. Let U_∞ be the free-stream velocity in a boundary layer or the centerline velocity in a channel and pipe. Let δ be the width of flow, which may be the width of the boundary layer, the channel half width, or the radius of the pipe. Assume that the wall is smooth, so that the height of the surface roughness elements is too small to affect the flow. Physical considerations suggest that the velocity profile near the wall depends only on the parameters that are relevant near the wall and does not depend on the free-stream velocity U_∞ or the thickness of the flow δ. Very near a smooth surface, then, we expect that

$$U = U(\rho, \tau_0, \nu, y), \tag{13.47}$$

where τ_0 is the shear stress at the wall. To express equation (13.47) in terms of dimensionless variables, note that only ρ and τ_0 involve the dimension of mass, so that these two variables must always occur together in any nondimensional group. The important ratio

$$u_* \equiv \sqrt{\frac{\tau_0}{\rho}}, \tag{13.48}$$

has the dimension of velocity and is called the *friction velocity*. Equation (13.47) can then be written as

$$U = U(u_*, \nu, y). \tag{13.49}$$

This relates four variables involving only the two dimensions of length and time. According to the pi theorem (Chapter 8, Section 4) there must be only $4 - 2 = 2$ nondimensional groups U/u_* and yu_*/ν, which should be related by some universal functional form

$$\frac{U}{u_*} = f\left(\frac{yu_*}{\nu}\right) = f(y_+) \qquad \text{(law of the wall)}, \tag{13.50}$$

where $y_+ \equiv yu_*/\nu$ is the distance nondimensionalized by the *viscous scale* ν/u_*. Equation (13.50) is called the *law of the wall*, and states that U/u_* must be a universal function of yu_*/ν near a smooth wall.

The inner part of the wall layer, right next to the wall, is dominated by viscous effects (Figure 13.18) and is called the *viscous sublayer*. It used to be called the "laminar sublayer," until experiments revealed the presence of considerable fluctuations within the layer. In spite of the fluctuations, the Reynolds stresses are still small here because of the dominance of viscous effects. Because of the thinness of the viscous sublayer, the stress can be taken as uniform within the layer and equal to the wall shear stress τ_0. Therefore the velocity gradient in the viscous sublayer is given by

$$\mu \frac{dU}{dy} = \tau_0,$$

which shows that the velocity distribution is linear. Integrating, and using the no-slip boundary condition, we obtain

$$U = \frac{y\tau_0}{\mu}.$$

In terms of nondimensional variables appropriate for a wall layer, this can be written as

$$\frac{U}{u_*} = y_+ \qquad \text{(viscous sublayer)}. \tag{13.51}$$

Experiments show that the linear distribution holds up to $yu_*/\nu \sim 5$, which may be taken to be the limit of the viscous sublayer.

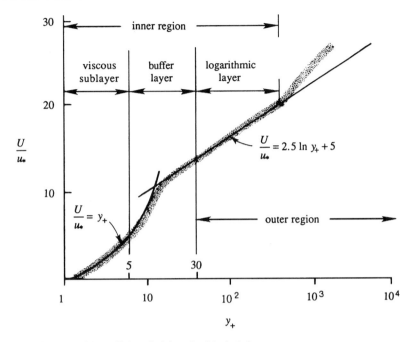

Figure 13.18   Law of the wall. A typical data cloud is shaded.

Outer Layer: Velocity Defect Law

We now explore the form of the velocity distribution in the outer part of a turbulent layer. The gross characteristics of the turbulence in the outer region are inviscid and resemble those of a wall-free turbulent flow. The existence of Reynolds stresses in the outer region results in a drag on the flow and generates a *velocity defect* $(U_\infty - U)$, which is expected to be proportional to the wall friction characterized by u_*. It follows that the velocity distribution in the outer region must have the form

$$\frac{U - U_\infty}{u_*} = F\left(\frac{y}{\delta}\right) = F(\xi) \qquad \text{(velocity defect law)}, \qquad (13.52)$$

where $\xi \equiv y/\delta$. This is called the *velocity defect law*.

Overlap Layer: Logarithmic Law

The velocity profiles in the inner and outer parts of the boundary layer are governed by different laws (13.50) and (13.52), in which the independent variable y is scaled differently. Distances in the outer part are scaled by δ, whereas those in the inner part are measured by the much smaller *viscous scale* v/u_*. In other words, the small distances in the inner layer are magnified by expressing them as yu_*/v. This is the typical behavior in singular perturbation problems (see Chapter 10, Sections 14 and 16). In these problems the inner and outer solutions are matched together in a *region of overlap* by taking the limits $y_+ \to \infty$ and $\xi \to 0$ *simultaneously*. Instead of matching velocity, in this case it is more convenient to match their gradients. (The derivation given here closely follows Tennekes and Lumley (1972).) From equations (13.50) and (13.52), the velocity gradients in the inner and outer regions are given by

$$\frac{dU}{dy} = \frac{u_*^2}{\nu}\frac{df}{dy_+},$$ (13.53)

$$\frac{dU}{dy} = \frac{u_*}{\delta}\frac{dF}{d\xi}.$$ (13.54)

Equating (13.53) and (13.54) and multiplying by y/u_*, we obtain

$$\xi\frac{dF}{d\xi} = y_+\frac{df}{dy_+} = \frac{1}{k},$$ (13.55)

valid for large y_+ and small ξ. As the left-hand side can only be a function of ξ and the right-hand side can only be a function of y_+, both sides must be equal to the same universal constant, say $1/k$, where k is called the *von Karman constant*. Experiments show that $k \simeq 0.41$. Integration of equation (13.55) gives

$$f(y_+) = \frac{1}{k}\ln y_+ + A,$$

$$F(\xi) = \frac{1}{k}\ln \xi + B.$$ (13.56)

Experiments show that $A = 5.0$ and $B = -1.0$ for a smooth flat plate, for which equations (13.56) become

$$\frac{U}{u_*} = \frac{1}{k}\ln\frac{yu_*}{\nu} + 5.0,$$ (13.57)

$$\frac{U - U_\infty}{u_*} = \frac{1}{k}\ln\frac{y}{\delta} - 1.0.$$ (13.58)

These are the velocity distributions in the *overlap layer*, also called the *inertial sub-layer* or simply the *logarithmic layer*. As the derivation shows, these laws are only valid for large y_+ and small y/δ.

The foregoing method of justifying the logarithmic velocity distribution near a wall was first given by Clark B. Millikan in 1938, before the formal theory of singular perturbation problems was fully developed. The logarithmic law, however, was known from experiments conducted by the German researchers, and several derivations based on semiempirical theories were proposed by Prandtl and von Karman. One such derivation by the so-called mixing length theory is presented in the following section.

The logarithmic velocity distribution near a surface can be derived solely on dimensional grounds. In this layer the velocity gradient dU/dy can only depend on the local distance y and on the only relevant velocity scale near the surface, namely u_*. (The layer is far enough from the wall so that the direct effect of ν is not relevant and far enough from the outer part of the turbulent layer so that the effect of δ is not relevant.) A dimensional analysis gives

$$\frac{dU}{dy} = \frac{u_*}{ky},$$

where the von Karman constant k is introduced for consistency with the preceding formulas. Integration gives

$$U = \frac{u_*}{k} \ln y + \text{const.} \tag{13.59}$$

It is therefore apparent that dimensional considerations alone lead to the logarithmic velocity distribution near a wall. In fact, the constant of integration can be adjusted to reduce equation (13.59) to equation (13.57) or (13.58). For example, matching the profile to the edge of the viscous sublayer at $y = 10.7\nu/u_*$ reduces equation (13.59) to equation (13.57) (Exercise 8). The logarithmic velocity distribution also applies to rough walls, as discussed later in the section.

The experimental data on the velocity distribution near a wall is sketched in Figure 13.18. It is a semilogarithmic plot in terms of the inner variables. It shows that the linear velocity distribution (13.51) is valid for $y_+ < 5$, so that we can take the *viscous sublayer thickness* to be

$$\delta_\nu \simeq \frac{5\nu}{u_*} \qquad \text{(viscous sublayer thickness)}.$$

The logarithmic velocity distribution (13.57) is seen to be valid for $30 < y_+ < 300$. The upper limit on y_+, however, depends on the Reynolds number and becomes larger as Re increases. There is therefore a large logarithmic overlap region in flows at large Reynolds numbers. The close analogy between the overlap region in physical space and inertial subrange in spectral space is evident. In both regions, there is little production or dissipation; there is simply an "inertial" transfer across the region by inviscid nonlinear processes. It is for this reason that the logarithmic layer is called the *inertial sublayer*.

As equation (13.58) suggests, a logarithmic velocity distribution in the overlap region can also be plotted in terms of the outer variables of $(U - U_\infty)/u_*$ vs y/δ. Such plots show that the logarithmic distribution is valid for $y/\delta < 0.2$. The logarithmic law, therefore, holds accurately in a rather small percentage ($\sim20\%$) of the total boundary layer thickness. The general defect law (13.52), where $F(\xi)$ is not necessarily logarithmic, holds almost everywhere except in the inner part of the wall layer.

The region $5 < y_+ < 30$, where the velocity distribution is neither linear nor logarithmic, is called the *buffer layer*. Neither the viscous stress nor the Reynolds stress is negligible here. This layer is dynamically very important, as the turbulence production $-\overline{uv}(dU/dy)$ reaches a maximum here due to the large velocity gradients.

Wosnik *et al.* (2000) very carefully reexamined turbulent pipe and channel flows and compared their results with superpipe data and scalings developed by Zagarola and Smits (1998), and others. Very briefly, Figure 13.18 is split into more regions in that a "mesolayer" is required between the buffer layer and the inertial sublayer. Proper description of the velocity in this mesolayer requires an offset parameter in the logarithm of equations (13.56). This is obtained by generalizing equation (13.55) to

$$(\xi + \bar{a})\frac{dF}{d(\xi + \bar{a})} = (y_+ + a_+)\frac{df}{d(y_+ + a_+)} = \frac{1}{k},$$

where $\bar{a} = a/\delta, a_+ = au_*/\nu$.

Equations (13.56) become

$$f(y_+) = k^{-1}\ln(y_+ + a_+) + A,$$
$$F(\xi) = k^{-1}\ln(\xi + \bar{a}) + B.$$

The value for a_+ suggested by Wosnik *et al.* that best fits the superpipe data is $a_+ = -8$.

A more rational asymptotic treatment was given by Buschmann and Gad-el-Hak (2003a) in terms of an expansion for large Karman number $\delta^+ = \sqrt{(C_f/2)}\cdot(\delta/\theta)\mathrm{Re}_\theta$ in the case of a zero pressure gradient turbulent boundary layer. Here C_f is the skin friction coefficient defined in (10.38) and θ is the momentum thickness defined in (10.17). Re_θ is the Reynolds number based on the local momentum thickness of the boundary layer. The second author had previously found $\delta^+ = 1.168(\mathrm{Re}_\theta)^{.875}$ empirically over a wide range of Re. U/u_* is expanded in both the inner layer (y^+) and the outer layer $(\eta = y/\delta)$ in negative powers of δ^+. To lowest order we recover the simple log velocity profile [(13.59)]. Higher-order terms include powers of the inner and outer variables. After matching in an overlap region, the remaining coefficients are ultimately determined by comparison with experiments. Comparing with alternative forms for the turbulent velocity profiles, Buschmann and Gad-el-Hak (2003b) conclude that the generalized log law gives a better fit over an extended range of y^+ than any alternative velocity profile. Also, as Re_θ increases, the higher-order terms in the Karman number expansion become asymptotically small.

The outer region of turbulent boundary layers $(y_+ > 100)$ is the subject of a similarity analysis by Castillo and George (2001). They found that 90% of a turbulent flow under all pressure gradients is characterized by a single pressure gradient parameter,

$$\Lambda = \frac{\delta}{\rho U_\infty^2\, d\delta/dx}\frac{dp_\infty}{dx}.$$

A requirement for "equilibrium" turbulent boundary layer flows, to which their analysis is restricted, is that $\Lambda =$ const., and this leads to similarity. Examination of data from many sources led them to conclude that "... there appear to be almost no flows that are not in equilibrium ...." Their most remarkable result is that only three values of Λ correlate the data for all pressure gradients: $\Lambda = 0.22$ (adverse pressure gradients); $\Lambda = -1.92$ (favorable pressure gradients); and $\Lambda = 0$ (zero pressure gradient). A direct consequence of $\Lambda =$ const. is that $\delta(x) \sim U_\infty^{-1/\Lambda}$. Data was well correlated by this result for both favorable and adverse pressure gradients. Walker and Castillo (2002) then correlated velocity defect profiles for favorable, zero, and adverse pressure gradients by plotting $[(U_\infty - U)/U_\infty]\,(\delta^*/\delta_{99})$ vs. y/δ_{99}. (See Section 10.3 for the definitions of δ_{99} and δ^*). Remarkably, only three distinct turbulent velocity profiles resulted. This correlation of data with only three values of Λ was contested by Maciel, Rossignol, and Lemay (2006) in their examination of data bases for adverse pressure gradient turbulent boundary layers. They found that scalings developed by Zagarola and Smits (1998) worked best but that Λ varied by a factor of 2 (from 0.16 to 0.33), while in each of the flows Λ was held constant and the flow was observed to be self-similar. The value of $\Lambda = 0.22$ held for only two of the nine adverse pressure gradient data sets listed by Maciel et al. Moreover, the velocity defect profiles for adverse pressure gradient flows did not collapse onto a single limiting profile, as asserted by Walker and Castillo.

Very close to separation, the boundary layer assumptions that $\partial/\partial x \ll \partial/\partial y$ and $v \ll u$ break down and new scalings become necessary as discussed in Indinger, Buschmann, and Gad-el-Hak (2006).

A review paper by W. K. George (2006) puts much of the analysis and discussion into a wide-view perspective. Three scalings for the outer 90% of the zero pressure gradient turbulent velocity deficit profiles are written, the first a generalization of Eq. (13.52) to include behavior with $\delta u_*/\nu$. The second is of the same form but the normalization is by U_∞ instead of u_*. The third scaling is of the form of the second with the functional behavior prefaced by δ^*/δ, where the displacement thickness δ^* is defined in Eq. (10.16). The first scaling gives the logarithmic behavior in the overlap between the inner and outer regions of the turbulent boundary layer whereas the second gives a power law.

Professor George pointed out that although the momentum integral equation for constant pressure turbulent boundary layers [Eq. (10.44)] holds, $d\theta/dx = (u_*/U_\infty)^2$, and the main contribution to θ [momentum thickness, Eq. (10.17)] comes from near wall regions, almost the entire value of $d\theta/dx$ comes from distances far from the wall.

The third scaling is due to Zagarola and Smits (1998) and can reduce to either of the first two depending on the Re $\to \infty$ asymptotic behavior. If $\delta^*/\delta \to u_*/U_\infty$, then the first scaling leading to a logarithmic overlap is obtained. If, on the other hand, $\delta^*/\delta \to const.$, then the second scaling leading to a power law is found. Here the limit Re $\to \infty$ must be taken as $x \to \infty$ downstream with fixed upstream and external conditions. It is also shown here that the logarithmic overlap is the lowest order in an expansion of the power law. The two results are very close largely because $du/dy \sim 1/y^1$ yields a logarithm (upon integration) but $du/dy \sim 1/y^\gamma$ where γ is anything other than $= 1$ but may be very close to 1, yields a power upon integration. Since the latter is more general and results from a difference in turbulence scales in the two regions involved in the overlap, it is likely to be correct, but more detailed and careful data is required to distinguish the two forms.

It is easier to discuss the details of the balance among regions of turbulence for a planar fully developed pressure driven turbulent shear flow. Consider a flow in the x-direction between two parallel plates at $y = 0$ and $y = 2\delta$ that is no longer evolving in the x-direction. The convective acceleration terms are then identically zero and the x-momentum equation reduces to $-\partial P/\partial x + d/dy(\mu \partial U/\partial y - \rho_0 \overline{uv}) = 0$. Since the pressure gradient is constant in x, it must be balanced by the shear stress on the two walls over any distance L, which yields the equality $\tau_0 = -\delta \cdot \partial P / \partial x$. Then with the dimensionless variables, $U/u_* = u_+, yu_*/\nu = y_+, \delta u_*/\nu = \delta_+, (\overline{uv})/u_*^2 = (\overline{uv})_+$, the x-momentum equation becomes $\delta_+^{-1} + d^2 u_+/dy_+^2 - d/dy(\overline{uv})_+ = 0$, representing a balance among the pressure gradient, viscous stress gradient, and Reynolds stress gradient. This is the starting point of the analysis by Fife, Wei, Klewicki, and McMurtry (2005). The authors find distinct scalings representing different balances of terms. These are described by the following in order of increasing distance from the wall. For $y_+ < 3$, there is the sublayer where the viscous stress gradient balances the mean pressure gradient, and the Reynolds stress gradient is small by comparison. As we go outwards from the wall, the viscous stress gradient balances the Reynolds stress gradient. For large enough Re this layer extends into the logarithmic region of the earlier models. Near the location of maximum Reynolds stress, since the Reynolds stress gradient is small, the viscous stress gradient and pressure gradient are again in balance. This is true despite the Reynolds stress being much larger than the viscous stress. This region is called the viscous/pressure gradient mesolayer. Still further from the wall the Reynolds stress gradient and pressure gradient are in balance.

Wei, Fife, and Klewicki (2007) codified this analysis for any combination of Couette and Poiseuille flows. We will concentrate on the latter. Explicitly, the inner normalized equations are written in terms of u_+ and y_+. The innermost viscous layer is then a simple generalization of Eq. (13.51) to include the pressure gradient. The balance of the inner normalized equations is between the viscous and Reynolds stress gradients. The outer normalization retains u_+ but uses $\xi = y/\delta$ as the outer length scale. In the outermost region, the Reynolds stress and pressure gradients balance giving: $1 - d/d\xi (\overline{uv})_+ = O(\delta_+^{-1}) \to 0$. Between these, in the neighborhood of maximum Reynolds stress, there is a mesolayer in which *all terms are in balance*. This is scaled by:

$$\hat{y} = (y_+ - y_{m+})\delta_+^{-1/2}, (\widehat{uv}) = \delta_+^{1/2}\left[(\overline{uv})_+ - (\overline{uv})_{m+}\right], \hat{u}(\hat{y})$$
$$= u_+ - u_{m+} - (du_+/dy_+)_m \cdot (y_+ - y_{m+}).$$

Here, all $\hat{\ }$ variables are finite in the limit $\delta_+ \equiv \mathrm{Re}_* \to \infty$. Although the location y_{m+} and the value of $(\overline{uv})_{m+}$ are not known *a priori*, good approximations are obtained if $y_{m+}\delta_+^{-1/2} = 1$ and $(\overline{uv})_{m+} = 1$.

Fife et al. (2005) and Wei et al. (2007) introduced the notion of scaling patches such that a patch is ". . . defined to be a region in the flow field specified by an interval of distance from the wall, together with a scaling or non-dimensionalization of the variables which is natural for that region." Here "natural" means that in appropriately normalized variables, the data within that region are independent of Reynolds number in the turbulent limit, $\mathrm{Re} \to \infty$. The authors assert that knowledge of the local scaling properties of the mean momentum and Reynolds stress profiles is essential to understanding wall bounded turbulent flows. The scaling to render all three terms in the mean momentum balance of the same order (in Re or δ_+) in the neighborhood of the Reynolds stress maximum is indeterminate, leaving one free parameter. All coefficients are $= 1$ in the $\mathrm{Re} \to \infty$ limit in this patch centered about the Reynolds stress maximum. A very special choice of this free parameter leads to a logarithm for u_+; any other choice gives power law growth. A different scaling patch may be found in the neighborhood of the channel centerline, again with one free parameter. A very special choice of the free parameter results in the classical defect law for the outermost layer.

Rough Surface

In deriving the logarithmic law (13.57), we assumed that the flow in the inner layer is determined by viscosity. This is true only in *hydrodynamically smooth* surfaces, for which the average height of the surface roughness elements is smaller than the thickness of the viscous sublayer. For a hydrodynamically rough surface, on the other hand, the roughness elements protrude out of the viscous sublayer. An example is the flow near the surface of the earth, where the trees and buildings act as roughness elements. This causes a wake behind each roughness element, and the stress is transmitted to the wall by the "pressure drag" on the roughness elements. Viscosity becomes irrelevant for determining either the velocity distribution or the overall drag on the surface. This is why the drag coefficients for a rough pipe and a rough flat surface become independent of the Reynolds number as Re → ∞.

The velocity distribution near a rough surface is again logarithmic, although it cannot be represented by equation (13.57). To find its form, we start with the general logarithmic law (13.59). The constant of integration can be determined by noting that the mean velocity U is expected to be negligible somewhere within the roughness elements (Figure 13.19b). We can therefore assume that (13.59) applies for $y > y_0$, where y_0 is a measure of the roughness heights and is defined as the value of y at which the logarithmic distribution gives $U = 0$. Equation (13.59) then gives

$$\frac{U}{u_*} = \frac{1}{k} \ln \frac{y}{y_0}.$$ (13.60)

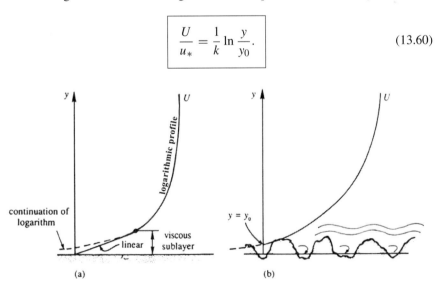

Figure 13.19 Logarithmic velocity distributions near smooth and rough surfaces: (a) smooth wall; and (b) rough wall.

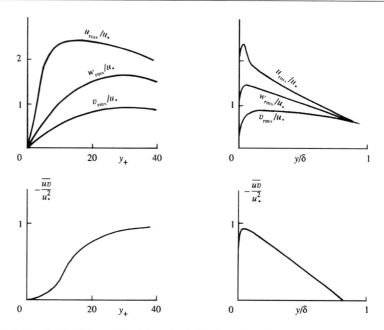

Figure 13.20 Sketch of observed variation of turbulent intensity and Reynolds stress across a channel of half-width δ. The left panels are plots as functions of the inner variable y_+, while the right panels are plots as functions of the outer variable y/δ.

Variation of Turbulent Intensity

The experimental data of turbulent intensity and Reynolds stress in a channel flow are given in Townsend (1976). Figure 13.20 shows a schematic representation of these data, plotted both in terms of the outer and the inner variables. It is seen that the turbulent velocity fluctuations are of order u_*. The longitudinal fluctuations are the largest because the shear production initially feeds the energy into the u-component; the energy is subsequently distributed into the lateral components v and w. (Incidentally, in a convectively generated turbulence the turbulent energy is initially fed to the *vertical component*.) The turbulent intensity initially rises as the wall is approached, but goes to zero right at the wall in a very thin wall layer. As expected from physical considerations, the normal component v_{rms} starts to feel the wall effect earlier. Figure 13.20 also shows that the distribution of each variable very close to the wall becomes clear only when the distances are magnified by the viscous scaling ν/u_*. The Reynolds stress profile in terms of the inner variable shows that the stresses are negligible within the viscous sublayer ($y_+ < 5$), beyond which the Reynolds stress is nearly constant throughout the wall layer. This is why the logarithmic layer is also called the *constant stress* layer.

12. Eddy Viscosity and Mixing Length

The equations for mean motion in a turbulent flow, given by equation (13.24), cannot be solved for $U_i(\mathbf{x})$ unless we have an expression relating the Reynolds stresses

$\overline{u_i u_j}$ in terms of the mean velocity field. Prandtl and von Karman developed certain semiempirical theories that attempted to provide this relationship.

These theories are based on an analogy between the momentum exchanges both in turbulent and in laminar flows. Consider first a unidirectional laminar flow $U(y)$, in which the shear stress is

$$\frac{\tau_{\text{lam}}}{\rho} = \nu \frac{dU}{dy}, \tag{13.61}$$

where ν is a property of the fluid. According to the kinetic theory of gases, the diffusive properties of a gas are due to the molecular motions, which tend to mix momentum and heat throughout the flow. It can be shown that the viscosity of a gas is of order

$$\nu \sim a\lambda, \tag{13.62}$$

where a is the rms speed of molecular motion, and λ is the mean free path defined as the average distance traveled by a molecule between collisions. The proportionality constant in equation (13.62) is of order 1.

One is tempted to speculate that the diffusive behavior of a turbulent flow may be qualitatively similar to that of a laminar flow and may simply be represented by a much larger diffusivity. By analogy with (13.61), Boussinesq proposed to represent the turbulent stress as

$$-\overline{uv} = \nu_e \frac{dU}{dy}, \tag{13.63}$$

where ν_e is the *eddy viscosity*. Note that, whereas ν is a known property of the fluid, ν_e in (13.63) depends on the conditions of the *flow*. We can always divide the turbulent stress by the mean velocity gradient and call it ν_e, but this is not progress unless we can formulate a rational method for finding the eddy viscosity from other known parameters of a turbulent flow.

The eddy viscosity relation (13.63) implies that the local gradient determines the flux. However, this cannot be valid if the eddies happen to be larger than the scale of curvature of the profile. Following Panofsky and Dutton (1984), consider the atmospheric concentration profile of carbon monoxide (CO) shown in Figure 13.21. An eddy viscosity relation would have the form

$$-\overline{wc} = \kappa_e \frac{dC}{dz}, \tag{13.64}$$

where C is the mean concentration (kilograms of CO per kilogram of air), c is its fluctuation, and κ_e is the eddy diffusivity. A positive κ_e requires that the flux of CO at P be downward. However, if the thermal convection is strong enough, the large eddies so generated can carry large amounts of CO from the ground to point P, and result in an upward flux there. The direction of flux at P in this case is not determined by the local gradient at P, but by the concentration difference between the surface and point P. In this case, the eddy diffusivity found from equation (13.64) would be negative and, therefore, not very meaningful.

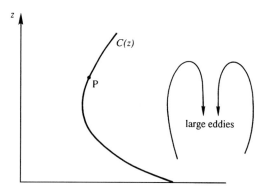

Figure 13.21 An illustration of breakdown of an eddy diffusivity type relation. The eddies are larger than the scale of curvature of the concentration profile $C(z)$ of carbon monoxide.

In cases where the concept of eddy viscosity may work, we may use the analogy with equation (13.62), and write

$$\nu_e \sim u' l_m, \tag{13.65}$$

where u' is a typical scale of the fluctuating velocity, and l_m is the *mixing length*, defined as the cross-stream distance traveled by a fluid particle before it gives up its momentum and loses identity. The concept of mixing length was first introduced by Taylor (1915), but the approach was fully developed by Prandtl and his coworkers. As with the eddy viscosity approach, little progress has been made by introducing the mixing length, because u' and l_m are just as unknown as ν_e is. Experience shows that in many situations u' is of the order of either the local mean speed U or the friction velocity u_*. However, there does not seem to be a rational approach for relating l_m to the mean flow field.

Prandtl derived the logarithmic velocity distribution near a solid surface by using the mixing length theory in the following manner. The scale of velocity fluctuations in a wall-bounded flow can be taken as $u' \sim u_*$. Prandtl also argued that the mixing length must be proportional to the distance y. Then equation (13.65) gives

$$\nu_e = k u_* y.$$

For points outside the viscous sublayer but still near the wall, the Reynolds stress can be taken equal to the wall stress ρu_*^2. This gives

$$\rho u_*^2 = \rho k u_* y \frac{dU}{dy},$$

which can be written as

$$\frac{dU}{dy} = \frac{u_*}{ky}. \tag{13.66}$$

This integrates to

$$\frac{U}{u_*} = \frac{1}{k} \ln y + \text{const.}$$

In recent years the mixing length theory has fallen into disfavor, as it is incorrect in principle (Tennekes and Lumley, 1972). It only works when there is a single length scale and a single time scale; for example in the overlap layer in a wall-bounded flow the only relevant length scale is y and the only time scale is y/u_*. However, its validity is then solely a consequence of dimensional necessity and not of any other fundamental physics. Indeed it was shown in the preceding section that the logarithmic velocity distribution near a solid surface can be derived from dimensional considerations alone. (Since u_* is the only characteristic velocity in the problem, the local velocity gradient dU/dy can only be a function of u_* and y. This leads to equation (13.66) merely on dimensional grounds.) Prandtl's derivation of the empirically known logarithmic velocity distribution has only historical value.

However, the relationship (13.65) is useful for estimating the order of magnitude of the eddy diffusivity in a turbulent flow, if we interpret the right-hand side as simply the product of typical velocity and length scales of large eddies. Consider the thermal convection between two horizontal plates in air. The walls are separated by a distance $L = 3$ m, and the lower layer is warmer by $\Delta T = 1\,°C$. The equation of motion (13.33) for the fluctuating field gives the vertical acceleration as

$$\frac{Dw}{Dt} \sim g\alpha T' \sim \frac{g\Delta T}{T}, \tag{13.67}$$

where we have used the fact that the temperature fluctuations are expected to be of order ΔT and that $\alpha = 1/T$ for a perfect gas. The time to rise through a height L is $t \sim L/w$, so that equation (13.67) gives a characteristic velocity fluctuation of

$$w \sim \sqrt{gL\Delta T/T} \approx \sqrt{0.1}\,\text{m/s} \approx 0.316\,\text{m/s}.$$

It is fair to assume that the largest eddies are as large as the separation between the plates. The eddy diffusivity is therefore

$$\kappa_e \sim wL \sim 0.95\,\text{m}^2/\text{s},$$

which is much larger than the molecular value of $2 \times 10^{-5}\,\text{m}^2/\text{s}$.

As noted in the preceding, the Reynolds averaged Navier–Stokes equations do not form a closed system. In order for them to be predictive and useful in solving problems of scientific and engineering interest, closures must be developed. Reynolds stresses or higher correlations must be expressed in terms of themselves or lower correlations with empirically determined constants. An excellent review of an important class of closures is provided by Speziale (1991). Critical discussions of various closures together with comparisons with each other, with experiments, or with numerical simulations are given for several idealized problems. Very tragically, Charles Speziale died while at the peak of his intellectual productivity. He contributed numerous papers on turbulence modeling and other subjects on the foundations of fluid mechanics. A memorial tribute and a number of papers on turbulence in his honor by some of the prominent authorities may be found in the May 2006 issue of *Journal of Applied Mechanics* (v. 73, no. 3).

A different approach to turbulence modeling is represented by renormalization group (RNG) theories. Rather than use the Reynolds averaged equations, turbulence is simulated by a solenoidal isotropic random (body) force field **f** (force/mass). Here **f** is chosen to generate the velocity field described by the Kolmogorov spectrum in the limit of large wavenumber K. For very small eddies (larger wavenumbers beyond the inertial subrange), the energy decays exponentially by viscous dissipation. The spectrum in Fourier space (K) is truncated at a cutoff wavenumber and the effect of these very small scales is represented by a modified viscosity. Then an iteration is performed successively moving back the cutoff into the inertial range. Smith and Reynolds (1992) provide a tutorial on the RNG method developed several years earlier by Yakhot and Orszag. Lam (1992) develops results in a different way and offers insights and plausible explanations for the various artifices in the theory.

13. Coherent Structures in a Wall Layer

The large-scale identifiable structures of turbulent events, called *coherent structures*, depend on the type of flow. A possible structure of large eddies found in the outer parts of a boundary layer, and in a wall-free shear flow, was illustrated in Figure 13.10. In this section we shall discuss the coherent structures observed within the *inner layer* of a wall-bounded shear flow. This is one of the most active areas of current turbulent research, and reviews of the subject can be found in Cantwell (1981) and Landahl and Mollo-Christensen (1986).

These structures are deduced from spatial correlation measurements, a certain amount of imagination, and plenty of flow visualization. The flow visualization involves the introduction of a marker, one example of which is dye. Another involves the "hydrogen bubble technique," in which the marker is generated electrically. A thin wire is stretched across the flow, and a voltage is applied across it, generating a line of hydrogen bubbles that travel with the flow. The bubbles produce white spots in the photographs, and the shapes of the white regions indicate where the fluid is traveling faster or slower than the average.

Flow visualization experiments by Kline *et al.* (1967) led to one of the most important advances in turbulence research. They showed that the inner part of the wall layer in the range $5 < y_+ < 70$ is not at all passive, as one might think. In fact, it is perhaps dynamically the most active, in spite of the fact that it occupies only about 1% of the total thickness of the boundary layer. Figure 13.22 is a photograph from Kline *et al.* (1967), showing the top view of the flow within the viscous sublayer at a distance $y_+ = 2.7$ from the wall. (Here x is the direction of flow, and z is the "spanwise" direction.) The wire producing the hydrogen bubbles in the figure was parallel to the z-axis. The streaky structures seen in the figure are generated by regions of fluid moving downstream faster or slower than the average. The figure reveals that the *streaks* of low-speed fluid are quasi-periodic in the spanwise direction. From

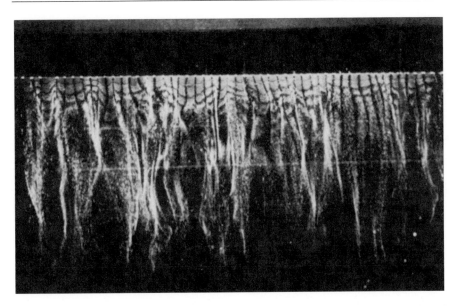

Figure 13.22   Top view of near-wall structure (at $y_+ = 2.7$) in a turbulent boundary layer on a horizontal flat plate. The flow is visualized by hydrogen bubbles. S. J. Kline *et al., Journal of Fluid Mechanics* **30**: 741–773, 1967 and reprinted with the permission of Cambridge University Press.

time to time these slowly moving streaks lift up into the buffer region, where they undergo a characteristic oscillation. The oscillations end violently and abruptly as the lifted fluid breaks up into small-scale eddies. The whole cycle is called *bursting*, or eruption, and is essentially an ejection of slower fluid into the flow above. The flow into which the ejection occurs decelerates, causing a point of inflection in the profile $u(y)$ (Figure 13.23). The secondary flow associated with the eruption motion causes a stretching of the spanwise vortex lines, as sketched in the figure. These vortex lines amplify due to the inherent instability of an inflectional profile, and readily break up, producing a source of small-scale turbulence. The strengths of the eruptions vary, and the stronger ones can go right through to the edge of the boundary layer.

It is clear that the bursting of the slow fluid associates a positive v with a negative u, generating a positive Reynolds stress $-\overline{uv}$. In fact, measurements show that most of the Reynolds stress is generated by either the bursting or its counterpart, called the *sweep* (or inrush) during which high-speed fluid moves toward the wall. The Reynolds stress generation is therefore an intermittent process, occurring perhaps 25% of the time.

Largely due to numerical simulations of turbulent flows, it is now understood that the very large turbulent wall shear stress (as compared with that in laminar flow) is due to streamwise vorticity in the buffer or inner wall layer ($y_+ = 10$–50). Kim (2003) reports on the history of discovery by computation and experimental verification of insight into the details of turbulent flows. This insight led to strategies to reduce the wall shear stress by active or passive controls. The availability of microsensors and MEMS actuators creates the possibility of actively modifying

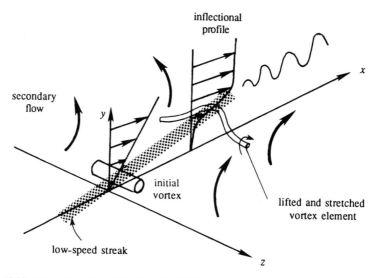

Figure 13.23   Mechanics of streak break up. S. J. Kline *et al.*, *Journal of Fluid Mechanics* **30**: 741–773, 1967 and reprinted with the permission of Cambridge University Press.

the flow near the wall to significantly reduce the shear stress. Passive modification is exemplified by adding riblets to the surface. These are fine streamwise corruga- tions that interfere with the interaction between the streamwise vortices and the wall. Much smaller drag reduction is achieved this way. An example of active modification of the near-wall flow is blowing and suctioning alternately on the surface to counter the streamwise vorticity. A surprising result of these studies is that linear control theory (for the Navier–Stokes equation linearized about a mean flow) provides excel- lent results for a strategy for reducing wall shear stress, provided that function to be extremized (which cannot be drag) is carefully chosen. All of these results apply only for small turbulence Reynolds number ($\mathrm{Re}_* = u_* \delta / \nu$). However, there has been a history of success in applying insights gained for small Re_* to larger, more realistic values.

14. Turbulence in a Stratified Medium

Effects of stratification become important in such laboratory flows as heat transfer from a heated plate and in geophysical flows such as those in the atmosphere and in the ocean. Some effects of stratification on turbulent flows will be considered in this section. Further discussion can be found in Tennekes and Lumley (1972), Phillips (1977), and Panofsky and Dutton (1984).

As is customary in geophysical literature, we shall take the z-direction as upward, and the shear flow will be denoted by $U(z)$. For simplicity the flow will be assumed homogeneous in the horizontal plane, that is independent of x and y. The turbulence in a stratified medium depends critically on the static stability. In the neutrally stable state of a compressible environment the density decreases upward, because of the decrease of pressure, at a rate $d\rho_a/dz$ called the *adiabatic density gradient*. This is discussed further in Chapter 1, Section 10. A medium is statically stable if the density decreases faster than the adiabatic decrease. The effective density gradient that determines the stability of the environment is then determined by the sign of $d(\rho - \rho_a)/dz$, where $\rho - \rho_a$ is called the *potential density*. In the following discussion, we shall assume that the adiabatic variations of density have been subtracted out, so that when we talk about density or temperature, we shall really mean potential density or potential temperature.

The Richardson Numbers

Let us first examine the equation for turbulent kinetic energy (13.34). Omitting the viscous transport and assuming that the flow is independent of x and y, it reduces to

$$\frac{D}{Dt}(q^2) = -\frac{\partial}{\partial z}\left(\frac{1}{\rho_0}\overline{pw} + \overline{q^2 w}\right) - \overline{uw}\frac{dU}{dz} + g\alpha\overline{wT'} - \varepsilon,$$

where $q^2 = (\overline{u^2} + \overline{v^2} + \overline{w^2})/2$. The first term on the right-hand side is the transport of turbulent kinetic energy by fluctuating w. The second term $-\overline{uw}(dU/dz)$ is the production of turbulent energy by the interaction of Reynolds stress and the mean shear; this term is almost always positive. The third term $g\alpha\overline{wT'}$ is the production of turbulent kinetic energy by the vertical heat flux; it is called the *buoyant production*, and was discussed in more detail in Section 7. In an unstable environment, in which the mean temperature $\bar{T}$ decreases upward, the heat flux $\overline{wT'}$ is positive (upward), signifying that the turbulence is generated convectively by upward heat fluxes. In the opposite case of a stable environment, the turbulence is suppressed by stratification. The ratio of the buoyant destruction of turbulent kinetic energy to the shear production is called the *flux Richardson number*:

$$\mathrm{Rf} = \frac{-g\alpha\overline{wT'}}{-\overline{uw}(dU/dz)} = \frac{\text{buoyant destruction}}{\text{shear production}}, \tag{13.68}$$

where we have oriented the x-axis in the direction of flow. As the shear production is positive, the sign of Rf depends on the sign of $\overline{wT'}$. For an unstable environment in which the heat flux is upward Rf is negative and for a stable environment it is positive. For Rf > 1, the buoyant destruction removes turbulence at a rate larger than the rate at which it is produced by shear production. However, the critical value of Rf at which the turbulence ceases to be self-supporting is less than unity, as dissipation is necessarily a large fraction of the shear production. Observations indicate that the critical value is $\mathrm{Rf}_{cr} \simeq 0.25$ (Panofsky and Dutton, 1984, p. 94). If measurements indicate the presence of turbulent fluctuations, but at the same time a value of Rf much larger than 0.25, then a fair conclusion is that the turbulence is decaying. When Rf is negative, a large $-\mathrm{Rf}$ means strong convection and weak mechanical turbulence.

    Instead of Rf, it is easier to measure the *gradient Richardson number*, defined as

$$\mathrm{Ri} \equiv \frac{N^2}{(dU/dz)^2} = \frac{\alpha g(d\bar{T}/dz)}{(dU/dz)^2}, \tag{13.69}$$

where N is the buoyancy frequency. If we make the eddy coefficient assumptions

$$-\overline{wT'} = \kappa_e \frac{d\bar{T}}{dz},$$

$$-\overline{uw} = \nu_e \frac{dU}{dz},$$

then the two Richardson numbers are related by

$$\mathrm{Ri} = \frac{\nu_e}{\kappa_e}\mathrm{Rf}. \tag{13.70}$$

The ratio ν_e/κ_e is the *turbulent Prandtl number*, which determines the relative efficiency of the vertical turbulent exchanges of momentum and heat. The presence of a stable stratification damps the vertical transports of both heat and momentum; however, the momentum flux is reduced less because the internal waves in a stable environment can transfer momentum (by moving vertically from one region to another) but not heat. Therefore, $\nu_e/\kappa_e > 1$ for a stable environment. Equation (13.70) then shows that turbulence can persist even when $\mathrm{Ri} > 1$, if the critical value of 0.25 applies on the *flux* Richardson number (Turner, 1981; Bradshaw and Woods, 1978). In an unstable environment, on the other hand, ν_e/κ_e becomes small. In a neutral environment it is usually found that $\nu_e \simeq \kappa_e$; the idea of equating the eddy coefficients of heat and momentum is called the *Reynolds analogy*.

Monin–Obukhov Length

The Richardson numbers are ratios that compare the relative importance of mechanical and convective turbulence. Another parameter used for the same purpose is not a ratio, but has the unit of length. It is the *Monin–Obukhov length*, defined as

$$L_M \equiv -\frac{u_*^3}{k\alpha g\overline{wT'}}, \tag{13.71}$$

where u_* is the friction velocity, $\overline{wT'}$ is the heat flux, α is the coefficient of thermal expansion, and k is the von Karman constant introduced for convenience. Although $\overline{wT'}$ is a function of z, the parameter L_M is effectively a constant for the flow, as it is used only in the logarithmic surface layer in which both the stress and the heat flux $\overline{wT'}$ are nearly constant. The Monin–Obukhov length then becomes a parameter determined from the boundary conditions of drag and the heat flux at the surface. Like Rf, it is positive for stable conditions and negative for unstable conditions.

The significance of L_M within the surface layer becomes clearer if we write Rf in terms of L_M, using the logarithmic velocity distribution (13.60), from which $dU/dz = u_*/kz$. (Note that we are now using z for distances perpendicular to the surface.) Using $\overline{uw} = u_*^2$ because of the near uniformity of stress in the logarithmic layer, equation (13.68) becomes

$$\text{Rf} = \frac{z}{L_M}.$$ (13.72)

As Rf is the ratio of buoyant destruction to shear production of turbulence, (13.72) shows that L_M is the height at which these two effects are of the same order. For both stable and unstable conditions, the effects of stratification are slight if $z \ll |L_M|$. At these small heights, then, the velocity profile is logarithmic, as in a neutral environment. This is called a *forced convection* region, because the turbulence is mechanically forced. For $z \gg |L_M|$, the effects of stratification dominate. In an unstable environment, it follows that the turbulence is generated mainly by buoyancy at heights $z \gg -L_M$, and the shear production is negligible. The region beyond the forced convecting layer is therefore called a zone of *free convection* (Figure 13.24), containing thermal plumes (columns of hot rising gases) characteristic of free convection from heated plates in the absence of shear flow.

Observations as well as analysis show that the effect of stratification on the velocity distribution in the surface layer is given by the log-linear profile (Turner, 1973)

$$U = \frac{u_*}{k}\left[\ln\frac{z}{z_0} + 5\frac{z}{L_M}\right].$$

The form of this profile is sketched in Figure 13.25 for stable and unstable conditions. It shows that the velocity is more uniform than $\ln z$ in the unstable case because of the enhanced vertical mixing due to buoyant convection.

Spectrum of Temperature Fluctuations

An equation for the intensity of temperature fluctuations $\overline{T'^2}$ can be obtained in a manner identical to that used for obtaining the turbulent kinetic energy. The procedure is therefore to obtain an equation for DT'/Dt by subtracting those for $D\tilde{T}/Dt$ and

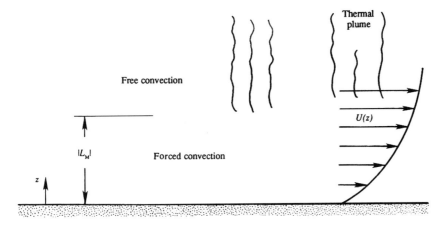

Figure 13.24 Forced and free convection zones in an unstable atmosphere.

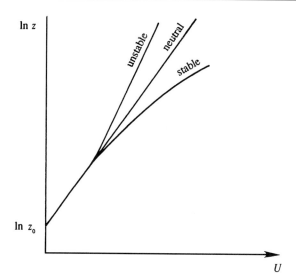

Figure 13.25   Effect of stability on velocity profiles in the surface layer.

$D\bar{T}/Dt$, and then to multiply the resulting equation for DT'/Dt by T' and take the average. The result is

$$\frac{1}{2}\frac{D\overline{T'^2}}{Dt} = -\overline{wT'}\frac{d\bar{T}}{dz} - \frac{\partial}{\partial z}\left(\frac{1}{2}\overline{T'^2 w} - \kappa\frac{\overline{dT'^2}}{dz}\right) - \varepsilon_\mathrm{T},$$

where $\varepsilon_\mathrm{T} \equiv \kappa\overline{(\partial T'/\partial x_j)^2}$ is the *dissipation of temperature fluctuation*, analogous to the dissipation of turbulent kinetic energy $\varepsilon = 2\nu\overline{e_{ij}\,e_{ij}}$. The first term on the right-hand side is the generation of $\overline{T'^2}$ by the mean temperature gradient, $\overline{wT'}$ being positive if $d\bar{T}/dz$ is negative. The second term on the right-hand side is the turbulent transport of $\overline{T'^2}$.

A wavenumber spectrum of temperature fluctuations can be defined such that

$$\overline{T'^2} \equiv \int_0^\infty \Gamma(K)\,dK.$$

As in the case of the kinetic energy spectrum, an inertial range of wavenumbers exists in which neither the production by large-scale eddies nor the dissipation by conductive and viscous effects are important. As the temperature fluctuations are intimately associated with velocity fluctuations, $\Gamma(K)$ in this range must depend not only on ε_T but also on the variables that determine the velocity spectrum, namely ε and K. Therefore

$$\Gamma(K) = \Gamma(\varepsilon_\mathrm{T}, \varepsilon, K) \qquad l^{-1} \ll K \ll \eta^{-1}.$$

The unit of Γ is $^\circ C^2$ m, and the unit of ε_T is $^\circ C^2/s$. A dimensional analysis gives

$$\Gamma(K) \propto \varepsilon_T \varepsilon^{-1/3} K^{-5/3} \qquad l^{-1} \ll K \ll \eta^{-1}, \qquad (13.73)$$

which was first derived by Obukhov in 1949. Comparing with equation (13.40), it is apparent that the spectra of both velocity and temperature fluctuations in the inertial subrange have the same $K^{-5/3}$ form.

The spectrum beyond the inertial subrange depends on whether the Prandtl number ν/κ of the fluid is smaller or larger than one. We shall only consider the case of $\nu/\kappa \gg 1$, which applies to water for which the Prandtl number is 7.1. Let η_T be the scale responsible for smearing out the temperature gradients and η be the Kolmogorov microscale at which the velocity gradients are smeared out. For $\nu/\kappa \gg 1$ we expect that $\eta_T \ll \eta$, because then the conductive effects are important at scales smaller than the viscous scales. In fact, Batchelor (1959) showed that $\eta_T \simeq \eta(\kappa/\nu)^{1/2} \ll \eta$. In such a case there exists a range of wavenumbers $\eta^{-1} \ll K \ll \eta_T^{-1}$, in which the scales are not small enough for the thermal diffusivity to smear out the temperature fluctuation. Therefore, $\Gamma(K)$ continues farther up to η_T^{-1}, although $S(K)$ drops off sharply. This is called the *viscous convective subrange*, because the spectrum is dominated by viscosity but is still actively convective. Batchelor (1959) showed that the spectrum in the viscous convective subrange is

$$\Gamma(K) \propto K^{-1} \qquad \eta^{-1} \ll K \ll \eta_T^{-1}. \qquad (13.74)$$

Figure 13.26 shows a comparison of velocity and temperature spectra, observed in a tidal flow through a narrow channel. The temperature spectrum shows that the spectral slope increases from $-\frac{5}{3}$ in the inertial subrange to -1 in the viscous convective subrange.

15. Taylor's Theory of Turbulent Dispersion

The large mixing rate in a turbulent flow is due to the fact that the fluid particles gradually wander away from their initial location. Taylor (1921) studied this problem and calculated the rate at which a particle disperses (i.e., moves away) from its initial location. The presentation here is directly adapted from his classic paper. He considered a point source emitting particles, say a chimney emitting smoke. The particles are emitted into a stationary and homogeneous turbulent medium in which the mean velocity is zero. Taylor used Lagrangian coordinates $\mathbf{X}(\mathbf{a}, t)$, which is the present location at time t of a particle that was at location $\mathbf{a}$ at time $t = 0$. We shall take the point source to be the origin of coordinates and consider an ensemble of experiments in which we measure the location $\mathbf{X}(\mathbf{0}, t)$ at time t of all the particles that started from the origin (Figure 13.27). For simplicity we shall suppress the first argument in $\mathbf{X}(\mathbf{0}, t)$ and write $\mathbf{X}(t)$ to mean the same thing.

Rate of Dispersion of a Single Particle

Consider the behavior of a single component of $\mathbf{X}$, say X_α ($\alpha = 1, 2,$ or 3). (We are using a Greek subscript α because we shall *not* imply the summation convention.)

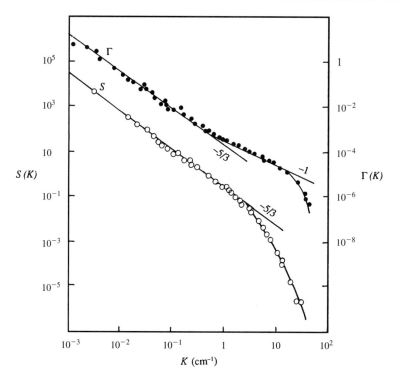

Figure 13.26 Temperature and velocity spectra measured by Grant *et al.* (1968). The measurements were made at a depth of 23 m in a tidal passage through islands near the coast of British Columbia, Canada. Wavenumber K is in cm^{-1}. Solid points represent Γ in $(°C)^2/cm^{-1}$, and open points represent S in $(cm/s)^2/cm^{-1}$. Powers of K that fit the observation are indicated by straight lines. O. M. Phillips, *The Dynamics of the Upper Ocean*, 1977 and reprinted with the permission of Cambridge University Press.

The average rate at which the *magnitude* of X_α increases with time can be found by finding $\overline{d(X_\alpha^2)}/dt$, where the overbar denotes ensemble average and not time average. We can write

$$\frac{d}{dt}(\overline{X_\alpha^2}) = 2\overline{X_\alpha \frac{dX_\alpha}{dt}}, \tag{13.75}$$

where we have used the commutation rule (13.3) of averaging and differentiation. Defining

$$u_\alpha = \frac{dX_\alpha}{dt},$$

as the *Lagrangian* velocity component of a fluid particle at time t, equation (13.75) becomes

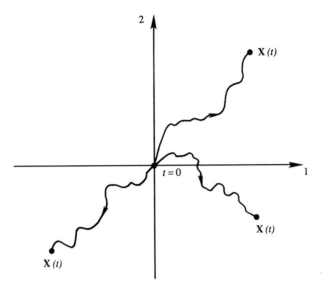

Figure 13.27 Three experimental outcomes of $\mathbf{X}(t)$, the current positions of particles from the origin at time $t = 0$.

$$\frac{d}{dt}(\overline{X_\alpha^2}) = 2\overline{X_\alpha u_\alpha} = 2\overline{\left[\int_0^t u_\alpha(t')\,dt'\right] u_\alpha}$$

$$= 2\int_0^t \overline{u_\alpha(t')u_\alpha(t)}\,dt', \qquad (13.76)$$

where we have used the commutation rule (13.4) of averaging and integration. We have also written

$$X_\alpha = \int_0^t u_\alpha(t')\,dt',$$

which is valid as X_α and u_α are associated with the same particle. Because the flow is assumed to be stationary, $\overline{u_\alpha^2}$ is independent of time, and the autocorrelation of $u_\alpha(t)$ and $u_\alpha(t')$ is only a function of the time difference $t - t'$. Defining

$$r_\alpha(\tau) \equiv \frac{\overline{u_\alpha(t)u_\alpha(t+\tau)}}{\overline{u_\alpha^2}},$$

to be the autocorrelation of Lagrangian velocity components of a particle, equation (13.76) becomes

$$\frac{d}{dt}(\overline{X_\alpha^2}) = 2\overline{u_\alpha^2}\int_0^t r_\alpha(t'-t)\,dt'$$

$$= 2\overline{u_\alpha^2}\int_0^t r_\alpha(\tau)\,d\tau, \qquad (13.77)$$

where we have changed the integration variable from t' to $\tau = t - t'$. Integrating, we obtain

$$\overline{X_\alpha^2}(t) = 2\overline{u_\alpha^2} \int_0^t dt' \int_0^{t'} r_\alpha(\tau)\,d\tau, \tag{13.78}$$

which shows how the variance of the particle position changes with time.

Another useful form of equation (13.78) is obtained by integrating it by parts. We have

$$\int_0^t dt' \int_0^{t'} r_\alpha(\tau)\,d\tau = \left[t' \int_0^{t'} r_\alpha(\tau)\,d\tau \right]_{t'=0}^{t} - \int_0^t t' r_\alpha(t')\,dt'$$

$$= t \int_0^t r_\alpha(\tau)\,d\tau - \int_0^t t' r_\alpha(t')\,dt'$$

$$= t \int_0^t \left(1 - \frac{\tau}{t} \right) r_\alpha(\tau)\,d\tau.$$

Equation (13.78) then becomes

$$\overline{X_\alpha^2}(t) = 2\overline{u_\alpha^2} t \int_0^t \left(1 - \frac{\tau}{t} \right) r_\alpha(\tau)\,d\tau. \tag{13.79}$$

Two limiting cases are examined in what follows.

Behavior for small t: If t is small compared to the correlation scale of $r_\alpha(\tau)$, then $r_\alpha(\tau) \simeq 1$ throughout the integral in equation (13.78) (Figure 13.28). This gives

$$\overline{X_\alpha^2}(t) \simeq \overline{u_\alpha^2} t^2. \tag{13.80}$$

Taking the square root of both sides, we obtain

$$\boxed{X_\alpha^{\mathrm{rms}} = u_\alpha^{\mathrm{rms}} t \qquad t \ll \mathcal{T},} \tag{13.81}$$

which shows that the rms displacement increases linearly with time and is proportional to the intensity of turbulent fluctuations in the medium.

Behavior for large t: If t is large compared with the correlation scale of $r_\alpha(\tau)$, then τ/t in equation (13.79) is negligible, giving

$$\overline{X_\alpha^2}(t) \simeq 2\overline{u_\alpha^2} \mathcal{T} t, \tag{13.82}$$

where

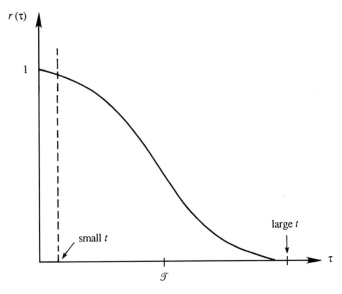

Figure 13.28 Small and large values of time on a plot of the correlation function.

$$\mathcal{T} \equiv \int_0^\infty r_\alpha(\tau)\, d\tau,$$

is the integral time scale determined from the Lagrangian correlation $r_\alpha(\tau)$. Taking the square root, equation (13.82) gives

$$X_\alpha^{\text{rms}} = u_\alpha^{\text{rms}} \sqrt{2\mathcal{T}t} \qquad t \gg \mathcal{T}. \tag{13.83}$$

The $t^{1/2}$ behavior of equation (13.83) at large times is similar to the behavior in a *random walk*, in which the distance traveled in a series of random (i.e., uncorrelated) steps increases as $t^{1/2}$. This similarity is due to the fact that for large t the fluid particles have "forgotten" their initial behavior at $t = 0$. In contrast, the small time behavior $X_\alpha^{\text{rms}} = u_\alpha^{\text{rms}} t$ is due to complete correlation, with *each experiment* giving $X_\alpha \simeq u_\alpha t$. The concept of random walk is discussed in what follows.

Random Walk

The following discussion is adapted from Feynman *et al.* (1963, pp. 6–5 and 41–8). Imagine a person walking in a random manner, by which we mean that there is no correlation between the directions of two consecutive steps. Let the vector $\mathbf{R}_n$ represent the distance from the origin after n steps, and the vector $\mathbf{L}$ represent the nth step (Figure 13.29). We assume that each step has the same magnitude L. Then

$$\mathbf{R}_n = \mathbf{R}_{n-1} + \mathbf{L},$$

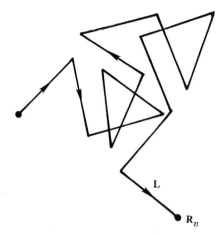

Figure 13.29   Random walk.

which gives

$$R_n^2 = \mathbf{R}_n \cdot \mathbf{R}_n = (\mathbf{R}_{n-1} + \mathbf{L}) \cdot (\mathbf{R}_{n-1} + \mathbf{L})$$
$$= R_{n-1}^2 + L^2 + 2\mathbf{R}_{n-1} \cdot \mathbf{L}.$$

Taking the average, we get

$$\overline{R_n^2} = \overline{R_{n-1}^2} + L^2 + 2\overline{\mathbf{R}_{n-1} \cdot \mathbf{L}}. \tag{13.84}$$

The last term is zero because there is no correlation between the direction of the nth step and the location reached after $n - 1$ steps. Using rule (13.84) successively, we get

$$\overline{R_n^2} = \overline{R_{n-1}^2} + L^2 = R_{n-2}^2 + 2L^2$$
$$= \overline{R_1^2} + (n - 1)L^2 = nL^2.$$

The rms distance traveled after n uncorrelated steps, each of length L, is therefore

$$\boxed{R_n^{\mathrm{rms}} = L\sqrt{n},} \tag{13.85}$$

which is called a random walk.

Behavior of a Smoke Plume in the Wind

Taylor's analysis can be adapted to account for the presence of mean velocity. Consider the dispersion of smoke into a wind blowing in the x-direction (Figure 13.30). Then a photograph of the smoke plume, in which the film is exposed for a long time, would outline the average width Y^{rms}. As the x-direction in this problem is similar to time in

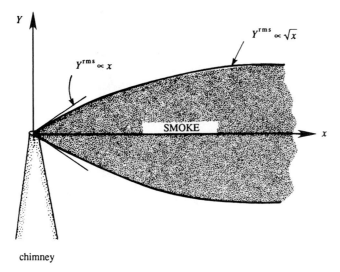

Figure 13.30 Average shape of a smoke plume in a wind blowing uniformly along the x-axis. G. I. Taylor, *Proc. London Mathematical Society* **20**: 196–211, 1921.

Taylor's problem, the limiting behavior in equations (13.81) and (13.83) shows that the smoke plume is parabolic with a *pointed* vertex.

Effective Diffusivity

An equivalent eddy diffusivity can be estimated from Taylor's analysis. The equivalence is based on the following idea: Consider the spreading of a concentrated source, say of heat or vorticity, in a fluid of *constant* diffusivity. What should the diffusivity be in order that the spreading rate equals that predicted by equation (13.77)? The problem of the sudden introduction of a line vortex of strength Γ, considered in Chapter 9, Section 9, is such a problem of diffusion of a concentrated source. It was shown there that the tangential velocity in this flow is given by

$$u_\theta = \frac{\Gamma}{2\pi r} e^{-r^2/4\nu t}.$$

The solution is therefore proportional to $\exp(-r^2/4\nu t)$, which has a Gaussian shape in the radial direction r, with a characteristic width ("standard deviation") of $\sigma = \sqrt{2\nu t}$. It follows that the momentum diffusivity ν in this problem is related to the variance σ^2 as

$$\nu = \frac{1}{2}\frac{d\sigma^2}{dt}, \tag{13.86}$$

which can be calculated if $\sigma^2(t)$ is known. Generalizing equation (13.86), we can say that the effective diffusivity in a problem of turbulent dispersion of a patch of particles issuing from a point is given by

$$\kappa_e \equiv \frac{1}{2}\frac{d}{dt}(\overline{X_\alpha^2}) = \overline{u_\alpha^2} \int_0^t r_\alpha(\tau)\,d\tau, \tag{13.87}$$

where we have used equation (13.77). From equations (13.80) and (13.82), the two limiting cases of equation (13.87) are

$$\kappa_e \simeq \overline{u_\alpha^2}t \qquad t \ll \mathcal{T}, \tag{13.88}$$

$$\kappa_e \simeq \overline{u_\alpha^2}\mathcal{T} \qquad t \gg \mathcal{T}. \tag{13.89}$$

Equation (13.88) shows the interesting fact that the eddy diffusivity initially increases with time, a behavior different from that in molecular diffusion with constant diffusivity. This can be understood as follows. The dispersion (or separation) of particles in a patch is caused by eddies with scales less than or equal to the scale of the patch, since the larger eddies simply advect the patch and do not cause any separation of the particles. As the patch size becomes larger, an *increasing* range of eddy sizes is able to cause dispersion, giving $\kappa_\alpha \propto t$. This behavior shows that *it is frequently impossible to represent turbulent diffusion by means of a large but constant eddy diffusivity*. Turbulent diffusion does not behave like molecular diffusion. For large times, on the other hand, the patch size becomes larger than the largest eddies present, in which case the diffusive behavior becomes similar to that of molecular diffusion with a constant diffusivity given by equation (13.89).

16. Concluding Remarks

Turbulence is an area of classical fluid mechanics that is the subject of continued intensive research. Frequent symposia are held to summarize and bring the research community up-to-date on new results and focus on promising approaches. Some noteworthy proceedings are listed in the **Supplementary Reading** section at the end of this chapter.

Exercises

1. Let $R(\tau)$ and $S(\omega)$ be a Fourier transform pair. Show that $S(\omega)$ is real and symmetric if $R(\tau)$ is real and symmetric.

2. Calculate the mean, standard deviation, and rms value of the periodic time series

$$u(t) = U_0 \cos \omega t + \bar{U}.$$

3. Show that the autocorrelation function $\overline{u(t)u(t+\tau)}$ of a periodic series $u = U \cos \omega t$ is itself periodic.

4. Calculate the zero-lag cross-correlation $\overline{u(t)v(t)}$ between two periodic series $u(t) = \cos \omega t$ and $v(t) = \cos (\omega t + \phi)$. For values of $\phi = 0, \pi/4$, and $\pi/2$, plot the

scatter diagrams of u vs v at different times, as in Figure 13.6. Note that the plot is a straight line if $\phi = 0$, an ellipse if $\phi = \pi/4$, and a circle if $\phi = \pi/2$; the straight line, as well as the axes of the ellipse, are inclined at $45°$ to the uv-axes. Argue that the straight line signifies a perfect correlation, the ellipse a partial correlation, and the circle a zero correlation.

5. Measurements in an atmosphere at $20\,°C$ show an rms vertical velocity of $w_{rms} = 1$ m/s and an rms temperature fluctuation of $T_{rms} = 0.1\,°C$. If the correlation coefficient is 0.5, calculate the heat flux $\rho C_p \overline{wT'}$.

6. A mass of $10\,kg$ of water is stirred by a mixer. After one hour of stirring, the temperature of the water rises by $1.0\,°C$. What is the power output of the mixer in watts? What is the size η of the dissipating eddies?

7. A horizontal smooth pipe $20\,cm$ in diameter carries water at a temperature of $20\,°C$. The drop of pressure is $dp/dx = 8\,N/m^2$ per meter. Assuming turbulent flow, verify that the thickness of the viscous sublayer is ≈ 0.25 mm. [*Hint*: Use $dp/dx = 2\tau_0/R$, as given in equation (9.12). This gives $\tau_0 = 0.4\,N/m^2$, and therefore $u_* = 0.02$ m/s.]

8. Derive the logarithmic velocity profile for a smooth wall

$$\frac{U}{u_*} = \frac{1}{k}\ln\frac{yu_*}{v} + 5.0,$$

by starting from

$$U = \frac{u_*}{k}\ln y + \text{const.}$$

and matching the profile to the edge of the viscous sublayer at $y = 10.7\,v/u_*$.

9. Estimate the Monin–Obukhov length in the atmospheric boundary layer if the surface stress is $0.1\,N/m^2$ and the upward heat flux is $200\,W/m^2$.

10. Consider a one-dimensional turbulent diffusion of particles issuing from a point source. Assume a Gaussian Lagrangian correlation function of particle velocity

$$r(\tau) = e^{-\tau^2/t_c^2},$$

where t_c is a constant. By integrating the correlation function from $\tau = 0$ to ∞, find the integral time scale $\mathcal{T}$ in terms of t_c. Using the Taylor theory, estimate the eddy diffusivity at large times $t/\mathcal{T} \gg 1$, given that the rms fluctuating velocity is 1 m/s and $t_c = 1$ s.

11. Show by dimensional reasoning as outlined in Section 10 that for self-preserving flows far downstream, $U_\infty - U_e \sim x^{-1/2}, \delta \sim \sqrt{x}$, for a wake, and $U_1 - U_2 = \text{const.}, \delta \sim x$, for a shear layer.

Literature Cited

Batchelor, G. K. (1959). "Small scale variation of convected quantities like temperature in turbulent fluid. Part I: General discussion and the case of small conductivity." *Journal of Fluid Mechanics* **5**: 113–133.

Bradshaw, P. and J. D. Woods (1978). "Geophysical turbulence and buoyant flows," *in: Turbulence,* P. Bradshaw, ed., New York: Springer-Verlag.

Buschmann, M. H. and M. Gad-el-Hak (2003a). "Generalized logarithmic law and its consequences." *AIAA Journal* **41**: 40–48.

Buschmann, M. H. and M. Gad-el-Hak (2003b). "Debate concerning the mean velocity profiles of a turbulent boundary layer." *AIAA Journal* **41**: 565–572.

Cantwell, B. J. (1981). "Organized motion in turbulent flow." *Annual Review of Fluid Mechanics* **13**: 457–515.

Castillo, L. and W. K. George (2001). "Similarity analysis for turbulent boundary layer with pressure gradient: Outer flow." *AIAA Journal* **39**: 41–47.

Feynman, R. P., R. B. Leighton, and M. Sands (1963). *The Feynman Lectures on Physics,* New York: Addison-Wesley.

Fife, P., T. Wei, J. Klewicki, and P. McMurtry (2005). "Stress gradient balance layers and scale hierarchies in wall-bounded turbulent flows." *Journal of Fluid Mechanics* **532**: 165–189.

George, W. K. (2006). "Recent advancements toward the understanding of turbulent boundary layers." *AIAA Journal* **44**: 2435–2449.

Grant, H. L., R. W. Stewart, and A. Moilliet (1962). "The spectrum of a cross-stream component of turbulence in a tidal stream." *Journal of Fluid Mechanics* **13**: 237–240.

Grant, H. L., B. A. Hughes, W. M. Vogel, and A. Moilliet (1968). "The spectrum of temperature fluctuation in turbulent flow." *Journal of Fluid Mechanics* **34**: 423–442.

Indinger, T., M. H. Buschmann, and M. Gad-el-Hak (2006). "Mean velocity profile of turbulent boundary layers approaching separation." *AIAA Journal* **44**: 2465–2474.

Kim, John (2003). "Control of turbulent boundary layers." *Physics of Fluids* **15**: 1093–1105.

Kline, S. J., W. C. Reynolds, F. A. Schraub, and P. W. Runstadler (1967). "The structure of turbulent boundary layers." *Journal of Fluid Mechanics* **30**: 741–773.

Lam, S. H. (1992). "On the RNG theory of turbulence." *The Physics of Fluids A* **4**: 1007–1017.

Landahl, M. T. and E. Mollo-Christensen (1986). *Turbulence and Random Processes in Fluid Mechanics,* London: Cambridge University Press.

Lesieur, M. (1987). *Turbulence in Fluids,* Dordrecht, Netherlands: Martinus Nijhoff Publishers.

Maciel, Y., K. S. Rossignol, and J. Lemay (2006). "Self-similarity in the outer region of adverse-pressure-gradient turbulent boundary layers." *AIAA Journal* **44**: 2450–2464.

Monin, A. S. and A. M. Yaglom (1971). *Statistical Fluid Mechanics,* Cambridge, MA: MIT Press.

Panofsky, H. A. and J. A. Dutton (1984). *Atmospheric Turbulence,* New York: Wiley.

Phillips, O. M. (1977). *The Dynamics of the Upper Ocean,* London: Cambridge University Press.

Smith, L. M. and W. C. Reynolds (1992). "On the Yakhot-Orszag renormalization group method for deriving turbulence statistics and models." *The Physics of Fluids A* **4**: 364–390.

Speziale, C. G. (1991). "Analytical methods for the development of Reynolds-stress closures in turbulence." *Annual Review of Fluid Mechanics* **23**: 107–157.

Taylor, G. I. (1915). "Eddy motion in the atmosphere." *Philosophical Transactions of the Royal Society of London* **A215**: 1–26.

Taylor, G. I. (1921). "Diffusion by continuous movements." *Proceedings of the London Mathematical Society* **20**: 196–211.

Tennekes, H. and J. L. Lumley (1972). *A First Course in Turbulence,* Cambridge, MA: MIT Press.

Townsend, A. A. (1976). *The Structure of Turbulent Shear Flow,* London: Cambridge University Press.

Turner, J. S. (1973). *Buoyancy Effects in Fluids,* London: Cambridge University Press.

Turner, J. S. (1981). "Small-scale mixing processes," *in: Evolution of Physical Oceanography,* B. A. Warren and C. Wunch, eds, Cambridge, MA: MIT Press.

Walker, D. J. and L. Castillo (2002). "Effect of initial conditions on turbulent boundary layers." *AIAA Journal* **40**: 2540–2542.

Wei, T., P. Fife, and J. Klewicki (2007). "On scaling the mean momentum balance and its solutions in turbulent Couette-Poiseuille flow." *Journal of Fluid Mechanics* **573**: 371–398.

Wosnik, M., L. Castillo, and W. K. George (2000). "A theory for turbulent pipe and channel flows." *Journal of Fluid Mechanics* **421**: 115–145.

Zagarola, M. V. and A. J. Smits (1998). "Mean-flow scaling of turbulent pipe flow." *Journal of Fluid Mechanics* **373**: 33–79.

Supplemental Reading

Hinze, J. O. (1975). *Turbulence*, 2nd ed., New York: McGraw-Hill.

Hunt, J. C. R., N. D. Sandham, J. C. Vassilicos, B. E. Launder, P. A. Monkewitz, and G. F. Hewitt(2001). Developments in turbulence research: are view based on the 1999 Programme of the Isaac Newton Institute, Cambridge. Published in *Journal of Fluid Mechanics*, **436**: 353–391.

Yakhot, V. and S. A. Orszag (1986). "Renormalization group analysis of turbulence. I. Basic theory." *Journal of Scientific Computing* **1**: 3–51.

Proceedings of the Boeing Symposium on Turbulence. Published in *Journal of Fluid Mechanics* (1970), **41**: Parts 1 (March) and 2 (April).

Symposium on Fluid Mechanics of Stirring and Mixing, IUTAM. Published in *Physics of Fluids* A (1991), **3** (5), May, Part 2.

The Turbulent Years: John Lumley at 70, A Symposium in Honor of John L. Lumley on his 70th Birthday. Published in *Physics of Fluids* (2002), **14**: 2424–2557.

Geophysical Fluid Dynamics

1. Introduction

The subject of geophysical fluid dynamics deals with the dynamics of the atmosphere and the ocean. It has recently become an important branch of fluid dynamics due to our increasing interest in the environment. The field has been largely developed by meteorologists and oceanographers, but non-specialists have also been interested in the subject. Taylor was not a geophysical fluid dynamicist, but he held the position of

©2010 Elsevier Inc. All rights reserved.
DOI: 10.1016/B978-0-12-381399-2.50014-9

a meteorologist for some time, and through this involvement he developed a special interest in the problems of turbulence and instability. Although Prandtl was mainly interested in the engineering aspects of fluid mechanics, his well-known textbook (Prandtl, 1952) contains several sections dealing with meteorological aspects of fluid mechanics. Notwithstanding the pressure for specialization that we all experience these days, it is worthwhile to learn something of this fascinating field even if one's primary interest is in another area of fluid mechanics.

The importance of the study of atmospheric dynamics can hardly be overemphasized. We live within the atmosphere and are almost helplessly affected by the weather and its rather chaotic behavior. The motion of the atmosphere is intimately connected with that of the ocean, with which it exchanges fluxes of momentum, heat and moisture, and this makes the dynamics of the ocean as important as that of the atmosphere. The study of ocean currents is also important in its own right because of its relevance to navigation, fisheries, and pollution disposal.

The two features that distinguish geophysical fluid dynamics from other areas of fluid dynamics are the rotation of the earth and the vertical density stratification of the medium. We shall see that these two effects dominate the dynamics to such an extent that entirely new classes of phenomena arise, which have no counterpart in the laboratory scale flows we have studied in the preceding chapters. (For example, we shall see that the dominant mode of flow in the atmosphere and the ocean is *along* the lines of constant pressure, not from high to low pressures.) The motion of the atmosphere and the ocean is naturally studied in a coordinate frame rotating with the earth. This gives rise to the Coriolis force, which is discussed in Chapter 4. The density stratification gives rise to buoyancy force, which is introduced in Chapter 4 (Conservation Laws) and discussed in further detail in Chapter 7 (Gravity Waves). In addition, important relevant material is discussed in Chapter 5 (Vorticity), Chapter 10 (Boundary Layer), Chapter 12 (Instability), and Chapter 13 (Turbulence). The reader should be familiar with these before proceeding further with the present chapter.

Because Coriolis forces and stratification effects play dominating roles in both the atmosphere and the ocean, there is a great deal of similarity between the dynamics of these two media; this makes it possible to study them together. There are also significant differences, however. For example the effects of lateral boundaries, due to the presence of continents, are important in the ocean but not in the atmosphere. The intense currents (like the Gulf Stream and the Kuroshio) along the western boundaries of the ocean have no atmospheric analog. On the other hand phenomena like cloud formation and latent heat release due to moisture condensation are typically atmospheric phenomena. Processes are generally slower in the ocean, in which a typical horizontal velocity is 0.1 m/s, although velocities of the order of 1–2 m/s are found within the intense western boundary currents. In contrast, typical velocities in the atmosphere are 10–20 m/s. The nomenclature can also be different in the two fields. Meteorologists refer to a flow directed to the west as an "easterly wind" (i.e., *from* the east), while oceanographers refer to such a flow as a "westward current." Atmospheric scientists refer to vertical positions by "heights" measured upward from the earth's surface, while oceanographers refer to "depths" measured downward from the sea surface. However, we shall always take the vertical coordinate z to be upward, so no confusion should arise.

We shall see that rotational effects caused by the presence of the Coriolis force have opposite signs in the two hemispheres. Note that *all figures and descriptions given here are valid for the northern hemisphere*. In some cases the sense of the rotational effect for the southern hemisphere has been explicitly mentioned. When the sense of the rotational effect is left unspecified for the southern hemisphere, it has to be assumed as opposite to that in the northern hemisphere.

2. *Vertical Variation of Density in Atmosphere and Ocean*

An important variable in the study of geophysical fluid dynamics is the density stratification. In equation (1.38) we saw that the static stability of a fluid medium is determined by the sign of the potential density gradient

$$\frac{d\rho_{\text{pot}}}{dz} = \frac{d\rho}{dz} + \frac{g\rho}{c^2}, \tag{14.1}$$

where c is the speed of sound. A medium is statically stable if the potential density decreases with height. The first term on the right-hand side corresponds to the *in situ* density change due to all sources such as pressure, temperature, and concentration of a constituent such as the salinity in the sea or the water vapor in the atmosphere. The second term on the right-hand side is the density gradient due to the pressure decrease with height in an adiabatic environment and is called the *adiabatic density gradient*. The corresponding temperature gradient is called the *adiabatic temperature gradient*. For incompressible fluids $c = \infty$ and the adiabatic density gradient is zero.

As shown in Chapter 1, Section 10, the temperature of a dry adiabatic atmosphere decreases upward at the rate of $\approx 10\,^{\circ}\text{C/km}$; that of a moist atmosphere decreases at the rate of ≈ 5–$6\,^{\circ}\text{C/km}$. In the ocean, the adiabatic density gradient is $g\rho/c^2$ $\sim 4 \times 10^{-3}\,\text{kg/m}^4$, taking a typical sonic speed of $c = 1520\,\text{m/s}$. The potential density in the ocean increases with depth at a much smaller rate of $0.6 \times 10^{-3}\,\text{kg/m}^4$, so that the two terms on the right-hand side of equation (14.1) are nearly in balance. It follows that most of the *in situ* density increase with depth in the ocean is due to the compressibility effects and not to changes in temperature or salinity. As potential density is the variable that determines the static stability, oceanographers take into account the compressibility effects by referring all their density measurements to the sea level pressure. Unless specified otherwise, throughout the present chapter potential density will simply be referred to as "density," omitting the qualifier "potential."

The mean vertical distribution of the *in situ* temperature in the lower 50 km of the atmosphere is shown in Figure 14.1. The lowest 10 km is called the *troposphere*, in which the temperature decreases with height at the rate of 6.5 °C/km. This is close to the moist adiabatic lapse rate, which means that the troposphere is close to being neutrally stable. The neutral stability is expected because turbulent mixing due to frictional and convective effects in the lower atmosphere keeps it well-stirred and therefore close to the neutral stratification. Practically all the clouds, weather changes, and water vapor of the atmosphere are found in the troposphere. The layer is capped by the *tropopause*, at an average height of 10 km, above which the temperature increases. This higher layer is called the *stratosphere*, because it is very stably stratified. The

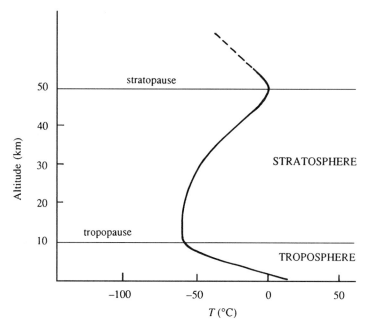

Figure 14.1   Vertical distribution of temperature in the lower 50 km of the atmosphere.

increase of temperature with height in this layer is caused by the absorption of the sun's ultraviolet rays by ozone. The stability of the layer inhibits mixing and consequently acts as a lid on the turbulence and convective motion of the troposphere. The increase of temperature stops at the *stratopause* at a height of nearly 50 km.

The vertical structure of density in the ocean is sketched in Figure 14.2, showing typical profiles of potential density and temperature. Most of the temperature increase with height is due to the absorption of solar radiation within the upper layer of the ocean. The density distribution in the ocean is also affected by the salinity. However, there is no characteristic variation of salinity with depth, and a decrease with depth is found to be as common as an increase with depth. In most cases, however, the vertical structure of density in the ocean is determined mainly by that of temperature, the salinity effects being secondary. The upper 50–200 m of ocean is well-mixed, due to the turbulence generated by the wind, waves, current shear, and the convective overturning caused by surface cooling. The temperature gradients decrease with depth, becoming quite small below a depth of 1500 m. There is usually a large temperature gradient in the depth range of 100–500 m. This layer of high stability is called the *thermocline*. Figure 14.2 also shows the profile of *buoyancy frequency N*, defined by

$$N^2 \equiv -\frac{g}{\rho_0}\frac{d\rho}{dz},$$

where ρ of course stands for the potential density and ρ_0 is a constant reference density. The buoyancy frequency reaches a typical maximum value of $N_{\max} \sim 0.01\,\mathrm{s}^{-1}$ (period $\sim 10\,\mathrm{min}$) in the thermocline and decreases both upward and downward.

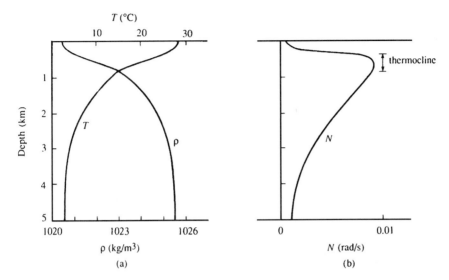

Figure 14.2 Typical vertical distributions of: (a) temperature and density; and (b) buoyancy frequency in the ocean.

3. Equations of Motion

In this section we shall review the relevant equations of motion, which are derived and discussed in Chapter 4. The equations of motion for a stratified medium, observed in a system of coordinates rotating at an angular velocity $\mathbf{\Omega}$ with respect to the "fixed stars," are

$$\nabla \cdot \mathbf{u} = 0,$$

$$\frac{D\mathbf{u}}{Dt} + 2\mathbf{\Omega} \times \mathbf{u} = -\frac{1}{\rho_0}\nabla p - \frac{g\rho}{\rho_0}\mathbf{k} + \mathbf{F}, \qquad (14.2)$$

$$\frac{D\rho}{Dt} = 0,$$

where $\mathbf{F}$ is the friction force per unit mass. The diffusive effects in the density equation are omitted in set (14.2) because they will not be considered here.

Set (14.2) makes the so-called *Boussinesq approximation*, discussed in Chapter 4, Section 18, in which the density variations are neglected everywhere except in the gravity term. Along with other restrictions, it assumes that the vertical scale of the motion is less than the "scale height" of the medium c^2/g, where c is the speed of sound. This assumption is very good in the ocean, in which $c^2/g \sim 200\,\mathrm{km}$. In the atmosphere it is less applicable, because $c^2/g \sim 10\,\mathrm{km}$. Under the Boussinesq approximation, the principle of mass conservation is expressed by $\nabla \cdot \mathbf{u} = 0$. In contrast, the density equation $D\rho/Dt = 0$ follows from the nondiffusive heat equation $DT/Dt = 0$ and an incompressible equation of state of the form $\delta\rho/\rho_0 = -\alpha\delta T$. (If the density is determined by the concentration S of a constituent, say the water vapor in the atmosphere or the salinity in the ocean, then $D\rho/Dt = 0$ follows from the nondiffusive conservation equation for the constituent in the form $DS/Dt = 0$, plus the incompressible equation of state $\delta\rho/\rho_0 = \beta\delta S$.)

The equations can be written in terms of the pressure and density *perturbations* from a state of rest. In the absence of any motion, suppose the density and pressure have the vertical distributions $\bar{\rho}(z)$ and $\bar{p}(z)$, where the z-axis is taken vertically upward. As this state is hydrostatic, we must have

$$\frac{d\bar{p}}{dz} = -\bar{\rho}g. \tag{14.3}$$

In the presence of a flow field $\mathbf{u}(\mathbf{x}, t)$, we can write the density and pressure as

$$\begin{aligned}
\rho(\mathbf{x}, t) &= \bar{\rho}(z) + \rho'(\mathbf{x}, t), \\
p(\mathbf{x}, t) &= \bar{p}(z) + p'(\mathbf{x}, t),
\end{aligned} \tag{14.4}$$

where ρ' and p' are the changes from the state of rest. With this substitution, the first two terms on the right-hand side of the momentum equation in (14.2) give

$$\begin{aligned}
-\frac{1}{\rho_0}\nabla p - \frac{g\rho}{\rho_0}\mathbf{k} &= -\frac{1}{\rho_0}\nabla(\bar{p} + p') - \frac{g(\bar{\rho} + \rho')}{\rho_0}\mathbf{k} \\
&= -\frac{1}{\rho_0}\left[\frac{d\bar{p}}{dz}\mathbf{k} + \nabla p'\right] - \frac{g(\bar{\rho} + \rho')}{\rho_0}\mathbf{k}.
\end{aligned}$$

Subtracting the hydrostatic state (14.3), this becomes

$$-\frac{1}{\rho_0}\nabla p - \frac{g\rho}{\rho_0}\mathbf{k} = -\frac{1}{\rho_0}\nabla p' - \frac{g\rho'}{\rho_0}\mathbf{k},$$

which shows that *we can replace p and ρ in* equation (14.2) *by the perturbation quantities p' and ρ'.*

Formulation of the Frictional Term

The friction force per unit mass $\mathbf{F}$ in equation (14.2) needs to be related to the velocity field. From Chapter 4, Section 7, the friction force is given by

$$F_i = \frac{\partial\tau_{ij}}{\partial x_j},$$

where τ_{ij} is the viscous stress tensor. The stress in a laminar flow is caused by the molecular exchanges of momentum. From equation (4.41), the viscous stress tensor in an isotropic incompressible medium in laminar flow is given by

$$\tau_{ij} = \rho\nu\left(\frac{\partial u_i}{\partial x_j} + \frac{\partial u_j}{\partial x_i}\right).$$

In large-scale geophysical flows, however, the frictional forces are provided by turbulent mixing, and the molecular exchanges are negligible. The complexity of turbulent behavior makes it impossible to relate the stress to the velocity field in a simple way. To proceed, then, we adopt the eddy viscosity hypothesis, assuming that the turbulent stress is proportional to the velocity gradient field.

Geophysical media are in the form of shallow stratified layers, in which the vertical velocities are much smaller than horizontal velocities. This means that the exchange of momentum across a horizontal surface is much weaker than that across a vertical surface. We expect then that the vertical eddy viscosity ν_v is much smaller than the horizontal eddy viscosity ν_H, and we assume that the turbulent stress components have the form

$$
\begin{aligned}
\tau_{xz} = \tau_{zx} &= \rho \nu_v \frac{\partial u}{\partial z} + \rho \nu_H \frac{\partial w}{\partial x}, \\
\tau_{yz} = \tau_{zy} &= \rho \nu_v \frac{\partial v}{\partial z} + \rho \nu_H \frac{\partial w}{\partial y}, \\
\tau_{xy} = \tau_{yx} &= \rho \nu_H \left(\frac{\partial u}{\partial y} + \frac{\partial v}{\partial x} \right), \\
\tau_{xx} = 2\rho \nu_H \frac{\partial u}{\partial x}, \quad \tau_{yy} &= 2\rho \nu_H \frac{\partial v}{\partial y}, \quad \tau_{zz} = 2\rho \nu_v \frac{\partial w}{\partial z}.
\end{aligned}
\tag{14.5}
$$

The difficulty with set (14.5) is that the expressions for τ_{xz} and τ_{yz} depend on the fluid *rotation* in the vertical plane and not just the deformation. In Chapter 4, Section 10, we saw that a requirement for a constitutive equation is that the stresses should be independent of fluid rotation and should depend only on the deformation. Therefore, τ_{xz} should depend only on the combination $(\partial u/\partial z + \partial w/\partial x)$, whereas the expression in equation (14.5) depends on both deformation and rotation. A tensorially correct geophysical treatment of the frictional terms is discussed, for example, in Kamenkovich (1967). However, the assumed form (14.5) leads to a simple formulation for viscous effects, as we shall see shortly. As the eddy viscosity assumption is of questionable validity (which Pedlosky (1971) describes as a "rather disreputable and desperate attempt"), there does not seem to be any purpose in formulating the stress–strain relation in more complicated ways merely to obey the requirement of invariance with respect to rotation.

With the assumed form for the turbulent stress, the components of the frictional force $F_i = \partial \tau_{ij}/\partial x_j$ become

$$
\begin{aligned}
F_x &= \frac{\partial \tau_{xx}}{\partial x} + \frac{\partial \tau_{xy}}{\partial y} + \frac{\partial \tau_{xz}}{\partial z} = \nu_H \left(\frac{\partial^2 u}{\partial x^2} + \frac{\partial^2 u}{\partial y^2} \right) + \nu_v \frac{\partial^2 u}{\partial z^2}, \\
F_y &= \frac{\partial \tau_{yx}}{\partial x} + \frac{\partial \tau_{yy}}{\partial y} + \frac{\partial \tau_{yz}}{\partial z} = \nu_H \left(\frac{\partial^2 v}{\partial x^2} + \frac{\partial^2 v}{\partial y^2} \right) + \nu_v \frac{\partial^2 v}{\partial z^2}, \\
F_z &= \frac{\partial \tau_{zx}}{\partial x} + \frac{\partial \tau_{zy}}{\partial y} + \frac{\partial \tau_{zz}}{\partial z} = \nu_H \left(\frac{\partial^2 w}{\partial x^2} + \frac{\partial^2 w}{\partial y^2} \right) + \nu_v \frac{\partial^2 w}{\partial z^2}.
\end{aligned}
\tag{14.6}
$$

Estimates of the eddy coefficients vary greatly. Typical suggested values are $\nu_v \sim 10 \, \mathrm{m^2/s}$ and $\nu_H \sim 10^5 \, \mathrm{m^2/s}$ for the lower atmosphere, and $\nu_v \sim 0.01 \, \mathrm{m^2/s}$

and $\nu_H \sim 100\,\text{m}^2/\text{s}$ for the upper ocean. In comparison, the molecular values are $\nu = 1.5 \times 10^{-5}\,\text{m}^2/\text{s}$ for air and $\nu = 10^{-6}\,\text{m}^2/\text{s}$ for water.

4. Approximate Equations for a Thin Layer on a Rotating Sphere

The atmosphere and the ocean are very thin layers in which the depth scale of flow is a few kilometers, whereas the horizontal scale is of the order of hundreds, or even thousands, of kilometers. The trajectories of fluid elements are very shallow and the vertical velocities are much smaller than the horizontal velocities. In fact, the continuity equation suggests that the scale of the vertical velocity W is related to that of the horizontal velocity U by

$$\frac{W}{U} \sim \frac{H}{L},$$

where H is the depth scale and L is the horizontal length scale. Stratification and Coriolis effects usually constrain the vertical velocity to be even smaller than UH/L.

Large-scale geophysical flow problems should be solved using spherical polar coordinates. If, however, the horizontal length scales are much smaller than the radius of the earth ($= 6371\,\text{km}$), then the curvature of the earth can be ignored, and the motion can be studied by adopting a *local* Cartesian system on a *tangent plane* (Figure 14.3). On this plane we take an xyz coordinate system, with x increasing eastward, y northward, and z upward. The corresponding velocity components are u (eastward), v (northward), and w (upward).

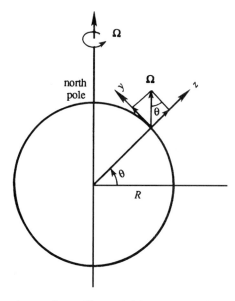

Figure 14.3   Local Cartesian coordinates. The x-axis is into the plane of the paper.

The earth rotates at a rate

$$\Omega = 2\pi \text{ rad/day} = 0.73 \times 10^{-4} \text{ s}^{-1},$$

around the polar axis, in a counterclockwise sense looking from above the north pole. From Figure 14.3, the components of angular velocity of the earth in the local Cartesian system are

$$\Omega_x = 0,$$
$$\Omega_y = \Omega \cos \theta,$$
$$\Omega_z = \Omega \sin \theta,$$

where θ is the latitude. The Coriolis force is therefore

$$2\mathbf{\Omega} \times \mathbf{u} = \begin{vmatrix} \mathbf{i} & \mathbf{j} & \mathbf{k} \\ 0 & 2\Omega \cos \theta & 2\Omega \sin \theta \\ u & v & w \end{vmatrix}$$

$$= 2\Omega[\mathbf{i}(w \cos \theta - v \sin \theta) + \mathbf{j}u \sin \theta - \mathbf{k}u \cos \theta].$$

In the term multiplied by $\mathbf{i}$ we can use the condition $w \cos \theta \ll v \sin \theta$, because the thin sheet approximation requires that $w \ll v$. The three components of the Coriolis force are therefore

$$(2\mathbf{\Omega} \times \mathbf{u})_x = -(2\Omega \sin \theta)v = -fv,$$
$$(2\mathbf{\Omega} \times \mathbf{u})_y = (2\Omega \sin \theta)u = fu, \tag{14.7}$$
$$(2\mathbf{\Omega} \times \mathbf{u})_z = -(2\Omega \cos \theta)u,$$

where we have defined

$$\boxed{f = 2\Omega \sin \theta,} \tag{14.8}$$

to be twice the *vertical* component of $\mathbf{\Omega}$. As vorticity is twice the angular velocity, f is called the *planetary vorticity*. More commonly, f is referred to as the *Coriolis parameter*, or the *Coriolis frequency*. It is positive in the northern hemisphere and negative in the southern hemisphere, varying from $\pm 1.45 \times 10^{-4} \text{ s}^{-1}$ at the poles to zero at the equator. This makes sense, since a person standing at the north pole spins around himself in an counterclockwise sense at a rate Ω, whereas a person standing at the equator does not spin around himself but simply translates. The quantity

$$T_i = 2\pi/f,$$

is called the *inertial period*, for reasons that will be clear in Section 11.

The vertical component of the Coriolis force, namely $-2\Omega u \cos\theta$, is generally negligible compared to the dominant terms in the vertical equation of motion, namely $g\rho'/\rho_0$ and $\rho_0^{-1}(\partial p'/\partial z)$. Using equations (14.6) and (14.7), the equations of motion (14.2) reduce to

$$\frac{Du}{Dt} - fv = -\frac{1}{\rho_0}\frac{\partial p}{\partial x} + \nu_H\left(\frac{\partial^2 u}{\partial x^2} + \frac{\partial^2 u}{\partial y^2}\right) + \nu_v\frac{\partial^2 u}{\partial z^2},$$

$$\frac{Dv}{Dt} + fu = -\frac{1}{\rho_0}\frac{\partial p}{\partial y} + \nu_H\left(\frac{\partial^2 v}{\partial x^2} + \frac{\partial^2 v}{\partial y^2}\right) + \nu_v\frac{\partial^2 v}{\partial z^2}, \qquad (14.9)$$

$$\frac{Dw}{Dt} = -\frac{1}{\rho_0}\frac{\partial p}{\partial z} - \frac{g\rho}{\rho_0} + \nu_H\left(\frac{\partial^2 w}{\partial x^2} + \frac{\partial^2 w}{\partial y^2}\right) + \nu_v\frac{\partial^2 w}{\partial z^2}.$$

These are the equations of motion for a thin shell on a rotating earth. Note that only the *vertical* component of the earth's angular velocity appears as a consequence of the flatness of the fluid trajectories.

f-Plane Model

The Coriolis parameter $f = 2\Omega \sin\theta$ varies with latitude θ. However, we shall see later that this variation is important only for phenomena having very long time scales (several weeks) or very long length scales (thousands of kilometers). For many purposes we can assume f to be a constant, say $f_0 = 2\Omega \sin\theta_0$, where θ_0 is the central latitude of the region under study. A model using a constant Coriolis parameter is called an *f-plane model*.

β-Plane Model

The variation of f with latitude can be approximately represented by expanding f in a Taylor series about the central latitude θ_0:

$$f = f_0 + \beta y, \qquad (14.10)$$

where we defined

$$\beta \equiv \left(\frac{df}{dy}\right)_{\theta_0} = \left(\frac{df}{d\theta}\frac{d\theta}{dy}\right)_{\theta_0} = \frac{2\Omega \cos\theta_0}{R}.$$

Here, we have used $f = 2\Omega \sin\theta$ and $d\theta/dy = 1/R$, where the radius of the earth is nearly

$$R = 6371 \text{ km}.$$

A model that takes into account the variation of the Coriolis parameter in the simplified form $f = f_0 + \beta y$, with β as constant, is called a *β-plane model*.

5. Geostrophic Flow

Consider quasi-steady large-scale motions in the atmosphere or the ocean, away from boundaries. For these flows an excellent approximation for the horizontal equilibrium is a balance between the Coriolis force and the pressure gradient:

$$
\begin{aligned}
-fv &= -\frac{1}{\rho_0}\frac{\partial p}{\partial x}, \\
fu &= -\frac{1}{\rho_0}\frac{\partial p}{\partial y}.
\end{aligned}
\tag{14.11}
$$

Here we have neglected the nonlinear acceleration terms, which are of order U^2/L, in comparison to the Coriolis force $\sim fU$ (U is the horizontal velocity scale, and L is the horizontal length scale.) The ratio of the nonlinear term to the Coriolis term is called the *Rossby number*:

$$
\text{Rossby number} = \frac{\text{Nonlinear acceleration}}{\text{Coriolis force}} \sim \frac{U^2/L}{fU} = \frac{U}{fL} = Ro.
$$

For a typical atmospheric value of $U \sim 10\,\text{m/s}$, $f \sim 10^{-4}\,\text{s}^{-1}$, and $L \sim 1000\,\text{km}$, the Rossby number turns out to be 0.1. The Rossby number is even smaller for many flows in the ocean, so that the neglect of nonlinear terms is justified for many flows.

The balance of forces represented by equation (14.11), in which the horizontal pressure gradients are balanced by Coriolis forces, is called a *geostrophic balance*. In such a system the velocity distribution can be determined from a measured distribution of the pressure field. The geostrophic equilibrium breaks down near the equator (within a latitude belt of $\pm 3°$), where f becomes small. It also breaks down if the frictional effects or unsteadiness become important.

Velocities in a geostrophic flow are perpendicular to the horizontal pressure gradient. This is because equation (14.11) implies that $\mathbf{v} \cdot \nabla p = 0$, i.e., ...

$$
(\mathbf{i}u + \mathbf{j}v) \cdot \nabla p = \frac{1}{\rho_0 f}\left(-\mathbf{i}\frac{\partial p}{\partial y} + \mathbf{j}\frac{\partial p}{\partial x}\right) \cdot \left(\mathbf{i}\frac{\partial p}{\partial x} + \mathbf{j}\frac{\partial p}{\partial y}\right) = 0.
$$

Thus, the horizontal velocity is *along*, and not across, the lines of constant pressure. If f is regarded as constant, then the geostrophic balance (14.11) shows that $p/f\rho_0$ can be regarded as a streamfunction. The isobars on a weather map are therefore nearly the streamlines of the flow.

Figure 14.4 shows the geostrophic flow around low and high pressure centers in the northern hemisphere. Here the Coriolis force acts to the right of the velocity vector. This requires the flow to be counterclockwise (viewed from above) around a low pressure region and clockwise around a high pressure region. The sense of circulation is opposite in the southern hemisphere, where the Coriolis force acts to the left of the velocity vector. (Frictional forces become important at lower levels in the atmosphere and result in a flow partially *across* the isobars. This will be discussed

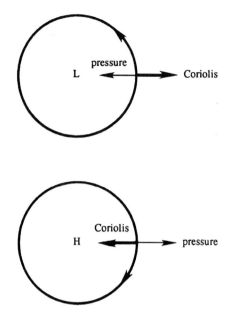

Figure 14.4 Geostrophic flow around low and high pressure centers. The pressure force $(-\nabla p)$ is indicated by a thin arrow, and the Coriolis force is indicated by a thick arrow.

in Section 7, where we will see that the flow around a low pressure center spirals *inward* due to frictional effects.)

The flow along isobars at first surprises a reader unfamiliar with the effects of the Coriolis force. A question commonly asked is: How is such a motion set up? A typical manner of establishment of such a flow is as follows. Consider a horizontally converging flow in the surface layer of the ocean. The convergent flow sets up the sea surface in the form of a gentle "hill," with the sea surface dropping away from the center of the hill. A fluid particle starting to move down the "hill" is deflected to the right in the northern hemisphere, and a steady state is reached when the particle finally moves *along* the isobars.

Thermal Wind

In the presence of a *horizontal* gradient of density, the geostrophic velocity develops a *vertical* shear. This is easy to demonstrate from an analysis of the geostrophic and hydrostatic balance

$$-fv = -\frac{1}{\rho_0}\frac{\partial p}{\partial x}, \tag{14.12}$$

$$fu = -\frac{1}{\rho_0}\frac{\partial p}{\partial y}, \tag{14.13}$$

$$0 = -\frac{\partial p}{\partial z} - g\rho. \tag{14.14}$$

Eliminating p between equations (14.12) and (14.14), and also between equations (14.13) and (14.14), we obtain, respectively,

$$\frac{\partial v}{\partial z} = -\frac{g}{\rho_0 f}\frac{\partial \rho}{\partial x},$$
$$\frac{\partial u}{\partial z} = \frac{g}{\rho_0 f}\frac{\partial \rho}{\partial y}. \tag{14.15}$$

Meteorologists call these the *thermal wind* equations because they give the vertical variation of wind from measurements of horizontal temperature gradients. The thermal wind is a baroclinic phenomenon, because the surfaces of constant p and ρ do not coincide.

Taylor–Proudman Theorem

A striking phenomenon occurs in the geostrophic flow of a *homogeneous* fluid. It can only be observed in a laboratory experiment because stratification effects cannot be avoided in natural flows. Consider then a laboratory experiment in which a tank of fluid is steadily rotated at a high angular speed Ω and a solid body is moved slowly along the bottom of the tank. The purpose of making Ω large and the movement of the solid body slow is to make the Coriolis force much larger than the acceleration terms, which must be made negligible for geostrophic equilibrium. Away from the frictional effects of boundaries, the balance is therefore geostrophic in the horizontal and hydrostatic in the vertical:

$$-2\Omega v = -\frac{1}{\rho}\frac{\partial p}{\partial x}, \tag{14.16}$$

$$2\Omega u = -\frac{1}{\rho}\frac{\partial p}{\partial y}, \tag{14.17}$$

$$0 = -\frac{1}{\rho}\frac{\partial p}{\partial z} - g. \tag{14.18}$$

It is useful to define an Ekman number as the ratio of viscous to Coriolis forces (per unit volume):

$$\text{Ekman number} = \frac{\text{viscous force}}{\text{Coriolis force}} = \frac{\rho \nu U/L^2}{\rho f U} = \frac{\nu}{f L^2} = E.$$

Under the circumstances already described here, both Ro and E are small.

Elimination of p by cross differentiation between the horizontal momentum equations gives

$$2\Omega\left(\frac{\partial v}{\partial y} + \frac{\partial u}{\partial x}\right) = 0.$$

Using the continuity equation, this gives

$$\frac{\partial w}{\partial z} = 0. \tag{14.19}$$

Also, differentiating equations (14.16) and (14.17) with respect to z, and using equation (14.18), we obtain

$$\frac{\partial v}{\partial z} = \frac{\partial u}{\partial z} = 0. \tag{14.20}$$

Equations (14.19) and (14.20) show that

$$\boxed{\frac{\partial \mathbf{u}}{\partial z} = 0,} \tag{14.21}$$

showing that the velocity vector cannot vary in the direction of $\boldsymbol{\Omega}$. In other words, steady slow motions in a rotating, homogeneous, inviscid fluid are two dimensional. This is the *Taylor–Proudman theorem*, first derived by Proudman in 1916 and demonstrated experimentally by Taylor soon afterwards.

In Taylor's experiment, a tank was made to rotate as a solid body, and a small cylinder was slowly dragged along the bottom of the tank (Figure 14.5). Dye was introduced from point A above the cylinder and directly ahead of it. In a nonrotating fluid the water would pass over the top of the moving cylinder. In the rotating experiment, however, the dye divides at a point S, as if it had been blocked by an upward extension of the cylinder, and flows around this *imaginary* cylinder, called the *Taylor column*. Dye released from a point B within the Taylor column remained there and moved with the cylinder. The conclusion was that the flow outside the upward extension of the cylinder is the same as if the cylinder extended across the entire water depth and that a column of water directly above the cylinder moves with it. The motion is two dimensional, although the solid body does not extend across the entire water depth. Taylor did a second experiment, in which he dragged a solid body *parallel* to the axis of rotation. In accordance with $\partial w/\partial z = 0$, he observed that a column of fluid is pushed ahead. The lateral velocity components u and v were zero. In both of these experiments, there are shear layers at the edge of the Taylor column.

In summary, Taylor's experiment established the following striking fact for steady inviscid motion of homogeneous fluid in a strongly rotating system: Bodies moving either parallel or perpendicular to the axis of rotation carry along with their motion a so-called Taylor column of fluid, oriented parallel to the axis. The phenomenon is analogous to the horizontal *blocking* caused by a solid body (say a mountain) in a strongly stratified system, shown in Figure 7.33.

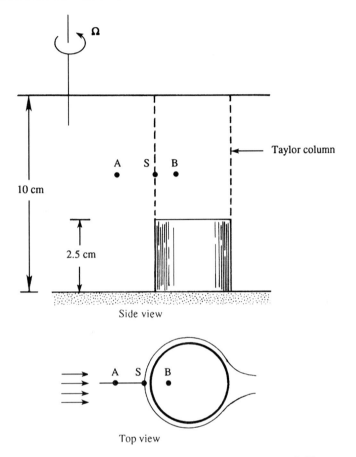

Figure 14.5 Taylor's experiment in a strongly rotating flow of a homogeneous fluid.

6. *Ekman Layer at a Free Surface*

In the preceding section, we discussed a steady linear inviscid motion expected to be valid away from frictional boundary layers. We shall now examine the motion within frictional layers over horizontal surfaces. In viscous flows unaffected by Coriolis forces and pressure gradients, the only term which can balance the viscous force is either the time derivative $\partial u/\partial t$ or the advection $\mathbf{u} \cdot \nabla \mathbf{u}$. The balance of $\partial u/\partial t$ and the viscous force gives rise to a viscous layer whose thickness increases with time, as in the suddenly accelerated plate discussed in Chapter 9, Section 7. The balance of $\mathbf{u} \cdot \nabla \mathbf{u}$ and the viscous force give rise to a viscous layer whose thickness increases in the direction of flow, as in the boundary layer over a semi-infinite plate discussed in Chapter 10, Sections 5 and 6. In a rotating flow, however, we can have a balance between the Coriolis and the viscous forces, and the thickness of the viscous layer can be invariant in time and space. Two examples of such layers are given in this and the following sections.

Consider first the case of a frictional layer near the free surface of the ocean, which is acted on by a wind stress τ in the x-direction. We shall not consider how the flow adjusts to the steady state but examine only the steady solution. We shall assume that the horizontal pressure gradients are zero and that the field is horizontally homogeneous. From equation (14.9), the horizontal equations of motion are

$$-fv = \nu_v \frac{d^2u}{dz^2}, \qquad (14.22)$$

$$fu = \nu_v \frac{d^2v}{dz^2}. \qquad (14.23)$$

Taking the z-axis vertically upward from the surface of the ocean, the boundary conditions are

$$\rho \nu_v \frac{du}{dz} = \tau \quad \text{at } z = 0, \qquad (14.24)$$

$$\frac{dv}{dz} = 0 \quad \text{at } z = 0, \qquad (14.25)$$

$$u, v \to 0 \quad \text{as } z \to -\infty. \qquad (14.26)$$

Multiplying equation (14.23) by $i = \sqrt{-1}$ and adding equation (14.22), we obtain

$$\frac{d^2V}{dz^2} = \frac{if}{\nu_v} V, \qquad (14.27)$$

where we have defined the "complex velocity"

$$V \equiv u + iv.$$

The solution of equation (14.27) is

$$V = A\,e^{(1+i)z/\delta} + B\,e^{-(1+i)z/\delta}, \qquad (14.28)$$

where we have defined

$$\delta \equiv \sqrt{\frac{2\,\nu_v}{f}}. \qquad (14.29)$$

We shall see shortly that δ is the thickness of the Ekman layer. The constant B is zero because the field must remain finite as $z \to -\infty$. The surface boundary conditions (14.24) and (14.25) can be combined as $\rho\nu_v(dV/dz) = \tau$ at $z = 0$, from which equation (14.28) gives

$$A = \frac{\tau \delta (1 - i)}{2 \rho \nu_v}.$$

Substitution of this into equation (14.28) gives the velocity components

$$u = \frac{\tau/\rho}{\sqrt{f \nu_v}} \, e^{z/\delta} \cos\left(-\frac{z}{\delta} + \frac{\pi}{4}\right),$$

$$v = -\frac{\tau/\rho}{\sqrt{f \nu_v}} \, e^{z/\delta} \sin\left(-\frac{z}{\delta} + \frac{\pi}{4}\right).$$

The Swedish oceanographer Ekman worked out this solution in 1905. The solution is shown in Figure 14.6 for the case of the northern hemisphere, in which f is positive. The velocities at various depths are plotted in Figure 14.6a, where each arrow represents the velocity vector at a certain depth. Such a plot of v vs u is sometimes called a "hodograph" plot. The vertical distributions of u and v are shown in Figure 14.6b. The hodograph shows that the surface velocity is deflected 45° to the right of the applied wind stress. (In the southern hemisphere the deflection is to the left of the surface stress.) The velocity vector rotates clockwise (looking down) with depth, and the magnitude exponentially decays with an e-folding scale of δ, which is called the *Ekman layer thickness*. The tips of the velocity vector at various depths form a spiral, called the *Ekman spiral*.

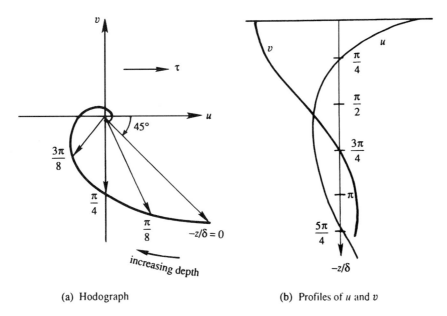

(a) Hodograph   (b) Profiles of u and v

Figure 14.6 Ekman layer at a free surface. The left panel shows velocity at various depths; values of $-z/\delta$ are indicated along the curve traced out by the tip of the velocity vectors. The right panel shows vertical distributions of u and v.

The components of the volume transport in the Ekman layer are

$$\int_{-\infty}^{0} u \, dz = 0,$$

$$\int_{-\infty}^{0} v \, dz = -\frac{\tau}{\rho f}. \tag{14.30}$$

This shows that the *net transport is to the right of the applied stress and is independent of $v_{\rm v}$*. In fact, the result $\int v \, dz = -\tau/f\rho$ follows directly from a vertical integration of the equation of motion in the form $-\rho f v = d(\text{stress})/dz$, so that the result does not depend on the eddy viscosity assumption. The fact that the transport is to the right of the applied stress makes sense, because then the net (depth-integrated) Coriolis force, directed to the right of the depth-integrated transport, can balance the wind stress.

The horizontal uniformity assumed in the solution is not a serious limitation. Since Ekman layers near the ocean surface have a thickness ($\sim$50 m) much smaller than the scale of horizontal variation ($L > 100$ km), the solution is still locally applicable. The absence of horizontal pressure gradient assumed here can also be relaxed easily. Because of the thinness of the layer, any imposed horizontal pressure gradient remains constant across the layer. The presence of a horizontal pressure gradient merely adds a depth-independent geostrophic velocity to the Ekman solution. Suppose the sea surface slopes down to the north, so that there is a pressure force acting northward throughout the Ekman layer and below (Figure 14.7). This means that at the bottom of the Ekman layer ($z/\delta \to -\infty$) there is a geostrophic velocity U to the right of the pressure force. The surface Ekman spiral forced by the wind stress joins smoothly to this geostrophic velocity as $z/\delta \to -\infty$.

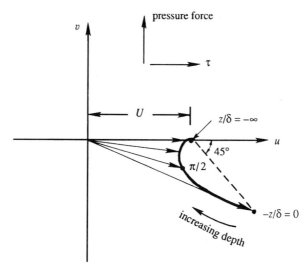

Figure 14.7   Ekman layer at a free surface in the presence of a pressure gradient. The geostrophic velocity forced by the pressure gradient is U.

Pure Ekman spirals are not observed in the surface layer of the ocean, mainly because the assumptions of constant eddy viscosity and steadiness are particularly restrictive. When the flow is averaged over a few days, however, several instances have been found in which the current does look like a spiral. One such example is shown in Figure 14.8.

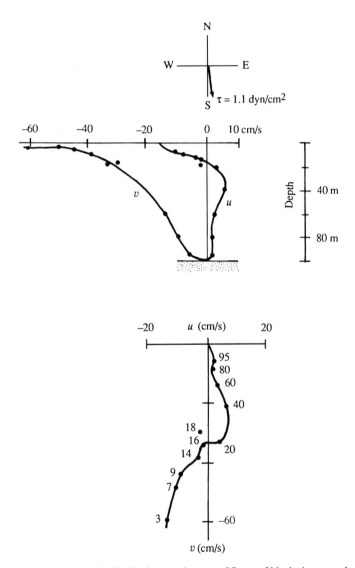

Figure 14.8 An observed velocity distribution near the coast of Oregon. Velocity is averaged over 7 days. Wind stress had a magnitude of 1.1 dyn/cm² and was directed nearly southward, as indicated at the top of the figure. The upper panel shows vertical distributions of *u* and *v*, and the lower panel shows the hodograph in which depths are indicated in meters. The hodograph is similar to that of a surface Ekman layer (of depth 16 m) lying over the bottom Ekman layer (extending from a depth of 16 m to the ocean bottom). P. Kundu, in *Bottom Tubulence*, J. C. J. Nihoul, ed., Elsevier, 1977 and reprinted with the permission of Jacques C. J. Nihoul.

Explanation in Terms of Vortex Tilting

We have seen in previous chapters that the thickness of a viscous layer usually grows in a nonrotating flow, either in time or in the direction of flow. The Ekman solution, in contrast, results in a viscous layer that does not grow either in time or space. This can be explained by examining the vorticity equation (Pedlosky, 1987). The vorticity components in the x- and y-directions are

$$\omega_x = \frac{\partial w}{\partial y} - \frac{\partial v}{\partial z} = -\frac{dv}{dz},$$

$$\omega_y = \frac{\partial u}{\partial z} - \frac{\partial w}{\partial x} = \frac{du}{dz},$$

where we have used $w = 0$. Using these, the z-derivative of the equations of motion (14.22) and (14.23) gives

$$-f\frac{dv}{dz} = \nu_v\frac{d^2\omega_y}{dz^2},$$
$$\tag{14.31}$$
$$-f\frac{du}{dz} = \nu_v\frac{d^2\omega_x}{dz^2}.$$

The right-hand side of these equations represent diffusion of vorticity. Without Coriolis forces this diffusion would cause a thickening of the viscous layer. The presence of planetary rotation, however, means that vertical fluid lines coincide with the planetary vortex lines. The tilting of vertical fluid lines, represented by terms on the left-hand sides of equations (14.31), then causes a rate of change of horizontal component of vorticity that just cancels the diffusion term.

7. *Ekman Layer on a Rigid Surface*

Consider now a horizontally independent and steady viscous layer on a solid surface in a rotating flow. This can be the atmospheric boundary layer over the solid earth or the boundary layer over the ocean bottom. We assume that at large distances from the surface the velocity is toward the x-direction and has a magnitude U. Viscous forces are negligible far from the wall, so that the Coriolis force can be balanced only by a pressure gradient:

$$fU = -\frac{1}{\rho}\frac{dp}{dy}.\tag{14.32}$$

This simply states that the flow outside the viscous layer is in geostrophic balance, U being the geostrophic velocity. For our assumed case of positive U and f, we must have $dp/dy < 0$, so that the pressure falls with y—that is, the pressure force is directed along the positive y direction, resulting in a geostrophic flow U to the right of the pressure force in the northern hemisphere. The horizontal pressure gradient remains constant within the thin boundary layer.

Near the solid surface the viscous forces are important, so that the balance within the boundary layer is

$$-fv = \nu_v \frac{d^2u}{dz^2}, \tag{14.33}$$

$$fu = \nu_v \frac{d^2v}{dz^2} + fU, \tag{14.34}$$

where we have replaced $-\rho^{-1}(dp/dy)$ by fU in accordance with equation (14.32). The boundary conditions are

$$u = U, \quad v = 0 \quad \text{as } z \to \infty, \tag{14.35}$$

$$u = 0, \quad v = 0 \quad \text{at } z = 0, \tag{14.36}$$

where z is taken vertically upward from the solid surface. Multiplying equation (14.34) by i and adding equation (14.33), the equations of motion become

$$\frac{d^2V}{dz^2} = \frac{if}{\nu_v}(V - U), \tag{14.37}$$

where we have defined the complex velocity $V \equiv u + iv$. The boundary conditions (14.35) and (14.36) in terms of the complex velocity are

$$V = U \quad \text{as } z \to \infty, \tag{14.38}$$

$$V = 0 \quad \text{at } z = 0. \tag{14.39}$$

The particular solution of equation (14.37) is $V = U$. The total solution is, therefore,

$$V = A e^{-(1+i)z/\delta} + B e^{(1+i)z/\delta} + U, \tag{14.40}$$

where $\delta \equiv \sqrt{2\nu_v/f}$. To satisfy equation (14.38), we must have $B = 0$. Condition (14.39) gives $A = -U$. The velocity components then become

$$u = U[1 - e^{-z/\delta} \cos(z/\delta)],$$
$$v = U e^{-z/\delta} \sin(z/\delta). \tag{14.41}$$

According to equation (14.41), the tip of the velocity vector describes a spiral for various values of z (Figure 14.9a). As with the Ekman layer at a free surface, the frictional effects are confined within a layer of thickness $\delta = \sqrt{2\nu_v/f}$, which increases with ν_v and decreases with the rotation rate f. Interestingly, the layer thickness is independent of the magnitude of the free-stream velocity U; this behavior is quite different from that of a steady nonrotating boundary layer on a semi-infinite plate (the Blasius solution of Section 10.5) in which the thickness is proportional to $1/\sqrt{U}$.

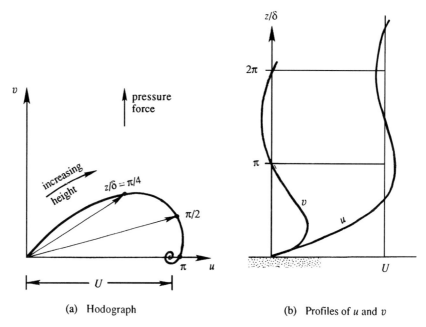

(a) Hodograph   (b) Profiles of u and v

Figure 14.9 Ekman layer at a rigid surface. The left panel shows velocity vectors at various heights; values of z/δ are indicated along the curve traced out by the tip of the velocity vectors. The right panel shows vertical distributions of u and v.

Figure 14.9b shows the vertical distribution of the velocity components. Far from the wall the velocity is entirely in the x-direction, and the Coriolis force balances the pressure gradient. As the wall is approached, retarding effects decrease u and the associated Coriolis force, so that the pressure gradient (which is independent of z) forces a component v in the direction of the pressure force. Using equation (14.41), the net transport in the Ekman layer normal to the uniform stream outside the layer is

$$\int_0^\infty v \, dz = U \left[\frac{\nu_v}{2f} \right]^{1/2} = \frac{1}{2} U \delta,$$

which is directed to the *left* of the free-stream velocity, in the direction of the pressure force.

If the atmosphere were in laminar motion, ν_v would be equal to its molecular value for air, and the Ekman layer thickness at a latitude of 45° (where $f \simeq 10^{-4}\,\mathrm{s}^{-1}$) would be $\approx \delta \sim 0.4\,\mathrm{m}$. The observed thickness of the atmospheric boundary layer is of order 1 km, which implies an eddy viscosity of order $\nu_v \sim 50\,\mathrm{m}^2/\mathrm{s}$. In fact, Taylor (1915) tried to estimate the eddy viscosity by matching the predicted velocity distributions (14.41) with the observed wind at various heights.

The Ekman layer solution on a solid surface demonstrates that the three-way balance among the Coriolis force, the pressure force, and the frictional force within the boundary layer results in a component of flow directed toward the lower pressure.

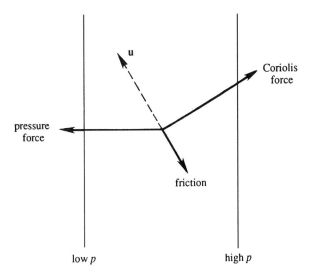

Figure 14.10 Balance of forces within an Ekman layer, showing that velocity u has a component toward low pressure.

The balance of forces within the boundary layer is illustrated in Figure 14.10. The net frictional force on an element is oriented approximately opposite to the velocity vector **u**. It is clear that a balance of forces is possible only if the velocity vector has a component from high to low pressure, as shown. Frictional forces therefore cause the flow around a low-pressure center to spiral *inward*. Mass conservation requires that the inward converging flow should rise over a low-pressure system, resulting in cloud formation and rainfall. This is what happens in a cyclone, which is a low-pressure system. In contrast, over a high-pressure system the air sinks as it spirals outward due to frictional effects. The arrival of high-pressure systems therefore brings in clear skies and fair weather, because the sinking air does not result in cloud formation.

Frictional effects, in particular the Ekman transport by surface winds, play a fundamental role in the theory of wind-driven ocean circulation. Possibly the most important result of such theories was given by Henry Stommel in 1948. He showed that the northward increase of the Coriolis parameter f is responsible for making the currents along the western boundary of the ocean (e.g., the Gulf Stream in the Atlantic and the Kuroshio in the Pacific) much stronger than the currents on the eastern side. These are discussed in books on physical oceanography and will not be presented here. Instead, we shall now turn our attention to the influence of Coriolis forces on inviscid wave motions.

8. Shallow-Water Equations

Both surface and internal gravity waves were discussed in Chapter 7. The effect of planetary rotation was assumed to be small, which is valid if the frequency ω of the wave is much larger than the Coriolis parameter f. In this chapter we are considering phenomena slow enough for ω to be comparable to f. Consider surface gravity waves on a shallow layer of homogeneous fluid whose mean depth is H. If we restrict ourselves to wavelengths λ much larger than H, then the vertical velocities are much smaller than the horizontal velocities. In Chapter 7, Section 6 we saw that the acceleration $\partial w / \partial t$ is then negligible in the vertical momentum equation, so that the pressure distribution is hydrostatic. We also demonstrated that the fluid particles execute a horizontal rectilinear motion that is independent of z. When the effects of planetary rotation are included, the horizontal velocity is still depth-independent, although the particle orbits are no longer rectilinear but elliptic on a horizontal plane, as we shall see in the following section.

Consider a layer of fluid over a flat horizontal bottom (Figure 14.11). Let z be measured upward from the bottom surface, and η be the displacement of the free surface. The pressure at height z from the bottom, which is hydrostatic, is given by

$$p = \rho g (H + \eta - z).$$

The horizontal pressure gradients are therefore

$$\frac{\partial p}{\partial x} = \rho g \frac{\partial \eta}{\partial x}, \quad \frac{\partial p}{\partial y} = \rho g \frac{\partial \eta}{\partial y}. \tag{14.42}$$

As these are independent of z, the resulting horizontal motion is also depth independent.

Now consider the continuity equation

$$\frac{\partial u}{\partial x} + \frac{\partial v}{\partial y} + \frac{\partial w}{\partial z} = 0.$$

As $\partial u / \partial x$ and $\partial v / \partial y$ are independent of z, the continuity equation requires that w vary linearly with z, from zero at the bottom to the maximum value at the free surface. Integrating vertically across the water column from $z = 0$ to $z = H + \eta$, and noting that u and v are depth independent, we obtain

$$(H + \eta) \frac{\partial u}{\partial x} + (H + \eta) \frac{\partial v}{\partial y} + w(\eta) - w(0) = 0, \tag{14.43}$$

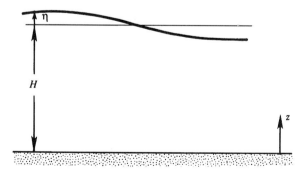

Figure 14.11   Layer of fluid on a flat bottom.

where $w(\eta)$ is the vertical velocity at the surface and $w(0) = 0$ is the vertical velocity at the bottom. The surface velocity is given by

$$w(\eta) = \frac{D\eta}{Dt} = \frac{\partial\eta}{\partial t} + u\frac{\partial\eta}{\partial x} + v\frac{\partial\eta}{\partial y}.$$

The continuity equation (14.43) then becomes

$$(H + \eta)\frac{\partial u}{\partial x} + (H + \eta)\frac{\partial v}{\partial y} + \frac{\partial\eta}{\partial t} + u\frac{\partial\eta}{\partial x} + v\frac{\partial\eta}{\partial y} = 0,$$

which can be written as

$$\frac{\partial\eta}{\partial t} + \frac{\partial}{\partial x}[u(H + \eta)] + \frac{\partial}{\partial y}[v(H + \eta)] = 0. \tag{14.44}$$

This says simply that the divergence of the horizontal transport depresses the free surface. For small amplitude waves, the quadratic nonlinear terms can be neglected in comparison to the linear terms, so that the divergence term in equation (14.44) simplifies to $H\nabla \cdot \mathbf{u}$.

The linearized continuity and momentum equations are then

$$\boxed{\begin{aligned} \frac{\partial\eta}{\partial t} + H\left(\frac{\partial u}{\partial x} + \frac{\partial v}{\partial y}\right) &= 0, \\ \frac{\partial u}{\partial t} - fv &= -g\frac{\partial\eta}{\partial x}, \\ \frac{\partial v}{\partial t} + fu &= -g\frac{\partial\eta}{\partial y}. \end{aligned}} \tag{14.45}$$

In the momentum equations of (14.45), the pressure gradient terms are written in the form (14.42) and the nonlinear advective terms have been neglected under the small amplitude assumption. Equations (14.45), called the *shallow water equations*, govern the motion of a layer of fluid in which the horizontal scale is much larger than the depth of the layer. These equations will be used in the following sections for studying various types of gravity waves.

Although the preceding analysis has been formulated for a layer of *homogeneous* fluid, equations (14.45) are applicable to internal waves in a stratified medium, if we replaced H by the *equivalent depth* H_e, defined by

$$c^2 = gH_e, \tag{14.46}$$

where c is the speed of long nonrotating *internal* gravity waves. This will be demonstrated in the following section.

9. *Normal Modes in a Continuously Stratified Layer*

In the preceding section we considered a homogeneous medium and derived the governing equations for waves of wavelength larger than the depth of the fluid layer. Now consider a continuously stratified medium and assume that the horizontal scale of motion is much larger than the vertical scale. The pressure distribution is therefore hydrostatic, and the equations of motion are

$$\frac{\partial u}{\partial x} + \frac{\partial v}{\partial y} + \frac{\partial w}{\partial z} = 0, \tag{14.47}$$

$$\frac{\partial u}{\partial t} - fv = -\frac{1}{\rho_0}\frac{\partial p}{\partial x}, \tag{14.48}$$

$$\frac{\partial v}{\partial t} + fu = -\frac{1}{\rho_0}\frac{\partial p}{\partial y}, \tag{14.49}$$

$$0 = -\frac{\partial p}{\partial z} - g\rho, \tag{14.50}$$

$$\frac{\partial \rho}{\partial t} - \frac{\rho_0 N^2}{g}w = 0, \tag{14.51}$$

where p and ρ represent *perturbations* of pressure and density from the state of rest. The advective term in the density equation is written in the linearized form $w(d\bar{\rho}/dz) = -\rho_0 N^2 w/g$, where $N(z)$ is the buoyancy frequency. In this form the rate of change of density at a point is assumed to be due only to the vertical advection of the background density distribution $\bar{\rho}(z)$, as discussed in Chapter 7, Section 18.

In a continuously stratified medium, it is convenient to use the method of separation of variables and write $q = \sum q_n(x, y, t)\psi_n(z)$ for some variable q. The solution is thus written as the sum of various vertical "modes," which are called *normal modes* because they turn out to be orthogonal to each other. The vertical structure of a mode is described by ψ_n and q_n describes the horizontal propagation of the mode. Although each mode propagates only horizontally, the *sum* of a number of modes can also propagate vertically if the various q_n are out of phase.

We assume separable solutions of the form

$$[u, v, p/\rho_0] = \sum_{n=0}^{\infty} [u_n, v_n, p_n]\psi_n(z), \tag{14.52}$$

$$w = \sum_{n=0}^{\infty} w_n \int_{-H}^{z} \psi_n(z)\,dz, \tag{14.53}$$

$$\rho = \sum_{n=0}^{\infty} \rho_n \frac{d\psi_n}{dz}, \tag{14.54}$$

where the amplitudes u_n, v_n, p_n, w_n, and ρ_n are functions of (x, y, t). The z-axis is measured from the upper free surface of the fluid layer, and $z = -H$ represents the bottom wall. The reasons for assuming the various forms of z-dependence in equations (14.52)–(14.54) are the following: Variables u, v, and p have the same vertical structure in order to be consistent with equations (14.48) and (14.49). Continuity equation (14.47) requires that the vertical structure of w should be the integral of $\psi_n(z)$. Equation (14.50) requires that the vertical structure of ρ must be the z-derivative of the vertical structure of p.

Subsititution of equations (14.53) and (14.54) into equation (14.51) gives

$$\sum_{n=0}^{\infty}\left[\frac{\partial\rho_n}{\partial t}\frac{d\psi_n}{dz} - \frac{\rho_0 N^2}{g}w_n\int_{-H}^{z}\psi_n\,dz\right] = 0.$$

This is valid for all values of z, and the modes are linearly independent, so the quantity within [ ] must vanish for each mode. This gives

$$\frac{d\psi_n/dz}{N^2\int_{-H}^{z}\psi_n\,dz} = \frac{\rho_0}{g}\frac{w_n}{\partial\rho_n/\partial t} \equiv -\frac{1}{c_n^2}. \tag{14.55}$$

As the first term is a function of z alone and the second term is a function of (x, y, t) alone, for consistency both terms must be equal to a constant; we take the "separation constant" to be $-1/c_n^2$. The vertical structure is then given by

$$\frac{1}{N^2}\frac{d\psi_n}{dz} = -\frac{1}{c_n^2}\int_{-H}^{z}\psi_n\,dz.$$

Taking the z-derivative,

$$\boxed{\frac{d}{dz}\left(\frac{1}{N^2}\frac{d\psi_n}{dz}\right) + \frac{1}{c_n^2}\psi_n = 0,} \tag{14.56}$$

which is the differential equation governing the vertical structure of the normal modes. Equation (14.56) has the so-called Sturm–Liouville form, for which the various solutions are orthogonal.

Equation (14.55) also gives

$$w_n = -\frac{g}{\rho_0 c_n^2}\frac{\partial\rho_n}{\partial t}.$$

Substitution of equations (14.52)–(14.54) into equations (14.47)–(14.51) finally gives the normal mode equations

$$\frac{\partial u_n}{\partial x} + \frac{\partial v_n}{\partial y} + \frac{1}{c_n^2}\frac{\partial p_n}{\partial t} = 0, \tag{14.57}$$

$$\frac{\partial u_n}{\partial t} - f v_n = -\frac{\partial p_n}{\partial x}, \tag{14.58}$$

$$\frac{\partial v_n}{\partial t} + f u_n = -\frac{\partial p_n}{\partial y}, \tag{14.59}$$

$$p_n = -\frac{g}{\rho_0} \rho_n, \tag{14.60}$$

$$w_n = \frac{1}{c_n^2} \frac{\partial p_n}{\partial t}. \tag{14.61}$$

Once equations (14.57)–(14.59) have been solved for u_n, v_n and p_n, the amplitudes ρ_n and w_n can be obtained from equations (14.60) and (14.61). The set (14.57)–(14.59) is identical to the set (14.45) governing the motion of a *homogeneous* layer, provided p_n is identified with $g\eta$ and c_n^2 is identified with gH. In a stratified flow each mode (having a fixed vertical structure) behaves, in the horizontal dimensions and in time, just like a homogeneous layer, with an *equivalent depth* H_e defined by

$$\boxed{c_n^2 \equiv g H_e.} \tag{14.62}$$

Boundary Conditions on ψ_n

At the layer bottom, the boundary condition is

$$w = 0 \quad \text{at } z = -H.$$

To write this condition in terms of ψ_n, we first combine the hydrostatic equation (14.50) and the density equation (14.51) to give w in terms of p:

$$w = \frac{g(\partial \rho / \partial t)}{\rho_0 N^2} = -\frac{1}{\rho_0 N^2} \frac{\partial^2 p}{\partial z \, \partial t} = -\frac{1}{N^2} \sum_{n=0}^{\infty} \frac{\partial p_n}{\partial t} \frac{d\psi_n}{dz}. \tag{14.63}$$

The requirement $w = 0$ then yields the bottom boundary condition

$$\frac{d\psi_n}{dz} = 0 \quad \text{at } z = -H. \tag{14.64}$$

We now formulate the surface boundary condition. The linearized surface boundary conditions are

$$w = \frac{\partial \eta}{\partial t}, \quad p = \rho_0 g \eta \quad \text{at } z = 0, \tag{14.65'}$$

where η is the free surface displacement. These conditions can be combined into

$$\frac{\partial p}{\partial t} = \rho_0 g w \quad \text{at } z = 0.$$

Using equation (14.63) this becomes

$$\frac{g}{N^2} \frac{\partial^2 p}{\partial z \, \partial t} + \frac{\partial p}{\partial t} = 0 \quad \text{at } z = 0.$$

Substitution of the normal mode decomposition (14.52) gives

$$\frac{d\psi_n}{dz} + \frac{N^2}{g} \psi_n = 0 \quad \text{at } z = 0. \tag{14.65}$$

The boundary conditions on ψ_n are therefore equations (14.64) and (14.65).

Solution of Vertical Modes for Uniform N

For a medium of uniform N, a simple solution can be found for ψ_n. From equations (14.56), (14.64), and (14.65), the vertical structure of the normal modes is given by

$$\frac{d^2\psi_n}{dz^2} + \frac{N^2}{c_n^2} \psi_n = 0, \tag{14.66}$$

with the boundary conditions

$$\frac{d\psi_n}{dz} + \frac{N^2}{g} \psi_n = 0 \quad \text{at } z = 0, \tag{14.67}$$

$$\frac{d\psi_n}{dz} = 0 \quad \text{at } z = -H. \tag{14.68}$$

The set (14.66)–(14.68) defines an eigenvalue problem, with ψ_n as the eigenfunction and c_n as the eigenvalue. The solution of equation (14.66) is

$$\psi_n = A_n \cos \frac{Nz}{c_n} + B_n \sin \frac{Nz}{c_n}. \tag{14.69}$$

Application of the surface boundary condition (14.67) gives

$$B_n = -\frac{c_n N}{g} A_n.$$

The bottom boundary condition (14.68) then gives

$$\tan \frac{NH}{c_n} = \frac{c_n N}{g}, \tag{14.70}$$

whose roots define the eigenvalues of the problem.

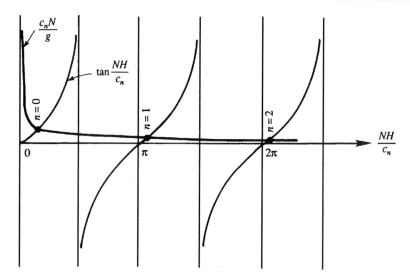

Figure 14.12 Calculation of eigenvalues c_n of vertical normal modes in a fluid layer of depth H and uniform stratification N.

The solution of equation (14.70) is indicated graphically in Figure 14.12. The first root occurs for $NH/c_n \ll 1$, for which we can write $\tan(NH/c_n) \simeq NH/c_n$, so that equation (14.70) gives (indicating this root by $n = 0$)

$$c_0 = \sqrt{gH}.$$

The vertical modal structure is found from equation (14.69). Because the magnitude of an eigenfunction is arbitrary, we can set $A_0 = 1$, obtaining

$$\psi_0 = \cos \frac{Nz}{c_0} - \frac{c_0 N}{g} \sin \frac{Nz}{c_0} \simeq 1 - \frac{N^2 z}{g} \simeq 1,$$

where we have used $N|z|/c_0 \ll 1$ (with $NH/c_0 \ll 1$), and $N^2 z/g \ll 1$ (with $N^2 H/g = (NH/c_0)(c_0 N/g) \ll 1$, both sides of equation (14.70) being much less than 1). For this mode the vertical structure of u, v, and p is therefore nearly depth-independent. The corresponding structure for w (given by $\int \psi_0 \, dz$, as indicated in equation (14.53)) is linear in z, with zero at the bottom and a maximum at the upper free surface. A stratified medium therefore has a mode of motion that behaves like that in an unstratified medium; this mode does not feel the stratification. The $n = 0$ mode is called the *barotropic mode*.

The remaining modes $n \geqslant 1$ are *baroclinic*. For these modes $c_n N/g \ll 1$ but NH/c_n is not small, as can be seen in Figure 14.12, so that the baroclinic roots of equation (14.70) are nearly given by

$$\tan \frac{NH}{c_n} = 0,$$

which gives

$$c_n = \frac{NH}{n\pi}, \qquad n = 1, 2, 3, \ldots. \qquad (14.71)$$

Taking a typical depth-average oceanic value of $N \sim 10^{-3}\,\text{s}^{-1}$ and $H \sim 5\,\text{km}$, the eigenvalue for the first baroclinic mode is $c_1 \sim 2\,\text{m/s}$. The corresponding equivalent depth is $H_e = c_1^2/g \sim 0.4\,\text{m}$.

An examination of the algebraic steps leading to equation (14.70) shows that neglecting the right-hand side is equivalent to replacing the upper boundary condition (14.65') by $w = 0$ at $z = 0$. This is called the rigid lid approximation. The *baroclinic modes are negligibly distorted by the rigid lid approximation.* In contrast, the rigid lid approximation applied to the *barotropic* mode would yield $c_0 = \infty$, as equation (14.71) shows for $n = 0$. Note that the rigid lid approximation does *not* imply that the free surface displacement corresponding to the baroclinic modes is negligible in the ocean. In fact, excluding the wind waves and tides, much of the free surface displacements in the ocean are due to baroclinic motions. The rigid lid approximation merely implies that, for baroclinic motions, the vertical displacements at the surface are much smaller than those within the fluid column. A valid baroclinic solution can therefore be obtained by setting $w = 0$ at $z = 0$. Further, the rigid lid approximation does not imply that the pressure is constant at the level surface $z = 0$; if a rigid lid were actually imposed at $z = 0$, then the pressure on the lid would vary due to the baroclinic motions.

The vertical mode shape under the rigid lid approximation is given by the cosine distribution

$$\psi_n = \cos\frac{n\pi z}{H}, \qquad n = 0, 1, 2, \ldots,$$

because it satisfies $d\psi_n/dz = 0$ at $z = 0, -H$. The nth mode ψ_n has n zero crossings within the layer (Figure 14.13).

A decomposition into normal modes is only possible in the absence of topographic variations and mean currents with shear. It is valid with or without Coriolis forces and with or without the β-effect. However, the hydrostatic approximation here means that the frequencies are much smaller than N. Under this condition the eigenfunctions are independent of the frequency, as equation (14.56) shows. Without the hydrostatic approximation the eigenfunctions ψ_n become dependent on the frequency ω. This is discussed, for example, in LeBlond and Mysak (1978).

Summary: Small amplitude motion in a frictionless continuously stratified ocean can be decomposed in terms of noninteracting vertical normal modes. The vertical structure of each mode is defined by an eigenfunction $\psi_n(z)$. If the horizontal scale of the waves is much larger than the vertical scale, then the equations governing the horizontal propagation of each mode are identical to those of a shallow *homogeneous* layer, with the layer depth H replaced by an equivalent depth H_e defined by $c_n^2 = gH_e$. For a medium of constant N, the baroclinic ($n \geqslant 1$) eigenvalues are

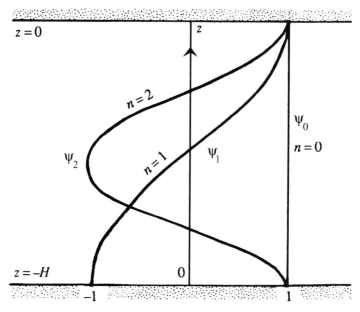

Figure 14.13   Vertical distribution of a few normal modes in a stratified medium of uniform buoyancy frequency.

given by $c_n = NH/\pi n$, while the barotropic eigenvalue is $c_0 = \sqrt{gH}$. The rigid lid approximation is quite good for the baroclinic modes.

10. High- and Low-Frequency Regimes in Shallow-Water Equations

We shall now examine what terms are negligible in the shallow-water equations for the various frequency ranges. Our analysis is valid for a single homogeneous layer or for a stratified medium. In the latter case H has to be interpreted as the *equivalent* depth, and c has to be interpreted as the speed of long nonrotating *internal* gravity waves. The β-effect will be considered in this section. As f varies only northward, horizontal isotropy is lost whenever the β-effect is included, and it becomes necessary to distinguish between the different horizontal directions. We shall follow the usual geophysical convention that the x-axis is directed eastward and the y-axis is directed northward, with u and v the corresponding velocity components.

The simplest way to perform the analysis is to examine the v-equation. A single equation for v can be derived by first taking the time derivatives of the momentum equations in (14.45) and using the continuity equation to eliminate $\partial \eta/\partial t$. This gives

$$\frac{\partial^2 u}{\partial t^2} - f\frac{\partial v}{\partial t} = gH\frac{\partial}{\partial x}\left(\frac{\partial u}{\partial x} + \frac{\partial v}{\partial y}\right), \tag{14.72}$$

$$\frac{\partial^2 v}{\partial t^2} + f\frac{\partial u}{\partial t} = gH\frac{\partial}{\partial y}\left(\frac{\partial u}{\partial x} + \frac{\partial v}{\partial y}\right). \tag{14.73}$$

Now take $\partial/\partial t$ of equation (14.73) and use equation (14.72), to obtain

$$\frac{\partial^3 v}{\partial t^3} + f\left[f\frac{\partial v}{\partial t} + gH\frac{\partial}{\partial x}\left(\frac{\partial u}{\partial x} + \frac{\partial v}{\partial y}\right)\right] = gH\frac{\partial^2}{\partial y\,\partial t}\left(\frac{\partial u}{\partial x} + \frac{\partial v}{\partial y}\right). \quad (14.74)$$

To eliminate u, we first obtain a vorticity equation by cross differentiating and subtracting the momentum equations in equation (14.45):

$$\frac{\partial}{\partial t}\left(\frac{\partial u}{\partial y} - \frac{\partial v}{\partial x}\right) - f_0\left(\frac{\partial u}{\partial x} + \frac{\partial v}{\partial y}\right) - \beta v = 0.$$

Here, we have made the customary β-plane approximation, valid if the y-scale is small enough so that $\Delta f/f \ll 1$. Accordingly, we have treated f as constant (and replaced it by an average value f_0) *except* when df/dy appears; this is why we have written f_0 in the second term of the preceding equation. Taking the x-derivative, multiplying by gH, and adding to equation (14.74), we finally obtain a vorticity equation in terms of v only:

$$\frac{\partial^3 v}{\partial t^3} - gH\frac{\partial}{\partial t}\nabla_H^2 v + f_0^2\frac{\partial v}{\partial t} - gH\beta\frac{\partial v}{\partial x} = 0, \quad (14.75)$$

where $\nabla_H^2 = \partial^2/\partial x^2 + \partial^2/\partial y^2$ is the horizontal Laplacian operator.

Equation (14.75) is Boussinesq, linear and hydrostatic, but otherwise quite general in the sense that it is applicable to both high and low frequencies. Consider wave solutions of the form

$$v = \hat{v}\, e^{i(kx+ly-\omega t)},$$

where k is the eastward wavenumber and l is the northward wavenumber. Then equation (14.75) gives

$$\omega^3 - c^2\omega K^2 - f_0^2\omega - c^2\beta k = 0, \quad (14.76)$$

where $K^2 = k^2 + l^2$ and $c = \sqrt{gH}$. It can be shown that all roots of equation (14.76) are real, two of the roots being superinertial ($\omega > f$) and the third being subinertial ($\omega \ll f$). Equation (14.76) is the complete dispersion relation for linear shallow-water equations. In various parametric ranges it takes simpler forms, representing simpler waves.

First, consider high-frequency waves $\omega \gg f$. Then the third term of equation (14.76) is negligible compared to the first term. Moreover, the fourth term is also negligible in this range. Compare, for example, the fourth and second terms:

$$\frac{c^2\beta k}{c^2\omega K^2} \sim \frac{\beta}{\omega K} \sim 10^{-3},$$

where we have assumed typical values of $\beta = 2 \times 10^{-11} \, \text{m}^{-1} \, \text{s}^{-1}$, $\omega = 3f$ $\sim 3 \times 10^{-4} \, \text{s}^{-1}$, and $2\pi/K \sim 100 \, \text{km}$. For $\omega \gg f$, therefore, the balance is between the first and second terms in equation (14.76), and the roots are $\omega = \pm K \sqrt{gH}$, which correspond to a propagation speed of $\omega/K = \sqrt{gH}$. The effects of both f and β are therefore negligible for high-frequency waves, as is expected as they are too fast to be affected by the Coriolis effects.

Next consider $\omega > f$, but $\omega \sim f$. Then the third term in equation (14.76) is not negligible, but the β-effect is. These are gravity waves influenced by Coriolis forces; gravity waves are discussed in the next section. However, the time scales are still too short for the motion to be affected by the β-effect.

Last, consider very slow waves for which $\omega \ll f$. Then the β-effect becomes important, and the first term in equation (14.76) becomes negligible. Compare, for example, the first and the last terms:

$$\frac{\omega^3}{c^2 \beta k} \ll 1.$$

Typical values for the ocean are $c \sim 200 \, \text{m/s}$ for the barotropic mode, $c \sim 2 \, \text{m/s}$ for the baroclinic mode, $\beta = 2 \times 10^{-11} \, \text{m}^{-1} \, \text{s}^{-1}$, $2\pi/k \sim 100 \, \text{km}$, and $\omega \sim 10^{-5} \, \text{s}^{-1}$. This makes the forementioned ratio about 0.2×10^{-4} for the barotropic mode and 0.2 for the baroclinic mode. The first term in equation (14.76) is therefore negligible for $\omega \ll f$.

Equation (14.75) governs the dynamics of a variety of wave motions in the ocean and the atmosphere, and the discussion in this section shows what terms can be dropped under various limiting conditions. An understanding of these limiting conditions will be useful in the following sections.

11. Gravity Waves with Rotation

In this chapter we shall examine several free-wave solutions of the shallow-water equations. In this section we shall study gravity waves with frequencies in the range $\omega > f$, for which the β-effect is negligible, as demonstrated in the preceding section. Consequently, the Coriolis frequency f is regarded as constant here. Consider progressive waves of the form

$$(u, v, \eta) = (\hat{u}, \hat{v}, \hat{\eta}) e^{i(kx + ly - \omega t)},$$

where $\hat{u}$, $\hat{v}$, and $\hat{\eta}$ are the complex amplitudes, and the real part of the right-hand side is meant. Then equation (14.45) gives

$$-i\omega \hat{u} - f\hat{v} = -ikg\hat{\eta}, \qquad (14.77)$$

$$-i\omega \hat{v} + f\hat{u} = -ilg\hat{\eta}, \qquad (14.78)$$

$$-i\omega \hat{\eta} + iH(k\hat{u} + l\hat{v}) = 0. \qquad (14.79)$$

Solving for $\hat{u}$ and $\hat{v}$ between equations (14.77) and (14.78), we obtain

$$\hat{u} = \frac{g\hat{\eta}}{\omega^2 - f^2}(\omega k + ifl),$$

$$\hat{v} = \frac{g\hat{\eta}}{\omega^2 - f^2}(-ifk + \omega l). \tag{14.80}$$

Substituting these in equation (14.79), we obtain

$$\omega^2 - f^2 = gH(k^2 + l^2). \tag{14.81}$$

This is the dispersion relation of gravity waves in the presence of Coriolis forces. (The relation can be most simply derived by setting the determinant of the set of linear homogeneous equations (14.77)–(14.79) to zero.) It can be written as

$$\omega^2 = f^2 + gHK^2, \tag{14.82}$$

where $K = \sqrt{k^2 + l^2}$ is the magnitude of the horizontal wavenumber. The dispersion relation shows that the waves can propagate in any horizontal direction and have $\omega > f$. Gravity waves affected by Coriolis forces are called *Poincaré waves, Sverdrup waves*, or simply *rotational gravity waves*. (Sometimes the name "Poincaré wave" is used to describe those rotational gravity waves that satisfy the boundary conditions in a channel.) In spite of their name, the solution was first worked out by Kelvin (Gill, 1982, p. 197). A plot of equation (14.82) is shown in Figure 14.14. It is seen that the waves are dispersive except for $\omega \gg f$ when equation (14.82) gives $\omega^2 \simeq gHK^2$, so that the propagation speed is $\omega/K = \sqrt{gH}$. The high-frequency limit agrees with our previous discussion of surface gravity waves unaffected by Coriolis forces.

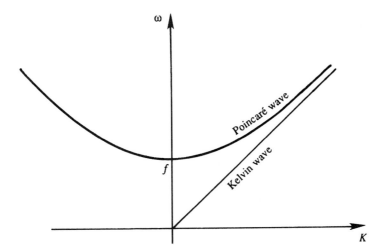

Figure 14.14 Dispersion relations for Poincaré and Kelvin waves.

Particle Orbit

The symmetry of the dispersion relation (14.81) with respect to k and l means that the x- and y-directions are not felt differently by the wavefield. The horizontal isotropy is a result of treating f as constant. (We shall see later that Rossby waves, which depend on the β-effect, are not horizontally isotropic.) We can therefore orient the x-axis along the wavenumber vector and set $l = 0$, so that the wavefield is invariant along the y-axis. To find the particle orbits, it is convenient to work with real quantities. Let the displacement be

$$\eta = \hat{\eta} \cos(kx - \omega t),$$

where $\hat{\eta}$ is real. The corresponding velocity components can be found by multiplying equation (14.80) by $\exp(ikx - i\omega t)$ and taking the real part of both sides. This gives

$$u = \frac{\omega \hat{\eta}}{kH} \cos(kx - \omega t),$$

$$v = \frac{f \hat{\eta}}{kH} \sin(kx - \omega t).$$

(14.83)

To find the particle paths, take $x = 0$ and consider three values of time corresponding to $\omega t = 0, \pi/2,$ and π. The corresponding values of u and v from equation (14.83) show that the velocity vector rotates clockwise (in the northern hemisphere) in elliptic paths (Figure 14.15). The ellipticity is expected, since the presence of Coriolis forces means that fu must generate $\partial v/\partial t$ according to the equation of motion (14.45). (In equation (14.45), $\partial \eta/\partial y = 0$ due to our orienting the x-axis along the direction of propagation of the wave.) Particles are therefore constantly deflected to the right by the Coriolis force, resulting in elliptic orbits. *The ellipses have an axis ratio of ω/f, and the major axis is oriented in the direction of wave propagation.* The ellipses become narrower as ω/f increases, approaching the rectilinear orbit of gravity waves

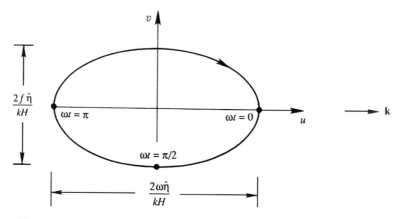

Figure 14.15 Particle orbit in a rotational gravity wave. Velocity components corresponding to $\omega t = 0$, $\pi/2$, and π are indicated.

unaffected by planetary rotation. However, the sea surface in a rotational gravity wave is no different than that for ordinary gravity waves, namely oscillatory in the direction of propagation and invariant in the perpendicular direction.

Inertial Motion

Consider the limit $\omega \rightarrow f$, that is when the particle paths are circular. The dispersion relation (14.82) then shows that $K \rightarrow 0$, implying a horizontal uniformity of the flow field. Equation (14.79) shows that $\hat{\eta}$ must tend to zero in this limit, so that there are no horizontal pressure gradients in this limit. Because $\partial u/\partial x = \partial v/\partial y = 0$, the continuity equation shows that $w = 0$. The particles therefore move on horizontal sheets, each layer decoupled from the one above and below it. The balance of forces is

$$\frac{\partial u}{\partial t} - fv = 0,$$

$$\frac{\partial v}{\partial t} + fu = 0.$$

The solution of this set is of the form

$$u = q \cos ft,$$

$$v = -q \sin ft,$$

where the speed $q = \sqrt{u^2 + v^2}$ is constant along the path. The radius r of the orbit can be found by adopting a Lagrangian point of view, and noting that the equilibrium of forces is between the Coriolis force fq and the centrifugal force $r\omega^2 = rf^2$, giving $r = q/f$. The limiting case of motion in circular orbits at a frequency f is called *inertial motion*, because in the absence of pressure gradients a particle moves by virtue of its inertia alone. The corresponding period $2\pi/f$ is called the *inertial period*. In the absence of planetary rotation such motion would be along straight lines; in the presence of Coriolis forces the motion is along circular paths, called *inertial circles*. Near-inertial motion is frequently generated in the surface layer of the ocean by sudden changes of the wind field, essentially because the equations of motion (14.45) have a natural frequency f. Taking a typical current magnitude of $q \sim 0.1 \, \text{m/s}$, the radius of the orbit is $r \sim 1 \, \text{km}$.

12. Kelvin Wave

In the preceding section we considered a shallow-water gravity wave propagating in a horizontally *unbounded* ocean. We saw that the crests are horizontal and oriented in a direction perpendicular to the direction of propagation. The *absence* of a transverse pressure gradient $\partial \eta/\partial y$ resulted in a transverse flow and elliptic orbits. This is clear from the third equation in (14.45), which shows that the presence of fu must result in $\partial v/\partial t$ if $\partial \eta/\partial y = 0$. In this section we consider a gravity wave propagating parallel to a wall, whose presence allows a pressure gradient $\partial \eta/\partial y$ that can decay away from the wall. We shall see that this allows a gravity wave in which fu is geostrophically balanced by $-g(\partial\eta/\partial y)$, and $v = 0$. Consequently the particle orbits are not elliptic but rectilinear.

Consider first a gravity wave propagating in a channel. From Figure 7.7 we know that the fluid velocity under a crest is "forward" (i.e., in the direction of propagation), and that under a trough it is backward. Figure 14.16 shows two transverse sections of the wave, one through a crest (left panel) and the other through a trough (right panel). The wave is propagating into the plane of the paper, along the x-direction. Then the fluid velocity under the crest is into the plane of the paper and that under the trough is out of the plane of the paper. The constraints of the side walls require that $v = 0$ at the walls, and we are exploring the possibility of a wave motion in which v is zero everywhere. Then the equation of motion along the y-direction requires that fu can only be geostrophically balanced by a transverse slope of the sea surface across the channel:

$$fu = -g\frac{\partial\eta}{\partial y}.$$

In the northern hemisphere, the surface must slope as indicated in the figure, that is downward to the left under the crest and upward to the left under the trough, so that the pressure force has the current directed to its right. The result is that the amplitude of the wave is larger on the right-hand side of the channel, looking into the direction of propagation, as indicated in Figure 14.16. The current amplitude, like the surface displacement, also decays to the left.

If the left wall in Figure 14.16 is moved away to infinity, we get a gravity wave trapped to the coast (Figure 14.17). A coastally trapped long gravity wave, in which the transverse velocity $v = 0$ everywhere, is called a *Kelvin wave*. It is clear that it can propagate only in a direction such that the coast is to the right (looking in the direction of propagation) in the northern hemisphere and to the left in the southern hemisphere. The opposite direction of propagation would result in a sea surface displacement increasing exponentially away from the coast, which is not possible.

An examination of the transverse momentum equation

$$\frac{\partial v}{\partial t} + fu = -g\frac{\partial\eta}{\partial y},$$

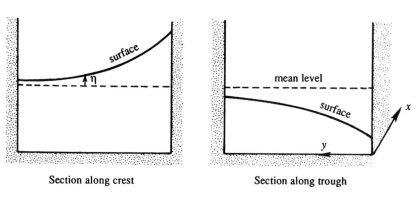

Section along crest   Section along trough

Figure 14.16   Free surface distribution in a gravity wave propagating through a channel into the plane of the paper.

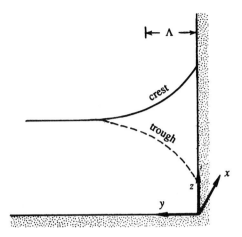

Figure 14.17 Coastal Kelvin wave propagating along the x-axis. Sea surface across a section through a crest is indicated by the continuous line, and that along a trough is indicated by the dashed line.

reveals fundamental differences between Poincaré waves and Kelvin waves. For a Poincaré wave the crests are horizontal, and the absence of a transverse pressure gradient requires a $\partial v/\partial t$ to balance the Coriolis force, resulting in elliptic orbits. In a Kelvin wave a transverse velocity is prevented by a geostrophic balance of fu and $-g(\partial \eta/\partial y)$.

From the shallow-water set (14.45), the equations of motion for a Kelvin wave propagating along a coast aligned with the x-axis (Figure 14.17) are

$$\frac{\partial \eta}{\partial t} + H\frac{\partial u}{\partial x} = 0,$$

$$\frac{\partial u}{\partial t} = -g\frac{\partial \eta}{\partial x}, \tag{14.84}$$

$$fu = -g\frac{\partial \eta}{\partial y}.$$

Assume a solution of the form

$$[u, \eta] = [\hat{u}(y), \hat{\eta}(y)]e^{i(kx-\omega t)}.$$

Then equation (14.84) gives

$$-i\omega\hat{\eta} + iHk\hat{u} = 0,$$

$$-i\omega\hat{u} = -igk\hat{\eta}, \tag{14.85}$$

$$f\hat{u} = -g\frac{d\hat{\eta}}{dy}.$$

The dispersion relation can be found solely from the first two of these equations; the third equation then determines the transverse structure. Eliminating $\hat{u}$ between the first two, we obtain

$$\hat{\eta}[\omega^2 - gHk^2] = 0.$$

A nontrivial solution is therefore possible only if $\omega = \pm k\sqrt{gH}$, so that the wave propagates with a nondispersive speed

$$\boxed{c = \sqrt{gH}.} \tag{14.86}$$

The propagation speed of a Kelvin wave is therefore identical to that of nonrotating gravity waves. Its dispersion equation is a straight line and is shown in Figure 14.14. All frequencies are possible.

To determine the transverse structure, eliminate $\hat{u}$ between the first and third of equation (14.85), giving

$$\frac{d\hat{\eta}}{dy} \pm \frac{f}{c}\hat{\eta} = 0.$$

The solution that decays away from the coast is

$$\hat{\eta} = \eta_0 \, e^{-fy/c},$$

where η_0 is the amplitude at the coast. Therefore, the sea surface slope and the velocity field for a Kelvin wave have the form

$$\eta = \eta_0 \, e^{-fy/c} \cos k(x - ct),$$
$$u = \eta_0 \sqrt{\frac{g}{H}} e^{-fy/c} \cos k(x - ct), \tag{14.87}$$

where we have taken the real parts, and have used equation (14.85) in obtaining the u field.

Equations (14.87) show that the transverse decay scale of the Kelvin wave is

$$\boxed{\Lambda \equiv \frac{c}{f},}$$

which is called the *Rossby radius of deformation*. For a deep sea of depth $H = 5\,\text{km}$, and a midlatitude value of $f = 10^{-4}\,\text{s}^{-1}$, we obtain $c = \sqrt{gH} = 220\,\text{m/s}$ and $\Lambda = c/f = 2200\,\text{km}$. Tides are frequently in the form of coastal Kelvin waves of semidiurnal frequency. The tides are forced by the periodic changes in the gravitational attraction of the moon and the sun. These waves propagate along the boundaries of an ocean basin and cause sea level fluctuations at coastal stations.

Analogous to the surface or "external" Kelvin waves discussed in the preceding, we can have *internal Kelvin waves* at the interface between two fluids of different densities (Figure 14.18). If the lower layer is very deep, then the speed of propagation is given by (see equation (7.126))

$$c = \sqrt{g'H},$$

where H is the thickness of the upper layer and $g' = g(\rho_2 - \rho_1)/\rho_2$ is the reduced gravity. For a continuously stratified medium of depth H and buoyancy frequency N, internal Kelvin waves can propagate at any of the normal mode speeds

$$c = NH/n\pi, \qquad n = 1, 2, \ldots.$$

The decay scale for *internal* Kelvin waves, $\Lambda = c/f$, is called the *internal Rossby radius of deformation*, whose value is much smaller than that for the external Rossby radius of deformation. For $n = 1$, a typical value in the ocean is $\Lambda = NH/\pi f \sim 50$ km; a typical atmospheric value is much larger, being of order $\Lambda \sim 1000$ km.

Internal Kelvin waves in the ocean are frequently forced by wind changes near coastal areas. For example, a southward wind along the west coast of a continent in the northern hemisphere (say, California) generates an Ekman layer at the ocean surface, in which the mass flow is *away* from the coast (to the right of the applied wind stress). The mass flux in the near-surface layer is compensated by the movement of deeper water toward the coast, which raises the thermocline. An upward movement of the thermocline, as indicated by the dashed line in Figure 14.18, is called *upwelling*. The vertical movement of the thermocline in the wind-forced region then propagates poleward along the coast as an internal Kelvin wave.

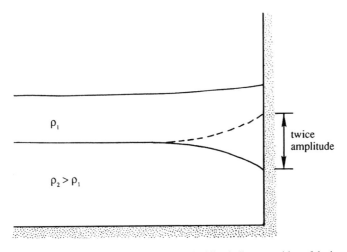

Figure 14.18 Internal Kelvin wave at an interface. Dashed line indicates position of the interface when it is at its maximum height. Displacement of the free surface is much smaller than that of the interface and is oppositely directed.

13. Potential Vorticity Conservation in Shallow-Water Theory

In this section we shall derive a useful conservation law for the vorticity of a shallow layer of fluid. From Section 8, the equations of motion for a shallow layer of homogeneous fluid are

$$\frac{\partial u}{\partial t} + u\frac{\partial u}{\partial x} + v\frac{\partial u}{\partial y} - fv = -g\frac{\partial \eta}{\partial x}, \tag{14.88}$$

$$\frac{\partial v}{\partial t} + u\frac{\partial v}{\partial x} + v\frac{\partial v}{\partial y} + fu = -g\frac{\partial \eta}{\partial y}, \tag{14.89}$$

$$\frac{\partial h}{\partial t} + \frac{\partial}{\partial x}(uh) + \frac{\partial}{\partial y}(vh) = 0, \tag{14.90}$$

where $h(x, y, t)$ is the depth of flow and η is the height of the sea surface measured from an arbitrary horizontal plane (Figure 14.19). The x-axis is taken eastward and the y-axis is taken northward, with u and v the corresponding velocity components. The Coriolis frequency $f = f_0 + \beta y$ is regarded as dependent on latitude. The nonlinear terms have been retained, including those in the continuity equation, which has been written in the form (14.44); note that $h = H + \eta$. We saw in Section 8 that the constant density of the layer and the hydrostatic pressure distribution make the horizontal pressure gradient depth-independent, so that only a depth-independent current can be generated. The vertical velocity is linear in z.

A vorticity equation can be derived by differentiating equation (14.88) with respect to y, equation (14.89) with respect to x, and subtracting. The pressure is eliminated, and we obtain

$$\frac{\partial}{\partial t}\left(\frac{\partial v}{\partial x} - \frac{\partial u}{\partial y}\right) + \frac{\partial}{\partial x}\left[u\frac{\partial v}{\partial x} + v\frac{\partial v}{\partial y}\right] - \frac{\partial}{\partial y}\left[u\frac{\partial u}{\partial x} + v\frac{\partial u}{\partial y}\right]$$

$$+ f_0\left(\frac{\partial u}{\partial x} + \frac{\partial v}{\partial y}\right) + \beta v = 0. \tag{14.91}$$

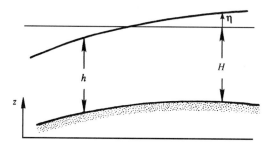

Figure 14.19 Shallow layer of instantaneous depth $h(x, y, t)$.

Following the customary β-plane approximation, we have treated f as constant (and replaced it by an average value f_0) *except* when df/dy appears. We now introduce

$$\zeta \equiv \frac{\partial v}{\partial x} - \frac{\partial u}{\partial y},$$

as the vertical component of *relative vorticity*, that is, the vorticity measured relative to the rotating earth. Then the nonlinear terms in equation (14.91) can easily be rearranged in the form

$$u\frac{\partial \zeta}{\partial x} + v\frac{\partial \zeta}{\partial y} + \left(\frac{\partial u}{\partial x} + \frac{\partial v}{\partial y}\right)\zeta.$$

Equation (14.91) then becomes

$$\frac{\partial \zeta}{\partial t} + u\frac{\partial \zeta}{\partial x} + v\frac{\partial \zeta}{\partial y} + \left(\frac{\partial u}{\partial x} + \frac{\partial v}{\partial y}\right)(\zeta + f_0) + \beta v = 0,$$

which can be written as

$$\frac{D\zeta}{Dt} + (\zeta + f_0)\left(\frac{\partial u}{\partial x} + \frac{\partial v}{\partial y}\right) + \beta v = 0, \qquad (14.92)$$

where D/Dt is the derivative following the horizontal motion of the layer:

$$\frac{D}{Dt} \equiv \frac{\partial}{\partial t} + u\frac{\partial}{\partial x} + v\frac{\partial}{\partial y}.$$

The horizontal divergence $(\partial u/\partial x + \partial v/\partial y)$ in equation (14.92) can be eliminated by using the continuity equation (14.90), which can be written as

$$\frac{Dh}{Dt} + h\left(\frac{\partial u}{\partial x} + \frac{\partial v}{\partial y}\right) = 0.$$

Equation (14.92) then becomes

$$\frac{D\zeta}{Dt} = \frac{\zeta + f_0}{h}\frac{Dh}{Dt} - \beta v.$$

This can be written as

$$\frac{D(\zeta + f)}{Dt} = \frac{\zeta + f_0}{h}\frac{Dh}{Dt}, \qquad (14.93)$$

where we have used

$$\frac{Df}{Dt} = \frac{\partial f}{\partial t} + u\frac{\partial f}{\partial x} + v\frac{\partial f}{\partial y} = v\beta.$$

Because of the absence of vertical shear, the vorticity in a shallow-water model is purely vertical and independent of depth. The relative vorticity measured with respect to the rotating earth is ζ, while f is the planetary vorticity, so that the *absolute vorticity* is $(\zeta + f)$. Equation (14.93) shows that the rate of change of absolute vorticity is proportional to the absolute vorticity times the vertical stretching Dh/Dt of the water column. It is apparent that $D\zeta/Dt$ can be nonzero even if $\zeta = 0$ initially. This is different from a nonrotating flow in which stretching a fluid line changes its vorticity only if the line has an *initial* vorticity. (This is why the process was called the *vortex stretching*; see Chapter 5, Section 7.) The difference arises because vertical lines in a rotating earth contain the planetary vorticity even when $\zeta = 0$. Note that the vortex *tilting* term, discussed in Chapter 5, Section 7, is absent in the shallow-water theory because the water moves in the form of vertical columns without ever tilting.

Equation (14.93) can be written in the compact form

$$\frac{D}{Dt}\left(\frac{\zeta + f}{h}\right) = 0, \tag{14.94}$$

where $f = f_0 + \beta y$, and we have assumed $\beta y \ll f_0$. The ratio $(\zeta + f)/h$ is called the *potential vorticity* in shallow-water theory. Equation (14.94) shows that the *potential vorticity is conserved along the motion*, an important principle in geophysical fluid dynamics. In the ocean, outside regions of strong current vorticity such as coastal boundaries, the magnitude of ζ is much smaller than that of f. In such a case $(\zeta + f)$ has the sign of f. The principle of conservation of potential vorticity means that an increase in h must make $(\zeta + f)$ more positive in the northern hemisphere and more negative in the southern hemisphere.

As an example of application of the potential vorticity equation, consider an eastward flow over a step (at $x = 0$) running north–south, across which the layer thickness changes discontinuously from h_0 to h_1 (Figure 14.20). The flow upstream of the step has a uniform speed U, so that the oncoming stream has no relative vorticity. To conserve the ratio $(\zeta + f)/h$, the flow must suddenly acquire negative (clockwise) relative vorticity due to the sudden decrease in layer thickness. The relative vorticity of a fluid element just after passing the step can be found from

$$\frac{f}{h_0} = \frac{\zeta + f}{h_1},$$

giving $\zeta = f(h_1 - h_0)/h_0 < 0$, where f is evaluated at the upstream latitude of the streamline. Because of the clockwise vorticity, the fluid starts to move south at $x = 0$. The southward movement decreases f, so that ζ must correspondingly increase so as to keep $(f + \zeta)$ constant. This means that the clockwise curvature of the stream reduces, and eventually becomes a counterclockwise curvature. In this manner an eastward flow over a step generates stationary undulatory flow on the downstream side. In Section 15 we shall see that the stationary oscillation is due to a Rossby wave generated at the step whose westward phase velocity is canceled by the eastward current. We shall see that the wavelength is $2\pi\sqrt{U/\beta}$.

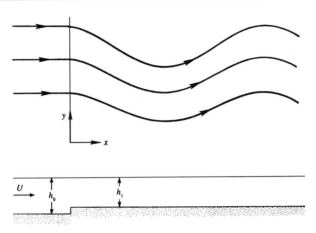

Figure 14.20   Eastward flow over a step, resulting in stationary oscillations of wavelength $2\pi\sqrt{U/\beta}$.

Suppose we try the same argument for a *westward* flow over a step. Then a particle should suddenly acquire clockwise vorticity as the depth of flow decreases at $x = 0$, which would require the particle to move north. It would then come into a region of larger f, which would require ζ to decrease further. Clearly, an exponential behavior is predicted, suggesting that the argument is not correct. Unlike an eastward flow, a westward current feels the *upstream* influence of the step so that it acquires a counterclockwise curvature *before* it encounters the step (Figure 14.21). The positive vorticity is balanced by a reduction in f, which is consistent with conservation of potential vorticity. At the location of the step the vorticity decreases suddenly. Finally, far downstream of the step a fluid particle is again moving westward at its original latitude. The westward flow over a topography is *not* oscillatory.

14.   Internal Waves

In Chapter 7, Section 19 we studied internal gravity waves unaffected by Coriolis forces. We saw that they are not isotropic; in fact the direction of propagation with respect to the vertical determines their frequency. We also saw that their frequency satisfies the inequality $\omega \leqslant N$, where N is the buoyancy frequency. Their phase-velocity vector $\mathbf{c}$ and the group-velocity vector $\mathbf{c}_g$ are perpendicular and have oppositely directed vertical components (Figure 7.32 and Figure 7.34). That is, phases propagate upward if the groups propagate downward, and vice versa. In this section we shall study the effect of Coriolis forces on internal waves, assuming that f is independent of latitude.

Internal waves are ubiquitous in the atmosphere and the ocean. In the lower atmosphere turbulent motions dominate, so that internal wave activity represents a minor component of the motion. In contrast, the stratosphere contains very little convective motion because of its stable density distribution, and consequently a great deal of internal wave activity. They generally propagate upward from the lower atmosphere, where they are generated. In the ocean they may be as common as the waves on

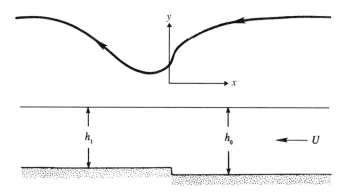

Figure 14.21 Westward flow over a step. Unlike the eastward flow, the westward flow is not oscillatory and feels the upstream influence of the step.

the surface, and measurements show that they can cause the isotherms to go up and down by as much as 50–100 m. Sometimes the internal waves break and generate small-scale turbulence, similar to the "foam" generated by breaking waves.

We shall now examine the nature of the fluid motion in internal waves. The equations of motion are

$$\frac{\partial u}{\partial x} + \frac{\partial v}{\partial y} + \frac{\partial w}{\partial z} = 0,$$

$$\frac{\partial u}{\partial t} - fv = -\frac{1}{\rho_0}\frac{\partial p}{\partial x},$$

$$\frac{\partial v}{\partial t} + fu = -\frac{1}{\rho_0}\frac{\partial p}{\partial y}, \tag{14.95}$$

$$\frac{\partial w}{\partial t} = -\frac{1}{\rho_0}\frac{\partial p}{\partial z} - \frac{\rho g}{\rho_0},$$

$$\frac{\partial \rho}{\partial t} - \frac{\rho_0 N^2}{g} w = 0.$$

We have not made the hydrostatic assumption because we are *not* assuming that the horizontal wavelength is long compared to the vertical wavelength. The advective term in the density equation is written in a linearized form $w(d\bar{\rho}/dz) = -\rho_0 N^2 w/g$. Thus the rate of change of density at a point is assumed to be due only to the vertical advection of the background density distribution $\bar{\rho}(z)$. Because internal wave activity is more intense in the thermocline where N varies appreciably (Figure 14.2), we shall be somewhat more general than in Chapter 7 and not assume that N is depth-independent.

An equation for w can be formed from the set (14.95) by eliminating all other variables. The algebraic steps of such a procedure are shown in Chapter 7, Section 18 without the Coriolis forces. This gives

$$\frac{\partial^2}{\partial t^2}\nabla^2 w + N^2 \nabla_H^2 w + f^2 \frac{\partial^2 w}{\partial z^2} = 0, \tag{14.96}$$

where

$$\nabla^2 \equiv \frac{\partial^2}{\partial x^2} + \frac{\partial^2}{\partial y^2} + \frac{\partial^2}{\partial z^2}$$

and

$$\nabla_H^2 \equiv \frac{\partial^2}{\partial x^2} + \frac{\partial^2}{\partial y^2}.$$

Because the coefficients of equation (14.96) are independent of the horizontal directions, equation (14.96) can have solutions that are trigonometric in x and y. We therefore assume a solution of the form

$$[u, v, w] = [\hat{u}(z), \hat{v}(z), \hat{w}(z)] \, e^{i(kx+ly-\omega t)}. \tag{14.97}$$

Substitution into equation (14.96) gives

$$(-i\omega)^2 \left[(ik)^2 + (il)^2 + \frac{d^2}{dz^2} \right] \hat{w} + N^2 [(ik)^2 + (il)^2]\hat{w} + f^2 \frac{d^2 \hat{w}}{dz^2} = 0,$$

from which we obtain

$$\frac{d^2 \hat{w}}{dz^2} + \frac{(N^2 - \omega^2)(k^2 + l^2)}{\omega^2 - f^2} \hat{w} = 0. \tag{14.98}$$

Defining

$$m^2(z) \equiv \frac{(k^2 + l^2)[N^2(z) - \omega^2]}{\omega^2 - f^2}, \tag{14.99}$$

Equation (14.98) becomes

$$\frac{d^2 \hat{w}}{dz^2} + m^2 \hat{w} = 0. \tag{14.100}$$

For $m^2 < 0$, the solutions of equation (14.100) are exponential in z signifying that the resulting motion is surface-trapped. It represents a surface wave propagating horizontally. For a positive m^2, on the other hand, solutions are trigonometric in z, giving internal waves propagating vertically as well as horizontally. From equation (14.99), therefore, internal waves are possible only in the frequency range:

$$\boxed{f < \omega < N,}$$

where we have assumed $N > f$, as is true for much of the atmosphere and the ocean.

WKB Solution

To proceed further, we assume that $N(z)$ is a slowly varying function in that its fractional change over a vertical wavelength is much less than unity. We are therefore considering only those internal waves whose vertical wavelength is short compared to the scale of variation of N. If H is a characteristic vertical distance over which N varies appreciably, then we are assuming that

$$Hm \gg 1.$$

For such slowly varying $N(z)$, we expect that $m(z)$ given by equation (14.99) is also a slowly varying function, that is, $m(z)$ changes by a small fraction in a distance $1/m$. Under this assumption the waves *locally* behave like plane waves, as if m is constant. This is the so-called *WKB approximation* (after Wentzel–Kramers–Brillouin), which applies when the properties of the medium (in this case N) are slowly varying.

To derive the approximate WKB solution of equation (14.100), we look for a solution in the form

$$\hat{w} = A(z)e^{i\phi(z)},$$

where the phase ϕ and the (slowly varying) amplitude A are real. (No generality is lost by assuming A to be real. Suppose it is complex and of the form $A = \bar{A} \exp(i\alpha)$, where $\bar{A}$ and α are real. Then $\hat{w} = \bar{A} \exp[i(\phi + \alpha)]$, a form in which $(\phi + \alpha)$ is the phase.) Substitution into equation (14.100) gives

$$\frac{d^2 A}{dz^2} + A\left[m^2 - \left(\frac{d\phi}{dz}\right)^2\right] + i2\frac{dA}{dz}\frac{d\phi}{dz} + iA\frac{d^2\phi}{dz^2} = 0.$$

Equating the real and imaginary parts, we obtain

$$\frac{d^2 A}{dz^2} + A\left[m^2 - \left(\frac{d\phi}{dz}\right)^2\right] = 0, \tag{14.101}$$

$$2\frac{dA}{dz}\frac{d\phi}{dz} + A\frac{d^2\phi}{dz^2} = 0. \tag{14.102}$$

In equation (14.101) the term $d^2 A/dz^2$ is negligible because its ratio with the second term is

$$\frac{d^2 A/dz^2}{Am^2} \sim \frac{1}{H^2 m^2} \ll 1.$$

Equation (14.101) then becomes approximately

$$\frac{d\phi}{dz} = \pm m, \tag{14.103}$$

whose solution is

$$\phi = \pm \int^z m \, dz,$$

the lower limit of the integral being arbitrary.

The amplitude is determined by writing equation (14.102) in the form

$$\frac{dA}{A} = -\frac{(d^2\phi/dz^2)\,dz}{2(d\phi/dz)} = -\frac{(dm/dz)\,dz}{2m} = -\frac{1}{2}\frac{dm}{m},$$

where equation (14.103) has been used. Integrating, we obtain $\ln A = -\frac{1}{2}\ln m + $ const., that is,

$$A = \frac{A_0}{\sqrt{m}},$$

where A_0 is a constant. The WKB solution of equation (14.100) is therefore

$$\hat{w} = \frac{A_0}{\sqrt{m}} e^{\pm i \int^z m\,dz}. \tag{14.104}$$

Because of neglect of the β-effect, the waves must behave similarly in x and y, as indicated by the symmetry of the dispersion relation (14.99) in k and l. Therefore, we lose no generality by orienting the x-axis in the direction of propagation, and taking

$$k > 0 \quad l = 0 \quad \omega > 0.$$

To find u and v in terms of w, use the continuity equation $\partial u/\partial x + \partial w/\partial z = 0$, noting that the y-derivatives are zero because of our setting $l = 0$. Substituting the wave solution (14.97) into the continuity equation gives

$$ik\hat{u} + \frac{d\hat{w}}{dz} = 0. \tag{14.105}$$

The z-derivative of $\hat{w}$ in equation (14.104) can be obtained by treating the denominator $\sqrt{m}$ as approximately constant because the variation of $\hat{w}$ is dominated by the wiggly behavior of the local plane wave solution. This gives

$$\frac{d\hat{w}}{dz} = \frac{A_0}{\sqrt{m}}(\pm im)e^{\pm i\int^z m\,dz} = \pm i A_0 \sqrt{m}\, e^{\pm i \int^z m\,dz},$$

so that equation (14.105) becomes

$$\hat{u} = \mp \frac{A_0 \sqrt{m}}{k} e^{\pm i \int^z m\,dz}. \tag{14.106}$$

An expression for $\hat{v}$ can now be obtained from the horizontal equations of motion in equation (14.95). Cross differentiating, we obtain the vorticity equation

$$\frac{\partial}{\partial t}\left(\frac{\partial u}{\partial y} - \frac{\partial v}{\partial x}\right) = f\left(\frac{\partial u}{\partial x} + \frac{\partial v}{\partial y}\right).$$

Using the wave solution equation (14.97), this gives

$$\frac{\hat{u}}{\hat{v}} = \frac{i\omega}{f}.$$

Equation (14.106) then gives

$$\hat{v} = \pm \frac{if}{\omega} \frac{A_0\sqrt{m}}{k} e^{\pm i \int^z m\, dz}. \qquad (14.107)$$

Taking real parts of equations (14.104), (14.106), and (14.107), we obtain the velocity field

$$u = \mp \frac{A_0\sqrt{m}}{k} \cos\left(kx \pm \int^z m\, dz - \omega t\right),$$

$$v = \mp \frac{A_0 f \sqrt{m}}{\omega k} \sin\left(kx \pm \int^z m\, dz - \omega t\right), \qquad (14.108)$$

$$w = \frac{A_0}{\sqrt{m}} \cos\left(kx \pm \int^z m\, dz - \omega t\right),$$

where the dispersion relation is

$$\boxed{m^2 = \frac{k^2(N^2 - \omega^2)}{\omega^2 - f^2}.} \qquad (14.109)$$

The meaning of $m(z)$ is clear from equation (14.108). If we call the argument of the trigonometric terms the "phase," then it is apparent that $\partial(\text{phase})/\partial z = m(z)$, so that $m(z)$ is the *local* vertical wavenumber. Because we are treating $k, m, \omega > 0$, it is also apparent that the *upper signs represent waves with upward phase propagation, and the lower signs represent downward phase propagation.*

Particle Orbit

To find the shape of the hodograph in the horizontal plane, consider the point $x = z = 0$. Then equation (14.108) gives

$$u = \mp \cos \omega t,$$

$$v = \pm \frac{f}{\omega} \sin \omega t, \qquad (14.110)$$

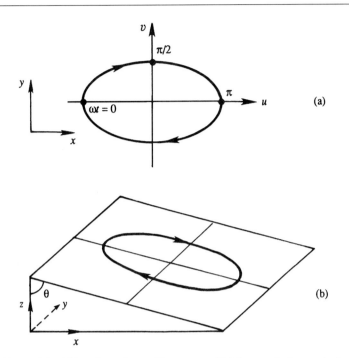

Figure 14.22 Particle orbit in an internal wave. The upper panel (a) shows projection on a horizontal plane; points corresponding to $\omega t = 0, \pi/2$, and π are indicated. The lower panel (b) shows a three-dimensional view. Sense of rotation shown is valid for the northern hemisphere.

where the amplitude of u has been arbitrarily set to one. Taking the upper signs in equation (14.110), the values of u and v are indicated in Figure 14.22a for three values of time corresponding to $\omega t = 0, \pi/2$, and π. It is clear that the horizontal hodographs are clockwise ellipses, with the major axis in the direction of propagation x, and the axis ratio is f/ω. The same conclusion applies for the lower signs in equation (14.110). The particle orbits in the horizontal plane are therefore identical to those of Poincaré waves (Figure 14.15).

However, the plane of the motion is no longer horizontal. From the velocity components equation (14.108), we note that

$$\frac{u}{w} = \mp \frac{m}{k} = \mp \tan \theta, \qquad (14.111)$$

where $\theta = \tan^{-1}(m/k)$ is the angle made by the wavenumber vector $\mathbf{K}$ with the horizontal (Figure 14.23). For upward phase propagation, equation (14.111) gives $u/w = -\tan \theta$, so that w is negative if u is positive, as indicated in Figure 14.23. A three-dimensional sketch of the particle orbit is shown in Figure 14.22b. It is easy to show (Exercise 6) that the phase velocity vector $\mathbf{c}$ is in the direction of $\mathbf{K}$, that $\mathbf{c}$ and $\mathbf{c}_g$ are perpendicular, and that the fluid motion $\mathbf{u}$ is parallel to $\mathbf{c}_g$; these facts are demonstrated in Chapter 7 for internal waves unaffected by Coriolis forces.

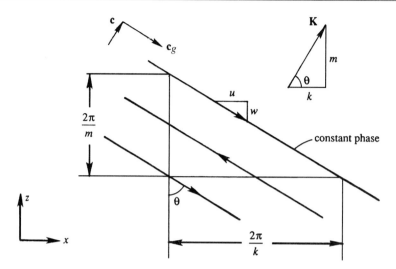

Figure 14.23 Vertical section of an internal wave. The three parallel lines are constant phase lines, with the arrows indicating fluid motion along the lines.

The velocity vector at any location rotates clockwise with time. Because of the vertical propagation of phase, the tips of the *instantaneous* vectors also turn with *depth*. Consider the turning of the velocity vectors with depth when the phase velocity is upward, so that the deeper currents have a phase lead over the shallower currents (Figure 14.24). Because the currents at all depths rotate clockwise in *time* (whether the vertical component of **c** is upward or downward), it follows that the tips of the instantaneous velocity vectors should fall on a helical spiral that turns clockwise with *depth*. Only such a turning in depth, coupled with a clockwise rotation of the velocity vectors with time, can result in a phase lead of the deeper currents. In the opposite case of a *downward* phase propagation, the helix turns *counterclockwise* with depth. The direction of turning of the velocity vectors can also be found from equation (14.108), by considering $x = t = 0$ and finding u and v at various values of z.

Discussion of the Dispersion Relation

The dispersion relation (14.109) can be written as

$$\omega^2 - f^2 = \frac{k^2}{m^2}(N^2 - \omega^2). \tag{14.112}$$

Introducing $\tan\theta = m/k$, equation (14.112) becomes

$$\omega^2 = f^2 \sin^2\theta + N^2 \cos^2\theta,$$

which shows that ω is a function of the angle made by the wavenumber with the horizontal and is not a function of the magnitude of **K**. For $f = 0$ the forementioned expression reduces to $\omega = N \cos\theta$, derived in Chapter 7, Section 19 without Coriolis forces.

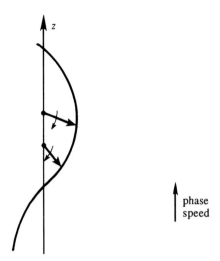

Figure 14.24 Helical spiral traced out by the tips of instantaneous velocity vectors in an internal wave with upward phase speed. Heavy arrows show the velocity vectors at two depths, and light arrows indicate that they are rotating clockwise with *time*. Note that the instantaneous vectors turn clockwise with *depth*.

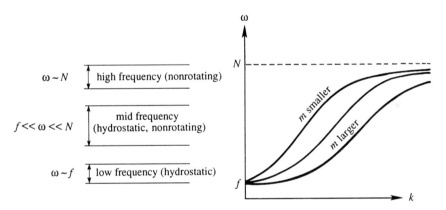

Figure 14.25 Dispersion relation for internal waves. The different regimes are indicated on the left-hand side of the figure.

A plot of the dispersion relation (14.112) is presented in Figure 14.25, showing ω as a function of k for various values of m. All curves pass through the point $\omega = f$, which represents inertial oscillations. Typically, $N \gg f$ in most of the atmosphere and the ocean. Because of the wide separation of the upper and lower limits of the internal wave range $f \leqslant \omega \leqslant N$, various limiting cases are possible, as indicated in Figure 14.25. They are

(1) *High-frequency regime* ($\omega \sim N$, *but* $\omega \leqslant N$): In this range f^2 is negligible in comparison with ω^2 in the denominator of the dispersion relation (14.109), which reduces to

$$m^2 \simeq \frac{k^2(N^2 - \omega^2)}{\omega^2}, \quad \text{that is,} \quad \omega^2 \simeq \frac{N^2 k^2}{m^2 + k^2}.$$

Using $\tan\theta = m/k$, this gives $\omega = N\cos\theta$. Thus, the high-frequency internal waves are the same as the nonrotating internal waves discussed in Chapter 7.

(2) *Low-frequency regime* ($\omega \sim f$, but $\omega \geqslant f$): In this range ω^2 can be neglected in comparison to N^2 in the dispersion relation (14.109), which becomes

$$m^2 \simeq \frac{k^2 N^2}{\omega^2 - f^2}, \quad \text{that is,} \quad \omega^2 \simeq f^2 + \frac{k^2 N^2}{m^2}.$$

The low-frequency limit is obtained by making the hydrostatic assumption, that is, neglecting $\partial w/\partial t$ in the vertical equation of motion.

(3) *Midfrequency regime* ($f \ll \omega \ll N$): In this range the dispersion relation (14.109) simplifies to

$$m^2 \simeq \frac{k^2 N^2}{\omega^2},$$

so that *both* the hydrostatic and the nonrotating assumptions are applicable.

Lee Wave

Internal waves are frequently found in the "lee" (that is, the downstream side) of mountains. In stably stratified conditions, the flow of air over a mountain causes a vertical displacement of fluid particles, which sets up internal waves as it moves downstream of the mountain. If the amplitude is large and the air is moist, the upward motion causes condensation and cloud formation.

Due to the effect of a mean flow, the lee waves are stationary with respect to the ground. This is shown in Figure 14.26, where the westward phase speed is canceled by the eastward mean flow. We shall determine what wave parameters make this cancellation possible. The frequency of lee waves is much larger than f, so that rotational effects are negligible. The dispersion relation is therefore

$$\omega^2 = \frac{N^2 k^2}{m^2 + k^2}. \tag{14.113}$$

However, we now have to introduce the effects of the mean flow. The dispersion relation (14.113) is still valid if ω is interpreted as the *intrinsic frequency*, that is, the frequency measured in a frame of reference moving with the mean flow. In a medium moving with a velocity $\mathbf{U}$, the *observed frequency* of waves at a fixed point is Doppler shifted to

$$\omega_0 = \omega + \mathbf{K} \cdot \mathbf{U},$$

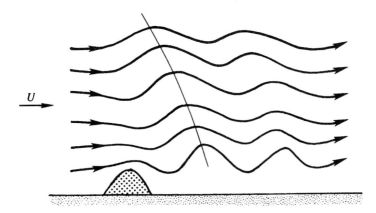

Figure 14.26 Streamlines in a lee wave. The thin line drawn through crests shows that the phase propagates downward and westward.

where ω is the intrinsic frequency; this is discussed further in Chapter 7, Section 3. For a stationary wave $\omega_0 = 0$, which requires that the intrinsic frequency is $\omega = -\mathbf{K} \cdot \mathbf{U} = kU$. (Here $-\mathbf{K} \cdot \mathbf{U}$ is positive because $\mathbf{K}$ is westward and $\mathbf{U}$ is eastward.) The dispersion relation (14.113) then gives

$$U = \frac{N}{\sqrt{k^2 + m^2}}.$$

If the flow speed U is given, and the mountain introduces a typical horizontal wavenumber k, then the preceding equation determines the vertical wavenumber m that generates stationary waves. Waves that do not satisfy this condition would radiate away.

The energy source of lee waves is at the surface. The energy therefore must propagate upward, and consequently the phases propagate downward. The intrinsic phase speed is therefore westward and downward in Figure 14.26. With this information, we can determine which way the constant phase lines should tilt in a stationary lee wave. Note that the wave pattern in Figure 14.26 would propagate to the left in the absence of a mean velocity, and only with the constant phase lines tilting backwards with height would the flow at larger height lead the flow at a lower height.

Further discussion of internal waves can be found in Phillips (1977) and Munk (1981); lee waves are discussed in Holton (1979).

15. Rossby Wave

To this point we have discussed wave motions that are possible with a constant Coriolis frequency f and found that these waves have frequencies larger than f. We shall now consider wave motions that owe their existence to the variation of f with latitude. With such a variable f, the equations of motion allow a very important type of wave motion called the *Rossby wave*. Their spatial scales are so large in the atmosphere that they usually have only a few wavelengths around the entire globe (Figure 14.27). This

Figure 14.27   Observed height (in decameters) of the 50 kPa pressure surface in the northern hemisphere. The center of the picture represents the north pole. The undulations are due to Rossby waves (dm = km/100). J. T. Houghton, *The Physics of the Atmosphere*, 1986 and reprinted with the permission of Cambridge University Press.

is why Rossby waves are also called *planetary waves*. In the ocean, however, their wavelengths are only about 100 km. Rossby-wave frequencies obey the inequality $\omega \ll f$. Because of this slowness the time derivative terms are an order of magnitude smaller than the Coriolis forces and the pressure gradients in the horizontal equations of motion. Such *nearly* geostrophic flows are called *quasi-geostrophic motions*.

Quasi-Geostrophic Vorticity Equation

We shall first derive the governing equation for quasi-geostrophic motions. For simplicity, we shall make the customary β-plane approximation valid for $\beta y \ll f_0$, keeping in mind that the approximation is not a good one for atmospheric Rossby waves, which have planetary scales. Although Rossby waves are frequently superposed on a mean flow, we shall derive the equations without a mean flow, and superpose a uniform mean flow at the end, assuming that the perturbations are small and that a linear superposition is valid. The first step is to simplify the vorticity equation for quasi-geostrophic motions, assuming that the *velocity is geostrophic to the lowest order*. The small departures from geostrophy, however, are important because they determine the *evolution* of the flow with time.

We start with the shallow-water potential vorticity equation

$$\frac{D}{Dt}\left(\frac{\zeta + f}{h}\right) = 0,$$

which can be written as

$$h\frac{D}{Dt}(\zeta + f) - (\zeta + f)\frac{Dh}{Dt} = 0.$$

We now expand the material derivative and substitute $h = H + \eta$, where H is the uniform undisturbed depth of the layer, and η is the surface displacement. This gives

$$(H + \eta)\left(\frac{\partial \zeta}{\partial t} + u\frac{\partial \zeta}{\partial x} + v\frac{\partial \zeta}{\partial y} + \beta v\right) - (\zeta + f_0)\left(\frac{\partial \eta}{\partial t} + u\frac{\partial \eta}{\partial x} + v\frac{\partial \eta}{\partial y}\right) = 0. \tag{14.114}$$

Here, we have used $Df/Dt = v(df/dy) = \beta v$. We have also replaced f by f_0 in the second term because the β-plane approximation neglects the variation of f *except* when it involves df/dy. For small perturbations we can neglect the quadratic nonlinear terms in equation (14.114), obtaining

$$H\frac{\partial \zeta}{\partial t} + H\beta v - f_0\frac{\partial \eta}{\partial t} = 0. \tag{14.115}$$

This is the linearized form of the potential vorticity equation. Its quasi-geostrophic version is obtained if we substitute the approximate geostrophic expressions for velocity:

$$u \simeq -\frac{g}{f_0}\frac{\partial \eta}{\partial y},$$
$$v \simeq \frac{g}{f_0}\frac{\partial \eta}{\partial x}. \tag{14.116}$$

From this the vorticity is found as

$$\zeta = \frac{g}{f_0}\left(\frac{\partial^2 \eta}{\partial x^2} + \frac{\partial^2 \eta}{\partial y^2}\right),$$

so that the vorticity equation (14.115) becomes

$$\frac{gH}{f_0}\frac{\partial}{\partial t}\left(\frac{\partial^2 \eta}{\partial x^2} + \frac{\partial^2 \eta}{\partial y^2}\right) + \frac{gH\beta}{f_0}\frac{\partial \eta}{\partial x} - f_0\frac{\partial \eta}{\partial t} = 0.$$

Denoting $c = \sqrt{gH}$, this becomes

$$\frac{\partial}{\partial t}\left(\frac{\partial^2 \eta}{\partial x^2} + \frac{\partial^2 \eta}{\partial y^2} - \frac{f_0^2}{c^2}\eta\right) + \beta\frac{\partial \eta}{\partial x} = 0. \tag{14.117}$$

This is the quasi-geostrophic form of the linearized vorticity equation, which governs the flow of large-scale motions. The ratio c/f_0 is recognized as the Rossby radius. Note that we have not set $\partial\eta/\partial t = 0$, in equation (14.115) during the derivation of equation (14.117), although a strict validity of the geostrophic relations (14.116) would require that the horizontal divergence, and hence $\partial\eta/\partial t$, be zero. This is because the *departure* from strict geostrophy determines the evolution of the flow described by equation (14.117). We can therefore use the geostrophic relations for velocity everywhere except in the horizontal divergence term in the vorticity equation.

Dispersion Relation

Assume solutions of the form

$$\eta = \hat{\eta}\, e^{i(kx+ly-\omega t)}.$$

We shall regard ω as positive; the signs of k and l then determine the direction of phase propagation. A substitution into the vorticity equation (14.117) gives

$$\boxed{\omega = -\frac{\beta k}{k^2 + l^2 + f_0^2/c^2}.} \tag{14.118}$$

This is the dispersion relation for *Rossby waves*. The asymmetry of the dispersion relation with respect to k and l signifies that the wave motion is not isotropic in the horizontal, which is expected because of the β-effect. Although we have derived it for a single homogeneous layer, it is equally applicable to stratified flows if c is replaced by the corresponding *internal* value, which is $c = \sqrt{g'H}$ for the reduced gravity model (see Chapter 7, Section 17) and $c = NH/n\pi$ for the nth mode of a continuously stratified model. For the barotropic mode c is very large, and f_0^2/c^2 is usually negligible in the denominator of equation (14.118).

The dispersion relation $\omega(k, l)$ in equation (14.118) can be displayed as a surface, taking k and l along the horizontal axes and ω along the vertical axis. The section of this surface along $l = 0$ is indicated in the upper panel of Figure 14.28, and sections of the surface for three values of ω are indicated in the bottom panel. The contours of constant ω are circles because the dispersion relation (14.118) can be written as

$$\left(k + \frac{\beta}{2\omega}\right)^2 + l^2 = \left(\frac{\beta}{2\omega}\right)^2 - \frac{f_0^2}{c^2}.$$

The definition of group velocity

$$\mathbf{c}_g = \mathbf{i}\frac{\partial\omega}{\partial k} + \mathbf{j}\frac{\partial\omega}{\partial l},$$

shows that the group velocity vector is the gradient of ω in the wavenumber space. The direction of $\mathbf{c}_g$ is therefore perpendicular to the ω contours, as indicated in the lower

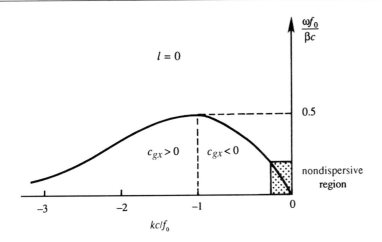

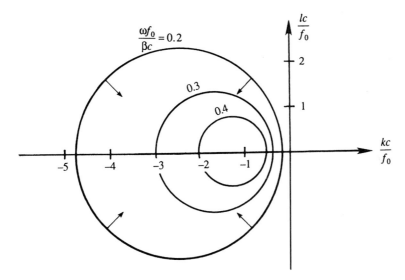

Figure 14.28 Dispersion relation $\omega(k, l)$ for a Rossby wave. The upper panel shows ω vs k for $l = 0$. Regions of positive and negative group velocity c_{gx} are indicated. The lower panel shows a plan view of the surface $\omega(k, l)$, showing contours of constant ω on a kl-plane. The values of $\omega f_0/\beta c$ for the three circles are 0.2, 0.3, and 0.4. Arrows perpendicular to ω contours indicate directions of group velocity vector $\mathbf{c}_g$. A. E. Gill, *Atmosphere–Ocean Dynamics*, 1982 and reprinted with the permission of Academic Press and Mrs. Helen Saunders-Gill.

panel of Figure 14.28. For $l = 0$, the maximum frequency and zero group speed are attained at $kc/f_0 = -1$, corresponding to $\omega_{max} f_0/\beta c = 0.5$. The maximum frequency is much smaller than the Coriolis frequency. For example, in the ocean the ratio $\omega_{max}/f_0 = 0.5\beta c/f_0^2$ is of order 0.1 for the barotropic mode, and of order 0.001 for a baroclinic mode, taking a typical midlatitude value of $f_0 \sim 10^{-4}\,\mathrm{s}^{-1}$, a barotropic gravity wave speed of $c \sim 200\,\text{m/s}$, and a baroclinic gravity wave speed of $c \sim 2\,\text{m/s}$. The shortest period of midlatitude baroclinic Rossby waves in the ocean can therefore be more than a year.

The eastward phase speed is

$$c_x = \frac{\omega}{k} = -\frac{\beta}{k^2 + l^2 + f_0^2/c^2}. \tag{14.119}$$

The negative sign shows that the *phase propagation is always westward*. The phase speed reaches a maximum when $k^2 + l^2 \to 0$, corresponding to very large wavelengths represented by the region near the origin of Figure 14.28. In this region the waves are nearly nondispersive and have an eastward phase speed

$$c_x \simeq -\frac{\beta c^2}{f_0^2}.$$

With $\beta = 2 \times 10^{-11}\,\text{m}^{-1}\,\text{s}^{-1}$, a typical baroclinic value of $c \sim 2\,\text{m/s}$, and a midlatitude value of $f_0 \sim 10^{-4}\,\text{s}^{-1}$, this gives $c_x \sim 10^{-2}\,\text{m/s}$. At these slow speeds the Rossby waves would take years to cross the width of the ocean at midlatitudes. The Rossby waves in the ocean are therefore more important at lower latitudes, where they propagate faster. (The dispersion relation (14.118), however, is not valid within a latitude band of 3° from the equator, for then the assumption of a near geostrophic balance breaks down. A different analysis is needed in the tropics. A discussion of the wave dynamics of the tropics is given in Gill (1982) and in the review paper by McCreary (1985).) In the atmosphere c is much larger, and consequently the Rossby waves propagate faster. A typical large atmospheric disturbance can propagate as a Rossby wave at a speed of several meters per second.

Frequently, the Rossby waves are superposed on a strong eastward mean current, such as the atmospheric jet stream. If U is the speed of this eastward current, then the observed eastward phase speed is

$$c_x = U - \frac{\beta}{k^2 + l^2 + f_0^2/c^2}. \tag{14.120}$$

Stationary Rossby waves can therefore form when the eastward current cancels the westward phase speed, giving $c_x = 0$. This is how stationary waves are formed downstream of the topographic step in Figure 14.20. A simple expression for the wavelength results if we assume $l = 0$ and the flow is barotropic, so that f_0^2/c^2 is negligible in equation (14.120). This gives $U = \beta/k^2$ for stationary solutions, so that the wavelength is $2\pi\sqrt{U/\beta}$.

Finally, note that we have been rather cavalier in deriving the quasi-geostrophic vorticity equation in this section, in the sense that we have substituted the approximate geostrophic expressions for velocity without a formal ordering of the scales. Gill (1982) has given a more precise derivation, expanding in terms of a small parameter. Another way to justify the dispersion relation (14.118) is to obtain it from the general dispersion relation (14.76) derived in Section 10:

$$\omega^3 - c^2\omega(k^2 + l^2) - f_0^2\omega - c^2\beta k = 0. \qquad (14.121)$$

For $\omega \ll f$, the first term is negligible compared to the third, reducing equation (14.121) to equation (14.118).

16. *Barotropic Instability*

In Chapter 12, Section 9 we discussed the inviscid stability of a shear flow $U(y)$ in a nonrotating system, and demonstrated that a necessary condition for its instability is that d^2U/dy^2 must change sign somewhere in the flow. This was called *Rayleigh's point of inflection criterion*. In terms of vorticity $\bar{\zeta} = -dU/dy$, the criterion states that $d\bar{\zeta}/dy$ must change sign somewhere in the flow. We shall now show that, on a rotating earth, the criterion requires that $d(\bar{\zeta} + f)/dy$ must change sign somewhere within the flow.

Consider a horizontal current $U(y)$ in a medium of uniform density. In the absence of horizontal density gradients only the barotropic mode is allowed, and $U(y)$ does not vary with depth. The vorticity equation is

$$\left(\frac{\partial}{\partial t} + \mathbf{u} \cdot \nabla\right)(\zeta + f) = 0. \qquad (14.122)$$

This is identical to the potential vorticity equation $D/Dt[(\zeta + f)/h] = 0$, with the added simplification that the layer depth is constant because $w = 0$. Let the total flow be decomposed into background flow plus a disturbance:

$$u = U(y) + u',$$
$$v = v'.$$

The total vorticity is then

$$\zeta = \bar{\zeta} + \zeta' = -\frac{dU}{dy} + \left(\frac{\partial v'}{\partial x} - \frac{\partial u'}{\partial y}\right) = -\frac{dU}{dy} + \nabla^2\psi,$$

where we have defined the perturbation streamfunction

$$u' = -\frac{\partial \psi}{\partial y}, \qquad v' = \frac{\partial \psi}{\partial x}.$$

Substituting into equation (14.122) and linearizing, we obtain the perturbation vorticity equation

$$\frac{\partial}{\partial t}(\nabla^2\psi) + U\frac{\partial}{\partial x}(\nabla^2\psi) + \left(\beta - \frac{d^2U}{dy^2}\right)\frac{\partial \psi}{\partial x} = 0. \qquad (14.123)$$

Because the coefficients of equation (14.123) are independent of x and t, there can be solutions of the form

$$\psi = \hat{\psi}(y) \, e^{ik(x-ct)}.$$

The phase speed c is complex and solutions are unstable if its imaginary part $c_i > 0$. The perturbation vorticity equation (14.123) then becomes

$$(U - c)\left[\frac{d^2}{dy^2} - k^2\right]\hat{\psi} + \left[\beta - \frac{d^2 U}{dy^2}\right]\hat{\psi} = 0.$$

Comparing this with equation (12.76) derived without Coriolis forces, it is seen that the effect of planetary rotation is the replacement of $-d^2 U/dy^2$ by $(\beta - d^2 U/dy^2)$. The analysis of the section therefore carries over to the present case, resulting in the following criterion: *A necessary condition for the inviscid instability of a barotropic current $U(y)$ is that the gradient of the absolute vorticity*

$$\frac{d}{dy}(\bar{\zeta} + f) = \beta - \frac{d^2 U}{dy^2}, \tag{14.124}$$

must change sign somewhere in the flow. This result was first derived by Kuo (1949).

Barotropic instability quite possibly plays an important role in the instability of currents in the atmosphere and in the ocean. The instability has no preference for any latitude, because the criterion involves β and not f. However, the mechanism presumably dominates in the tropics because midlatitude disturbances prefer the *baroclinic instability* mechanism discussed in the following section. An unstable distribution of westward tropical wind is shown in Figure 14.29.

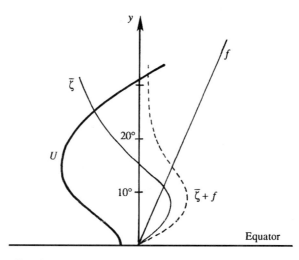

Figure 14.29 Profiles of velocity and vorticity of a westward tropical wind. The velocity distribution is barotropically unstable as $d(\bar{\zeta} + f)/dy$ changes sign within the flow. J. T. Houghton, *The Physics of the Atmosphere*, 1986 and reprinted with the permission of Cambridge University Press.

17. Baroclinic Instability

The weather maps at midlatitudes invariably show the presence of wavelike horizontal excursions of temperature and pressure contours, superposed on eastward mean flows such as the jet stream. Similar undulations are also found in the ocean on eastward currents such as the Gulf Stream in the north Atlantic. A typical wavelength of these disturbances is observed to be of the order of the internal Rossby radius, that is, about 4000 km in the atmosphere and 100 km in the ocean. They seem to be propagating as Rossby waves, but their erratic and unexpected appearance suggests that they are not forced by any external agency, but are due to an inherent *instability* of midlatitude eastward flows. In other words, the eastward flows have a spontaneous tendency to develop wavelike disturbances. In this section we shall investigate the instability mechanism that is responsible for the spontaneous relaxation of eastward jets into a meandering state.

The poleward decrease of the solar irradiation results in a poleward decrease of the temperature and a consequent increase of the density. An idealized distribution of the atmospheric density in the northern hemisphere is shown in Figure 14.30. The density increases northward due to the lower temperatures near the poles and decreases upward because of static stability. According to the thermal wind relation (14.15), an eastward flow (such as the jet stream in the atmosphere or the Gulf Stream in the Atlantic) in equilibrium with such a density structure must have a velocity that increases with height. A system with inclined density surfaces, such as the one in Figure 14.30, has more potential energy than a system with horizontal density surfaces, just as a system with an inclined free surface has more potential energy than a system with a horizontal free surface. It is therefore potentially unstable because it can release the stored potential energy by means of an instability that would cause the density surfaces to flatten out. In the process, vertical shear of the mean flow $U(z)$ would decrease, and perturbations would gain kinetic energy.

Instability of baroclinic jets that release potential energy by flattening out the density surfaces is called the *baroclinic instability*. Our analysis would show that the preferred scale of the unstable waves is indeed of the order of the Rossby radius, as observed for the midlatitude weather disturbances. The theory of baroclinic instability

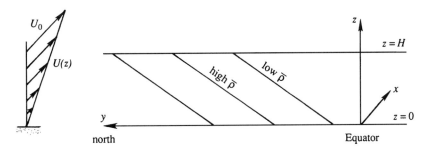

Figure 14.30 Lines of constant density in the northern hemispheric atmosphere. The lines are nearly horizontal and the slopes are greatly exaggerated in the figure. The velocity $U(z)$ is into the plane of paper.

was developed in the 1940s by Bjerknes *et al.* and is considered one of the major triumphs of geophysical fluid mechanics. Our presentation is essentially based on the review article by Pedlosky (1971).

Consider a basic state in which the density is stably stratified in the vertical with a *uniform* buoyancy frequency N, and increases northward at a *constant* rate $\partial \bar{\rho}/\partial y$. According to the thermal wind relation, the constancy of $\partial \bar{\rho}/\partial y$ requires that the vertical shear of the basic eastward flow $U(z)$ also be constant. The β-effect is neglected as it is not an essential requirement of the instability. (The β-effect does modify the instability, however.) This is borne out by the spontaneous appearance of undulations in laboratory experiments in a rotating annulus, in which the inner wall is maintained at a higher temperature than the outer wall. The β-effect is absent in such an experiment.

Perturbation Vorticity Equation

The equations for total flow are

$$\frac{\partial u}{\partial t} + u\frac{\partial u}{\partial x} + v\frac{\partial u}{\partial y} - fv = -\frac{1}{\rho_0}\frac{\partial p}{\partial x},$$

$$\frac{\partial v}{\partial t} + u\frac{\partial v}{\partial x} + v\frac{\partial v}{\partial y} + fu = -\frac{1}{\rho_0}\frac{\partial p}{\partial y},$$

$$0 = -\frac{\partial p}{\partial z} - \rho g, \tag{14.125}$$

$$\frac{\partial u}{\partial x} + \frac{\partial v}{\partial y} + \frac{\partial w}{\partial z} = 0,$$

$$\frac{\partial \rho}{\partial t} + u\frac{\partial \rho}{\partial x} + v\frac{\partial \rho}{\partial y} + w\frac{\partial \rho}{\partial z} = 0,$$

where ρ_0 is a constant reference density. We assume that the total flow is composed of a basic eastward jet $U(z)$ in geostrophic equilibrium with the basic density structure $\bar{\rho}(y, z)$ shown in Figure 14.30, plus perturbations. That is,

$$u = U(z) + u'(x, y, z),$$
$$v = v'(x, y, z),$$
$$w = w'(x, y, z), \tag{14.126}$$
$$\rho = \bar{\rho}(y, z) + \rho'(x, y, z),$$
$$p = \bar{p}(y, z) + p'(x, y, z).$$

The basic flow is in geostrophic and hydrostatic balance:

$$fU = -\frac{1}{\rho_0}\frac{\partial \bar{p}}{\partial y},$$

$$0 = -\frac{\partial \bar{p}}{\partial z} - \bar{\rho}g. \tag{14.127}$$

Eliminating the pressure, we obtain the thermal wind relation

$$\frac{dU}{dz} = \frac{g}{f\rho_0}\frac{\partial\bar\rho}{\partial y}, \tag{14.128}$$

which states that the eastward flow must increase with height because $\partial\bar\rho/\partial y > 0$. For simplicity, we assume that $\partial\bar\rho/\partial y$ is constant, and that $U = 0$ at the surface $z = 0$. Thus the background flow is

$$U = \frac{U_0 z}{H},$$

where U_0 is the velocity at the top of the layer at $z = H$.

We first form a vorticity equation by cross differentiating the horizontal equations of motion in equation (14.125), obtaining

$$\frac{\partial\zeta}{\partial t} + u\frac{\partial\zeta}{\partial x} + v\frac{\partial\zeta}{\partial y} - (\zeta + f)\frac{\partial w}{\partial z} = 0. \tag{14.129}$$

This is identical to equation (14.92), except for the exclusion of the β-effect here; the algebraic steps are therefore not repeated. Substituting the decomposition (14.126), and noting that $\zeta = \zeta'$ because the basic flow $U = U_0 z/H$ has no vertical component of vorticity, (14.129) becomes

$$\frac{\partial\zeta'}{\partial t} + U\frac{\partial\zeta'}{\partial x} - f\frac{\partial w'}{\partial z} = 0, \tag{14.130}$$

where the nonlinear terms have been neglected. This is the perturbation vorticity equation, which we shall now write in terms of p'.

Assume that the perturbations are large-scale and slow, so that the velocity is nearly geostrophic:

$$u' \simeq -\frac{1}{\rho_0 f}\frac{\partial p'}{\partial y}, \quad v' \simeq \frac{1}{\rho_0 f}\frac{\partial p'}{\partial x}, \tag{14.131}$$

from which the perturbation vorticity is found as

$$\zeta' = \frac{1}{\rho_0 f}\nabla_H^2 p'. \tag{14.132}$$

We now express w' in equation (14.130) in terms of p'. The density equation gives

$$\frac{\partial}{\partial t}(\bar\rho + \rho') + (U + u')\frac{\partial}{\partial x}(\bar\rho + \rho') + v'\frac{\partial}{\partial y}(\bar\rho + \rho') + w'\frac{\partial}{\partial z}(\bar\rho + \rho') = 0.$$

Linearizing, we obtain

$$\frac{\partial\rho'}{\partial t} + U\frac{\partial\rho'}{\partial x} + v'\frac{\partial\bar\rho}{\partial y} - \frac{\rho_0 N^2 w'}{g} = 0, \tag{14.133}$$

where $N^2 = -g\rho_0^{-1}(\partial\bar{\rho}/\partial z)$. The perturbation density ρ' can be written in terms of p' by using the hydrostatic balance in equation (14.125), and subtracting the basic state (14.127). This gives

$$0 = -\frac{\partial p'}{\partial z} - \rho' g, \qquad (14.134)$$

which states that the perturbations are hydrostatic. Equation (14.133) then gives

$$w' = -\frac{1}{\rho_0 N^2}\left[\left(\frac{\partial}{\partial t} + U\frac{\partial}{\partial x}\right)\frac{\partial p'}{\partial z} - \frac{dU}{dz}\frac{\partial p'}{\partial x}\right], \qquad (14.135)$$

where we have written $\partial\bar{\rho}/\partial y$ in terms of the thermal wind dU/dz. Using equations (14.132) and (14.135), the perturbation vorticity equation (14.130) becomes

$$\left(\frac{\partial}{\partial t} + U\frac{\partial}{\partial x}\right)\left[\nabla_{\mathrm{H}}^2 p' + \frac{f^2}{N^2}\frac{\partial^2 p'}{\partial z^2}\right] = 0. \qquad (14.136)$$

This is the equation that governs the quasi-geostrophic perturbations on an eastward current $U(z)$.

Wave Solution

We assume that the flow is confined between two horizontal planes at $z = 0$ and $z = H$ and that it is unbounded in x and y. Real flows are likely to be bounded in the y direction, especially in a laboratory situation of flow in an annular region, where the walls set boundary conditions parallel to the flow. The boundedness in y, however, simply sets up normal modes in the form $\sin(n\pi y/L)$, where L is the width of the channel. Each of these modes can be replaced by a periodicity in y. Accordingly, we assume wavelike solutions

$$p' = \hat{p}(z)\,e^{i(kx+ly-\omega t)}. \qquad (14.137)$$

The perturbation vorticity equation (14.136) then gives

$$\frac{d^2\hat{p}}{dz^2} - \alpha^2\hat{p} = 0, \qquad (14.138)$$

where

$$\alpha^2 \equiv \frac{N^2}{f^2}(k^2 + l^2). \qquad (14.139)$$

The solution of equation (14.138) can be written as

$$\hat{p} = A\cosh\alpha\left(z - \frac{H}{2}\right) + B\sinh\alpha\left(z - \frac{H}{2}\right). \qquad (14.140)$$

Boundary conditions have to be imposed on solution (14.140) in order to derive an instability criterion.

Boundary Conditions

The conditions are

$$w' = 0 \quad \text{at } z = 0, H.$$

The corresponding conditions on p' can be found from equation (14.135) and $U = U_0 z / H$. We obtain

$$-\frac{\partial^2 p'}{\partial t \, \partial z} - \frac{U_0 z}{H} \frac{\partial^2 p'}{\partial x \, \partial z} + \frac{U_0}{H} \frac{\partial p'}{\partial x} = 0 \quad \text{at } z = 0, H,$$

where we have also used $U = U_0 z / H$. The two boundary conditions are therefore

$$\frac{\partial^2 p'}{\partial t \, \partial z} - \frac{U_0}{H} \frac{\partial p'}{\partial x} = 0 \quad \text{at } z = 0,$$

$$\frac{\partial^2 p'}{\partial t \, \partial z} - \frac{U_0}{H} \frac{\partial p'}{\partial x} + U_0 \frac{\partial^2 p'}{\partial x \, \partial z} = 0 \quad \text{at } z = H.$$

Instability Criterion

Using equations (14.137) and (14.140), the foregoing boundary conditions require

$$A \left[\alpha c \sinh \frac{\alpha H}{2} - \frac{U_0}{H} \cosh \frac{\alpha H}{2} \right]$$

$$+ B \left[-\alpha c \cosh \frac{\alpha H}{2} + \frac{U_0}{H} \sinh \frac{\alpha H}{2} \right] = 0,$$

$$A \left[\alpha (U_0 - c) \sinh \frac{\alpha H}{2} - \frac{U_0}{H} \cosh \frac{\alpha H}{2} \right]$$

$$+ B \left[\alpha (U_0 - c) \cosh \frac{\alpha H}{2} - \frac{U_0}{H} \sinh \frac{\alpha H}{2} \right] = 0,$$

where $c = \omega / k$ is the eastward phase velocity.

This is a pair of homogeneous equations for the constants A and B. For nontrivial solutions to exist, the determinant of the coefficients must vanish. This gives, after some straightforward algebra, the phase velocity

$$c = \frac{U_0}{2} \pm \frac{U_0}{\alpha H} \sqrt{ \left(\frac{\alpha H}{2} - \tanh \frac{\alpha H}{2} \right) \left(\frac{\alpha H}{2} - \coth \frac{\alpha H}{2} \right) }. \qquad (14.141)$$

Whether the solution grows with time depends on the sign of the radicand. The behavior of the functions under the radical sign is sketched in Figure 14.31. It is apparent that the first factor in the radicand is positive because $\alpha H / 2 > \tanh(\alpha H / 2)$

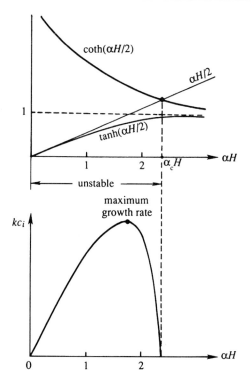

Figure 14.31 Baroclinic instability. The upper panel shows behavior of the functions in equation (14.141), and the lower panel shows growth rates of unstable waves.

for all values of αH. However, the second factor is negative for small values of αH for which $\alpha H/2 < \coth(\alpha H/2)$. In this range the roots of c are complex conjugates, with $c = U_0/2 \pm ic_i$. Because we have assumed that the perturbations are of the form $\exp(-ikct)$, the existence of a nonzero c_i implies the possibility of a perturbation that grows as $\exp(kc_i t)$, and the solution is unstable. The marginal stability is given by the critical value of α satisfying

$$\frac{\alpha_c H}{2} = \coth\left(\frac{\alpha_c H}{2}\right),$$

whose solution is

$$\alpha_c H = 2.4,$$

and the flow is unstable if $\alpha H < 2.4$. Using the definition of α in equation (14.139), it follows that the flow is unstable if

$$\frac{HN}{f} < \frac{2.4}{\sqrt{k^2 + l^2}}.$$

As all values of k and l are allowed, we can always find a value of $k^2 + l^2$ low enough to satisfy the forementioned inequality. *The flow is therefore always unstable (to low wavenumbers).* For a north–south wavenumber $l = 0$, instability is ensured if the east–west wavenumber k is small enough such that

$$\frac{HN}{f} < \frac{2.4}{k}. \tag{14.142}$$

In a continuously stratified ocean, the speed of a long internal wave for the $n = 1$ baroclinic mode is $c = NH/\pi$, so that the corresponding internal Rossby radius is $c/f = NH/\pi f$. It is usual to omit the factor π and define the Rossby radius in a continuously stratified fluid as

$$\Lambda \equiv \frac{HN}{f}.$$

The condition (14.142) for baroclinic instability is therefore that the east–west wavelength be large enough so that

$$\lambda > 2.6\Lambda.$$

However, the wavelength $\lambda = 2.6\Lambda$ does not grow at the fastest rate. It can be shown from equation (14.141) that the wavelength with the largest growth rate is

$$\boxed{\lambda_{\max} = 3.9\Lambda.}$$

This is therefore the wavelength that is observed when the instability develops. Typical values for f, N, and H suggest that $\lambda_{\max} \sim 4000\,\text{km}$ in the atmosphere and $200\,\text{km}$ in the ocean, which agree with observations. Waves much smaller than the Rossby radius do not grow, and the ones much larger than the Rossby radius grow very slowly.

Energetics

The foregoing analysis suggests that the existence of "weather waves" is due to the fact that small perturbations can grow spontaneously when superposed on an eastward current maintained by the sloping density surfaces (Figure 14.30). Although the basic current does have a vertical shear, the perturbations do not grow by extracting energy from the vertical shear field. Instead, they extract their energy from the *potential energy* stored in the system of sloping density surfaces. The energetics of the baroclinic instability is therefore quite different than that of the Kelvin–Helmholtz instability (which also has a vertical shear of the mean flow), where the perturbation Reynolds stress $\overline{u'w'}$ interacts with the vertical shear and extracts energy from the mean shear flow. The baroclinic instability is *not* a shear flow instability; the Reynolds stresses are too small because of the small w in quasi-geostrophic large-scale flows.

The energetics of the baroclinic instability can be understood by examining the equation for the perturbation kinetic energy. Such an equation can be derived by multiplying the equations for $\partial u'/\partial t$ and $\partial v'/\partial t$ by u' and v', respectively, adding the two, and integrating over the region of flow. Because of the assumed periodicity in x and y, the extent of the region of integration is chosen to be one wavelength in either direction. During this integration, the boundary conditions of zero normal flow on the walls and periodicity in x and y are used repeatedly. The procedure is similar to that for the derivation of equation (12.83) and is not repeated here. The result is

$$\frac{dK}{dt} = -g \int w'\rho'\, dx\, dy\, dz,$$

where K is the global perturbation kinetic energy

$$K \equiv \frac{\rho_0}{2} \int (u'^2 + v'^2)\, dx\, dy\, dz.$$

In unstable flows we must have $dK/dt > 0$, which requires that the volume integral of $w'\rho'$ must be negative. Let us denote the volume average of $w'\rho'$ by $\overline{w'\rho'}$. A negative $\overline{w'\rho'}$ means that on the average the lighter fluid rises and the heavier fluid sinks. By such an interchange the center of gravity of the system, and therefore its potential energy, is lowered. The interesting point is that this cannot happen in a stably stratified system with *horizontal* density surfaces; in that case an exchange of fluid particles *raises* the potential energy. Moreover, a basic state with inclined density surfaces (Figure 14.30) cannot have $\overline{w'\rho'} < 0$ if the particle excursions are vertical. If, however, the particle excursions fall within the wedge formed by the constant density lines and the horizontal (Figure 14.32), then an exchange of fluid particles takes lighter particles upward (and northward) and denser particles downward (and southward). Such an interchange would tend to make the density surfaces more horizontal, releasing potential energy from the mean density field with a consequent growth of the

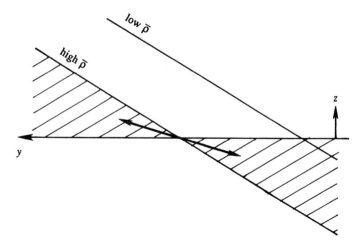

Figure 14.32 Wedge of instability (shaded) in a baroclinic instability. The wedge is bounded by constant density lines and the horizontal. Unstable waves have a particle trajectory that falls within the wedge.

perturbation energy. This type of convection is called *sloping convection*. According to Figure 14.32 the exchange of fluid particles within the *wedge of instability* results in a net poleward transport of heat from the tropics, which serves to redistribute the larger solar heat received by the tropics.

In summary, baroclinic instability draws energy from the potential energy of the mean density field. The resulting eddy motion has particle trajectories that are oriented at a small angle with the horizontal, so that the resulting heat transfer has a poleward component. The preferred scale of the disturbance is the Rossby radius.

18. Geostrophic Turbulence

Two common modes of instability of a large-scale current system were presented in the preceding sections. When the flow is strong enough, such instabilities can make a flow chaotic or turbulent. A peculiarity of large-scale turbulence in the atmosphere or the ocean is that it is essentially two dimensional in nature. The existence of the Coriolis force, stratification, and small thickness of geophysical media severely restricts the vertical velocity in large-scale flows, which tend to be quasi-geostrophic, with the Coriolis force balancing the horizontal pressure gradient to the lowest order. Because vortex stretching, a key mechanism by which ordinary three-dimensional turbulent flows transfer energy from large to small scales, is absent in two-dimensional flow, one expects that the dynamics of geostrophic turbulence are likely to be fundamentally different from that of three-dimensional laboratory-scale turbulence discussed in Chapter 13. However, we can still call the motion "turbulent" because it is unpredictable and diffusive.

A key result on the subject was discovered by the meteorologist Fjortoft (1953), and since then Kraichnan, Leith, Batchelor, and others have contributed to various aspects of the problem. A good discussion is given in Pedlosky (1987), to which the reader is referred for a fuller treatment. Here, we shall only point out a few important results.

An important variable in the discussion of two-dimensional turbulence is *enstrophy*, which is the mean square vorticity $\overline{\zeta^2}$. In an isotropic turbulent field we can define an energy spectrum $S(K)$, a function of the magnitude of the wavenumber K, as

$$\overline{u^2} = \int_0^\infty S(K)\, dK.$$

It can be shown that the enstrophy spectrum is $K^2 S(K)$, that is,

$$\overline{\zeta^2} = \int_0^\infty K^2 S(K)\, dK,$$

which makes sense because vorticity involves the spatial gradient of velocity.

We consider a freely evolving turbulent field in which the shape of the velocity spectrum changes with time. The large scales are essentially inviscid, so that both energy and enstrophy are nearly conserved:

$$\frac{d}{dt} \int_0^\infty S(K)\,dK = 0, \tag{14.143}$$

$$\frac{d}{dt} \int_0^\infty K^2 S(K)\,dK = 0, \tag{14.144}$$

where terms proportional to the molecular viscosity ν have been neglected on the right-hand sides of the equations. The enstrophy conservation is unique to two-dimensional turbulence because of the absence of vortex stretching.

Suppose that the energy spectrum initially contains all its energy at wavenumber K_0. Nonlinear interactions transfer this energy to other wavenumbers, so that the sharp spectral peak smears out. For the sake of argument, suppose that all of the initial energy goes to two neighboring wavenumbers K_1 and K_2, with $K_1 < K_0 < K_2$. Conservation of energy and enstrophy requires that

$$S_0 = S_1 + S_2,$$
$$K_0^2 S_0 = K_1^2 S_1 + K_2^2 S_2,$$

where S_n is the spectral energy at K_n. From this we can find the ratios of energy and enstrophy spectra before and after the transfer:

$$\frac{S_1}{S_2} = \frac{K_2 - K_0}{K_0 - K_1} \frac{K_2 + K_0}{K_1 + K_0},$$
$$\frac{K_1^2 S_1}{K_2^2 S_2} = \frac{K_1^2}{K_2^2} \frac{K_2^2 - K_0^2}{K_0^2 - K_1^2}. \tag{14.145}$$

As an example, suppose that nonlinear smearing transfers energy to wavenumbers $K_1 = K_0/2$ and $K_2 = 2K_0$. Then equations (14.145) show that $S_1/S_2 = 4$ and $K_1^2 S_1/K_2^2 S_2 = \frac{1}{4}$, so that more energy goes to lower wavenumbers (large scales), whereas more enstrophy goes to higher wavenumbers (smaller scales). This important result on two-dimensional turbulence was derived by Fjortoft (1953). Clearly, the constraint of enstrophy conservation in two-dimensional turbulence has prevented a symmetric spreading of the initial energy peak at K_0.

The unique character of two-dimensional turbulence is evident here. In small-scale three-dimensional turbulence studied in Chapter 13, the energy goes to smaller and smaller scales until it is dissipated by viscosity. In geostrophic turbulence, on the other hand, the energy goes to larger scales, where it is less susceptible to viscous dissipation. Numerical calculations are indeed in agreement with this behavior, which shows that the energy-containing eddies grow in size by coalescing. On the other hand, the vorticity becomes increasingly confined to thin shear layers on the eddy boundaries; these shear layers contain very little energy. The backward (or inverse) energy cascade and forward enstrophy cascade are represented schematically in Figure 14.33. It is clear that there are two "inertial" regions in the spectrum of a two-dimensional turbulent flow, namely, the energy cascade region and the enstrophy cascade region.

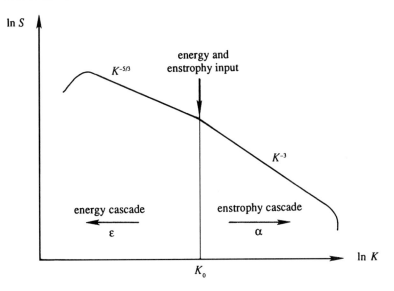

Figure 14.33 Energy and enstrophy cascade in two-dimensional turbulence.

If energy is injected into the system at a rate ε, then the energy spectrum in the energy cascade region has the form $S(K) \propto \varepsilon^{2/3} K^{-5/3}$; the argument is essentially the same as in the case of the Kolmogorov spectrum in three-dimensional turbulence (Chapter 13, Section 9), except that the transfer is backwards. A dimensional argument also shows that the energy spectrum in the enstrophy cascade region is of the form $S(K) \propto \alpha^{2/3} K^{-3}$, where α is the forward enstrophy flux to higher wavenumbers. There is negligible energy flux in the enstrophy cascade region.

As the eddies grow in size, they become increasingly immune to viscous dissipation, and the inviscid assumption implied in equation (14.143) becomes increasingly applicable. (This would not be the case in three-dimensional turbulence in which the eddies continue to decrease in size until viscous effects drain energy out of the system.) In contrast, the corresponding assumption in the enstrophy conservation equation (14.144) becomes less and less valid as enstrophy goes to smaller scales, where viscous dissipation drains enstrophy out of the system. At later stages in the evolution, then, equation (14.144) may not be a good assumption. However, it can be shown (see Pedlosky, 1987) that the dissipation of enstrophy actually *intensifies* the process of energy transfer to larger scales, so that the red cascade (that is, transfer to larger scales) of energy is a general result of two-dimensional turbulence.

The eddies, however, do not grow in size indefinitely. They become increasingly slower as their length scale l increases, while their velocity scale u remains constant. The slower dynamics makes them increasingly wavelike, and the eddies transform into Rossby-wave packets as their length scale becomes of order (Rhines, 1975)

$$l \sim \sqrt{\frac{u}{\beta}} \qquad \text{(Rhines length)},$$

where $\beta = df/dy$ and u is the rms fluctuating speed. The Rossby-wave propagation results in an anisotropic elongation of the eddies in the east–west ("zonal") direction, while the eddy size in the north–south direction stops growing at $\sqrt{u/\beta}$. Finally, the velocity field consists of zonally directed jets whose north–south extent is of order $\sqrt{u/\beta}$. This has been suggested as an explanation for the existence of zonal jets in the atmosphere of the planet Jupiter (Williams, 1979). The inverse energy cascade regime may not occur in the earth's atmosphere and the ocean at midlatitudes because the Rhines length (about 1000 km in the atmosphere and 100 km in the ocean) is of the order of the internal Rossby radius, where the energy is injected by baroclinic instability. (For the inverse cascade to occur, $\sqrt{u/\beta}$ needs to be larger than the scale at which energy is injected.)

Eventually, however, the kinetic energy has to be dissipated by molecular effects at the Kolmogorov microscale η, which is of the order of a few millimeters in the ocean and the atmosphere. A fair hypothesis is that processes such as internal waves drain energy out of the mesoscale eddies, and breaking internal waves generate three-dimensional turbulence that finally cascades energy to molecular scales.

A recent review of intense storm motion (lower atmosphere dynamics and thermodynamics) was published by Chan (2005), whereas upper atmospheric motion was discussed by Haynes (2005). Oceanic flow transport was treated by Wiggins (2005).

Exercises

1. The Gulf Stream flows northward along the east coast of the United States with a surface current of average magnitude 2 m/s. If the flow is assumed to be in geostrophic balance, find the average slope of the sea surface across the current at a latitude of 45° N. [*Answer*: 2.1 cm per km]

2. A plate containing water ($\nu = 10^{-6}$ m^2/s) above it rotates at a rate of 10 revolutions per minute. Find the depth of the Ekman layer, assuming that the flow is laminar.

3. Assume that the atmospheric Ekman layer over the earth's surface at a latitude of 45° N can be approximated by an eddy viscosity of $\nu_v = 10$ m^2/s. If the geostrophic velocity above the Ekman layer is 10 m/s, what is the Ekman transport across isobars? [*Answer*: 2203 m^2/s]

4. Find the axis ratio of a hodograph plot for a semidiurnal tide in the middle of the ocean at a latitude of 45° N. Assume that the midocean tides are rotational surface gravity waves of long wavelength and are unaffected by the proximity of coastal boundaries. If the depth of the ocean is 4 km, find the wavelength, the phase velocity, and the group velocity. Note, however, that the wavelength is comparable to the width of the ocean, so that the neglect of coastal boundaries is not very realistic.

5. An internal Kelvin wave on the thermocline of the ocean propagates along the west coast of Australia. The thermocline has a depth of 50 m and has a nearly discontinuous density change of 2 kg/m^3 across it. The layer below the thermocline

is deep. At a latitude of 30° S, find the direction and magnitude of the propagation speed and the decay scale perpendicular to the coast.

6. Using the dispersion relation $m^2 = k^2(N^2 - \omega^2)/(\omega^2 - f^2)$ for internal waves, show that the group velocity vector is given by

$$[c_{gx}, c_{gz}] = \frac{(N^2 - f^2)\,km}{(m^2 + k^2)^{3/2}(m^2 f^2 + k^2 N^2)^{1/2}}[m, -k]$$

[*Hint*: Differentiate the dispersion relation partially with respect to k and m.] Show that $\mathbf{c}_g$ and $\mathbf{c}$ are perpendicular and have oppositely directed vertical components. Verify that $\mathbf{c}_g$ is parallel to $\mathbf{u}$.

7. Suppose the atmosphere at a latitude of 45° N is idealized by a uniformly stratified layer of height 10 km, across which the potential temperature increases by 50 °C.

  (i)   What is the value of the buoyancy frequency N?

  (ii) Find the speed of a long gravity wave corresponding to the $n = 1$ baroclinic mode.

 (iii) For the $n = 1$ mode, find the westward speed of nondispersive (i. e., very large wavelength) Rossby waves. [*Answer*: $N = 0.01279\,\text{s}^{-1}$; $c_1 = 40.71$ m/s; $c_x = -3.12$ m/s]

8. Consider a steady flow rotating between plane parallel boundaries a distance L apart. The angular velocity is Ω and a small rectilinear velocity U is superposed. There is a protuberance of height $h \ll L$ in the flow. The Ekman and Rossby numbers are both small: $Ro \ll 1$, $E \ll 1$. Obtain an integral of the relevant equations of motion that relates the modified pressure and the streamfunction for the motion, and show that the modified pressure is constant on streamlines.

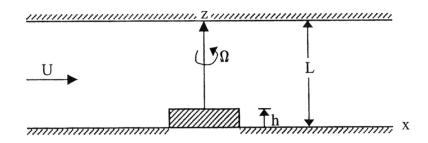

Literature Cited

Chan, J. C. L. (2005). "The physics of tropical cyclone motion." *Annual Review of Fluid Mechanics* **37**: 99–128.

Fjortoft, R. (1953). "On the changes in the spectral distributions of kinetic energy for two-dimensional non-divergent flow." *Tellus* **5**: 225–230.

Gill, A. E. (1982). *Atmosphere–Ocean Dynamics*, New York: Academic Press.

Haynes, P. (2005). "Stratospheric dynamics." *Annual Review of Fluid Mechanics* **37**: 263–293.

Holton, J. R. (1979). *An Introduction to Dynamic Meteorology*, New York: Academic Press.

Houghton, J. T. (1986). *The Physics of the Atmosphere*, London: Cambridge University Press.

Kamenkovich, V. M. (1967). "On the coefficients of eddy diffusion and eddy viscosity in large-scale oceanic and atmospheric motions." *Izvestiya, Atmospheric and Oceanic Physics* **3**: 1326–1333.

Kundu, P. K. (1977). "On the importance of friction in two typical continental waters: Off Oregon and Spanish Sahara," in *Bottom Turbulence*, J. C. J. Nihoul, ed., Amsterdam: Elsevier.

Kuo, H. L. (1949). "Dynamic instability of two-dimensional nondivergent flow in a barotropic atmosphere." *Journal of Meteorology* **6**: 105–122.

LeBlond, P. H. and L. A. Mysak (1978). *Waves in the Ocean*, Amsterdam: Elsevier.

McCreary, J. P. (1985). "Modeling equatorial ocean circulation." *Annual Review of Fluid Mechanics* **17**: 359–409.

Munk, W. (1981). "Internal waves and small-scale processes," in *Evolution of Physical Oceanography*, B. A. Warren and C. Wunch, eds., Cambridge, MA: MIT Press.

Pedlosky, J. (1971). "Geophysical fluid dynamics," in *Mathematical Problems in the Geophysical Sciences*, W. H. Reid, ed., Providence, Rhode Island: American Mathematical Society.

Pedlosky, J. (1987). *Geophysical Fluid Dynamics*, New York: Springer-Verlag.

Phillips, O. M. (1977). *The Dynamics of the Upper Ocean*, London: Cambridge University Press.

Prandtl, L. (1952). *Essentials of Fluid Dynamics*, New York: Hafner Publ. Co.

Rhines, P. B. (1975). "Waves and turbulence on a β-plane." *Journal of Fluid Mechanics* **69**: 417–443.

Taylor, G. I. (1915). "Eddy motion in the atmosphere." *Philosophical Transactions of the Royal Society of London* **A215**: 1–26.

Wiggins, S. (2005). "The dynamical systems approach to Lagrangian transport in oceanic flows." *Annual Review of Fluid Mechanics* **37**: 295–328.

Williams, G. P. (1979). "Planetary circulations: 2. The Jovian quasi-geostrophic regime." *Journal of Atmospheric Sciences* **36**: 932–968.

Aerodynamics

1. Introduction

Aerodynamics is the branch of fluid mechanics that deals with the determination of the flow past bodies of aeronautical interest. Gravity forces are neglected, and viscosity is regarded as small so that the viscous forces are confined to thin boundary layers (Figure 10.1). The subject is called *incompressible aerodynamics* if the flow speeds are low enough (Mach number < 0.3) for the compressibility effects to be negligible. At larger Mach numbers the subject is normally called *gas dynamics*, which deals with flows in which compressibility effects are important. In this chapter we shall study some elementary aspects of incompressible flow around aircraft wing shapes. The blades of turbomachines (such as turbines and compressors) have the same cross section as that of an aircraft wing, so that much of our discussion will also apply to the flow around the blades of a turbomachine.

©2010 Elsevier Inc. All rights reserved.
DOI: 10.1016/B978-0-12-381399-2.50015-0

Because the viscous effects are confined to thin boundary layers, the bulk of the flow is still irrotational. Consequently, a large part of our discussion of irrotational flows presented in Chapter 6 is relevant here. It is assumed that the reader is familiar with that chapter.

2.   *The Aircraft and Its Controls*

Although a book on fluid mechanics is not the proper place for describing an aircraft and its controls, we shall do this here in the hope that the reader will find it interesting. Figure 15.1 shows three views of an aircraft. The body of the aircraft, which houses the passengers and other payload, is called the *fuselage*. The engines (jets or propellers) are often attached to the wings; sometimes they may be mounted on the fuselage. Figure 15.2 shows the plan view of a wing. The outer end of each wing is called the *wing tip*, and the distance between the wing tips is called the *wing span s*. The distance between the leading and trailing edges of the wing is called the *chord length c*, which

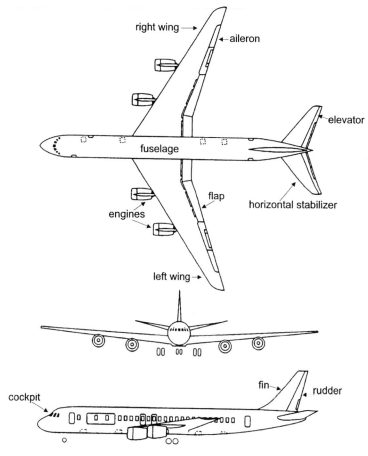

Figure 15.1   Three views of a transport aircraft and its control surfaces (NASA).

varies along the spanwise direction. The plan area of the wing is called the *wing area A*. The narrowness of the wing planform is measured by its *aspect ratio*

$$\Lambda \equiv \frac{s^2}{A} = \frac{s}{\bar{c}},$$

where $\bar{c}$ is the average chord length.

The various possible rotational motions of an aircraft can be referred to three axes, called the *pitch axis*, the *roll axis*, and the *yaw axis* (Figure 15.3).

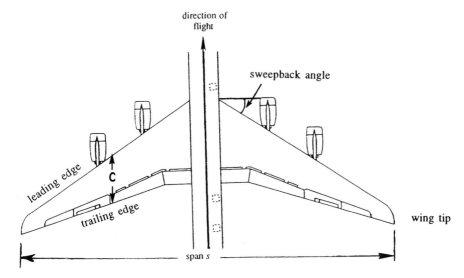

Figure 15.2 Wing planform geometry.

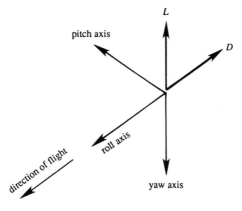

Figure 15.3 Aircraft axes.

Control Surfaces

The aircraft is controlled by the pilot by moving certain control surfaces described in the following paragraphs.

Aileron: These are portions of each wing near the wing tip (Figure 15.1), joined to the main wing by a hinged connection, as shown in Figure 15.4. They move differentially in the sense that one moves up while the other moves down. A depressed aileron increases the lift, and a raised aileron decreases the lift, so that a rolling moment results. The object of situating the ailerons near the wing tip is to generate a large rolling moment. The pilot generally controls the ailerons by moving a control stick, whose movement to the left or right causes a roll to the left or right. In larger aircraft the aileron motion is controlled by rotating a small wheel that resembles one half of an automobile steering wheel.

Elevator: The elevators are hinged to the trailing edge of the tail plane. Unlike ailerons they move together, and their movement generates a pitching motion of the aircraft. The elevator movements are imparted by the forward and backward movement of a control stick, so that a backward pull lifts the nose of the aircraft.

Rudder: The yawing motion of the aircraft is governed by the hinged rear portion of the tail fin, called the rudder. The pilot controls the rudder by pressing his feet against two rudder pedals so arranged that moving the left pedal forward moves the aircraft's nose to the left.

Flap: During take off, the speed of the aircraft is too small to generate enough lift to support the weight of the aircraft. To overcome this, a section of the rear of the wing is "split," so that it can be rotated downward to increase the lift (Figure 15.5). A further function of the flap is to increase both lift and drag during landing.

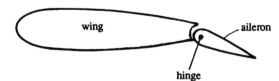

Figure 15.4   The aileron.

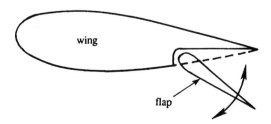

Figure 15.5   The flap.

Modern jet transports also have "spoilers" on the top surface of each wing. When raised slightly, they separate the boundary layer early on part of the top of the wing and this decreases its lift. They can be deployed together or individually. Reducing the lift on one wing will bank the aircraft so that it would turn in the direction of the lowered wing. Deployed together, lift would be decreased and the aircraft would descend to a new equilibrium altitude. Spoilers have another function as well. Upon touchdown during landing they are deployed fully as flat plates nearly perpendicular to the wing surface. As such they add greatly to the drag to slow the aircraft and shorten its roll down the runway.

An aircraft is said to be in trimmed flight when there are no moments about its center of gravity. Trim tabs are small adjustable surfaces within or adjacent to the major control surfaces described in the preceding: ailerons, elevators, and rudder. Deflections of these surfaces may be set and held to adjust for a change in the aircraft's center of gravity in flight due to consumption of fuel or a change in the direction of the prevailing wind with respect to the flight path. These are set for steady level flight on a straight path with minimum deflection of the major control surfaces.

3. Airfoil Geometry

Figure 15.6 shows the shape of the cross section of a wing, called an *airfoil* section (spelled aerofoil in the British literature). The leading edge of the profile is generally rounded, whereas the trailing edge is sharp. The straight line joining the centers of curvature of the leading and trailing edges is called the *chord*. The meridian line of the section passing midway between the upper and lower surfaces is called the *camber line*. The maximum height of the camber line above the chord line is called the *camber* of the section. Normally the camber varies from nearly zero for high-speed supersonic wings, to ≈5% of chord length for low-speed wings. The angle α between the chord line and the direction of flight (i.e., the direction of the undisturbed stream) is called the *angle of attack* or *angle of incidence*.

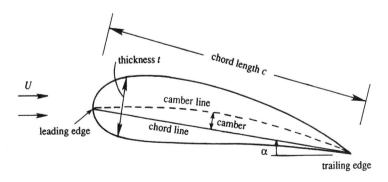

Figure 15.6 Airfoil geometry.

4. Forces on an Airfoil

The resultant aerodynamic force F on an airfoil can be resolved into a *lift force L* perpendicular to the direction of undisturbed flight and a *drag force D* in the direction of flight (Figure 15.7). In steady level flight the drag is balanced by the thrust of the engine, and the lift equals the weight of the aircraft. These forces are expressed nondimensionally by defining the coefficients of lift and drag:

$$C_L \equiv \frac{L}{(1/2)\rho U^2 A}, \qquad C_D \equiv \frac{D}{(1/2)\rho U^2 A}. \tag{15.1}$$

The drag results from the tangential stress and normal pressure distributions on the surface. These are called the *friction drag* and the *pressure drag*, respectively. The lift is almost entirely due to the pressure distribution. Figure 15.8 shows the distribution of the pressure coefficient $C_p = (p - p_\infty)/\frac{1}{2}\rho U^2$ at a moderate angle of attack. The outward arrows correspond to a negative C_p, while a positive C_p is represented by inward arrows. It is seen that the pressure coefficient is negative over most of the surface, except over small regions near the nose and the tail. However, the pressures over most of the upper surface are smaller than those over the bottom surface, which results in a lift force. The top and bottom surfaces of an airfoil are popularly referred to as the *suction side* and the *compression side*, respectively.

5. Kutta Condition

In Chapter 6, Section 11 we showed that the lift per unit span in an irrotational flow over a two-dimensional body of arbitrary cross section is

$$L = \rho U \Gamma, \tag{15.2}$$

where U is the free-stream velocity and Γ is the circulation around the body. Relation (15.2) is called the *Kutta–Zhukhovsky lift theorem*. The question is, how does a flow develop such a circulation? Obviously, a circular or elliptic cylinder does not develop

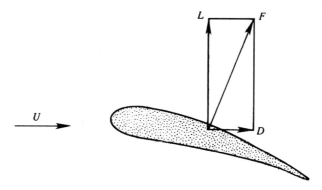

Figure 15.7   Forces on an airfoil.

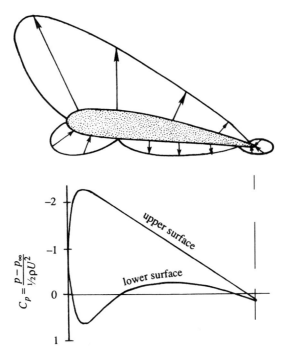

Figure 15.8 Distribution of the pressure coefficient over an airfoil. The upper panel shows C_p plotted normal to the surface and the lower panel shows C_p plotted normal to the chord line.

any circulation around it, unless it is rotated. It has been experimentally observed that only bodies having a sharp trailing edge, such as an airfoil, can generate circulation and lift.

Figure 15.9 shows the irrotational flow pattern around an airfoil for increasing values of clockwise circulation. For $\Gamma = 0$, there is a stagnation point A located just below the leading edge and a stagnation point B on the top surface near the trailing edge. When some clockwise circulation is superimposed, both stagnation points move slightly down. For a particular value of Γ, the stagnation point B coincides with the trailing edge. (If the circulation is further increased, the rear stagnation point moves to the *lower* surface.) As far as irrotational flow of an ideal fluid is concerned, all these flow patterns are possible solutions. A real flow, however, develops a specific amount of circulation, depending on the airfoil shape and the angle of attack.

Consider the irrotational flow around the trailing edge of an airfoil. It is shown in Chapter 6, Section 4 that, for flow in a corner of included angle γ, the velocity at the corner point is zero if $\gamma < 180°$ and infinite if $\gamma > 180°$ (see Figure 6.4). In the upper two panels of Figure 15.9 the fluid goes from the lower to the upper side by turning around the trailing edge, so that γ is slightly less than 360°. The resulting velocity at the trailing edge is therefore infinite in the upper two panels of Figure 15.9. In the bottom panel, on the other hand, the trailing edge is a *stagnation point* because γ is slightly less than 180°.

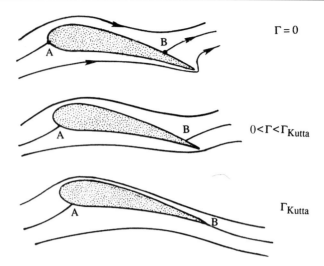

$\Gamma = 0$

$0 < \Gamma < \Gamma_{Kutta}$

Γ_{Kutta}

Figure 15.9   Irrotational flow pattern over an airfoil for various values of clockwise circulation.

Photographs of flow around airfoils reveal that the pattern sketched in the bottom panel of Figure 15.9 is the one developed in practice. The German aerodynamist Wilhelm Kutta proposed the following rule in 1902: *In flow over a two-dimensional body with a sharp trailing edge, there develops a circulation of magnitude just sufficient to move the rear stagnation point to the trailing edge.* This is called the *Kutta condition*, sometimes also called the *Zhukhovsky hypothesis*. At the beginning of the twentieth century it was merely an experimentally observed fact. Justification for this empirical rule became clear after the boundary layer concepts were understood. In the following section we shall see why a real flow should satisfy the Kutta condition.

Historical Notes

According to von Karman (1954, p. 34), the connection between the lift of airplane wings and the circulation around them was recognized and developed by three persons. One of them was the Englishman Frederick Lanchester (1887–1946). He was a multisided and imaginative person, a practical engineer as well as an amateur mathematician. His trade was automobile building; in fact, he was the chief engineer and general manager of the Lanchester Motor Company. He once took von Karman for a ride around Cambridge in an automobile that he built himself, but von Karman "felt a little uneasy discussing aerodynamics at such rather frightening speed." The second person is the German mathematician Wilhelm Kutta (1867–1944), well-known for the Runge–Kutta scheme used in the numerical integration of ordinary differential equations. He started out as a pure mathematician, but later became interested in aerodynamics. The third person is the Russian physicist Nikolai Zhukhovsky, who developed the mathematical foundations of the theory of lift for wings of infinite span, independently of Lanchester and Kutta. An excellent book on the history of flight and the science of aerodynamics was recently authored by Anderson (1998).

6. Generation of Circulation

We shall now discuss why a real flow around an airfoil should satisfy the Kutta condition. The explanation lies in the frictional and boundary layer nature of a real flow. Consider an airfoil starting from rest in a real fluid. The flow immediately after starting is irrotational everywhere, because the vorticity adjacent to the surface has not yet diffused outward. The velocity at this stage has a near discontinuity adjacent to the surface. The flow has no circulation, and resembles the pattern in the upper panel of Figure 15.9. The fluid goes around the trailing edge with a very high velocity and overcomes a steep deceleration and pressure rise from the trailing edge to the stagnation point.

Within a fraction of a second (in a time of the order of that taken by the flow to move one chord length), however, boundary layers develop on the airfoil, and the retarded fluid does not have sufficient kinetic energy to negotiate the steep pressure rise from the trailing edge toward the rear stagnation point. This generates a back-flow in the boundary layer and a separation of the boundary layer at the trailing edge. The consequence of all this is the generation of a shear layer, which rolls up into a spiral form under the action of its own induced vorticity (Figure 15.10). The rolled-up shear layer is carried downstream by the flow and is left at the location where the airfoil started its motion. This is called the *starting vortex*.

The sense of circulation of the starting vortex is counterclockwise in Figure 15.10, which means that it must leave behind a clockwise circulation around the airfoil. To see this, imagine that the fluid is stationary and the airfoil is moving to the left. Consider a material circuit ABCD, made up of the same fluid particles and large enough to enclose both the initial and final locations of the airfoil (Figure 15.11). Initially the trailing edge was within the region BCD, which now contains the starting vortex only. According to the Kelvin circulation theorem, the circulation around any material circuit remains constant, if the circuit remains in a region of inviscid flow (although viscous processes may go on *inside* the region enclosed by the circuit). The circulation around the large curve ABCD therefore remains zero, since it was zero initially. Consequently the counterclockwise circulation of the starting vortex around DBC is balanced by an equal clockwise circulation around ADB. The wing is therefore left with a circulation Γ equal and opposite to the circulation of the starting vortex.

It is clear from Figure 15.9 that a value of circulation other than the one that moves the rear stagnation point exactly to the trailing edge would result in a sequence of events as just described and would lead to a readjustment of the flow. The only value

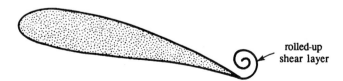

Figure 15.10 Formation of a spiral vortex sheet soon after an airfoil begins to move.

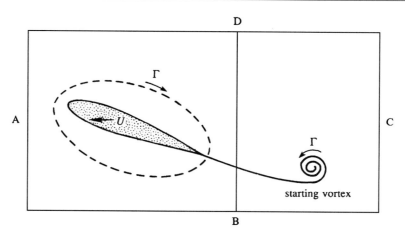

Figure 15.11   A material circuit ABCD in a stationary fluid and an airfoil moving to the left.

of the circulation that would not result in further readjustment is the one required by the Kutta condition. With every change in the speed of the airflow or in the angle of attack, a new starting vortex is cast off and left behind. A new value of circulation around the airfoil is established so as to place the rear stagnation point at the trailing edge in each case.

It is apparent that the *viscosity of the fluid is not only responsible for the drag, but also for the development of circulation and lift*. In developing the circulation, the flow leads to a steady state where a further boundary layer separation is prevented. The establishment of circulation around an airfoil-shaped body in a real fluid is a remarkable result.

7. Conformal Transformation for Generating Airfoil Shape

In the study of airfoils, one is interested in finding the flow pattern and pressure distribution. The *direct* solution of the Laplace equation for the prescribed boundary shape of the airfoil is quite straightforward using a computer, but analytically difficult. In general the analytical solutions are possible only when the airfoil is assumed thin. This is called *thin airfoil theory*, in which the airfoil is replaced by a vortex sheet coinciding with the camber line. An integral equation is developed for the local vorticity distribution from the condition that the camber line be a streamline (velocity tangent to the camber line). The velocity at each point on the camber line is the superposition (i.e., integral) of velocities induced at that point due to the vorticity distribution at all other points on the camber line plus that from the oncoming stream (at infinity). Since the maximum camber is small, this is usually evaluated on the x–y-plane. The Kutta condition is represented by the requirement that the strength of the vortex sheet at the trailing edge is zero. This is treated in detail in Kuethe and Chow (1998, chapter 5) and Anderson (2007, chapter 4). An *indirect* way of solving the problem involves the method of conformal transformation, in which a mapping function is determined such that the arbitrary airfoil shape is transformed into a circle. Then a study of the flow around the circle would determine the flow pattern around the airfoil. This is called *Theodorsen's method*, which is complicated and will not be discussed here.

Instead, we shall deal with a case in which a *given* transformation maps a circle into an airfoil-like shape and determines the properties of the airfoil generated thereby. This is the *Zhukhovsky transformation*

$$z = \zeta + \frac{b^2}{\zeta}, \tag{15.3}$$

where b is a constant. It maps regions of the ζ-plane into the z-plane, some examples of which are discussed in Chapter 6, Section 14. Here, we shall assume circles of different configurations in the ζ-plane and examine their transformed shapes in the z-plane. It will be seen that one of them will result in an airfoil shape.

Transformation of a Circle into a Straight Line

Consider a circle, centered at the origin in the ζ-plane, whose radius b is the same as the constant in the Zhukhovsky transformation (Figure 15.12). For a point $\zeta = b\,e^{i\theta}$ on the circle, the corresponding point in the z-plane is

$$z = b\,e^{i\theta} + b\,e^{-i\theta} = 2b\cos\theta.$$

As θ varies from 0 to π, z goes along the x-axis from $2b$ to $-2b$. As θ varies from π to 2π, z goes from $-2b$ to $2b$. The circle of radius b in the ζ-plane is thus transformed into a straight line of length $4b$ in the z-plane. It is clear that the region *outside* the circle in the ζ-plane is mapped into the *entire* z-plane. (It can be shown that the region inside the circle is also transformed into the entire z-plane. This, however, is of no concern to us, since we shall not consider the interior of the circle in the ζ-plane.)

Transformation of a Circle into a Circular Arc

Let us consider a circle of radius $a\,(>b)$ in the ζ-plane, the center of which is displaced along the η-axis and which cuts the ξ-axis at $(\pm b, 0)$, as shown in Figure 15.13. If a

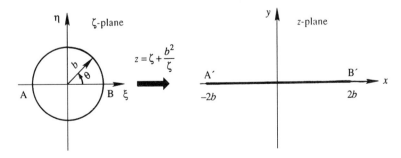

Figure 15.12   Transformation of a circle into a straight line.

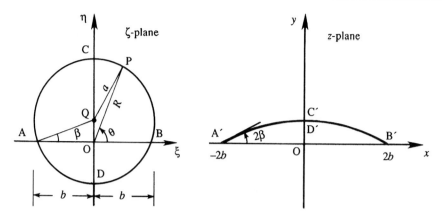

Figure 15.13 Transformation of a circle into a circular arc.

point on the circle in the ζ-plane is represented by $\zeta = Re^{i\theta}$, then the corresponding point in the z-plane is

$$z = Re^{i\theta} + \frac{b^2}{R}e^{-i\theta},$$

whose real and imaginary parts are

$$x = (R + b^2/R)\cos\theta,$$
$$y = (R - b^2/R)\sin\theta. \tag{15.4}$$

Eliminating R, we obtain

$$x^2\sin^2\theta - y^2\cos^2\theta = 4b^2\sin^2\theta\cos^2\theta. \tag{15.5}$$

To understand the shape of the curve represented by equation (15.5) we must express θ in terms of x, y, and the known constants. From triangle OQP, we obtain

$$QP^2 = OP^2 + OQ^2 - 2(OQ)(OP)\cos(Q\hat{O}P).$$

Using $QP = a = b/\cos\beta$ and $OQ = b\tan\beta$, this becomes

$$\frac{b^2}{\cos^2\beta} = R^2 + b^2\tan^2\beta - 2Rb\tan\beta\cos(90° - \theta),$$

which simplifies to

$$2b\tan\beta\sin\theta = R - b^2/R = y/\sin\theta, \tag{15.6}$$

where equation (15.4) has been used. We now eliminate θ between equations (15.5) and (15.6). First note from equation (15.6) that $\cos^2\theta = (2b\tan\beta - y)/2b\tan\beta$, and

$\cot^2\theta = (2b\tan\beta - y)/y$. Then divide equation (15.5) by $\sin^2\theta$, and substitute these expressions of $\cos^2\theta$ and $\cot^2\theta$. This gives

$$x^2 + (y + 2b\cot 2\beta)^2 = (2b\csc 2\beta)^2,$$

where β is known from $\cos\beta = b/a$. This is the equation of a circle in the z-plane, having the center at $(0, -2b\cot 2\beta)$ and a radius of $2b\csc 2\beta$. The Zhukhovsky transformation has thus mapped a complete circle into a circular arc.

Transformation of a Circle into a Symmetric Airfoil

Instead of displacing the center of the circle along the imaginary axis of the ζ-plane, suppose that it is displaced to a point Q on the real axis (Figure 15.14). The radius of the circle is $a\ (>b)$, and we assume that a is slightly larger than b:

$$a \equiv b(1 + e) \qquad e \ll 1. \tag{15.7}$$

A numerical evaluation of the Zhukhovsky transformation (15.3), with assumed values for a and b, shows that the corresponding shape in the z-plane is a streamlined body that is symmetrical about the x-axis. Note that the airfoil in Figure 15.14 has a rounded nose and thickness, while the one in Figure 15.13 has a camber but no thickness.

Transformation of a Circle into a Cambered Airfoil

As can be expected from Figures 15.13 and 15.14, the transformed figure in the z-plane will be a general airfoil with both camber and thickness if the circle in the ζ-plane is displaced in both η and ξ directions (Figure 15.15). The following relations can be proved for $e \ll 1$:

$$c \simeq 4b,$$
$$\text{camber} \simeq \tfrac{1}{2}\beta c, \tag{15.8}$$
$$t_{\max}/c \simeq 1.3\,e.$$

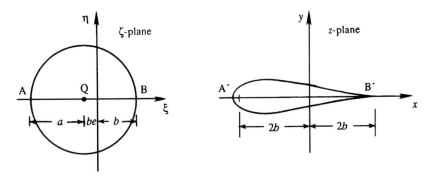

Figure 15.14 Transformation of a circle into a symmetric airfoil.

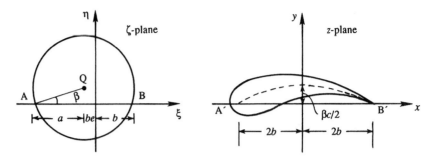

Figure 15.15   Transformation of a circle into a cambered airfoil.

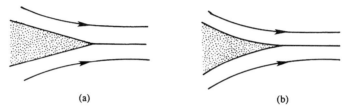

Figure 15.16   Shapes of the trailing edge: (a) trailing edge with finite angle; and (b) cusped trailing edge.

Here t_{max} is the maximum thickness, which is reached nearly at the quarter chord position $x = -b$. The "camber," defined in Figure 15.6, is indicated in Figure 15.15.

Such airfoils generated from the Zhukhovsky transformation are called *Zhukhovsky airfoils*. They have the property that the trailing edge is a *cusp*, which means that the upper and lower surfaces are tangent to each other at the trailing edge. Without the Kutta condition, the trailing edge is a point of infinite velocity, as discussed in Section 5. If the trailing edge angle is nonzero (Figure 15.16a), the coincidence of the stagnation point with the point of infinite velocity still makes the trailing edge a stagnation point, because of the following argument: The fluid velocity on the upper and lower surfaces is parallel to its respective surface. At the trailing edge this leads to normal velocities in different directions, which cannot be possible. The velocities on both sides of the airfoil must therefore be zero at the trailing edge. This is not true for the cusped trailing edge of a Zhukhovsky airfoil (Figure 15.16b). In that case the tangents to the upper and lower surfaces coincide at the trailing edge, and the fluid leaves the trailing edge smoothly. The trailing edge for the Zhukhovsky airfoil is simply an ordinary point where the velocity is neither zero nor infinite.

8. Lift of Zhukhovsky Airfoil

The preceding section has shown how a circle is transformed into an airfoil with the help of the Zhukhovsky transformation. We are now going to determine certain flow properties of such an airfoil. Consider flow around the circle with clockwise circulation Γ in the ζ-plane, in which the approach velocity is inclined at an angle α with the ξ-axis (Figure 15.17). The corresponding pattern in the z-plane is the flow

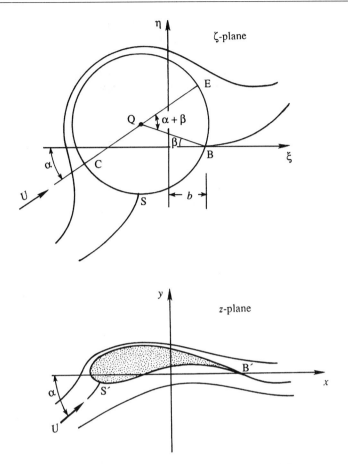

Figure 15.17 Transformation of flow around a circle into flow around an airfoil.

around an airfoil with circulation Γ and angle of attack α. It can be shown that the circulation does not change during a conformal transformation. If $w = \phi + i\psi$ is the complex potential, then the velocities in the two planes are related by

$$\frac{dw}{dz} = \frac{dw}{d\zeta}\frac{d\zeta}{dz}.$$

Using the Zhukhovsky transformation (15.3), this becomes

$$\frac{dw}{dz} = \frac{dw}{d\zeta}\frac{\zeta^2}{\zeta^2 - b^2}. \tag{15.9}$$

Here $dw/dz = u - iv$ is the complex velocity in the z-plane, and $dw/d\zeta$ is the complex velocity in the ζ-plane. Equation (15.9) shows that the velocities in the two planes become equal as $\zeta \to \infty$, which means that the free-stream velocities are inclined at the same angle α in the two planes.

Point B with coordinates $(b, 0)$ in the ζ-plane is transformed into the trailing edge B' of the airfoil. Because $\zeta^2 - b^2$ vanishes there, it follows from equation (15.9) that the velocity at the trailing edge will in general be infinite. If, however, we arrange that B is a stagnation point in the ζ-plane at which $dw/d\zeta = 0$, then dw/dz at the trailing edge will have the 0/0 form. Our discussion of Figure 15.16b has shown that this will in fact result in a finite velocity at B'.

From equation (6.39), the tangential velocity at the surface of the cylinder is given by

$$u_\theta = -2U \sin\theta - \frac{\Gamma}{2\pi a}, \qquad (15.10)$$

where θ is measured from the diameter CQE. At point B, we have $u_\theta = 0$ and $\theta = -(\alpha + \beta)$. Therefore equation (15.10) gives

$$\boxed{\Gamma = 4\pi U a \sin(\alpha + \beta),} \qquad (15.11)$$

which is the clockwise circulation required by the Kutta condition. It shows that the circulation around an airfoil depends on the speed U, the chord length c $(\simeq 4a)$, the angle of attack α, and the camber/chord ratio $\beta/2$. The coefficient of lift is

$$\boxed{C_L = \frac{L}{(1/2)\rho U^2 c} \simeq 2\pi(\alpha + \beta),} \qquad (15.12)$$

where we have used $4a \simeq c$, $L = \rho U \Gamma$, and $\sin(\alpha + \beta) \simeq (\alpha + \beta)$ for small angles of attack. Equation (15.12) shows that the lift can be increased by adding a certain amount of camber. The lift is zero at a negative angle of attack $\alpha = -\beta$, so that the angle $(\alpha + \beta)$ can be called the "absolute" angle of attack. The fact that the lift of an airfoil is proportional to the angle of attack is important, as it suggests that the pilot can control the lift simply by adjusting the attitude of the airfoil.

A comparison of the theoretical lift equation (15.12) with typical experimental results on a Zhukovsky airfoil is shown in Figure 15.18. The small disagreement can be attributed to the finite thickness of the boundary layer changing the effective shape of the airfoil. The sudden drop of the lift at $(\alpha + \beta) \simeq 20°$ is due to a severe boundary layer separation, at which point the airfoil is said to *stall*. This is discussed in Section 12.

Zhukovsky airfoils are not practical for two basic reasons. First, they demand a cusped trailing edge, which cannot be practically constructed or maintained. Second, the camber line in a Zhukovsky airfoil is nearly a circular arc, and therefore the maximum camber lies close to the center of the chord. However, a maximum camber within the forward portion of the chord is usually preferred so as to obtain a desirable pressure distribution. To get around these difficulties, other families of airfoils have been generated from circles by means of more complicated transformations. Nevertheless, the results for a Zhukovsky airfoil given here have considerable application as reference values.

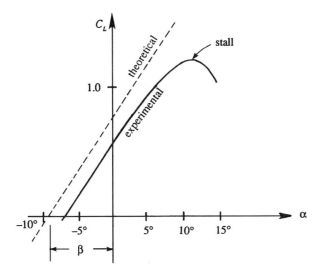

Figure 15.18 Comparison of theoretical and experimental lift coefficients for a cambered Zhukhovsky airfoil.

9. Wing of Finite Span

So far we have considered only two-dimensional flows around wings of infinite span. We shall now consider wings of finite span and examine how the lift and drag are modified. Figure 15.19 shows a schematic view of a wing, looking downstream from the aircraft. As the pressure on the lower surface of the wing is greater than that on the upper surface, air flows around the wing tips from the lower into the upper side. Therefore, there is a spanwise component of velocity toward the wing tip on the underside of the wing and toward the center on the upper side, as shown by the streamlines in Figure 15.20a. The spanwise momentum continues as the fluid goes over the wing and into the wake downstream of the trailing edge. On the stream surface extending downstream from the wing, therefore, the lateral component of the flow is outward (toward the wing tips) on the underside and inward on the upper side. On this surface, then, there is vorticity with axes oriented in the streamwise direction. The vortices have opposite signs on the two sides of the central axis OQ. The streamwise vortex filaments downstream of the wing are called *trailing vortices*, which form a *vortex sheet* (Figure 15.20b). As discussed in Chapter 5, Section 9, a vortex sheet is composed of closely spaced vortex filaments and generates a discontinuity in tangential velocity.

Downstream of the wing the vortex sheet rolls up into two distinct vortices, which are called *tip* or *trailing vortices*. The circulation around each of the tip vortices is equal to Γ_0, the circulation at the center of the wing (Figure 15.21). The existence of the tip vortices becomes visually evident when an aircraft flies in humid air. The decreased pressure (due to the high velocity) and temperature in the core of the tip vortices often cause atmospheric moisture to condense into droplets, which are seen in the form of *vapor trails* extending for kilometers across the sky.

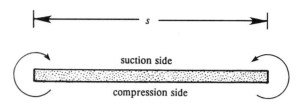

Figure 15.19 Flow around wind tips.

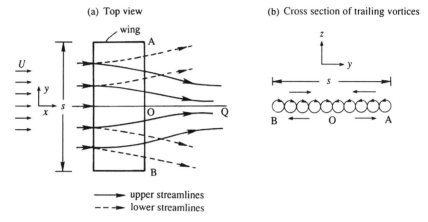

Figure 15.20 Flow over a wing of finite span: (a) top view of streamline patterns on the upper and lower surfaces of the wing; and (b) cross section of trailing vortices behind the wing.

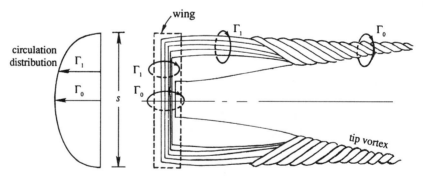

Figure 15.21 Rolling up of trailing vortices to form tip vortices.

One of Helmholtz's vortex theorems states that a vortex filament cannot end in the fluid, but must either end at a solid surface or form a closed loop or "vortex ring." In the case of the finite wing, the tip vortices start at the wing and are joined together at the other end by the starting vortices. The starting vortices are left behind at the point where the aircraft took off, and some of them may be left where the angle of attack was last changed. In any case, they are usually so far behind the wing that their effect on the wing may be neglected, and the tip vortices may be regarded as extending to an infinite distance behind the wing.

As the aircraft proceeds the tip vortices get longer, which means that kinetic energy is being constantly supplied to generate the vortices. It follows that an additional drag force is experienced by a wing of finite span. This is called the *induced drag*, which is explored in the following section.

10. Lifting Line Theory of Prandtl and Lanchester

In this section we shall formalize the concepts presented in the preceding section and derive an expression for the lift and induced drag of a wing of finite span. The basic assumption of the theory is that the value of the aspect ratio span/chord is large, so that the flow around a section is approximately two dimensional. Although a formal mathematical account of the theory was first published by Prandtl, many of the important underlying ideas were first conceived by Lanchester. The historical controversy regarding the credit for the theory is noted at the end of the section.

Bound and Trailing Vortices

It is known that a vortex, like an airfoil, experiences a lift force when placed in a uniform stream. In fact, the disturbance created by an airfoil in a uniform stream is in many ways similar to that created by a vortex filament. It therefore follows that a wing can be replaced by a vortex, with its axis parallel to the wing span. This hypothetical vortex filament replacing the wing is called the *bound vortex*, "bound" signifying that it moves with the wing. We say that the bound vortex is located on a *lifting line*, which is the core of the wing. Recall the discussion in Section 7 where the camber line was replaced by a vortex sheet in thin airfoil theory. This sheet may be regarded as the bound vorticity. According to one of the Helmholtz theorems (Chapter 5, Section 4), a vortex cannot begin or end in the fluid; it must end at a wall or form a closed loop. The bound vortex therefore bends downstream and forms the trailing vortices.

The strength of the circulation around the wing varies along the span, being maximum at the center and zero at the wing tips. A relation can be derived between the distribution of circulation along the wing span and the strength of the trailing vortex filaments. Suppose that the clockwise circulation of the bound vortex changes from Γ to $\Gamma - d\Gamma$ at a certain point (Figure 15.22a). Then another vortex AC of strength $d\Gamma$ must emerge from the location of the change. In fact, the strength and sign of the circulation around AC is such that, when AC is folded back onto AB, the circulation is uniform along the composite vortex tube. (Recall the vortex theorem of Helmholtz, which says that the strength of a vortex tube is constant along its length.)

Now consider the circulation distribution $\Gamma(y)$ over a wing (Figure 15.22b). The change in circulation in length dy is $d\Gamma$, which is a decrease if $dy > 0$. It follows that the magnitude of the trailing vortex filament of width dy is

$$-\frac{d\Gamma}{dy}\, dy,$$

The trailing vortices will be stronger near the wing tips where $d\Gamma/dy$ is the largest.

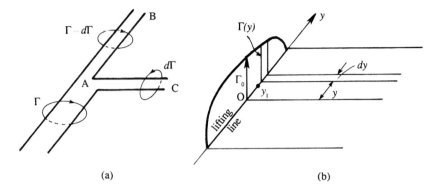

Figure 15.22 Lifting line theory: (a) change of vortex strength; and (b) nomenclature.

Downwash

Let us determine the velocity induced at a point y_1 on the lifting line by the trailing vortex sheet. Consider a semi-infinite trailing vortex filament, whose one end is at the lifting line. Such a vortex of width dy, having a strength $-(d\Gamma/dy)\,dy$, will induce a downward velocity of magnitude

$$dw(y_1) = \frac{-(d\Gamma/dy)\,dy}{4\pi(y - y_1)}.$$

Note that this is *half* the velocity induced by an infinitely long vortex, which equals (circulation)/$(2\pi r)$ where r is the distance from the axis of the vortex. The bound vortex makes no contribution to the velocity induced at the lifting line itself.

The total downward velocity at y_1 due to the entire vortex sheet is therefore

$$w(y_1) = \frac{1}{4\pi} \int_{-s/2}^{s/2} \frac{d\Gamma}{dy} \frac{dy}{(y_1 - y)}, \tag{15.13}$$

which is called the *downwash* at y_1 on the lifting line of the wing. The vortex sheet also induces a smaller downward velocity in front of the airfoil and a larger one behind the airfoil (Figure 15.23).

The effective incident flow on any element of the wing is the resultant of U and w (Figure 15.24). The downwash therefore changes the attitude of the airfoil, decreasing the "geometrical angle of attack" α by the angle

$$\varepsilon = \tan\frac{w}{U} \simeq \frac{w}{U},$$

so that the *effective angle of attack* is

$$\alpha_e = \alpha - \varepsilon = \alpha - \frac{w}{U}. \tag{15.14}$$

Because the aspect ratio is assumed large, ε is small. Each element dy of the finite wing may then be assumed to act as though it is an isolated two-dimensional section

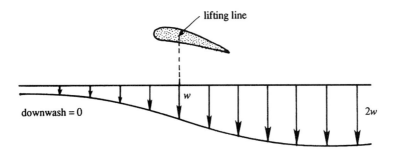

Figure 15.23   Variation of downwash ahead of and behind an airfoil.

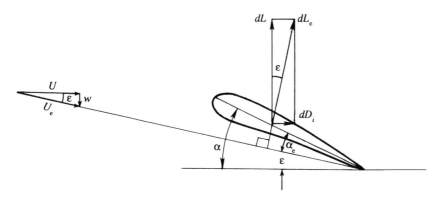

Figure 15.24   Lift and induced drag on a wing element dy.

set in a stream of uniform velocity U_e, at an angle of attack α_e. According to the Kutta–Zhukhovsky lift theorem, a circulation Γ superimposed on the actual resultant velocity U_e generates an elementary aerodynamic force $dL_e = \rho U_e \Gamma \, dy$, which acts normal to U_e. This force may be resolved into two components, the conventional lift force dL normal to the direction of flight and a component dD_i parallel to the direction of flight (Figure 15.24). Therefore

$$dL = dL_e \cos \varepsilon = \rho U_e \Gamma \, dy \cos \varepsilon \simeq \rho U \Gamma \, dy,$$
$$dD_i = dL_e \sin \varepsilon = \rho U_e \Gamma \, dy \sin \varepsilon \simeq \rho w \Gamma \, dy.$$

In general w, Γ, U_e, ε, and α_e are all functions of y, so that for the entire wing

$$L = \int_{-s/2}^{s/2} \rho U \Gamma \, dy,$$
$$D_i = \int_{-s/2}^{s/2} \rho w \Gamma \, dy.$$

$$(15.15)$$

These expressions have a simple interpretation: Whereas the interaction of U and Γ generates L, which acts normal to U, the interaction of w and Γ generates D_i, which acts normal to w.

Induced Drag

The drag force D_i induced by the trailing vortices is called the *induced drag*, which is zero for an airfoil of infinite span. It arises because a wing of finite span continuously creates trailing vortices and the rate of generation of the kinetic energy of the vortices must equal the rate of work done against the induced drag, namely $D_i U$. For this reason the induced drag is also known as the *vortex drag*. It is analogous to the *wave drag* experienced by a ship, which continuously radiates gravity waves during its motion. As we shall see, the induced drag is the largest part of the total drag experienced by an airfoil.

A basic reason why there must be a downward velocity behind the wing is the following: The fluid exerts an upward lift force on the wing, and therefore the wing exerts a downward force on the fluid. The fluid must therefore constantly gain downward momentum as it goes past the wing. (See the photograph of the spinning baseball (Figure 10.27), which exerts an upward force on the fluid.)

For a *given* $\Gamma(y)$, it is apparent that $w(y)$ can be determined from equation (15.13) and D_i can then be determined from equation (15.15). However, $\Gamma(y)$ itself depends on the distribution of $w(y)$, essentially because the effective angle of attack is changed due to $w(y)$. To see how $\Gamma(y)$ may be estimated, first note that the lift coefficient for a two-dimensional Zhukhovsky airfoil is nearly $C_L = 2\pi(\alpha + \beta)$. For a finite wing we may assume

$$C_L = K\left[\alpha - \frac{w(y)}{U} + \beta(y)\right], \tag{15.16}$$

where $(\alpha - w/U)$ is the effective angle of attack, $-\beta(y)$ is the angle of attack for zero lift (found from experimental data such as Figure 15.18), and K is a constant whose value is nearly 6 for most airfoils. ($K = 2\pi$ for a Zhukhovsky airfoil.) An expression for the circulation can be obtained by noting that the lift coefficient is related to the circulation as $C_L \equiv L/(\frac{1}{2}\rho U^2 c) = \Gamma/(\frac{1}{2}Uc)$, so that $\Gamma = \frac{1}{2}UcC_L$. The assumption equation (15.16) is then equivalent to the assumption that the circulation for a wing of finite span is

$$\Gamma(y) = \frac{K}{2}Uc(y)\left[\alpha - \frac{w(y)}{U} + \beta(y)\right]. \tag{15.17}$$

For a given U, α, $c(y)$, and $\beta(y)$, equations (15.13) and (15.17) define an integral equation for determining $\Gamma(y)$. (An integral equation is one in which the unknown function appears under an integral sign.) The problem can be solved numerically by iterative techniques. Instead of pursuing this approach, in the next section we shall assume that $\Gamma(y)$ is given.

Lanchester versus Prandtl

There is some controversy in the literature about who should get more credit for developing modern wing theory. Since Prandtl in 1918 first published the theory in a mathematical form, textbooks for a long time have called it the "Prandtl Lifting Line Theory." Lanchester was bitter about this, because he felt that his contributions were not adequately recognized. The controversy has been discussed by von Karman (1954, p. 50), who witnessed the development of the theory. He gives a lot of credit to Lanchester, but falls short of accusing his teacher Prandtl of being deliberately unfair. Here we shall note a few facts that von Karman brings up.

Lanchester was the first person to study a wing of finite span. He was also the first person to conceive that a wing can be replaced by a bound vortex, which bends backward to form the tip vortices. Last, Lanchester was the first to recognize that the minimum power necessary to fly is that required to generate the kinetic energy field of the downwash field. It seems, then, that Lanchester had conceived all of the basic ideas of the wing theory, which he published in 1907 in the form of a book called "Aerodynamics." In fact, a figure from his book looks very similar to our Figure 15.21.

Many of these ideas were explained by Lanchester in his talk at Göttingen, long before Prandtl published his theory. Prandtl, his graduate student von Karman, and Carl Runge were all present. Runge, well-known for his numerical integration scheme of ordinary differential equations, served as an interpreter, because neither Lanchester nor Prandtl could speak the other's language. As von Karman said, "both Prandtl and Runge learned very much from these discussions."

However, Prandtl did not want to recognize Lanchester for priority of ideas, saying that he conceived of them before he saw Lanchester's book. Such controversies cannot be settled. And great men have been involved in controversies before. For example, astrophysicist Stephen Hawking (1988), who occupied Newton's chair at Cambridge (after Lighthill), described Newton to be a rather mean man who spent much of his later years in unfair attempts at discrediting Leibniz, in trying to force the Royal astronomer to release some unpublished data that he needed to verify his predictions, and in heated disputes with his lifelong nemesis Robert Hooke.

In view of the fact that Lanchester's book was already in print when Prandtl published his theory, and the fact that Lanchester had all the ideas but not a formal mathematical theory, we have called it the "Lifting Line Theory of Prandtl and Lanchester."

11. *Results for Elliptic Circulation Distribution*

The induced drag and other properties of a finite wing depend on the distribution of $\Gamma(y)$. The circulation distribution, however, depends in a complicated way on the wing planform, angle of attack, and so on. It can be shown that, for a given total lift and wing area, the induced drag is a minimum when the circulation distribution is elliptic. (See, for e.g., Ashley and Landahl, 1965, for a proof.) Here we shall simply assume an elliptic distribution of the form (see Figure 15.22b)

$$\Gamma = \Gamma_0 \left[1 - \left(\frac{2y}{s} \right)^2 \right]^{1/2}, \tag{15.18}$$

and determine the resulting expressions for downwash and induced drag.

The total lift force on a wing is then

$$L = \int_{-s/2}^{s/2} \rho U \Gamma \, dy = \frac{\pi}{4} \rho U \Gamma_0 s. \tag{15.19}$$

To determine the downwash, we first find the derivative of equation (15.18):

$$\frac{d\Gamma}{dy} = -\frac{4\Gamma_0 y}{s\sqrt{s^2 - 4y^2}}.$$

From equation (15.13), the downwash at y_1 is

$$w(y_1) = \frac{1}{4\pi} \int_{-s/2}^{s/2} \frac{d\Gamma}{dy} \frac{dy}{y_1 - y} = \frac{\Gamma_0}{\pi s} \int_{-s/2}^{s/2} \frac{y \, dy}{(y - y_1)\sqrt{s^2 - 4y^2}}.$$

Writing $y = (y - y_1) + y_1$ in the numerator, we obtain

$$w(y_1) = \frac{\Gamma_0}{\pi s} \left[\int_{-s/2}^{s/2} \frac{dy}{\sqrt{s^2 - 4y^2}} + y_1 \int_{-s/2}^{s/2} \frac{dy}{(y - y_1)\sqrt{s^2 - 4y^2}} \right].$$

The first integral has the value $\pi/2$. The second integral can be reduced to a standard form (listed in any mathematical handbook) by substituting $x = y - y_1$. On setting limits the second integral turns out to be zero, although the integrand is not an odd function. The downwash at y_1 is therefore

$$w(y_1) = \frac{\Gamma_0}{2s}, \tag{15.20}$$

which shows that, for an elliptic circulation distribution, the induced velocity at the wing is constant along the span.

Using equations (15.18) and (15.20), the induced drag is found as

$$D_{\mathrm{i}} = \int_{-s/2}^{s/2} \rho w \Gamma \, dy = \frac{\pi}{8} \rho \Gamma_0^2.$$

In terms of the lift equation (15.19), this becomes

$$D_{\mathrm{i}} = \frac{2L^2}{\rho U^2 \pi s^2},$$

which can be written as

$$\boxed{C_{D_{\mathrm{i}}} = \frac{C_L^2}{\pi \Lambda},} \tag{15.21}$$

where we have defined the coefficients (here A is the wing planform area)

$$\Lambda \equiv \frac{s^2}{A} = \text{aspect ratio}$$

$$C_{D_i} \equiv \frac{D_i}{(1/2)\rho U^2 A}, \qquad C_L \equiv \frac{L}{(1/2)\rho U^2 A}.$$

Equation (15.21) shows that $C_{D_i} \to 0$ in the two-dimensional limit $\Lambda \to \infty$. More important, it shows that the *induced drag coefficient increases as the square of the lift coefficient*. We shall see in the following section that the induced drag generally makes the largest contribution to the total drag of an airfoil.

Since an elliptic circulation distribution minimizes the induced drag, it is of interest to determine the circumstances under which such a circulation can be established. Consider an element dy of the wing (Figure 15.25). The lift on the element is

$$dL = \rho U \Gamma \, dy = C_L \tfrac{1}{2} \rho U^2 c \, dy, \tag{15.22}$$

where $c \, dy$ is an elementary wing area. Now if the circulation distribution is elliptic, then the downwash is independent of y. In addition, if the wing profile is geometrically similar at every point along the span and has the same geometrical angle of attack α, then the effective angle of attack and hence the lift coefficient C_L will be independent of y. Equation (15.22) shows that the chord length c is then simply proportional to Γ, and so $c(y)$ is also elliptically distributed. Thus, an untwisted wing with elliptic planform, or composed of two semiellipses (Figure 15.25), will generate an elliptic circulation distribution. However, the same effect can also be achieved with nonelliptic planforms if the angle of attack varies along the span, that is, if the wing is given a "twist."

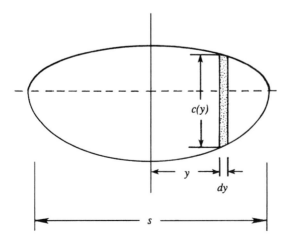

Figure 15.25 Wing of elliptic planform.

12. Lift and Drag Characteristics of Airfoils

Before an aircraft is built its wings are tested in a wind tunnel, and the results are generally given as plots of C_L and C_D vs the angle of attack. A typical plot is shown in Figure 15.26. It is seen that, in a range of incidence angle from $\alpha = -4°$ to $\alpha = 12°$, the variation of C_L with α is approximately linear, a typical value of $dC_L/d\alpha$ being ≈ 0.1 per degree. The lift reaches a maximum value at an incidence of $\approx 15°$. If the angle of attack is increased further, the steep adverse pressure gradient on the upper surface of the airfoil causes the flow to separate nearly at the leading edge, and a very large wake is formed (Figure 15.27). The lift coefficient drops suddenly, and the wing is said to *stall*. Beyond the stalling incidence the lift coefficient levels off again and remains at ≈ 0.7–0.8 for fairly large angles of incidence.

The maximum lift coefficient depends largely on the Reynolds number Re. At lower values of Re $\sim 10^5$–10^6, the flow separates before the boundary layer undergoes

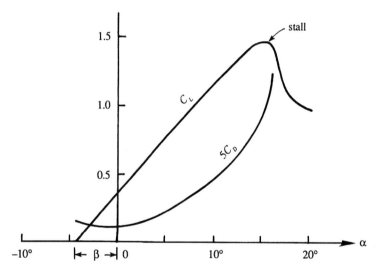

Figure 15.26   Lift and drag coefficients vs angle of attack.

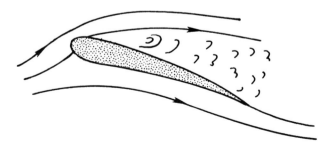

Figure 15.27   Stalling of an airfoil.

transition, and a very large wake is formed. This gives maximum lift coefficients <0.9. At larger Reynolds numbers, say Re $> 10^7$, the boundary layer undergoes transition to turbulent flow before it separates. This produces a somewhat smaller wake, and maximum lift coefficients of ≈ 1.4 are obtained.

The angle of attack at zero lift, denoted by $-\beta$ here, is a function of the section camber. (For a Zhukhovsky airfoil, $\beta = 2$(camber)/chord.) The effect of increasing the airfoil camber is to raise the entire graph of C_L vs α, thus increasing the maximum values of C_L without stalling. A cambered profile delays stalling essentially because its leading edge points into the airstream while the rest of the airfoil is inclined to the stream. Rounding the airfoil nose is very helpful, for an airfoil of zero thickness would undergo separation at the leading edge. Trailing edge flaps act to increase the camber when they are deployed. Then the maximum lift coefficient is increased, allowing for lower landing speeds.

Various terms are in common usage to describe the different components of the drag. The total drag of a body can be divided into a *friction drag* due to the tangential stresses on the surface and *pressure drag* due to the normal stresses. The pressure drag can be further subdivided into an *induced drag* and a *form drag*. The induced drag is the "drag due to lift" and results from the work done by the body to supply the kinetic energy of the downwash field as the trailing vortices increase in length. The form drag is defined as the part of the total pressure drag that remains after the induced drag is subtracted out. (Sometimes the skin friction and form drags are grouped together and called the *profile drag*, which represents the drag due to the "profile" alone and not due to the finiteness of the wing.) The form drag depends strongly on the shape and orientation of the airfoil and can be minimized by good design. In contrast, relatively little can be done about the induced drag if the aspect ratio is fixed.

Normally the induced drag constitutes the major part of the total drag of a wing. As C_{D_i} is nearly proportional to C_L^2, and C_L is nearly proportional to α, it follows that $C_{D_i} \propto \alpha^2$. This is why the drag coefficient in Figure 15.26 seems to increase quadratically with incidence.

For high-speed aircraft, the appearance of shock waves can adversely affect the behavior of the lift and drag characteristics. In such cases the maximum *flow* speeds can be close to or higher than the speed of sound even when the aircraft is flying at subsonic speeds. Shock waves can form when the local flow speed exceeds the local speed of sound. To reduce their effect, the wings are given a *sweepback angle*, as shown in Figure 15.2. The maximum flow speeds depend primarily on the component of the oncoming stream perpendicular to the leading edge; this component is reduced as a result of the sweepback. As a result, increased flight speeds are achievable with highly swept wings. This is particularly true when the aircraft flies at supersonic speeds, in which there is invariably a shock wave in front of the nose of the fuselage, extending downstream in the form of a cone. Highly swept wings are then used in order that the wing does not penetrate this shock wave. For flight speeds exceeding Mach numbers of order 2, the wings have such large sweepback angles that they resemble the Greek letter Δ; these wings are sometimes called *delta wings*.

13. *Propulsive Mechanisms of Fish and Birds*

The propulsive mechanisms of many animals utilize the aerodynamic principle of lift generation on winglike surfaces. We shall now describe some of the basic ideas of this interesting subject, which is discussed in more detail by Lighthill (1986).

Locomotion of Fish

First consider the case of a fish. It develops a *forward* thrust by horizontally oscillating its tail from *side to side*. The tail has a cross section resembling that of a symmetric airfoil (Figure 15.28a). One-half of the oscillation is represented in Figure 15.28b, which shows the top view of the tail. The sequence 1 to 5 represents the positions of the tail during the tail's motion to the left. A quick change of *orientation* occurs at one extreme position of the oscillation during 1 to 2; the tail then moves to the left during 2 to 4, and another quick change of orientation occurs at the other extreme during 4 to 5.

Suppose the tail is moving to the left at speed V, and the fish is moving forward at speed U. The fish controls these magnitudes so that the resultant fluid velocity U_r (relative to the tail) is inclined to the tail surface at a positive "angle of attack." The resulting lift L is perpendicular to U_r and has a forward component $L \sin \theta$. (It is easy to verify that there is a similar forward propulsive force when the tail moves from left to right.) This thrust, working at the rate $UL \sin \theta$, propels the fish. To achieve this propulsion, the tail of the fish pushes sideways on the water against a force of $L \cos \theta$, which requires work at the rate $VL \cos \theta$. As $V/U = \tan \theta$, ideally the conversion of energy is perfect—all of the oscillatory work done by the fish tail goes into the translational mode. In practice, however, this is not the case because of the presence of induced drag and other effects that generate a wake.

Most fish stay afloat by controlling the buoyancy of a swim bladder inside their stomach. In contrast, some large marine mammals such as whales and dolphins

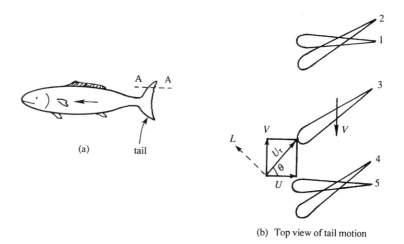

(b) Top view of tail motion

Figure 15.28 Propulsion of fish. (a) Cross section of the tail along AA is a symmetric airfoil. Five positions of the tail during its motion to the left are shown in (b). The lift force L is normal to the resultant speed U_r of water with respect to the tail.

develop *both* a forward thrust and a vertical lift by moving their tails *vertically*. They are able to do this because their tail surface is *horizontal*, in contrast to the vertical tail shown in Figure 15.28. A recent review by Fish and Lauder (2006) provided evidence that leading edge tubercles as seen on humpback whale flippers increase lift and reduce drag at high angles of attack. This is because separation is delayed due to the creation of streamwise vortices on the suction side. Cetacean flukes or flippers and fish tail fins as well as dorsal and pectoral fins are flexible and can vary their camber during a stroke. As a result they are very efficient propulsive devices.

Flight of Birds and Insects

Now consider the flight of birds, who flap their wings to generate *both* the lift to support their body weight and the forward thrust to overcome the drag. Figure 15.29 shows a vertical section of the wing positions during the upstroke and downstroke of the wing. (Birds have cambered wings, but this is not shown in the figure.) The angle of inclination of the wing with the airstream changes suddenly at the end of each stroke, as shown. The important point is that the upstroke is inclined at a greater angle to the airstream than the downstroke. As the figure shows, the downstroke develops a lift force L perpendicular to the resultant velocity of the air relative to the wing. Both

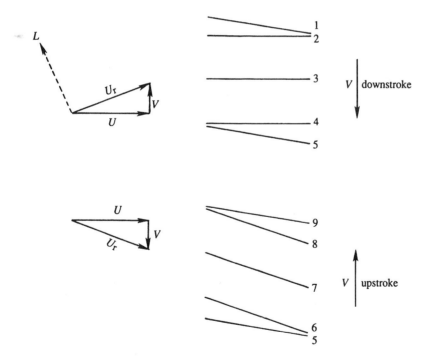

Figure 15.29 Propulsion of a bird. A cross section of the wing is shown during upstroke and downstroke. During the downstroke, a lift force L acts normal to the resultant speed U_r of air with respect to the wing. During the upstroke, U_r is nearly parallel to the wing and very little aerodynamic force is generated.

a forward thrust and an upward force result from the downstroke. In contrast, very little aerodynamic force is developed during the upstroke, as the resultant velocity is then nearly parallel to the wing. Birds therefore do most of the work during the downstroke, and the upstroke is "easy."

A recent study by Liu et al. (2006) provides the most complete description to date of wing planform, camber, airfoil section, and spanwise twist distribution of seagulls, mergansers, teals, and owls. Moreover, flapping as viewed by video images from free flight was digitized and modeled by a two-jointed wing at the quarter chord point. The data from this paper can be used to model the aerodynamics of bird flight.

Using previously measured kinematics and experiments on an approximately 100X upscaled model, Ramamurti and Sandberg (2001) calculated the flow about a Drosophila (fruit fly) in flight. They matched Reynolds number (based on wing-tip speed and average chord) and found that viscosity had negligible effect on thrust and drag at a flight Reynolds number of 120. The wings were near elliptical plates with axis ratio 3:1.2 and thickness about 1/80 of the span. Averaged over a cycle, the mean thrust coefficient (thrust/[dynamic pressure × wing surface]) was 1.3 and the mean drag coefficient close to 1.5.

14. Sailing against the Wind

People have sailed without the aid of an engine for thousands of years and have known how to arrive at a destination against the wind. Actually, it is not possible to sail exactly against the wind, but it is possible to sail at $\approx$40–45° to the wind. Figure 15.30 shows how this is made possible by the aerodynamic lift on the sail, which is a piece of large stretched cloth. The wind speed is U, and the sailing speed is V, so that the apparent wind speed relative to the boat is U_r. If the sail is properly oriented, this gives rise to a lift force perpendicular to U_r and a drag force parallel to U_r. The resultant force F can be resolved into a driving component (thrust) along the motion of the boat and a lateral component. The driving component performs work in moving the boat; most of this work goes into overcoming the frictional drag and in generating the gravity waves that radiate outward. The lateral component does not cause much sideways drift because of the shape of the hull. It is clear that the thrust decreases as the angle θ decreases and normally vanishes when θ is $\approx$40–45°. The energy for sailing comes from the wind field, which loses kinetic energy after passing through the sail.

In the foregoing discussion we have not considered the hydrodynamic forces exerted by the water on the hull. At constant sailing speed the net hydrodynamic force must be equal and opposite to the net aerodynamic force on the sail. The hydrodynamic force can be decomposed into a drag (parallel to the direction of motion) and a lift. The lift is provided by the "keel," which is a thin vertical surface extending downward from the bottom of the hull. For the keel to act as a lifting surface, the longitudinal axis of the boat points at a small angle to the direction of motion of the boat, as indicated near the bottom right part of Figure 15.30. This "angle of attack" is generally <3° and is not noticeable. The hydrodynamic lift developed by the keel opposes the aerodynamic lateral force on the sail. It is clear that without the keel the lateral aerodynamic force on the sail would topple the boat around its longitudinal axis.

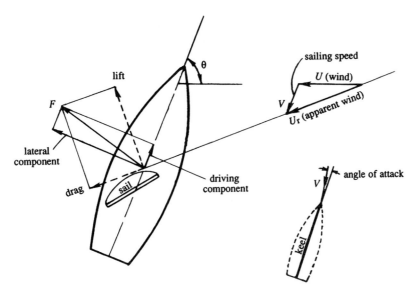

Figure 15.30 Principle of a sailboat.

To arrive at a destination directly against the wind, one has to sail in a zig-zag path, always maintaining an angle of $\approx 45°$ to the wind. For example, if the wind is coming from the east, we can first proceed northeastward as shown, then change the orientation of the sail to proceed southeastward, and so on. In practice, a combination of a number of sails is used for effective maneuvering. The mechanics of sailing yachts is discussed in Herreshoff and Newman (1966).

Exercises

1. Consider an airfoil section in the xy-plane, the x-axis being aligned with the chordline. Examine the pressure forces on an element $d\mathbf{s} = (dx, dy)$ on the surface, and show that the net force (per unit span) in the y-direction is

$$F_y = -\int_0^c p_u \, dx + \int_0^c p_l \, dx,$$

where p_u and p_l are the pressures on the upper and the lower surfaces and c is the chord length. Show that this relation can be rearranged in the form

$$C_y \equiv \frac{F_y}{(1/2)\rho U^2 c} = \oint C_p \, d\left(\frac{x}{c}\right),$$

where $C_p \equiv (p - p_\infty)/(\frac{1}{2}\rho U^2)$, and the integral represents the area enclosed in a C_p vs x/c diagram, such as Figure 15.8. Neglect shear stresses. [Note that C_y is not exactly the lift coefficient, since the airstream is inclined at a small angle α with the x-axis.]

2. The measured pressure distribution over a section of a two-dimensional airfoil at 4° incidence has the following form:

> *Upper Surface*: C_p is constant at -0.8 from the leading edge to a distance equal to 60% of chord and then increases linearly to 0.1 at the trailing edge.
> *Lower Surface*: C_p is constant at -0.4 from the leading edge to a distance equal to 60% of chord and then increases linearly to 0.1 at the trailing edge.

Using the results of Exercise 1, show that the lift coefficient is nearly 0.32.

3. The Zhukhovsky transformation $z = \zeta + b^2/\zeta$ transforms a circle of radius b, centered at the origin of the ζ-plane, into a flat plate of length $4b$ in the z-plane. The circulation around the cylinder is such that the Kutta condition is satisfied at the trailing edge of the flat plate. If the plate is inclined at an angle α to a uniform stream U, show that

(i) The complex velocity in the ζ-plane is

$$w = U \left(\zeta\, e^{-i\alpha} + \frac{1}{\zeta} b^2\, e^{i\alpha} \right) + \frac{i\Gamma}{2\pi} \ln\, (\zeta\, e^{-i\alpha}),$$

where $\Gamma = 4\pi U b \sin\alpha$. Note that this represents flow over a circular cylinder with circulation, in which the oncoming velocity is oriented at an angle α.

(ii) The velocity components at point P $(-2b, 0)$ in the ζ-plane are $[\frac{3}{4} U \cos\alpha, \frac{9}{4} U \sin\alpha]$.

(iii) The coordinates of the transformed point P$'$ in the xy-plane are $[-5b/2, 0]$.

(iv) The velocity components at $[-5b/2, 0]$ in the xy-plane are $[U \cos\alpha, 3U \sin\alpha]$.

4. In Figure 15.13, the angle at A$'$ has been marked 2β. Prove this. [*Hint*: Locate the center of the circular arc in the z-plane.]

5. Consider a cambered Zhukhovsky airfoil determined by the following parameters:

$$a = 1.1,$$
$$b = 1.0,$$
$$\beta = 0.1.$$

Using a computer, plot its contour by evaluating the Zhukhovsky transformation. Also plot a few streamlines, assuming an angle of attack of 5°.

6. A thin Zhukhovsky airfoil has a lift coefficient of 0.3 at zero incidence. What is the lift coefficient at 5° incidence?

7. An untwisted elliptic wing of 20-m span supports a weight of 80,000 N in a level flight at 300 km/hr. Assuming sea level conditions, find (i) the induced drag and (ii) the circulation around sections halfway along each wing.

8. The circulation across the span of a wing follows the parabolic law

$$\Gamma = \Gamma_0 \left(1 - \frac{4y^2}{s^2} \right)$$

Calculate the induced velocity w at midspan, and compare the value with that obtained when the distribution is elliptic.

Literature Cited

Anderson, John D., Jr. (1998). *A History of Aerodynamics*, London: Cambridge University Press.

Anderson, John D., Jr. (2007). *Fundamentals of Aerodynamics*, New York: McGraw-Hill.

Ashley, H. and M. Landahl (1965). *Aerodynamics of Wings and Bodies*, Reading, MA: Addison-Wesley.

Fish, F. E. and G. V. Lauder (2006). "Passive and Active Control by Swimming Fishes and Mammals." *Annual Rev. Fluid Mech.* **38**: 193–224.

Hawking, S. W. (1988). *A Brief History of Time*, New York: Bantam Books.

Herreshoff, H. C. and J. N. Newman (1986). "The study of sailing yachts." *Scientific American* **215** (August issue): 61–68.

von Karman, T. (1954). *Aerodynamics*, New York: McGraw-Hill. (A delightful little book, written for the nonspecialist, full of historical anecdotes and at the same time explaining aerodynamics in the easiest way.)

Kuethe, A. M. and C. Y. Chow (1998). *Foundations of Aerodynamics: Basis of Aerodynamic Design*, New York: Wiley.

Lighthill, M. J. (1986). *An Informal Introduction to Theoretical Fluid Mechanics*, Oxford, England: Clarendon Press.

Liu, T., K. Kuykendoll, R. Rhew, and S. Jones (2006). "Avian Wing Geometry and Kinematics." *AIAA J.* **44**: 954–963.

Ramamurti, R. and W. C. Sandberg (2001). "Computational Study of 3-D Flapping Foil Flows." AIAA Paper 2001–0605.

Supplemental Reading

Batchelor, G. K. (1967). *An Introduction to Fluid Dynamics*, London: Cambridge University Press.

Karamcheti, K. (1980). *Principles of Ideal-Fluid Aerodynamics*, Melbourne, FL: Krieger Publishing Co.

Prandtl, L. (1952). *Essentials of Fluid Dynamics*, London: Blackie & Sons Ltd. (This is the English edition of the original German edition. It is very easy to understand, and much of it is still relevant today.) Printed in New York by Hafner Publishing Co. If this is unavailable, see the following reprints in paperback that contain much if not all of this material:

Prandtl, L. and O. G. Tietjens (1934) [original publication date]. *Fundamentals of Hydro and Aero-mechanics*, New York: Dover Publ. Co.; and

Prandtl, L. and O. G. Tietjens (1934) [original publication date]. *Applied Hydro and Aeromechanics*, New York: Dover Publ. Co. This contains many original flow photographs from Prandtl's laboratory.

Compressible Flow

1. Introduction

To this point we have neglected the effects of density variations due to pressure changes. In this chapter we shall examine some elementary aspects of flows in which the compressibility effects are important. The subject of compressible flows is also

©2010 Elsevier Inc. All rights reserved.
DOI: 10.1016/B978-0-12-381399-2.50016-2

called *gas dynamics*, which has wide applications in high-speed flows around objects of engineering interest. These include *external flows* such as those around airplanes, and *internal flows* in ducts and passages such as nozzles and diffusers used in jet engines and rocket motors. Compressibility effects are also important in astrophysics. Two popular books dealing with compressibility effects in engineering applications are those by Liepmann and Roshko (1957) and Shapiro (1953), which discuss in further detail most of the material presented here.

Our study in this chapter will be rather superficial and elementary because this book is essentially about incompressible flows. However, this small chapter on compressible flows is added because a complete ignorance about compressibility effects is rather unsatisfying. Several startling and fascinating phenomena arise in compressible flows (especially in the supersonic range) that go against our intuition developed from a knowledge of incompressible flows. Discontinuities (shock waves) appear within the flow, and a rather strange circumstance arises in which an increase of flow area *accelerates* a (supersonic) stream. Friction can also make the flow go faster and adding heat can lower the temperature in subsonic duct flows. We will see this later in this chapter. Some understanding of these phenomena, which have no counterpart in low-speed flows, is desirable even if the reader may not make much immediate use of this knowledge. Except for our treatment of friction in constant area ducts, we shall limit our study to that of frictionless flows outside boundary layers. Our study will, however, have a great deal of practical value because the boundary layers are especially thin in high-speed flows. Gravitational effects, which are minor in high-speed flows, will be neglected.

Criterion for Neglect of Compressibility Effects

Compressibility effects are determined by the magnitude of the *Mach number* defined as

$$M \equiv \frac{u}{c},$$

where u is the speed of flow, and c is the speed of sound given by

$$c^2 = \left(\frac{\partial p}{\partial \rho}\right)_s,$$

where the subscript "s" signifies that the partial derivative is taken at constant entropy. To see how large the Mach number has to be for the compressibility effects to be appreciable in a steady flow, consider the one-dimensional steady flow version of the continuity equation $\nabla \cdot (\rho \mathbf{u}) = 0$, that is,

$$u\frac{\partial \rho}{\partial x} + \rho\frac{\partial u}{\partial x} = 0.$$

The incompressibility assumption requires that

$$u\frac{\partial \rho}{\partial x} \ll \rho\frac{\partial u}{\partial x}$$

or that

$$\frac{\delta\rho}{\rho} \ll \frac{\delta u}{u}. \qquad (16.1)$$

Pressure changes can be estimated from the definition of c, giving

$$\delta p \simeq c^2 \delta \rho. \qquad (16.2)$$

The Euler equation requires

$$u\,\delta u \simeq \frac{\delta p}{\rho}. \qquad (16.3)$$

By combining equations (16.2) and (16.3), we obtain

$$\frac{\delta\rho}{\rho} \simeq \frac{u^2}{c^2}\frac{\delta u}{u}.$$

From comparison with equation (16.1) we see that the density changes are negligible if

$$\frac{u^2}{c^2} = M^2 \ll 1.$$

The constant density *assumption is therefore valid if M < 0.3, but not at higher Mach numbers.*

Although the significance of the ratio u/c was known for a long time, the Swiss aerodynamist Jacob Ackeret introduced the term "Mach number," just as the term Reynolds number was introduced by Sommerfeld many years after Reynolds' experiments. The name of the Austrian physicist Ernst Mach (1836–1916) was chosen because of his pioneering studies on supersonic motion and his invention of the so-called *Schlieren method* for optical studies of flows involving density changes; see von Karman (1954, p. 106). (Mach distinguished himself equally well in philosophy. Einstein acknowledged that his own thoughts on relativity were influenced by "Mach's principle," which states that properties of space had no independent existence but are determined by the mass distribution within it. Strangely, Mach never accepted either the theory of relativity or the atomic structure of matter.)

Classification of Compressible Flows

Compressible flows can be classified in various ways, one of which is based on the Mach number M. A common way of classifying flows is as follows:

(i) *Incompressible flow*: $M < 0.3$ everywhere in the flow. Density variations due to pressure changes can be neglected. The gas medium is compressible but the density may be regarded as constant.

(ii) *Subsonic flow*: M exceeds 0.3 somewhere in the flow, but does not exceed 1 anywhere. Shock waves do not appear in the flow.

(iii) *Transonic flow*: The Mach number in the flow lies in the range 0.8–1.2. Shock waves appear and lead to a rapid increase of the drag. Analysis of transonic flows is difficult because the governing equations are inherently nonlinear, and also because a separation of the inviscid and viscous aspects of the flow is often impossible. (The word "transonic" was invented by von Karman and Hugh Dryden, although the latter argued in favor of having two s's in the word. von Karman (1954, p. 116) stated that "I first introduced the term in a report to the U.S. Air Force. I am not sure whether the general who read the word knew what it meant, but his answer contained the word, so it seemed to be officially accepted.")

(iv) *Supersonic flow*: M lies in the range 1–3. Shock waves are generally present. In many ways analysis of a flow that is supersonic everywhere is easier than an analysis of a subsonic or incompressible flow as we shall see. This is because information propagates along certain directions, called characteristics, and a determination of these directions greatly facilitates the computation of the flow field.

(v) *Hypersonic flow*: $M > 3$. The very high flow speeds cause severe heating in boundary layers, resulting in dissociation of molecules and other chemical effects.

Useful Thermodynamic Relations

As density changes are accompanied by temperature changes, thermodynamic principles will be constantly used here. Most of the necessary concepts and relations have been summarized in Sections 8 and 9 of Chapter 1, which may be reviewed before proceeding further. Some of the most frequently used relations, valid for a perfect gas with constant specific heats, are listed here for quick reference:

$$\textit{Equation of state} \quad p = \rho R T,$$

$$\textit{Internal energy} \quad e = C_v T,$$

$$\textit{Enthalpy} \quad h = C_p T,$$

$$\textit{Specific heats} \quad C_p = \frac{\gamma R}{\gamma - 1},$$

$$C_v = \frac{R}{\gamma - 1},$$

$$C_p - C_v = R,$$

$$\textit{Speed of sound} \quad c = \sqrt{\gamma R T},$$

$$\textit{Entropy change} \quad S_2 - S_1 = C_p \ln \frac{T_2}{T_1} - R \ln \frac{p_2}{p_1}, \tag{16.4}$$

$$= C_v \ln \frac{T_2}{T_1} - R \ln \frac{\rho_2}{\rho_1}. \tag{16.5}$$

An isentropic process of a perfect gas between states 1 and 2 obeys the following relations:

$$\frac{p_2}{p_1} = \left(\frac{\rho_2}{\rho_1}\right)^\gamma,$$

$$\frac{T_2}{T_1} = \left(\frac{\rho_2}{\rho_1}\right)^{\gamma-1} = \left(\frac{p_2}{p_1}\right)^{(\gamma-1)/\gamma}.$$

Some important properties of air at ordinary temperatures and pressures are

$$R = 287 \, \mathrm{m}^2/(\mathrm{s}^2 \, \mathrm{K}),$$
$$C_p = 1005 \, \mathrm{m}^2/(\mathrm{s}^2 \, \mathrm{K}),$$
$$C_v = 718 \, \mathrm{m}^2/(\mathrm{s}^2 \, \mathrm{K}),$$
$$\gamma = 1.4.$$

These values will be useful for solution of the exercises.

2. *Speed of Sound*

We know that a pressure pulse in an incompressible flow behaves in the same way as that in a rigid body, where a displaced particle simultaneously displaces *all* the particles in the medium. The effects of pressure or other changes are therefore instantly felt throughout the medium. A compressible fluid, in contrast, behaves similarly to an elastic solid, in which a displaced particle compresses and increases the density of adjacent particles that move and increase the density of the neighboring particles, and so on. In this way a disturbance in the form of an *elastic wave*, or a *pressure wave*, travels through the medium. The speed of propagation is faster when the medium is more rigid. If the amplitude of the elastic wave is infinitesimal, it is called an *acoustic wave*, or a *sound wave*.

We shall now find an expression for the speed of propagation of sound. Figure 16.1a shows an infinitesimal pressure pulse propagating to the left with speed c into a still fluid. The fluid properties ahead of the wave are p, T, and ρ, while the flow speed is $u = 0$. The properties behind the wave are $p + dp$, $T + dT$, and $\rho + d\rho$, whereas the flow speed is du directed to the left. We shall see that a "compression wave" (for which the fluid pressure *rises* after the passage of the wave) must move the fluid in the direction of propagation, as shown in Figure 16.1a. In contrast, an "expansion wave" moves the fluid "backwards."

To make the analysis steady, we superimpose a velocity c, directed to the right, on the entire system (Figure 16.1b). The wave is now stationary, and the fluid enters the wave with velocity c and leaves with a velocity $c - du$. Consider an area A on the wavefront. A mass balance gives

$$A\rho c = A(\rho + d\rho)(c - du).$$

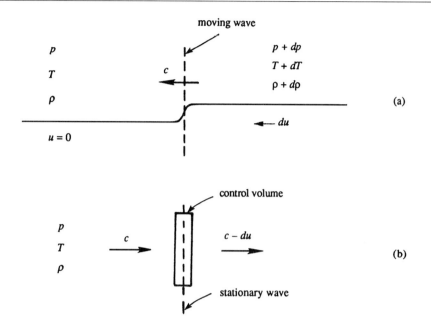

Figure 16.1 Propagation of a sound wave: (a) wave propagating into still fluid; and (b) stationary wave.

Because the amplitude is assumed small, we can neglect the second-order terms, obtaining

$$du = c(d\rho/\rho). \tag{16.6}$$

This shows that $du > 0$ if $d\rho$ is positive, thus passage of a compression wave leaves behind a fluid moving in the direction of the wave, as shown in Figure 16.1a.

Now apply the momentum equation, which states that the net force in the x-direction on the control volume equals the rate of outflow of x-momentum minus the rate of inflow of x-momentum. This gives

$$pA - (p + dp)A = (A\rho c)(c - du) - (A\rho c)c,$$

where viscous stresses have been neglected. Here, $A\rho c$ is the mass flow rate. The first term on the right-hand side represents the rate of outflow of x-momentum, and the second term represents the rate of inflow of x-momentum. Simplifying the momentum equation, we obtain

$$dp = \rho c\, du. \tag{16.7}$$

Eliminating du between equations (16.6) and (16.7), we obtain

$$c^2 = \frac{dp}{d\rho}. \tag{16.8}$$

If the amplitude of the wave is infinitesimal, then each fluid particle undergoes a nearly isentropic process as the wave passes by. The basic reason for this is that the irreversible entropy production is proportional to the *squares* of the velocity and temperature gradients (see Chapter 4, Section 15) and is therefore negligible for weak waves. The particles do undergo small temperature changes, but the changes are due to adiabatic expansion or compression and are not due to heat transfer from the neighboring particles. The entropy of a fluid particle then remains constant as a weak wave passes by. This will also be demonstrated in Section 6, where it will be shown that the entropy change across the wave is $dS \propto (dp)^3$, implying that dS goes to zero much faster than the rate at which the amplitude dp tends to zero.

It follows that the derivative $dp/d\rho$ in equation (16.8) should be replaced by the partial derivative at constant entropy, giving

$$c^2 = \left(\frac{\partial p}{\partial \rho}\right)_s. \tag{16.9}$$

For a perfect gas, the use of p/ρ^{γ} = const. and $p = \rho RT$ reduces the speed of sound (16.9) to

$$c = \sqrt{\frac{\gamma p}{\rho}} = \sqrt{\gamma RT}. \tag{16.10}$$

For air at $15\,°C$, this gives $c = 340\,\text{m/s}$. We note that the nonlinear terms that we have neglected do change the shape of a propagating wave depending on whether it is a compression or expansion, as follows. Because $\gamma > 1$, the isentropic relations show that if $dp > 0$ (compression), then $dT > 0$, and from equation (16.10) the sound speed c is increased. Therefore, the sound speed behind the front is greater than that at the front and the back of the wave catches up with the front of the wave. Thus the wave steepens as it travels. The opposite is true for an expansion wave, for which $dp < 0$ and $dT < 0$ so c decreases. The back of the wave falls farther behind the front so an expansion wave flattens as it travels.

Finite amplitude waves, across which there is a discontinuous change of pressure, will be considered in Section 6. These are called *shock waves*. It will be shown that the finite waves are not isentropic and that they propagate through a still fluid *faster* than the sonic speed.

The first approximate expression for c was found by Newton, who assumed that dp was proportional to $d\rho$, as would be true if the process undergone by a fluid particle was isothermal. In this manner Newton arrived at the expression $c = \sqrt{RT}$. He attributed the discrepancy of this formula with experimental measurements as due to "unclean air." The science of thermodynamics was virtually nonexistent at the time, so that the idea of an isentropic process was unknown to Newton. The correct expression for the sound speed was first given by Laplace.

To show explicitly that small disturbances in a compressible fluid obey a wave equation, we consider a slightly perturbed uniform flow in the x-direction so that

$$\mathbf{u} = U_\infty(\mathbf{i}_x + \mathbf{u}'), \quad p = p_\infty(1 + p'), \quad \rho = \rho_\infty(1 + \rho'), \quad \text{and so on}$$

where the perturbations $()'$ are all $<< 1$. We substitute this assumed flow into the equations for conservation of mass, momentum, and energy. We shall neglect the effects of viscous stresses and heat conduction here but we will include them at the end of Section 6, where they are determinative of shock structure. We may write the conservation laws in the form

$$D\rho/Dt + \rho\nabla \cdot \mathbf{u} = 0$$

$$\rho D\mathbf{u}/Dt + \nabla p = 0$$

$$\rho Dh/Dt - Dp/Dt = 0$$

where body forces have also been neglected and D/Dt denotes the derivative following the fluid particle, $D/Dt = \partial/\partial t + \mathbf{u} \cdot \nabla$. Substituting the assumed flow into mass conservation first,

$$\rho_\infty \partial\rho'/\partial t + \rho_\infty U_\infty \partial\rho'/\partial x + \rho_\infty U_\infty \mathbf{u}' \cdot \nabla\rho' + \rho_\infty U_\infty \nabla \cdot \mathbf{u}' + \rho_\infty U_\infty \rho'\nabla \cdot \mathbf{u}' = 0.$$

We neglect the squares and products of the perturbations, leaving

$$\partial\rho'/\partial t + U_\infty \partial\rho'/\partial x + U_\infty \nabla \cdot \mathbf{u}' = 0.$$

Similarly, the momentum equation yields

$$\partial\mathbf{u}'/\partial t + U_\infty \partial\mathbf{u}'/\partial x + [p_\infty/(\rho_\infty U_\infty)]\nabla p' = 0.$$

We may eliminate $\mathbf{u}'$ by taking the divergence of the momentum equation and substituting into mass conservation, giving,

$$(\partial/\partial t + U_\infty \partial/\partial x)\nabla \cdot \mathbf{u}' = -(1/U_\infty)(\partial/\partial t + U_\infty \partial/\partial x)^2\rho' = -[p_\infty/(\rho_\infty U_\infty)]\nabla^2 p'.$$

The energy equation is put in terms of p', ρ' for a perfect gas with constant specific heats, $h = (C_p/R)(p/\rho)$ and $C_p/R = \gamma/(\gamma - 1)$. This results in $D/Dt(p/\rho^\gamma) = 0$ but $p/\rho^\gamma = (p_\infty/\rho_\infty^\gamma)(1 + p' - \gamma\rho')$, with squares and products of the perturbations neglected. Then $(\partial/\partial t + U_\infty \partial/\partial x)p' - \gamma(\partial/\partial t + U_\infty \partial/\partial x)\rho' = 0$. Using this to eliminate ρ', $(\partial/\partial t + U_\infty \partial/\partial x)^2 p' = (\gamma p_\infty/\rho_\infty)\nabla^2 p' = c^2\nabla^2 p'$. This is a classical linear wave equation for p'. We can translate this back to a frame at rest by a Galilean transformation, $(x, y, z, t) \rightarrow (x', y', z', t')$ with $t' = t + x/U_\infty, x' = x, y' = y, z' = z$. Thus $\partial/\partial t' = \partial/\partial t + U_\infty \partial/\partial x$ and we are left with

$$\partial^2 p/\partial t^2 = c^2\nabla^2 p$$

(primes suppressed), as seen in Section 7.2, p.200. The solution in one dimension is given there and it is seen that c is the wave speed.

3. Basic Equations for One-Dimensional Flow

In this section we begin our study of certain compressible flows that can be analyzed by a one-dimensional approximation. Such a simplification is valid in flow through a duct whose centerline does not have a large curvature and whose cross section does not vary abruptly. The overall behavior in such flows can be studied by ignoring the variation of velocity and other properties across the duct and replacing the property distributions by their average values over the cross section (Figure 16.2). The area of the duct is taken as $A(x)$, and the flow properties are taken as $p(x)$, $\rho(x)$, $u(x)$, and so on. Unsteadiness can be introduced by including t as an additional independent variable. The forms of the basic equations in a one-dimensional compressible flow are discussed in what follows.

Continuity Equation

For steady flows, conservation of mass requires that

$$\rho u A = \text{independent of } x.$$

Differentiating, we obtain

$$\frac{d\rho}{\rho} + \frac{du}{u} + \frac{dA}{A} = 0. \tag{16.11}$$

Energy Equation

Consider a control volume within the duct, shown by the dashed line in Figure 16.2. The first law of thermodynamics for a control volume fixed in space is

$$\frac{d}{dt} \int \rho \left(e + \frac{u^2}{2}\right) dV + \int \left(e + \frac{u^2}{2}\right) \rho u_j \, dA_j = \int u_i \tau_{ij} \, dA_j - \int \mathbf{q} \cdot d\mathbf{A}, \tag{16.12}$$

where $u^2/2$ is the kinetic energy per unit mass. The first term on the left-hand side represents the rate of change of "stored energy" (the sum of internal and kinetic

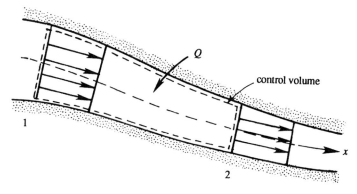

Figure 16.2 A one-dimensional flow.

energies) within the control volume, and the second term represents the flux of energy out of the control surface. The first term on the right-hand side represents the rate of work done on the control surface, and the second term on the right-hand side represents the heat *input* through the control surface. Body forces have been neglected in equation (16.12). (Here, $\mathbf{q}$ is the heat flux per unit area per unit time, and $d\mathbf{A}$ is directed along the outward normal, so that $\int \mathbf{q} \cdot d\mathbf{A}$ is the rate of *outflow* of heat.) Equation (16.12) can easily be derived by integrating the differential form given by equation (4.65) over the control volume.

Assume steady state, so that the first term on the left-hand side of equation (16.12) is zero. Writing $\dot{m} = \rho_1 u_1 A_1 = \rho_2 u_2 A_2$ (where the subscripts denote sections 1 and 2), the second term on the left-hand side in equation (16.12) gives

$$\int \left(e + \frac{1}{2} u^2 \right) \rho u_j \, dA_j = \dot{m} \left[e_2 + \frac{1}{2} u_2^2 - e_1 - \frac{1}{2} u_1^2 \right].$$

The work done on the control surfaces is

$$\int u_i \tau_{ij} \, dA_j = u_1 p_1 A_1 - u_2 p_2 A_2.$$

Here, we have assumed no-slip on the sidewalls and frictional stresses on the endfaces 1 and 2 are negligible. The rate of heat addition to the control volume is

$$-\int \mathbf{q} \cdot d\mathbf{A} = Q\dot{m},$$

where Q is the heat added per unit mass. (Checking units, Q is in J/kg, and $\dot{m}$ is in kg/s, so that $Q\dot{m}$ is in J/s.) Then equation (16.12) becomes, after dividing by $\dot{m}$,

$$e_2 + \frac{1}{2} u_2^2 - e_1 - \frac{1}{2} u_1^2 = \frac{1}{\dot{m}} [u_1 p_1 A_1 - u_2 p_2 A_2] + Q. \tag{16.13}$$

The first term on the right-hand side can be written in a simple manner by noting that

$$\frac{uA}{\dot{m}} = v,$$

where v is the specific volume. This must be true because $uA = \dot{m}v$ is the volumetric flow rate through the duct. (Checking units, $\dot{m}$ is the mass flow rate in kg/s, and v is the specific volume in m^3/kg, so that $\dot{m}v$ is the volume flow rate in m^3/s.) Equation (16.13) then becomes

$$e_2 + \tfrac{1}{2} u_2^2 - e_1 - \tfrac{1}{2} u_1^2 = p_1 v_1 - p_2 v_2 + Q. \tag{16.14}$$

It is apparent that $p_1 v_1$ is the work done (per unit mass) by the surroundings in pushing fluid into the control volume. Similarly, $p_2 v_2$ is the work done by the fluid inside the control volume on the surroundings in pushing fluid out of the control volume. Equation (16.14) therefore has a simple meaning. Introducing the enthalpy $h \equiv e + pv$, we obtain

$$h_2 + \tfrac{1}{2}u_2^2 = h_1 + \tfrac{1}{2}u_1^2 + Q. \qquad (16.15)$$

This is the energy equation, which is valid even if there are frictional or nonequilibrium conditions (e.g., shock waves) between sections 1 and 2. It is apparent that the *sum of enthalpy and kinetic energy remains constant in an adiabatic flow*. Therefore, enthalpy plays the same role in a flowing system that internal energy plays in a nonflowing system. The difference between the two types of systems is the *flow work* pv required to push matter across a section.

Bernoulli and Euler Equations

For inviscid flows, the steady form of the momentum equation is the Euler equation

$$u \, du + \frac{dp}{\rho} = 0. \qquad (16.16)$$

Integrating along a streamline, we obtain the Bernoulli equation for a compressible flow:

$$\frac{1}{2}u^2 + \int \frac{dp}{\rho} = \text{const.}, \qquad (16.17)$$

which agrees with equation (4.78).

  For adiabatic frictionless flows the Bernoulli equation is identical to the energy equation. To see this, note that this is an isentropic flow, so that the $T \, dS$ equation

$$T \, dS = dh - v \, dp,$$

gives

$$dh = dp/\rho.$$

Then the Euler equation (16.16) becomes

$$u \, du + dh = 0,$$

which is identical to the adiabatic form of the energy equation (16.15). The collapse of the momentum and energy equations is expected because the constancy of entropy has eliminated one of the flow variables.

Momentum Principle for a Control Volume

If the centerline of the duct is straight, then the steady form of the momentum principle for a finite control volume, which cuts across the duct at sections 1 and 2, gives

$$p_1 A_1 - p_2 A_2 + F \equiv \rho_2 u_2^2 A_2 - \rho_1 u_1^2 A_1, \qquad (16.18)$$

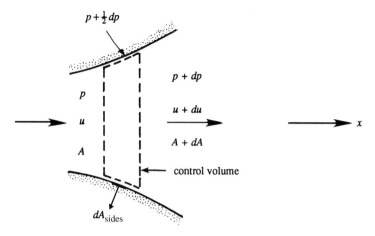

Figure 16.3 Application of the momentum principle to an infinitesimal control volume in a duct.

where F is the x-component of the resultant force exerted on the fluid by the walls. The momentum principle (16.18) is applicable even when there are frictional and dissipative processes (such as shock waves) within the control volume:

$$F = \left[\int_{\text{sides}} (-p\delta_{ij} + \sigma_{ij}) \, dA_j \right]_x = \int_{x_1}^{x_2} p \, dA(x) - (f_\sigma)_x,$$

$$f_{\sigma,x} = -\left[\int_{\text{sides}} \sigma_{ij} \, dA_j \right]_x.$$

If frictional processes are absent, then equation (16.18) reduces to the Euler equation (16.16). To see this, consider an infinitesimal area change between sections 1 and 2 (Figure 16.3). Then the average pressure exerted by the walls on the control surface is $(p + \frac{1}{2}dp)$, so that $F = dA(p + \frac{1}{2}dp)$. Then equation (16.18) becomes

$$pA - (p + dp)(A + dA) + \left(p + \tfrac{1}{2}dp\right) dA = \rho u A(u + du) - \rho u^2 A,$$

where by canceling terms and neglecting second-order terms, this reduces to the Euler equation (16.16).

4. Stagnation and Sonic Properties

A very useful reference state for computing compressible flows is the stagnation state in which the velocity is zero. Suppose the properties of the flow (such as h, ρ, u) are known at a certain point. The stagnation properties at a point are defined as those that would be obtained if the local flow were *imagined* to slow down to zero velocity *isentropically*. The stagnation properties are denoted by a subscript zero. Thus the *stagnation enthalpy* is defined as

$$h_0 \equiv h + \tfrac{1}{2}u^2.$$

For a perfect gas, this gives

$$C_p T_0 = C_p T + \tfrac{1}{2} u^2, \tag{16.19}$$

which defines the *stagnation temperature*.

It is useful to express the ratios such as T_0/T in terms of the local Mach number. From equation (16.19), we obtain

$$\frac{T_0}{T} = 1 + \frac{u^2}{2 C_p T} = 1 + \frac{\gamma - 1}{2} \frac{u^2}{\gamma R T},$$

where we have used $C_p = \gamma R / (\gamma - 1)$. Therefore

$$\boxed{\frac{T_0}{T} = 1 + \frac{\gamma - 1}{2} M^2,} \tag{16.20}$$

from which the stagnation temperature T_0 can be found for a given T and M. The isentropic relations can then be used to obtain the *stagnation pressure* and *stagnation density*:

$$\frac{p_0}{p} = \left(\frac{T_0}{T}\right)^{\gamma/(\gamma-1)} = \left[1 + \frac{\gamma - 1}{2} M^2\right]^{\gamma/(\gamma-1)}, \tag{16.21}$$

$$\frac{\rho_0}{\rho} = \left(\frac{T_0}{T}\right)^{1/(\gamma-1)} = \left[1 + \frac{\gamma - 1}{2} M^2\right]^{1/(\gamma-1)}. \tag{16.22}$$

In a general flow the stagnation properties can vary throughout the flow field. If, however, the flow is adiabatic (but not necessarily isentropic), then $h + u^2/2$ is constant throughout the flow as shown in equation (16.15). It follows that h_0, T_0, *and* c_0 ($= \sqrt{\gamma R T_0}$) *are constant throughout an adiabatic flow, even in the presence of friction. In contrast, the stagnation pressure* p_0 *and density* ρ_0 *decrease if there is friction.* To see this, consider the entropy change in an adiabatic flow between sections 1 and 2, with 2 being the downstream section. Let the flow at both sections hypothetically be brought to rest by isentropic processes, giving the local stagnation conditions p_{01}, p_{02}, T_{01}, and T_{02}. Then the entropy change between the two sections can be expressed as

$$S_2 - S_1 = S_{02} - S_{01} = -R \ln \frac{p_{02}}{p_{01}} + C_p \ln \frac{T_{02}}{T_{01}},$$

where we have used equation (16.4) for computing entropy changes. The last term is zero for an adiabatic flow in which $T_{02} = T_{01}$. As the second law of thermodynamics requires that $S_2 > S_1$, it follows that

$$p_{02} < p_{01},$$

which shows that the stagnation pressure falls due to friction.

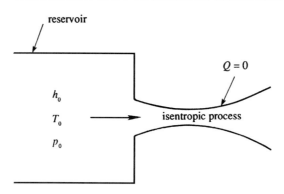

Figure 16.4 An isentropic process starting from a reservoir. Stagnation properties are uniform everywhere and are equal to the properties in the reservoir.

It is apparent that all stagnation properties are constant along an isentropic flow. If such a flow happens to start from a large reservoir where the fluid is practically at rest, then the properties in the reservoir equal the stagnation properties everywhere in the flow (Figure 16.4).

In addition to the stagnation properties, there is another useful set of reference quantities. These are called *sonic* or *critical* conditions and are denoted by an asterisk. Thus, p^*, ρ^*, c^*, and T^* are properties attained if the local fluid is imagined to expand or compress isentropically until it reaches $M = 1$. It is easy to show (Exercise 1) that the area of the passage A^*, at which the sonic conditions are attained, is given by

$$\frac{A}{A^*} = \frac{1}{M}\left[\frac{2}{\gamma+1}\left(1+\frac{\gamma-1}{2}M^2\right)\right]^{(1/2)(\gamma+1)/(\gamma-1)}. \tag{16.23}$$

We shall see in the following section that sonic conditions can only be reached at the *throat* of a duct, where the area is minimum. Equation (16.23) shows that we can find the throat area A^* of an isentropic duct flow if we know the Mach number M and the area A at some point of the duct. Note that it is not necessary that a throat actually should exist in the flow; the sonic variables are simply reference values that are reached *if* the flow were brought to the sonic state isentropically. From its definition it is clear that the value of A^* in a flow remains constant along an isentropic flow. The presence of shock waves, friction, or heat transfer changes the value of A^* along the flow.

The values of T_0/T, p_0/p, ρ_0/ρ, and A/A^* at a point can be determined from equations (16.20)–(16.23) if the local Mach number is known. For $\gamma = 1.4$, these ratios are tabulated in Table 16.1. The reader should examine this table at this point. Examples 16.1 and 16.2 given later will illustrate the use of this table.

TABLE 16.1　Isentropic Flow of a Perfect Gas ($\gamma = 1.4$)

M	p/p_0	ρ/ρ_0	T/T_0	A/A^*	M	p/p_0	ρ/ρ_0	T/T_0	A/A^*
0.0	1.0	1.0	1.0	∞	0.52	0.8317	0.8766	0.9487	1.3034
0.02	0.9997	0.9998	0.9999	28.9421	0.54	0.8201	0.8679	0.9449	1.2703
0.04	0.9989	0.9992	0.9997	14.4815	0.56	0.8082	0.8589	0.9410	1.2403
0.06	0.9975	0.9982	0.9993	9.6659	0.58	0.7962	0.8498	0.9370	1.2130
0.08	0.9955	0.9968	0.9987	7.2616	0.6	0.7840	0.8405	0.9328	1.1882
0.1	0.9930	0.9950	0.9980	5.8218	0.62	0.7716	0.8310	0.9286	1.1656
0.12	0.9900	0.9928	0.9971	4.8643	0.64	0.7591	0.8213	0.9243	1.1451
0.14	0.9864	0.9903	0.9961	4.1824	0.66	0.7465	0.8115	0.9199	1.1265
0.16	0.9823	0.9873	0.9949	3.6727	0.68	0.7338	0.8016	0.9153	1.1097
0.18	0.9776	0.9840	0.9936	3.2779	0.7	0.7209	0.7916	0.9107	1.0944
0.2	0.9725	0.9803	0.9921	2.9635	0.72	0.7080	0.7814	0.9061	1.0806
0.22	0.9668	0.9762	0.9904	2.7076	0.74	0.6951	0.7712	0.9013	1.0681
0.24	0.9607	0.9718	0.9886	2.4956	0.76	0.6821	0.7609	0.8964	1.0570
0.26	0.9541	0.9670	0.9867	2.3173	0.78	0.6690	0.7505	0.8915	1.0471
0.28	0.9470	0.9619	0.9846	2.1656	0.8	0.6560	0.7400	0.8865	1.0382
0.3	0.9395	0.9564	0.9823	2.0351	0.82	0.6430	0.7295	0.8815	1.0305
0.32	0.9315	0.9506	0.9799	1.9219	0.84	0.6300	0.7189	0.8763	1.0237
0.34	0.9231	0.9445	0.9774	1.8229	0.86	0.6170	0.7083	0.8711	1.0179
0.36	0.9143	0.9380	0.9747	1.7358	0.88	0.6041	0.6977	0.8659	1.0129
0.38	0.9052	0.9313	0.9719	1.6587	0.9	0.5913	0.6870	0.8606	1.0089
0.4	0.8956	0.9243	0.9690	1.5901	0.92	0.5785	0.6764	0.8552	1.0056
0.42	0.8857	0.9170	0.9659	1.5289	0.94	0.5658	0.6658	0.8498	1.0031
0.44	0.8755	0.9094	0.9627	1.4740	0.96	0.5532	0.6551	0.8444	1.0014
0.46	0.8650	0.9016	0.9594	1.4246	0.98	0.5407	0.6445	0.8389	1.0003
0.48	0.8541	0.8935	0.9559	1.3801	1.0	0.5283	0.6339	0.8333	1.0000
0.5	0.8430	0.8852	0.9524	1.3398	1.02	0.5160	0.6234	0.8278	1.0003
1.04	0.5039	0.6129	0.8222	1.0013	2.04	0.1201	0.2200	0.5458	1.7451
1.06	0.4919	0.6024	0.8165	1.0029	2.06	0.1164	0.2152	0.5409	1.7750
1.08	0.4800	0.5920	0.8108	1.0051	2.08	0.1128	0.2104	0.5361	1.8056
1.1	0.4684	0.5817	0.8052	1.0079	2.1	0.1094	0.2058	0.5313	1.8369
1.12	0.4568	0.5714	0.7994	1.0113	2.12	0.1060	0.2013	0.5266	1.8690
1.14	0.4455	0.5612	0.7937	1.0153	2.14	0.1027	0.1968	0.5219	1.9018
1.16	0.4343	0.5511	0.7879	1.0198	2.16	0.0996	0.1925	0.5173	1.9354
1.18	0.4232	0.5411	0.7822	1.0248	2.18	0.0965	0.1882	0.5127	1.9698
1.2	0.4124	0.5311	0.7764	1.0304	2.2	0.0935	0.1841	0.5081	2.0050
1.22	0.4017	0.5213	0.7706	1.0366	2.22	0.0906	0.1800	0.5036	2.0409
1.24	0.3912	0.5115	0.7648	1.0432	2.24	0.0878	0.1760	0.4991	2.0777
1.26	0.3809	0.5019	0.7590	1.0504	2.26	0.0851	0.1721	0.4947	2.1153
1.28	0.3708	0.4923	0.7532	1.0581	2.28	0.0825	0.1683	0.4903	2.1538
1.3	0.3609	0.4829	0.7474	1.0663	2.3	0.0800	0.1646	0.4859	2.1931
1.32	0.3512	0.4736	0.7416	1.0750	2.32	0.0775	1.1609	0.4816	2.2333
1.34	0.3417	0.4644	0.7358	1.0842	2.34	0.0751	0.1574	0.4773	2.2744
1.36	0.3323	0.4553	0.7300	1.0940	2.36	0.0728	0.1539	0.4731	2.3164
1.38	0.3232	0.4463	0.7242	1.1042	2.38	0.0706	0.1505	0.4688	2.3593
1.4	0.3142	0.4374	0.7184	1.1149	2.4	0.0684	0.1472	0.4647	2.4031
1.42	0.3055	0.4287	0.7126	1.1262	2.42	0.0663	0.1439	0.4606	2.4479
1.44	0.2969	0.4201	0.7069	1.1379	2.44	0.0643	0.1408	0.4565	2.4936

TABLE 16.1   (*Continued*)

M	p/p_0	ρ/ρ_0	T/T_0	A/A^*	M	p/p_0	ρ/ρ_0	T/T_0	A/A^*
1.46	0.2886	0.4116	0.7011	1.1501	2.46	0.0623	0.1377	0.4524	2.5403
1.48	0.2804	0.4032	0.6954	1.1629	2.48	0.0604	0.1346	0.4484	2.5880
1.5	0.2724	0.3950	0.6897	1.1762	2.5	0.0585	0.1317	0.4444	2.6367
1.52	0.2646	0.3869	0.6840	1.1899	2.52	0.0567	0.1288	0.4405	2.6865
1.54	0.2570	0.3789	0.6783	1.2042	2.54	0.0550	0.1260	0.4366	2.7372
1.56	0.2496	0.3710	0.6726	1.2190	2.56	0.0533	0.1232	0.4328	2.7891
1.58	0.2423	0.3633	0.6670	1.2344	2.58	0.0517	0.1205	0.4289	2.8420
1.6	0.2353	0.3557	0.6614	1.2502	2.6	0.0501	0.1179	0.4252	2.8960
1.62	0.2284	0.3483	0.6558	1.2666	2.62	0.0486	0.1153	0.4214	2.9511
1.64	0.2217	0.3409	0.6502	1.2836	2.64	0.0471	0.1128	0.4177	3.0073
1.66	0.2151	0.3337	0.6447	1.3010	2.66	0.0457	0.1103	0.4141	3.0647
1.68	0.2088	0.3266	0.6392	1.3190	2.68	0.0443	0.1079	0.4104	3.1233
1.7	0.2026	0.3197	0.6337	1.3376	2.7	0.0430	0.1056	0.4068	3.1830
1.72	0.1966	0.3129	0.6283	1.3567	2.72	0.0417	0.1033	0.4033	3.2440
1.74	0.1907	0.3062	0.6229	1.3764	2.74	0.0404	0.1010	0.3998	3.3061
1.76	0.1850	0.2996	0.6175	1.3967	2.76	0.0392	0.0989	0.3963	3.3695
1.78	0.1794	0.2931	0.6121	1.4175	2.78	0.0380	0.0967	0.3928	3.4342
1.8	0.1740	0.2868	0.6068	1.4390	2.8	0.0368	0.0946	0.3894	3.5001
1.82	0.1688	0.2806	0.6015	1.4610	2.82	0.0357	0.0926	0.3860	3.5674
1.84	0.1637	0.2745	0.5963	1.4836	2.84	0.0347	0.0906	0.3827	3.6359
1.86	0.1587	0.2686	0.5910	1.5069	2.86	0.0336	0.0886	0.3794	3.7058
1.88	0.1539	0.2627	0.5859	1.5308	2.88	0.0326	0.0867	0.3761	3.7771
1.9	0.1492	0.2570	0.5807	1.5553	2.9	0.0317	0.0849	0.3729	3.8498
1.92	0.1447	0.2514	0.5756	1.5804	2.92	0.0307	0.0831	0.3696	3.9238
1.94	0.1403	0.2459	0.5705	1.6062	2.94	0.0298	0.0813	0.3665	3.9993
1.96	0.1360	0.2405	0.5655	1.6326	2.96	0.0289	0.0796	0.3633	4.0763
1.98	0.1318	0.2352	0.5605	1.6597	2.98	0.0281	0.0779	0.3602	4.1547
2.0	0.1278	0.2300	0.5556	1.6875	3.0	0.0272	0.0762	0.3571	4.2346
2.02	0.1239	0.2250	0.5506	1.7160	3.02	0.0264	0.0746	0.3541	4.3160
3.04	0.0256	0.0730	0.3511	4.3990	4.04	0.0062	0.0266	0.2345	11.1077
3.06	0.0249	0.0715	0.3481	4.4835	4.06	0.0061	0.0261	0.2327	11.3068
3.08	0.0242	0.0700	0.3452	4.5696	4.08	0.0059	0.0256	0.2310	11.5091
3.1	0.0234	0.0685	0.3422	4.6573	4.1	0.0058	0.0252	0.2293	11.7147
3.12	0.0228	0.0671	0.3393	4.7467	4.12	0.0056	0.0247	0.2275	11.9234
3.14	0.0221	0.0657	0.3365	4.8377	4.14	0.0055	0.0242	0.2258	12.1354
3.16	0.0215	0.0643	0.3337	4.9304	4.16	0.0053	0.0238	0.2242	12.3508
3.18	0.0208	0.0630	0.3309	5.0248	4.18	0.0052	0.0234	0.2225	12.5695
3.2	0.0202	0.0617	0.3281	5.1210	4.2	0.0051	0.0229	0.2208	12.7916
3.22	0.0196	0.0604	0.3253	5.2189	4.22	0.0049	0.0225	0.2192	13.0172
3.24	0.0191	0.0591	0.3226	5.3186	4.24	0.0048	0.0221	0.2176	13.2463
3.26	0.0185	0.0579	0.3199	5.4201	4.26	0.0047	0.0217	0.2160	13.4789
3.28	0.0180	0.0567	0.3173	5.5234	4.28	0.0046	0.0213	0.2144	13.7151
3.3	0.0175	0.0555	0.3147	5.6286	4.3	0.0044	0.0209	0.2129	13.9549
3.32	0.0170	0.0544	0.3121	5.7358	4.32	0.0043	0.0205	0.2113	14.1984
3.34	0.0165	0.0533	0.3095	5.8448	4.34	0.0042	0.0202	0.2098	14.4456
3.36	0.0160	0.0522	0.3069	5.9558	4.36	0.0041	0.0198	0.2083	14.6965
3.38	0.0156	0.0511	0.3044	6.0687	4.38	0.0040	0.0194	0.2067	14.9513
3.4	0.0151	0.0501	0.3019	6.1837	4.4	0.0039	0.0191	0.2053	15.2099
3.42	0.0147	0.0491	0.2995	6.3007	4.42	0.0038	0.0187	0.2038	15.4724

TABLE 16.1   (*Continued*)

M	p/p_0	ρ/ρ_0	T/T_0	A/A^*	M	p/p_0	ρ/ρ_0	T/T_0	A/A^*
3.44	0.0143	0.0481	0.2970	6.4198	4.44	0.0037	0.0184	0.2023	15.7388
3.46	0.0139	0.0471	0.2946	6.5409	4.46	0.0036	0.0181	0.2009	16.0092
3.48	0.0135	0.0462	0.2922	6.6642	4.48	0.0035	0.0178	0.1994	16.2837
3.5	0.0131	0.0452	0.2899	6.7896	4.5	0.0035	0.0174	0.1980	16.5622
3.52	0.0127	0.0443	0.2875	6.9172	4.52	0.0034	0.0171	0.1966	16.8449
3.54	0.0124	0.0434	0.2852	7.0471	4.54	0.0033	0.0168	0.1952	17.1317
3.56	0.0120	0.0426	0.2829	7.1791	4.56	0.0032	0.0165	0.1938	17.4228
3.58	0.0117	0.0417	0.2806	7.3135	4.58	0.0031	0.0163	0.1925	17.7181
3.6	0.0114	0.0409	0.2784	7.4501	4.6	0.0031	0.0160	0.1911	18.0178
3.62	0.0111	0.0401	0.2762	7.5891	4.62	0.0030	0.0157	0.1898	18.3218
3.64	0.0108	0.0393	0.2740	7.7305	4.64	0.0029	0.0154	0.1885	18.6303
3.66	0.0105	0.0385	0.2718	7.8742	4.66	0.0028	0.0152	0.1872	18.9433
3.68	0.0102	0.0378	0.2697	8.0204	4.68	0.0028	0.0149	0.1859	19.2608
3.7	0.0099	0.0370	0.2675	8.1691	4.7	0.0027	0.0146	0.1846	19.5828
3.72	0.0096	0.0363	0.2654	8.3202	4.72	0.0026	0.0144	0.1833	19.9095
3.74	0.0094	0.0356	0.2633	8.4739	4.74	0.0026	0.0141	0.1820	20.2409
3.76	0.0091	0.0349	0.2613	8.6302	4.76	0.0025	0.0139	0.1808	20.5770
3.78	0.0089	0.0342	0.2592	8.7891	4.78	0.0025	0.0137	0.1795	20.9179
3.8	0.0086	0.0335	0.2572	8.9506	4.8	0.0024	0.0134	0.1783	21.2637
3.82	0.0084	0.0329	0.2552	9.1148	4.82	0.0023	0.0132	0.1771	21.6144
3.84	0.0082	0.0323	0.2532	0.2817	4.84	0.0023	0.0130	0.1759	21.9700
3.86	0.0080	0.0316	0.2513	9.4513	4.86	0.0022	0.0128	0.1747	22.3306
3.88	0.0077	0.0310	0.2493	9.6237	4.88	0.0022	0.0125	0.1735	22.6963
3.9	0.0075	0.0304	0.2474	9.7990	4.9	0.0021	0.0123	0.1724	23.0671
3.92	0.0073	0.0299	0.2455	9.9771	4.92	0.0021	0.0121	0.1712	23.4431
3.94	0.0071	0.0293	0.2436	10.1581	4.94	0.0020	0.0119	0.1700	23.8243
3.96	0.0069	0.0287	0.2418	10.3420	4.96	0.0020	0.0117	0.1689	24.2109
3.98	0.0068	0.0282	0.2399	10.5289	4.98	0.0019	0.0115	0.1678	24.6027
4.0	0.0066	0.0277	0.2381	10.7188	5.0	0.0019	0.0113	0.1667	25.0000
4.02	0.0064	0.0271	0.2363	10.9117					

5. Area–Velocity Relations in One-Dimensional Isentropic Flow

Some surprising consequences of compressibility are dramatically demonstrated by considering an isentropic flow in a duct of varying area. Before we demonstrate this effect, we shall make some brief comments on two common devices of varying area in which the flow can be approximately isentropic. One of them is the *nozzle* through which the flow expands from high to low pressure to generate a high-speed jet. An example of a nozzle is the exit duct of a rocket motor. The second device is called the *diffuser*, whose function is opposite to that of a nozzle. (Note that the diffuser has nothing to do with heat diffusion.) In a diffuser a high-speed jet is decelerated and compressed. For example, air enters the jet engine of an aircraft after passing through a diffuser, which raises the pressure and temperature of the air. In incompressible flow, a nozzle profile converges in the direction of flow to increase the velocity, while a diffuser profile diverges. We shall see that this conclusion is true for subsonic flows, but not for supersonic flows.

Consider two sections of a duct (Figure 16.3). The continuity equation gives

$$\frac{d\rho}{\rho} + \frac{du}{u} + \frac{dA}{A} = 0. \tag{16.24}$$

In a constant density flow $d\rho = 0$, for which the continuity equation requires that a decreasing area leads to an increase of velocity.

As the flow is assumed to be frictionless, we can use the Euler equation

$$u\,du = -\frac{dp}{\rho} = -\frac{dp}{d\rho}\frac{d\rho}{\rho} = -c^2\frac{d\rho}{\rho}, \tag{16.25}$$

where we have used the fact that $c^2 = dp/d\rho$ in an isentropic flow. The Euler equation requires that an increasing speed ($du > 0$) in the direction of flow must be accompanied by a fall of pressure ($dp < 0$). In terms of the Mach number, equation (16.25) becomes

$$\frac{d\rho}{\rho} = -M^2\frac{du}{u}. \tag{16.26}$$

This shows that for $M \ll 1$, the percentage change of density is much smaller than the percentage change of velocity. The density changes in the continuity equation (16.24) can therefore be neglected in low Mach number flows, a fact also demonstrated in Section 1.

Substituting equation (16.26) into equation (16.24), we obtain

$$\frac{du}{u} = \frac{-dA/A}{1 - M^2}. \tag{16.27}$$

This relation leads to the following important conclusions about compressible flows:

(i) At subsonic speeds ($M < 1$) a decrease of area increases the speed of flow. A subsonic nozzle therefore must have a convergent profile, and a subsonic diffuser must have a divergent profile (upper row of Figure 16.5). The behavior is therefore qualitatively the same as in incompressible flows.

(ii) At supersonic speeds ($M > 1$) the denominator in equation (16.27) is negative, and we arrive at the surprising conclusion that an *increase* in area leads to an increase of speed. The reason for such a behavior can be understood from equation (16.26), which shows that for $M > 1$ the density decreases faster than the velocity increases, thus the area must increase in an accelerating flow in order that the product $A\rho u$ is constant.

The supersonic portion of a nozzle therefore must have a divergent profile, while the supersonic part of a diffuser must have a convergent profile (bottom row of Figure 16.5).

Suppose a nozzle is used to generate a supersonic stream, starting from low speeds at the inlet (Figure 16.6). Then the Mach number must increase continuously from $M = 0$ near the inlet to $M > 1$ at the exit. The foregoing discussion shows

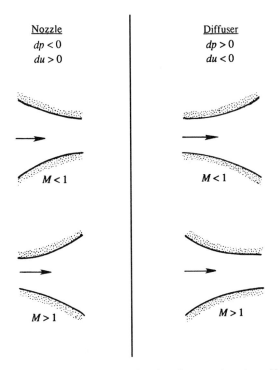

Figure 16.5 Shapes of nozzles and diffusers in subsonic and supersonic regimes. Nozzles are shown in the left column and diffusers are shown in the right column.

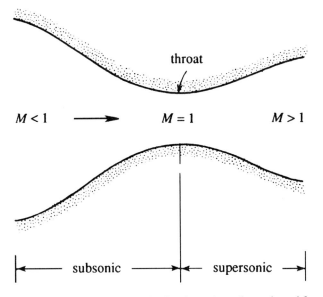

Figure 16.6 A convergent–divergent nozzle. The flow is continuously accelerated from low speed to supersonic Mach number.

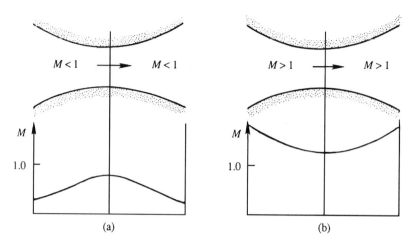

Figure 16.7   Convergent–divergent passages in which the condition at the throat is not sonic.

that the nozzle must converge in the subsonic portion and diverge in the supersonic portion. Such a nozzle is called a *convergent–divergent nozzle*. From Figure 16.6 it is clear that the Mach number must be unity at the *throat*, where the area is neither increasing nor decreasing. This is consistent with equation (16.27), which shows that du can be nonzero at the throat only if $M = 1$. It follows that the *sonic velocity can be achieved only at the throat of a nozzle or a diffuser and nowhere else.*

It does not, however, follow that M must necessarily be unity at the throat. According to equation (16.27), we may have a case where $M \neq 1$ at the throat if $du = 0$ there. As an example, note that the flow in a convergent–divergent tube may be subsonic everywhere, with M increasing in the convergent portion and decreasing in the divergent portion, with $M \neq 1$ at the throat (Figure 16.7a). The first half of the tube here is acting as a nozzle, whereas the second half is acting as a diffuser. Alternatively, we may have a convergent–divergent tube in which the flow is supersonic everywhere, with M decreasing in the convergent portion and increasing in the divergent portion, and again $M \neq 1$ at the throat (Figure 16.7b).

Example 16.1

The nozzle of a rocket motor is designed to generate a thrust of 30,000 N when operating at an altitude of 20 km. The pressure inside the combustion chamber is 1000 kPa while the temperature is 2500 K. The gas constant of the fluid in the jet is $R = 280 \, \text{m}^2/(\text{s}^2 \, \text{K})$, and $\gamma = 1.4$. Assuming that the flow in the nozzle is isentropic, calculate the throat and exit areas. Use the isentropic table (Table 16.1).

Solution: At an altitude of 20 km, the pressure of the standard atmosphere (Section A4 in Appendix A) is 5467 Pa. If subscripts "0" and "e" refer to the stagnation and exit conditions, then a summary of the information given is as follows:

$$p_e = 5467 \, \text{Pa},$$
$$p_0 = 1000 \, \text{kPa},$$

$$T_0 = 2500\,\text{K},$$
$$\text{Thrust} = \rho_e A_e u_e^2 = 30{,}000\,\text{N}.$$

Here, we have used the facts that the thrust equals mass flow rate times the exit velocity, and the pressure inside the combustion chamber is nearly equal to the stagnation pressure. The pressure ratio at the exit is

$$\frac{p_e}{p_0} = \frac{5467}{(1000)(1000)} = 5.467 \times 10^{-3}.$$

For this ratio of p_e/p_0, the isentropic table (Table 16.1) gives

$$M_e = 4.15,$$
$$\frac{A_e}{A^*} = 12.2,$$
$$\frac{T_e}{T_0} = 0.225.$$

The exit temperature and density are therefore

$$T_e = (0.225)(2500) = 562\,\text{K},$$
$$\rho_e = p_e/RT_e = 5467/(280)(562) = 0.0347\,\text{kg/m}^3.$$

The exit velocity is

$$u_e = M_e\sqrt{\gamma R T_e} = 4.15\sqrt{(1.4)(280)(562)} = 1948\,\text{m/s}.$$

The exit area is found from the expression for thrust:

$$A_e = \frac{\text{Thrust}}{\rho_e u_e^2} = \frac{30{,}000}{(0.0347)(1948)^2} = 0.228\,\text{m}^2.$$

Because $A_e/A^* = 12.2$, the throat area is

$$A^* = \frac{0.228}{12.2} = 0.0187\,\text{m}^2.$$

6. Normal Shock Wave

A shock wave is similar to a sound wave except that it has finite strength. The thickness of such a wavefront is of the order of micrometers, so that the properties vary almost discontinuously across a shock wave. The high gradients of velocity and temperature result in entropy production within the wave, due to which the isentropic relations cannot be used across the shock. In this section we shall derive the relations between properties of the flow on the two sides of a *normal shock*, where the wavefront is perpendicular to the direction of flow. We shall treat the shock wave as a discontinuity; a treatment of Navier-Strokes shock structure is given at the end of this section.

To derive the relationships between the properties on the two sides of the shock, consider a control volume shown in Figure 16.8, where the sections 1 and 2 can be taken arbitrarily close to each other because of the discontinuous nature of the wave. The area change between the upstream and the downstream sides can then be neglected. The basic equations are

$$\text{Continuity:} \qquad \rho_1 u_1 = \rho_2 u_2, \tag{16.28}$$

$$\text{x-momentum:} \qquad p_1 - p_2 = \rho_2 u_2^2 - \rho_1 u_1^2, \tag{16.29}$$

$$\text{Energy:} \qquad h_1 + \tfrac{1}{2}u_1^2 = h_2 + \tfrac{1}{2}u_2^2.$$

In the application of the momentum theorem, we have neglected any frictional drag from the walls because such forces go to zero as the wave thickness goes to zero. Note that we cannot use the Bernoulli equation because the process inside the wave is dissipative. We have written down four unknowns (h_2, u_2, p_2, ρ_2) and three equations. The additional relation comes from the perfect gas relationship

$$h = C_p T = \frac{\gamma R}{\gamma - 1} \frac{p}{\rho R} = \frac{\gamma p}{(\gamma - 1)\rho},$$

so that the energy equation becomes

$$\frac{\gamma}{\gamma - 1} \frac{p_1}{\rho_1} + \frac{1}{2}u_1^2 = \frac{\gamma}{\gamma - 1} \frac{p_2}{\rho_2} + \frac{1}{2}u_2^2. \tag{16.30}$$

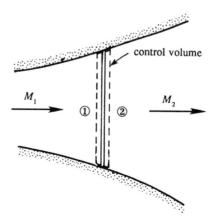

Figure 16.8 Normal shock wave.

We now have three unknowns (u_2, p_2, ρ_2) and three equations (16.28)–(16.30). Elimination of ρ_2 and u_2 from these gives, after some algebra,

$$\frac{p_2}{p_1} = 1 + \frac{2\gamma}{\gamma+1}\left[\frac{\rho_1 u_1^2}{\gamma p_1} - 1\right].$$

This can be expressed in terms of the upstream Mach number M_1 by noting that $\rho u^2/\gamma p = u^2/\gamma RT = M^2$. The pressure ratio then becomes

$$\boxed{\frac{p_2}{p_1} = 1 + \frac{2\gamma}{\gamma+1}(M_1^2 - 1).} \tag{16.31}$$

Let us now derive a relation between M_1 and M_2. Because $\rho u^2 = \rho c^2 M^2 = \rho(\gamma p/\rho)M^2 = \gamma p M^2$, the momentum equation (16.29) gives

$$p_1 + \gamma p_1 M_1^2 = p_2 + \gamma p_2 M_2^2.$$

Using equation (16.31), this gives

$$\boxed{M_2^2 = \frac{(\gamma-1)M_1^2 + 2}{2\gamma M_1^2 + 1 - \gamma},} \tag{16.32}$$

which is plotted in Figure 16.9. Because $M_2 = M_1$ (state 2 = state 1) is a solution of equations (16.28)–(16.30), that is shown as well indicating two possible solutions for M_2 for all $M_1 > [(\gamma-1)/2\gamma]^{1/2}$. We show in what follows that $M_1 \geqslant 1$ to avoid violation of the second law of thermodynamics. The two possible solutions are: (a) no change of state; and (b) a sudden transition from supersonic to subsonic flow with consequent increases in pressure, density, and temperature. The density, velocity, and temperature ratios can be similarly obtained. They are

$$\frac{\rho_2}{\rho_1} = \frac{u_1}{u_2} = \frac{(\gamma+1)M_1^2}{(\gamma-1)M_1^2 + 2}, \tag{16.33}$$

$$\frac{T_2}{T_1} = 1 + \frac{2(\gamma-1)}{(\gamma+1)^2}\frac{\gamma M_1^2 + 1}{M_1^2}(M_1^2 - 1). \tag{16.34}$$

The normal shock relations (16.31)–(16.34) were worked out independently by the British engineer W. J. M. Rankine (1820–1872) and the French ballistician Pierre Henry Hugoniot (1851–1887). These equations are sometimes known as the *Rankine–Hugoniot relations*.

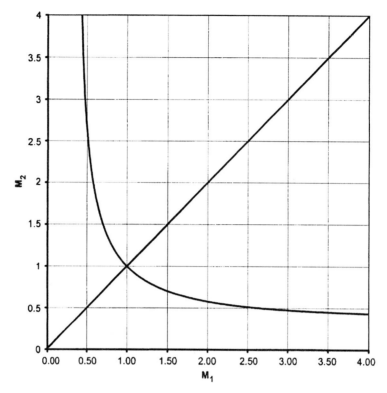

Figure 16.9 Normal shock-wave solution $M_2(M_1)$ for $\gamma = 1.4$. Trivial (no change) solution is also shown. Asymptotes are $[(\gamma - 1)/2\gamma]^{1/2} = 0.378$.

An important quantity is the change of entropy across the shock. Using equation (16.4), the entropy change is

$$\frac{S_2 - S_1}{C_v} = \ln\left[\frac{p_2}{p_1}\left(\frac{\rho_1}{\rho_2}\right)^\gamma\right]$$

$$= \ln\left\{\left[1 + \frac{2\gamma}{\gamma + 1}(M_1^2 - 1)\right]\left[\frac{(\gamma - 1)M_1^2 + 2}{(\gamma + 1)M_1^2}\right]^\gamma\right\}, \quad (16.35)$$

which is plotted in Figure 16.10. This shows that the entropy across an expansion shock would decrease, which is impermissible for the perfect gas equation of state. However, for a heavy fluorocarbon gas (FC-70), Fergason et al. (2001), using different equations of state, have numerically simulated a rarefaction (expansion) shock in a shock tube type flow. Equation (16.36) calculates the entropy change for a perfect gas with constant specific heats explicitly in the neighborhood of $M_1 = 1$. Now assume that the upstream Mach number M_1 is only slightly larger than 1, so that $M_1^2 - 1$ is

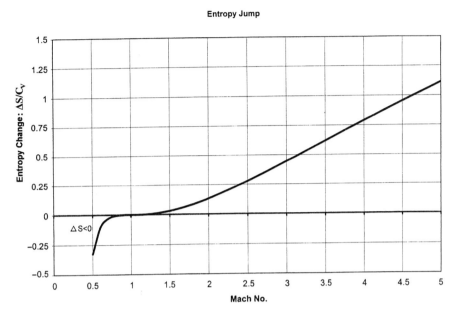

Figure 16.10 Entropy change $(S_2 - S_1)/C_v$ as a function of M_1 for $\gamma = 1.4$. Note higher-order contact at $M = 1$.

a small quantity. It is straightforward to show that equation (16.35) then reduces to (Exercise 2)

$$\frac{S_2 - S_1}{C_v} \simeq \frac{2\gamma(\gamma - 1)}{3(\gamma + 1)^2}(M_1^2 - 1)^3. \tag{16.36}$$

This shows that we must have $M_1 > 1$ because the entropy of an adiabatic process cannot decrease. Equation (16.32) then shows that $M_2 < 1$. Thus, the *Mach number changes from supersonic to subsonic values across a normal shock*; a discontinuous change from subsonic to supersonic conditions would lead to a violation of the second law of thermodynamics. (A shock wave is therefore analogous to a hydraulic jump (Chapter 7, Section 12) in a gravity current, in which the Froude number jumps from supercritical to subcritical values; see Figure 7.23.) Equations (16.31), (16.33), and (16.34) then show that the jumps in p, ρ, and T are also from low to high values, so that a shock wave compresses and heats a fluid.

Note that the terms involving the first two powers of $(M_1^2 - 1)$ do not appear in equation (16.36). Using the pressure ratio (16.31), equation (16.36) can be written as

$$\frac{S_2 - S_1}{C_v} \simeq \frac{\gamma^2 - 1}{12\gamma^2}\left(\frac{\Delta p}{p_1}\right)^3.$$

This shows that as the wave amplitude decreases, the entropy jump goes to zero much faster than the rate at which the pressure jump (or the jumps in velocity or temperature) goes to zero. Weak shock waves are therefore nearly isentropic.

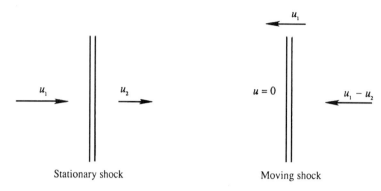

Figure 16.11   Stationary and moving shocks.

This is why we argued that the propagation of sound waves is an isentropic process.

Because of the adiabatic nature of the process, the stagnation properties T_0 and h_0 are constant across the shock. In contrast, the stagnation properties p_0 and ρ_0 decrease across the shock due to the dissipative processes inside the wavefront.

Normal Shock Propagating in a Still Medium

Frequently, one needs to calculate the properties of flow due to the propagation of a shock wave through a still medium, for example, due to an explosion. The transformation necessary to analyze this problem is indicated in Figure 16.11. The left panel shows a stationary shock, with incoming and outgoing velocities u_1 and u_2, respectively. On this flow we add a velocity u_1 directed to the left, so that the fluid entering the shock is stationary, and the fluid downstream of the shock is moving to the *left* at a speed $u_1 - u_2$, as shown in the right panel of the figure. This is consistent with our remark in Section 2 that the passage of a compression wave "pushes" the fluid forward in the direction of propagation of the wave. The shock speed is therefore u_1, with a supersonic Mach number $M_1 = u_1/c_1 > 1$. It follows that a *finite pressure disturbance propagates through a still fluid at supersonic speed*, in contrast to infinitesimal waves that propagate at the sonic speed. The expressions for all the thermodynamic properties of the flow, such as those given in equations (16.31)–(16.36), are still applicable.

Shock Structure

We shall now note a few points about the structure of a shock wave. The viscous and heat conductive processes within the shock wave result in an entropy increase across the front. However, the magnitude of the viscosity μ and thermal conductivity k only determines the thickness of the front and not the magnitude of the entropy increase. The entropy increase is determined solely by the upstream Mach number as shown by equation (16.36). We shall also see later that the *wave drag* experienced by a body due to the appearance of a shock wave is independent of viscosity or thermal conductivity. (The situation here is analogous to the viscous dissipation in fully turbulent flows (Chapter 13, Section 8), in which the dissipation rate ε is determined by the velocity and length scales of a large-scale turbulence field ($\varepsilon \sim u^3/l$) and not by the magnitude of the viscosity; a change in viscosity merely changes the scale at which the dissipation takes place (namely, the Kolmogorov microscale).)

The shock wave is in fact a very thin boundary layer. However, the velocity gradient du/dx is entirely longitudinal, in contrast to the lateral velocity gradient involved in a viscous boundary layer near a solid surface. Analysis shows that the thickness δ of a shock wave is given by

$$\frac{\delta \Delta u}{\nu} \sim 1,$$

where the left-hand side is a Reynolds number based on the velocity change across the shock, its thickness, and the average value of viscosity. Taking a typical value for air of $\nu \sim 10^{-5}\,\mathrm{m^2/s}$, and a velocity jump of $\Delta u \sim 100\,\mathrm{m/s}$, we obtain a shock thickness of

$$\delta \sim 10^{-7}\,\mathrm{m}.$$

This is not much larger than the mean free path (average distance traveled by a molecule between collisions), which suggests that the continuum hypothesis becomes of questionable validity in analyzing shock structure.

To gain some insight into the structure of shock waves, we shall consider the one-dimensional steady Navier–Stokes equations, including heat conduction and Newtonian viscous stresses. Despite the fact that the significant length scale for the structure pushes the limits of validity of the continuum formulation, the solution we obtain provides a smooth transition between upstream and downstream states, looks reasonable, and agrees with experiments and kinetic theory models for upstream Mach numbers less than about 2. The equations for conservation of mass, momentum, and energy are, respectively,

$$d(\rho u)/dx = 0$$
$$\rho u \, du/dx + dp/dx = d(\mu'' du/dx)/dx, \quad \mu'' = 2\mu + \lambda$$
$$\rho u \, dh/dx - u \, dp/dx = \mu''(du/dx)^2 + d(kdT/dx)/dx.$$

By adding to the energy equation the product of u with the momentum equation, these can be integrated once to yield,

$$\rho u = m$$
$$mu + p - \mu'' du/dx = mV$$
$$m(h + u^2/2) - \mu'' u \, du/dx - kdT/dx = mI,$$

where m, V, I are the constants of integration. These are evaluated upstream (state 1) and downstream (state 2) where gradients vanish and yield the Rankine-Hugoniot relations derived above. We also need the equations of state for a perfect gas with constant specific heats to solve for the structure: $h = C_p T$, $p = \rho R T$. Multiplying the energy equation by C_p/k we obtain the form

$$(mC_p/k)(C_p T + u^2/2) - (\mu'' C_p/k)d(u^2/2)/dx - d(C_p T)/dx = mC_p I/k.$$

This has an exact integral in the special case $\text{Pr}'' \equiv \mu'' C_p/k = 1$. This was found by Becker in 1922. If Stokes relation is assumed [(4.42)], $3\lambda + 2\mu = 0$ then $\mu'' = 4\mu/3$ and $\text{Pr} = \mu C_p/k = 3/4$, which is quite close to the actual value for air. The Becker integral is $C_p T + u^2/2 = I$. Eliminating all variables but u from the momentum equation, using the equations of state, mass conservation, and the energy integral,

$$mu + (m/u)(R/C_p)(I - u^2/2) - \mu''du/dx = mV.$$

With $C_p/R = \gamma/(\gamma - 1)$, multiplying by u/m, we obtain

$$-[2\gamma/(\gamma + 1)](\mu''/m)udu/dx = -u^2 + [2\gamma/(\gamma + 1)]uV - 2I(\gamma - 1)/(\gamma + 1)$$
$$\equiv (U_1 - U)(U - U_2).$$

Divide by V^2 and let $u/V = U$. The equation for the structure becomes

$$-U(U_1 - U)^{-1}(U - U_2)^{-1}dU = [(\gamma + 1)/2\gamma](m/\mu'')dx,$$

where the roots of the quadratic are

$$U_{1,2} = [\gamma/(\gamma + 1)]\{1 \pm [1 - 2(\gamma^2 - 1)I/(\gamma^2 V^2)]^{1/2}\},$$

the dimensionless speeds far up- and downstream of the shock. The left-hand side of the equation for the structure is rewritten in terms of partial fractions and then integrated to obtain

$$[U_1 \ln(U_1 - U) - U_2 \ln(U - U_2)]/(U_1 - U_2)$$
$$= [(\gamma + 1)/(2\gamma)]m \int dx/\mu'' \equiv [(\gamma + 1)/(2\gamma)]\eta$$

The structure is shown in Figure 16.12 in terms of the stretched coordinate $\eta = \int (m/\mu'')dx$ where μ'' is often a strong function of temperature and thus of x. A similar structure is obtained for all except quite small values of Pr''. In the limit $\text{Pr}'' \to 0$, Hayes (1958) points out that there must be a "shock within a shock" because

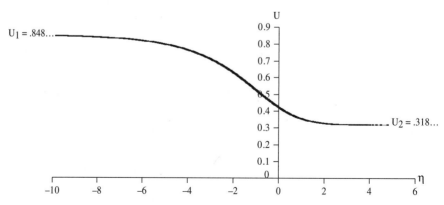

Figure 16.12 Shock structure velocity profile for the case $U_1 = 0.848485$, $U_2 = 0.31818$, corresponding to $M_1 = 2.187$.

heat conduction alone cannot provide the entire structure. In fact, Becker (1922) (footnote, p. 341) credits Prandtl for originating this idea. Cohen and Moraff (1971) provided the structure of both the outer (heat conducting) and inner (isothermal viscous) shocks. The variable η is a dimensionless length scale measured very roughly in units of mean free paths. We see that a measure of shock thickness is of the order of 5 mean free paths.

7. *Operation of Nozzles at Different Back Pressures*

Nozzles are used to accelerate a fluid stream and are employed in such systems as wind tunnels, rocket motors, and steam turbines. A pressure drop is maintained across it. In this section we shall examine the behavior of a nozzle as the exit pressure is varied. It will be assumed that the fluid is supplied from a large reservoir where the pressure is maintained at a constant value p_0 (the stagnation pressure), while the "back pressure" p_B in the exit chamber is varied. In the following discussion, we need to note that the pressure p_{exit} at the exit plane of the nozzle must equal the back pressure p_B if the flow at the exit plane is subsonic, but *not* if it is supersonic. This must be true because sharp pressure changes are only allowed in a supersonic flow.

Convergent Nozzle

Consider first the case of a convergent nozzle shown in Figure 16.13, which examines a sequence of states *a* through *c* during which the back pressure is gradually lowered. For curve *a*, the flow throughout the nozzle is subsonic. As p_B is lowered, the Mach number increases everywhere and the mass flux through the nozzle also increases. This continues until sonic conditions are reached at the exit, as represented by curve *b*. Further lowering of the back pressure has no effect on the flow inside the nozzle. This is because the fluid at the exit is now moving downstream at the velocity at which no pressure changes can propagate upstream. Changes in p_B therefore cannot propagate upstream after sonic conditions are reached at the exit. We say that the nozzle at this stage is *choked* because the mass flux cannot be increased by further lowering of back pressure. If p_B is lowered further (curve *c* in Figure 16.13), supersonic flow is generated *outside* the nozzle, and the jet pressure adjusts to p_B by means of a series of "oblique expansion waves," as schematically indicated by the oscillating pressure distribution for curve *c*. (The concepts of oblique expansion waves and oblique shock waves will be explained in Sections 10 and 11. It is only necessary to note here that they are oriented at an angle to the direction of flow, and that the pressure decreases through an oblique expansion wave and increases through an oblique shock wave.)

Convergent–Divergent Nozzle

Now consider the case of a convergent–divergent passage (Figure 16.14). Completely subsonic flow applies to curve *a*. As p_B is lowered to p_b, sonic condition is reached at the throat. On further reduction of the back pressure, the flow upstream of the throat does not respond, and the nozzle has "choked" in the sense that it is allowing the maximum mass flow rate for the given values of p_0 and throat area. There is a

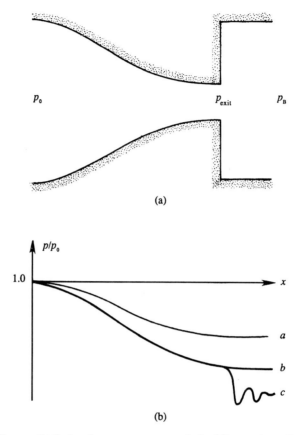

Figure 16.13 Pressure distribution along a convergent nozzle for different values of back pressure p_B: (a) diagram of nozzle; and (b) pressure distributions.

range of back pressures, shown by curves c and d, in which the flow initially becomes supersonic in the divergent portion, but then adjusts to the back pressure by means of a normal shock standing inside the nozzle. The flow downstream of the shock is, of course, subsonic. In this range the position of the shock moves downstream as p_B is decreased, and for curve d the normal shock stands right at the exit plane. The flow in the entire divergent portion up to the exit plane is now supersonic and remains so on further reduction of p_B. When the back pressure is further reduced to p_e, there is no normal shock anywhere within the nozzle, and the jet pressure adjusts to p_B by means of oblique shock waves outside the exit plane. These oblique shock waves vanish when $p_B = p_f$. On further reduction of the back pressure, the adjustment to p_B takes place outside the exit plane by means of oblique expansion waves.

Example 16.2

A convergent–divergent nozzle is operating under off-design conditions, resulting in the presence of a shock wave in the diverging portion. A reservoir containing air at

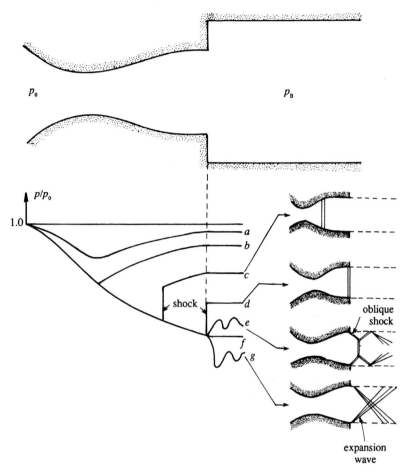

Figure 16.14 Pressure distribution along a convergent–divergent nozzle for different values of back pressure p_B. Flow patterns for cases c, d, e, and g are indicated schematically on the right. H. W. Liepmann and A. Roshko, *Elements of Gas Dynamics*, Wiley, New York 1957 and reprinted with the permission of Dr. Anatol Roshko.

400 kPa and 800 K supplies the nozzle, whose throat area is $0.2 \, \text{m}^2$. The upstream Mach number of the shock is $M_1 = 2.44$. The area at the exit is $0.7 \, \text{m}^2$. Find the area at the location of the shock and the exit temperature.

Solution: Figure 16.15 shows the profile of the nozzle, where sections 1 and 2 represent conditions across the shock. As a shock wave can exist only in a supersonic stream, we know that sonic conditions are reached at the throat, and the throat area equals the critical area A^*. The values given are therefore

$$p_0 = 400 \, \text{kPa},$$
$$T_0 = 800 \, \text{K},$$

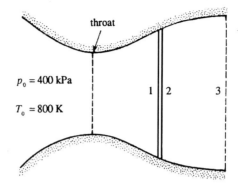

Figure 16.15   Example 16.2.

$$A_{\text{throat}} = A_1^* = 0.2\,\text{m}^2,$$
$$M_1 = 2.44,$$
$$A_3 = 0.7\,\text{m}^2.$$

Note that A^* is constant upstream of the shock, up to which the process is isentropic; this is why we have set $A_{\text{throat}} = A_1^*$.

The technique of solving this problem is to proceed downstream from the given stagnation conditions. Corresponding to the Mach number $M_1 = 2.44$, the isentropic table Table 16.1 gives

$$\frac{A_1}{A_1^*} = 2.5,$$

so that

$$A_1 = A_2 = (2.5)(0.2) = 0.5\,\text{m}^2.$$

This is the area at the location of the shock. Corresponding to $M_1 = 2.44$, the normal shock Table 16.2 gives

$$M_2 = 0.519,$$

$$\frac{p_{02}}{p_{01}} = 0.523.$$

There is no loss of stagnation pressure up to section 1, so that

$$p_{01} = p_0,$$

which gives

$$p_{02} = 0.523\,p_0 = 0.523(400) = 209.2\,\text{kPa}.$$

The value of A^* changes across a shock wave. The ratio A_2/A_2^* can be found from the *isentropic* table (Table 16.1) corresponding to a Mach number of $M_2 = 0.519$. (Note that A_2^* simply denotes the area that would be reached if the flow from state 2 were accelerated isentropically to sonic conditions.) Corresponding to $M_2 = 0.519$, Table 16.1 gives

$$\frac{A_2}{A_2^*} = 1.3,$$

TABLE 16.2   One-Dimensional Normal-Shock Relations ($\gamma = 1.4$)

M_1	M_2	p_2/p_1	T_2/T_1	$(p_0)_2/(p_0)_1$	M_1	M_2	p_2/p_1	T_2/T_1	$(p_0)_2/(p_0)_1$
1.00	1.000	1.000	1.000	1.000	1.96	0.584	4.315	1.655	0.740
1.02	0.980	1.047	1.013	1.000	1.98	0.581	4.407	1.671	0.730
1.04	0.962	1.095	1.026	1.000	2.00	0.577	4.500	1.688	0.721
1.06	0.944	1.144	1.039	1.000	2.02	0.574	4.594	1.704	0.711
1.08	0.928	1.194	1.052	0.999	2.04	0.571	4.689	1.720	0.702
1.10	0.912	1.245	1.065	0.999	2.06	0.567	4.784	1.737	0.693
1.12	0.896	1.297	1.078	0.998	2.08	0.564	4.881	1.754	0.683
1.14	0.882	1.350	1.090	0.997	2.10	0.561	4.978	1.770	0.674
1.16	0.868	1.403	1.103	0.996	2.12	0.558	5.077	1.787	0.665
1.18	0.855	1.458	1.115	0.995	2.14	0.555	5.176	1.805	0.656
1.20	0.842	1.513	1.128	0.993	2.16	0.553	5.277	1.822	0.646
1.22	0.830	1.570	1.140	0.991	2.18	0.550	5.378	1.837	0.637
1.24	0.818	1.627	1.153	0.988	2.20	0.547	5.480	1.857	0.628
1.26	0.807	1.686	1.166	0.986	2.22	0.544	5.583	1.875	0.619
1.28	0.796	1.745	1.178	0.983	2.24	0.542	5.687	1.892	0.610
1.30	0.786	1.805	1.191	0.979	2.26	0.539	5.792	1.910	0.601
1.32	0.776	1.866	1.204	0.976	2.28	0.537	5.898	1.929	0.592
1.34	0.766	1.928	1.216	0.972	2.30	0.534	6.005	1.947	0.583
1.36	0.757	1.991	1.229	0.968	2.32	0.532	6.113	1.965	0.575
1.38	0.748	2.055	1.242	0.963	2.34	0.530	6.222	1.984	0.566
1.40	0.740	2.120	1.255	0.958	2.36	0.527	6.331	2.003	0.557
1.42	0.731	2.186	1.268	0.953	2.38	0.525	6.442	2.021	0.549
1.44	0.723	2.253	1.281	0.948	2.40	0.523	6.553	2.040	0.540
1.46	0.716	2.320	1.294	0.942	2.42	0.521	6.666	2.060	0.532
1.48	0.708	2.389	1.307	0.936	2.44	0.519	6.779	2.079	0.523
1.50	0.701	2.458	1.320	0.930	2.46	0.517	6.894	2.098	0.515
1.52	0.694	2.529	1.334	0.923	2.48	0.515	7.009	2.118	0.507
1.54	0.687	2.600	1.347	0.917	2.50	0.513	7.125	2.138	0.499
1.56	0.681	2.673	1.361	0.910	2.52	0.511	7.242	2.157	0.491
1.58	0.675	2.746	1.374	0.903	2.54	0.509	7.360	2.177	0.483
1.60	0.668	2.820	1.388	0.895	2.56	0.507	7.479	2.198	0.475
1.62	0.663	2.895	1.402	0.888	2.58	0.506	7.599	2.218	0.468
1.64	0.657	2.971	1.416	0.880	2.60	0.504	7.720	2.238	0.460
1.66	0.651	3.048	1.430	0.872	2.62	0.502	7.842	2.260	0.453

TABLE 16.2   (*Continued*)

M_1	M_2	p_2/p_1	T_2/T_1	$(p_0)_2/(p_0)_1$	M_1	M_2	p_2/p_1	T_2/T_1	$(p_0)_2/(p_0)_1$
1.68	0.646	3.126	1.444	0.864	2.64	0.500	7.965	2.280	0.445
1.70	0.641	3.205	1.458	0.856	2.66	0.499	8.088	2.301	0.438
1.72	0.635	3.285	1.473	0.847	2.68	0.497	8.213	2.322	0.431
1.74	0.631	3.366	1.487	0.839	2.70	0.496	8.338	2.343	0.424
1.76	0.626	3.447	1.502	0.830	2.72	0.494	8.465	2.364	0.417
1.78	0.621	3.530	1.517	0.821	2.74	0.493	8.592	2.386	0.410
1.80	0.617	3.613	1.532	0.813	2.76	0.491	8.721	2.407	0.403
1.82	0.612	3.698	1.547	0.804	2.78	0.490	8.850	2.429	0.396
1.84	0.608	3.783	1.562	0.795	2.80	0.488	8.980	2.451	0.389
1.86	0.604	3.869	1.577	0.786	2.82	0.487	9.111	2.473	0.383
1.88	0.600	3.957	1.592	0.777	2.84	0.485	9.243	2.496	0.376
1.90	0.596	4.045	1.608	0.767	2.86	0.484	9.376	2.518	0.370
1.92	0.592	4.134	1.624	0.758	2.88	0.483	9.510	2.541	0.364
1.94	0.588	4.224	1.639	0.749	2.90	0.481	9.645	2.563	0.358
2.92	0.480	9.781	2.586	0.352	2.98	0.476	10.194	2.656	0.334
2.94	0.479	9.918	2.609	0.346	3.00	0.475	10.333	2.679	0.328
2.96	0.478	10.055	2.632	0.340					

which gives

$$A_2^* = \frac{A_2}{1.3} = \frac{0.5}{1.3} = 0.3846 \, \text{m}^2.$$

The flow from section 2 to section 3 is isentropic, during which A^* remains constant. Thus

$$\frac{A_3}{A_3^*} = \frac{A_3}{A_2^*} = \frac{0.7}{0.3846} = 1.82.$$

We should now find the conditions at the exit from the isentropic table (Table 16.1). However, we could locate the value of $A/A^* = 1.82$ either in the supersonic or the subsonic branch of the table. As the flow downstream of a normal shock can only be subsonic, we should use the subsonic branch. Corresponding to $A/A^* = 1.82$, Table 16.1 gives

$$\frac{T_3}{T_{03}} = 0.977.$$

The stagnation temperature remains constant in an adiabatic process, so that $T_{03} = T_0$. Thus

$$T_3 = 0.977(800) = 782 \, \text{K}.$$

8. Effects of Friction and Heating in Constant-Area Ducts

In a duct of constant area, the equations of mass, momentum, and energy reduced to one-dimensional steady form become

$$\rho_1 u_1 = \rho_2 u_2,$$

$$p_1 + \rho_1 u_1^2 = p_2 + \rho_2 u_2^2 + p_1 f,$$

$$h_1 + \tfrac{1}{2}u_1^2 + h_1 q = h_2 + \tfrac{1}{2}u_2^2.$$

Here, $f = (f_\sigma)_x/(p_1 A)$ is a dimensionless friction parameter and $q = Q/h_1$ is a dimensionless heating parameter. In terms of Mach number, for a perfect gas with constant specific heats, the momentum and energy equations become, respectively,

$$p_1(1 + \gamma M_1^2 - f) = p_2(1 + \gamma M_2^2),$$

$$h_1\left(1 + \frac{\gamma-1}{2}M_1^2 + q\right) = h_2\left(1 + \frac{\gamma-1}{2}M_2^2\right).$$

Using mass conservation, the equation of state $p = \rho RT$, and the definition of Mach number, all thermodynamic variables can be eliminated resulting in

$$\frac{M_2}{M_1} = \frac{1 + \gamma M_2^2}{1 + \gamma M_1^2 - f}\left[\frac{1 + ((\gamma-1)/2)M_1^2 + q}{1 + ((\gamma-1)/2)M_2^2}\right]^{1/2}.$$

Bringing the unknown M_2 to the left-hand side and assuming q and f are specified along with M_1,

$$\frac{M_2^2(1 + ((\gamma-1)/2)M_2^2)}{(1 + \gamma M_2^2)^2} = \frac{M_1^2(1 + ((\gamma-1)/2)M_1^2 + q)}{(1 + \gamma M_1^2 - f)^2} \equiv \mathcal{A}.$$

This is a biquadratic equation for M_2 with the solution

$$M_2^2 = \frac{-(1 - 2\mathcal{A}\gamma) \pm [1 - 2\mathcal{A}(\gamma + 1)]^{1/2}}{(\gamma - 1) - 2\mathcal{A}\gamma^2}. \tag{16.37}$$

Figures 16.16 and 16.17 are plots of equation (16.37), $M_2 = F(M_1)$ first with f as a parameter (16.16) and $q = 0$ and then with q as a parameter and $f = 0$ (16.17). Generally, we specify the properties of the flow at the inlet station (station 1) and wish to calculate the properties at the outlet (station 2). Here, we will regard the dimensionless friction and heat transfer f and q as specified. Then we see that once M_2 is calculated from (16.37), all of the other properties may be obtained from the dimensionless formulation of the conservation laws. When q and $f = 0$, two solutions are possible: the trivial solution $M_1 = M_2$ and the normal shock solution that we obtained in Section 6 in the preceding. We also showed that the upper left branch of the solution $M_2 > 1$ when $M_1 < 1$ is inaccessible because it violates the second law of thermodynamics, that is, it results in a spontaneous decrease of entropy.

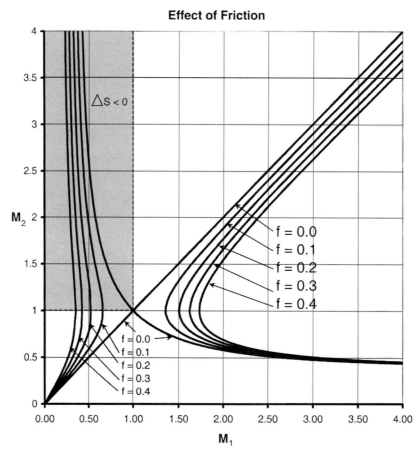

Figure 16.16 Flow in a constant-area duct with friction f as parameter; $q = 0$. Upper left quadrant is inaccessible because $\Delta S < 0$. $\gamma = 1.4$.

Effect of Friction

Referring to the left branch of Figure 16.16, the solution indicates that for $M_2 > M_1$ so that friction accelerates a subsonic flow. Then the pressure, density, and temperature are all diminished with respect to the entrance values. How can friction make the flow go faster? Friction is manifested by boundary layers at the walls. The boundary layer displacement thickness grows downstream so that the flow behaves as if it is in a convergent duct, which, as we have seen, is a subsonic nozzle. We will discuss in what follows what actually happens when there is no apparent solution for M_2. When M_1 is supersonic, two solutions are generally possible—one for which $1 < M_2 < M_1$ and the other where $M_2 < 1$. They are connected by a normal shock. Whether or not a shock occurs depends on the downstream pressure. There is also the possibility of M_1 insufficiently large or f too large so that no solution is indicated. We will discuss that in the following but note that the two solutions coalesce when $M_2 = 1$ and the

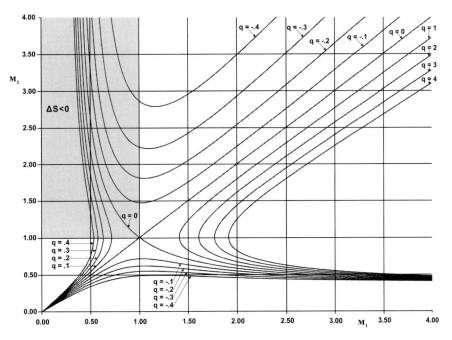

Figure 16.17 Flow in a constant-area duct with heating/cooling q as parameter; $f = 0$. Upper left quadrant is inaccessible because $\Delta S < 0$. $\gamma = 1.4$.

flow is said to be choked. At this condition the maximum mass flow is passed by the duct. In the case $1 < M_2 < M_1$, the flow is decelerated and the pressure, density, and temperature all increase in the downstream direction. The stagnation pressure is always decreased by friction as the entropy is increased.

Effect of Heat Transfer

The range of solutions is twice as rich in this case as q may take both signs. Figure 16.17 shows that for $q > 0$ solutions are similar in most respects to those with friction ($f > 0$). Heating accelerates a subsonic flow and lowers the pressure and density. However, heating generally increases the fluid temperature except in the limited range $1/\sqrt{\gamma} < M_1 < 1$ in which the tendency to accelerate the fluid is greater than the ability of the heat flux to raise the temperature. The energy from heat addition goes preferentially into increasing the kinetic energy of the fluid. The fluid temperature is decreased by heating in this limited range of Mach number. The supersonic branch $M_2 > 1$ when $M_1 < 1$ is inaccessible because those solutions violate the second law of thermodynamics. Again, as with f too large or M_1 too close to 1, there is a possibility with q too large of no solution indicated; this is discussed in what follows. When $M_1 > 1$, two solutions for M_2 are generally possible and they are connected by a normal shock. The shock is absent if the downstream pressure is low and present if the downstream pressure is high. Although $q > 0$ (and $f > 0$) does not always indicate a solution (if the flow has been choked), there will always be a solution for $q < 0$. Cooling a supersonic flow accelerates it, thus decreasing its pressure, temperature, and density. If no shock occurs, $M_2 > M_1$. Conversely, cooling a subsonic flow decelerates it so that the pressure and density increase. The temperature decreases when heat is removed from the flow except in the limited range $1/\sqrt{\gamma} < M_1 < 1$ in which the heat removal decelerates the flow so rapidly that the temperature increases.

For high molecular weight gases, near critical conditions (high pressure, low temperature), gasdynamic relationships as developed here for perfect gases may be completely different. Cramer and Fry (1993) found that such gases may support expansion shocks, accelerated flow through "antithroats," and generally behave in unfamiliar ways.

Choking by Friction or Heat Addition

We can see from Figures 16.16 and 16.17 that heating a flow or accounting for friction in a constant-area duct will make that flow tend towards sonic conditions. For any given M_1, the maximum f or $q > 0$ that is permissible is the one for which $M = 1$ at the exit station. The flow is then said to be choked, and no more mass/time can flow through that duct. This is analogous to flow in a convergent duct. Imagine pouring liquid through a funnel from one container into another. There is a maximum volumetric flow rate that can be passed by the funnel, and beyond that flow rate, the funnel overflows. The same thing happens here. If f or q is too large, such that no (steady-state) solution is possible, there is an external adjustment that reduces the mass flow rate to that for which the exit speed is just sonic. Both for $M_1 < 1$ and $M_1 > 1$ the limiting curves for f and q indicating choked flow intersect $M_2 = 1$ at right angles. Qualitatively, the effect is the same as choking by area contraction.

9. Mach Cone

So far in this chapter we have considered one-dimensional flows in which the flow properties varied only in the direction of flow. In this section we begin our study of wave motions in more than one dimension. Consider a point source emitting infinitesimal pressure disturbances in a still fluid in which the speed of sound is c. If the point disturbance is stationary, then the wavefronts are concentric spheres. This is shown in Figure 16.18a, where the wavefronts at intervals of Δt are shown.

Now suppose that the source propagates to the left at speed $U < c$. Figure 16.18b shows four locations of the source, that is, 1 through 4, at equal intervals of time Δt, with point 4 being the present location of the source. At point 1, the source emitted a wave that has spherically expanded to a radius of $3c\,\Delta t$ in an interval of time $3\,\Delta t$. During this time the source has moved to location 4, at a distance of $3U\,\Delta t$ from point 1. The figure also shows the locations of the wavefronts emitted while the source was at points 2 and 3. It is clear that the wavefronts do not intersect because $U < c$. As in the case of the stationary source, the wavefronts propagate everywhere in the flow field, upstream and downstream. It therefore follows that *a body moving at a subsonic speed influences the entire flow field*; information propagates upstream as well as downstream of the body.

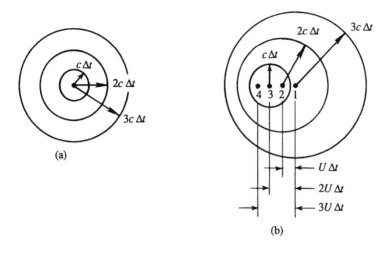

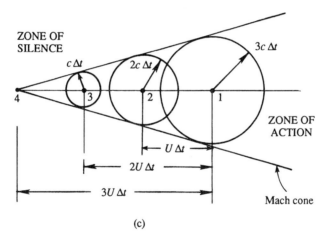

Figure 16.18 Wavefronts emitted by a point source in a still fluid when the source speed U is: (a) $U = 0$; (b) $U < c$; and (c) $U > c$.

Now consider a case where the disturbance moves supersonically at $U > c$ (Figure 16.18c). In this case the spherically expanding wavefronts cannot catch up with the faster moving disturbance and form a conical tangent surface called the *Mach cone*. In plane two-dimensional flow, the tangent surface is in the form of a wedge, and the tangent lines are called *Mach lines*. An examination of the figure shows that the half-angle of the Mach cone (or wedge), called the *Mach angle* μ, is given by $\sin \mu = (c \, \Delta t)/(U \, \Delta t)$, so that

$$\sin \mu = \frac{1}{M}. \tag{16.38}$$

The Mach cone becomes wider as M decreases and becomes a plane front (that is, $\mu = 90°$) when $M = 1$.

The point source considered here could be part of a solid body, which sends out pressure waves as it moves through the fluid. Moreover, Figure 16.18c applies equally if the point source is stationary and the fluid is approaching at a supersonic speed U. It is clear that in a supersonic flow an observer outside the Mach cone would not "hear" a signal emitted by a point disturbance, hence this region is called the *zone of silence*. In contrast, the region inside the Mach cone is called the *zone of action*, within which the effects of the disturbance are felt. This explains why the sound of a supersonic airplane does not reach an observer until the Mach cone arrives, *after* the plane has passed overhead.

At every point in a planar supersonic flow there are two Mach lines, oriented at $\pm\mu$ to the local direction of flow. Information propagates along these lines, which are the *characteristics* of the governing differential equation. It can be shown that the nature of the governing differential equation is hyperbolic in a supersonic flow and elliptic in a subsonic flow.

10. Oblique Shock Wave

In Section 6 we examined the case of a normal shock wave, oriented perpendicular to the direction of flow, in which the velocity changes from supersonic to subsonic values. However, a shock wave can also be oriented obliquely to the flow (Figure 16.19a), the velocity changing from V_1 to V_2. The flow can be analyzed by considering a normal shock across which the normal velocity varies from u_1 to u_2 and superposing a velocity v parallel to it (Figure 16.19b). By considering conservation of momentum in a direction tangential to the shock, we may show that v is unchanged across a shock (Exercise 12). The magnitude and direction of the velocities on the two sides of the shock are

$$V_1 = \sqrt{u_1^2 + v^2} \qquad \text{oriented at } \sigma = \tan^{-1}(u_1/v),$$

$$V_2 = \sqrt{u_2 + v^2} \qquad \text{oriented at } \sigma - \delta = \tan^{-1}(u_2/v).$$

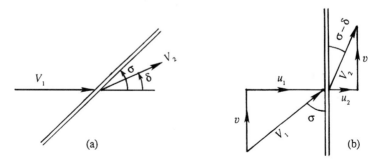

Figure 16.19 (a) Oblique shock wave in which δ = deflection angle and σ = shock angle; and (b) analysis by considering a normal shock and superposing a velocity v parallel to the shock.

The normal Mach numbers are

$$M_{n1} = u_1/c_1 = M_1 \sin \sigma > 1,$$
$$M_{n2} = u_2/c_2 = M_2 \sin(\sigma - \delta) < 1.$$

Because $u_2 < u_1$, there is a sudden change of direction of flow across the shock; in fact the flow turns *toward* the shock by an amount δ. The angle σ is called the *shock angle* or *wave angle* and δ is called the *deflection angle*.

Superposition of the tangential velocity v does not affect the *static* properties, which are therefore the same as those for a normal shock. The expressions for the ratios $p_2/p_1, \rho_2/\rho_1, T_2/T_1$, and $(S_2 - S_1)/C_v$ are therefore those given by equations (16.31), (16.33)–(16.35), if M_1 is replaced by the normal component of the upstream Mach number $M_1 \sin \sigma$. For example,

$$\frac{p_2}{p_1} = 1 + \frac{2\gamma}{\gamma + 1}(M_1^2 \sin^2 \sigma - 1), \tag{16.39}$$

$$\frac{\rho_2}{\rho_1} = \frac{(\gamma + 1)M_1^2 \sin^2 \sigma}{(\gamma - 1)M_1^2 \sin^2 \sigma + 2} = \frac{u_1}{u_2} = \frac{\tan \sigma}{\tan (\sigma - \delta)}. \tag{16.40}$$

The normal shock table, Table 16.2, is therefore also applicable to oblique shock waves if we use $M_1 \sin \sigma$ in place of M_1.

The relation between the upstream and downstream Mach numbers can be found from equation (16.32) by replacing M_1 by $M_1 \sin \sigma$ and M_2 by $M_2 \sin (\sigma - \delta)$. This gives

$$M_2^2 \sin^2(\sigma - \delta) = \frac{(\gamma - 1)M_1^2 \sin^2 \sigma + 2}{2\gamma M_1^2 \sin^2 \sigma + 1 - \gamma}. \tag{16.41}$$

An important relation is that between the deflection angle δ and the shock angle σ for a given M_1, given in equation (16.40). Using the trigonometric identity for $\tan (\sigma - \delta)$, this becomes

$$\tan \delta = 2 \cot \sigma \frac{M_1^2 \sin^2 \sigma - 1}{M_1^2(\gamma + \cos 2\sigma) + 2}. \tag{16.42}$$

A plot of this relation is given in Figure 16.20. The curves represent δ vs σ for constant M_1. The value of M_2 varies along the curves, and the locus of points corresponding to $M_2 = 1$ is indicated. It is apparent that there is a maximum deflection angle δ_{max} for oblique shock solutions to be possible; for example, $\delta_{max} = 23°$ for $M_1 = 2$. For a given M_1, δ becomes zero at $\sigma = \pi/2$ corresponding to a normal shock, and at $\sigma = \mu = \sin^{-1}(1/M_1)$ corresponding to the Mach angle. For a fixed M_1 and $\delta < \delta_{max}$, there are two possible solutions: a *weak shock* corresponding to a smaller σ, and a *strong shock* corresponding to a larger σ. It is clear that the flow downstream of a strong shock is always subsonic; in contrast, the flow downstream of a weak shock is generally supersonic, except in a small range in which δ is slightly smaller than δ_{max}.

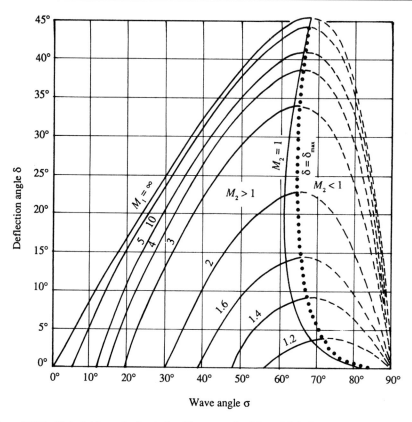

Figure 16.20 Plot of oblique shock solution. The strong shock branch is indicated by dashed lines, and the heavy dotted line indicates the maximum deflection angle δ_{max}. (From NACA Report 1135.)

Generation of Oblique Shock Waves

Consider the supersonic flow past a wedge of half-angle δ, or the flow over a wall that turns inward by an angle δ (Figure 16.21). If M_1 and δ are given, then σ can be obtained from Figure 16.20, and M_{n2} (and therefore $M_2 = M_{n2}/\sin(\sigma - \delta)$) can be obtained from the shock table, Table 16.2. An attached shock wave, corresponding to the weak solution, forms at the nose of the wedge, such that the flow is parallel to the wedge after turning through an angle δ. The shock angle σ decreases to the Mach angle $\mu_1 = \sin^{-1}(1/M_1)$ as the deflection δ tends to zero. It is interesting that the corner velocity in a supersonic flow is finite. In contrast, the corner velocity in a subsonic (or incompressible) flow is either zero or infinite, depending on whether the wall shape is concave or convex. Moreover, the streamlines in Figure 16.21 are straight, and computation of the field is easy. By contrast, the streamlines in a subsonic flow are curved, and the computation of the flow field is not easy. The basic reason for this is that, in a supersonic flow, the disturbances do not propagate upstream of Mach lines or shock waves emanating from the disturbances, hence the flow field can be constructed step by step, *proceeding downstream*. In contrast, the disturbances

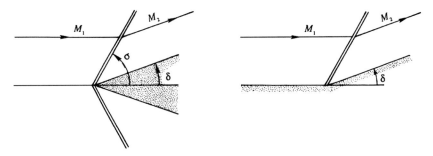

Figure 16.21 Oblique shocks in supersonic flow.

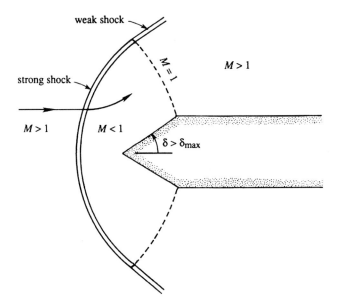

Figure 16.22 Detached shock.

propagate both upstream and downstream in a subsonic flow, so that all features in the entire flow field are related to each other.

As δ is increased beyond δ_{max}, attached oblique shocks are not possible, and a detached curved shock stands in front of the body (Figure 16.22). The central streamline goes through a normal shock and generates a subsonic flow in front of the wedge. The *strong* shock solution of Figure 16.20 therefore holds near the nose of the body. Farther out, the shock angle decreases, and the weak shock solution applies. If the wedge angle is not too large, then the curved detached shock in Figure 16.22 becomes an oblique attached shock as the Mach number is increased. In the case of a blunt-nosed body, however, the shock at the leading edge is always detached, although it moves closer to the body as the Mach number is increased.

We see that shock waves may exist in supersonic flows and their location and orientation adjust to satisfy boundary conditions. In external flows, such as those just described, the boundary condition is that streamlines at a solid surface must be tangent to that surface. In duct flows the boundary condition locating the shock is usually the downstream pressure.

The Weak Shock Limit

A simple and useful expression can be derived for the pressure change across a weak shock by considering the limiting case of a small deflection angle δ. We first need to simplify equation (16.42) by noting that as $\delta \to 0$, the shock angle σ tends to the Mach angle $\mu_1 = \sin^{-1}(1/M_1)$.

Also from equation (16.39) we note that $(p_2 - p_1)/p_1 \to 0$ as $M_1^2 \sin^2 \sigma - 1 \to 0$, (as $\sigma \to \mu$ and $\delta \to 0$). Then from equations (16.39) and (16.42)

$$\tan \delta = 2 \cot \sigma \frac{\gamma + 1}{2\gamma} \left(\frac{p_2 - p_1}{p_1} \right) \frac{1}{M_1^2(\gamma + 1 - 2\sin^2 \sigma) + 2}. \qquad (16.43)$$

As $\delta \to 0$, $\tan \delta \approx \delta$, $\cot \mu = \sqrt{M_1^2 - 1}$, $\sin \sigma \approx 1/M_1$, and

$$\boxed{\frac{p_2 - p_1}{p_1} \simeq \frac{\gamma M_1^2}{\sqrt{M_1^2 - 1}} \delta.} \qquad (16.44)$$

The interesting point is that relation (16.44) is also applicable to a weak *expansion* wave and not just a weak compression wave. By this we mean that the pressure increase due to a small deflection of the wall toward the flow is the same as the pressure *decrease* due to a small deflection of the wall *away* from the flow. This is because the entropy change across a shock goes to zero much faster than the rate at which the pressure difference across the wave decreases as our study of normal shock waves has shown. Very weak "shock waves" are therefore approximately isentropic or reversible. Relationships for a weak shock wave can therefore be applied to a weak expansion wave, except for some sign changes. In Section 12, equation (16.44) will be applied in estimating the lift and drag of a thin airfoil in supersonic flow.

11. Expansion and Compression in Supersonic Flow

Consider the supersonic flow over a gradually curved wall (Figure 16.23). The wavefronts are now Mach lines, inclined at an angle of $\mu = \sin^{-1}(1/M)$ to the *local* direction of flow. The flow orientation and Mach number are constant on each Mach line. In the case of compression, the Mach number decreases along the flow, so that the Mach angle increases. The Mach lines therefore coalesce and form an oblique shock. In the case of gradual expansion, the Mach number increases along the flow and the Mach lines diverge.

If the wall has a sharp deflection away from the approaching stream, then the pattern of Figure 16.23b takes the form of Figure 16.24. The flow expands through a "fan" of Mach lines centered at the corner, called the *Prandtl–Meyer expansion*

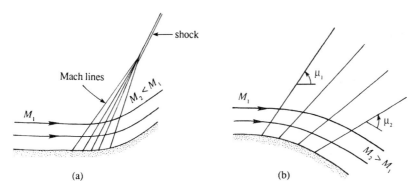

Figure 16.23 Gradual compression and expansion in supersonic flow: (a) gradual compression, resulting in shock formation; and (b) gradual expansion.

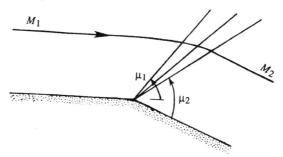

Figure 16.24 The Prandtl–Meyer expansion fan.

fan. The Mach number increases through the fan, with $M_2 > M_1$. The first Mach line is inclined at an angle of μ_1 to the local flow direction, while the last Mach line is inclined at an angle of μ_2 to the *local* flow direction. The pressure falls gradually along a streamline through the fan. (Along the wall, however, the pressure remains constant along the upstream wall, falls discontinuously at the corner, and then remains constant along the downstream wall.) Figure 16.24 should be compared with Figure 16.21, in which the wall turns *inward* and generates a shock wave. By contrast, the expansion in Figure 16.24 is gradual and isentropic.

The flow through a Prandtl–Meyer fan is calculated as follows. From Figure 16.19b, conservation of momentum tangential to the shock shows that the tangential velocity is unchanged, or

$$V_1 \cos \sigma = V_2 \cos(\sigma - \delta) = V_2(\cos \sigma \cos \delta + \sin \sigma \sin \delta).$$

We are concerned here with very small deflections, $\delta \to 0$ so $\sigma \to \mu$. Here, $\cos \delta \approx 1$, $\sin \delta \approx \delta$, $V_1 \approx V_2(1 + \delta \tan \sigma)$, so $(V_2 - V_1)/V_1 \approx \delta \tan \sigma \approx -\delta/\sqrt{M_1^2 - 1}$.

Regarding this as appropriate for infinitesimal change in V for an infinitesimal deflection, we can write this as $d\delta = -dV \sqrt{M^2 - 1}/V$ (first quadrant deflection). Because $V = Mc$, $dV/V = dM/M + dc/c$. With $c = \sqrt{\gamma RT}$ for a perfect gas,

$dc/c = dT/2T$. Using equation (16.20) for adiabatic flow of a perfect gas, $dT/T = -(\gamma - 1)M\, dM/[1 + ((\gamma - 1)/2)M^2]$.

Then

$$d\delta = -\frac{\sqrt{M^2 - 1}}{M}\frac{dM}{1 + ((\gamma - 1)/2)M^2}.$$

Integrating δ from 0 (radians) and M from 1 gives

$$\delta + \nu(M) = \text{const.},$$

where

$$\nu(M) = \int_1^M \frac{\sqrt{M^2 - 1}}{1 + ((\gamma - 1)/2)M^2}\frac{dM}{M}$$

$$= \sqrt{\frac{\gamma + 1}{\gamma - 1}}\tan^{-1}\sqrt{\frac{\gamma - 1}{\gamma + 1}(M^2 - 1)} - \tan^{-1}\sqrt{M^2 - 1}, \qquad (16.45)$$

is called the Prandtl–Meyer function. The sign of $\sqrt{M^2 - 1}$ originates from the identification of $\tan\sigma = \tan\mu = 1/\sqrt{M^2 - 1}$ for a first quadrant deflection (upper half-plane). For a fourth quadrant deflection (lower half-plane), $\tan\mu = -1/\sqrt{M^2 - 1}$. For example, in Figure 16.23 we would write

$$\delta_1 + \nu_1(M_1) = \delta_2 + \nu_2(M_2),$$

where, for example, δ_1, δ_2, and M_1 are given. Then

$$\nu_2(M_2) = \delta_1 - \delta_2 + \nu_1(M_1).$$

In panel (a), $\delta_1 - \delta_2 < 0$, so $\nu_2 < \nu_1$ and $M_2 < M_1$. In panel (b), $\delta_1 - \delta_2 > 0$, so $\nu_2 > \nu_1$ and $M_2 > M_1$.

12. Thin Airfoil Theory in Supersonic Flow

Simple expressions can be derived for the lift and drag coefficients of an airfoil in supersonic flow if the thickness and angle of attack are small. The disturbances caused by a thin airfoil are small, and the total flow can be built up by superposition of small disturbances emanating from points on the body. Such a linearized theory of lift and drag was developed by Ackeret. Because all flow inclinations are small, we can use the relation (16.44) to calculate the pressure changes due to a change in flow direction. We can write this relation as

$$\frac{p - p_\infty}{p_\infty} = \frac{\gamma M_\infty^2 \delta}{\sqrt{M_\infty^2 - 1}}, \qquad (16.46)$$

where p_∞ and M_∞ refer to the properties of the free stream, and p is the pressure at a point where the flow is inclined at an angle δ to the free-stream direction. The sign of δ determines the sign of $(p - p_\infty)$.

To see how the lift and drag of a thin body in a supersonic stream can be estimated, consider a flat plate inclined at a small angle α to a stream (Figure 16.25). At the leading edge there is a weak expansion fan on the top surface and a weak oblique shock on the bottom surface. The streamlines ahead of these waves are straight. The streamlines above the plate turn through an angle α by expanding through a centered fan, downstream of which they become parallel to the plate with a pressure $p_2 < p_\infty$. The upper streamlines then turn sharply across a shock emanating from the trailing edge, becoming parallel to the free stream once again. Opposite features occur for the streamlines below the plate. The flow first undergoes compression across a shock coming from the leading edge, which results in a pressure $p_3 > p_\infty$. It is, however, not important to distinguish between shocks and expansion waves in Figure 16.25, because the linearized theory treats them the same way, except for the sign of the pressure changes they produce.

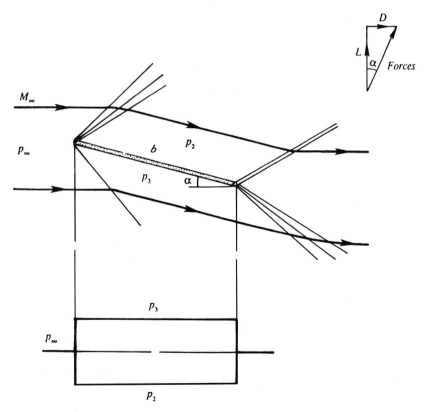

Figure 16.25 Inclined flat plate in a supersonic stream. The upper panel shows the flow pattern and the lower panel shows the pressure distribution.

The pressures above and below the plate can be found from equation (16.46), giving

$$\frac{p_2 - p_\infty}{p_\infty} = -\frac{\gamma M_\infty^2 \alpha}{\sqrt{M_\infty^2 - 1}},$$

$$\frac{p_3 - p_\infty}{p_\infty} = \frac{\gamma M_\infty^2 \alpha}{\sqrt{M_\infty^2 - 1}}.$$

The pressure difference across the plate is therefore

$$\frac{p_3 - p_2}{p_\infty} = \frac{2\alpha\gamma M_\infty^2}{\sqrt{M_\infty^2 - 1}}.$$

If b is the chord length, then the lift and drag forces per unit span are

$$L = (p_3 - p_2)b \cos\alpha \simeq \frac{2\alpha\gamma M_\infty^2 p_\infty b}{\sqrt{M_\infty^2 - 1}},$$

$$D = (p_3 - p_2)b \sin\alpha \simeq \frac{2\alpha^2\gamma M_\infty^2 p_\infty b}{\sqrt{M_\infty^2 - 1}}. \tag{16.47}$$

The lift coefficient is defined as

$$C_L \equiv \frac{L}{(1/2)\rho_\infty U_\infty^2 b} = \frac{L}{(1/2)\gamma p_\infty M_\infty^2 b},$$

where we have used the relation $\rho U^2 = \gamma p M^2$. Using equation (16.47), the lift and drag coefficients for a flat lifting surface are

$$C_L \simeq \frac{4\alpha}{\sqrt{M_\infty^2 - 1}},$$

$$C_D \simeq \frac{4\alpha^2}{\sqrt{M_\infty^2 - 1}}. \tag{16.48}$$

These expressions do not hold at transonic speeds $M_\infty \to 1$, when the process of linearization used here breaks down. The expression for the lift coefficient should be compared to the incompressible expression $C_L \simeq 2\pi\alpha$ derived in the preceding chapter. Note that the flow in Figure 16.25 does have a circulation because the velocities at the upper and lower surfaces are parallel but have different magnitudes. However, in a supersonic flow it is not necessary to invoke the Kutta condition (discussed in the preceding chapter) to determine the magnitude of the circulation. The flow in Figure 16.25 does leave the trailing edge smoothly.

The drag in equation (16.48) is the *wave drag* experienced by a body in a supersonic stream, and exists even in an inviscid flow. The d'Alembert paradox therefore does not apply in a supersonic flow. The supersonic wave drag is analogous to the gravity wave drag experienced by a ship moving at a speed greater than the velocity of surface gravity waves, in which a system of bow waves is carried with the ship. The magnitude of the supersonic wave drag is independent of the value of the viscosity, although the energy spent in overcoming this drag is finally dissipated through viscous effects within the shock waves. In addition to the wave drag, additional drags due to viscous and finite-span effects, considered in the preceding chapter, act on a real wing.

In this connection, it is worth noting the difference between the airfoil shapes used in subsonic and supersonic airplanes. Low-speed airfoils have a streamlined shape, with a rounded nose and a sharp trailing edge. These features are not helpful in supersonic airfoils. The most effective way of reducing the drag of a supersonic airfoil is to reduce its thickness. Supersonic wings are characteristically thin and have a sharp leading edge.

Exercises

1. The critical area A^* of a duct flow was defined in Section 4. Show that the relation between A^* and the actual area A at a section, where the Mach number equals M, is that given by equation (16.23). This relation was not proved in the text. [*Hint*: Write

$$\frac{A}{A^*} = \frac{\rho^* c^*}{\rho u} = \frac{\rho^*}{\rho_0} \frac{\rho_0}{\rho} \frac{c^*}{c} \frac{c}{u} = \frac{\rho^*}{\rho_0} \frac{\rho_0}{\rho} \sqrt{\frac{T^*}{T_0} \frac{T_0}{T}} \frac{1}{M}.$$

Then use the relations given in Section 4.]

2. The entropy change across a normal shock is given by equation (16.35). Show that this reduces to expression (16.36) for weak shocks. [*Hint*: Let $M_1^2 - 1 \ll 1$. Write the terms within the two brackets [ ] [ ] in equation (16.35) in the form $[1 + \varepsilon_1][1 + \varepsilon_2]^\gamma$, where ε_1 and ε_2 are small quantities. Then use series expansion $\ln(1 + \varepsilon) = \varepsilon - \varepsilon^2/2 + \varepsilon^3/3 + \cdots$. This gives equation (16.36) times a function of M_1 in which we can set $M_1 = 1$.]

3. Show that the maximum velocity generated from a reservoir in which the stagnation temperature equals T_0 is

$$u_{\max} = \sqrt{2 C_p T_0}.$$

What are the corresponding values of T and M?

4. In an adiabatic flow of air through a duct, the conditions at two points are

$$u_1 = 250 \, \text{m/s},$$
$$T_1 = 300 \, \text{K},$$
$$p_1 = 200 \, \text{kPa},$$
$$u_2 = 300 \, \text{m/s},$$
$$p_2 = 150 \, \text{kPa}.$$

Show that the loss of stagnation pressure is nearly 34.2 kPa. What is the entropy increase?

5. A shock wave generated by an explosion propagates through a still atmosphere. If the pressure downstream of the shock wave is 700 kPa, estimate the shock speed and the flow velocity downstream of the shock.

6. A wedge has a half-angle of 50°. Moving through air, can it ever have an attached shock? What if the half-angle were 40°? [*Hint*: The argument is based entirely on Figure 16.20.]

7. Air at standard atmospheric conditions is flowing over a surface at a Mach number of $M_1 = 2$. At a downstream location, the surface takes a sharp inward turn by an angle of 20°. Find the wave angle σ and the downstream Mach number. Repeat the calculation by using the weak shock assumption and determine its accuracy by comparison with the first method.

8. A flat plate is inclined at 10° to an airstream moving at $M_\infty = 2$. If the chord length is $b = 3$ m, find the lift and wave drag per unit span.

9. A perfect gas is stored in a large tank at the conditions specified by p_o, T_o. Calculate the maximum mass flow rate that can exhaust through a duct of cross-sectional area A. Assume that A is small enough that during the time of interest p_o and T_o do not change significantly and that the flow is adiabatic.

10. For flow of a perfect gas entering a constant area duct at Mach number M_1, calculate the maximum admissable values of f, q for the same mass flow rate. Case (a) $f = 0$; Case (b) $q = 0$.

11. Using thin airfoil theory calculate the lift and drag on the airfoil shape given by $y_u = t \sin(\pi x/c)$ for the upper surface and $y_l = 0$ for the lower surface. Assume a supersonic stream parallel to the x-axis. The thickness ratio $t/c \ll 1$.

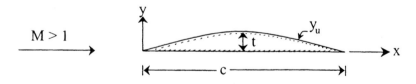

12. Write momentum conservation for the volume of the small "pill box" shown in Figure 4.22 (p. 121) where the interface is a shock with flow from side 1 to side 2. Let the two end faces approach each other as the shock thickness → 0 and assume viscous stresses may be neglected on these end faces (outside the structure). Show that the **n** component of momentum conservation yields (16.29) and the **t** component gives **u** · **t** is conserved or v is continuous across the shock.

Literature Cited

Ames Research Staff (1953). NACA Report 1135: "Equations, Tables, and Charts for Compressible Flow."

Becker, R. (1922). "Stosswelle und Detonation." *Z. Physik* **8**: 321–362.

Cohen, I. M. and C. A. Moraff (1971). "Viscous inner structure of zero Prandtl number shocks." *Phys. Fluids* **14**: 1279–1280.

Cramer, M. S. and R. N. Fry (1993). "Nozzle flows of dense gases." *The Physics of Fluids A* **5**: 1246–1259.

Fergason, S. H., T. L. Ho, B. M. Argrow, and G. Emanuel (2001). "Theory for producing a single-phase rarefaction shock wave in a shock tube." *Journal of Fluid Mechanics* **445**: 37–54.

Hayes, W. D. (1958). "The basic theory of gasdynamic discontinuities," Sect. D of *Fundamentals of Gasdynamics*, Edited by H. W. Emmons, Vol. III of *High Speed Aerodynamics and Jet Propulsion*, Princeton, NJ: Princeton University Press.

Liepmann, H. W. and A. Roshko (1957). *Elements of Gas Dynamics*, New York: Wiley.

Shapiro, A. H. (1953). *The Dynamics and Thermodynamics of Compressible Fluid Flow,* 2 volumes. New York: Ronald.

von Karman, T. (1954). *Aerodynamics,* New York: McGraw-Hill.

Supplemental Reading

Courant, R. and K. O. Friedrichs (1977). *Supersonic Flow and Shock Waves*, New York: Springer-Verlag.

Yahya, S. M. (1982). *Fundamentals of Compressible Flow*, New Delhi: Wiley Eastern.

Chapter 17

Introduction to Biofluid Mechanics

Portonovo S. Ayyaswamy
University of Pennsylvania
Philadelphia, PA

1. Introduction

This chapter is intended to be of an introductory nature to the vast field of biofluid mechanics. Here, we shall consider the ideas and principles of the preceding chapters in the context of fluid motion in biological systems. First we will learn about some aspects of the fluid motion in the human body, and later we will learn about some aspects of fluid mechanics of plants.

The human body is a complex system that requires materials such as air, water, minerals and nutrients for survival and function. Upon intake, these materials have to be transported and distributed around the body as required. The associated biotransport and distribution processes involve interactions with membranes, cells, tissues, and organs comprising the body. Subsequent to cellular metabolism in the tissues, waste

©2010 Elsevier Inc. All rights reserved.
DOI: 10.1016/B978-0-12-381399-2.50017-4

by products have to be transported to the excretory organs for synthesis and removal. In addition to these functions, biotransport systems and processes are required for homeostasis (physiological regulation–for example, maintenance of pH and of body temperature), and for enabling the movement of immune substances to aid in body's defense and recovery from infection and injury. Furthermore, in certain other specialized systems such as the cochlea in the ear, fluid transport enables hearing and motion sensing. Evidently, in the human body, there are multiple types of fluid dynamic systems that operate at multiple and widely disparate scales. These scales are at various levels such as macro, micro, nano, pico and so on. Systems at the micro and macro levels, for example, include cells (micro), tissue (micro–macro), and organs (macro). Transport at the micro, nano and pico levels would include ion channeling, binding, signaling, endocytosis, and so on. Tissues constitute organs, and organs as systems perform various functions. For example, the cardiovascular system consists of the heart, blood vessels (arteries, arterioles, venules, veins, capillaries), lymphatic vessels, and the lungs. Its function is to provide adequate blood flow and regulate the flow as required by the various organs of the body. In this chapter, as related to the human body, we shall restrict attention to some aspects of the cardiovascular system for blood circulation.

2. *The Circulatory System in the Human Body*

The primary functions of the cardiovascular system are: (i) to pick up oxygen and nutrients from the lungs and the intestine, respectively, and deliver them to tissues (cells) of various parts, (ii) to remove waste and carbon dioxide from the body for excretion through the kidneys and the lungs, respectively, and (iii) to regulate body temperature by convecting the heat generated and dissipating it through transport across the skin. The circulatory system in the normal human body (as in all vertebrates and some other select group of species) can be considered as a closed system, meaning that the blood never leaves the system of blood vessels. The driving potential for blood flow is the prevailing pressure gradient.

The circulations associated with the cardiovascular system may be considered under three subsystems. These are the (i) systemic circulation, (ii) pulmonary circulation, and (iii) coronary circulation. (See Fig. 17.1.) In the systemic circulation, blood flows to all of the tissues in the body except the lungs. Contraction of the left ventricle of the heart pumps oxygen-rich blood to a relatively high pressure and ejects it through the aortic valve into the aorta. Branches from the aorta supply blood to the various organs via systemic arteries and arterioles. These, in turn, carry blood to the capillaries in the tissues of various organs. Oxygen and nutrients are transported by diffusion across the walls of the capillaries to the tissues. Cellular metabolism in the tissues generates carbon dioxide and byproducts (waste). Carbon dioxide dissolves in the blood and waste is carried by the blood stream. Blood drains into venules and veins. These vessels ultimately empty into two large veins called the superior vena cava (SVC) and and inferior vena cava (IVC) that return carbon dioxide rich blood to the right atrium. The mean blood pressure of the systemic circulation ranges from a high of 93 mmHg in the arteries to a low of few mmHg in the venae cavae. Fig. 17.2

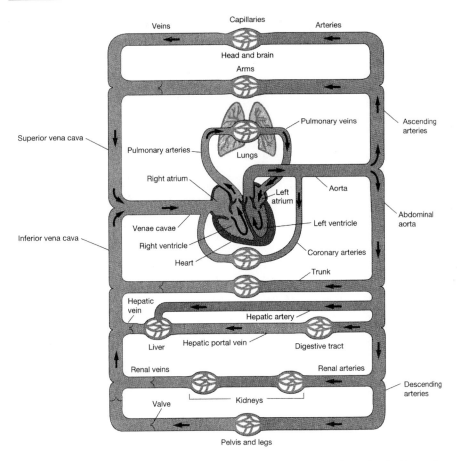

Figure 17.1 Schematic of blood flow in systemic and pulmonary circulation. (Reproduced with permission from Silverthorn, D.U. (2001) *Human Physiology: An Integrated Approach*, 2nd ed., Prentice Hall, Upper Saddle River, NJ.).

shows that pressure falls continuously as blood moves farther from the heart. The highest pressure in the vessels of the circulatory system is in the aorta and in the systemic arteries while the lowest pressure is in the venae cavae.

In pulmonary circulation, contraction of the right atrium ejects carbon dioxide rich blood through the tricuspid valve into the right ventricle. Contraction of the right ventricle pumps the blood through the pulmonic valve (also called semilunar valve) into the pulmonary arteries. These arteries bifurcate and transport blood into the complex network of pulmonary capillaries in the lungs. These capillaries lie between and around the alveoli walls. During respiratory inhalation, the concentration of oxygen in the air is greater in the air sacs of the alveolar region than in the capillary blood. Oxygen diffuses across capillary walls into blood. Simultaneously, the concentration of carbon dioxide in the blood is higher than in the air and carbon dioxide diffuses from the blood into the alveoli. Carbon dioxide exits through the mouth and nostrils. Oxygenated blood leaves the lungs through the pulmonary veins and enters the left

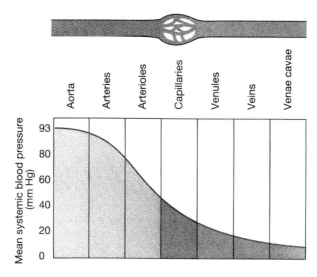

Figure 17.2 Pressure gradient in the blood vessels. (Reproduced with permission from Silverthorn, D.U. (2001) *Human Physiology: An Integrated Approach*, 2nd ed., Prentice Hall, Upper Saddle River, NJ.).

atrium. When the left atrium contracts, it pumps blood through the bicuspid (mitral) valve into the left ventricle. Figs. 17.3 and 17.4 provide an overview of external and cellular respiration and the branching of the airways, respectively.

Blood is pumped through the systemic and pulmonary circulations at a rate of about 5.2 liters per minute under normal conditions. The systemic and pulmonary circulations described above constitute one cardiac cycle. The cardiac cycle denotes any one or all of such events related to the flow of blood that occur from the beginning of one heartbeat to the beginning of the next. Throughout the cardiac cycle, the blood pressure increases and decreases. The frequency of the cardiac cycle is the heart rate. The cardiac cycle is controlled by a portion of the autonomic nervous system (that part of the nervous system which does not require the brain's involvement in order to function).

In coronary circulation, blood is supplied to and from the heart muscle itself. The muscle tissue of the heart, or myocardium, is thick and it requires coronary blood vessels to deliver blood deep into the myocardium. The vessels that supply blood with a high concentration of oxygen to the myocardium are known as coronary arteries. The main coronary artery arises from the root of the aorta and branches into the left and right coronary arteries. Up to about seventy five percent of the coronary blood supply goes to the left coronary artery, the remainder going to the right coronary artery. Blood flows through the capillaries of the heart and returns through the cardiac veins which remove the deoxygenated blood from the heart muscle. The coronary arteries that run on the surface of the heart are relatively narrow vessels and are commonly affected by atherosclerosis and can become blocked, causing angina or a heart attack. The coronary arteries are classified as "end circulation," since they represent the only source of blood supply to the myocardium.

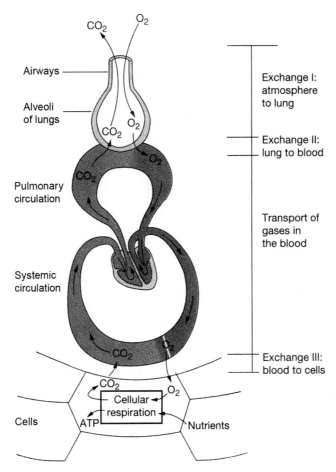

Figure 17.3 Overview of external and cellular respiration. (Reproduced with permission from Silverthorn, D.U. (2001) *Human Physiology: An Integrated Approach*, 2nd ed., Prentice Hall, Upper Saddle River, NJ.).

The Heart as a Pump

The heart has four pumping chambers–two atria (upper) and two ventricles (lower). The left and right parts of the heart are separated by a muscle called the septum which keeps the blood volumes in each part separate. The upper chambers interact with the lower chambers via the heart valves. The heart has four valves which ensure that blood flows only in the desired direction. The atrio-ventricular valves (AV) consist of the tricuspid (three flaps) valve between the right atrium and the right ventricle, and the bicuspid (two flaps, also called the mitral) valve between the left atrium and the left ventricle. The pulmonary valve is between the right ventricle and the pulmonary artery, and the aortic valve is between the left ventricle and the aorta. Both the pulmonary and aortic valves have three symmetrical half moon shaped valve flaps (cusps), and are called the semilunar valves. The function of the four chambers in

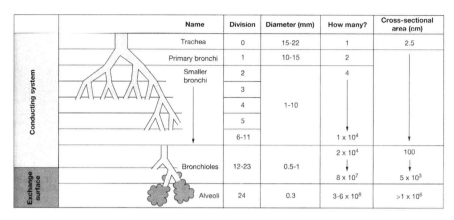

	Name	Division	Diameter (mm)	How many?	Cross-sectional area (cm)
	Trachea	0	15-22	1	2.5
	Primary bronchi	1	10-15	2	
	Smaller bronchi	2		4	
		3			
		4	1-10		
		5			
		6-11		1×10^4	
	Bronchioles	12-23	0.5-1	2×10^4	100
				8×10^7	5×10^3
	Alveoli	24	0.3	$3\text{-}6 \times 10^8$	$>1 \times 10^6$

Figure 17.4 Branching of the airways. Areas have units of cm^2. (Reproduced with permission from Silverthorn, D.U. (2001) *Human Physiology: An Integrated Approach*, 2nd ed., Prentice Hall, Upper Saddle River, NJ.).

the heart is to pump blood through pulmonary and systemic circulations. The atria receive blood from the veins–right atrium receives carbon dioxide rich blood from the SVC and IVC, and the left atrium receives oxygen rich blood from the pulmonary veins. The heart is controlled by a single electrical impulse and both sides of the heart act synchronously. Electrical activity stimulates the heart muscle (myocardium) of the chambers of the heart to make them contract. This is immediately followed by mechanical contraction of the heart. Both atria contract at the same time. The contraction of the atria moves the blood from the upper chambers through the valves into the ventricles. The atrial muscles are electrically separated from the ventricular muscles except for one pathway through which an electrical impulse is conducted from the atria to the ventricles. The impulse reaching the ventricles is delayed by about 110 ms while the conduction occurs through the pathway. This delay allows the ventricles to be filled before they contract. The left ventricle is a high pressure pump and its contraction supplies systemic circulation while the right ventricle is a low pressure pump supplying pulmonary circulation (Lungs offer much less resistance to flow than systemic organs).

From the above discussions, we see that the pumping action of the heart can be regarded as a two phase process–a contraction phase (systole) and a filling (relaxation) phase (diastole). Systole describes that portion of the heartbeat during which contraction of the heart muscle and hence ejection of blood takes place. A single "beat" of the heart involves three operations: atrial systole, ventricular systole and complete cardiac diastole. Atrial systole is the contraction of the heart muscle of the left and right atria, and occurs over a period of 0.1s. As the atria contract, the blood pressure in each atrium increases, which forces the mitral and tricuspid valves to open forcing blood into the ventricles. The AV valves remain open during atrial systole. Following atrial systole, ventricular systole which is the contraction of the muscles of the left and right ventricles occurs over a period of 0.3s. The ventricular systole generates enough pressure to force the AV valves to close, and the aortic and pulmonic valves open. (The aortic and pulmonic valves are always closed except for the short period of ventricular systole when the pressure in the ventricle rises above the pressure in the aorta for the left ventricle and above the pressure in the pulmonary artery for the right ventricle.) During systole, the typical pressures in the aorta and the pulmonary artery rise to 120 mmHg and 24 mmHg, respectively, (note conversion, 1 mmHg = 133 Pa). In normal adults, blood flow through the aortic valve begins at the start of ventricular systole, and rapidly accelerates to a peak value of approximately 1.35 m/s during the first one-third of systole. Thereafter, the blood flow begins to decelerate. Pulmonic valve peak velocities are lower and in normal adults, they are about 0.75 m/s. Contraction of the ventricles in systole ejects about two thirds of the blood from these chambers. As the left ventricle empties, its pressure falls below the pressure in the aorta, and the aortic valve closes. Similarly, as the pressure in the right ventricle falls below the pressure in the pulmonary artery, the pulmonic valve closes. Thus, at the end of the the ventricular systole, the aortic and pulmonic valves close, with the aortic valve closing a little earlier than the pulmonic valve. Diastole describes that portion of the heart beat during which the chamber refilling takes place. The cardiac diastole is the period of time when the heart relaxes after contraction in preparation for refilling with circulating blood. The ventricles refill or ventricular diastole occurs during atrial systole. When the ventricle is filled and ventricular systole begins, then the AV valves are closed and the atria begin refilling with blood or atrial diastole occurs. About a period of 0.4s following ventricular systole, both the atria and the ventricles begin refilling and both chambers are in diastole. During this period, both AV valves are open and aortic and pulmonic valves are closed. The typical diastolic pressure in the aorta is 80 mmHg and, in the pulmonary artery, it is 8 mmHg. Thus, the typical systolic and diastolic pressure ratios are 120/80 mmHg for the aorta and 24/8 mmHg for the pulmonary artery. The systolic pressure minus the diastolic pressure is called the pressure pulse, and for the aorta (left ventricle) it is 40 mmHg. The pulse pressure is a measure of the strength of the pressure wave. It increases with increased stroke volume (say, due to activity or exercise). Pressure waves created by the ventricular contraction diminish in amplitude with the distance and are not perceptible in the capillaries. Fig. 17.5 shows the pressure throughout the systemic circulation.

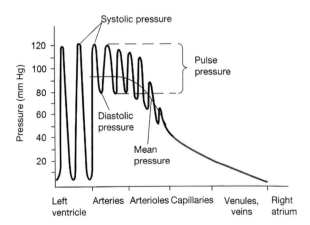

Figure 17.5 Pressure throughout the systemic circulation. (Reproduced with permission from Silverthorn, D.U. (2001) *Human Physiology: An Integrated Approach*, 2nd ed., Prentice Hall, Upper Saddle River, NJ.).

Net Work Done by the Ventricle on Blood During One Cardiac Cycle

The work done by the ventricle on blood may be calculated from the area enclosed by the pressure–volume curve for the ventricle. Consider, for example, the left ventricle (LV). Fig. 17.6 shows the pressure–volume curve for the LV.

Blood pressure is measured in mm of Hg, and the volume in *ml*. At A, the ventricular pressure and volume are at their lowest values. With the increase of atrial pressure, the bicuspid valve will open and let blood flow into the ventricle. AB represents diastolic ventricular filling. During AB work is being done by the blood in the LV to increase the volume. At B, the ventricular volume is filled to its maximum and this volume is called the end diastolic volume (*EDV*). The ventricular muscles begin to contract, pressure increases, and the bicuspid valve closes. BC is the constant volume contraction of the ventricle. No work is done during BC but energy is stored as elastic energy in the muscles. At C, ventricular pressure is greater than that in the aorta, the aortic valve opens and blood is ejected into the aorta. Ventricular volume decreases, but the ventricle continues to contract and the pressure increases. However, at D, pressure in the aorta exceeds that of the ventricular pressure and the aortic valve closes. During CD, work is done by the heart muscles on blood. The volume in the

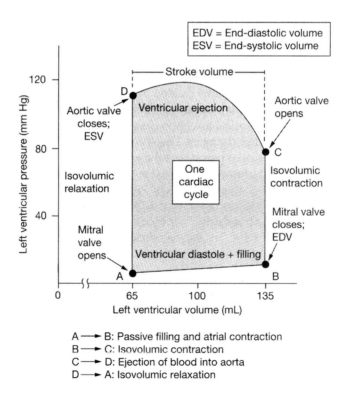

Figure 17.6 Left ventricular pressure–volume curve. (Reproduced with permission from Silverthorn, D.U. (2001) *Human Physiology: An Integrated Approach*, 2nd ed., Prentice Hall, Upper Saddle River, NJ.).

LV at D is at its lowest value, and this is called the end systolic volume (*ESV*). DA is the constant volume pressure decrease in the ventricle due to muscle relaxation and no work is done during this process. Ventricular pressure falls below that in the aorta causing the aortic valve to close. ABCD constitutes one cardiac cycle, and the area within the pressure-volume diagram represents the net work done by the LV on blood. The energy required to perform this work is derived from the oxygen in the blood. A similar development applies for the right ventricle.

Typically, work done by the heart is only about 10–15% of the total input energy, and the remainder is dissipated as heat.

The volume of blood pumped by the LV into the systemic circulation in a cardiac cycle is called the stroke volume (*SV*), and it is expressed in ml/beat. The normal stroke volume is 70 ml/beat.

$$SV = EDV - ESV \qquad (17.1)$$

A parameter that is related to stroke volume is ejection fraction (*EF*). *EF* is the fraction of blood ejected by the LV during systole. At the start of systole, the LV is filled with blood to the *EDV*. During systole, the LV contracts and ejects blood until it reaches *ESV*. *EF* is given by

$$EF = (SV/EDV) \times 100\% \qquad (17.2)$$

Cardiac output (*CO*) is the volume of blood being pumped by the heart (in particular, by a ventricle) in a minute. It is the time averaged flow rate. It is equal to the heart rate multiplied by the stroke volume. Thus,

$$CO = SV \times HR, \qquad (17.3)$$

where *HR* is the heart rate in beats/min. For a normal adult, the typical *HR* is between 70 and 75 beats per minute. With 70 beats per minute, and 70 ml blood ejection with each beat of the heart, the *CO* is 4900 ml/m. This value is typical for a normal adult at rest, although *CO* may reach up to 30 l/m during extreme activity (say, exercise). Heart rate can vary by a factor of approximately 3, between 60 and 180 beats per minute, while the stroke volume can vary between 70 and 120 ml, a factor of only 1.7. The cardiac index (*CI*) relates *CO* with the body surface area, *BSA* as given by,

$$CI = CO/BSA = SV \times HR/BSA, \qquad (17.4)$$

where, *BSA* is in square meters.

Nature of Blood

Composition of Blood

Blood is about 7% of the human body weight. Its density is approximately 1054 kg/m^3. The pH of normal blood is in the range $7.35 < \text{pH} < 7.45$. The normal adult has a blood volume of about 5 liters. At any given time, about 13% of the total blood volume resides in the arteries and about 7% resides in the capillaries. Blood is a complex circulating liquid tissue consisting of several types of formed elements (corpuscles or cells) (about 45% by volume) suspended in a fluid medium known as plasma (about 55% by volume; 2.7−3.0 liters in a normal human). The plasma is a dilute electrolyte solution (almost 92% water) containing, about 8% by weight, three major types of blood proteins—fibrinogen (5%), globulin (45%), and albumin (50%) in water. Beta lipoprotein and lipalbumin are also present in trace amounts. Plasma proteins are large molecules with high molecular weight and do not pass through the capillary wall. The formed elements (cells) consist of red blood cells (erythrocytes; about 45% of blood volume), white blood cells (leukocytes; about 1% of blood volume), and platelets (thrombocytes; <1% of blood volume). Thus, the formed elements in blood consist of 95% red blood cells, 0.13% white blood cells, and about 4.9% platelets. The specific gravity of red blood cells is about 1.06. The white blood cells further consist of monocytes, lymphocytes, neutrophils, eosinophils, and basophils.

In humans, mature red blood cells lack a nucleus and organelles. They are produced in the bone marrow, and the cell life span is about 125 days. The red blood cell is biconcave in shape. It consists of a concentrated solution of hemoglobin, an oxygen carrying protein, surrounded by a flexible membrane. The hemoglobin transports oxygen from the lungs to capillaries in various tissues, and some carbon dioxide. The cell is about 8.5μm in diameter with transverse dimensions of 2.5μm at the thickest portion and about 1μm at the thinnest portion. However, its flexibility is such that it can bend and pass through capillaries as small as 5μm in diameter. The surface area of the cell is about $163(\mu\text{m})^2$, and the intracellular fluid volume is about $87(\mu\text{m})^3$. There are approximately 5×10^6 red blood cells in each mm^3 of blood. The biconcave shape of the cell provides it with a very large ratio of surface area to volume. This enables efficient gas exchange in the capillaries. The percentage of blood volume made up by red blood cells is referred to as the hematocrit. Hematocrit ranges from 42 to 45 in normal blood, and plays a major role in determining the rheological properties of blood. White blood cells or leukocytes are cells of the immune system which defend the body against both infectious disease and foreign materials. Several different and diverse types of leukocytes exist and they are all produced in the bone marrow. There are normally about 10^4 white blood cells in each mm^3 of blood. Platelets or thrombocytes are cell fragments circulating in blood that are involved in the cellular mechanisms of hemostasis leading to the formation of blood clots. They are smaller in size than red or white blood cells. Low levels of platelets predisposes to bleeding, while high levels increase the risk of thrombosis (coagulation of blood in the heart or a blood vessel).

Viscosity of Blood

An important property of a flowing fluid is its viscosity. The viscosity of blood depends on the viscosity of the plasma and its protein content, the hematocrit, the temperature, the shear rate (also called the rate of shearing strain), and the narrowness of the vessel in which it is flowing (for example, a narrow diameter capillary). The dependence on the narrowness of the vessel diameter is called the Fahraeus-Lindqvist effect. The dependence on the prevailing shear rate and the Fahraeus-Lindqvist effect, each classify blood as a non-Newtonian fluid.

The presence of white cells and platelets do not significantly affect the viscosity since they are in such small proportions. We will briefly discuss these various features of blood viscosity.

The viscosity of plasma and blood are often given in terms of relative viscosity as compared to that of water (viscosity of water is about 0.8 centipoise at room temperature; 1 centipoise (1 cP) $= 0.01$ Poise, conversion: 1 Poise $= 1 dyne$ s/cm^2 $=$ 0.1 N s/m^2). The viscosity of plasma depends on the protein composition of the plasma and ranges between 1.1 and 1.6 centipoise. The viscosity of whole blood at a physiological hematocrit of 45% is about 3.2 cP. Higher hematocrit results in higher viscosity. At a hematocrit of 60%, the relative viscosity of blood is about 8. Viscosity of blood increases with decreasing temperature, and the increase is approximately 2% for each °C. The dependence of viscosity on flow rate in vessels is complicated. As noted in earlier chapters, viscous flow rates in vessels are significantly influenced by the shear stress, τ, and the associated rate of shearing strain (or shear rate), $\dot{\gamma}$. For Newtonian fluids, τ is linearly related to $\dot{\gamma}$, and the slope of this characteristic is the viscosity, μ. For whole blood, this relationship is complicated due to the following reasons. In a blood volume at rest, above a minimum hematocrit of about 5–8%, blood cells form a continuous structure. A finite stress (called the yield stress), τ_y, is required to break this continuous structure into a suspension of aggregates in the plasma. This yield stress depends also on the concentration of plasma proteins, in particular, fibrinogen. An empirical correlation for the yield stress is given by the expression:

$$\sqrt{\tau_y} = (H - 0.1)(C_F + 0.5), \qquad (17.5)$$

where H is the hematocrit expressed as a fraction and it is >0.1, and C_F is the fibrinogen content in grams per 100 ml and $0.21 < C_F < 0.46$. For 45% hematocrit blood, the yield stress is in the range $0.01 < \tau_y < 0.06$ $dynes$/cm^2, (conversion: 1 $dyne$/cm$^2 = 0.1$ N/m^2). Beyond the yield, when sheared in the bulk, up to about $\dot{\gamma} < 50$ sec^{-1}, the aggregates in blood break into smaller units called rouleaux formations. For shear rates up to about 200 sec^{-1}, the rouleaux progressively break into individual cells. Beyond this, no further reduction in structure is noted to occur with an increase in the shearing rate.

For whole blood, at low shear rates, $\dot{\gamma} < 200$ sec^{-1}, the variation of τ with $\dot{\gamma}$ is noted to be nonlinear. This behavior at low $\dot{\gamma}$ is non-Newtonian. Low $\dot{\gamma}$ values are associated with flows in small arteries and capillaries (microcirculation). At higher shear rates, $\dot{\gamma} > 200$ sec^{-1}, the relationship between τ and $\dot{\gamma}$ is linear, and the viscosity approaches an asymptotic value of about 3.5 cP. Blood flows in large arteries have such high shear rates, and the viscosity in such cases may be assumed as constant and equal to 3.5 cP. Since whole blood behaves like a non-Newtonian yield stress fluid, the slope of the shear stress—rate of strain characteristic at any given point on the curve is defined as the apparent viscosity of blood at that point, μ_{app}. Clearly, μ_{app} is not a constant but depends on the prevailing $\dot{\gamma}$ at that point. (See Fig. 17.7.) There

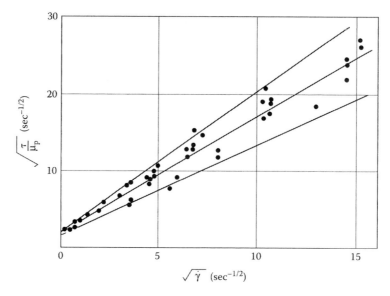

Figure 17.7 Shear stress vs. Shear rate for blood flow. (Reproduced with permission from Whitmore, R. L. (1968) *Rheology of Circulation*, Pergamon Press, New York).

are a number of constitutive equations available in the literature that attempt to model the relationship between shear stress and shear rate of flowing blood. A commonly used one is called the Casson model and it is expressed as follows:

$$\sqrt{\frac{\tau}{\mu_p}} = k_c\sqrt{\dot{\gamma}} + \sqrt{\frac{\tau_y}{\mu_p}}, \tag{17.6}$$

where μ_p is plasma viscosity and k_c is the Casson viscosity coefficient (a dimensionless number). An expression based on a least square fit of the experimental data and expressed in Casson form is that of Whitmore (1968),

$$\sqrt{\frac{\tau}{\mu_p}} = 1.53\sqrt{\dot{\gamma}} + 2.0. \tag{17.7}$$

This expression is plotted in Fig. 17.8. Apparent viscosity significantly increases at low rates of shear. It must be noted that although the Casson model is suitable at low shear rates, it still assumes that blood can be modelled as a homogeneous fluid.

In blood vessels of less than about 500 μm in diameter, the inhomogeneous nature of blood starts to have an effect on the apparent viscosity. This feature will be discussed next.

Fahraeus-Lindqvist Effect

When blood flows through narrow tubes of decreasing radii, approximately in the range (15 μm $< d <$ 500 μm), the apparent viscosity, μ_{app}, decreases with

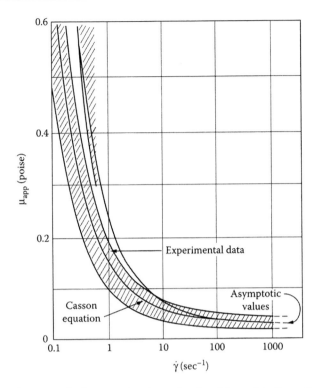

Figure 17.8 A least square fit of apparent viscosity as a function of shear rate in Casson form. (Reproduced with permission from Whitmore, R. L. (1968) *Rheology of Circulation*, Pergamon Press, New York).

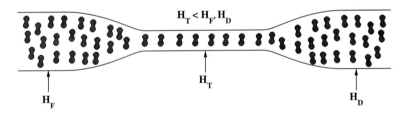

Figure 17.9 The Fahraeus effect.

decreasing radius of the vessel. This is a second non-Newtonian characteristic of blood and is called the Fahraeus-Lindqvist (FL) effect. The reduced viscosity in narrow tubes is beneficial to the pumping action of the heart.

The basis for the FL effect is the Fahraeus effect.

When blood of constant hematocrit (feed hematocrit or bulk hematocrit, H_F) flows from a large vessel into a small vessel (vessel sizes in the ranges cited above), the hematocrit in the small vessel (dynamic or tube hematocrit, H_T) decreases as the tube diameter decreases. (See Fig. 17.9.) This phenomenon is called Fahraeus effect and must not be confused with a diminution of particle concentration in the smaller vessel because of an entrance effect whereby particle entry into the smaller vessel is hindered (see, Goldsmith et al. (1989) for detailed discussions). To separate this possible "screening effect" and confirm the Fahraeus effect, H_T may be compared with the hematocrit in the blood flowing out (discharge hematocrit, H_D) from the smaller tube into a discharge vessel of comparable size to the feed vessel. In the steady state, $H_F = H_D$. In vivo and in vitro experiments show that, $H_T < H_D$ in tubes up to about 15 μm in diameter. The H_T/H_D ratio decreases from about 1 to about 0.46 as the capillary diameter decreases from about 600 μm to about 15 μm. While the discharge hematocrit value may be 45%, the corresponding dynamic hematocrit in a narrow sized vessel such as an arteriole may just be 20%. As a consequence, the apparent viscosity decreases in the diameter range 15 μm $< d <$ 500 μm. However, for tubes less than about 15 μm in diameter, the ratio H_T/H_D starts to increase.

Why does the hematocrit decrease in small blood vessels? The reason for this effect is not fully understood at this time. In blood vessel flow, there seems to be a tendency for the red cells to move toward the axis of the tube, leaving a layer of plasma, whose width, usually designated by δ, increases with increase in the shear rate. This tendency to move away from the wall is not observed with rigid particles; thus, the deformability of the red cell appears to be the reason for lateral migration. Deformable particles are noted to experience a net radial hydrodynamic force even at low Reynolds numbers and tend to migrate towards the tube axis. (see, Fung (1993) for detailed discussions). Chandran et al. (2007) state that as the blood flows through a tube, the blood cells (with their deformable biconcave shape) rotate (spin)in the shear field. Due to this spinning, they tend to move away from the wall and toward the center of the tube. The cell free plasma layer reduces the tube hematocrit. As the size of the vessel gets smaller, the fraction of the volume occupied by the cell-free layer increases, and the tube hematocrit is further lowered. A numerical validation of this reasoning is available in a recent paper by Liu and Liu (2006). There is yet another reason. Blood vessels have many smaller sized branches. If a branching daughter vessel is so located that it draws blood from the larger parent vessel mainly from the cell free layer, the hematocrit in the branch will end up being lower. This is called *plasma skimming*. In all these circumstances, the tube hematocrit is lowered. The viscosity of blood at the core may be higher due to a higher core hematocrit, H_c, there, but the overall apparent viscosity in the tube flow is lower.

As the tube diameter becomes less than about 6 μm, the apparent viscosity increases dramatically. The erythrocyte is about 8 μm in diameter and can enter tubes somewhat smaller in size, and a tube of about 2.7 microns is about the smallest size that an RBC can enter, Fournier (2007), Fung (1993). When the tube diameter becomes very small, the pressure drop associated with the flow increases greatly and there is increase in apparent viscosity.

If we consider laminar blood flow in straight, horizontal, circular, feed and capillary tubes, a number of straightforward relationships between Q_F, Q_c, Q_p, H_F, H_T, H_c, δ, and a may be established based on the law of conservation of blood cells. Here, Q denotes flow rate, subscripts c and p denote core and plasma regions, respectively, and a is the radius of the capillary tube. Thus,

$$Q_F H_F = Q_c H_c, \quad Q_c + Q_p = Q_F, \quad \text{and,} \quad H_T a^2 = H_c (a - \delta)^2, \quad (17.8)$$

where a is the radius of the capillary tube. Equation (17.8) will be useful in modelling the FL phenomenon. A simple mathematical model for the FL effect is included in a subsequent section.

Nature of the Blood Vessels

All blood vessels other than capillaries are usually composed of three layers: the tunica intima, tunica media, and tunica adventitia. The tunica intima consists of a layer of endothelial cells lining the lumen of the vessel (the hollow internal cavity in which the blood flows), as well as a subendothelial layer made up of mostly loose connective tissue. The endothelial cells are in direct contact with the blood flow. An internal elastic lamina often separates the tunica intima from the tunica media. The tunica media is composed chiefly of circumferentially arranged smooth muscle cells. Again, an external elastic lamina often separates the tunica media from the tunica adventitia. The tunica adventitia is primarily composed of loose connective tissue made up of fibroblasts and associated collagen fibers. In the largest arteries, such as the aorta, the amount of elastic tissue is very considerable. Veins have the same three layers as arteries, but boundaries are indistinct, walls are thinner, and elastic components are not as well developed.

Blood flows under high pressure in the aorta (about 120 mmHg systolic, 80 mmHg diastolic, pressure pulse of 40 mm Hg at the root) and the major arteries. These vessels have strong walls. The aorta is an elastic artery, about 25 mm in diameter with a wall thickness of about 2 mm, and is quite distensible. During left ventricular systole (about 1/3 of the cardiac cycle), the aorta expands. This stretching gives the potential energy that will help maintain blood pressure during diastole. During the diastole (about 2/3 of the cardiac cycle), the pressure pulse decays exponentially and the aorta contracts passively. Medium arteries are about 4 mm in diameter with a wall thickness of about 1 mm. Arterioles are about 50 μm in diameter and have thin muscular walls (usually only one to two layers of smooth muscle) of about 20 μm thickness. Their vascular tone is controlled by regulatory mechanisms, and they constrict or relax as needed to maintain blood pressure. Arterioles are the primary site of vascular resistance and blood flow distribution to various regions is controlled by changes in resistance offered by various arterioles. True capillaries average from 9 to 12 μm in diameter, just large enough to permit passage of cellular components of blood. The thin wall consists of extremely attenuated endothelial cells. In cross section, the lumen of small capillaries may be encircled by a single endothelial cell, while larger capillaries may be made up of portions of 2 or 3 cells. No smooth muscle is present. Venules are about 20 μm in diameter and allow deoxygenated blood to return from the capillary beds to the larger veins. They have three layers. An inner endothelium layer which acts a membrane, a middle layer of muscle and elastic tissue, and an outer layer of fibrous connective tissue. The middle layer is poorly developed. The walls of venules are about 2 μm in thickness, and thus are very much thinner than those of arterioles. Veins are thin walled, distensible, and collapsible tubes. Some of them may be collapsed in normal function. They transport blood at a lower pressure than the arteries. They are about 5 mm in diameter and the wall thickness is about 500 μm. They are surrounded by helical bands of smooth muscles which help maintain blood flow to the right atrium. Most veins have one-way flaps called venous valves. These valves prevent gravity from causing blood to flow back and collect in the lower extremities. Veins more distal to the heart have more valves. Pulmonary veins and the smallest venules have no valves. Veins also have a thick collagen outer layer, which helps maintain blood pressure. In the venous system, a large increase in the blood volume results in a relatively small increase in pressure compared to the arterial system (see, Chandran et al. (2007)). The veins act as the main reservoir for blood in the circulatory system and the total capacity of the veins is more than sufficient to hold the entire blood volume of the body. This capacity is reduced through the constriction of smooth muscles, minimizing the cross-sectional area (and hence volume) of the individual veins and therefore the total venous system. The superior vena cava is a large, yet short vein that carries de-oxygenated blood from the upper half of the body to the heart's right atrium. The inferior vena cava is the large vein that carries de-oxygenated blood from the lower half of the body into the heart. The vena cava is about 30 mm in diameter with a wall thickness of about 1.5 mm. The venae cavae have no valves. Fig. 17.10 shows the cross-sectional areas of different parts of the systemic circulation with velocity of blood flow in each part. The fastest flow is in the arterial system. The slowest flow is in the capillaries and venules.

Total Peripheral Resistance Concept

As stated earlier, arterioles are the primary site of vascular resistance, and blood flow distribution to various regions is controlled by changes in resistance offered by various arterioles. To quantify the resistance of the arterioles in an averaged sense, the concept of *total peripheral resistance* is introduced. Total peripheral resistance essentially refers to the cumulative resistance of the thousands of arterioles involved in the systemic or pulmonary circulation, respectively. For systemic circulation, with time averaging of quantities over a cardiac cycle,

$$\text{Total Peripheral Resistance} = R = \frac{\Delta \bar{p}}{Q}, \tag{17.9}$$

where R denotes resistance, $\Delta \bar{p}$ is the difference between the time averaged pressure at the aortic valve and the time averaged venous pressure at the right atrium, and Q is the time averaged flow rate (cardiac output). The units of peripheral resistance would therefore be in mmHg per cm^3/s. This unit of measuring resistance is called the Peripheral Resistance Unit (*PRU*). Letting $\bar{p}_A$ and $\bar{p}_V$ to denote the time averaged pressures at the aortic valve and at the right atrium, respectively,

$$\Delta p = \bar{p}_A - \bar{p}_V, \tag{17.10}$$

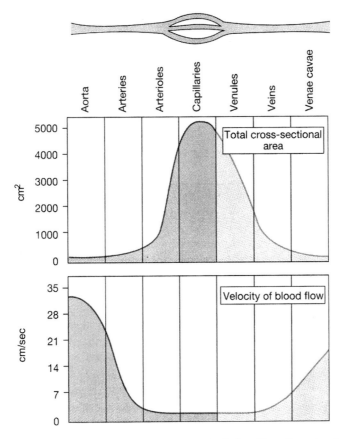

Figure 17.10 Vessel diameter, total cross-sectional area, and velocity of flow. (Reproduced with permission from Silverthorn, D. U. (2001) *Human Physiology: An Integrated Approach*, 2nd ed., Prentice Hall, Upper Saddle River, NJ.).

and, with $\bar{p}_V = 0$, $\Delta \bar{p} = \bar{p}_A$, the time averaged arterial pressure. Then, $\bar{p}_A = QR$. The average pressure, $\bar{p}_A$, may be estimated as:

$$\bar{p}_A = \frac{1}{3} p_S + \frac{2}{3} p_D = p_D + \frac{1}{3}(p_S - p_D), \tag{17.11}$$

where, p_S is the systolic pressure, p_D is the diastolic pressure, and $(p_S - p_D)$ is the pressure pulse (see, Kleinstreuer (2006)). For a normal person at rest, with $\bar{p}_A = 100$ mmHg, $Q = 86.6$ cm^3/s, $R = 1.2$ *PRU*. An expression similar to that in equation (17.9) would apply for pulmonary circulation and would involve the difference between time averaged pressures at the pulmonary artery and at the left atrium, and the flow rate in pulmonary circulation (same as that in systemic circulation). Since the difference between time averaged pressures in pulmonary circulation is about an order of magnitude smaller than in the systemic circulation, the corresponding *PRU* would be an order of magnitude smaller.

3. Modelling of Flow in Blood Vessels

There are approximately 100 000 km of blood vessels in the adult human body (Brown et al. (1999)). In this section, we will examine several models for describing blood flow in some important vessels.

General Introduction

Blood flow in the circulatory system is in general unsteady. In most regions it is pulsatile due to the systolic and diastolic pumping. In pulsatile flow, the flow has a periodic behavior and a net directional motion over the cycle. Pressure and velocity profiles vary periodically with time, over the duration of a cardiac cycle. A dimensionless parameter called the Womersley number, α, is used to characterize the pulsatile nature of blood flow, and it is defined by:

$$\alpha = a\sqrt{\frac{\omega}{\nu}}, \tag{17.12}$$

where, a is the radius of the tube, ω is the frequency of the pulse wave (heart rate expressed in radians/sec), and ν is the kinematic viscosity. This definition shows that Womersley number is a composite parameter of the Reynolds number, $Re = u\,2a/\nu$, and the Strouhal number, $St = \omega\,2a/u$. The square of the Womersley number is called the Stokes number. The Womersley number denotes the ratio of unsteady inertial forces to viscous forces in the flow. It ranges from as large as about 20 in the aorta, significantly greater than 1 in all large arteries, to as small as about 10^{-3} in the capillaries. Let us estimate the Womersley number for an illustration. With a normal heart rate of 72 beats per minute, $\omega = (2\,\pi\,72/60) \approx 8$ rad/s. Take $\rho = 1.05$ g cm^{-3}, $\mu = 0.04$ g cm^{-1} s^{-1} and an artery of radius $a = 0.5$ cm. Then $\alpha \approx 7$. Decreasing α values correspond to increasing role of viscous forces and, for $\alpha < 1$, viscous effects are dominant. In that highly viscous regime, the flow may be regarded as quasi-steady. With increasing α, inertial forces become important. In pulsatile flows, flow separation may occur both by a geometric adverse pressure gradient and/or by time varying changes in the driving pressure. Geometric adverse pressure gradients may arise due to varying cross sectional areas through which the flow occurs. On the other hand, time varying changes in a cardiac cycle result in acceleration and deceleration during the cycle. An adverse pressure gradient during the deceleration phase may result in flow separation.

Blood vessel walls are viscoelastic in their behavior. The ability of a blood vessel wall to expand and contract passively with changes in pressure is an important function of large arteries and veins. This ability of a vessel to distend and increase volume with increasing transmural pressure difference (inside minus outside pressure) is quantified as vessel compliance. During systole, pressure from the left ventricle is transmitted as a wave due to the elasticity of the arteries. Due to the compliant nature of the arteries and their finite thickness, the pressure travels like a wave at a speed much faster than the flow velocity. Since blood vessels may have many branches, the reflection and transmission of waves in such branching vessels significantly complicate the understanding of such flows. In this chapter, a reasonably simplified picture of these various complex features will be presented. Further reading in advanced treatments such as the book by Fung (1997) will be necessary to obtain a comprehensive understanding.

First we start with the study of laminar, steady flow of blood in circular tubes, and in subsequent sections, we shall consider more realistic models.

Hagen-Poiseuille Flow

In the simplest model, blood flow in a vessel is modelled as a laminar, steady, incompressible, fully developed flow of a Newtonian fluid through a straight, rigid, cylindrical, horizontal tube of constant circular cross section (see Fig. 17.11). Such a flow is called the circular Poiseuille or more commonly as the Hagen-Poiseuille flow. This flow has been treated in Chapter 9, Section 5.

How valid is the Hagen-Poiseuille model?

In the normal body, blood flow in vessels is generally laminar. However, at high flow rates, particularly in the ascending aorta, the flow may become turbulent at or near to peak systole. Disturbed flow may occur during the deceleration phase of the cardiac cycle (Chandran et al. (2007)). Turbulent flow may also occur in large arteries at branch points. However, under normal conditions, the critical Reynolds number, Re_c, for blood flow in long, straight, smooth blood vessels is relatively high, and the flow remains laminar. Let us consider some estimates. The aorta is about 40 cm long and the average velocity of flow in it is about 40 cm/s. The lumen diameter at the root of the aorta is $d = 25$ mm, and the corresponding $Re = \rho\, u\, d/\mu$ is 3000. The maximum Reynolds number may be as high as 9000. The average value for Re in the vena cava is about 3000. Arteries have varying sizes and the maximum Re is about 1000. For Newtonian fluid flow in a straight cylindrical rigid tube, the critical Re_c is about 2300. However, aorta and arteries are distensible tubes, and the Re_c of 2300 criterion does not apply. In the case of blood flow, laminar flow conditions generally prevail even at a high Reynolds number of 10,000 (Mazumdar (2004)). In summary, the laminar flow assumption is reasonable in many cases.

Blood flow in the circulatory system is in general unsteady and pulsatile. The large arteries have elastic walls and are subject to substantially pulsatile flow. The steady flow assumption is inapplicable until the flow has reached smaller muscular arteries and arterioles in the circulatory system. Blood flow in arteries has been described by several authors (see, McDonald (1974), Pedley (1980), Ku (1997)). In the heart chambers and blood vessels, blood may be considered incompressible. In the walls

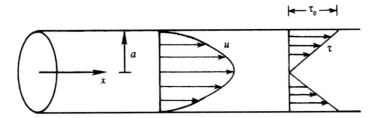

Figure 17.11　Poiseuille flow.

of the heart and in the blood vessel walls, it may not be considered as incompressible (Fung (1997)). Since blood flow remains laminar at very high Reynolds numbers, the entry length is very large in many cases. Branches and curved vessels hinder flow development. The fully developed flow assumption is very restrictive in describing blood flow in vessels.

Flow in large blood vessels may be generally regarded as Newtonian. The Newtonian fluid assumption is inapplicable at low shear rates such as those that would occur in arterioles and capillaries.

Many blood vessels are not straight but are curved and have branches. However, flow may be regarded to occur in straight sections in many cases of interest.

Arterial walls are not rigid but are viscoelastic and distensible. The pressure pulse generated during left ventricular contraction travels through the arterial wall. The speed of wave propagation depends upon the elastic properties of the wall and the fluid—structure interaction. Arterial branches and curves may cause reflections of the wave.

Gravitational and hydrostatic effects become very important for orientations of the body other than the supine position.

Systemic arteries are generally circular tubes but may have tapering cross sections, while the veins and pulmonary arteries tend to be elliptical.

Since there are many situations where the Hagen-Poiseuille model is reasonably applicable, we will now start with the recapitulation of the flow results provided in Chapter 9. The pertinent results are:

Axial flow velocity, $u = u(r)$, in a pipe of radius, a (see, equation (9.10)):

$$u = \frac{r^2 - a^2}{4\mu} \left(\frac{dp}{dx} \right).$$
(17.13)

Pressure drop: In a fully developed flow, the pressure gradient, (dp/dx), is a constant, and, it may be expressed in terms of the pressure gradient along the entire tube:

$$\left(\frac{dp}{dx} \right) = -\frac{\Delta p}{L} = -\frac{(p_1 - p_2)}{L},$$
(17.14)

where, Δp is the imposed pressure difference, subscripts 1 and 2 denote inlet and exit ends, respectively, and L is the length of the entire tube. With equation (17.14), equation (17.13) becomes,

$$u = \frac{r^2 - a^2}{4\mu} \left(\frac{dp}{dx} \right) = \frac{p_1 - p_2}{4\mu L} (a^2 - r^2) = \left[\frac{\Delta p \, a^2}{4\mu L} \right] \left[1 - \left(\frac{r}{a} \right)^2 \right].$$
(17.15)

The maximum velocity occurs at the center of the tube, $r = 0$, and is given by

$$u_{\max} = \left[\frac{\Delta p \, a^2}{4\mu L} \right]$$
(17.16)

The volume flow rate is:

$$Q = \int_0^a u 2\pi r\, dr = -\frac{\pi a^4}{8\mu}\left(\frac{dp}{dx}\right) = \frac{\pi a^4}{8\mu}\frac{(p_1 - p_2)}{L} = \frac{\pi\ a^4}{8\mu}\frac{\Delta p}{L} = \frac{u_{max}}{2}\pi a^2 \tag{17.17}$$

This equation (17.17) is called the Poiseuille formula. The average velocity over the cross section is :

$$V = \frac{Q}{A} = \frac{Q}{\pi a^2} = \frac{u_{max}}{2}, \tag{17.18}$$

where A is the cross section of the tube. The shear stress at tube wall is:

$$\tau_{xr}|_{r=a} = \tau = -\mu\left(\frac{du}{dr}\right)\Big|_{r=a} = -\frac{a}{2}\left(\frac{dp}{dx}\right) = -\frac{a}{2}\frac{\Delta p}{L}, \tag{17.19}$$

where the negative sign has been included to give $\tau > 0$ with $\left(\frac{du}{dr}\right) < 0$ (the velocity decreases from the tube centerline to the tube wall). The maximum shear stress occurs at the walls, and the stress decreases towards the center of the vessel.

The Hagen-Poiseuille equation and its derivatives are most applicable to flow in the muscular arteries, but modifications are likely to be required outside this range (see Brown et al. (1999)). For an application of Poiseuille flow relationships in the context of perfused tissue heat transfer and thermally significant blood vessels, see Baish et al. (1986a, 1986b).

With the results for the Hagen-Poiseuille flow, we have from Eq. (17.9),

$$\text{Total Peripheral Resistance} = R = \frac{\Delta\bar{p}}{Q} = \frac{8\mu l}{\pi a^4}. \tag{17.20}$$

Equation (17.20) shows that peripheral resistance to the flow of blood is inversely proportional to the fourth power of vessel diameter.

Hagen-Poiseuille Flow and the Fahraeus-Lindqvist Effect

Consider laminar, steady flow of blood through a straight, rigid, cylindrical, horizontal tube of constant circular cross section and radius a, as shown in Fig. 17.12.

Let the flow be divided into two regions: a central core containing RBCs and a cell free plasma layer of thickness δ surrounding the core. Let the viscosities of the core and the plasma layer be μ_c and μ_p, respectively. Let the shear rates be such that each region can be considered Newtonian, and that we could employ Hagen-Poiseuille theory.

The shear stress distribution in the core region is governed by,

$$\tau_{xr} = -\mu_c\frac{du^c}{dr} = -\frac{r}{2}\frac{\Delta p}{L}, \tag{17.21}$$

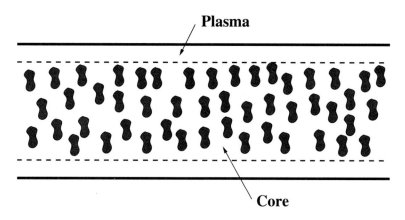

Figure 17.12 Fahraeus-Lindqvist effect.

subject to conditions,

$$\frac{du^c}{dr} = 0, \text{ at } r = 0, \tag{17.22}$$

$$\tau_{xr}|_c = \tau_{xr}|_p, \text{ at } r = (a - \delta). \tag{17.23}$$

The shear stress distribution in the plasma region is governed by,

$$\tau_{xr} = -\mu_p \frac{du^p}{dr} = -\frac{r}{2} \frac{\Delta p}{L}, \tag{17.24}$$

subject to conditions,

$$u^c = u^p, \text{ at } r = (a - \delta), \tag{17.25}$$

$$u^p = 0, \text{ at } r = a. \tag{17.26}$$

Integration of equations (17.21) and (17.24) subject to the indicated conditions yield the following expressions for the axial velocities in the plasma and core regions:

$$u^p = \frac{a^2}{4\mu_p} \frac{\Delta p}{L} \left[1 - \left(\frac{r}{a}\right)^2 \right], \text{ for } a - \delta \leq r \leq a, \tag{17.27}$$

and,

$$u^c = \frac{a^2}{4\mu_p} \frac{\Delta p}{L} \left[1 - \left(\frac{a-\delta}{a}\right)^2 - \frac{\mu_p}{\mu_c} \left(\frac{r}{a}\right)^2 + \frac{\mu_p}{\mu_c} \left(\frac{a-\delta}{a}\right)^2 \right], \text{ for } 0 \leq r \leq a-\delta. \tag{17.28}$$

The volume flow rates in the plasma, Q_p, and core region, Q_c, are:

$$Q_p = 2\pi \int_{a-\delta}^{a} u^p r dr = \frac{\pi \Delta p}{8\mu_p L} \left[a^2 - (a - \delta)^2 \right]^2, \tag{17.29}$$

and,

$$Q_c = 2\pi \int_0^{a-\delta} u^c r \, dr$$

$$= \frac{\pi a^2 \Delta p}{4\mu_p L} \left[a^2 - \left(1 - \frac{\mu_p}{2\mu_c}\right) \frac{(a-\delta)^4}{a^2} \right]. \tag{17.30}$$

The total flow rate of blood within the tube, Q, is the sum of the flow rates in the plasma and core regions. Therefore,

$$Q = Q_p + Q_c = \frac{\pi a^4 \Delta p}{8\mu_p L} \left[1 - \left(1 - \frac{\delta}{a}\right)^4 \left(1 - \frac{\mu_p}{\mu_c}\right) \right]. \tag{17.31}$$

From the equation (17.31), we could calculate the apparent viscosity of the two region fluid by measuring Q, and $\Delta p/L$. Define μ_{app}, by analogy with Hagen-Poiseuille flow, as given by,

$$Q = \frac{\pi a^4 \Delta p}{8\mu_{app} L}. \tag{17.32}$$

From equations (17.31) and (17.32), the apparent viscosity, μ_{app}, may be expressed in terms of μ_p as,

$$\mu_{app} = \mu_p \left[1 - \left(1 - \frac{\delta}{a}\right)^4 \left(1 - \frac{\mu_p}{\mu_c}\right) \right]^{-1}. \tag{17.33}$$

In the limit $(\delta/a) \ll 1$, $\left(1 - \frac{\delta}{a}\right)^4 \approx (1 - 4\delta/a)$. Then, equation (17.33) reduces to:

$$\mu_{app} = \mu_c \left[1 + 4\frac{\delta}{a} \left(\frac{\mu_c}{\mu_p} - 1\right) \right]^{-1} \rightarrow \mu_c \rightarrow \mu. \tag{17.34}$$

In equations (17.31) and (17.33), δ and μ_c are unknown. From equation (17.8), we have $H_c/H_F = 1 + (Q_p/Q_c)$. We still need input from experimental data to set up a modelling procedure for FL. Fournier (2007) recommends the use of Charm and Kurland's equation for this purpose, (see Charm and Kurland (1974) for details),

$$\mu_c = \mu_p \frac{1}{1 - \alpha_c H_c}, \tag{17.35}$$

where,

$$\alpha_c = 0.070 \exp\left[2.49 H_c + \frac{1107}{T} \exp(-1.69 H_c) \right], \tag{17.36}$$

where T is temperature in K. Equation (17.36) may be used to a hematocrit of 0.60.

With this input, a modelling procedure can be developed for various flow and tube parameters.

Effect of Developing Flow

When we discussed Poiseuille flow, we noted that the fully developed flow assumption that is often invoked in the study of blood flow in vessels is very restrictive. We will now learn about some of the limitations of this assumption.

When a fluid under the action of a pressure gradient enters a cylindrical tube, it takes a certain distance called the *inlet or entrance length*, ℓ, before the flow in the tube becomes steady and fully developed. When the flow is fully developed and laminar, the velocity profile is parabolic that is characteristic of Poiseuille flow. Within the inlet length, the velocity profile changes in the direction of the flow and the fluid accelerates or decelerates as it flows. There is a balance among pressure, viscous, and inertia (acceleration) forces. Compared to fully developed flow, the entrance region is subject to large velocity gradients near the wall and these result in high wall shear stresses. The entry of blood from the ventricular reservoir into the aortic tube or from a large artery into a smaller branch will involve an entrance length. It must be understood, however, that the inlet length with pulsating flow (say, in the proximal aorta) is different from that for a steady flow.

If we assumed that the fluid enters the tube from a reservoir, the profile at the inlet is virtually flat. The transition from a flat velocity distribution, at the entrance of a tube, to the fully developed parabolic velocity profile is illustrated in Fig. 17.13. Once inside the tube, the layer of fluid immediately in contact with the wall will become stationary (no-slip condition) and the laminae adjacent to it slide on it subject to viscous forces and a boundary layer is formed. The presence of the endothelial lining on the inside of a blood vessel wall does not negate the no-slip condition. The motion of the bulk of fluid in the central region of the tube will not be affected by the viscous forces and will have a flat velocity profile. As flow progresses down the tube, the boundary layer will grow in thickness as the viscous drag involves more and more of the fluid.

Eventually, the boundary layer fills the whole of the tube and the steady viscous flow is established or the flow is fully developed. In the literature (see, for example, Mohanty and Asthana (1979)), there are discussions which divide the entrance region into two parts, the inlet region and the filled region. At the end of the inlet region, the boundary layers meet at the tube axis but the velocity profiles are not yet similar. In the filled region, adjustment of the completely viscous profile takes place until the Poiseuille similar profile is attained at the end of it. In our discussion here, we will treat the entrance region as a region with a potential core and a developing

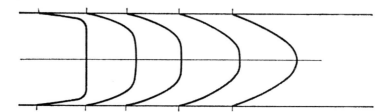

Figure 17.13   Developing velocity profile in a tube flow.

boundary layer at the wall. The shape of the velocity profile in the tube depends on whether the flow is laminar or turbulent, as does the length of the entrance region, ℓ. This is a direct consequence of the differences in the nature of the shear stress in laminar and turbulent flows. The magnitude of the pressure gradient, $\partial p/\partial x$, is larger in the entrance region than in the fully developed region. There is also an expenditure of kinetic energy involved in transition from a flat to a parabolic profile. For steady flow of a Newtonian fluid in a rigid walled horizontal circular tube, the entrance length may be estimated from,

$$\frac{\ell}{d} = 0.06 \; Re \;\; \text{for} \;\; \text{laminar flow} \;\; \text{and} \;\; Re > 50,$$

$$\frac{\ell}{d} = 0.693 \; Re^{1/4} \;\; \text{for} \;\; \text{turbulent flow} \tag{17.37}$$

For steady flow at low Reynolds number, the entrance region is approximately one tube radius long (for $Re \leq 0.01$, say in capillaries, $\ell/d = 0.65$). In large arteries, the entrance length is relatively long and over a significant length of the artery the velocity gradients are high near the wall. This affects the mass exchange of gas and nutrient molecules between the blood and artery wall.

Unsteady flow through the entrance region with a pulsating flow depends on the Womersley and Reynolds numbers. For a medium sized artery, the Reynolds number is typically on the order of 100 to 1000, and the Womersley number ranges from 1 to 10. Pedley (1980) has estimated the wall shear stress in the entrance region for pulsatile flow using asymptotic boundary layer theory while He and Ku (1994) have employed a spectral element simulation to investigate unsteady entrance flow in a straight tube. For a mean Re of 200 and α varying from 1.8 *to* 12.5 and an inlet wave form $1 + \sin \omega t$, He and Ku have computed variations in entrance length during the pulsatile cycle. The amplitude of the entrance length variation decreases with an increase in α. The phase lag between the entrance length and the inlet flow waveform increases for α up to 5.0 and decreases for larger values of α. For low α, the maximum entrance length during pulsatile flow is approximately the same as the steady entrance length for the peak flow and is primarily dependent on the Reynolds number. For high α, the Stokes boundary layer growth is faster and the entrance length is more uniform during the cycle. For $\alpha \geq 12.5$, the pulsatile entrance length is approximately the same length as the entrance length of the mean flow. At all α, the wall shear rate converges to its fully developed value at about half the length at which the centerline velocity converges to its fully developed value. This leads to the conclusion that the upstream flow conditions leading to a specific artery may or may not be fully developed and can be predicted only by the magnitudes of the Reynolds number and Womersley number.

Effect of Tube Wall Elasticity on Poiseuille Flow

Here, we will include the elastic behavior of the vessel wall and examine the effect on the Hagen-Poiseuille model.

Consider a pressure gradient driven, laminar, steady flow of a Newtonian fluid in a long, circular, cylindrical, thin walled, elastic tube. Let the initial radius of the tube be a_0, and h be the wall thickness and it is small compared to a_0. Because the tube is elastic, it will distend more at the high pressure end (inlet) than at the outlet end. The tube radius, a, will now be a function of x.

The variation in tube radius due to wall elasticity has to be ascertained. The difference between the pressure external to the tube (on the outside of the tube), p_e, and the pressure inside the tube, $p(x)$, at any cross section of the tube (the negative of transmural pressure difference), is $(p_e - p(x))$. This pressure difference acts across h at every cross section, and will induce a circumferential stress. There will be a corresponding circumferential strain. This strain is the ratio of the change in radius to the original radius of the tube. In this way, we can ascertain the cross section at x.

Consider the static force equilibrium on a cylindrical segment of the blood vessel consisting of the top half cross section and of unit length. Let $\sigma_{\theta\theta}$ denote the average circumferential (hoop) stress in the tube wall. The net downward force due to the pressure difference will be balanced by the net upward force, and this balance is,

$$2\sigma_{\theta\theta}h = \int_0^\pi (p(x) - p_e)\, a(x) \sin\theta\, d\theta, \tag{17.38}$$

which results in,

$$\sigma_{\theta\theta} = \frac{(p(x) - p_e)\, a(x)}{h}. \tag{17.39}$$

From Hooke's law, the circumferential strain, $e_{\theta\theta}$ is given by,

$$e_{\theta\theta} = \frac{\sigma_{\theta\theta}}{E} = \frac{(a(x) - a_0)}{a_0} = \left(\frac{a(x)}{a_0}\right) - 1, \tag{17.40}$$

where, E is the Young's modulus of the tube wall material, and we have neglected the radial stress σ_{rr} as compared to $\sigma_{\theta\theta}$ in the thin walled tube. The wall is considered thin if $(h/a) \ll 1$. From equations (17.39) and (17.40), we get, the pressure—radius relationship,

$$a(x) = a_0 \left[1 - \frac{a_0}{Eh}(p(x) - p_e)\right]^{-1} \tag{17.41}$$

Now since the flow is laminar and steady, we can still apply Hagen-Poiseuille formula, equation (17.17), to the flow. Thus,

$$Q = -\frac{\pi}{8\mu}\left(\frac{dp}{dx}\right)(a(x))^4 \tag{17.42}$$

Therefore,

$$\frac{dp}{dx} = -\frac{8\mu Q}{\pi(a(x))^4} \tag{17.43}$$

With equation (17.41),

$$\left[1 - \frac{a_0}{Eh}(p(x) - p_e)\right]^{-4} dp = -\frac{8\mu}{\pi a_0{}^4}Q\, dx \tag{17.44}$$

This is subject to the conditions, $p = p_1$ at $x = 0$, and $p = p_2$ at $x = L$. By integration of equation (17.44) and from the boundary conditions,

$$\frac{Eh}{3a_0} \left\{ \left[1 - \frac{a_0}{Eh} (p_2 - p_e) \right]^{-3} - \left[1 - \frac{a_0}{Eh} (p_1 - p_e) \right]^{-3} \right\} = -\frac{8\mu}{\pi a_0^4} L \, Q.$$

$$(17.45)$$

Solving for Q,

$$Q = \frac{\pi a_0^3 Eh}{24\mu L} \left\{ \left[1 - \frac{a_0}{Eh} (p_1 - p_e) \right]^{-3} - \left[1 - \frac{a_0}{Eh} (p_2 - p_e) \right]^{-3} \right\} \quad (17.46)$$

From equation (17.46), we see that the flow is a nonlinear function of pressure drop if wall elasticity is taken into account. In the above development, we have assumed Hookean behavior for the stress-strain relationship. However, blood vessels do not necessarily obey Hooke's law, their zero-stress states are open sectors, and their constitutive equations may be non-linear (see, Zhou and Fung (1997)).

Pulsatile Flow Theory

As stated earlier, blood flow in the arteries is pulsatile in nature. One of the earliest attempts to model pulsatile flow was carried out by Otto Frank in 1899 (see Fung (1997)).

Elasticity of the Aorta and the Windkessel Theory

Recall that when the left ventricle contracts during systole, pressure within the chamber increases until it is greater than the pressure in the aorta, leading to the opening of the aortic valve. The ventricular muscles continue to contract increasing the chamber pressure while ejecting blood into the aorta. As a result, the ventricular volume decreases. The pressure in the aorta starts to build up and the aorta begins to distend due to wall elasticity. At the end of the systole, ventricular muscles start to relax, the ventricular pressure rapidly falls below that of the aorta and the aortic valve closes. Not all of the blood pumped into the aorta, however, immediately goes into systemic circulation. A part of the blood is used to distend the aorta and a part of the blood is sent to peripheral vessels. The distended aorta acts as an elastic reservoir or a Windkessel (the name in German for an elastic reservoir), the rate of outflow from which is determined by the total peripheral resistance of the system (systemic). As the distended aorta contracts, the pressure diminishes in the aorta. The rate of pressure decrease in the aorta is much slower compared to that in the heart chamber. In other words, during the systole part of the heart pumping cycle, the large fluctuation of blood pressure in the left ventricle is converted to a pressure wave with a high mean value and a smaller fluctuation in the distended aorta (Fung (1997)). This behavior of the distended aorta was thought to be analogous to the high-pressure air chamber (Windkessel) of 19th century fire engines in Germany, and hence the name Windkessel theory was used by Otto Frank to describe this phenomenon.

In the Windkessel theory, blood flow at a rate $Q(t)$ from the left ventricle enters an elastic chamber (the aorta) and a part of this flows out into a single rigid tube representative of all of the peripheral vessels. The rigid tube offers constant resistance, R, equal to the total peripheral resistance that was evaluated in the Hagen-Poiseuille model, equation (17.9). From the law of conservation of mass, assuming blood is incompressible,

Rate of Inflow into Aorta $=$ Rate of change of volume of elastic chamber

$+$ Rate of outflow into rigid tube.   (17.47)

Let the instantaneous blood pressure in the elastic chamber be $p(t)$, and its volume be $v(t)$. The pressure on the outside of the aorta is taken to be zero. The rate of change of volume of an elastic chamber is given by,

$$\frac{dv}{dt} = \left(\frac{dv}{dp}\right)\left(\frac{dp}{dt}\right). \tag{17.48}$$

In equation (17.48), the quantity $\left(\frac{dv}{dp}\right)$ is the compliance, $\mathcal{K}$, of the vessel and is a measure of the distensibility. Compliance at a given pressure is the rate of change in volume with respect to a change in pressure. Here pressures are always understood to be transmural pressure differences. Compliance essentially represents the distensibility of the vascular walls in response to a certain pressure. Also, from equation (17.9), rate of flow into peripherals is given by $(p(t)/R)$, where we have assumed $\bar{p}_V = 0$. Therefore, equation (17.47) becomes,

$$Q(t) = \mathcal{K}\left(\frac{dp}{dt}\right) + \left(\frac{p(t)}{R}\right). \tag{17.49}$$

The equation (17.49) is a linear equation of the form,

$$Q = \frac{dy}{dx} + Py, \tag{17.50}$$

whose solution is,

$$ye^{\int Pdx} = A + \int Qe^{\int Pdx}dx. \tag{17.51}$$

From equations (17.49) and (17.51), with p_0 denoting p at $t = 0$, the instantaneous pressure p in the aorta as a function of the left ventricular ejection rate $Q(t)$ is given by,

$$p(t) = \frac{1}{\mathcal{K}}e^{-t/R\mathcal{K}}\int_0^t Q(\tau)e^{\tau/R\mathcal{K}}d\tau + p_0 e^{-t/R\mathcal{K}}. \tag{17.52}$$

In equation (17.52), p_0 would be the aortic pressure at the end of diastolic phase.

A fundamental assumption in the Windkessel theory is that the pressure pulse wave generated by the heart is transmitted instantaneously throughout the arterial system and disappears before the next cardiac cycle. In reality, pressure waves require finite but small transmission times, and are modified by reflection at bifurcations, bends, tapers, and at the end of short tubes of finite length, and so on. We will now account for some of these features.

Pulse Wave Propagation in an Elastic Tube: Inviscid Theory

Consider a homogeneous, incompressible, and inviscid fluid in an infinitely long, horizontal, cylindrical, thin walled, elastic tube. Let the fluid be initially at rest. The propagation of a disturbance wave of small amplitude and long wave length compared to the tube radius is of interest to us. In particular, we wish to calculate the wave speed. Since the disturbance wave length is much greater than the tube diameter, the time dependent internal pressure can be taken to be a function only of (x, t).

Before we embark on developing the solution, we need to understand the inviscid approximation. For flow in large arteries, the Reynolds and Womersley numbers are large, the wall boundary layers are very thin compared to the radius of the vessel. The inviscid approximation may be useful in giving us insights in understanding such flows. Clearly, this will not be the case with arterioles, venules and capillaries. However, the inviscid analysis is strictly of limited use since it is the viscous stress that is dominant in determining flow stability in large arteries.

Under the various conditions prescribed, the resulting flow may be treated as one dimensional.

Let $A(x, t)$ and $u(x, t)$ denote the the cross sectional area of the tube and the longitudinal velocity component, respectively. The continuity equation is:

$$\frac{\partial A}{\partial t} + \frac{\partial (Au)}{\partial x} = 0, \tag{17.53}$$

and, the equation for the conservation of momentum is:

$$\rho A \left(\frac{\partial u}{\partial t} + u \frac{\partial u}{\partial x} \right) = -\frac{\partial ((p - p_e) A)}{\partial x}, \tag{17.54}$$

where $(p - p_e)$ is the transmural pressure difference. Since the tube wall is assumed to be elastic (not viscoelastic), under the further assumption that A depends on the transmural pressure difference $(p - p_e)$ alone, and the material obeys Hooke's law, we have from equation (17.41), the pressure—radius relationship (referred to as "tube law"),

$$p - p_e = \frac{Eh}{a_0} \left(1 - \frac{a_0}{a} \right) = \frac{Eh}{a_0} \left[1 - \left(\frac{A_0}{A} \right)^{\frac{1}{2}} \right], \tag{17.55}$$

where $A = \pi a^2$, and $A_0 = \pi a_0^2$. The equations (17.53), (17.54), and (17.55) govern the wave propagation. We may simplify this equation system further by linearizing it. This is possible if the pressure amplitude $(p - p_e)$ compared to p_0, the induced fluid speed u, and $(A - A_0)$ compared to A_0, and their derivatives are all small. If the pulse is moving slowly relative to the speed of sound in the fluid, the wave amplitude is much smaller than the wave length, and the distension at one cross section has no effect on the distension elsewhere, the assumptions are reasonable. As discussed by Pedley (2000), in normal human beings, the mean blood pressure, relative to atmospheric, at the level of the heart is about 100 mmHg, and there is a cyclical variation between 80 and 120 mmHg, so the amplitude-to-mean ratio is 0.2, which is reasonably small. Also, in the ascending aorta, the pulse wave speed, c, is about 5 m/s, and the maximum value of u is about 1 m/s, and (u/c) is also around 0.2. In that case, the system of equations reduce to

$$\frac{\partial A}{\partial t} + A_0 \frac{\partial u}{\partial x} = 0, \tag{17.56}$$

and,

$$\rho \frac{\partial u}{\partial t} = -\frac{\partial p}{\partial x}, \tag{17.57}$$

and,

$$p - p_e = \frac{Eh}{2a_0 A_0} (A - A_0), \quad \text{and} \quad \frac{\partial p}{\partial A} = \frac{Eh}{2a_0 A_0} \tag{17.58}$$

Differentiating equation (17.56) with respect to t and equation (17.57) with respect to x, and subtracting the resulting equations, we get,

$$\frac{\partial^2 A}{\partial t^2} = \frac{A_0}{\rho} \frac{\partial^2 p}{\partial x^2}, \tag{17.59}$$

and with equation (17.58), we obtain,

$$\frac{\partial^2 p}{\partial t^2} = \frac{Eh}{2a_0 A_0} \frac{\partial^2 A}{\partial t^2} = \frac{\partial p}{\partial A} \frac{A_0}{\rho} \frac{\partial^2 p}{\partial x^2}. \tag{17.60}$$

Combining equations (17.59) and (17.60), we produce,

$$\frac{\partial^2 p}{\partial x^2} = \frac{1}{c^2} \frac{\partial^2 p}{\partial t^2}, \quad \text{or,} \quad \frac{\partial^2 p}{\partial t^2} = c^2(A_0) \frac{\partial^2 p}{\partial x^2}, \tag{17.61}$$

where, $c^2 = \frac{Eh}{2\rho a_0} = \frac{A}{\rho} \frac{dp}{dA}$. Equation (17.61) is the wave equation, and the quantity,

$$c = \sqrt{\frac{Eh}{2\rho a_0}} = \sqrt{\frac{A}{\rho} \frac{dp}{dA}}, \tag{17.62}$$

is the speed of propagation of the pressure pulse. This is known as the Moens-Korteweg wave speed. If the thin wall assumption is not made, following Fung (1997), by evaluating the strain on the midwall of the tube,

$$c = \sqrt{\frac{Eh}{2\rho (a_0 + h/2)}}, \tag{17.63}$$

Next, similar to equation (17.61), we can develop,

$$\frac{\partial^2 u}{\partial x^2} = \frac{1}{c^2} \frac{\partial^2 u}{\partial t^2}, \tag{17.64}$$

for the velocity component u. The wave equation (17.61) has the general solution,

$$p = f_1 \left(t - \frac{x}{c} \right) + f_2 \left(t + \frac{x}{c} \right), \tag{17.65}$$

where f_1 and f_2 are arbitrary functions; f_2 is zero if the wave propagates only in the $+x$ direction. This result states that the small amplitude disturbance can propagate along the tube, in either direction, without change of shape of the wave form, at speed $c(a_0)$. Also, the velocity wave form is predicted to be of the same shape as the pressure wave form.

In principle, the Moens-Korteweg wave speed given in equation (17.63) must enable the determination of the arterial modulus E as a function of a by noninvasive measurement of the values of arterial dimensions (a, h), the wave forms of the arterial inner radius at two sites, the transit time (as the time interval between the wave form peaks), and hence the pulse wave velocity. More details in this regard are available in the book by Mazumdar (1999).

Next, consider the solutions of wave equations (17.61) and (17.64),

$$p = \hat{p}_1 f(x - ct) + \hat{p}_2 g(x + ct), \tag{17.66}$$

and,

$$u = \hat{u}_1 f(x - ct) + \hat{u}_2 g(x + ct), \tag{17.67}$$

where $\hat{p}_1$, $\hat{u}_1$, $\hat{p}_2$, and, $\hat{u}_2$, are the pressure and velocity amplitudes for waves travelling in the positive x-direction and negative x-direction, respectively. From equation (17.57),

$$\hat{p}_1 = \rho c \hat{u}_1, \quad \text{and,} \quad \hat{p}_2 = -\rho c \hat{u}_2. \tag{17.68}$$

This equation (17.68) relates the amplitudes of the pressure and velocity waves.

The above analysis would equally apply if the inviscid fluid in the tube was initially in steady motion, say from left to right. In that case, u would have to be regarded as a small perturbation superposed on the steady flow, and c would be the speed of the perturbation wave relative to the undisturbed flow.

Let us now examine the limitations of the above model. For typical flow in the aorta, the speed of propagation of the pulse is about 4 m/s (Brown et al. (1999)), about 5 m/s in the ascending aorta, rising to about 8 m/s in more peripheral arteries. These predictions are very close to measured values in normal subjects, either dogs or humans (Pedley (2000)). The peak flow speed is about 1 m/s. The speed of propagation in a collapsible vein might be as low as 1 m/s, and this may lead to phenomena analogous to sonic flow (Brown et al. (1999)). From equation (17.62), for given E, h, ρ, and size of vessel, the wave speed is a constant. Experimental studies indicate, however, that the wave speed is a function of frequency. The shape of the wave form does not remain the same. The theory must be modified to account for peaking of the pressure pulse due to wave reflection from arterial junctions, wave front steepening due to nonlinear dispersion effects (Lighthill (1978)), and observed velocity wave form by including dissipative effects due to viscosity (Lighthill (1975), Pedley (2000)). The neglect of the inertial terms and the effects of viscosity have therefore to be examined to address these concerns and to develop a systematic understanding. These issues will be addressed in later sections in the following order. First, we will learn about pulsatile viscous flow in a rigid walled single straight tube. This would imply the assumption of an infinite wave speed. Subsequent to that, we will examine the effects of wall elasticity on pulsatile viscous flow in a single tube to gain a more realistic understanding. This would allow us to understand wave transmission at finite speed. Following this, we will study blood vessel bifurcation. This will be extended to understand the effects of wave reflection from arterial junctions under the inviscid flow approximation.

Pulsatile Flow in a Rigid Cylindrical Tube: Viscous Effects Included, Infinite Wave Speed Assumption

Consider the axi-symmetric flow of a Newtonian incompressible fluid in a long, thin, circular, cylindrical, horizontal, rigid walled tube. Clearly, the assumption of a rigid wall implies that the speed of wave propagation is infinite and unrealistic. However, the development presented here will provide us with useful insights and these will be helpful in formulating a much improved theory in the next section.

We shall employ the cylindrical coordinates (r, θ, x) with velocity components $(u_r, u_\theta,$ and, $u_x)$, respectively. Let λ be the wave length of the pulse. This is long, and $a \ll \lambda$. Since the wave speed is infinite, all the velocity components are very much smaller than the wave speed. These assumptions would enable us to drop the inertial terms in the momentum equations. With the additional assumptions of axi-symmetry $(u_\theta = 0,$ and $\frac{\partial}{\partial \theta} = 0)$, and rigid tube wall, $(u_r = 0)$, and omitting the subscript x in u_x for convenience, the continuity equation may be written:

$$\frac{\partial u}{\partial x} = 0, \tag{17.69}$$

and the r-momentum equation is:

$$0 = -\frac{\partial p}{\partial r} \tag{17.70}$$

and the x momentum equation is:

$$\rho \frac{\partial u}{\partial t} = -\frac{\partial p}{\partial x} + \mu \left[\frac{\partial^2 u}{\partial r^2} + \frac{1}{r} \frac{\partial u}{\partial r} \right] \tag{17.71}$$

We see that $u = u(r, t)$ and $p = p(x, t)$. Therefore, we are left with just one equation:

$$\mu \left[\frac{\partial^2 u}{\partial r^2} + \frac{1}{r} \frac{\partial u}{\partial r} \right] - \rho \frac{\partial u}{\partial t} = \frac{\partial p}{\partial x}. \tag{17.72}$$

In equation (17.72), since $p = p(x, t)$, $\frac{\partial p}{\partial x}$ will be a function only of t. Since the pressure wave form is periodic, it is convenient to express the partial derivative of pressure using a Fourier series. Such a periodic function depends on the fundamental frequency of the signal, ω, heart rate (unit, rad/s), and the time t. Recall that ω is also called the circular frequency, $\omega/2\pi$ is the frequency (unit, Hz), and λ is the wave length, (unit, m). Also, $\lambda = c/(\omega/2\pi)$, where c is wave speed. The wave length is the wave speed divided by frequency, or the distance travelled per cycle.

We set

$$\frac{\partial p}{\partial x} = -Ge^{i\omega t}, \tag{17.73}$$

where G is a constant denoting the amplitude of the pressure gradient pulse and $e^{i\omega t} = \cos \omega t + i \sin \omega t$. With this representation for $p(t)$, equation (17.72) becomes:

$$\mu \left[\frac{\partial^2 u}{\partial r^2} + \frac{1}{r} \frac{\partial u}{\partial r} \right] - \rho \left[\frac{\partial u}{\partial t} \right] = \frac{\partial p}{\partial x} = -Ge^{i\omega t} \tag{17.74}$$

This is a linear, second order, partial differential equation with a forcing function. For $\omega = 0$, the flow is described by the Hagen-Poiseuille model. Womersley (1955a, 1955b), has solved this problem, and we will provide essential details.

For $\omega \neq 0$, we may try solutions of the form,

$$u(r, t) = U(r)e^{i\omega t}, \tag{17.75}$$

where, $U(r)$ is the velocity profile in any cross section of the tube. The real part in equation (17.75) gives the velocity for the pressure gradient $G \cos \omega t$ and the imaginary part gives the velocity for the pressure gradient $G \sin \omega t$. Assume that the flow is identical at each cross section along the tube. From equations (17.74) and (17.75), we get:

$$\frac{d^2U}{dr^2} + \frac{1}{r} \frac{dU}{dr} - \frac{i\omega\rho}{\mu} U = \frac{G}{\mu}. \tag{17.76}$$

This is a Bessel's differential equation, and the solution would involve Bessel functions of zeroth order and complex arguments. Thus,

$$U(r) = C_1 J_0 \left(i\sqrt{(i\omega\rho/\mu)}\, r \right) + C_2 Y_0 \left(i\sqrt{(i\omega\rho/\mu)}\, r \right) + \frac{G}{\omega\rho i}, \tag{17.77}$$

where C_1 and C_2 are constants. In equation (17.77), from the requirement that U is finite at $r = 0$, $C_2 = 0$. For a rigid walled tube, $U = 0$ at $r = a$. Therefore,

$$C_1 J_0 \left(i^{3/2}\sqrt{(\omega\rho/\mu)}\, a \right) + \frac{G}{\omega\rho i} = 0. \tag{17.78}$$

From equation (17.12), the Womersley number is defined by $\alpha = a\sqrt{\omega/\nu}$. Therefore, from equation (17.78), we may write,

$$C_1 = \frac{iG}{\omega\rho} \frac{1}{J_0 \left(i^{3/2}\alpha \right)}. \tag{17.79}$$

Therefore, from equation (17.77),

$$U(r) = -\frac{iG}{\omega\rho}\left(1 - \frac{J_0\left(i^{3/2}\alpha\, r/a\right)}{J_0\left(i^{3/2}\alpha\right)}\right). \tag{17.80}$$

Introduce, for convenience,

$$F_1(\alpha) = \left(\frac{J_0\left(i^{3/2}\alpha\, r/a\right)}{J_0\left(i^{3/2}\alpha\right)}\right). \tag{17.81}$$

Now, from equation (17.75),

$$u(r,t) = U(r)e^{i\omega t} = -\frac{iG}{\omega\rho}\left(1 - F_1(\alpha)\right)e^{i\omega t} = \frac{Ga^2}{i\mu\alpha^2}\left(1 - F_1(\alpha)\right)e^{i\omega t}. \tag{17.82}$$

In the above development, we have found the velocity as a function of radius r and time t for the entire driving pressure gradient. Since we have represented both $\frac{\partial p}{\partial x}$ and $u(r,t)$ in terms of Fourier modes, we could also express the solution for both these quantities in terms of individual Fourier modes or harmonics explicitly as,

$$\frac{\partial p}{\partial x} = -\sum_{n=0}^{N} G_n e^{in\omega t}, \tag{17.83}$$

where N is the number of modes (harmonics), and the $n = 0$ term represents the mean pressure gradient. Similarly, for velocity,

$$u(r,t) = u_0(r) + \sum_{1}^{N} u_n(r)e^{in\omega t} \tag{17.84}$$

In equation (17.84),

$$u_0(r) = \frac{G_0 a^2}{4\mu}\left(1 - \frac{r^2}{a^2}\right), \tag{17.85}$$

is the mean flow and is recognized as the steady Hagen-Poiseuille flow with G_0 as the mean pressure gradient, and, for each harmonic,

$$u_n(r) = \frac{G_n a^2}{i\mu\alpha_n^2}\left(1 - F_1(\alpha_n)\right) \tag{17.86}$$

We can now write down the expressions for $u_n(r)$ in the limits of α_n small and large. These are, for α_n small,

$$u_n(r) \approx \frac{G_n a^2}{4\mu}\left(1 - \frac{r^2}{a^2}\right), \tag{17.87}$$

which represents a quasi-steady flow, and for α_n large,

$$u_n(r) \approx \frac{G_n a^2}{i\mu\alpha_n^2} \left\{1 - \exp\left[-\sqrt{\frac{\omega}{2\nu}}(1+i)(a-r)\right]\right\}, \qquad (17.88)$$

which is the velocity boundary layer on a plane wall in an oscillating flow. This flow was discussed in Chapter 9 (Stokes second problem).

The volume flow rate, $Q(t)$, may be obtained by integrating the velocity profile across the cross section. Thus, from equations (17.85) and (17.86)

$$Q(t) = \int_0^a u \, 2\pi r dr = \pi a^2 \left\{\frac{G_0 a^2}{8\mu} + \frac{a^2}{i\mu} \sum_1^\infty \frac{G_n}{\alpha_n^2}[1 - F_2(\alpha_n)] \, e^{in\omega t}\right\}, \qquad (17.89)$$

or equivalently, with equation (17.82),

$$Q(t) = \int_0^a 2\pi e^{i\omega t} \frac{Ga^2}{i\mu\alpha^2}(1 - F_1(\alpha)) \, rdr = \frac{\pi a^4}{i\mu\alpha^2} G \, (1 - F_2(\alpha)) \, e^{i\omega t}, \quad (17.90)$$

where,

$$F_2(\alpha) = \frac{2J_1\left(i^{3/2}\alpha\right)}{i^{3/2}\alpha J_0\left(i^{3/2}\alpha\right)}. \qquad (17.91)$$

The real part of equation (17.90) gives the volume flow rate when the pressure gradient is $G \cos \omega t$ and the imaginary part gives the rate when the pressure gradient is $G \sin \omega t$.

Next, the wall shear rate, $\tau(t)|_{r=a}$ is given by,

$$\tau(t)|_{r=a} = \frac{\partial u}{\partial r}\bigg|_{r=a} = \frac{G_0 a}{2} + \frac{a}{2} \sum_1^N G_n F(\alpha_n) e^{in\omega t} \qquad (17.92)$$

We may now examine the flow rates in the limit cases of $\alpha \to 0$ and $\alpha \to \infty$. As $\alpha \to 0$, by Taylor's expansion,

$$F_2(\alpha) \approx 1 - \frac{i\alpha^2}{8} - O(\alpha^4), \qquad (17.93)$$

and, from equation (17.90), in the limit as $\alpha \to 0$,

$$Q = \frac{\pi Ga^4}{8\mu} e^{in\omega t}, \qquad (17.94)$$

and the magnitude of the volumetric flow rate, Q_0, in the limit as $\alpha \to 0$ is,

$$|Q_0| = \frac{\pi Ga^4}{8\mu}, \qquad (17.95)$$

as would be expected (Hagen-Poiseuille result). As $\alpha \to \infty$,

$$F_2(\alpha) \approx \frac{2}{i^{1/2}\alpha}\left(1 + \frac{1}{2\alpha}\right), \qquad (17.96)$$

Next, in Hagen-Poiseuille flow, the steady flow rate is the maximum attainable and there is no phase lag between the applied pressure gradient and the flow. To understand the phase difference between the applied pressure gradient pulse and the flow rate in the present flow model, we set,

$$(1 - F_2(\alpha)) = Z(\alpha), \quad Z(\alpha) = X(\alpha) + iY(\alpha) \qquad (17.97)$$

Then from equation (17.90),

$$Q = \frac{\pi G a^4}{\mu \alpha^2} \{[Y \cos(\omega t) + X \sin(\omega t)] - i [X \cos(\omega t) - Y \sin(\omega t)]\} \qquad (17.98)$$

The magnitude of Q is,

$$|Q| = \frac{\pi G a^4}{\mu \alpha^2} \sqrt{X^2 + Y^2}. \qquad (17.99)$$

The phase angle between the applied pressure gradient $Ge^{i\omega t}$ and the flow rate (equation (17.90)) is now noted to be,

$$\tan \phi = \frac{X}{Y}. \qquad (17.100)$$

With increasing ω, the phase lag between the pressure gradient and the flow rate increases, and the flow rate decreases. Thus, the magnitude of the volumetric flow rate, $|Q|$, given by equation (17.99) will be considerably less than the magnitude $|Q_0|$ given by equation (17.95) as would be expected. For an arterial flow, with $\alpha = 8$, $X \approx 0.85$, $Y \approx 0.16$, the pulsed volumetric flow rate, $|Q|$ would be just about one tenth of the steady value, $|Q_0|$. For more detailed discussions and comparisons with measured values of pressure gradients and flow rates in blood vessels, see Nichols and O'Rourke (1998).

The above analysis assumed an infinite wave speed of propagation. In order to accommodate the requirement of wave transmission at a finite wave speed, we need to account for vessel wall elasticity. This will be discussed in the next section.

Wave Propagation in a Viscous Liquid Contained in an Elastic Cylindrical Tube

Blood vessel walls are viscoelastic. But in large arteries the effect of nonlinear viscoelasticity on wave propagation is not so severe (Fung, 1997). Even where viscoelastic effects are important, an understanding based on elastic walls will be useful. In this section, we will first study the effects of elastic walls. Then, we will briefly discuss the effects of wall viscoelasticity.

Consider a long, thin, circular, cylindrical, horizontal elastic tube containing a Newtonian, incompressible fluid. Let this system be set in motion solely due to a pressure wave, and the amplitude of the disturbance be small enough so that quadratic terms in the amplitude are negligible compared with linear ones.

In the formulation, we have to consider the fluid flow equations together with the equations of motion governing tube wall displacements. Assume that the tube wall material obeys Hooke's law. Since the tube is thin, membrane theory for modelling the wall displacements is adequately accurate, and we will neglect bending stresses.

The primary question is, how does viscosity attenuate velocity and pressure in this flow?

We shall employ the cylindrical coordinates (r, θ, x) with velocity components $(u_r, u_\theta, \text{ and, } u_x)$, respectively. With the assumption of axi-symmetry, $u_\theta = 0$ and $\frac{\partial}{\partial \theta} = 0$. For convenience, we write the u_r component as v, and we omit the subscript x in u_x.

Restricting the analysis to small disturbances, the governing equations for the fluid are:

$$\frac{\partial u}{\partial x} + \frac{1}{r}\frac{\partial (rv)}{\partial r} = 0, \tag{17.101}$$

$$\rho \frac{\partial u}{\partial t} = -\frac{\partial p}{\partial x} + \mu \left(\frac{\partial^2 u}{\partial r^2} + \frac{1}{r}\frac{\partial u}{\partial r} + \frac{\partial^2 u}{\partial x^2} \right), \tag{17.102}$$

$$\rho \frac{\partial v}{\partial t} = -\frac{\partial p}{\partial r} + \mu \left(\frac{\partial^2 v}{\partial r^2} + \frac{1}{r}\frac{\partial v}{\partial r} + \frac{\partial^2 v}{\partial x^2} - \frac{v}{r^2} \right), \tag{17.103}$$

where u and v are the velocity components in the axial and radial directions, respectively.

These have to be supplemented with the tube wall displacement equations. Let the tube wall displacements in the (r, θ, x) directions be $(\eta, \zeta, \text{ and } \xi)$, respectively, and the tube material density be ρ_w. The initial radius of the tube is a_0, and the wall thickness is h.

For this thin elastic tube, the circumferential (hoop) tension and the tension in the axial direction are related by Hooke's law as follows:

$$T_\theta = \frac{Eh}{1 - \hat{v}^2} \left(\frac{\eta}{a_0} + \hat{v}\frac{\partial \xi}{\partial x} \right), \tag{17.104}$$

and

$$T_x = \frac{Eh}{1 - \hat{v}^2} \left(\frac{\partial \xi}{\partial x} + \hat{v}\frac{\eta}{a_0} \right), \tag{17.105}$$

where $\hat{v}$ is Poisson's ratio.

By a force balance on a wall element of volume $(h \; rd\theta \; dx)$, the equations governing wall displacements may be written as:

- r-direction

$$\rho_w h \frac{\partial^2 \eta}{\partial t^2} = \sigma_{rr}|_{r=a} - \frac{T_\theta}{a_0}, \tag{17.106}$$

and

- x-direction

$$\rho_w h \frac{\partial^2 \xi}{\partial t^2} = +\frac{\partial T_x}{\partial x} - \sigma_{rx}|_{r=a}. \tag{17.107}$$

There is no displacement equation for the θ direction. In equations (17.106) and (17.107), $\sigma_{rr}|_{r=a}$ and $\sigma_{rx}|_{r=a}$ refer to radial and shear stresses, respectively, which the fluid exerts on the tube wall. These equations are based on the assumptions that shear and bending stresses in the tube wall material are negligible and the slope of the disturbed tube wall ($\partial a/\partial x$) is small. These also imply that the ratios (a/λ) and (h/λ), where λ is the wavelength of disturbance, are small.

From equations (17.104), (17.105), (17.106), (17.107), we obtain,

$$\rho_w h \frac{\partial^2 \eta}{\partial t^2} = \sigma_{rr}|_{r=a} - \frac{Eh}{1-\hat{v}^2}\left(\frac{\eta}{a_0^2} + \frac{\hat{v}}{a_0}\frac{\partial \xi}{\partial x}\right), \tag{17.108}$$

and

$$\rho_w h \frac{\partial^2 \xi}{\partial t^2} = -\mu\left(\frac{\partial u}{\partial r} + \frac{\partial v}{\partial x}\right)\Bigg|_{r=a} + \frac{Eh}{1-\hat{v}^2}\left(\frac{\partial^2 \xi}{\partial x^2} + \frac{\hat{v}}{a_0}\frac{\partial \eta}{\partial x}\right). \tag{17.109}$$

In the above equations, from the theory of fluid flow, the normal compressive stress due to fluid flow on element of area perpendicular to the radius is given by,

$$\sigma_{rr} = +p - 2\mu\frac{\partial v}{\partial r}, \tag{17.110}$$

and the shear stress due to fluid flow acting in a direction parallel to the axis of the tube on an element of area perpendicular to a radius is,

$$\sigma_{rx} = \mu\left(\frac{\partial u}{\partial r} + \frac{\partial v}{\partial x}\right). \tag{17.111}$$

These are the radial and shear stresses exerted by the fluid on the wall of the vessel. With equations (17.110) and (17.111), equations (17.108), and (17.109) become,

$$\rho_w h \frac{\partial^2 \eta}{\partial t^2} = +p|_{r=a} - 2\mu\frac{\partial v}{\partial r}\Bigg|_{r=a} - \frac{Eh}{1-\hat{v}^2}\left(\frac{\eta}{a_0^2} + \frac{\hat{v}}{a_0}\frac{\partial \xi}{\partial x}\right), \tag{17.112}$$

and

$$\rho_w h \frac{\partial^2 \xi}{\partial t^2} = -\mu\left(\frac{\partial u}{\partial r} + \frac{\partial v}{\partial x}\right)\Bigg|_{r=a} + \frac{Eh}{1-\hat{v}^2}\left(\frac{\partial^2 \xi}{\partial x^2} + \frac{\hat{v}}{a_0}\frac{\partial \eta}{\partial x}\right). \tag{17.113}$$

We have to solve equations (17.101), (17.102), (17.103), together with (17.112), and (17.113) subject to prescribed conditions. The boundary conditions at the wall are that the velocity components of the fluid be equal to those of the wall. Thus,

$$u|_{r=a_0} = \frac{\partial \xi}{\partial t}\Bigg|_{r=a_0}, \tag{17.114}$$

and

$$v|_{r=a_0} = \frac{\partial \eta}{\partial t}\Bigg|_{r=a_0}. \tag{17.115}$$

We note that the boundary conditions given in equations (17.114) and (17.115) are linearized conditions, since we are evaluating u and v at the undisturbed radius a_0.

We now represent the various quantities in terms of Fourier modes. Thus,

$$u(x, r, t) = \hat{u}(r)e^{i(kx-\omega t)}, \quad v(x, r, t) = \hat{v}(r)e^{i(kx-\omega t)},$$
$$p(x, t) = \hat{p}e^{i(kx-\omega t)}, \quad \xi(x, t) = \hat{\xi}e^{i(kx-\omega t)},$$
$$\eta(x, t) = \hat{\eta}e^{i(kx-\omega t)}, \tag{17.116}$$

where $\hat{u}(r)$, $\hat{v}(r)$, $\hat{p}$, $\hat{\xi}$, and $\hat{\eta}$ are the amplitudes, $\omega = 2\pi/T$ is a real constant, the frequency of the forced disturbance, T is the period of the heart cycle, and $k = k_1 + ik_2$ is a complex constant, k_1 being the wave number and k_2 is a measure of the decay of the disturbance as it travels along the vessel (damping constant), $|k| = \sqrt{k_1^2 + k_2^2} = 2\pi/\lambda$ where λ is the wave length of disturbance, and $c = \omega/k_1$ is the wave speed.

The above formulation has been solved by Morgan and Kiely (1954) and by Womersley (1957a, 1957b), and we will provide the essential details here. The analysis will be restricted to disturbances of long wavelength, that is, $a/\lambda \ll 1$, and large Womersley number, $\alpha \gg 1$.

From equation (17.101),

$$\left|\frac{v}{u}\right| = \left|\frac{\hat{v}(r)}{\hat{u}(r)}\right| = O(|ak|). \tag{17.117}$$

For small damping, we note that $|k| \approx k_1 = 2\pi/\lambda$, and $c = \omega/k_1$ is the wave speed.

From equations (17.102) and (17.103), we may make the following observations. In equation (17.102), $\frac{\partial^2 u}{\partial x^2}$ may be neglected in comparison with the other terms since $a/\lambda \ll 1$ and $\lambda\alpha \gg 1$. In equation (17.103), $\frac{\partial p}{\partial r}$ is of a higher order of magnitude in a/λ than is $\frac{\partial p}{\partial x}$. In fact, we may neglect all terms that are of order a/λ. In effect, we are neglecting radial acceleration and damping terms and taking the pressure to be uniform over each cross section. The fluid equations become,

$$\frac{\partial u}{\partial x} + \frac{1}{r}\frac{\partial (rv)}{\partial r} = 0, \tag{17.118}$$

$$\rho\frac{\partial u}{\partial t} = -\frac{\partial p}{\partial x} + \mu\left(\frac{\partial^2 u}{\partial r^2} + \frac{1}{r}\frac{\partial u}{\partial r}\right), \tag{17.119}$$

$$\frac{\partial p}{\partial r} = 0, \tag{17.120}$$

$$p = \hat{p}e^{i(kx-\omega t)}. \tag{17.121}$$

Now substitute the assumed forms given in equation (17.116) into equations (17.118) and (17.119) to produce,

$$\frac{d(r\hat{v})}{dr} = -ikr\hat{u}, \tag{17.122}$$

$$\frac{d^2\hat{u}}{dr^2} + \frac{1}{r}\frac{d\hat{u}}{dr} + \frac{i\omega\rho}{\mu}\hat{u} = \frac{ik\hat{p}}{\mu}, \tag{17.123}$$

The boundary conditions given by equations (17.114) and (17.115) become:

$$\hat{u}(a_0)e^{i(kx-\omega t)} = -i\omega\hat{\xi}e^{i(kx-\omega t)}, \tag{17.124}$$

$$\hat{v}(a_0)e^{i(kx-\omega t)} = -i\omega\hat{\eta}e^{i(kx-\omega t)}. \tag{17.125}$$

We may now note that the linearization of the boundary conditions will involve an error of the same order as that caused by neglecting the nonlinear terms in the equations. The error would be small if $\hat{\xi}$ and $\hat{\eta}$ are very small compared to a.

Next, introduce the assumed form given in equation (17.116), and use equation (17.120) in the displacement equations (17.112) and (17.113) to develop,

$$-\rho_w h\omega^2\hat{\eta} = \hat{p} - 2\mu \left(\frac{d\hat{v}}{dr}\right)\Bigg|_{r=a_0} - \frac{Eh}{1-\hat{v}^2}\left(\frac{\hat{\eta}}{a_0{}^2} + \frac{i\hat{v}k}{a_0}\hat{\xi}\right), \tag{17.126}$$

$$-\rho_w h\omega^2\hat{\xi} = -\mu \left(\frac{d\hat{u}}{dr} + ik\hat{v}\right)\Bigg|_{r=a_0} + \frac{Eh}{1-\hat{v}^2}\left(-k^2\hat{\xi} + \frac{i\hat{v}k}{a_0}\hat{\eta}\right). \tag{17.127}$$

Now invoke the assumptions that $h/a \ll 1$, ρ is of the same order of magnitude as ρ_w, and $a^2/\lambda^2 \ll 1$ in equations (17.126) and (17.127). This amounts to neglecting the terms which represent tube inertia, and approximating σ_{rx} in equation (17.111) by $\mu\left(\frac{\partial v}{\partial x}\right)$ and σ_{rr} in (17.110) by p. After considerable algebra, equations (17.126) and (17.127) reduce to:

$$\hat{p} = \frac{Eh}{a_0{}^2}\hat{\eta} - \frac{i\hat{v}}{a_0 k}\mu\frac{d\hat{u}}{dr}\Bigg|_{r=a_0}, \tag{17.128}$$

$$\hat{\xi} = \frac{i\hat{v}}{ka_0}\hat{\eta} - \frac{1-\hat{v}^2}{Ehk^2}\mu\frac{d\hat{u}}{dr}\Bigg|_{r=a_0} \tag{17.129}$$

We are now left with equations (17.122), (17.123), (17.128), and (17.129), subject to boundary conditions given by (17.124) and (17.125) and the pseudo boundary condition that $u(r)$ be nonsingular at $r = 0$.

Equations (17.123) and (17.128) can be combined to give:

$$\frac{d^2\hat{u}}{dr^2} + \frac{1}{r}\frac{d\hat{u}}{dr} + \frac{i\omega\rho}{\mu}\hat{u} = \frac{ik}{\mu}\frac{Eh}{a_0{}^2}\hat{\eta} + \frac{\hat{v}}{a_0}\frac{d\hat{u}}{dr}\Bigg|_{r=a_0}. \tag{17.130}$$

Satisfying the pseudo boundary condition, the solution to this Bessel's differential equation is given by:

$$\hat{u}(r) = AJ_0(\beta r) + \frac{k}{\omega}\frac{Eh}{\rho a_0{}^2}\hat{\eta} - \frac{\hat{v}}{\beta a_0} A J_1(\beta a_0), \tag{17.131}$$

where, $\beta = \sqrt{i\omega/\nu}$, and A is an arbitrary constant. Next, from equation (17.122),

$$\hat{v} = -\frac{ik}{r}\int_0^r r\hat{u}(r)dr. \tag{17.132}$$

From equation (17.131) and equation (17.132),

$$\hat{v}(r) = -\frac{ikA}{\beta} J_1(\beta r) - \frac{ik^2}{\omega} \frac{Eh\hat{\eta}}{\rho a_0^2} \frac{r}{2} + \frac{ik\hat{v}}{\beta a_0} A \frac{r}{2} J_1(\beta a_0). \tag{17.133}$$

Equations (17.131) and (17.133) give the expressions for $\hat{u}(r)$ and $\hat{v}(r)$, respectively. Subjecting them to the boundary conditions given in equations (17.124) and (17.125), introducing $\hat{\beta} = \beta a_0$, and eliminating $\hat{\xi}$ by the use of equation (17.129), the following two linear homogeneous equations for $\hat{\eta}$ are developed:

$$\hat{\eta} \left[\frac{\omega}{k} \frac{\hat{v}}{a_0} - \frac{kEh}{\omega \rho a_0^2} \right] = A \left[J_0(\hat{\beta}) + J_1(\hat{\beta}) \left\{ \frac{i\beta \omega \mu (1 - \hat{v}^2)}{Ehk^2} - \frac{\hat{v}}{\hat{\beta}} \right\} \right], \tag{17.134}$$

$$\hat{\eta} \left[1 - \frac{k^2 Eh}{\omega^2 2 \rho a_0} \right] = A J_1(\hat{\beta}) \left[\frac{k}{\omega \beta} - \frac{k\hat{v}}{2\omega \beta} \right]. \tag{17.135}$$

For non zero solutions, the determinant of the above set of linear algebraic equations in $\hat{\eta}$ and A must be zero. As a result, the following characteristic equation is developed:

$$\left(\frac{k^2}{\omega^2} \frac{Eh}{2\rho a_0} \right)^2 \left[2\hat{\beta} \frac{J_0(\hat{\beta})}{J_1(\hat{\beta})} - 4 \right] + \left(\frac{k^2}{\omega^2} \frac{Eh}{2\rho a_0} \right) \left[4\hat{v} - 1 - 2\hat{\beta} \frac{J_0(\hat{\beta})}{J_1(\hat{\beta})} \right]$$
$$+ \left(1 - \hat{v}^2 \right) = 0. \tag{17.136}$$

The solution to this quadratic equation will give k^2/ω^2 in terms of known quantities. Then we can find, $k/\omega = (k_1 + ik_2)/\omega$. The wave speed, ω/k_1, and the damping factor may be evaluated by determining the real and imaginary parts of k/ω.

Morgan and Kiely (1954) have provided explicit results for the wave speed, c, and the damping constant, k_2, in the limits of small and large α. Mazumdar (1999) has indicated that by an *in vivo* study, the wave speed, ω/k_1, can be evaluated non-invasively by monitoring the transit time as the time interval between the peaks of ultrasonically measured waveforms of the arterial diameter at two arterial sites at a known distance apart. Then from equation (17.136), E can be calculated. From either of the equations (17.134) or (17.135), A can be expressed in terms of $\hat{\eta}$, and with that $\hat{u}(r)$ can be related to $\hat{p}$. Mazumdar gives details as to how the cardiac output may be calculated with the information so developed in conjunction with pulsed Doppler flowmetry.

Figure 17.14 shows velocity profiles at intervals of $\Delta \omega t = 15°$, of the flow resulting from a pressure gradient varying as $\cos(\omega t)$ in a tube. As this is harmonic motion, only half cycle is illustrated and for $\omega t > 180°$, the velocity profiles are of the same form but opposite in sign. α is the Womersley number. The reversal of flow starts in the laminae near the wall. As the Womersley number increases, the profiles become flatter in the central region, there is a reduction in the amplitudes of the flow, and the rate of reversal of flow increases close to the wall. At $\alpha = 6.67$, the central mass of the fluid is seen to reciprocate like a solid core.

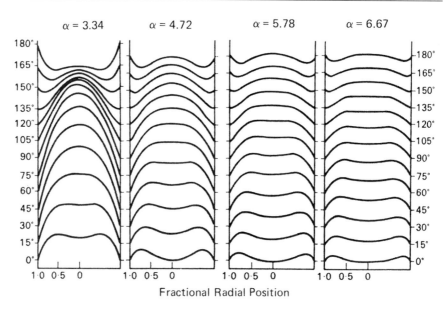

Figure 17.14 Velocity profiles of a sinusoidally oscillating flow in a pipe. (Reproduced from McDonald, D. A. (1974) *Blood Flow in Arteries*, The Williams & Wilkins Company, Baltimore).

Effect of Viscoelasticity of Tube Material

In general, the wall of a blood vessel must be treated as viscoelastic. This means that the relations given in equations (17.104) and (17.105) must be replaced by corresponding relations for a tube of viscoelastic material. In this problem, all the stresses and strains in the problem are assumed to vary as $e^{i(kx - \omega t)}$, and we will further assume that the effect of the strain rates on the stresses is small compared to the effect of the strains. For the purely elastic case, only two real elastic constants were needed. Morgan and Kiely (1954) have shown that by substituting suitable complex quantities for the elastic modulus and the Poisson's ratio the viscoelastic behavior of the tube wall may be accommodated. They introduce,

$$E^* = E - i\omega E', \quad \text{and,} \quad \hat{v}^* = \hat{v} - i\omega \hat{v}', \quad (17.137)$$

where, E' and $\hat{v}'$ are new constants. In equations (17.104) and (17.105), E^* and $\hat{v}^*$ will replace E and $\hat{v}$, respectively. The formulation will otherwise remain the same. An equation for k/ω will arise as before. The fact that E^* and $\hat{v}^*$ are complex has to be taken into account while evaluating the wave velocity and the damping factor. Morgan and Kiely provide results appropriate for small and large α.

Morgan and Ferrante (1955) have extended the study by Morgan and Kiely (1954) discussed above to the situation for small α values where there is Poiseuille like flow in the thin, elastic walled tube. The flow oscillations are small and they are superimposed on a large steady stream velocity. The steady flow modifies the wave velocity. The wave velocity in the presence of a steady flow is the algebraic sum of the normal wave velocity and the steady flow velocity. Morgan and Ferrante predict a decrease in the damping of a wave propagated in the direction of the stream and an increase in the damping when propagated upstream. However, the steady flow component in arteries is so small in comparison with the pulse wave velocity that its role in damping is of little importance (see McDonald (1974)). Womersley (1957a) has considered the situation where the flow oscillations are large in amplitude compared to the mean stream velocity. This is similar to the situation in an artery. He predicts that the presence of a steady stream velocity would produce a small increase in the damping.

Next, we will study blood flow in branching tubes.

Blood Vessel Bifurcation: An Application of Poiseuille's Formula and Murray's Law

Blood vessels bifurcate into smaller daughter vessels which in turn bifurcate to even smaller ones. On the basis that the flow satisfies Poiseuille's formula in the parent and all the daughter vessels, and by invoking the principle of minimization of energy dissipation in the flow, we can determine the optimal size of the vessels and the geometry of bifurcation. We recall that Hagen-Poiseuille flow involves established (fully developed) flow in a long tube. Here, for simplicity, we will assume that established Poiseuille flow exists in all the vessels. This is obviously a drastic assumption but the analysis will provide us with some useful insights.

Let the parent and daughter vessels be straight, circular in cross section, and lie in a plane.

Consider a parent vessel AB of length L_0 of radius a_0 in which the flow rate is Q which bifurcates into two daughter vessels BC and BD with lengths L_1, and L_2, radii a_1 and a_2, and flow rates Q_1 and Q_2, respectively. The axes of vessels BC and BD are inclined at angles θ and ϕ with respect to the axis of AB, as shown in Fig. (17.15). Points A,C,D are fixed. The optimal sizes of the vessels and the optimal location of B have to determined from the principle of minimization of energy dissipation.

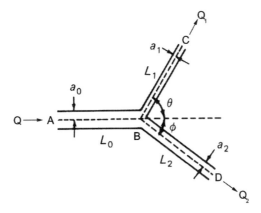

Figure 17.15 Schematic of an arterial bifurcation.

The total rate of energy dissipation by flow rate Q in a blood vessel of length of L and radius a is equal to sum of the rate at which work is done on the blood, $Q\Delta p$, and the rate at which energy is used up by the blood vessel by metabolism, $K\pi a^2 L$, where K is a constant. For Hagen-Poiseuille flow, from equation (17.17), $Q = \frac{\pi a^4}{8\mu}\frac{\Delta p}{L}$. Therefore,

$$\text{Total energy dissipation} = \frac{8\mu L}{\pi a^4}Q^2 + K\pi a^2 L = \hat{E}_1, \text{(say).} \qquad (17.138)$$

To obtain the optimal size of a vessel for transport, for a given length of vessel, we need to minimize this quantity with respect to radius of the vessel. Thus,

$$\frac{\partial \hat{E}_1}{\partial a} = -\frac{32\mu L}{\pi}Q^2 a^{-5} + 2K\pi La = 0. \qquad (17.139)$$

Solving for a,

$$a = \left[\frac{16\mu}{\pi^2}K\right]^{1/6} Q^{1/3}. \qquad (17.140)$$

The equation (17.140) gives the optimal radius for the blood vessel indicating that minimum energy dissipation occurs under this condition. The optimal relationship, $Q \sim a^3$, is called Murray's Law.

With equation (17.140), the minimum value for energy dissipation is

$$\hat{E}_{1,\min} = \frac{3\pi}{2}KLa^2. \qquad (17.141)$$

Next, consider the flow with the branches. The minimum value for energy dissipation with branches is

$$\hat{E}_{2,\min} = \frac{3\pi}{2}K\left(L_0 a_0^2 + L_1 a_1^2 + L_2 a_2^2\right). \qquad (17.142)$$

Also,

$$a_0 = \left[\frac{16\mu}{\pi^2}K\right]^{1/6} Q_0^{1/3}, \quad a_1 = \left[\frac{16\mu}{\pi^2}K\right]^{1/6} Q_1^{1/3}, \quad \text{and,} \quad a_2 = \left[\frac{16\mu}{\pi^2}K\right]^{1/6} Q_2^{1/3}, \qquad (17.143)$$

and, from mass conservation,

$$Q = Q_1 + Q_2 \; \rightarrow \; a_0^3 = a_1^3 + a_2^3. \qquad (17.144)$$

The lengths L_0, L_1, L_2 depend on the location of point B. The optimum location of the point B is determined by examining associated variational problems (see Fung (1997)).

Any small movement of B changes $\hat{E}_{2,\min}$ by $\delta\hat{E}_{2,\min}$, and,

$$\delta\hat{E}_{2,\min} = \frac{3\pi}{2}K\left(\delta L_0\, a_0^2 + \delta L_1\, a_1^2 + \delta L_2\, a_2^2\right) \qquad (17.145)$$

The optimal location of B would be such as to make $\delta\hat{E}_{2,\min} = 0$ for arbitrary small movement δL of point B. By making such displacements of B, one at a time, in the direction of AB, in the direction of BC, and finally in the direction of DB, and setting the value of the corresponding $\delta\hat{E}_{2,\min}$ to zero, we develop a set of three conditions governing optimization. These are:

$$\cos\theta = \frac{a_0^4 + a_1^4 - a_2^4}{2a_0^2 a_1^2}, \quad \cos\phi = \frac{a_0^4 - a_1^4 + a_2^4}{2a_0^2 a_2^2}, \quad \cos(\theta + \phi) = \frac{a_0^4 - a_1^4 - a_2^4}{2a_1^2 a_2^2}.$$
$$(17.146)$$

Together with equation (17.144), equation set (17.146) may be solved for the optimum angle θ as,

$$\cos\theta = \frac{a_0^4 + a_1^4 - \left(a_0^3 - a_1^3\right)^{4/3}}{2a_0^2 a_1^2}, \qquad (17.147)$$

and a similar equation for ϕ. Comparison of these optimization results with experimental data are noted to be excellent (see Fung (1997)).

Reflection of Waves at Arterial Junctions: Inviscid Flow and Long Wave Length Approximation

Arteries have branches. When a pressure or a velocity wave reaches a junction where the parent artery 1 bifurcates into daughter tubes 2 and 3 as shown in the Figure 17.16, the incident wave is partially reflected at the junction into the parent tube and partially transmitted down the daughters. In the long wave length approximation, we may neglect the flow at the junction. Let the longitudinal coordinate in each tube be x,

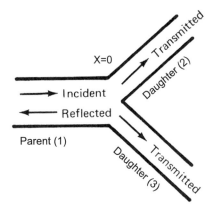

Figure 17.16   Schematic of an arterial bifurcation: Reflection.

with $x = 0$ at the bifurcation. the incident wave in the parent tube comes from $x = \infty$.

Let p_I be the oscillatory pressure associated with the incident wave, p_R that associated with the reflected wave, and p_{T1} and p_{T2}, those associated with the transmitted waves. Let the pressure be a single valued and continuous function at the junction for all time t. The continuity requirement ensures that there are no local accelerations. Under these conditions, at the junction,

$$p_I + p_R = p_{T1} = p_{T2}. \tag{17.148}$$

Next, let Q_I be the flow rate associated with the incident wave, Q_R that associated with the reflected wave, and Q_{T1} and Q_{T2}, those associated with the transmitted waves. The flow rate is also taken to be single valued and continuous at the junction for all time t. The continuity requirement ensures conservation of mass. At the junction,

$$Q_I - Q_R = Q_{T1} + Q_{T2}. \tag{17.149}$$

Let the undisturbed cross sectional areas of the tubes be A_1, A_2, and A_3, and the intrinsic wave speeds be c_1, c_2, c_3, respectively. In general, for a fluid of density ρ flowing under the influence of a wave with intrinsic wave speed c, through a tube of cross sectional area A, the flow rate Q is related to the mean velocity u by,

$$Q = Au = \pm \frac{A}{\rho c} p, \tag{17.150}$$

where we have employed the relationship given in equation (17.68). The plus or the minus sign applies depending on whether the wave is going in the positive x direction or in the negative x direction. The quantity $A/\rho c$ is called the characteristic admittance of tube and is denoted by Y, while, $\rho c/A$, is called the characteristic impedance of the tube and is denoted by Z. Admittance is seen to be the ratio of the oscillatory flow to the oscillatory pressure when the wave goes in the direction of $+x$ axis. With these definitions,

$$Q = Au = \pm Yp = \pm \frac{p}{Z}. \tag{17.151}$$

The equation (17.149) may be written in terms of admittances or impedances as:

$$Y_1(p_I - p_R) = \sum_{j=2}^{3} Y_j p_{Tj}, \text{ or } \frac{(p_I - p_R)}{Z_1} = \sum_{j=2}^{3} \frac{p_{Tj}}{Z_j}. \tag{17.152}$$

We can simultaneously solve equations (17.148) and (17.152) to produce,

$$\frac{p_R}{p_I} = \frac{Y_1 - \sum Y_j}{Y_1 + \sum Y_j} = \mathcal{R}, \text{ and } \frac{p_{Tj}}{p_I} = \frac{2Y_1}{Y_1 + \sum Y_j} = \mathcal{T}, \tag{17.153}$$

or,

$$\frac{p_R}{p_I} = \frac{Z_1^{-1} - \sum Z_j^{-1}}{Z_1^{-1} + \sum Z_j^{-1}}, \text{ and } \frac{p_{Tj}}{p_I} = \frac{2Z_1^{-1}}{Z_1^{-1} + \sum Z_j^{-1}}. \tag{17.154}$$

In equation (17.153), $\mathcal{R}$ and $\mathcal{T}$ are called the reflection and transmission coefficients, respectively. From equation (17.153), the amplitudes of the reflected and transmitted pressure waves are $\mathcal{R}$ and $\mathcal{T}$ times the amplitude of the incident pressure wave, respectively. These relations can be written in more explicit manner as follows (see Lighthill (1978)):

The contribution of the incident wave to the pressure in the parent tube is given by,

$$p_I = P_I \, f \left(t - \frac{x}{c_1} \right), \tag{17.155}$$

where, P_I is an amplitude parameter, and f is a continuous, periodic function whose maximum value is 1. The corresponding contribution to the flow rate is,

$$Q_I = A_1 u = Y_1 P_I f \left(t - \frac{x}{c_1} \right). \tag{17.156}$$

The contributions to pressure from the reflected and transmitted waves to the parent and daughter tubes, respectively, are:

$$p_R = P_R \, g \left(t + \frac{x}{c_1} \right), \quad \text{and,} \quad p_{Tj} = P_{Tj} h_j \left(t - \frac{x}{c_j} \right), \quad (j = 2, 3). \tag{17.157}$$

where P_R and P_T are amplitude parameters, and g and h are are continuous, periodic functions. The corresponding contributions to the flow rates are:

$$Q_R = -Y_1 P_R \, g \left(t + \frac{x}{c_1} \right), \quad \text{and,} \quad Q_{Tj} = Y_j P_{Tj} h_j \left(t - \frac{x}{c_j} \right), \quad (j = 2, 3). \tag{17.158}$$

Therefore, the pressure perturbation in the parent tube is given by equation (17.155) and (17.157) to be:

$$\frac{p}{P_I} = f \left(t - \frac{x}{c_1} \right) + \frac{P_R}{P_I} f \left(t + \frac{x}{c_1} \right), \tag{17.159}$$

and the flow rate, from equations (17.156) and (17.158), is:

$$Q = Y_1 P_I \left[f \left(t - \frac{x}{c_1} \right) - \frac{P_R}{P_I} f \left(t + \frac{x}{c_1} \right) \right]. \tag{17.160}$$

The transmission of energy by the pressure waves is of interest. The rate of work done by the wave motion through the cross section of the tube or equivalently, the rate of transmission of energy by the wave is clearly, $p \, A \, u$ or pQ, which is the same as p^2/Z from equation (17.151). Now we could calculate the incident, reflected and transmitted quantities at the junction. Thus,

$$\text{Rate of energy transmission by incident wave} = \frac{p_I^2}{Z_1}, \qquad (17.161)$$

$$\text{Rate of energy transmission by reflected wave} = \frac{(\mathcal{R}p_I)^2}{Z_1}$$

$$= \mathcal{R}^2 \frac{p_I^2}{Z_1} \qquad (17.162)$$

The quantity $\mathcal{R}^2$ is called the energy reflection coefficient. Similarly, the energy transmission coefficient which is the rate of energy transfer in the two transmitted waves compared with that in the incident wave may be defined by,

$$\frac{\frac{p_{T2}^2}{Z_2} + \frac{p_{T3}^2}{Z_3}}{\frac{p_I^2}{Z_1}} = \frac{Z_2^{-1} + Z_3^{-1}}{Z_1^{-1}} \left(\frac{p_{T2}}{p_I}\right)^2 = \frac{Z_2^{-1} + Z_3^{-1}}{Z_1^{-1}} \mathcal{T}^2, \qquad (17.163)$$

where we have noted that in our case, $p_{T2} = p_{T3}$.

A comparison of equations (17.159) and (17.160) shows that, if we include reflection at bifurcations, the pressure and flow wave forms are no longer of the same shape. Pedley (1980) has offered interesting discussions about the behavior of the waves at the junction. From equation (17.153), for real values of c_j and Y_j, if $\sum Y_j < Y_1$, then the reflected wave has the same sign as the incident wave, and the pressures in the two waves are in phase at $x = 0$. They combine additively to form a large-amplitude fluctuation at the junction, and the effect of the junction is similar to that of a closed end ($P_R = P_I$). If $\sum Y_j > Y_1$, there is a phase change at $x = 0$, the smallest-amplitude pressure fluctuation occurs there, and the junction resembles an open end ($P_R = -P_I$). If $\sum Y_j = Y_1$, there is no reflected wave, and the junction is said to be perfectly matched. Pedley (2000) has noted that the increase in the pressure wave amplitude in the aorta with distance down the vessel may indicate that there is a closed end type of reflection at (or beyond) the iliac bifurcation. Peaking of the pressure pulse is a consequence of closed end type of reflection in a blood vessel.

Waves in more complex systems consisting of many branches may be analyzed by repeated application of the results presented above.

Next, we will study blood flow in curved tubes. Almost all blood vessels have curvature and the curvature affects both the nature (stability) and volume rate of flow.

Flow in a Rigid Walled Curved Tube

Blood vessels are typically curved and the curvature effects have to be accounted for in modelling in order to get a realistic understanding. The aortic arch is a 3D bend twisting through more than $180°$, Ku (1997). In a curved tube, fluid motion is not everywhere parallel to the curved axis of the tube (see Fig. 17.17), secondary motions are generated, the velocity profile is distorted, and there is increased energy

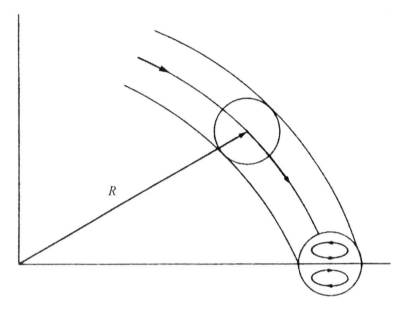

Figure 17.17   Schematic of flow in a curved tube.

dissipation. However, curving of a tube increases the stability of flow, and the critical Reynolds number increases significantly, and a critical Reynolds number of 5,000 is easily obtained (see, McDonald (1974)). Flows in curved tubes are discussed in detail by McConalogue and Srivastava (1968), Singh (1974), Pedley (1980) and by Berger et al. (1983). In this Section, we will concentrate on some of the most important aspects and will focus on the flow in a uniformly curved vessel of small curvature. The wall is considered to be rigid. Pulsatile flow through a curved tube can induce complicated secondary flows with flow reversals and is very difficult to analyze. It may be noted that steady viscous flow in a symmetrical bifurcation resembles that in two curved tubes stuck together. Thus, an understanding gained in studying curved flows will be beneficial in that regard as well.

Consider fully developed, steady, laminar, viscous flow in a curved tube of radius a and a uniform radius of curvature R. Let us employ the toroidal coordinate system (r', α, θ), where r' denotes the distance from the center of the circular cross section of the pipe, α is the angle between the radius vector and the plane of symmetry, and θ is the angular distance of the cross section from the entry of the pipe (see Fig. 17.18). Let the corresponding dimensional velocity components be (u', v', w'). As a fluid particle traverses a curved path of radius R (radius of curvature) with a (longitudinal) speed w' along the θ direction, it will experience a lateral (centrifugal) acceleration of w'^2/R, and a lateral force equal to $m_p \, w'^2/R$, where m_p is the mass of the particle. The radii of curvature of the particle paths near the inner bend, the central axis, and the outer bend will be of increasing magnitude as we move away from the inner bend. Also, due to the no-slip condition, the velocities, w', of particles near the inner and outer bends will be lower, while that of the particle at the central axis will be the highest. The particle at the central axis will experience the highest centrifugal force

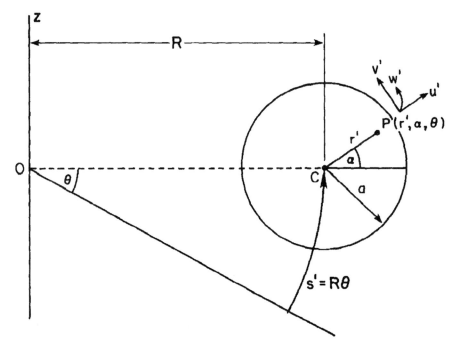

Figure 17.18   Toroidal coordinate system.

while that near the outer bend will experience the least. A lateral pressure gradient will cause the faster flowing fluid near the center to be swept towards the outside of the bend and be replaced at the inside by the slower moving fluid near the wall. In effect, a secondary circulation will be set up resulting in two vortices, called Dean vortices because Dean (1928) was the first to systematically study these secondary motions in curved tubes (see Fig. 17.17). Dean vortices significantly influence the axial flow. The wall shear near the outside of the bend is relatively higher than the (much reduced) wall shear on the inside of the bend. Fully developed flow upstream of or through curved tubes exhibits velocity that skews toward the outer wall of the bend. For most arterial flows, skewing will be toward the outer wall. If the flow into the entrance region of a curved tube is not developed, then the inviscid core of the fluid in the curve can act like a potential vortex with velocity skewing toward the inner wall.

Secondary flow in curved tubes is utilized in heart-lung machines to promote oxygenation of blood (Fung (1997)). In the machine, blood flows inside the curved tube and oxygen flows on the outside. The tube is permeable to oxygen. The secondary flow in the tube stirs up the blood and results in faster oxygenation.

Let us now analyze the flow in a curved tube so as to understand the salient features. Introduce non-dimensional variables, $r = r'/a$, $s = R\theta/a$, $\mathbf{u} = \mathbf{u}'/\bar{W}_0$, and $p = p'/\rho \bar{W}_0^2$, where $\mathbf{u} = (u, v, w)$ is the velocity vector, p is the pressure, ρ is the density, and $\bar{W}_0$ is the mean axial velocity in the pipe. Restrict consideration to the case where the flow is fully developed ($\partial \mathbf{u}/\partial s = 0$). Introduce the dimensionless ratio,

$$\delta = \frac{\text{radius of tube cross section}}{\text{radius of curvature of the centerline}} = \frac{a}{R}, \qquad (17.164)$$

We restrict consideration to a uniformly curved tube, $\delta = $ constant, and with a slight curvature (weakly curved), $\delta \ll 1$. Since δ is a constant, the velocity field is independent of s, the components are functions only of r and θ, and the pressure gradient $\partial p/\partial s$ is independent of s. With δ constant, the only way that the transverse velocities are affected by the axial velocity is through the centrifugal force, and it is the centrifugal force that drives the secondary motion. This means that the centrifugal force terms must be of the same order of magnitude as the viscous and inertial terms in the momentum equation, and this requires rescaling the velocities. The transformation that accomplishes this is $(u, v, w) \rightarrow (\sqrt{\delta}\hat{u}, \ \sqrt{\delta}\hat{v}, \ \hat{w})$. We will also let $s = R\theta/a = \sqrt{1/\delta}\ \tilde{s}$ for convenience.

In the following, we shall omit writing the "^" on u, v, w, and the "~" on s for convenience. When $\delta \ll 1$, the major contribution to the axial pressure gradient may be separated from the transverse component, and we may write,

$$p = p_0(s) + \delta p_1(r, \alpha, s) + \dots, \qquad (17.165)$$

Under all these restrictions, the governing equations become,

$$\frac{\partial u}{\partial r} + \frac{u}{r} + \frac{1}{r}\frac{\partial v}{\partial \alpha} = 0, \qquad (17.166)$$

$$u\frac{\partial u}{\partial r} + \frac{v}{r}\frac{\partial u}{\partial \alpha} - \frac{v^2}{r} - w^2 \cos \alpha = -\frac{\partial p_1}{\partial r} - \frac{2}{\kappa}\frac{1}{r}\frac{\partial}{\partial \alpha}\left(\frac{\partial v}{\partial r} + \frac{v}{r} - \frac{1}{r}\frac{\partial u}{\partial \alpha}\right), \quad (17.167)$$

$$u\frac{\partial v}{\partial r} + \frac{v}{r}\frac{\partial v}{\partial \alpha} + \frac{uv}{r} + w^2 \sin \alpha = -\frac{1}{r}\frac{\partial p_1}{\partial \alpha} + \frac{2}{\kappa}\frac{\partial}{\partial r}\left(\frac{\partial v}{\partial r} + \frac{v}{r} - \frac{1}{r}\frac{\partial u}{\partial \alpha}\right), \quad (17.168)$$

$$u\frac{\partial w}{\partial r} + \frac{v}{r}\frac{\partial w}{\partial \alpha} = -\frac{\partial p_0}{\partial s} + \frac{2}{\kappa}\left(\frac{\partial^2 w}{\partial r^2} + \frac{1}{r}\frac{\partial w}{\partial r} + \frac{1}{r^2}\frac{\partial^2 w}{\partial \alpha^2}\right). \qquad (17.169)$$

The boundary conditions are:

$$u = v = w = 0 \text{ at } r = 1, \quad \text{no singularity at } r = 0. \qquad (17.170)$$

The flow is governed by just one parameter κ in the equations, and it is called the Dean number. It is given by,

$$\kappa = \sqrt{\delta}\ \frac{2\,a\,\bar{W}_0}{\nu} = \sqrt{\delta}\ 2\ Re, \qquad (17.171)$$

where $\bar{W}_0$ is the mean axial velocity in the pipe. The Dean number is the Reynolds number modified by the pipe curvature. The appearance of the numerical constant 2 in the definition of the Dean number is by convention. At higher Dean numbers, the flow can separate along the inner boundary curve.

There are many different definitions of Dean number in the literature and the reader must be careful to see which particular form is being used in any given discussion.

From equation (17.169), $\partial p_0 / \partial s$ is independent of s, and p_0 can be written as $p_0(s) = -G\, s$, where G is a constant. The equation (17.166) admits the existence of a stream function for the secondary flow, ψ, defined by

$$u = \frac{1}{r} \frac{\partial \psi}{\partial \alpha}, \quad v = -\frac{\partial \psi}{\partial r}. \tag{17.172}$$

Substitution of equation (17.172) into equation (17.169) yields,

$$\nabla_1^2 w - \frac{\kappa}{2} \frac{\partial p_0}{\partial s} = \frac{\kappa}{2r} \left(\frac{\partial \psi}{\partial \alpha} \frac{\partial w}{\partial r} - \frac{\partial \psi}{\partial r} \frac{\partial w}{\partial \alpha} \right), \tag{17.173}$$

while elimination of pressure from equations (17.167) and (17.168) yields,

$$\frac{2}{\kappa} \nabla_1^4 \psi - \frac{1}{r} \left(\frac{\partial \psi}{\partial r} \frac{\partial}{\partial \alpha} - \frac{\partial \psi}{\partial \alpha} \frac{\partial}{\partial r} \right) \nabla_1^2 \psi = -2w \left(\sin\alpha \frac{\partial w}{\partial r} + \frac{\cos\alpha}{r} \frac{\partial w}{\partial \alpha} \right), \tag{17.174}$$

where,

$$\nabla_1^2 \psi = \frac{\partial^2}{\partial r^2} + \frac{1}{r} \frac{\partial}{\partial r} + \frac{1}{r^2} \frac{\partial^2}{\partial \alpha^2}. \tag{17.175}$$

The boundary conditions are:

$$\psi = \frac{\partial \psi}{\partial r} = w = 0, \quad at \ \ r = 1. \tag{17.176}$$

The equations (17.173) and (17.174) subject to conditions (17.176) have to be solved.

For small values of Dean number, following Dean (1928), we expand w and ψ in terms of a series in powers of the Dean number as follows:

$$w = \sum_{n=0}^{\infty} \kappa^{2n} w_n(r, \alpha), \quad \text{and,} \quad \psi = \kappa \sum_{n=0}^{\infty} \kappa^{2n} \psi_n(r, \alpha). \tag{17.177}$$

The w_0 term corresponds to Poiseuille flow in a straight tube with rigid walls. The ψ_0 term is $O(\kappa)$. The series expansion in κ is equivalent to the successive approximation of inertia terms in lubrication theory. The leading term in the secondary flow takes the form of a pair of counter rotating helical vortices, placed symmetrically with respect to the plane of symmetry. This flow pattern arises because of a centrifugally induced pressure gradient, approximately uniform over the cross section. The dimensionless volume flux is,

$$\frac{Q}{\pi a^2 \bar{W}} = 1 - 0.0306 \left(\frac{K}{576} \right)^2 + 0.0120 \left(\frac{K}{576} \right)^4 + O\left(K^6 \right), \tag{17.178}$$

where $K = (2a/R)(W_{max}a/\nu)^2 = 2(\kappa)^2$, is another frequently used definition of Dean's number. Here, $W_{max} = 2\bar{W}$; W_{max} and $\bar{W}$ are the maximum and mean velocities, respectively, in a straight pipe of the same radius under the same axial pressure gradient and under fully developed flow conditions. The first term corresponds to the Poiseuille straight pipe solution. The effect of curvature is seen to reduce the flux.

Many other authors define Dean's number by,

$$D = \sqrt{2\delta}\frac{\hat{G}a^2}{\mu}\frac{a}{\nu}, \qquad (17.179)$$

where, $-\hat{G}$ is the dimensional pressure gradient,

$$\hat{G} = -\frac{8\mu\bar{W}}{a^2}. \qquad (17.180)$$

In terms of D, equation (17.178) becomes,

$$\frac{Q}{\pi a^2 \bar{W}} = 1 - 0.0306\left(\frac{D}{96}\right)^4 + 0.0120\left(\frac{D}{96}\right)^8 + O\left(D^{12}\right), \qquad (17.181)$$

Next, consider the friction factor for flow in a curved tube. Let λ_c and λ_s denote the flow resistance in a curved and a straight pipe, respectively, while the flows in them are subject to pressure gradients equal in magnitude. The ratio λ is,

$$\lambda = \frac{\lambda_c}{\lambda_s} = \left(\frac{Q_c}{Q_s}\right)^{-1} = 1 + 0.0306\left(\frac{K}{576}\right)^2 - 0.0110\left(\frac{K}{576}\right)^4 + \ldots, \qquad (17.182)$$

where, Q_c and Q_s are the fluxes in straight and curved pipes, respectively. The flow resistance in a curved tube is not affected by the first order terms and is increased only by higher order terms. With regard to shear stress, the curvature increases axial wall shear on the outside wall and decreases it on the inside, and it also generates a positive secondary shear in the α direction.

The size of the coefficients suggests that the small D expansion is valid for values of D up to about 100 or $K \approx 600$, and the results here are useful only for smaller blood vessels. Pedley points out that in the canine aorta, where $\delta \approx 0.2$, the mean D is greater than 2000. As mentioned earlier, flow in a curved tube is much more stable than that in a straight tube and the critical Reynolds number could be as high as 5000 which corresponds to $K \approx 1.6 \times 10^6$.

For intermediate values of D, only numerical solutions are possible due to the importance of non-linear terms. Numerical results of Collins and Dennis (1975) for developed flow up to a D of 5000 are stated to compare very well with experimental results. At intermediate values of D, a boundary layer develops on the outside wall of the bend where the axial shear is high. The secondary flow in the core is approximately uniform and continues to manifest a two vortex structure. At higher values of D, there is greater distortion of the secondary streamlines. The wall shear at $r = 1$, $\alpha = 0$, is proportional to D ($\approx 0.85D$); see Pedley (2000).

At large Dean numbers, the centers of the two vortices move toward the outer bend, $\alpha = 0$, and the flow is very much reduced compared with a straight pipe for equal magnitude pressure gradients. Detailed studies using advanced computational methods are required to resolve the flow structure at large D. They are as yet unavailable in the published literature.

Pedley (2000) discusses nonuniqueness of curved tube flow results. When D is sufficiently small, the steady flow equations have just one solution and there is a single secondary flow vortex in each half of the tube. However, there is a critical value of D, above which more than one steady solution exists and these may correspond to four vortices, two in each half. Again, detailed computational studies are necessary to resolve these features.

We will next study the flow of blood in collapsible tubes. The role of pressure difference, $(p_e - p(x))$, on the vessel wall will be significant in such flows.

Flow in Collapsible Tubes

At large negative values of the transmural pressure difference (the difference between the pressure inside and the pressure outside), the cross sectional area of a blood vessel is either very small, the lumen being reduced to two narrow channels separated by a flat region of contact between the opposite walls or it may even fall to zero. There is an intermediate range of values of transmural pressure difference in which the cross section is very compliant and even the small viscous or inertial pressure drop of the flow may be enough to cause a large reduction in area, that is, collapse. Collapse occurs in a number of situations and a listing is given by Kamm and Pedley (1989). Collapse occurs, for example, in systemic veins above the heart (and outside the skull), as a result of the gravitational decrease in internal pressure with height; intramyocardial coronary blood vessels during systole; systemic arteries compressed by a sphygmomanometer cuff, or within the chest during cardiopulmonary resuscitation; pulmonary blood vessels in the upper levels of the lung; large intrathoracic airways during forced expiration or coughing; the urethra during micturition and in the ureter during peristaltic pumping. Collapse, therefore occurs both in small and large blood vessels, and as a result both at low and high Reynolds numbers. In certain cases, at high Reynolds number, collapse is accompanied by self-excited, flow-induced oscillations. There is audible sound. For example, Korotkoff sounds heard during sphygmomanometry are associated with this.

A Note on Korotkoff Sounds

Korotkoff sounds, named after Dr. Nikolai Korotkoff, a physician who described them in 1905, are sounds that physicians listen for when they are taking blood pressure. When the cuff of a sphygmomanometer is placed around the upper arm and inflated to a pressure above the systolic pressure, there will be no sound audible because the pressure in the cuff would be high enough to completely occlude the blood flow. If the pressure is now dropped, the first Korotkoff sound will be heard. As the pressure in the cuff is the same as the pressure produced by the heart, some blood will be able to pass through the upper arm when the pressure in the artery rises during systole.

This blood flows in spurts as the pressure in the artery rises above the pressure in the cuff and then drops back down, resulting in turbulence that results in audible sound. As the pressure in the cuff is allowed to fall further, thumping sounds continue to be heard as long as the pressure in the cuff is between the systolic and diastolic pressures, as the arterial pressure keeps on rising above and dropping back below the pressure in the cuff. Eventually, as the pressure in the cuff drops further, the sounds change in quality, then become muted, then disappear altogether when the pressure in the cuff drops below the diastolic pressure. Korotkoff described 5 types of Korotkoff sounds. The first Korotkoff sound is the snapping sound first heard at the systolic pressure. The second sounds are the murmurs heard for most of the area between the systolic and diastolic pressures. The third and the fourth sounds appear at pressures within 10 mmHg above the diastolic blood pressure, and are described as "thumping" and "muting". The fifth Korotkoff sound is silence as the cuff pressure drops below the diastolic pressure. Traditionally, the systolic blood pressure is taken to be the pressure at which the first Korotkoff sound is first heard and the diastolic blood pressure is the pressure at which the fourth Korotkoff sound is just barely audible. There has recently been a move towards the use of the 5th Korotkoff sound (i.e. silence) as the diastolic pressure, as this has been felt to be more reproducible.

Starling Resistor: A Motivating Experiment for Flow in Collapsible Tubes

The study of flows in collapsible tubes is facilitated by a well known experiment carried out under varying conditions by different researchers. In the experiment, a length of uniform collapsible tube is mounted at each end to a shorter length of rigid tube and is enclosed in a chamber whose pressure p_e can be adjusted. Fluid, say water, flows through the tube. The inlet and outlet pressures at the ends of the collapsible tube are p_1 and p_2. The volume rate of flow is Q. The pressures and the flow rate are next varied in a systematic way and the results are noted. The set up described is called a Starling resistor after physiologist Starling (see Fung (1997)). This experiment will enable us understand some aspects of actual flows in physiological systems. There are many different versions of the description of the Starling resistor experiment in the literature. The experiments have been carried out under both steady flow and unsteady flow conditions. We will describe the experiments as reported by Kamm and Pedley (1989).

Case (1): $(p_1 - p_2)$ is increased while $(p_1 - p_e)$ is held constant.

This is accomplished either by reducing p_2 with p_1 and p_e fixed, or by simultaneously increasing p_1 and p_e while p_2 is held constant. With either procedure, Q at first increases, but above a critical value it levels off and the condition of *flow limitation* is reached. In this condition, however much the driving pressure is increased the flow rate remains constant, or may even fall as a result of increasingly severe tube collapse. This experiment is relevant to forced expiration from the lung, to venous return, and to micturition.

Case (2): $(p_1 - p_2)$ or Q is increased while $(p_2 - p_e)$ is held constant at some negative value.

In this case, the tube is collapsed at low flow rates, but starts to open up from the upstream end as Q increases above a critical value, so that the resistance falls and $(p_1 - p_2)$ ceases to rise. This is termed *pressure-drop limitation*. This experiment does not seem to apply to any particular physiological condition.

Case (3): $(p_1 - p_2)$ is held constant while $(p_2 - p_e)$ is decreased from a large positive value.

In this case, the tube first behaves as though it were rigid and the flow rate is nearly constant. Then as $(p_2 - p_e)$ becomes sufficiently negative to produce partial collapse, the resistance rises and Q begins to fall. This experiment is relevant to pulmonary capillary flows.

Case (4): p_e fixed. The outlet end is connected to a flow resistor. The pressure downstream of the flow resistor is fixed (flow is exposed to atmosphere). Thus p_2 is equal to atmospheric pressure plus Q times the fixed resistance. p_1 is varied.

In this case, p_2 varies with Q due to the presence of a fixed downstream resistance. The degree of tube collapse (progressive collapse) also varies with Q for the same reason. At high flow rates, the tube is distended and its resistance is low. As the flow rate is reduced below a critical value the tube starts to collapse. Its resistance and $(p_1 - p_2)$ both increase as Q is decreased. Only when the tube is severely collapsed along most of its length does $(p_1 - p_2)$ start to decrease again as Q approaches zero. When p_1 is approximately equal to p_e, virtually the entire tube is collapsed (Fung (1997)). The tube often flutters in Case 4 (see discussions in Fung).

Case (5): Unsteady flow experiments

Excepting at small Reynolds numbers, there is always some parameter range where flow oscillations occur. The oscillations have a wide variety of modes.

The experiments reveal the importance of a tube law relating transmural pressure difference with the area of cross section of the collapsible tube and the flow and pressure drop limitations when analyzing collapsible tubes. Shapiro (1977a, 1977b) has developed a comprehensive one dimensional theory for steady flow based on a suitable tube law. Kamm and Shapiro (1979) have extended it to unsteady flow in a collapsible tube. In the following, we shall discuss the steady flow theory.

One-Dimensional Flow Treatment

The equations describing one-dimensional flow in a collapsible tube are similar to those in gas dynamics or channel flow of a liquid with a free surface (see,Shapiro (1977a)). Here, we will study the one dimensional steady flow formulation for the collapsible tube. However, before that let us recapitulate the traditional basic equations for one dimensional flow in a smoothly varying elastic tube. (see Section 3.3.2 on Pulse wave propagation in an elastic tube: Inviscid theory).

We had studied flow in an elastic tube with cross section $A(x, t)$ and longitudinal velocity $u(x, t)$. The constant external pressure on the tube was set at p_e. The primary mechanism of unsteady flow in the tube was wave propagation. The transmural pressure difference, $(p - p_e)$ was related to the local cross sectional area by a "tube law" which involved hoop tension, and it may be expressed as,

$$(p - p_e) = \hat{P}(A) \qquad (17.183)$$

where the functional form $\hat{P}$ depends on data. For disturbances of small amplitude and long wave length compared to the tube diameter,

$$A = A_0 + A', \quad p - p_e = \hat{P}(A_0) + p', \quad |A'| \ll A_0, \quad |p'| \ll \hat{P}(A_0) \quad (17.184)$$

and the wave speed is given by,

$$c^2 = \frac{A}{\rho} \frac{d\hat{P}}{dA} = \frac{A}{\rho} \frac{d(p - p_e)}{dA}. \quad (17.185)$$

Tube collapse is associated with negative transmural pressure difference and the pressure difference is supported by bending stiffness of the tube wall (see Fig. 17.19). Contrast this with positive transmural pressure difference discussed earlier which was supported by hoop tension. Following Shapiro (1977a), introduce,

$$\mathcal{P} = \frac{(p - p_e)}{K_p}, \quad \text{and,} \quad \alpha = \frac{A}{A_0}, \quad (17.186)$$

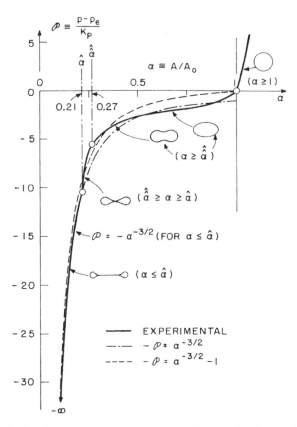

Figure 17.19 Behavior of a collapsible tube. (Reproduced with permission from the American Society of Mechanical Engineers, NY.).

where K_p is a parameter proportional to the bending stiffness of the wall material, and A_0 is the reference area of the tube for zero transmural pressure difference. The pressure difference is supported primarily by the bending stiffness of the tube wall. For a linear elastic tube wall material, K_p is proportional to the modulus of elasticity E, and the bending moment I, as in,

$$K_p \propto EI, \quad I = (h/a_0)^3/(1 - \hat{v}^2), \tag{17.187}$$

where h is wall thickness and $\hat{v}$ is Poisson's ratio.

From a fit of experimental data (see, Shapiro (1977a)), the tube law for flow in a collapsible tube is taken to be,

$$-\mathcal{P} \approx \alpha^{-n} - 1, \quad \text{and} \quad n = \frac{3}{2}. \tag{17.188}$$

For $\mathcal{P} < 0$, the tube is partially collapsed. If the tube is in longitudinal tension, say, T_L, then there will be a local curvature R_L in the longitudinal plane. The effect of T_L is to change p_e by T_L/R_L, and the tube law equation (17.188) will not hold (see, Cancelli and Pedley (1985)). We will here assume that $T_L/R_L \ll (p - p_e)$. Now, if the tube law equation (17.188) and transmural pressure difference are assumed to be uniform along the length of the tube, then with equation (17.185), at any location x, the phase velocity of long area waves is given by

$$c^2 = \frac{A}{\rho} \frac{\partial(p - p_e)}{\partial A} = \left[\frac{nK_p\alpha^{-n}}{\rho}\right], \tag{17.189}$$

for the square of the wave speed.

The assumptions of uniformity of tube law and transmural pressure difference are not valid under most physiological circumstances and these have to be relaxed. The physical causes that negate uniformity include: friction, gravity, variations of external pressure or of muscular tone, longitudinal variations in A_0, and longitudinal changes in the mechanical properties of the tube. To address some of these issues, we consider a more general formulation given by Shapiro.

The flow will still be considered steady, one dimensional, and incompressible.

The governing equations now are:

$$\frac{dA}{A} + \frac{du}{u} = 0, \tag{17.190}$$

and,

$$-Adp - \tau_w s dx - \rho g A dz = \rho A u du = \rho A u^2 \frac{du}{u}, \tag{17.191}$$

where, τ_w is the wall shear stress, s is the perimeter of the tube, z is the elevation in the gravity field g. For the shear stress, Shapiro (1977a) considers the cases of fully developed turbulent flow and fully developed laminar Poiseuille flow in the tube. For turbulent flow,

$$\frac{\tau_w s dx}{A} = \frac{1}{2}\rho u^2 \frac{4 f_T dx}{d_e}, \tag{17.192}$$

where $d_e = 4A/s$ is the equivalent hydraulic diameter and f_T is skin friction co-efficient for turbulent flow, while for laminar flow,

$$\frac{\tau_w \, s \, dx}{A} = \frac{\mu \, u}{d_0} \frac{1}{\alpha} \frac{4 \, f'_L \, dx}{d_0}, \quad \text{where,} \quad f'_L(\alpha) = \left(\frac{A}{A_e}\right) f_L, \qquad (17.193)$$

and d_0 is the diameter for A_0, and f_L is laminar skin friction coefficient.

With equation (17.190), the equation (17.191) may be written,

$$d(p + \rho g z) + \frac{\tau_w s dx}{A} - \rho u^2 \frac{dA}{A} = 0, \qquad (17.194)$$

where the appropriate expression for the shear stress must be introduced depending on the nature of the flow.

Shapiro (1977a) introduces a dimensionless speed-index, S,

$$S = \frac{u}{c}, \quad \text{so that} \quad \left(\frac{dS^2}{S^2}\right) = 2\frac{du}{u} - 2\frac{dc}{c}. \qquad (17.195)$$

This index facilitates in the development of the theory and in the interpretation of results. Its role is comparable in significance to that of Mach number and Froude number in gas dynamics and in free-surface channel flow, respectively (Shapiro (1977a)). By analogy with results of gas dynamics, in steady flow, when $S < 1$ (subcritical), friction causes the area and pressure to decrease in the downstream direction, and the velocity to increase. When $S > 1$ (supercritical), the area and pressure increase along the tube, while the velocity decreases. In general, whatever the effect of changes of A_0, p_e, z, etc. in a subcritical flow, the effect is of opposite sign in supercritical flow. For example, let p_e be increased while all other independent variables such as A_0, elasticity, etc. are held constant. Then A and p will decrease for $S < 1$, but they will increase for $S > 1$. When $S = 1$, choking of flow and flow limitation as at the throat of a Laval nozzle will occur. Again, as in gas dynamics, there is the possibility of continuous transitions from supercritical to subcritical flow, and also rapid transitions from supercritical to subcritical as in shock waves.

In the steady flow problem, the known quantities are dA_0, dp_e, gdz, $f dx$, dK_p, $\partial \mathcal{P}/\partial x$, $\partial \mathcal{P}/\partial \alpha$, while the unknowns are du, dA, dp, $d\alpha$, dS and so on.

In order to develop the final set of equations relating the dependent and independent variables, a number of useful relationships may be established between the differential quantities.

The external pressure is $p_e(x)$, $dp_e = (dp_e/dx) \, dx$, the area $A_0 = A_0(x)$, and $dA_0 = (dA_0/dx) \, dx$. Since $\alpha = A/A_0$,

$$\frac{d\alpha}{\alpha} = \left(\frac{dA}{A} - \frac{dA_0}{A_0}\right). \qquad (17.196)$$

The bending stiffness parameter is $K_p = K_p(x)$, $dK_p = (dK_p/dx) \, dx$, and the tube law is,

$$\mathcal{P} = \frac{p - p_e}{K_p(x)} = \mathcal{P}(\alpha, x), \ \rightarrow \ dp = dp_e + K_p d\mathcal{P} + \mathcal{P} dK_p. \qquad (17.197)$$

The appropriate form of equation (17.185) is,

$$c^2(A, x) = \frac{A}{\rho} \left[\frac{\partial(p - p_e)}{\partial A} \right]_x \ \rightarrow \ c^2(\alpha, x) = \frac{\alpha}{\rho} K_p \left. \frac{\partial \mathcal{P}}{\partial \alpha} \right|_{x=\text{constant}}. \qquad (17.198)$$

In equation (17.197),

$$d\mathcal{P} = \frac{\partial \mathcal{P}}{\partial \alpha} d\alpha + \frac{\partial \mathcal{P}}{\partial x} dx. \qquad (17.199)$$

With equations (17.198) and (17.199), equation (17.197) becomes,

$$dp = dp_e + \rho \, c^2 \frac{d\alpha}{\alpha} + K_p \frac{\partial \mathcal{P}}{\partial x} dx + \mathcal{P} dK_p. \qquad (17.200)$$

With equations (17.198) and (17.197), we get,

$$2 \frac{dc}{c} = \left(1 + \frac{\alpha \, \partial^2 \mathcal{P}/\partial \alpha^2}{\partial \mathcal{P}/\partial \alpha} \right) \frac{d\alpha}{\alpha} + \frac{dK_p}{K_p} + \frac{\alpha K_p}{\rho c^2} \frac{\partial}{\partial x} \left(\frac{\partial \mathcal{P}}{\partial x} \right) dx, \qquad (17.201)$$

and, with equation (17.195), equation (17.196) becomes,

$$\left(\frac{dS^2}{S^2} \right) = -2 \frac{d\alpha}{\alpha} - 2 \frac{dA_0}{A_0} - 2 \frac{dc}{c}. \qquad (17.202)$$

We now have equations (17.194), (17.196), (17.200), (17.201), and (17.202). With these, Shapiro (1977a) has developed a series of equations that relate each dependent variable as a linear sum of terms each containing an independent variable multiplied by appropriate coefficients (influence coefficients by analogy with one-dimensional gas dynamics). A comprehensive listing of equations is provided in the paper by Shapiro. From the listing, the most important dependent variables turn out to be $d\alpha/dx$ and dS^2/dx. Once these are known, other dependent quantities such as $\mathcal{P}, u, c$ etc. may be calculated easily. We now list these equations.

Let us consider cases where $\mathcal{P}$ is just a function of α alone, that is, $\mathcal{P}(\alpha)$. For the tube law,

$$p - p_e(x) = K_p(x)\mathcal{P}(\alpha), \qquad (17.203)$$

the equation governing the variation in α is,

$$\left(1 - S^2 \right) \frac{1}{\alpha} \frac{d\alpha}{dx} = \frac{S^2}{A_0} \frac{dA_0}{dx} - \frac{1}{\rho c^2} \left[\frac{dp_e}{dx} + \rho g \frac{dz}{dx} + \mathcal{R}Q + \mathcal{P} \frac{dK_p}{dx} \right], \qquad (17.204)$$

where $\mathcal{R}$ is viscous resistance per unit length (laminar or turbulent) and Q is flow rate, and the equation governing the speed index is,

$$\left(1 - S^2\right) \frac{1}{S^2} \frac{dS^2}{dx} = \frac{1}{A_0} \frac{dA_0}{dx} \left[-2 + (2 - \mathcal{M}) S^2\right]$$

$$+ \frac{\mathcal{M}}{\rho c^2} \left[\frac{dp_e}{dx} + \rho g \frac{dz}{dx} + \mathcal{R}Q \right]$$

$$+ \frac{1}{\rho c^2} \frac{dK_p}{dx} \left[\mathcal{M}\mathcal{P} - \left(1 - S^2\right) \alpha \frac{d\mathcal{P}}{d\alpha} \right], \quad (17.205)$$

where

$$\mathcal{M} = 3 + \frac{\alpha \partial^2 \mathcal{P}/\partial \alpha^2}{\partial \mathcal{P}/\partial \alpha}. \quad (17.206)$$

The equations for $d\alpha/dx$ and dS^2/dx are coupled and must be solved simultaneously by using numerical procedures. Shapiro (1977a) has included results for several limit cases. These include several examples in which a smooth transition through the critical condition $S = 1$ is possible, that is, continuous passage of flow from regime $S < 1$ through $S = 1$ into $S > 1$ might occur. Fig. 17.20 shows the transition from subcritical to supercritical flow by means of a minimum in the neutral area A_0. The pressure decreases continuously in the axial direction, and the area A of the deformed cross section would also decrease continuously in the axial direction. Fig. 17.20 shows the transition through $S = 1$ caused by a weight or clamp, a sphincter or pressurized cuff, due to changing p_e. The fluid pressure and the area, both decrease continuously in the axial direction. $S = 1$ occurs in the region where a sharp constriction exists.

Pedley (2000) points out that when $S = 1$, the right hand side of equation (17.205) is $-\mathcal{M}$ times that of equation (17.204). Therefore, at $S = 1$, it is possible for $d\alpha/dx$ or dS^2/dx to be non-zero as long as the right hand sides are zero. Of the terms on the right hand side, RQ is associated with friction and is always positive. This means that at least one of $d(p_e + \rho g z)/dx$, dK_p/dx or $-dA_0/dx$ should be negative, that is, the external pressure, the height or the stiffness should decrease with x or the undisturbed cross-sectional area should increase. An example where dz/dx in a vertical collapsible tube is negative ($= -1$) is the jugular vein of an upright giraffe and this problem has been discussed in detail by Pedley. Apparently, the giraffe jugular vein is normally partially collapsed!

In the next section, we shall learn about the modelling of a Casson fluid flow in a tube. We recall that blood behaves as a non-Newtonian fluid at low shear rates below about 200/s, and the apparent viscosity increases to relatively large magnitudes at low rates of shear. The modelling of such a fluid flow is important and will enable us understand blood flow at various shear rates.

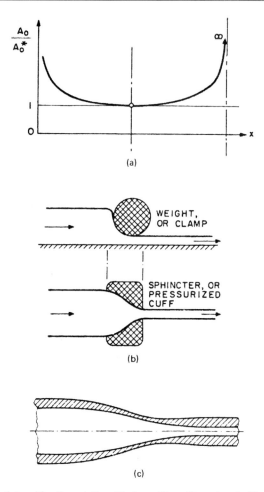

Figure 17.20 Smooth transition through the critical condition. (Reproduced with permission from the American Society of Mechanical Engineers, NY.).

Laminar Flow of a Casson Fluid in a Rigid Walled Tube

As shear rates decrease below about $200/s$, the apparent viscosity of blood rapidly increases. (See Fig. 17.7). As mentioned earlier, the variation of shear stress in blood flow with shear rate is accurately expressed by the equation (17.6):

$$\tau^{1/2} = \tau_y^{1/2} + K_c \dot{\gamma}^{1/2}, \quad \text{for} \ \ \tau \geq \tau_y, \ \ \text{and} \ \ \dot{\gamma} = 0, \ \ \text{for} \ \ \tau < \tau_y, \qquad (17.207)$$

where τ_y and K_c are determined from viscometer data. The yield stress τ_y for normal blood at 37 °C is about 0.04 $dynes/\text{cm}^2$. In modelling the flow, this behavior must be included.

Consider the steady laminar axi-symmetric flow of a Casson fluid in a rigid walled, horizontal, cylindrical tube under the action of an imposed pressure gradient,

$(p_1 - p_2)/L$. We shall employ cylindrical coordinates (r, θ, x) with velocity components $(u_r, u_\theta$, and $u_x)$, respectively. With the assumption of axi-symmetry, $\left(u_\theta = 0, \text{ and } \frac{\partial}{\partial \theta} = 0\right)$. For convenience, we write u_r component as v, and we omit the subscript x in u_x.

The maximum shear stress in the flow, τ_w, would be at the vessel wall. If the magnitude of τ_w is equal to or greater than the yield stress, τ_y, then there will be flow. We may estimate the minimum pressure gradient required to cause flow of a yield stress fluid in a cylindrical tube by a straightforward force balance on a cylindrical volume of fluid of radius r and length Δx. For steady flow, the viscous force opposing motion must be balanced by the force due to the applied pressure gradient. Thus,

$$\tau_{rx} \, 2\pi r \Delta x = -\pi r^2 (p|_{x+\Delta x} - p|_x), \qquad (17.208)$$

and, as $\Delta x \to 0$,

$$\tau_{rx}(r) = \frac{r}{2} \frac{dp}{dx} = \frac{(p_1 - p_2)r}{2L}. \qquad (17.209)$$

The shear stress at the wall, $\tau_w = -(a/2)(dp/dx) = (p_1 - p_2)a/2L$. When τ_y is equal to or less than τ_w, there will be fluid motion. The minimum pressure differential to cause flow is given by $(p_1 - p_2)|_{\min} = 2L\tau_y/a$. With $\tau_y = 0.04 \ dynes/cm^2$, for a blood vessel of $L/a = 500$, the minimum pressure drop required for flow is $40 \ dynes/cm^2$ or 0.03 mmHg. Recall that during systole, the typical pressures in the aorta and the pulmonary artery rise to 120 mm Hg and 24 mm Hg, respectively.

For axi-symmetric blood flow in a cylindrical tube, at low shear rates, the fully developed flow is noted to consist of a central core region where the shear rate is zero and the velocity profile is flat, surrounded by a region where the flow has a varying velocity profile (see Fig. 17.21). In the core, the fluid moves as if it were a solid body (also called, plug flow).

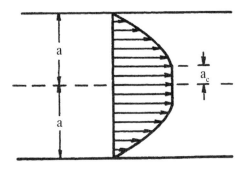

Figure 17.21   Velocity profile for axi-symmetric blood flow in a circular tube.

Let the radius of this core region be a_c. Then,

$$\tau = \tau_y \text{ at } r = a_c, \text{ and } \dot{\gamma} = 0 \text{ for } 0 \le r < a_c,$$

$$a_c = 2L\tau_y/(p_1 - p_2) = a\left(\frac{\tau_y}{\tau_w}\right),$$

$$\tau^{1/2} = \tau_y^{1/2} + K_c\dot{\gamma}^{1/2} \text{ for } a_c < r \le a. \tag{17.210}$$

In the core region, $\dot{\gamma} = 0 \Rightarrow (du/dr) = 0 \Rightarrow u = \text{constant} = u_c$ (say). Outside the core region, the velocity is a function of r only, and,

$$\dot{\gamma} = -\frac{du}{dr} = \frac{\left[\tau + \tau_y - 2\sqrt{\tau\tau_y}\right]}{K_c^2} \tag{17.211}$$

Let $(p_1 - p_2) = \Delta p$, $\tau = \Delta p\, r/2L$, and $\tau_y = \Delta p\, a_c/2L$. From equation (17.211),

$$-\frac{du}{dr} = \frac{1}{2K_c^2}\frac{\Delta p}{L}\left(r + a_c - 2\sqrt{ra_c}\right) \tag{17.212}$$

By integration,

$$u = \frac{1}{2K_c^2}\frac{\Delta p}{L}\left(\frac{4}{3}\sqrt{a_c r^3} - \frac{r^2}{2} - a_c r + C\right), \tag{17.213}$$

where C is the integration constant. With the no-slip boundary condition at the wall of the vessel, $u = 0$ at $r = a$,

$$C = -\left(\frac{4}{3}\sqrt{a_c a^3} - \frac{a^2}{2} - a_c a\right). \tag{17.214}$$

Therefore,

$$u = \frac{1}{4K_c^2}\frac{\Delta p}{L}\left[(a^2 - r^2) - \frac{8}{3}\sqrt{a_c}\left(\sqrt{a^3} - \sqrt{r^3}\right) + 2a_c(a - r)\right], \tag{17.215}$$

in $(a_c \le r \le a)$. With $u = u_c$ at $r = r_c$, in terms of τ_w and τ_y, the equation (17.215) becomes,

$$u = \frac{a\tau_w}{2K_c^2}\left\{\left[1 - \left(\frac{r}{a}\right)^2\right] - \frac{8}{3}\sqrt{\frac{\tau_y}{\tau_w}}\left[1 - \left(\frac{r}{a}\right)^{3/2}\right] + 2\left(\frac{\tau_y}{\tau_w}\right)\left(1 - \frac{r}{a}\right)\right\}, \tag{17.216}$$

in $(a_c \le r \le a)$. We get the velocity in the core, u_c, by setting,

$$\left(\frac{r}{a}\right) = \left(\frac{a_c}{a}\right) = \left(\frac{\tau_y}{\tau_w}\right), \tag{17.217}$$

in equation (17.216). In terms of pressure gradient, a, and a_c, u_c becomes,

$$u_c = \frac{1}{4K_c^2} \frac{\Delta p}{L} \left(\sqrt{a} - \sqrt{a_c}\right)^3 \left(\sqrt{a} + \frac{1}{3}\sqrt{a_c}\right). \qquad (17.218)$$

The volume rate of flow is given by,

$$Q = \pi a_c^2 u_c + \int_{a_c}^{a} 2\pi r u \, dr. \qquad (17.219)$$

After considerable algebra,

$$Q = \frac{\pi}{8} \frac{1}{K_c^2} \frac{\Delta p}{L} a^4 \left[1 - \frac{16}{7}\left(\frac{a_c}{a}\right)^{1/2} + \frac{4}{3}\left(\frac{a_c}{a}\right) - \frac{1}{21}\left(\frac{a_c}{a}\right)^4\right]. \qquad (17.220)$$

The Casson model predicts results that are in very good agreement with experimental results for blood flow over a large range of shear rates (see, Charm and Kurland (1974)).

Pulmonary Circulation

Pulmonary circulation is the movement of blood from the heart, to the lungs, and back to the heart again. The veins bring oxygen depleted blood back to the right atrium. The contraction of the right ventricle ejects blood into the pulmonary artery. In the human heart, the main pulmonary artery begins at the base of the right ventricle. It is short and wide—approximately 5 cm in length and 3 cm in diameter, and extends about 4 cm before it branches into the right and left pulmonary arteries that feed the two lungs. The pulmonary arteries are larger in size and more distensible than the systemic arteries and the resistance in pulmonary circulation is lower. In the lungs, red blood cells release carbon dioxide and pick up oxygen during respiration. The oxygenated blood then leaves the lungs through the pulmonary veins, which return it to the left heart, completing the pulmonary cycle. The pulmonary veins, like the pulmonary arteries, are also short, but their distensibility characteristics are similar to those of the systemic circulation (Guyton (1968)). The blood is then distributed to the body through the systemic circulation before returning again to the pulmonary circulation. The pulmonary circulation loop is virtually bypassed in fetal circulation. The fetal lungs are collapsed, and blood passes from the right atrium directly into the left atrium through the foramen ovale, an open passage between the two atria. When the lungs expand at birth, the pulmonary pressure drops and blood is drawn from the right atrium into the right ventricle and through the pulmonary circuit.

The rate of blood flow through the lungs is equal to the cardiac output except for the 1 to 2% that goes through the bronchial circulation (Guyton (1968)). Since almost the entire cardiac output flows through the lungs, the flow rate is very high. However, the low pulmonic pressures generated by the right ventricle are still sufficient to maintain this flow rate because pulmonary circulation involves a much shorter flow path than systemic circulation, and the pulmonary arteries are, as noted earlier, are larger and more distensible.

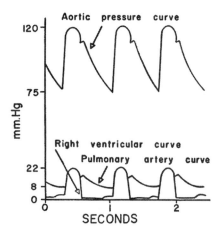

Figure 17.22 Pressure pulse contours in the right ventricle, and pulmonary artery. (Reproduced with permission from Guyton, A. C. and Hall, J. E. (2000) *Textbook of Medical Physiology*, W. B. Saunders Company, Philadelphia).

The nutrition to lungs themselves are supplied by bronchial arteries which are a part of systemic circulation. The bronchial circulation empties into pulmonary veins and returns to the left atrium by passing alveoli.

The Pressure Pulse Purve in the Right Ventricle

The pressure pulse curves of the right ventricle and pulmonary artery are illustrated in the Fig. (17.22). As described by Guyton (1968), approximately 0.16 second prior to ventricular systole, the atrium contracts, pumping a small quantity of blood into the right ventricle, and thereby causing about 4 mmHg initial rise in the right ventricular diastolic pressure even before the ventricle contracts. Following this, the right ventricle contracts, and the right ventricular pressure rises rapidly until it equals the pressure in the pulmonary artery. The pulmonary valve opens, and for approximately 0.3 second blood flows from the right ventricle into the pulmonary artery. When the right ventricle relaxes, the pulmonary valve closes, and the right ventricular pressure falls to its diastolic level of about zero. The systolic pressure in the right ventricle of the normal human being averages approximately 22 mmHg, and the diastolic pressure averages about 0 to 1 mmHg.

Effect of Pulmonary Arterial Pressure on Pulmonary Resistance

At the end of systole, the ventricular pressure falls while the pulmonary arterial pressure remains elevated, then falls gradually as blood flows through the capillaries of the lungs. The pulse pressure in the pulmonary arteries averages 14 mmHg which is almost two thirds as much as the systolic pressure. Fig. (17.23) shows the variation in pulmonary resistance with pulmonary arterial pressure. At low arterial pressures, pulmonary resistance is very high and at high pressures the resistance falls to low values. The rapid fall is due to the high distensibility of the pulmonary vessels.

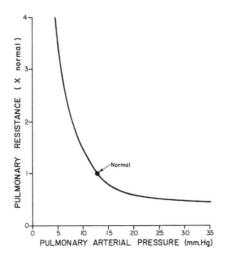

Figure 17.23 Effect of pulmonary arterial pressure on pulmonary resistance. (Reproduced with permission from Guyton, A. C. and Hall, J. E. (2000) *Textbook of Medical Physiology*, W. B. Saunders Company, Philadelphia).

The ability of lungs to accommodate greatly increased blood flow with little increase in pulmonary arterial pressure helps to conserve the energy of the heart. As described by Guyton, the only reason for flow of blood through the lungs is to pick up oxygen and to release carbon dioxide. The ability of pulmonary vessels to accommodate greatly increased blood flow without an increase in pulmonary arterial pressure accomplishes the required gaseous exchange without overworking the right ventricle.

In the earlier sections, we have discussed several modelling procedures in relation to systemic blood circulation. The modelling of the blood flow in pulmonary vessels are similar to what we studied in those sections.

A discussion of gas and material exchange in the capillary beds is beyond the scope of this introductory chapter and a reference may be made to the article by Grotberg (1994) for details.

4. *Introduction to the Fluid Mechanics of Plants*

Plant life comprises 99% of the earth's biomass (Bidwell (1974), Rand (1983)).

The basic unit of a plant is a plant cell. Plant cells are formed at meristems, and then develop into cell types which are grouped into tissues. Plants have three tissue types: 1) Dermal; 2) Ground; and 3) Vascular. Dermal tissue covers the outer surface and is composed of closely packed epidermal cells that secrete a waxy material that aids in the prevention of water loss. The ground tissue comprises the bulk of the primary plant body. Parenchyma, collenchyma, and sclerenchyma cells are common in the ground tissue. Vascular tissue transports food, water, hormones and minerals within the plant.

Basically, a plant has two organ systems: 1) the shoot system, and 2) the root system. The shoot system is above ground and includes the organs such as leaves, buds, stems, flowers and fruits. The root system includes those parts of the plant below ground, such as the roots, tubers, and rhizomes. There is transport between the roots and the shoots (see Fig. (17.24)).

Transport in plants occurs on three levels: (1). The uptake and loss of water and solutes by individual cells, (2). Short-distance transport of substances from cell to cell at the level of tissues or organs, and, (3). Long-distance transport of sap within xylem and phloem at the level of the whole plant.

The transport occurs as a result of gradients in chemical concentration (Fickian diffusion), hydrostatic pressure, and gravitational potential. These three driving potentials are grouped under one single quantity, the water potential. The water potential is designated ψ, and

$$\psi = p - RTc + \rho g z, \tag{17.221}$$

where p is hydrostatic pressure (*bar*), R is gas constant ($= 83.141 \text{cm}^3 - \text{bar}/\text{mole } K$), T is temperature (K), c is the concentration of all solutes in assumed

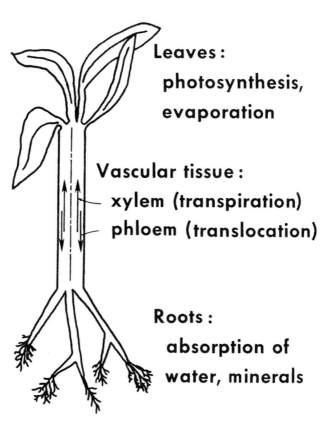

Figure 17.24 Overview of plant fluid mechanics. (Reproduced with permission from *Annual Review of Fluid Mechanics*, Vol. 15 ©1983 *Annual Reviews www.AnnualReviews.org.*).

dilute solution (mole/cm^3), ρ is density of water (g/cm^3), g is acceleration due to gravity (= 980 cm/sec^2), and z is height (cm). ψ is in *bars* (Conversion: 1 *bar* = 10^6 *dyne*/cm^2).

Transport at the cellular level in a plant depends on the selective permeability of plasma membranes which controls the movement of solutes between the cell and the extracellular solution. Molecules move down their concentration gradient across a membrane without the direct expenditure of metabolic energy (Fickian diffusion). Transport proteins embedded in the membrane speed up the movement across the membrane. Differences in water potential, ψ, drive water transport in plant cells. Uptake or loss of water by a cell occurs by osmosis across a membrane. Water moves across a membrane from a higher water potential to a lower water potential. If a plant cell is introduced into a solution with a higher water potential than that of the cell, osmotic uptake of water will cause the cell to swell. As the cell swells, it will push against the elastic wall, creating a "turgor" pressure inside the cell. Loss of water causes loss of turgor pressure and may result in wilting.

In contrast to the human circulatory system, the vascular system of plants is open. Unlike the blood vessels of human physiology, the vessels (conduits) of plants are formed of individual plant cells placed adjacent to one another. During cell differentiation the common walls of two adjacent cells develop pores which permit fluid to pass between them. Vascular tissue includes xylem, phloem, parenchyma, and cambium cells. Xylem and phloem make up the big transportation system of vascular plants. Long distance transport of materials (such as nutrients) in plants is driven by the prevailing pressure gradient.

In this section we will restrict attention to the vascular system that includes xylem and phloem cells.

Xylem

The term Xylem applies to woody walls of certain cells of plants. Xylem cells tend to conduct water and minerals from roots to leaves. Generally speaking, the xylem of a plant is the system of tubes and transport cells that circulates water and dissolved minerals. Xylem is made of vessels that are connected end to end to enable efficient transport. The xylem contains tracheids and vessel elements (see Fig. (17.25) from Rand (1983)). Xylem tissue dies after one year and then develops a new (rings in the tree trunk).

Water and mineral salts from soil enter the plant through the epidermis of roots, cross the root cortex, pass into the stele, and then flow up xylem vessels to the shoot system. The xylem flow is also called transpirational flow. Perforated end walls of xylem vessel elements enhance the bulk flow.

The movement of water and solutes through xylem vessels occurs due to a pressure gradient. In xylem, it is actually tension (negative pressure) that drives long-distance transport. Transpiration (evaporation of water from a leaf) reduces pressure in the leaf xylem and creates a tension that pulls xylem sap upward from the roots. While transpiration enables the pull, the cohesion of water due to hydrogen bonding transmits the upward pull along the entire length of the xylem from the leaves to the root tips. The pull extends down only through an unbroken chain of

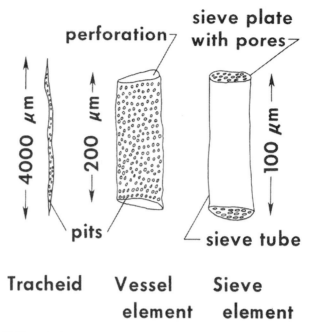

Figure 17.25   Fluid-conducting cells in the vascular tissue of plants. (Reproduced with permission from *Annual Review of Fluid Mechanics*, Vol. 15 ⓒ 1983 *Annual Reviews www.AnnualReviews.org.*).

water molecules. Cavitation, formation of water vapor pockets in the xylem vessel, may break the chain. Cavitation will occur when xylem sap freezes in water and as a result the vessel function will be compromised. Absorption of solar energy drives transpiration by causing water to evaporate from the moist walls of mesophyll cells of a leaf and by maintaining a high humidity in the air spaces within the leaf. To facilitate gas exchange between the inner parts of leaves, stems, and fruits, plants have a series of openings known as stomata. These enable exchange of water vapor, oxygen and carbon dioxide.

The pressure gradient for transpiration flow is essentially created by solar power, and in principle, a plant expends no energy in transporting xylem sap up to the leaves by bulk flow. The detailed mechanism of transpiration from a leaf is very complicated and depends on the interplay of adhesive and cohesive forces of water molecules at mesophyll cell—air space interfaces resulting in surface tension gradients and capillary forces. This will not be discussed in this section.

Xylem sap flows upward to veins that branch throughout each leaf, providing each with water. Plants lose a huge amount of water by transpiration—an average-sized maple tree loses more than about 200 liters of water per hour during the summer. Flow of water up the xylem replaces water lost by transpiration and carries minerals to the shoots. At night, when transpiration is very low, root cells are still expending energy to pump mineral ions into the xylem, accumulation of minerals in the stele lowers water potential, generating a positive pressure, called root pressure, that forces fluid up the xylem. It is the root pressure that is responsible for guttation, the exudation of water droplets that can be seen in the morning on tips of grass blades or leaf margins of some plants. Root pressure is not the main mechanism driving the ascent of xylem sap. It can force water upward by only a few meters, and many plants generate no root pressure at all. Small plants may use root pressure to refill xylem vessels in spring. Thus, for the most part, xylem sap is not pushed from below but pulled upward by the leaves.

Xylem Flow

Water and minerals absorbed in the roots are brought up to the leaves through the xylem. The upward flow in the xylem (also called the transpiration flow) is driven by evaporation at the leaves. In the xylem, the flow may be treated as quasi-steady. The rigid tube model for flow description is appropriate because plant cells have stiff walls. The xylem is about 0.02 mm in radius and the typical values for flow are, velocity 0.1 cm/s, the kinematic viscosity of the fluid 0.01 cm^2/s and the Reynolds number, $Re = ud/v$ is 0.04. In view of the low Reynolds number, the Stokes flow in a rigid tube approximation is appropriate.

Phloem

Phloem cells are usually located outside the xylem and conduct food from leaves to rest of the plant. The two most common cells in the phloem are the companion cells and sieve cells. Phloem cells are laid out end-to-end throughout the plant to form long tubes with porous cross walls between cells. These tubes enable translocation of the sugars and other molecules created by the plant during photosynthesis. Phloem flow is also called translocation flow. Phloem sap is an aqueous solution with sucrose as the most prevalent solute. It also contains minerals, amino acids, and hormones. Dissolved food, such as sucrose, flows through the sieve cells. In general, sieve tubes carry food from a sugar source (for example, mature leaves) to a sugar sink (roots, shoots or fruits). A tuber or a bulb, may be either a source or a sink, depending on the season. Sugar must be loaded into sieve-tube members before it can be exported to sugar sinks. Companion cells pass sugar they accumulate into the sieve-tube members via plasmodesmata. Translocation through the phloem is dependent on metabolic activity of the phloem cells (in contrast to transport in the xylem).

Unlike the xylem, phloem is always alive. In contrast to xylem sap, the direction that phloem sap travels is variable depending on locations of source and sink.

The pressure-flow hypothesis is employed to explain the movement of nutrients through the phloem. It proposes that water containing nutrient molecules flows under pressure through the phloem. The pressure is created by the difference in water concentration of the solution in the phloem and the relatively pure water in the nearby xylem ducts.

At their "source"—the leaves—sugars are pumped by active transport into the companion cells and sieve elements of the phloem. The exact mechanism of sugar transport in the phloem is not known, but it cannot be simple diffusion. As sugars and other products of photosynthesis accumulate in the phloem, the water potential in the leaf phloem is decreased and water diffuses from the neighboring xylem vessels by osmosis. This increases the hydrostatic pressure in the phloem. Turgor pressure builds up in the sieve tubes (similar to the creation of root pressure). Water and dissolved solutes are forced downwards to relieve the pressure. As the fluid is pushed down (and up) the phloem, sugars are removed by the cortex cells of both stem and root (the "sinks") and consumed or converted into starch. Starch is insoluble and exerts no osmotic effect. Therefore, the osmotic pressure of the contents of the phloem decreases. Finally, relatively pure water is left in the phloem. At the same time, ions are being pumped into the xylem from the soil by active transport, reducing the water potential in the xylem. The xylem now has a lower water potential than the phloem, so water diffuses by osmosis from the phloem to the xylem. Water and its dissolved ions are pulled up the xylem by tension from the leaves. Thus it is the pressure gradient between "source" (leaves) and "sink" (shoot and roots) that drives the contents of the phloem up and down through the sieve tubes.

Phloem Flow

Phloem flow occurs mainly through cells called sieve tubes which are arranged end to end and are joined by perforated cell walls called sieve plates (see Fig. (17.26) from Rand and Cooke (1978)). As a model of Phloem flow, Rand et al. (1980) have derived an approximate formula for the pressure drop for flow through a series of sieve tubes with periodically placed sieve sieve plates with pores (see Fig. (17.27) from Rand et al. (1980)). The approximation arises from treating the transport through the pore as creeping conical flow (see (Happel and Brenner, 1983)).

The approximate formula given by Rand et al. (1980) is:

$$\Delta p = \frac{8\mu Q}{\pi a^4}\left[L + \frac{\ell}{N}\left(\frac{a}{r}\right)^4\right] + 2\,\Delta p', \quad \text{where,}$$

$$\Delta p' = \frac{8\mu Q}{\pi a^3}\left(\frac{a_e}{r}\right)\left\{0.57N\left[\left(\frac{a_e}{r}\right)^3 - 1\right] - 1.5\left(1 - \frac{r}{a_e}\right)\right\}. \quad (17.222)$$

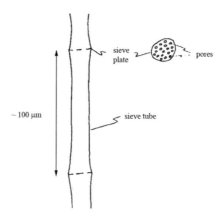

Figure 17.26   Sieve tube with sieve plate. (Reproduced with permission from the American Society of Agricultural Engineers, MI.).

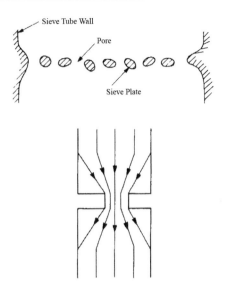

Figure 17.27 Sieve tube with pores and stream lines for conical flow through one pore. (Reproduced with permission from the American Society of Agricultural Engineers, MI.).

In equation (17.222), Δp is the pressure drop due to one sieve tube and one sieve plate, μ is the viscosity of fluid in $(g/cm - s)$, Q is the flow rate in (cm^3/s), N is the number of pores in sieve plate, a is sieve tube radius in cm, r is average radius of sieve pore in cm, L is the sieve tube length in cm, ℓ is sieve plate thickness in cm, and the effective tube radius $a_e = a/\sqrt{N}$.

Rand et al. (1980) note that the approximate formula has not been tested for $N \neq 1$.

Exercises

1. Consider steady laminar flow of a Newtonian fluid in a long, cylindrical, elastic tube of length L. The radius of the tube at any cross section is $a = a(x)$. Poiseuille's formula for the flow rate is a good approximation in this case.

(a) Develop an expression for the outlet pressure $p(L)$ in terms of the higher inlet pressure, the flow rate $\dot{Q}$, fluid viscosity μ, and $a(x)$.

(b) For a pulmonary blood vessel, we may assume that the pressure-radius relationship is linear: $a(x) = a_0 + \dfrac{\alpha p}{2}$, where a_0 is the tube radius when the transmural pressure is zero and α is the compliance of the tube. For a tube of length L, show that

$$\dot{Q} = \frac{\pi}{20\mu\alpha L}\left\{[a(0)]^5 - [a(L)]^5\right\},$$

where $a(0)$ and $a(L)$ are the values of $a(x)$ at $x = 0$ and $x = L$, respectively.

2. For pulsatile flow in a rigid cylindrical tube of length L, the pressure drop Δp may be expressed as: $\Delta p = f(L, a, \rho, \mu, \omega, U)$, where a is tube radius, ρ is

density, μ is viscosity, ω is frequency, and U is the average velocity of flow. Using dimensional analysis, show that

$$\frac{\Delta p}{\rho U^2} = C_1 \left(\frac{L}{a}\right)^{C_2} (Re)^{C_3} (St)^{C_4},$$

where C_1, C_2, C_3, and C_4 are constants, Re is Reynolds number, and St is Strouhal number defined as $a\omega/U$.

3. Localized narrowing of an artery may be caused by the formation of arthero sclerotic plaque in that region. Such localized narrowing is called stenosis. It is important to understand the flow characteristics in the vicinity of a stenosis. Flow in a tube with mild stenosis may be approximated by axi-symmetric flow through a converging-diverging tube. In this context, follow the details described in the paper by B.E. Morgan and D.F. Young, Bull. Math. Biol. 36, (1974), pp. 39–53, and obtain expressions for the velocity profile and wall shear stress.

4. Shapiro in his analysis of the steady flow in collapsible tubes (Trans. ASME, August 1977, pp. 126–147) has developed a series of equations that relate the dependent variables $du, dA, dp, d\alpha, dS$ etc., with the independent variables such as dA_0, dp_e, $g\ dz$, $f_T\ dx$ etc. In Section IV of that study, explicit calculations of certain simple flows are presented. In particular, consider pure pressure-gravity flows. Discuss the flow behavior patterns in this case.

5. Consider the Power-law model to describe the non-Newtonian behavior of blood. In this model, $\tau = \mu\dot{\gamma}^n$, where τ is the shear stress and the $\dot{\gamma}$ is the rate of shearing strain. Determine the flux for the flow of such a fluid in a rigid cylindrical tube of radius R. Show that when $n = 1$, the results correspond to the Poiseuille flow.

6. Consider the Herschel-Bulkely model to describe the non-Newtonian behavior of blood. In this model,

$$\tau = \mu\dot{\gamma}^n + \tau_0, \quad \tau \geq \tau_0$$
$$\dot{\gamma} = 0, \quad \tau < \tau_0$$

Determine the flux for the flow of such a fluid in a rigid cylindrical tube of radius R. Show that in the limit $\tau_0 = 0$, the results for the Herschel-Bulkley model coincide with those for the Power-law model.

Acknowledgment

The help received from Dr. K. Mukundakrishnan and Mrs. Olivia Brubaker during the development of this chapter is gratefully acknowledged.

Literature Cited

Baish, J. W., P. S. Ayyaswamy and K.R. Foster (1986a). "Small scale temperature fluctuations in perfused tissue during local hyperthermia." *J. BioMech. Eng.* **108**: 246–250.
Baish, J. W., P. S. Ayyaswamy and K. R. Foster (1986b). "Heat transport mechanisms in vascular tissues: A model comparison." *J. BioMech. Eng.* **108**: 324–331.

Berger, S. A., L. Talbot and L.-S. Yao (1983). "Flow in curved pipes." *Annual Review of Fluid Mechanics* **15**: 461–512.

Bidwell, R. G. S. (1974). *Plant Physiology*, New York: MacMillan.

Brown, B. H., R. H. Smallwood, D. C. Barber, P. V. Lawford, and D. R. Hose (1999). *Medical Physics and Biomedical Engineering*, London: Institute of Physics Publishing.

Cancelli, C. and T. J. Pedley (1985). "A separated flow model for collapsible tube oscillations." *Journal of Fluid Mechanics* **157**: 375–404.

Chandran, K. B., A. P. Yoganathan and S.E. Rittgers (2007). *Biofluid Mechanics—The Human Circulation*, Boca Raton, FL: Taylor & Francis.

Charm, S. E. and G. S. Kurland (1974). *Blood Flow and Microcirculation*, New York: John Wiley & Sons.

Collins, W. M. and S. C. R. Dennis (1975). "The steady motion of a viscous fluid in a curved tube." *Q. J. Mech. Appl. Math.* **28**: 133–156.

Dean, W. R. (1928). "The streamline motion of fluid in a curved pipe." *Philosophical Magazine,* Series 7, **30**: 673–693.

Fournier, R. L. (2007). *Basic Transport Phenomena in Biomedical Engineering*, New York: Taylor & Francis.

Fung, Y. C. (1993). *Biomechanics: Mechanical Properties of Living Tissues, Second Edition*, Boca Raton, FL: Springer.

Fung, Y. C. (1997). *Biomechanics: Circulation, Second Edition*, Boca Raton, FL: Springer.

Goldsmith, H. L., G. R. Cokelet and P. Gaehtgens (1989). "Robin Fahraeus: Evolution of his concepts in cardiovascular physiology." *Am. J of Physiology - Heart and Circulatory Physiology* **257(3)**: H1005–H1015.

Grotberg, J. B. (1994). "Pulmonary flow and transport phenomena." *Annual Review of Fluid Mechanics* **26**: 529–571.

Guyton, A. C. (1968). *Textbook of Medical Physiology*, Philadelphia: W. B. Saunders Company.

Happel, J. and H. Brenner (1983). *Low Reynolds Number Hydrodynamics*, New York: McGraw-Hill.

He, X. and D. N. Ku (1994). "Unsteady entrance flow development in a straight tube." *Journal of Biomechanical Engineering* **116**: 355–360.

Kamm, R. D. and T. J. Pedley (1989). "Flow in collapsible tubes: A brief review." *Journal of Biomechanical Engineering* **111**: 177–179.

Kamm, R. D. and A. H. Shapiro (1979). "Unsteady flow in a collapsible tube subjected to external pressure or body forces." *Journal of Fluid Mechanics* **95**: 1–78.

Kleinstreuer, C. (2006). *Biofluid Dynamics: Principles and Selected Applications*, Boca Raton, FL: Taylor & Francis.

Ku, D. N. (1997). "Blood flow in arteries." *Annual Review of Fluid Mechanics* **29**: 399–434.

Lighthill, M. J. (1975). *Mathematical Biofluiddynamics*, Philadelphia: Soc. Ind. Appl. Math.

Lighthill, M. J. (1978). *Waves in Fluids*, Cambridge: Cambridge University Press.

Liu, Y. and W. K. Liu (2006). "Rheology of red blood cell aggregation by computer simulation." *J. Comput. Phys.* **220(1)**: 139–154.

Mazumdar, J. N. (1999). *An Introduction to Mathematical Physiology and Biology,* Second Edition, Cambridge: Cambridge University Press.

Mazumdar, J. N. (2004). *Biofluid Mechanics,* Third Edition, Singapore: World Scientific.

McConalogue, D. J. and R.S. Srivastava (1968). "Motion of a fluid in a curved tube." *Proc. Roy. Soc. A.* **307**: 37–53.

McDonald, D. A. (1974). *Blood Flow in Arteries, Second Edition*, Baltimore: The Williams & Wilkins Company.

Mohanty, A. K. and S. B. L. Asthana (1979). "Laminar flow in the entrance region of a smooth pipe." *Journal of Fluid Mechanics* **90**: 433–447.

Morgan, G. W. and W. R. Ferrante (1955). "Wave propagation in elastic tubes filled with streaming liquid." *The Journal of the Acoustical Society of America* **27(4)**: 715–725.

Morgan, G. W. and J. P. Kiely (1954). "Wave propagation in a viscous liquid contained in a flexible tube." *The Journal of the Acoustical Society of America* **26(3)**: 323–328.

Nichols, W. W. and M. F. O'Rourke (1998). *McDonald's Blood Flow in Arteries: Theoretical, Experimental and Clinical Principles*, London: Arnold.

Pedley, T. J. (1980). *The Fluid Mechanics of Large Blood Vessels*, Cambridge: Cambridge University Press.

Pedley, T. J. (2000). "Blood flow in arteries and veins." *in: Perspectives in Fluid Dynamics*, G. K. Batchelor, H. K. Moffat and M. G. Worster, eds, Cambridge: Cambridge University Press.

Rand, R. H. (1983). "Fluid mechanics of green plants." *Annual Review of Fluid Mechanics* **15**: 29–45.

Rand, R. H. and J. R. Cooke (1978). "Fluid dynamics of phloem flow: an axi-symmetric model." *Trans. ASAE.* **21**: 898–900, 906.

Rand, R. H., S. K. Upadhyaya and J.R. Cooke (1980). "Fluid dynamics of phloem flow. II. An approximate formula." *Trans. ASAE.* **23**: 581–584.

Shapiro, A. H. (1977a). "Steady flow in collapsible tubes." *Journal of Biomechanical Engineering* **99**: 126–147.

Shapiro, A. H. (1977b). "Physiologic and medical aspects of flow in collapsible tubes." *Proc. 6th Can. Congr. Appl. Mech.* pp. 883–906.

Singh, M. P. (1974). "Entry flow in a curved pipe." *Journal of Fluid Mechanics* **65**: 517–539.

Whitmore, R. L. (1968). *Rheology of the Circulation,* Oxford: Pergamon Press.

Womersley, J. R. (1955a). "Method for the calculation of velocity, rate of flow and viscous drag in arteries when the pressure gradient is known." *Journal of Physiology* **127**: 553–563.

Womersley, J. R. 1955b). "Oscillatory motion of a viscous liquid in a thin-walled elastic tube. I. The linear approximation for long waves." *Philosophical Magazine* **46**: 199–221.

Womersley, J. R. (1957a). "The mathematical analysis of arterial circulation in a state of oscillatory motion." Technical Report WADC-TR-56-614, Wright Air Development Center, Dayton, OH

Womersley, J. R. (1957b). "Oscillatory flow in arteries: the constrained elastic tube as a model of arterial flow and pulse transmission." *Physics in Medicine and Biology* **2**: 178–187.

Zhou, J. and Y. C. Fung (1997). "The degree of nonlinearity and anisotropy of blood vessel elasticity." *Proc. Natl. Acad. Sci. USA* **94**: 14255–14260.

Some Properties of Common Fluids

A1. Useful Conversion Factors

Length: $1\,\text{m} = 3.2808\,\text{ft}$
$1\,\text{in.} = 2.540\,\text{cm}$
$1\,\text{mile} = 1.609\,\text{km}$
$1\,\text{nautical mile} = 1.852\,\text{km}$

Mass: $1\,\text{kg} = 2.2046\,\text{lb}$
$1\,\text{metric ton} = 1000\,\text{kg}$

Time: $1\,\text{day} = 86{,}400\,\text{s}$

Density: $1\,\text{kg/m}^3 = 0.062428\,\text{lb/ft}^3$

Velocity: $1\,\text{knot} = 0.5144\,\text{m/s}$

Force: $1\,\text{N} = 10^5\,\text{dyn}$

Pressure: $1\,\text{dyn/cm}^2 = 0.1\,\text{N/m}^2 = 0.1\,\text{Pa}$
$1\,\text{bar} = 10^5\,\text{Pa}$

Energy: $1\,\text{J} = 10^7\,\text{erg} = 0.2389\,\text{cal}$
$1\,\text{cal} = 4.186\,\text{J}$

Energy flux: $1\,\text{W/m}^2 = 2.39 \times 10^{-5}\,\text{cal}\,\text{cm}^{-2}\,\text{s}^{-1}$

©2010 Elsevier Inc. All rights reserved.
DOI: 10.1016/B978-0-12-381399-2.50018-6

A2. Properties of Pure Water at Atmospheric Pressure

Here, ρ = density, α = coefficient of thermal expansion, μ = shear viscosity, ν = kinematic viscosity = μ/ρ, κ = thermal diffusivity = $k/(\rho C_p)$, [k is first defined on p.6] Pr = Prandtl number, and 1.0×10^{-n} is written as $1.0\text{E} - n$

T °C	ρ kg/m^3	α K^{-1}	μ kg m^{-1} s^{-1}	ν m^2/s	κ m^2/s	C_p J kg^{-1} K^{-1}	Pr ν/κ
0	1000	−0.6E − 4	1.787E − 3	1.787E − 6	1.33E − 7	4217	13.4
10	1000	+0.9E − 4	1.307E − 3	1.307E − 6	1.38E − 7	4192	9.5
20	997	2.1E − 4	1.002E − 3	1.005E − 6	1.42E − 7	4182	7.1
30	995	3.0E − 4	0.799E − 3	0.802E − 6	1.46E − 7	4178	5.5
40	992	3.8E − 4	0.653E − 3	0.658E − 6	1.52E − 7	4178	4.3
50	988	4.5E − 4	0.548E − 3	0.555E − 6	1.58E − 7	4180	3.5

Latent heat of vaporization at $100\,°C = 2.257 \times 10^6$ J/kg.
Latent heat of melting of ice at $0\,°C = 0.334 \times 10^6$ J/kg.
Density of ice $= 920\,$kg/m^3.
Surface tension between water and air at $20\,°C = 0.0728$ N/m.
Sound speed at $25\,°C \simeq 1500$ m/s.

A3. Properties of Dry Air at Atmospheric Pressure

T °C	ρ kg/m^3	μ kg m^{-1} s^{-1}	ν m^2/s	κ m^2/s	Pr ν/κ
0	1.293	1.71E − 5	1.33E − 5	1.84E − 5	0.72
10	1.247	1.76E − 5	1.41E − 5	1.96E − 5	0.72
20	1.200	1.81E − 5	1.50E − 5	2.08E − 5	0.72
30	1.165	1.86E − 5	1.60E − 5	2.25E − 5	0.71
40	1.127	1.87E − 5	1.66E − 5	2.38E − 5	0.71
60	1.060	1.97E − 5	1.86E − 5	2.65E − 5	0.71
80	1.000	2.07E − 5	2.07E − 5	2.99E − 5	0.70
100	0.946	2.17E − 5	2.29E − 5	3.28E − 5	0.70

At $20\,°C$ and 1 atm,

$$C_p = 1012\,\text{J}\,\text{kg}^{-1}\,\text{K}^{-1}$$
$$C_v = 718\,\text{J}\,\text{kg}^{-1}\,\text{K}^{-1}$$
$$\gamma = 1.4$$
$$\alpha = 3.38 \times 10^{-3}\,\text{K}^{-1}$$
$$c = 340.6\,\text{m/s} \quad \text{(velocity of sound)}$$

Constants for dry air:

$$\text{Gas constant } R = 287.04\,\text{J}\,\text{kg}^{-1}\,\text{K}^{-1}$$
$$\text{Molecular mass } m = 28.966\,\text{kg/kmol}$$

A4. Properties of Standard Atmosphere

The following average values are accepted by international agreement. Here, z is the height above sea level.

z km	T °C	p kPa	ρ kg/m^3
0	15.0	101.3	1.225
0.5	11.5	95.5	1.168
1	8.5	89.9	1.112
2	2.0	79.5	1.007
3	−4.5	70.1	0.909
4	−11.0	61.6	0.819
5	−17.5	54.0	0.736
6	−24.0	47.2	0.660
8	−37.0	35.6	0.525
10	−50.0	26.4	0.413
12	−56.5	19.3	0.311
14	−56.5	14.1	0.226
16	−56.5	10.3	0.165
18	−56.5	7.5	0.120
20	−56.5	5.5	0.088

Curvilinear Coordinates

B1. *Cylindrical Polar Coordinates*

The coordinates are (R, θ, x), where θ is the azimuthal angle (see Figure 3.1b, where φ is used instead of θ). The equations are presented assuming ψ is a scalar, and

$$\mathbf{u} = \mathbf{i}_R u_R + \mathbf{i}_\theta u_\theta + \mathbf{i}_x u_x,$$

where $\mathbf{i}_R$, $\mathbf{i}_\theta$, and $\mathbf{i}_x$ are the local unit vectors at a point.

Gradient of a scalar

$$\nabla \psi = \mathbf{i}_R \frac{\partial \psi}{\partial R} + \frac{\mathbf{i}_\theta}{R} \frac{\partial \psi}{\partial \theta} + \mathbf{i}_x \frac{\partial \psi}{\partial x}.$$

Laplacian of a scalar

$$\nabla^2 \psi = \frac{1}{R} \frac{\partial}{\partial R} \left(R \frac{\partial \psi}{\partial R} \right) + \frac{1}{R^2} \frac{\partial^2 \psi}{\partial \theta^2} + \frac{\partial^2 \psi}{\partial x^2}.$$

Divergence of a vector

$$\nabla \cdot \mathbf{u} = \frac{1}{R} \frac{\partial (R u_R)}{\partial R} + \frac{1}{R} \frac{\partial u_\theta}{\partial \theta} + \frac{\partial u_x}{\partial x}.$$

Curl of a vector

$$\nabla \times \mathbf{u} = \mathbf{i}_R \left(\frac{1}{R} \frac{\partial u_x}{\partial \theta} - \frac{\partial u_\theta}{\partial x} \right) + \mathbf{i}_\theta \left(\frac{\partial u_R}{\partial x} - \frac{\partial u_x}{\partial R} \right) + \mathbf{i}_x \left[\frac{1}{R} \frac{\partial (R u_\theta)}{\partial R} - \frac{1}{R} \frac{\partial u_R}{\partial \theta} \right].$$

Laplacian of a vector

$$\nabla^2 \mathbf{u} = \mathbf{i}_R \left(\nabla^2 u_R - \frac{u_R}{R^2} - \frac{2}{R^2} \frac{\partial u_\theta}{\partial \theta} \right) + \mathbf{i}_\theta \left(\nabla^2 u_\theta + \frac{2}{R^2} \frac{\partial u_R}{\partial \theta} - \frac{u_\theta}{R^2} \right) + \mathbf{i}_x \nabla^2 u_x.$$

©2010 Elsevier Inc. All rights reserved.
DOI: 10.1016/B978-0-12-381399-2.50019-8

Strain rate and viscous stress (for incompressible form $\sigma_{ij} = 2\mu e_{ij}$)

$$e_{RR} = \frac{\partial u_R}{\partial R} = \frac{1}{2\mu}\sigma_{RR},$$

$$e_{\theta\theta} = \frac{1}{R}\frac{\partial u_\theta}{\partial \theta} + \frac{u_R}{R} = \frac{1}{2\mu}\sigma_{\theta\theta},$$

$$e_{xx} = \frac{\partial u_x}{\partial x} = \frac{1}{2\mu}\sigma_{xx},$$

$$e_{R\theta} = \frac{R}{2}\frac{\partial}{\partial R}\left(\frac{u_\theta}{R}\right) + \frac{1}{2R}\frac{\partial u_R}{\partial \theta} = \frac{1}{2\mu}\sigma_{R\theta},$$

$$e_{\theta x} = \frac{1}{2R}\frac{\partial u_x}{\partial \theta} + \frac{1}{2}\frac{\partial u_\theta}{\partial x} = \frac{1}{2\mu}\sigma_{\theta x},$$

$$e_{xR} = \frac{1}{2}\frac{\partial u_R}{\partial x} + \frac{1}{2}\frac{\partial u_x}{\partial R} = \frac{1}{2\mu}\sigma_{xR}.$$

Vorticity ($\boldsymbol{\omega} = \nabla \times \mathbf{u}$)

$$\omega_R = \frac{1}{R}\frac{\partial u_x}{\partial \theta} - \frac{\partial u_\theta}{\partial x},$$

$$\omega_\theta = \frac{\partial u_R}{\partial x} - \frac{\partial u_x}{\partial R},$$

$$\omega_x = \frac{1}{R}\frac{\partial}{\partial R}(Ru_\theta) - \frac{1}{R}\frac{\partial u_R}{\partial \theta}.$$

Equation of continuity

$$\frac{\partial \rho}{\partial t} + \frac{1}{R}\frac{\partial}{\partial R}(\rho R u_R) + \frac{1}{R}\frac{\partial}{\partial \theta}(\rho u_\theta) + \frac{\partial}{\partial x}(\rho u_x) = 0.$$

Navier–Stokes equations with constant ρ and ν, and no body force

$$\frac{\partial u_R}{\partial t} + (\mathbf{u}\cdot\nabla)u_R - \frac{u_\theta^2}{R} = -\frac{1}{\rho}\frac{\partial p}{\partial R} + \nu\left(\nabla^2 u_R - \frac{u_R}{R^2} - \frac{2}{R^2}\frac{\partial u_\theta}{\partial \theta}\right),$$

$$\frac{\partial u_\theta}{\partial t} + (\mathbf{u}\cdot\nabla)u_\theta + \frac{u_R u_\theta}{R} = -\frac{1}{\rho R}\frac{\partial p}{\partial \theta} + \nu\left(\nabla^2 u_\theta + \frac{2}{R^2}\frac{\partial u_R}{\partial \theta} - \frac{u_\theta}{R^2}\right),$$

$$\frac{\partial u_x}{\partial t} + (\mathbf{u}\cdot\nabla)u_x = -\frac{1}{\rho}\frac{\partial p}{\partial x} + \nu\nabla^2 u_x,$$

where

$$\mathbf{u}\cdot\nabla = u_R\frac{\partial}{\partial R} + \frac{u_\theta}{R}\frac{\partial}{\partial \theta} + u_x\frac{\partial}{\partial x},$$

$$\nabla^2 = \frac{1}{R}\frac{\partial}{\partial R}\left(R\frac{\partial}{\partial R}\right) + \frac{1}{R^2}\frac{\partial^2}{\partial \theta^2} + \frac{\partial^2}{\partial x^2}.$$

B2. Plane Polar Coordinates

The plane polar coordinates are (r, θ), where r is the distance from the origin (Figure 3.1a). The equations for plane polar coordinates can be obtained from those of the cylindrical coordinates presented in Section B1, replacing R by r and suppressing all components and derivatives in the axial direction x. Some of the expressions are repeated here because of their frequent occurrence.

Strain rate and viscous stress (for incompressible form $\sigma_{ij} = 2\mu e_{ij}$)

$$e_{rr} = \frac{\partial u_r}{\partial r} = \frac{1}{2\mu}\sigma_{rr},$$

$$e_{\theta\theta} = \frac{1}{r}\frac{\partial u_\theta}{\partial \theta} + \frac{u_r}{r} = \frac{1}{2\mu}\sigma_{\theta\theta},$$

$$e_{r\theta} = \frac{r}{2}\frac{\partial}{\partial r}\left(\frac{u_\theta}{r}\right) + \frac{1}{2r}\frac{\partial u_r}{\partial \theta} = \frac{1}{2\mu}\sigma_{r\theta}.$$

Vorticity

$$\omega_z = \frac{1}{r}\frac{\partial}{\partial r}(ru_\theta) - \frac{1}{r}\frac{\partial u_r}{\partial \theta}.$$

Equation of continuity

$$\frac{\partial \rho}{\partial t} + \frac{1}{r}\frac{\partial}{\partial r}(\rho r u_r) + \frac{1}{r}\frac{\partial}{\partial \theta}(\rho u_\theta) = 0.$$

Navier–Stokes equations with constant ρ and ν, and no body force

$$\frac{\partial u_r}{\partial t} + u_r\frac{\partial u_r}{\partial r} + \frac{u_\theta}{r}\frac{\partial u_r}{\partial \theta} - \frac{u_\theta^2}{r} = -\frac{1}{\rho}\frac{\partial p}{\partial r} + \nu\left(\nabla^2 u_r - \frac{u_r}{r^2} - \frac{2}{r^2}\frac{\partial u_\theta}{\partial \theta}\right),$$

$$\frac{\partial u_\theta}{\partial t} + u_r\frac{\partial u_\theta}{\partial r} + \frac{u_\theta}{r}\frac{\partial u_\theta}{\partial \theta} + \frac{u_r u_\theta}{r} = -\frac{1}{\rho r}\frac{\partial p}{\partial \theta} + \nu\left(\nabla^2 u_\theta + \frac{2}{r^2}\frac{\partial u_r}{\partial \theta} - \frac{u_\theta}{r^2}\right),$$

where

$$\nabla^2 = \frac{1}{r}\frac{\partial}{\partial r}\left(r\frac{\partial}{\partial r}\right) + \frac{1}{r^2}\frac{\partial^2}{\partial \theta^2}.$$

B3. Spherical Polar Coordinates

The spherical polar coordinates used are (r, θ, φ), where φ is the azimuthal angle (Figure 3.1c). Equations are presented assuming ψ is a scalar, and

$$\mathbf{u} = \mathbf{i}_r u_r + \mathbf{i}_\theta u_\theta + \mathbf{i}_\varphi u_\varphi,$$

where $\mathbf{i}_r$, $\mathbf{i}_\theta$, and $\mathbf{i}_\varphi$ are the local unit vectors at a point.

Gradient of a scalar

$$\nabla \psi = \mathbf{i}_r \frac{\partial \psi}{\partial r} + \mathbf{i}_\theta \frac{1}{r} \frac{\partial \psi}{\partial \theta} + \mathbf{i}_\varphi \frac{1}{r \sin \theta} \frac{\partial \psi}{\partial \varphi}.$$

Laplacian of a scalar

$$\nabla^2 \psi = \frac{1}{r^2} \frac{\partial}{\partial r} \left(r^2 \frac{\partial \psi}{\partial r} \right) + \frac{1}{r^2 \sin \theta} \frac{\partial}{\partial \theta} \left(\sin \theta \frac{\partial \psi}{\partial \theta} \right) + \frac{1}{r^2 \sin^2 \theta} \frac{\partial^2 \psi}{\partial \varphi^2}.$$

Divergence of a vector

$$\nabla \cdot \mathbf{u} = \frac{1}{r^2} \frac{\partial (r^2 u_r)}{\partial r} + \frac{1}{r \sin \theta} \frac{\partial (u_\theta \sin \theta)}{\partial \theta} + \frac{1}{r \sin \theta} \frac{\partial u_\theta}{\partial \varphi}.$$

Curl of a vector

$$\nabla \times \mathbf{u} = \frac{\mathbf{i}_r}{r \sin \theta} \left[\frac{\partial (u_\varphi \sin \theta)}{\partial \theta} - \frac{\partial u_\theta}{\partial \varphi} \right] + \frac{\mathbf{i}_\theta}{r} \left[\frac{1}{\sin \theta} \frac{\partial u_r}{\partial \varphi} - \frac{\partial (r u_\varphi)}{\partial r} \right]$$
$$+ \frac{\mathbf{i}_\varphi}{r} \left[\frac{\partial (r u_\theta)}{\partial r} - \frac{\partial u_r}{\partial \theta} \right].$$

Laplacian of a vector

$$\nabla^2 \mathbf{u} = \mathbf{i}_r \left[\nabla^2 u_r - \frac{2 u_r}{r^2} - \frac{2}{r^2 \sin \theta} \frac{\partial (u_\theta \sin \theta)}{\partial \theta} - \frac{2}{r^2 \sin \theta} \frac{\partial u_\varphi}{\partial \varphi} \right]$$
$$+ \mathbf{i}_\theta \left[\nabla^2 u_\theta + \frac{2}{r^2} \frac{\partial u_r}{\partial \theta} - \frac{u_\theta}{r^2 \sin^2 \theta} - \frac{2 \cos \theta}{r^2 \sin^2 \theta} \frac{\partial u_\varphi}{\partial \varphi} \right]$$
$$+ \mathbf{i}_\varphi \left[\nabla^2 u_\varphi + \frac{2}{r^2 \sin \theta} \frac{\partial u_r}{\partial \varphi} + \frac{2 \cos \theta}{r^2 \sin^2 \theta} \frac{\partial u_\theta}{\partial \varphi} - \frac{u_\varphi}{r^2 \sin^2 \theta} \right].$$

Strain rate and viscous stress (for incompressible form $\sigma_{ij} = 2\mu e_{ij}$)

$$e_{rr} = \frac{\partial u_r}{\partial r} = \frac{1}{2\mu} \sigma_{rr},$$

$$e_{\theta\theta} = \frac{1}{r} \frac{\partial u_\theta}{\partial \theta} + \frac{u_r}{r} = \frac{1}{2\mu} \sigma_{\theta\theta},$$

$$e_{\varphi\varphi} = \frac{1}{r \sin \theta} \frac{\partial u_\varphi}{\partial \varphi} + \frac{u_r}{r} + \frac{u_\theta \cot \theta}{r} = \frac{1}{2\mu} \sigma_{\varphi\varphi},$$

$$e_{\theta\varphi} = \frac{\sin \theta}{2r} \frac{\partial}{\partial \theta} \left(\frac{u_\varphi}{\sin \theta} \right) + \frac{1}{2r \sin \theta} \frac{\partial u_\theta}{\partial \varphi} = \frac{1}{2\mu} \sigma_{\theta\varphi},$$

$$e_{\varphi r} = \frac{1}{2r \sin \theta} \frac{\partial u_r}{\partial \varphi} + \frac{r}{2} \frac{\partial}{\partial r} \left(\frac{u_\varphi}{r} \right) = \frac{1}{2\mu} \sigma_{\varphi r},$$

$$e_{r\theta} = \frac{r}{2} \frac{\partial}{\partial r} \left(\frac{u_\theta}{r} \right) + \frac{1}{2r} \frac{\partial u_r}{\partial \theta} = \frac{1}{2\mu} \sigma_{r\theta}.$$

Vorticity

$$\omega_r = \frac{1}{r \sin \theta} \left[\frac{\partial}{\partial \theta} (u_\varphi \sin \theta) - \frac{\partial u_\theta}{\partial \varphi} \right],$$

$$\omega_\theta = \frac{1}{r} \left[\frac{1}{\sin \theta} \frac{\partial u_r}{\partial \varphi} - \frac{\partial (r u_\varphi)}{\partial r} \right],$$

$$\omega_\varphi = \frac{1}{r} \left[\frac{\partial}{\partial r} (r u_\theta) - \frac{\partial u_r}{\partial \theta} \right].$$

Equation of continuity

$$\frac{\partial \rho}{\partial t} + \frac{1}{r^2} \frac{\partial}{\partial r} (\rho r^2 u_r) + \frac{1}{r \sin \theta} \frac{\partial}{\partial \theta} (\rho u_\theta \sin \theta) + \frac{1}{r \sin \theta} \frac{\partial}{\partial \varphi} (\rho u_\varphi) = 0.$$

Navier–Stokes equations with constant ρ and v, and no body force

$$\frac{\partial u_r}{\partial t} + (\mathbf{u} \cdot \nabla) u_r - \frac{u_\theta^2 + u_\varphi^2}{r}$$
$$= -\frac{1}{\rho} \frac{\partial p}{\partial r} + v \left[\nabla^2 u_r - \frac{2 u_r}{r^2} - \frac{2}{r^2 \sin \theta} \frac{\partial (u_\theta \sin \theta)}{\partial \theta} - \frac{2}{r^2 \sin \theta} \frac{\partial u_\varphi}{\partial \varphi} \right],$$

$$\frac{\partial u_\theta}{\partial t} + (\mathbf{u} \cdot \nabla) u_\theta + \frac{u_r u_\theta}{r} - \frac{u_\varphi^2 \cot \theta}{r}$$
$$= -\frac{1}{\rho r} \frac{\partial p}{\partial \theta} + v \left[\nabla^2 u_\theta + \frac{2}{r^2} \frac{\partial u_r}{\partial \theta} - \frac{u_\theta}{r^2 \sin^2 \theta} - \frac{2 \cos \theta}{r^2 \sin^2 \theta} \frac{\partial u_\varphi}{\partial \varphi} \right],$$

$$\frac{\partial u_\varphi}{\partial t} + (\mathbf{u} \cdot \nabla) u_\varphi + \frac{u_\varphi u_r}{r} + \frac{u_\theta u_\varphi \cot \theta}{r}$$
$$= -\frac{1}{\rho r \sin \theta} \frac{\partial p}{\partial \varphi} + v \left[\nabla^2 u_\varphi + \frac{2}{r^2 \sin \theta} \frac{\partial u_r}{\partial \varphi} + \frac{2 \cos \theta}{r^2 \sin^2 \theta} \frac{\partial u_\theta}{\partial \varphi} - \frac{u_\varphi}{r^2 \sin^2 \theta} \right],$$

where

$$\mathbf{u} \cdot \nabla = u_r \frac{\partial}{\partial r} + \frac{u_\theta}{r} \frac{\partial}{\partial \theta} + \frac{u_\varphi}{r \sin \theta} \frac{\partial}{\partial \varphi},$$

$$\nabla^2 = \frac{1}{r^2} \frac{\partial}{\partial r} \left(r^2 \frac{\partial}{\partial r} \right) + \frac{1}{r^2 \sin \theta} \frac{\partial}{\partial \theta} \left(\sin \theta \frac{\partial}{\partial \theta} \right) + \frac{1}{r^2 \sin^2 \theta} \frac{\partial^2}{\partial \varphi^2}.$$

Founders of Modern Fluid Dynamics

Ludwig Prandtl (1875–1953)

Ludwig Prandtl was born in Freising, Germany, in 1875. He studied mechanical engineering in Munich. For his doctoral thesis he worked on a problem on elasticity under August Föppl, who himself did pioneering work in bringing together applied and theoretical mechanics. Later, Prandtl became Föppl's son-in-law, following the good German academic tradition in those days. In 1901, he became professor of mechanics at the University of Hanover, where he continued his earlier efforts to provide a sound theoretical basis for fluid mechanics. The famous mathematician Felix Klein, who stressed the use of mathematics in engineering education, became interested in Prandtl and enticed him to come to the University of Göttingen. Prandtl was a great admirer of Klein and kept a large portrait of him in his office. He served as professor of applied mechanics at Göttingen from 1904 to 1953; the quiet university town of Göttingen became an international center of aerodynamic research.

In 1904, Prandtl conceived the idea of a boundary layer, which adjoins the surface of a body moving through a fluid, and is perhaps the greatest single discovery in the history of fluid mechanics. He showed that frictional effects in a slightly viscous fluid are confined to a thin layer near the surface of the body; the rest of the flow can be considered inviscid. The idea led to a rational way of simplifying the equations of motion in the different regions of the flow field. Since then the boundary layer technique has been generalized and has become a most useful tool in many branches of science.

His work on wings of finite span (the Prandtl–Lanchester wing theory) eluci-dated the generation of induced drag. In compressible fluid motions he contributed the Prandtl–Glauert rule of subsonic flow, the Prandtl–Meyer expansion fan in supersonic flow around a corner, and published the first estimate of the thickness of a shock wave.

©2010 Elsevier Inc. All rights reserved.
DOI: 10.1016/B978-0-12-381399-2.50020-4

He made notable innovations in the design of wind tunnels and other aerodynamic equipment. His advocacy of monoplanes greatly advanced heavier-than-air aviation. In experimental fluid mechanics he designed the Pitot-static tube for measuring velocity. In turbulence theory he contributed the mixing length theory.

Prandtl liked to describe himself as a plain mechanical engineer. So naturally he was also interested in solid mechanics; for example, he devised a soap-film analogy for analyzing the torsion stresses of structures with noncircular cross sections. In this respect he was like G. I. Taylor, and his famous student von Karman; all three of them did a considerable amount of work on solid mechanics. Toward the end of his career Prandtl became interested in dynamic meteorology and published a paper generalizing the Ekman spiral for turbulent flows.

Prandtl was endowed with rare vision for understanding physical phenomena. His mastery of mathematical tricks was limited; indeed many of his collaborators were better mathematicians. However, Prandtl had an unusual ability of putting ideas in simple mathematical forms. In 1948, Prandtl published a simple and popular textbook on fluid mechanics, which has been referred to in several places here. His varied interest and simplicity of analysis is evident throughout this book. Prandtl died in Göttingen 1953.

Geoffrey Ingram Taylor (1886–1975)

Geoffrey Ingram Taylor's name almost always includes his initials G. I. in references, and his associates and friends simply refer to him as "G. I." He was born in 1886 in London. He apparently inherited a bent toward mathematics from his mother, who was the daughter of George Boole, the originator of "Boolean algebra." After graduation from the University of Cambridge, Taylor started to work with J. J. Thomson in pure physics.

He soon gave up pure physics and changed his interest to mechanics of fluids and solids. At this time a research position in dynamic meteorology was created at Cambridge and it was awarded to Taylor, although he had no knowledge of meteorology! At the age of 27 he was invited to serve as meteorologist on a British ship that sailed to Newfoundland to investigate the sinking of the *Titanic*. He took the opportunity to make measurements of velocity, temperature, and humidity profiles up to 2000 m by flying kites and releasing balloons from the ship. These were the very first measurements on the turbulent transfers of momentum and heat in the frictional layer of the atmosphere. This activity started his lifelong interest in turbulent flows.

During World War I he was commissioned as a meteorologist by the British Air Force. He learned to fly and became interested in aeronautics. He made the first measurements of the pressure distribution over a wing in full-scale flight. Involvement in aeronautics led him to an analysis of the stress distribution in propeller shafts. This work finally resulted in a fundamental advance in solid mechanics, the "Taylor dislocation theory."

Taylor had a extraordinarily long and productive research career (1909–1972). The amount and versatility of his work can be illustrated by the size and range of his *Collected Works* published in 1954: Volume I contains "Mechanics of Solids" (41 papers, 593 pages); Volume II contains "Meteorology, Oceanography, and

Turbulent Flow" (45 papers, 515 pages); Volume III contains "Aerodynamics and the Mechanics of Projectiles and Explosions" (58 papers, 559 pages); and Volume IV contains "Miscellaneous Papers on Mechanics of Fluids" (49 papers, 579 pages). Perhaps G. I. Taylor is best known for his work on turbulence. When asked, however, what gave him maximum *satisfaction*, Taylor singled out his work on the stability of Couette flow.

Professor George Batchelor, who has encountered many great physicists at Cambridge, described G. I. Taylor as one of the greatest physicists of the century. He combined a remarkable capacity for analytical thought with physical insight by which he knew "how things worked." He loved to conduct simple experiments, not to gather data to understand a phenomenon, but to demonstrate his theoretical calculations; in most cases he already knew what the experiment would show. Professor Batchelor has stated that Taylor was a thoroughly lovable man who did not suffer from the maladjustment and self-concern that many of today's institutional scientists seem to suffer (because of pressure!), and this allowed his creative energy to be used to the fullest extent.

He thought of himself as an amateur, and worked for pleasure alone. He did not take up a regular faculty position at Cambridge, had no teaching responsibilities, and did not visit another institution to pursue his research. He never had a secretary or applied for a research grant; the only facility he needed was a one-room laboratory and one technical assistant. He did not "keep up with the literature," tended to take up problems that were entirely new, and chose to work alone. Instead of mastering tensor notation, electronics, or numerical computations, G. I. Taylor chose to do things his own way, and did them better than anybody else.

Supplemental Reading

Batchelor, G. K. (1976). "Geoffrey Ingram Taylor, 1886–1975." *Biographical Memoirs of Fellows of the Royal Society* **22**: 565–633.

Batchelor, G. K. (1986). "Geoffrey Ingram Taylor, 7 March 1886–27 June 1975." *Journal of Fluid Mechanics* **173**: 1–14.

Oswatitsch, K. and K. Wieghardt (1987). "Ludwig Prandtl and his Kaiser-Wilhelm-Institute." *Annual Review of Fluid Mechanics* **19**: 1–25.

Von Karman, T. (1954). *Aerodynamics*, New York: McGraw-Hill.

Visual Resources

Following is a list of films, all but the first by the National Committee for Fluid Mechanics Films (NCFMF), founded in 1961 by the late Ascher H. Shapiro, then Professor of Mechanical Engineering at M.I.T. Descriptive text for the films was published separately as described below.

The Fluid Dynamics of Drag, Parts I, II, III, IV (1960)
    Text: Ascher H. Shapiro, *Shape and Flow:The Fluid Dynamics of Drag*,
    Doubleday and Co., New York (1961).
Vorticity, Parts I, II (1961)
    The text for this and all following films is: NCFMF, *Illustrated Experiments in*
    Fluid Mechanics, MIT Press, Cambridge, MA (1972).
Deformation of Continuous Media (1963)
Flow Visualization (1963)
Pressure Fields and Fluid Acceleration (1963)
Surface Tension in Fluid Mechanics (1964)
Waves in Fluids (1964)
*Boundary Layer Control (1965)
Rheological Behavior of Fluids (1965)
Secondary Flow (1965)
Channel Flow of a Compressible Fluid (1967)
Low-Reynolds-Number Flows (1967)
Magnetohydrodynamics (1967)
Cavitation (1968)
Eulerian and Lagrangian Descriptions in Fluid Mechanics (1968)
Flow Instabilities (1968)
Fundamentals of Boundary Layers (1968)
Rarefied Gas Dynamics (1968)
Stratified Flow (1968)
Aerodynamic Generation of Sound (1969)

©2010 Elsevier Inc. All rights reserved.
DOI: 10.1016/B978-0-12-381399-2.50021-6

Rotating Flows (1969)
Turbulence (1969)

Although these films are decades old, they remain excellent visualizations of the principles of fluid mechanics. All but the one marked "*" are available for viewing on the MIT web site as follows: *http://web.mit.edu/fluids/www/Shapiro/ncfmf.html* It would be a very good idea to view the film appropriate to the corresponding section of the text.

Index